2/11/93

Texts in Applied Mathematics 12

Texts in Applied Mathematics

1. *Sirovich:* Introduction to Applied Mathematics.
2. *Wiggins:* Introduction to Applied Nonlinear Dynamical Systems and Chaos.
3. *Hale/Koçak:* Dynamics and Bifurcations.
4. *Chorin/Marsden:* A Mathematical Introduction to Fluid Mechanics, 3rd ed.
5. *Hubbard/West:* Differential Equations: A Dynamical Systems Approach, Part I: Ordinary Differential Equations.
6. *Sontag:* Mathematical Control Theory: Deterministic Finite Dimensional Systems.
7. *Perko:* Differential Equations and Dynamical Systems.
8. *Seaborn:* Hypergeometric Functions and Their Applications.
9. *Pipkin:* A Course on Integral Equations.
10. *Hoppensteadt/Peskin:* Mathematics in Medicine and the Life Sciences.
11. *Braun:* Differential Equations and Their Applications: An Introduction to Applied Mathematics, 4th ed.
12. *Stoer/Bulirsch:* Introduction to Numerical Analysis, 2nd ed.

J. Stoer R. Bulirsch

Introduction to Numerical Analysis

Second Edition

Translated by R. Bartels, W. Gautschi, and C. Witzgall

With 35 Illustrations

Springer-Verlag
New York Berlin Heidelberg London Paris
Tokyo Hong Kong Barcelona Budapest

J. Stoer
Institut für Angewandte Mathematik
Universität Würzburg
am Hubland
8700 Würzburg, FRG

R. Bulirsch
Institut für Mathematik
Technische Universität
8000 München, FRG

R. Bartels
Department of Computer Science
University of Waterloo
Waterloo, Ontario N2L 3G1
Canada

W. Gautschi
Department of Computer Sciences
Purdue University
West Lafayette, IN 47907
USA

C. Witzgall
Center for Applied Mathematics
National Bureau of Standards
Washington, DC 20234
USA

Editors

F. John
Courant Institute of Mathematical Sciences
New York University
New York, NY 10012, USA

J.E. Marsden
Department of Mathematics
University of California
Berkeley, CA 94720, USA

L. Sirovich
Division of Applied Mathematics
Brown University
Providence, RI 02912, USA

M. Golubitsky
Department of Mathematics
University of Houston
Houston, TX 77004
USA

W. Jäger
Department of Applied Mathematics
Universtität Heidelberg
Im Neuenheimer Feld 294
6900 Heidelberg, FRG

Mathematics Subject Classification (1991): 65–01

Library of Congress Cataloging-in-Publication Data
Stoer, Josef.
[Einführung in die Numerische Mathematik. English]
Introduction to numerical analysis / J. Stoer, R. Bulirsch; translated by R. Bartels, W. Gautschi, and C. Witzgall.—2nd ed.
p. cm.
Includes bibliographical references and index.
ISBN 0-387-97878-X
1. Numerical analysis. I. Bulirsch, Roland. II. Title.
QA297.S8213 1992
519.4—dc20 92-20536

Printed on acid-free paper.

Title of the German Original Edition: Einführung in die Numerische Mathematik, I, II.
Publisher: Springer-Verlag Berlin Heidelberg, 1972, 1976.

Printed and bound by R.R. Donnelley & Sons, Harrisonburg, VA.
Printed in the United States of America.

9 8 7 6 5 4 3 2 1

ISBN 0-387-97878-X Springer-Verlag New York Berlin Heidelberg
ISBN 3-540-97878-X Springer-Verlag Berlin Heidelberg New York

Preface to the Second Edition

On the occasion of this new edition, the text was enlarged by several new sections. Two sections on B-splines and their computation were added to the chapter on spline functions: Due to their special properties, their flexibility, and the availability of well-tested programs for their computation, B-splines play an important role in many applications.

Also, the authors followed suggestions by many readers to supplement the chapter on elimination methods with a section dealing with the solution of large sparse systems of linear equations. Even though such systems are usually solved by iterative methods, the realm of elimination methods has been widely extended due to powerful techniques for handling sparse matrices. We will explain some of these techniques in connection with the Cholesky algorithm for solving positive definite linear systems.

The chapter on eigenvalue problems was enlarged by a section on the Lanczos algorithm; the sections on the LR and QR algorithm were rewritten and now contain a description of implicit shift techniques.

In order to some extent take into account the progress in the area of ordinary differential equations, a new section on implicit differential equations and differential-algebraic systems was added, and the section on stiff differential equations was updated by describing further methods to solve such equations.

The last chapter on the iterative solution of linear equations was also improved. The modern view of the conjugate gradient algorithm as an iterative method was stressed by adding an analysis of its convergence rate and a description of some preconditioning techniques. Finally, a new section on multigrid methods was incorporated: It contains a description of their basic ideas in the context of a simple boundary value problem for ordinary differential equations.

Many of the changes were suggested by several colleagues and readers. In particular, we would like to thank R. Seydel, P. Rentrop, and A. Neumaier for detailed proposals and our translators R. Bartels, W. Gautschi, and C. Witzgall for their valuable work and critical commentaries. The original German version was handled by F. Jarre, and I. Brugger was responsible for the expert typing of the many versions of the manuscript.

Finally we thank Springer-Verlag for the encouragement, patience, and close cooperation leading to this new edition.

Würzburg, München J. Stoer
May 1991 R. Bulirsch

Preface to the First Edition

This book is based on a one-year introductory course on numerical analysis given by the authors at several universities in Germany and the United States. The authors concentrate on methods which can be worked out on a digital computer. For important topics, algorithmic descriptions (given more or less formally in ALGOL 60), as well as thorough but concise treatments of their theoretical foundations, are provided. Where several methods for solving a problem are presented, comparisons of their applicability and limitations are offered. Each comparison is based on operation counts, theoretical properties such as convergence rates, and, more importantly, the intrinsic numerical properties that account for the reliability or unreliability of an algorithm. Within this context, the introductory chapter on error analysis plays a special role because it precisely describes basic concepts, such as the numerical stability of algorithms, that are indispensable in the thorough treatment of numerical questions.

The remaining seven chapters are devoted to describing numerical methods in various contexts. In addition to covering standard topics, these chapters encompass some special subjects not usually found in introductions to numerical analysis. Chapter 2, which discusses interpolation, gives an account of modern fast Fourier transform methods. In Chapter 3, extrapolation techniques for speeding up the convergence of discretization methods in connection with Romberg integration are explained at length.

The following chapter on solving linear equations contains a description of a numerically stable realization of the simplex method for solving linear programming problems. Further minimization algorithms for solving unconstrained minimization problems are treated in Chapter 5, which is devoted to solving nonlinear equations.

After a long chapter on eigenvalue problems for matrices, Chapter 7 is devoted to methods for solving ordinary differential equations. This chapter contains a broad discussion of modern multiple shooting techniques for solving two-point boundary-value problems. In contrast, methods for partial differential equations are not treated systematically. The aim is only to point out analogies to certain methods for solving ordinary differential equations, e.g., difference methods and variational techniques. The final chapter is devoted to discussing special methods for solving large sparse systems of linear equations resulting primarily from the application of difference or finite element techniques to partial differential equations. In addition to iteration methods, the conjugate gradient algorithm of Hestenes and Stiefel and the Buneman algorithm (which provides an example of a modern direct method for solving the discretized Poisson problem) are described.

Within each chapter numerous examples and exercises illustrate the numerical and theoretical properties of the various methods. Each chapter concludes with an extensive list of references.

The authors are indebted to many who have contributed to this introduction into numerical analysis. Above all, we gratefully acknowledge the deep influence of the early lectures of F.L. Bauer on our presentation. Many colleagues have helped us with their careful reading of manuscripts and many useful suggestions. Among others we would like to thank are C. Reinsch, M.B. Spijker, and, in particular, our indefatigable team of translators, R. Bartels, W. Gautschi, and C. Witzgall. Our co-workers K. Butendeich, G. Schuller, J. Zowe, and I. Brugger helped us to prepare the original German edition. Last but not least we express our sincerest thanks to Springer-Verlag for their good cooperation during the past years.

Würzburg, München J. Stoer
August 1979 R. Bulirsch

Contents

1 Error Analysis

Assessing the accuracy of the results of calculations is a paramount goal in numerical analysis. One distinguishes several kinds of errors which may limit this accuracy:

(1) *errors in the input data,*
(2) *roundoff errors,*
(3) *approximation errors.*

Input or data errors are beyond the control of the calculation. They may be due, for instance, to the inherent imperfections of physical measurements. Roundoff errors arise if one calculates with numbers whose representation is restricted to a finite number of digits, as is usually the case.

As for the third kind of error, many methods will not yield the exact solution of the given problem P, even if the calculations are carried out without rounding, but rather the solution of another simpler problem $\tilde{P}$ which approximates P. For instance, the problem P of summing an infinite series, e.g.,

$$e = 1 + \frac{1}{1!} + \frac{1}{2!} + \frac{1}{3!} + \cdots,$$

may be replaced by the simpler problem $\tilde{P}$ of summing only up to a finite number of terms of the series. The resulting approximation error is commonly called a *truncation error* (however, this term is also used for the roundoff related error committed by deleting any last digit of a number representation). Many approximating problems P are obtained by "discretizing" the original problem P: definite integrals are approximated by finite sums, differential quotients by a difference quotients, etc. In such cases, the approximation error is often referred to as *discretization error.*

Some authors extend the term "truncation error" to cover discretization errors.

In this chapter, we will examine the general effect of input and roundoff errors on the result of a calculation. Approximation errors will be discussed in later chapters as we deal with individual methods. For a comprehensive treatment of roundoff errors in floating-point computation see Sterbenz (1974).

1.1 Representation of Numbers

Based on their fundamentally different ways of representing numbers, two categories of computing machinery can be distinguished:

(1) *analog computers*,
(2) *digital computers*.

Examples of analog computers are slide rules and mechanical integrators as well as electronic analog computers. When using these devices one replaces numbers by physical quantities, e.g., the length of a bar or the intensity of a voltage, and simulates the mathematical problem by a physical one, which is solved through measurement, yielding a solution for the original mathematical problem as well. The scales of a slide rule, for instance, represent numbers x by line segments of length $k \ln x$. Multiplication is simulated by positioning line segments contiguously and measuring the combined length for the result.

It is clear that the accuracy of analog devices is directly limited by the physical measurements they employ.

Digital computers express the digits of a number representation by a sequence of discrete physical quantities. Typical instances are desk calculators and electronic digital computers.

Example

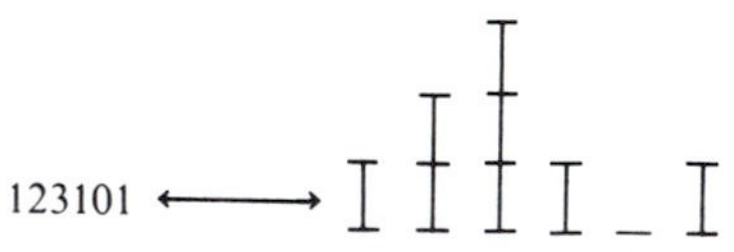

Each digit is represented by a specific physical quantity. Since only a small finite number of different digits have to be encoded—in the decimal number system, for instance, there are only 10 digits—the representation of digits in digital computers need not be quite as precise as the representation of numbers in analog computers. Thus one might tolerate voltages between, say, 7.8 and 8.2 when aiming at a representation of the digit 8 by 8 volts.

Consequently, the accuracy of digital computers is not directly limited by the precision of physical measurements.

For technical reasons, most modern electronic digital computers represent numbers internally in *binary* rather than decimal form. Here the coefficients or *bits* α_i of a decomposition by powers of 2 play the role of digits in the representation of a number x:

$$x = \pm(\alpha_n 2^n + \alpha_{n-1} 2^{n-1} + \cdots + \alpha_0 2^0 + \alpha_{-1} 2^{-1} + \alpha_{-2} 2^{-2} + \cdots),$$
$$\alpha_i = 0 \quad \text{or} \quad 1.$$

In order not to confuse decimal and binary representations of numbers, we denote the bits of a binary number representation by **O** and **L**, respectively.

EXAMPLE. The number $x = 18.5$ admits the decomposition

$$18.5 = 1 \times 2^4 + 0 \times 2^3 + 0 \times 2^2 + 1 \times 2^1 + 0 \times 2^0 + 1 \times 2^{-1}$$

and has therefore the binary representation

LOOLO.L.

We will use mainly the decimal system, pointing out differences between the two systems whenever it is pertinent to the examination at hand.

As the example $3.999\ldots = 4$ shows, the decimal representation of a number may not be unique. The same holds for binary representations. To exclude such ambiguities, we will always refer to the finite representation unless otherwise stated.

In general, digital computers must make do with a fixed finite number of places, the *word length*, when internally representing a number. This number n is determined by the make of the machine, although some machines have built-in extensions to integer multiples $2n$, $3n$, ... (double word length, triple word length, ...) of n to offer greater precision if needed. A word length of n places can be used in several different fashions to represent a number.

Fixed-point representation specifies a fixed number n_1 of places before and a fixed number n_2 after the decimal (binary) point, so that $n = n_1 + n_2$ (usually $n_1 = 0$ or $n_1 = n$).

EXAMPLE. For $n = 10$, $n_1 = 4$, $n_2 = 6$

	n_1	n_2
30.421 →	0030	421000
0.0437 →	0000	043700

In this representation, the position of the decimal (binary) point is fixed. A few simple digital devices, mainly for accounting purposes, are still restricted to fixed-point representation. Much more important, in particular for scientific calculations, are digital computers featuring *floating-point representation* of numbers. Here the decimal (binary) point is not fixed at the outset; rather its position with respect to the first digit is indicated for each number separately. This is done by specifying a so-called *exponent*. In other words, each real number can be represented in the form

$$(1.1.1) \qquad x = a \times 10^b (x = a \times 2^b) \quad \text{with } |a| < 1, b \text{ integer}$$

(say, 30.421 by 0.30421×10^2), where the exponent b indicates the position of the decimal point with respect to the *mantissa* a. Rutishauser proposed the following "semilogarithmic" notation, which displays the basis of the number system at the subscript level and moves the exponent down to the level of the mantissa:

$$0.30421_{10}2$$

Analogously,

$$\mathbf{O.LOOLOL_2LOL}$$

denotes the number 18.5 in the binary system. On any digital computer there are, of course, only fixed finite numbers t and e, $n = t + e$, of places available for the representation of mantissa and exponent, respectively.

EXAMPLE. For $t = 4$, $e = 2$ one would have the floating-point representation

$$0\ \boxed{5420}_{10}\ \boxed{04} \quad \text{or more concisely} \quad \boxed{5420 \mid 04}$$

for the number 5420 in the decimal system.

The floating-point representation of a number need not be unique. Since $5420 = 0.542_{10}4 = 0.0542_{10}5$, one could also have the floating-point representation

$$0\ \boxed{0542}_{10}\ \boxed{05} \quad \text{or} \quad \boxed{0542 \mid 05}$$

instead of the one given in the above example.

A floating-point representation is *normalized* if the first digit (bit) of the mantissa is different from 0 (**O**). Then $|a| \geqslant 10^{-1}$ ($|a| \geqslant 2^{-1}$) holds in (1.1.1). The *significant digits* (*bits*) of a number are the digits of the mantissa not counting leading zeros.

In what follows, we will only consider normalized floating-point representations and the corresponding floating-point arithmetic. The numbers t and e determine—together with the basis $B = 10$ or $B = 2$ of the number representation—the set $A \subseteq \mathbb{R}$ of real numbers which can be represented exactly within a given machine. The elements of A are called the *machine numbers*.

While normalized floating-point arithmetic is prevalent on current electronic digital computers, unnormalized arithmetic has been proposed to ensure that only truly significant digits are carried [Ashenhurst and Metropolis, (1959)].

1.2 Roundoff Errors and Floating-Point Arithmetic

The set A of numbers which are representable in a given machine is only finite. The question therefore arises of how to approximate a number $x \notin A$ which is not a machine number by a number $g \in A$ which is. This problem is encountered not only when reading data into a computer, but also when representing intermediate results within the computer during the course of a calculation. Indeed, straightforward examples show that the results of elementary arithmetic operations $x \pm y$, $x \times y$, x/y need not belong to A, even if both operands $x, y \in A$ are machine numbers.

It is natural to postulate that the approximation of any number $x \notin A$ by a machine number $\operatorname{rd}(x) \in A$ should satisfy

$$(1.2.1) \qquad |x - \operatorname{rd}(x)| \leqslant |x - g| \quad \text{for all } g \in A.$$

Such a machine-number approximation $\operatorname{rd}(x)$ can be obtained in most cases by *rounding*.

Example 1 ($t = 4$)

$$\operatorname{rd}(0.14285_{10}0) = 0.1429_{10}0,$$
$$\operatorname{rd}(3.14159_{10}0) = 0.3142_{10}1,$$
$$\operatorname{rd}(0.142842_{10}2) = 0.1428_{10}2.$$

In general, one can proceed as follows in order to find $\operatorname{rd}(x)$ for a t-digit computer: $x \notin A$ is first represented in normalized form $x = a \times 10^b$, so that $|a| \geqslant 10^{-1}$. Suppose the decimal representation of $|a|$ is given by

$$|a| = 0.\alpha_1\alpha_2 \ldots \alpha_i\alpha_{i+1} \ldots, \qquad 0 \leqslant \alpha_i \leqslant 9, \quad \alpha_1 \neq 0.$$

Then one forms

$$a' := \begin{cases} 0.\alpha_1\alpha_2 \ldots \alpha_t & \text{if } 0 \leqslant \alpha_{t+1} \leqslant 4, \\ 0.\alpha_1\alpha_2 \ldots \alpha_t + 10^{-t} & \text{if } \alpha_{t+1} \geqslant 5, \end{cases}$$

that is, one increases α_t by 1 if the $(t+1)$st digit $\alpha_{t+1} \geqslant 5$, and deletes all digits after the tth one. Finally one puts

$$\widetilde{\operatorname{rd}}(x) := \operatorname{sign}(x) \cdot a' \times 10^b.$$

Since $|a| \geqslant 10^{-1}$, the "relative error" of $\widetilde{\mathrm{rd}}(x)$ admits the following bound (Scarborough, 1950):

$$\left|\frac{\widetilde{\mathrm{rd}}(x) - x}{x}\right| \leqslant \frac{5 \times 10^{-(t+1)}}{|a|} \leqslant 5 \times 10^{-t}.$$

With the abbreviation $\mathrm{eps} := 5 \times 10^{-t}$, this can be written as

(1.2.2) $$\widetilde{\mathrm{rd}}(x) = x(1 + \varepsilon), \quad \text{where } |\varepsilon| \leqslant \mathrm{eps}.$$

The quantity $\mathrm{eps} = 5 \times 10^{-t}$ is called the *machine precision*. In the binary system, $\widetilde{\mathrm{rd}}(x)$ is defined analogously: Starting with a decomposition $x = a \times 2^b$ satisfying $2^{-1} \leqslant |a| < 1$ and the binary representation of $|a|$,

$$|a| = 0.\alpha_1 \ldots \alpha_t \alpha_{t+1} \ldots, \qquad \alpha_i = \mathbf{O} \text{ or } \mathbf{L}, \quad \alpha_1 = \mathbf{L},$$

one forms

$$a' := \begin{cases} 0.\alpha_1 \ldots \alpha_t & \text{if } \alpha_{t+1} = \mathbf{O}, \\ 0.\alpha_1 \ldots \alpha_t + 2^{-t} & \text{if } \alpha_{t+1} = \mathbf{L}, \end{cases}$$

$$\widetilde{\mathrm{rd}}(x) := \mathrm{sign}(x) \cdot a' \times 2^b.$$

Again (1.2.2) holds, provided one defines the machine precision by $\mathrm{eps} := 2^{-t}$.

Whenever $\widetilde{\mathrm{rd}}(x) \in A$ is a machine number, then $\widetilde{\mathrm{rd}}$ has the property (1.2.1) of a correct rounding process, and we may define

$$\widetilde{\mathrm{rd}}(x) := \mathrm{rd}(x) \quad \text{for all } x \text{ with } \widetilde{\mathrm{rd}}(x) \in A.$$

Because only a finite number e of places are available to express the exponent in a floating-point representation, there are unfortunately always numbers $x \notin A$ with $\widetilde{\mathrm{rd}}(x) \notin A$.

EXAMPLE 2 ($t = 4$, $e = 2$).

$$\begin{aligned}
&\text{(a)}\ \widetilde{\mathrm{rd}}(0.31794_{10}110) &&= 0.3179_{10}110 &&\notin A.\\
&\text{(b)}\ \widetilde{\mathrm{rd}}(0.99997_{10}99) &&= 0.1000_{10}100 &&\notin A.\\
&\text{(c)}\ \widetilde{\mathrm{rd}}(0.012345_{10}-99) &&= 0.1235_{10}-100 &&\notin A.\\
&\text{(d)}\ \widetilde{\mathrm{rd}}(0.54321_{10}-110) &&= 0.5432_{10}-110 &&\notin A.
\end{aligned}$$

In cases (a) and (b) the exponent is too greatly positive to fit the allotted space: These are instances of *exponent overflow*. Case (b) is particularly pathological: exponent overflow happens only after rounding. Cases (c) and (d) are instances of *exponent underflow*, i.e., the exponent of the number represented is too greatly negative. In cases (c) and (d) exponent underflow may be prevented by defining

(1.2.3) $$\begin{aligned}\mathrm{rd}(0.012345_{10}-99) &= 0.0123_{10}-99 \in A,\\ \mathrm{rd}(0.54321_{10}-110) &= 0 \in A.\end{aligned}$$

But then rd does not satisfy (1.2.2), that is, the relative error of $\mathrm{rd}(x)$ may exceed eps. Digital computers treat occurrences of exponent overflow and underflow as irregularities of the calculation. In the case of exponent underflow, $\mathrm{rd}(x)$ may be formed as indicated in (1.2.3). Exponent overflow may cause a halt in calculations. In the remaining regular cases (but not for all makes of computers), rounding is defined by

$$\mathrm{rd}(x) := \widetilde{\mathrm{rd}}(x).$$

Exponent overflow and underflow can be avoided to some extent by suitable scaling of the input data and by incorporating special checks and rescalings during computations. Since each different numerical method will require its own special protection techniques, and since overflow and underflow do not happen very frequently, we will make the idealized assumption that $e = \infty$ in our subsequent discussions, so that $\mathrm{rd} := \widetilde{\mathrm{rd}}$ does indeed provide a rule for rounding which ensures

$$\text{(1.2.4)} \qquad \begin{aligned} &\mathrm{rd} : \mathbb{R} \to A, \\ &\mathrm{rd}(x) = x(1+\varepsilon) \quad \text{with } |\varepsilon| \leqslant \mathrm{eps} \qquad \text{for all } x \in \mathbb{R}. \end{aligned}$$

In further examples we will, correspondingly, give the length t of the mantissa only. The reader must bear in mind, however, that subsequent statements regarding roundoff errors may be invalid if overflows or underflows are allowed to happen.

We have seen that the results of arithmetic operations $x \pm y$, $x \times y$, x/y need not be machine numbers, even if the operands x and y are. Thus one cannot expect to reproduce the arithmetic operations exactly on a digital computer. One will have to be content with substitute operations $+^*$, $-^*$, $\times^*$, $/^*$, so-called *floating-point operations*, which approximate the arithmetic operations as well as possible [v.Neumann and Goldstein (1947)]. Such operations may be defined, for instance, with the help of the rounding map rd as follows

$$\text{(1.2.5)} \qquad \left.\begin{aligned} x +^* y &:= \mathrm{rd}(x+y), \\ x -^* y &:= \mathrm{rd}(x-y) \\ x \times^* y &:= \mathrm{rd}(x \times y), \\ x /^* y &:= \mathrm{rd}(x / y), \end{aligned}\right. \quad \text{for } x, y \in A,$$

so that (1.2.4) implies

$$\text{(1.2.6)} \qquad \left.\begin{aligned} x +^* y &= (x+y)(1+\varepsilon_1) \\ x -^* y &= (x-y)(1+\varepsilon_2) \\ x \times^* y &= (x \times y)(1+\varepsilon_3) \\ x /^* y &= (x / y)(1+\varepsilon_4) \end{aligned}\right\} \quad |\varepsilon_i| \leqslant \mathrm{eps}.$$

On many modern computer installations, the floating-point operations $\pm *, \ldots$ are not defined by (1.2.5), but instead in such a way that (1.2.6) holds with only a somewhat weaker bound, say, $|\varepsilon_i| \leqslant k \cdot \text{eps}$, $k \geqslant 1$ being a small integer. Since these small deviations from (1.2.6) are not significant for our examinations, we will assume for simplicity that the floating-point operations are in fact defined by (1.2.5) and hence satisfy (1.2.6).

It should be pointed out that the floating-point operations do not satisfy the well-known laws for arithmetic operations. For instance,

$$x +^* y = x \quad \text{if } |y| < \frac{\text{eps}}{B}|x|, \qquad x, y \in A,$$

where B is the basis of the number system. The machine precision eps could indeed be defined as the smallest positive machine number g for which $1 +^* g > 1$:

$$\text{eps} = \min\{g \in A \mid 1 +^* g > 1 \text{ and } g > 0\}.$$

Furthermore, floating-point operations need not be associative or distributive.

EXAMPLE 3 ($t = 8$). With

$$\begin{aligned} a &:= 0.23371258_{10}-4, \\ b &:= 0.33678429_{10}2, \\ c &:= -0.33677811_{10}2, \end{aligned}$$

one has

$$\begin{aligned} a +^* (b +^* c) &= 0.23371258_{10}-4 +{}^* 0.61800000_{10}-3 \\ &= 0.64137126_{10}-3, \\ (a +^* b) +^* c &= 0.33678452_{10}2 -{}^* 0.33677811_{10}2 \\ &= 0.64100000_{10}-3. \end{aligned}$$

The exact result is

$$a + b + c = 0.641371258_{10}-3.$$

When subtracting two numbers $x, y \in A$ of the same sign, one has to watch out for *cancellation*. This occurs if x and y agree in one or more leading digits with respect to the same exponent, e.g.,

$$\begin{aligned} x &= 0.315876_{10}1, \\ y &= 0.314289_{10}1. \end{aligned}$$

The subtraction causes the common leading digits to disappear. The exact result $x - y$ is consequently a machine number, so that no *new* roundoff error $x -^* y = x - y$ arises. In this sense, subtraction in the case of cancellation is a quite harmless operation. We will see in the next section, however, that cancellation is extremely dangerous concerning the propagation of *old* errors, which stem from the calculations of x and y *prior* to carrying out the subtraction $x - y$.

For expressing the result of floating-point calculations, a convenient but slightly imprecise notation has been widely accepted, and we will use it frequently ourselves: If it is clear from the context how to evaluate an arithmetic expression E (if need be this can be specified by inserting suitable parentheses), then $\mathrm{fl}(E)$ denotes the value of the expression as obtained by floating-point arithmetic.

EXAMPLE 4

$$\begin{aligned} \mathrm{fl}(x \times y) &:= x \times^* y, \\ \mathrm{fl}(a + (b + c)) &:= a +^* (b +^* c), \\ \mathrm{fl}((a + b) + c) &:= (a +^* b) +^* c. \end{aligned}$$

We will also use the notation $\mathrm{fl}(\sqrt{x})$, $\mathrm{fl}(\cos(x))$, etc., whenever the digital computer approximates functions $\sqrt{\ }$, cos, etc., by substitutes $\sqrt{\ }^*$, cos*, etc. Thus $\mathrm{fl}(\sqrt{x}) := \sqrt{x}^*$, and so on.

The arithmetic operations $+$, $-$, $\times$, $/$, together with those basic functions like $\sqrt{\ }$, cos, for which floating-point substitutes $\sqrt{\ }^*$, cos*, etc., have been specified, will be called *elementary operations.*

1.3 Error Propagation

We have seen in the previous section (Example 3) that two different but mathematically equivalent methods $(a + b) + c$, $a + (b + c)$ for evaluating the same expression $a + b + c$ may lead to different results if floating-point arithmetic is used. For numerical purposes it is therefore important to distinguish between different evaluation schemes even if they are mathematically equivalent. Thus we call a finite sequence of elementary operations (as given for instance by consecutive computer instructions) which prescribes how to calculate the solution of a problem from given input data, an *algorithm.*

We will formalize the notion of an algorithm somewhat. Suppose a problem consists of calculating desired result numbers $y_1, \ldots, y_m$ from input numbers $x_1, \ldots, x_n$. If we introduce the vectors

$$x = \begin{bmatrix} x_1 \\ \vdots \\ x_n \end{bmatrix}, \qquad y = \begin{bmatrix} y_1 \\ \vdots \\ y_m \end{bmatrix},$$

then solving the above problem means determining the value $y = \varphi(x)$ of a certain multivariate vector function $\varphi\colon D \to \mathbb{R}^m$, $D \subseteq \mathbb{R}^n$, where φ is given by m real functions φ_i,

$$y_i = \varphi_i(x_1, \ldots, x_n), \qquad i = 1, \ldots, m.$$

At each stage of a calculation there is an *operand set* of numbers, which either are original input numbers x_i or have resulted from previous operations. A single operation calculates a new number from one or more elements of the operand set. The new number is either an intermediate or a final result. In any case, it is adjoined to the operand set, which then is purged of all entries that will not be needed as operands during the remainder of the calculation. The final operand set will consist of the desired results $y_1, \ldots, y_m$.

Therefore, an operation corresponds to a transformation of the operand set. Writing consecutive operand sets as vectors,

$$x^{(i)} = \begin{bmatrix} x_1^{(i)} \\ \vdots \\ x_{n_i}^{(i)} \end{bmatrix} \in \mathbb{R}^{n_i},$$

we can associate with an elementary operation an *elementary map*

$$\varphi^{(i)}\colon D_i \to \mathbb{R}^{n_{i+1}}, \qquad D_i \subseteq \mathbb{R}^{n_i},$$

so that

$$\varphi^{(i)}(x^{(i)}) = x^{(i+1)},$$

where $x^{(i+1)}$ is a vector representation of the transformed operand set. The elementary map $\varphi^{(i)}$ is uniquely defined except for inconsequential permutations of $x^{(i)}$ and $x^{(i+1)}$ which stem from the arbitrariness involved in arranging the corresponding operand sets in the form of vectors.

Given an algorithm, then its sequence of elementary operations gives rise to a decomposition of φ into a sequence of elementary maps

$$\text{(1.3.1)} \quad \begin{aligned} &\varphi^{(i)}\colon D_i \to D_{i+1}, \quad i = 0, 1, \ldots, r, \qquad D_j \subseteq \mathbb{R}^{n_j}, \\ &\varphi = \varphi^{(r)} \circ \varphi^{(r-1)} \circ \cdots \circ \varphi^{(0)}, \qquad D_0 = D, \quad D_{r+1} \subseteq \mathbb{R}^{n_{r+1}} = \mathbb{R}^m \end{aligned}$$

which characterize the algorithm.

EXAMPLE 1. For $\varphi(a, b, c) = a + b + c$, consider the two algorithms $\eta := a + b$, $y := c + \eta$ and $\eta := b + c$, $y := a + \eta$. The decompositions (1.3.1) are

$$\varphi^{(0)}(a, b, c) := \begin{bmatrix} a + b \\ c \end{bmatrix} \in \mathbb{R}^2, \qquad \varphi^{(1)}(u, v) := u + v \in \mathbb{R}$$

and

$$\varphi^{(0)}(a, b, c) := \begin{bmatrix} a \\ b + c \end{bmatrix} \in \mathbb{R}^2, \qquad \varphi^{(1)}(u, v) := u + v \in \mathbb{R}.$$

EXAMPLE 2. Since $a^2 - b^2 = (a + b)(a - b)$, one has for the calculation of $\varphi(a, b) := a^2 - b^2$ the two algorithms

$$\begin{array}{llll} \textit{Algorithm 1:} & \eta_1 := a \times a, & \textit{Algorithm 2:} & \eta_1 := a + b, \\ & \eta_2 := b \times b, & & \eta_2 := a - b, \\ & y := \eta_1 - \eta_2, & & y := \eta_1 \times \eta_2. \end{array}$$

Corresponding decompositions (1.3.1) are

$$\textit{Algorithm 1:} \quad \varphi^{(0)}(a, b) := \begin{bmatrix} a^2 \\ b \end{bmatrix}, \quad \varphi^{(1)}(u, v) := \begin{bmatrix} u \\ v^2 \end{bmatrix}, \quad \varphi^{(2)}(u, v) := u - v,$$

$$\textit{Algorithm 2:} \quad \varphi^{(0)}(a, b) := \begin{bmatrix} a \\ b \\ a + b \end{bmatrix}, \quad \varphi^{(1)}(a, b, u) := \begin{bmatrix} u \\ a - b \end{bmatrix}, \quad \varphi^{(2)}(u, v) := u \cdot v.$$

Note that the decomposition of $\varphi(a, b) := a^2 - b^2$ corresponding to Algorithm 1 above can be telescoped into a simpler decomposition:

$$\bar{\varphi}^{(0)}(a, b) = \begin{bmatrix} a^2 \\ b^2 \end{bmatrix}, \qquad \bar{\varphi}^{(1)}(u, v) := u - v.$$

Strictly speaking, however, map $\bar{\varphi}^{(0)}$ is not elementary. Moreover the decomposition does not determine the algorithm uniquely, since there is still a choice, however numerically insignificant, of what to compute first, a^2 or b^2.

Hoping to find criteria for judging the quality of algorithms, we will now examine the reasons why different algorithms for solving the same problem generally yield different results. Error propagation, for one, plays a decisive role, as the example of the sum $y = a + b + c$ shows (see Example 3 in Section 1.2). Here floating-point arithmetic yields an approximation $\tilde{y} = \mathrm{fl}((a + b) + c)$ to y which, according to (1.2.6), satisfies

$$\begin{aligned} \eta &:= \mathrm{fl}(a + b) = (a + b)(1 + \varepsilon_1), \\ \tilde{y} &:= \mathrm{fl}(\eta + c) = (\eta + c)(1 + \varepsilon_2) \\ &= [(a + b)(1 + \varepsilon_1) + c](1 + \varepsilon_2) \\ &= (a + b + c)\left[1 + \frac{a + b}{a + b + c}\varepsilon_1(1 + \varepsilon_2) + \varepsilon_2\right]. \end{aligned}$$

For the relative error $\varepsilon_y := (\tilde{y} - y)/y$ of $\tilde{y}$,

$$\varepsilon_y = \frac{a + b}{a + b + c}\varepsilon_1(1 + \varepsilon_2) + \varepsilon_2$$

or disregarding terms of order higher than 1 in ε's such as $\varepsilon_1\varepsilon_2$,

$$\varepsilon_y \doteq \frac{a + b}{a + b + c}\varepsilon_1 + 1 \cdot \varepsilon_2.$$

The amplification factors $(a+b)/(a+b+c)$ and 1, respectively, measure the effect of the roundoff errors ε_1, ε_2 on the error ε_y of the result. The factor $(a+b)/(a+b+c)$ is critical: depending on whether $|a+b|$ or $|b+c|$ is the smaller of the two, it is better to proceed via $(a+b)+c$ rather than $a+(b+c)$ for computing $a+b+c$.

In Example 3 of the previous section,

$$\frac{a+b}{a+b+c} = \frac{0.33\ldots{}_{10}2}{0.64\ldots{}_{10}-3} \approx \tfrac{1}{2}\times 10^5,$$

$$\frac{b+c}{a+b+c} = \frac{0.618\ldots{}_{10}-3}{0.64\ldots{}_{10}-3} \approx 0.97,$$

which explains the higher accuracy of $\mathrm{fl}(a+(b+c))$.

The above method of examining the propagation of particular errors while disregarding higher-order terms can be extended systematically to provide a *differential error analysis* of an algorithm for computing $\varphi(x)$ if this function is given by a decomposition (1.3.1):

$$\varphi = \varphi^{(r)} \circ \varphi^{(r-1)} \circ \cdots \circ \varphi^{(0)}.$$

To this end we must investigate how the input errors Δx of x as well as the roundoff errors accumulated during the course of the algorithm affect the final result $y = \varphi(x)$. We start this investigation by considering the input errors Δx alone, and we will apply any insights we gain to the analysis of the propagation of roundoff errors. We suppose that the function

$$\varphi\colon D \to \mathbb{R}^m, \qquad \varphi(x) = \begin{bmatrix} \varphi_1(x_1, \ldots, x_n) \\ \vdots \\ \varphi_m(x_1, \ldots, x_n) \end{bmatrix},$$

is defined on an open subset D of $\mathbb{R}^n$, and that its component functions φ_i, $i = 1, \ldots, n$, have continuous first derivatives on D. Let $\tilde{x}$ be an approximate value for x. Then we denote by

$$\Delta x_i := \tilde{x}_i - x_i, \qquad \Delta x := \tilde{x} - x$$

the *absolute error* of $\tilde{x}_i$ and $\tilde{x}$, respectively. The *relative error* of $\tilde{x}_i$ is defined as the quantity

$$\varepsilon_{\tilde{x}_i} := \frac{\tilde{x}_i - x_i}{x_i} \quad \text{if } x_i \neq 0.$$

Replacing the input data x by $\tilde{x}$ leads to the result $\tilde{y} := \varphi(\tilde{x})$ instead of $y = \varphi(x)$. Expanding in a Taylor series and disregarding higher-order terms gives

$$\begin{aligned} \Delta y_i := \tilde{y}_i - y_i = \varphi_i(\tilde{x}) - \varphi_i(x) &\doteq \sum_{j=1}^{n} (\tilde{x}_j - x_j)\frac{\partial \varphi_i(x)}{\partial x_j} \\ &= \sum_{j=1}^{n} \frac{\partial \varphi_i(x)}{\partial x_j}\Delta x_j, \qquad i = 1, \ldots, m, \end{aligned} \tag{1.3.2}$$

or in matrix notation,

$$(1.3.3)\quad \Delta y = \begin{bmatrix} \Delta y_1 \\ \vdots \\ \Delta y_m \end{bmatrix} \doteq \begin{bmatrix} \dfrac{\partial \varphi_1}{\partial x_1} & \cdots & \dfrac{\partial \varphi_1}{\partial x_n} \\ \vdots & & \vdots \\ \dfrac{\partial \varphi_m}{\partial x_1} & \cdots & \dfrac{\partial \varphi_m}{\partial x_n} \end{bmatrix} \begin{bmatrix} \Delta x_1 \\ \vdots \\ \Delta x_n \end{bmatrix} = D\varphi(x) \cdot \Delta x$$

with the Jacobian matrix $D\varphi(x)$.

The notation " $\doteq$ " instead of " $=$ ", which has been used occasionally before, is meant to indicate that the corresponding equations are only a first order approximation, i.e., they do not take quantities of higher order (in ε's or Δ's) into account.

The quantity $\partial\varphi_i(x)/\partial x_j$ in (1.3.3) represents the sensitivity with which y_i reacts to absolute perturbations Δx_j of x_j. If $y_i \neq 0$ for $i = 1, \ldots, m$ and $x_j \neq 0$ for $j = 1, \ldots, n$, then a similar error propagation formula holds for relative errors:

$$(1.3.4)\qquad \varepsilon_{y_i} \doteq \sum_{j=1}^{n} \frac{x_j}{\varphi_i(x)} \frac{\partial \varphi_i(x)}{\partial x_j} \varepsilon_{x_j}.$$

Again the quantity $(x_j/\varphi_i)\,\partial\varphi_i/\partial x_j$ indicates how strongly a relative error in x_j affects the relative error in y_i. The amplification factors $(x_j/\varphi_i)\,\partial\varphi_i/\partial x_j$ for the relative error have the advantage of not depending on the scales of y_i and x_j. The amplification factors for relative errors are generally called *condition numbers*. If any condition numbers are present which have large absolute values, then one speaks of an *ill-conditioned* problem; otherwise, of a *well-conditioned* problem. For ill-conditioned problems, small relative errors in the input data x can cause large relative errors in the results $y = \varphi(x)$.

The above concept of condition number suffers from the fact that it is meaningful only for nonzero y_i, x_j. Moreover, it is impractical for many purposes, since the condition of φ is described by mn numbers. For these reasons, the conditions of special classes of problems are frequently defined in a more convenient fashion. In linear algebra, for example, it is customary to call numbers c condition numbers if, in conjunction with a suitable norm $\|\cdot\|$,

$$\frac{\|\varphi(\tilde{x}) - \varphi(x)\|}{\|\varphi(x)\|} \leqslant c \frac{\|\tilde{x} - x\|}{\|x\|}$$

(see Section 4.4).

EXAMPLE 3. For $y = \varphi(a, b, c) := a + b + c$, (1.3.4) gives

$$\varepsilon_y \doteq \frac{a}{a+b+c}\varepsilon_a + \frac{b}{a+b+c}\varepsilon_b + \frac{c}{a+b+c}\varepsilon_c.$$

The problem is well conditioned if every summand a, b, c is small compared to $a + b + c$.

EXAMPLE 4. Let $y = \varphi(p, q) := -p + \sqrt{p^2 + q}$. Then

$$\frac{\partial\varphi}{\partial p} = -1 + \frac{p}{\sqrt{p^2+q}} = \frac{-y}{\sqrt{p^2+q}}, \qquad \frac{\partial\varphi}{\partial q} = \frac{1}{2\sqrt{p^2+q}},$$

so that

$$\varepsilon_y \doteq \frac{-p}{\sqrt{p^2+q}}\varepsilon_p + \frac{q}{2y\sqrt{p^2+q}}\varepsilon_q = -\frac{p}{\sqrt{p^2+q}}\varepsilon_p + \frac{p+\sqrt{p^2+q}}{2\sqrt{p^2+q}}\varepsilon_q.$$

Since

$$\left|\frac{p}{\sqrt{p^2+q}}\right| \leqslant 1, \quad \left|\frac{p+\sqrt{p^2+q}}{2\sqrt{p^2+q}}\right| \leqslant 1 \qquad \text{for } q > 0,$$

φ is well conditioned if $q > 0$, and badly conditioned if $q \approx -p^2$.

For the arithmetic operations (1.3.4) specializes to $(x \neq 0,\ y \neq 0)$

(1.3.5a) $\varphi(x, y) := x \cdot y$: $\quad \varepsilon_{xy} \doteq \varepsilon_x + \varepsilon_y$

(1.3.5b) $\varphi(x, y) := x \,/\, y$: $\quad \varepsilon_{x/y} \doteq \varepsilon_x - \varepsilon_y$

(1.3.5c) $\varphi(x, y) := x \pm y$: $\quad \varepsilon_{x\pm y} = \dfrac{x}{x\pm y}\varepsilon_x \pm \dfrac{y}{x\pm y}\varepsilon_y \quad$ if $x \pm y \neq 0$.

(1.3.5d) $\varphi(x) := \sqrt{x}$: $\quad \varepsilon_{\sqrt{x}} \doteq \frac{1}{2}\varepsilon_x$

It follows that the multiplication, division, and square root are not dangerous: The relative errors of the operands don't propagate strongly into the result. This is also the case for the addition, provided the operands x and y have the same sign. Indeed, the condition numbers $x/(x + y)$, $y/(x + y)$ then lie between 0 and 1, and they add up to 1, whence

$$|\varepsilon_{x+y}| \leqslant \max\{|\varepsilon_x|, |\varepsilon_y|\}.$$

If one operand is small compared to the other, but carries a large relative error, the result $x + y$ will still have a small relative error so long as the other operand has only a small relative error: *error damping* results. If, however, two operands of different sign are to be added, then at least one of the factors

$$\left|\frac{x}{x+y}\right|, \quad \left|\frac{y}{x+y}\right|$$

is bigger than 1, and at least one of the relative errors ε_x, ε_y will be *amplified*. This amplification is drastic if $x \approx -y$ holds and therefore *cancellation* occurs.

We will now employ the formula (1.3.3) to describe the propagation of roundoff errors for a given algorithm. An algorithm for computing the function $\varphi: D \to \mathbb{R}^m$, $D \subseteq \mathbb{R}^n$, for a given $x = (x_1, \ldots, x_n)^T \in D$ corresponds to a decomposition of the map φ into elementary maps $\varphi^{(i)}$ [see (1.3.1)], and leads from $x^{(0)} := x$ via a chain of intermediate results

(1.3.6) $\quad x = x^{(0)} \to \varphi^{(0)}(x^{(0)}) = x^{(1)} \to \cdots \to \varphi^{(r)}(x^{(r)}) = x^{(r+1)} = y$

to the result y. Again we assume that every $\varphi^{(i)}$ is continuously differentiable on D_i.

Now let us denote by $\psi^{(i)}$ the "remainder map"

$$\psi^{(i)} = \varphi^{(r)} \circ \varphi^{(r-1)} \circ \cdots \circ \varphi^{(i)}: D_i \to \mathbb{R}^m, \qquad i = 0, 1, 2, \ldots, r.$$

Then $\psi^{(0)} \equiv \varphi$. $D\varphi^{(i)}$ and $D\psi^{(i)}$ are the Jacobian matrices of the maps $\varphi^{(i)}$ and $\psi^{(i)}$. Since Jacobian matrices are multiplicative with respect to function composition,

$$D(f \circ g)(x) = Df(g(x)) \cdot Dg(x),$$

we note for further reference that

(1.3.7a) $$D\varphi(x) = D\varphi^{(r)}(x^{(r)}) \cdot D\varphi^{(r-1)}(x^{(r-1)}) \cdot \cdots \cdot D\varphi^{(0)}(x),$$

(1.3.7b) $$D\psi^{(i)}(x^{(i)}) = D\varphi^{(r)}(x^{(r)}) \cdot D\varphi^{(r-1)}(x^{(r-1)}) \cdot \cdots \cdot D\varphi^{(i)}(x^{(i)}),$$

$$i = 0, 1, \ldots, r.$$

With floating-point arithmetic, input and roundoff errors will perturb the intermediate (exact) results $x^{(i)}$ so that approximate values $\tilde{x}^{(i)}$ with $\tilde{x}^{(i+1)} = \mathrm{fl}(\varphi^{(i)}(\tilde{x}^{(i)}))$ will be obtained instead. For the absolute errors $\Delta x^{(i)} = \tilde{x}^{(i)} - x^{(i)}$,

(1.3.8) $$\Delta x^{(i+1)} = [\mathrm{fl}(\varphi^{(i)}(\tilde{x}^{(i)})) - \varphi^{(i)}(\tilde{x}^{(i)})] + [\varphi^{(i)}(\tilde{x}^{(i)}) - \varphi^{(i)}(x^{(i)})].$$

By (1.3.3) (disregarding higher-order error terms),

(1.3.9) $$\varphi^{(i)}(\tilde{x}^{(i)}) - \varphi^{(i)}(x^{(i)}) \doteq D\varphi^{(i)}(x^{(i)})\, \Delta x^{(i)}.$$

If $\varphi^{(i)}$ is an elementary map, or if it involves only independent elementary operations, the floating-point evaluation of $\varphi^{(i)}$ will yield the rounding of the exact value:

(1.3.10) $$\mathrm{fl}(\varphi^{(i)}(u)) = \mathrm{rd}(\varphi^{(i)}(u)).$$

Note, in this context, that the map $\varphi^{(i)}: D_i \to D_{i+1} \subseteq \mathbb{R}^{n_{i+1}}$ is actually a vector of component functions $\varphi_j^{(i)}: D_i \to \mathbb{R}$,

$$\varphi^{(i)}(u) = \begin{bmatrix} \varphi_1^{(i)}(u) \\ \vdots \\ \varphi_{n_i+1}^{(i)}(u) \end{bmatrix}.$$

Thus (1.3.10) must be interpreted componentwise:

(1.3.11) $$\begin{gathered} \mathrm{fl}(\varphi_j^{(i)}(u)) = \mathrm{rd}(\varphi_j^{(i)}(u)) = (1 + \varepsilon_j)\varphi_j^{(i)}(u), \\ |\varepsilon_j| \leqslant \mathrm{eps}, \qquad j = 1, 2, \ldots, n_{i+1}. \end{gathered}$$

Here ε_j is the new relative roundoff error generated during the calculation of the jth component of $\varphi^{(i)}$ in floating-point arithmetic. Plainly, (1.3.10) can also be written in the form

$$\mathrm{fl}(\varphi^{(i)}(u)) = (I + E_{i+1}) \cdot \varphi^{(i)}(u)$$

with the identity matrix I and the diagonal error matrix

$$E_{i+1} := \begin{bmatrix} \varepsilon_1 & & & 0 \\ & \varepsilon_2 & & \\ 0 & & \ddots & \\ & & & \varepsilon_{n_i+1} \end{bmatrix}, \qquad |\varepsilon_j| \leqslant \text{eps}.$$

This yields the following expression for the first bracket in (1.3.8):

$$\text{fl}(\varphi^{(i)}(\tilde{x}^{(i)})) - \varphi^{(i)}(\tilde{x}^{(i)}) = E_{i+1} \cdot \varphi^{(i)}(\tilde{x}^{(i)}).$$

Furthermore $E_{i+1} \cdot \varphi^{(i)}(\tilde{x}^{(i)}) \doteq E_{i+1} \cdot \varphi^{(i)}(x^{(i)})$, since the error terms by which $\varphi^{(i)}(\tilde{x}^{(i)})$ and $\varphi^{(i)}(x^{(i)})$ differ are multiplied by the error terms on the diagonal of E_{i+1}, giving rise to higher-order error terms. Therefore

$$\text{(1.3.12)} \qquad fl(\varphi^{(i)}(\tilde{x}^{(i)})) - \varphi^{(i)}(\tilde{x}^{(i)}) \doteq E_{i+1} \cdot \varphi^{(i)}(x^{(i)}) = E_{i+1} \cdot x^{(i+1)} =: \alpha_{i+1}.$$

The quantity α_{i+1} can be interpreted as the absolute roundoff error newly created when $\varphi^{(i)}$ is evaluated in floating-point arithmetic, and the diagonal elements of E_{i+1} can be similarly interpreted as the corresponding relative roundoff errors. Thus by (1.3.8), (1.3.9) and (1.3.12), $\Delta x^{(i+1)}$ can be expressed in first-order approximation as follows

$$\Delta x^{(i+1)} \doteq \alpha_{i+1} + D\varphi^{(i)}(x^{(i)}) \cdot \Delta x^{(i)} = E_{i+1} \cdot x^{(i+1)} + D\varphi^{(i)}(x^{(i)}) \cdot \Delta x^{(i)},$$
$$i \geqslant 0, \qquad \Delta x^{(0)} := \Delta x.$$

Consequently

$$\Delta x^{(1)} \doteq D\varphi^{(0)}(x)\, \Delta x + \alpha_1,$$
$$\Delta x^{(2)} \doteq D\varphi^{(1)}(x^{(1)})[D\varphi^{(0)}(x) \cdot \Delta x + \alpha_1] + \alpha_2,$$
$$\vdots$$
$$\Delta y = \Delta x^{(r+1)} \doteq D\varphi^{(r)} \ldots D\varphi^{(0)} \cdot \Delta x + D\varphi^{(r)} \ldots D\varphi^{(1)} \cdot \alpha_1 + \cdots + \alpha_{r+1}.$$

In view of (1.3.7), we finally arrive at the following formulas which describe the effect of the input errors Δx and the roundoff errors α_i on the result $y = x^{(r+1)} = \varphi(x)$:

$$\begin{aligned} \Delta y &\doteq D\varphi(x) \cdot \Delta x + D\psi^{(1)}(x^{(1)}) \cdot \alpha_1 + \cdots + D\psi^{(r)}(x^{(r)}) \cdot \alpha_r + \alpha_{r+1} \\ &= D\varphi(x) \cdot \Delta x + D\psi^{(1)}(x^{(1)}) \cdot E_1 x^{(1)} + \cdots + D\psi^{(r)}(x^{(r)}) \cdot E_r \cdot x^{(r)} \qquad \text{(1.3.13)} \\ &\quad + E_{r+1} \cdot y. \end{aligned}$$

It is therefore the size of the Jacobian matrix $D\psi^{(i)}$ of the remainder map $\psi^{(i)}$ which is critical for the effect of the intermediate roundoff errors α_i or E_i on the final result.

EXAMPLE 5. For the two algorithms for computing $y = \varphi(a, b) = a^2 - b^2$ given in Example 2 we have for *Algorithm 1:*

$$x = x^{(0)} = \begin{bmatrix} a \\ b \end{bmatrix}, \qquad x^{(1)} = \begin{bmatrix} a^2 \\ b \end{bmatrix}, \qquad x^{(2)} = \begin{bmatrix} a^2 \\ b^2 \end{bmatrix}, \qquad x^{(3)} = y = a^2 - b^2,$$

$$\psi^{(1)}(u, v) = u - v^2, \qquad \psi^{(2)}(u, v) = u - v,$$

$$D\varphi(x) = (2a, \ -2b),$$

$$D\psi^{(1)}(x^{(1)}) = (1, \ -2b), \qquad D\psi^{(2)}(x^{(2)}) = (1, \ -1)$$

$$\alpha_1 = \begin{bmatrix} \varepsilon_1 a^2 \\ 0 \end{bmatrix}, \quad E_1 = \begin{bmatrix} \varepsilon_1 & 0 \\ 0 & 0 \end{bmatrix}, \qquad \text{since } \mathrm{fl}(\varphi^{(0)}(x^{(0)})) - \varphi^{(0)}(x^{(0)}) = \begin{bmatrix} a \times^* a \\ b \end{bmatrix} - \begin{bmatrix} a^2 \\ b \end{bmatrix}$$

$$\alpha_2 = \begin{bmatrix} 0 \\ \varepsilon_2 b^2 \end{bmatrix}, \quad E_2 = \begin{bmatrix} 0 & 0 \\ 0 & \varepsilon_2 \end{bmatrix}$$

$$\alpha_3 = \varepsilon_3(a^2 - b^2), \qquad |\varepsilon_i| \leqslant \text{eps} \quad \text{for } i = 1, 2, 3.$$

From (1.1.13) with $\Delta x = \begin{bmatrix} \Delta a \\ \Delta b \end{bmatrix}$,

(1.3.14) $$\Delta y \doteq 2a\,\Delta a - 2b\,\Delta b + a^2\varepsilon_1 - b^2\varepsilon_2 + (a^2 - b^2)\varepsilon_3.$$

Analogously for Algorithm 2:

$$x = x^{(0)} = \begin{bmatrix} a \\ b \end{bmatrix}, \qquad x^{(1)} = \begin{bmatrix} a + b \\ a - b \end{bmatrix}, \qquad x^{(2)} = y = a^2 - b^2,$$

$$\psi^{(1)}(u, v) = u \cdot v,$$

$$D\varphi(x) = (2a, \ -2b), \qquad D\psi^{(1)}(x^{(1)}) = (a - b, \ a + b),$$

$$\alpha_1 = \begin{bmatrix} \varepsilon_1(a + b) \\ \varepsilon_2(a - b) \end{bmatrix}, \quad \alpha_2 = \varepsilon_3(a^2 - b^2), \quad E_1 = \begin{bmatrix} \varepsilon_1 & 0 \\ 0 & \varepsilon_2 \end{bmatrix}, \qquad |\varepsilon_i| \leqslant \text{eps},$$

and therefore (1.1.13) again yields

(1.3.15) $$\Delta y \doteq 2a\,\Delta a - 2b\,\Delta b + (a^2 - b^2)(\varepsilon_1 + \varepsilon_2 + \varepsilon_3).$$

If one selects a different algorithm for calculating the same result $\varphi(x)$ (in other words, a different decomposition of φ into elementary maps), then $D\varphi$ remains unchanged; the Jacobian matrices $D\psi^{(i)}$, which measure the propagation of roundoff, will be different, however, and so will be the total effect of rounding,

(1.3.16) $$D\psi^{(1)}\alpha_1 + \cdots + D\psi^{(r)}\alpha_r + \alpha_{r+1}.$$

An algorithm is called *numerically more trustworthy* than another algorithm for calculating $\varphi(x)$ if, for a given set of data x, the total effect of rounding, (1.3.16), is less for the first algorithm than for the second one.

EXAMPLE 6. The total effect of rounding using Algorithm 1 in Example 2 is, by (1.3.14),

(1.3.17) $$|a^2\varepsilon_1 - b^2\varepsilon_2 + (a^2 - b^2)\varepsilon_3| \leqslant (a^2 + b^2 + |a^2 - b^2|)\text{eps},$$

and that of Algorithm 2, by (1.3.15),

(1.3.18) $$|(a^2 - b^2)(\varepsilon_1 + \varepsilon_2 + \varepsilon_3)| \leqslant 3|a^2 - b^2|\text{eps}.$$

Algorithm 2 is numerically more trustworthy than algorithm 1 whenever $\frac{1}{3} < |a/b|^2 < 3$; otherwise algorithm 1 is more trustworthy. This follows from the equivalence of the two relations $\frac{1}{3} \leqslant |a/b|^2 \leqslant 3$ and $3|a^2 - b^2| \leqslant a^2 + b^2 + |a^2 - b^2|$.

For $a := 0.3237$, $b := 0.3134$, using four places ($t = 4$), we obtain the following results.

Algorithm 1: $a \times^* a = 0.1048$, $b \times^* b = 0.9822_{10}-1$,
$(a \times^* a) -^* (b \times^* b) = 0.6580_{10} - 2$.

Algorithm 2: $a +^* b = 0.6371$, $a -^* b = 0.1030_{10}-1$,
$(a +^* b) \times^* (a -^* b) = 0.6562_{10}-2$.

Exact result: $a^2 - b^2 = 0.656213_{10}-2$.

In the error propagation formula (1.3.13), the last term admits the following bound:[1]

$$|E_{r+1}y| \leqslant |y|\text{eps},$$

no matter what algorithm had been used for computing $y = \varphi(x)$. Hence an error Δy of magnitude $|y|\text{eps}$ has to be expected for any algorithm. Note, moreover, when using mantissas of t places, that the rounding of the input data $x = (x_1, \ldots, x_n)^T$ will cause an input error $\Delta^{(0)}x$ with

$$|\Delta^{(0)}x| \leqslant |x|\text{eps},$$

unless the input data are already machine numbers and therefore representable exactly. Since the latter cannot be counted on, any algorithm for computing $y = \varphi(x)$ will have to be assumed to incur the error $D\varphi(x) \cdot \Delta^{(0)}x$, so that altogether for *every* such algorithm an error of magnitude

(1.3.19) $$\Delta^{(0)}y := [|D\varphi(x)| \cdot |x| + |y|]\text{eps}$$

must be expected. We call $\Delta^{(0)}y$ the *inherent error* of y. Since this error will have to be reckoned with in any case, it would be unreasonable to ask that the influence of intermediate roundoff errors on the final result be considerably smaller than $\Delta^{(0)}y$. We therefore call roundoff errors α_i or E_i *harmless* if their contribution in (1.3.13) towards the total error Δy is of at most the same order of magnitude as the inherent error $\Delta^{(0)}y$ from (1.3.19):

$$|D\psi^{(i)}(x^{(i)}) \cdot \alpha_i| = |D\psi^{(i)}(x^{(i)}) \cdot E_i x^{(i)}| \approx \Delta^{(0)}y.$$

If all roundoff errors of an algorithm are harmless, then the algorithm is said to be *well behaved* or *numerically stable*. This particular notion of numerical stability has been promoted by Bauer et al. (1965); Bauer also uses the term

[1] The absolute values of vectors and matrices are to be understood componentwise, e.g., $|y| = (|y_1|, \ldots, |y_m|)^T$.

benign (1974). Finding numerically stable algorithms is a primary task of numerical analysis.

EXAMPLE 7. Both algorithms of Example 2 are numerically stable. Indeed, the inherent error $\Delta^{(0)}y$ is as follows:

$$\Delta^{(0)}y = \left([2|a|, 2|b|] \cdot \begin{bmatrix} |a| \\ |b| \end{bmatrix} + |a^2 - b^2|\right)\text{eps} = (2(a^2 + b^2) + |a^2 - b^2|)\text{eps}.$$

Comparing this with (1.3.17) and (1.3.18) even shows that the total roundoff error of each of the two algorithms cannot exceed $\Delta^{(0)}y$.

Let us pause to review our usage of terms. Numerical trustworthiness, which we will use as a comparative term, relates to the roundoff errors associated with two or more algorithms for the same problem. Numerical stability, which we will use as an absolute term, relates to the inherent error and the corresponding harmlessness of the roundoff errors associated with a single algorithm. Thus one algorithm may be numerically more trustworthy than another, yet neither may be numerically stable. If both are numerically stable, the numerically more trustworthy algorithm is to be preferred. We attach the qualifier "numerically" because of the widespread use of the term "stable" without that qualifier in other contexts such as the terminology of differential equations, economic models, and linear multistep iterations, where it has different meanings. Further illustrations of the concepts which we have introduced above will be found in the next section.

A general technique for establishing the numerical stability of an algorithm, the so-called *backward analysis*, has been introduced by Wilkinson (1960) for the purpose of examining algorithms in linear algebra. He tries to show that the floating-point result $\tilde{y} = y + \Delta y$ of an algorithm for computing $y = \varphi(x)$ may be written in the form $\tilde{y} = \varphi(x + \Delta x)$, that is, as the result of an exact calculation based on perturbed input data $x + \Delta x$. If Δx turns out to have the same order of magnitude as $|\Delta^{(0)}x| \leqslant |x|\text{eps}$, then the algorithm is indeed numerically stable.

Bauer (1974) associates graphs with algorithms in order to illuminate their error patterns. For instance, Algorithms 1 and 2 of example 2 give rise to the graphs in Figure 1. The nodes of these graphs correspond to the intermediate results. Node i is linked to node j by a directed arc if the intermediate result corresponding to node i is an operand of the elementary operation which produces the result corresponding to node j. At each node there arises a new relative roundoff error, which is written next to its node. Amplification factors for the relative errors are similarly associated with, and written next to, the arcs of the graph. Tracing through the graph of Algorithm 1, for instance, one obtains the following error relations:

$$\varepsilon_{\eta_1} = 1 \cdot \varepsilon_a + 1 \cdot \varepsilon_a + \varepsilon_1, \qquad \varepsilon_{\eta_2} = 1 \cdot \varepsilon_b + 1 \cdot \varepsilon_b + \varepsilon_2,$$

$$\varepsilon_y = \frac{\eta_1}{\eta_1 - \eta_2} \cdot \varepsilon_{\eta_1} - \frac{\eta_2}{\eta_1 - \eta_2} \cdot \varepsilon_{\eta_2} + \varepsilon_3.$$

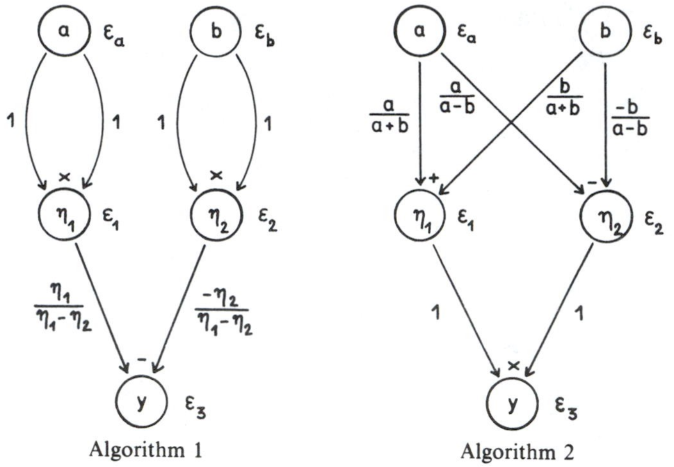

Figure 1 Graphs Representing Algorithms and Their Error Propagation.

To find the factor by which to multiply the roundoff error at node i in order to get its contribution to the error at node j, one multiplies all arc factors for each directed path from i to j and adds these products. The graph of Algorithm 2 thus indicates that the input error ε_a contributes

$$\left(\frac{a}{a+b}\cdot 1+\frac{a}{a-b}\cdot 1\right)\cdot \varepsilon_a$$

to the error ε_y.

1.4 Examples

Example 1. This example follows up Example 4 of the previous section: given $p > 0$, $q > 0$, $p \gg q$, determine the root

$$y = -p + \sqrt{p^2 + q}$$

with smallest absolute value of the quadratic equation

$$y^2 + 2py - q = 0.$$

Input data: p, q. Result: $y = \varphi(p, q) = -p + \sqrt{p^2 + q}$.

The problem was seen to be well conditioned for $p > 0$, $q > 0$. It was also shown that the relative input errors ε_p, ε_q make the following contribution to the relative error of the result $y = \varphi(p, q)$:

$$\frac{-p}{\sqrt{p^2+q}}\varepsilon_p + \frac{q}{2y\sqrt{p^2+q}}\varepsilon_q = \frac{-p}{\sqrt{p^2+q}}\varepsilon_p + \frac{p+\sqrt{p^2+q}}{2\sqrt{p^2+q}}\varepsilon_q.$$

Since

$$\left| \frac{p}{\sqrt{p^2 + q}} \right| \leqslant 1, \qquad \left| \frac{p + \sqrt{p^2 + q}}{2\sqrt{p^2 + q}} \right| \leqslant 1,$$

the inherent error $\Delta^{(0)} y$ satisfies

$$\text{eps} \leqslant \varepsilon_y^{(0)} := \frac{\Delta^{(0)} y}{y} \leqslant 3 \text{ eps}.$$

We will now consider two algorithms for computing $y = \varphi(p, q)$.

$$\textit{Algorithm 1:} \quad \begin{aligned} s &:= p^2, \\ t &:= s + q, \\ u &:= \sqrt{t}, \\ y &:= -p + u. \end{aligned}$$

Obviously, $p \gg q$ causes cancellation when $y := -p + u$ is evaluated, and it must therefore be expected that the roundoff error

$$\Delta u := \varepsilon \sqrt{t} = \varepsilon \sqrt{p^2 + q},$$

generated during the floating-point calculation of the square root

$$\mathrm{fl}(\sqrt{t}) = \sqrt{t}(1 + \varepsilon), \qquad |\varepsilon| \leqslant \text{eps},$$

will be greatly amplified. Indeed, the above error contributes the following term to the error of y:

$$\begin{aligned} \frac{1}{y} \Delta u &= \frac{\sqrt{p^2 + q}}{-p + \sqrt{p^2 + q}} \cdot \varepsilon \\ &= \frac{1}{q} (p\sqrt{p^2 + q} + p^2 + q)\varepsilon = k \cdot \varepsilon. \end{aligned}$$

Since $p, q > 0$, the amplification factor k admits the following lower bound:

$$k > \frac{2p^2}{q} > 0,$$

which is large, since $p \gg q$ by hypothesis. Therefore, the proposed algorithm is not numerically stable, because the influence of rounding $\sqrt{p^2 + q}$ alone exceeds that of the inherent error $\varepsilon_y^{(0)}$ by an order of magnitude.

$$\textit{Algorithm 2:} \quad \begin{aligned} s &:= p^2, \\ t &:= s + q, \\ u &:= \sqrt{t}, \\ v &:= p + u, \\ y &:= q/v. \end{aligned}$$

This algorithm does not cause cancellation when calculating $v := p + u$. The roundoff error $\Delta u = \varepsilon\sqrt{p^2 + q}$, which stems from rounding $\sqrt{p^2 + q}$, will be amplified according to the remainder map $\psi(u)$:

$$u \to p + u \to \frac{q}{p + u} = \psi(u).$$

Thus it contributes the following term to the relative error of y:

$$\begin{aligned} \frac{1}{y}\frac{\partial\psi}{\partial u} \cdot \Delta u &= \frac{-q}{y(p + u)^2} \cdot \Delta u \\ &= \frac{-q\sqrt{p^2 + q}}{(-p + \sqrt{p^2 + q})(p + \sqrt{p^2 + q})^2} \cdot \varepsilon \\ &= -\frac{\sqrt{p^2 + q}}{p + \sqrt{p^2 + q}} \cdot \varepsilon = k\varepsilon \end{aligned}$$

The amplification factor k remains small; indeed, $|k| < 1$, and Algorithm 2 is therefore numerically stable.

The following numerical results illustrate the difference between Algorithms 1 and 2. They were obtained using floating-point arithmetic of 40 binary mantissa places—about 13 decimal places—as will be the case in subsequent numerical examples.

$p = 1000$, $q = 0.018\ 000\ 000\ 081$.

Result y according to Algorithm 1: $0.900\ 030\ 136\ 108_{10}-5$,

Result y according to Algorithm 2: $0.899\ 999\ 999\ 999_{10}-5$,

Exact value of y: $0.900\ 000\ 000\ 000_{10}-5$.

EXAMPLE 2. For given fixed x and integer k, the value of $\cos kx$ may be computed recursively using for $m = 1, 2, \ldots, k - 1$ the formula

$$\cos(m + 1)x = 2 \cos x \cos mx - \cos(m - 1)x.$$

In this case, a trigonometric-function evaluation has to be carried out only once, to find $c = \cos x$. Now let $|x| \neq 0$ be a small number. The calculation of c causes a small roundoff error:

$$\tilde{c} = (1 + \varepsilon) \cos x, \qquad |\varepsilon| \leqslant \text{eps}.$$

How does this roundoff error affect the calculation of $\cos kx$?

$\cos kx$ can be expressed in terms of c: $\cos kx = \cos(k \arccos c) =: f(c)$. Since

$$\frac{df}{dc} = \frac{k \sin kx}{\sin x},$$

the error $\varepsilon \cos x$ of c causes, to first approximation, an absolute error

$$(1.4.1) \qquad \Delta \cos kx \doteq \varepsilon \frac{\cos x}{\sin x} k \sin kx = \varepsilon \cdot k \cot x \sin kx$$

in $\cos kx$.

On the other hand, the inherent error $\Delta^{(0)}c_k$ (1.3.19) of the result $c_k := \cos kx$ is

$$\Delta^{(0)}c_k = [k|x \sin kx| + |\cos kx|]\text{eps}.$$

Comparing this with (1.4.1) shows that $\Delta \cos kx$ may be considerably larger than $\Delta^{(0)}c_k$ for small $|x|$; hence the algorithm is not numerically stable.

EXAMPLE 3. For given x and a "large" positive integer k, the numbers $\cos kx$ and $\sin kx$ are to be computed recursively using

$$\cos mx := \cos x \cos(m-1)x - \sin x \sin(m-1)x,$$
$$\sin mx := \sin x \cos(m-1)x + \cos x \sin(m-1)x, \qquad m = 1, 2, \ldots, k$$

How do small errors $\varepsilon_c \cos x$, $\varepsilon_s \sin x$ in the calculation of $\cos x$, $\sin x$ affect the final results $\cos kx$, $\sin kx$? Abbreviating $c_m := \cos mx$, $s_m := \sin mx$, $c := \cos x$, $s := \sin x$, and putting

$$U := \begin{bmatrix} c & -s \\ s & c \end{bmatrix},$$

we have

$$\begin{bmatrix} c_m \\ s_m \end{bmatrix} = U \begin{bmatrix} c_{m-1} \\ s_{m-1} \end{bmatrix}, \qquad m = 1, \ldots, k$$

Here U is a unitary matrix, which corresponds to a rotation by the angle x. Repeated application of the formula above gives

$$\begin{bmatrix} c_k \\ s_k \end{bmatrix} = U^k \cdot \begin{bmatrix} c_o \\ s_o \end{bmatrix} = U^k \cdot \begin{bmatrix} 1 \\ 0 \end{bmatrix}.$$

Now

$$\frac{\partial U}{\partial c} = \begin{bmatrix} 1 & 0 \\ 0 & 1 \end{bmatrix}, \qquad \frac{\partial U}{\partial s} = \begin{bmatrix} 0 & -1 \\ 1 & 0 \end{bmatrix} =: A,$$

and therefore

$$\frac{\partial}{\partial c} U^k = kU^{k-1},$$

$$\frac{\partial}{\partial s} U^k = AU^{k-1} + UAU^{k-2} + \cdots + U^{k-1}A$$
$$= kAU^{k-1},$$

because A commutes with U. Since U describes a rotation in $\mathbb{R}^2$ by the angle x,

$$\frac{\partial}{\partial c} U^k = k \begin{bmatrix} \cos(k-1)x & -\sin(k-1)x \\ \sin(k-1)x & \cos(k-1)x \end{bmatrix},$$

$$\frac{\partial}{\partial s} U^k = k \begin{bmatrix} -\sin(k-1)x & -\cos(k-1)x \\ \cos(k-1)x & -\sin(k-1)x \end{bmatrix}.$$

The relative errors ε_c, ε_s of $c = \cos x$, $s = \sin x$ effect the following absolute errors of $\cos kx$, $\sin kx$:

$$
\begin{aligned}
\begin{bmatrix}\Delta c_k\\ \Delta s_k\end{bmatrix} &\doteq \left[\frac{\partial}{\partial c}U^k\right]\begin{bmatrix}1\\0\end{bmatrix}\cdot\varepsilon_c\cos x + \left[\frac{\partial}{\partial s}U^k\right]\begin{bmatrix}1\\0\end{bmatrix}\cdot\varepsilon_s\sin x \\
&= \varepsilon_c k\cos x\begin{bmatrix}\cos(k-1)x\\ \sin(k-1)x\end{bmatrix} + \varepsilon_s k\sin x\begin{bmatrix}-\sin(k-1)x\\ \cos(k-1)x\end{bmatrix}.
\end{aligned}
\tag{1.4.2}
$$

The inherent errors $\Delta^{(0)}c_k$ and $\Delta^{(0)}s_k$ of $c_k = \cos kx$ and $s_k = \sin kx$, respectively, are given by

$$
\begin{aligned}
\Delta^{(0)}c_k &= [k|x\sin kx| + |\cos kx|]\text{eps},\\
\Delta^{(0)}s_k &= [k|x\cos kx| + |\sin kx|]\text{eps}.
\end{aligned}
\tag{1.4.3}
$$

Comparison of (1.4.2) and (1.4.3) reveals that for big k and $|kx| \approx 1$ the influence of the roundoff error ε_c is considerably bigger than the inherent errors, while the round-off error ε_s is harmless. The algorithm is not numerically stable, albeit numerically more trustworthy than the algorithm of Example 2 as far as the computation of c_k alone is concerned.

EXAMPLE 4. For small $|x|$, the recursive calculation of

$$c_m = \cos mx, \quad s_m = \sin mx, \qquad m = 1, 2, \ldots,$$

based on

$$
\begin{aligned}
\cos(m+1)x &= \cos x\cos mx - \sin x\sin mx,\\
\sin(m+1)x &= \sin x\cos mx + \cos x\sin mx,
\end{aligned}
$$

as in Example 3, may be further improved numerically. To this end, we express the differences ds_{m+1} and dc_{m+1} of subsequent sine and cosine values as follows:

$$
\begin{aligned}
dc_{m+1} &:= \cos(m+1)x - \cos mx\\
&= 2(\cos x - 1)\cos mx - \sin x\sin mx - \cos x\cos mx + \cos mx\\
&= -4\left(\sin^2\frac{x}{2}\right)\cos mx + [\cos mx - \cos(m-1)x]\\
ds_{m+1} &:= \sin(m+1)x - \sin mx\\
&= 2(\cos x - 1)\sin mx + \sin x\cos mx - \cos x\sin mx + \sin mx\\
&= -4\left(\sin^2\frac{x}{2}\right)\sin mx + [\sin mx - \sin(m-1)x].
\end{aligned}
$$

This leads to a more elaborate recursive algorithm for computing c_k, s_k in the case $x > 0$:

$$
\begin{aligned}
dc_1 &:= -2\sin^2\frac{x}{2}, \qquad t := 2\,dc_1,\\
ds_1 &:= \sqrt{-dc_1(2 + dc_1)},\\
s_0 &:= 0, \qquad c_0 := 1,
\end{aligned}
$$

and for $m := 1, 2, \ldots, k$:

$$c_m := c_{m-1} + dc_m, \qquad dc_{m+1} := t \cdot c_m + dc_m,$$

$$s_m := s_{m-1} + ds_m, \qquad ds_{m+1} := t \cdot s_m + ds_m.$$

For the error analysis, note that c_k and s_k are functions of $s = \sin(x/2)$:

$$c_k = \cos(2k \arcsin s) =: \varphi_1(s),$$

$$s_k = \sin(2k \arcsin s) =: \varphi_2(s).$$

An error $\Delta s = \varepsilon_s \sin(x/2)$ in the calculation of s therefore causes—to a first-order approximation—the following errors in c_k:

$$\frac{\partial \varphi_1}{\partial s} \varepsilon_s \sin \frac{x}{2} = \varepsilon_s \cdot \frac{-2k \sin kx}{\cos(x/2)} \sin \frac{x}{2}$$

$$= -2k \tan \frac{x}{2} \sin kx \cdot \varepsilon_s,$$

and in s_k:

$$\frac{\partial \varphi_2}{\partial s} \varepsilon_s \sin \frac{x}{2} = 2k \tan \frac{x}{2} \cos kx \cdot \varepsilon_s.$$

Comparison with the inherent errors (1.4.3) shows these errors to be harmless for small $|x|$. The algorithm is then numerically stable, at least as far as the influence of the roundoff error ε_s is concerned.

Again we illustrate our analytical considerations with some numerical results. Let $x = 0.001$, $k = 1000$.

Algorithm	Result for $\cos kx$	Relative error
Example 2	0.540 302 121 124	$-0.34_{10}-6$
Example 3	0.540 302 305 776	$-0.17_{10}-9$
Example 4	0.540 302 305 865	$-0.58_{10}-11$
Exact value	0.540 302 305 868 140...	

Example 5. We will derive some results which will be useful for the analysis of algorithms for solving linear equations in Section 4.5. Given the quantities $c, a_1, \ldots, a_n, b_1, \ldots, b_{n-1}$ with $a_n \neq 0$, we want to find the solution β_n of the linear equation

(1.4.4) $$c - a_1 b_1 - \cdots - a_{n-1} b_{n-1} - a_n \beta_n = 0.$$

Floating-point arithmetic yields the approximate solution

(1.4.5) $$b_n = \mathrm{fl}\left(\frac{c - a_1 b_1 - \cdots - a_{n-1} b_{n-1}}{a_n}\right)$$

as follows:

$$s_0 := c;$$

(1.4.6) for $j = 1, 2, \ldots, n-1$,

$$s_j := \mathrm{fl}(s_{j-1} - a_j b_j) = (s_{j-1} - a_j b_j(1 + \mu_j))(1 + \alpha_j);$$

$$b_n := \mathrm{fl}(s_{n-1}/a_n) = (1 + \delta)s_{n-1}/a_n,$$

with $|\mu_j|$, $|\alpha_j|$, $|\delta| \leqslant \text{eps}$. If $a_n = 1$, as is frequently the case in applications, then $\delta = 0$, since $b_n := s_{n-1}$.

We will now describe two useful estimates for the residual

$$r := c - a_1 b_1 - \cdots - a_n b_n.$$

From (1.4.6) follow the equations

$$s_0 - c = 0,$$

$$s_j - (s_{j-1} - a_j b_j) = s_j - \left(\frac{s_j}{1 + \alpha_j} + a_j b_j \mu_j\right)$$

$$= s_j \frac{\alpha_j}{1 + \alpha_j} - a_j b_j \mu_j, \qquad j = 1, 2, \ldots, n-1,$$

$$a_n b_n - s_{n-1} = \delta\, s_{n-1}.$$

Summing these equations yields

$$r = c - \sum_{i=1}^{n} a_i b_i = \sum_{j=1}^{n-1} \left(-s_j \frac{\alpha_j}{1 + \alpha_j} + a_j b_j \mu_j\right) - \delta\, s_{n-1}$$

and thereby the first one of the promised estimates:

$$\text{(1.4.7)} \qquad |r| \leqslant \frac{\text{eps}}{1 - \text{eps}} [\delta' \cdot |s_{n-1}| + \sum_{j=1}^{n-1} (|s_j| + |a_j b_j|)],$$

$$\delta' := \begin{cases} 0 & \text{if } a_n = 1, \\ 1 & \text{otherwise.} \end{cases}$$

The second estimate is cruder than (1.4.7). (1.4.6) gives

$$\text{(1.4.8)} \qquad b_n = \left[c \prod_{k=1}^{n-1} (1 + \alpha_k) - \sum_{j=1}^{n-1} a_j b_j (1 + \mu_j) \prod_{k=j}^{n-1} (1 + \alpha_k)\right] \frac{1 + \delta}{a_n},$$

which can be solved for c:

$$\text{(1.4.9)} \qquad c = \sum_{j=1}^{n-1} a_j b_j (1 + \mu_j) \prod_{k=1}^{j-1} (1 + \alpha_k)^{-1} + a_n b_n (1 + \delta)^{-1} \prod_{k=1}^{n-1} (1 + \alpha_k)^{-1}.$$

A simple induction argument over m shows that

$$(1 + \sigma) = \prod_{k=1}^{m} (1 + \sigma_k)^{\pm 1}, \qquad |\sigma_k| \leqslant \text{eps}, \qquad m \cdot \text{eps} < 1$$

implies

$$|\sigma| \leqslant \frac{m \cdot \text{eps}}{1 - m \cdot \text{eps}}.$$

In view of (1.4.9) this ensures the existence of quantities ε_j with

$$c = \sum_{j=1}^{n-1} a_j b_j (1 + j \cdot \varepsilon_j) + a_n b_n (1 + (n - 1 + \delta')\varepsilon_n), \tag{1.4.10}$$

$$|\varepsilon_j| \leqslant \frac{\text{eps}}{1 - n \cdot \text{eps}}, \qquad \delta' := \begin{cases} 0 & \text{if } a_n = 1, \\ 1 & \text{otherwise.} \end{cases}$$

For $r = c - a_1 b_1 - a_2 b_2 - \cdots - a_n b_n$ we have consequently

$$|r| \leqslant \frac{\text{eps}}{1 - n \cdot \text{eps}} \left[\sum_{j=1}^{n-1} j |a_j b_j| + (n - 1 + \delta') |a_n b_n| \right]. \tag{1.4.11}$$

In particular, (1.4.8) reveals the numerical stability of our algorithm for computing β_n. The roundoff error α_m contributes the amount

$$\frac{c - a_1 b_1 - a_2 b_2 - \cdots - a_m b_m}{a_n} \alpha_m .$$

to the absolute error in β_n. This, however, is at most equal to

$$\left| \frac{c \cdot \varepsilon_c - a_1 b_1 \varepsilon_{a_1} - \cdots - a_m b_m \varepsilon_{a_m}}{a_n} \right|$$

$$\leqslant \frac{\left(|c| + \sum_{i=1}^{m} |a_i b_i| \right) \text{eps}}{|a_n|},$$

which represents no more than the influence of the input errors ε_c and ε_{a_i} of c and a_i, $i = 1, \ldots, m$, respectively, provided $|\varepsilon_c|, |\varepsilon_{a_i}| \leqslant$ eps. The remaining roundoff errors μ_k and δ are similarly shown to be harmless.

The numerical stability of the above algorithm is often shown by interpreting (1.4.10) in the sense of backward analysis: The computed approximate solution b_n is the exact solution of the equation

$$c - \bar{a}_1 b_1 - \cdots - \bar{a}_n b_n = 0,$$

whose coefficients

$$\bar{a}_j := a_j (1 + j\varepsilon_j), \qquad 1 \leqslant j \leqslant n - 1,$$

$$\bar{a}_n := a_n (1 + (n - 1 + \delta')\varepsilon_n)$$

have been changed only slightly from their original values a_j. This kind of analysis, however, involves the difficulty of having to define how large n can be so that errors of the form $n\varepsilon$, $|\varepsilon| \leqslant$ eps can still be considered as being of the same order of magnitude as the machine precision eps.

1.5 Interval Arithmetic; Statistical Roundoff Estimation

The effect of a *few* roundoff errors can be quite readily estimated, to a first-order approximation, by the methods of Section 1.3. For a typical numerical method, however, the number of arithmetic operations, and consequently the number of individual roundoff errors, is very large, and the corresponding algorithm is too complicated to permit the estimation of the total effect of *all* roundoff errors in this fashion.

A technique known as *interval arithmetic* offers an approach to determining exact upper bounds for the absolute error of an algorithm, taking into account all roundoff and data errors. Interval arithmetic is based on the realization that the exact values for all real numbers $a \in R$ which either enter an algorithm or are computed as intermediate or final results are usually not known. At best one knows small intervals wich contain a. For this reason, the interval-arithmetic approach is to calculate systematically in terms of such intervals

$$\tilde{a} = [a', a''],$$

bounded by machine numbers $a', a'' \in A$, rather than in terms of single real numbers a. Each unknown number a is represented by an interval $\tilde{a} = [a', a'']$ with $a \in \tilde{a}$. The arithmetic operations $\circledcirc{\tau} \in \{\oplus, \ominus, \otimes, \oslash\}$ between intervals are defined so as to be *compatible* with the above interpretation. That is, $\tilde{c} = \tilde{a} \circledcirc{\tau} \tilde{b}$ is defined as an interval (as small as possible) satisfying

$$\tilde{c} \supset \{a \,\tau\, b \mid a \in \tilde{a} \text{ and } b \in \tilde{b}\}$$

and having machine number endpoints.

In the case of addition, for instance, this holds if $\oplus$ is defined as follows:

$$[c', c''] = [a', a''] \oplus [b', b''],$$

where

$$c' := \max\{\gamma' \in A \mid \gamma' \leqslant a' + b'\},$$

$$c'' := \min\{\gamma'' \in A \mid \gamma'' \geqslant a'' + b''\},$$

with A denoting again the set of machine numbers. In the case of multiplication $\otimes$, assuming, say, $a' > 0$, $b' > 0$,

$$[c', c''] = [a', a''] \otimes [b', b'']$$

can be defined by letting

$$c' := \max\{\gamma' \in A \mid \gamma' \leqslant a' \times b'\},$$

$$c'' := \min\{\gamma'' \in A \mid \gamma'' \geqslant a'' \times b''\}.$$

Replacing, in these and similar fashions, every quantity by an interval and every arithmetic operation by its corresponding interval operation—this is readily implemented on computers—we obtain interval algorithms which produce intervals guaranteed to contain the desired exact solutions. The data for these interval algorithms will be again intervals, chosen to allow for data errors.

It has been found, however, that an uncritical utilization of interval arithmetic techniques leads to error bounds which, while certainly reliable, are in most cases much too pessimistic. It is not enough to simply substitute interval operations for arithmetic operations without taking into account *how* the particular roundoff or data errors enter into the respective results. For example, it happens quite frequently that a certain roundoff error ε impairs some *intermediate* results $u_1, \ldots, u_n$ of an algorithm considerably,

$$\left|\frac{\partial u_i}{\partial \varepsilon}\right| \gg 1 \quad \text{for } i = 1, \ldots, n,$$

while the *final* result $y = f(u_1, \ldots, u_n)$ is not strongly affected,

$$\left|\frac{\partial y}{\partial \varepsilon}\right| \leqslant 1,$$

even though it is calculated from the highly inaccurate intermediate values $u_1, \ldots, u_n$: the algorithm shows error damping.

EXAMPLE 1. Evaluate $y = \varphi(x) = x^3 - 3x^2 + 3x = ((x - 3) \times x + 3) \times x$ using Horner's scheme:

$$u := x - 3,$$

$$v := u \times x,$$

$$w := v + 3,$$

$$y := w \times x.$$

The value x is known to lie in the interval

$$x \in \tilde{x} = [0.9, 1.1].$$

Starting with this interval and using straight interval arithmetic, we find

$$\tilde{u} = \tilde{x} \ominus [3, 3] = [-2.1, -1.9],$$

$$\tilde{v} = \tilde{u} \otimes \tilde{x} = [-2.31, -1.71],$$

$$\tilde{w} = \tilde{v} \ominus [3, 3] = [0.69, 1.29],$$

$$\tilde{y} = \tilde{w} \otimes \tilde{x} = [0.621, 1.419].$$

The interval $\tilde{y}$ is much too large compared to the interval

$$\{\varphi(x) \mid x \in \tilde{x}\} = [0.999, 1.001],$$

which describes the actual effect of an error in x on $\varphi(x)$.

EXAMPLE 2. Using just ordinary 2-digit arithmetic gives considerably more accurate results than the interval arithmetic suggests:

	$x = 0.9$	$x = 1.1$
u	-2.1	-1.9
v	-1.9	-2.1
w	1.1	0.9
y	0.99	0.99

For the successful application of interval arithmetic, therefore, it is not sufficient merely to replace the arithmetic operations of commonly used algorithms by interval operations: It is necessary to develop new algorithms producing the same final results but having an improved error-dependence pattern for the intermediate results.

EXAMPLE 3. In Example 1 a simple transformation of $\varphi(x)$ suffices:

$$y = \varphi(x) = 1 + (x - 1)^3.$$

When applied to the corresponding evaluation algorithm and the same starting interval $\tilde{x} = [0.9, 1.1]$, interval arithmetic now produces the optimal result:

$$\begin{aligned}
\tilde{u}_1 &:= \tilde{x} \ominus [1, 1] = [-0.1, 0.1],\\
\tilde{u}_2 &:= \tilde{u}_1 \otimes \tilde{u}_1 = [-0.01, 0.01],\\
\tilde{u}_3 &:= \tilde{u}_2 \otimes \tilde{u}_1 = [-0.001, 0.001],\\
\tilde{y} &:= \tilde{u}_3 \oplus [1, 1] = [0.999, 1.001].
\end{aligned}$$

As far as ordinary arithmetic is concerned, there is not much difference between the two evaluation algorithms of Example 1 and Example 3. Using two digits again, the results are practically identical to those in Example 2:

	$x = 0.9$	$x = 1.1$
u_1	-0.1	0.1
u_2	0.01	0.01
u_3	-0.001	0.001
y	1.0	1.0

For an in-depth treatment of interval arithmetic the reader should consult, for instance, Moore (1966) or Kulisch (1969).

In order to obtain *statistical roundoff estimates* [Rademacher (1948)], we assume that the relative roundoff error [see (1.2.6)] which is caused by an elementary operation is a random variable with values in the interval [−eps, eps]. Furthermore we assume that the roundoff errors ε attributable to different operations are independent random variables. By μ_ε we denote the expected value and by σ_ε^2 the variance of the above round-off distribution. They satisfy the general relationship

$$\sigma_\varepsilon^2 = E(\varepsilon - E(\varepsilon))^2 = E(\varepsilon^2) - (E(\varepsilon))^2 = \mu_{\varepsilon^2} - \mu_\varepsilon^2 .$$

Assuming a uniform distribution in the interval [−eps, eps], we get

$$(1.5.1) \quad \mu_\varepsilon := E(\varepsilon) = 0, \qquad \sigma_\varepsilon^2 = E(\varepsilon^2) = \frac{1}{2\,\text{eps}} \int_{-\text{eps}}^{\text{eps}} t^2\,dt = \tfrac{1}{3}\,\text{eps}^2 =: \bar{\varepsilon}^2.$$

Closer examinations show the roundoff distribution to be not quite uniform [see Sterbenz (1974), Exercise 22, p. 122]. It should also be kept in mind that the ideal roundoff pattern is only an approximation to the roundoff patterns observed in actual computing machinery, so that the quantities μ_ε and σ_ε^2 may have to be determined empirically.

The results x of algorithms subjected to random roundoff errors become random variables themselves with expected values μ_x and variances σ_x^2, connected again by the basic relation

$$\sigma_x^2 = E(x - E(x))^2 = E(x^2) - (E(x))^2 = \mu_{x^2} - \mu_x^2.$$

The propagation of previous roundoff effects through elementary operations is described by the following formulas for arbitrary independent random variables x, y and constants α, $\beta \in \mathbb{R}$:

$$\begin{aligned} \mu_{\alpha x \pm \beta y} &= E(\alpha x \pm \beta y) = \alpha E(x) \pm \beta E(y) = \alpha\mu_x \pm \beta\mu_y, \\ \sigma_{\alpha x \pm \beta y}^2 &= E((\alpha x \pm \beta y)^2) - (E(\alpha x \pm \beta y))^2 \\ &= \alpha^2 E(x - E(x))^2 + \beta^2 E(y - E(y))^2 = \alpha^2\sigma_x^2 + \beta^2\sigma_y^2. \end{aligned} \tag{1.5.2}$$

The first of the above formulas follows by the linearity of the expected-value operator. It holds for arbitrary random variables x, y. The second formula is based on the relation $E(xy) = E(x)E(y)$, which holds whenever x and y are independent. Similarly, we obtain for independent x and y

$$\begin{aligned} \mu_{x \times y} &= E(x \times y) = E(x)E(y) = \mu_x\mu_y, \\ \sigma_{x \times y}^2 &= E(x \times y - E(x)E(y))^2 = \mu_{x^2}\mu_{y^2} - \mu_x^2\mu_y^2 \\ &= \sigma_x^2\sigma_y^2 + \mu_x^2\sigma_y^2 + \mu_y^2\sigma_x^2. \end{aligned} \tag{1.5.3}$$

EXAMPLE. For calculating $y = a^2 - b^2$ (see Example 2 in Section 1.3) we find, under the assumptions (1.5.1), $E(a) = a$, $\sigma_a^2 = 0$, $E(b) = b$, $\sigma_b^2 = 0$, and using (1.5.2) and (1.5.3), that

$$\begin{aligned} \eta_1 &= a^2(1 + \varepsilon_1), & E(\eta_1) &= a^2, \quad \sigma_{\eta_1}^2 = a^4\bar{\varepsilon}^2, \\ \eta_2 &= b^2(1 + \varepsilon_2), & E(\eta_2) &= b^2, \quad \sigma_{\eta_2}^2 = b^4\bar{\varepsilon}^2, \\ y &= (\eta_1 - \eta_2)(1 + \varepsilon_3), & E(y) &= E(\eta_1 - \eta_2)E(1 + \varepsilon_3) = a^2 - b^2 \end{aligned}$$

(η_1, η_2, ε_3 are assumed to be independent),

$$\begin{aligned} \sigma_y^2 &= \sigma_{\eta_1 - \eta_2}^2\sigma_{1+\varepsilon_3}^2 + \mu_{\eta_1 - \eta_2}^2\sigma_{1+\varepsilon_3}^2 + \mu_{1+\varepsilon_3}^2\sigma_{\eta_1 - \eta_2}^2 \\ &= (\sigma_{\eta_1}^2 + \sigma_{\eta_2}^2)\bar{\varepsilon}^2 + (a^2 - b^2)^2\bar{\varepsilon}^2 + 1(\sigma_{\eta_1}^2 + \sigma_{\eta_2}^2) \\ &= (a^4 + b^4)\bar{\varepsilon}^4 + [(a^2 - b^2)^2 + a^4 + b^4]\bar{\varepsilon}^2. \end{aligned}$$

Neglecting $\bar{\varepsilon}^4$ compared to $\bar{\varepsilon}^2$ yields

$$\sigma_y^2 \doteq ((a^2 - b^2)^2 + a^4 + b^4)\bar{\varepsilon}^2.$$

For $a := 0.3237$, $b := 0.3134$, $\text{eps} := 5 \times 10^{-4}$ (see Example 5 in Section 1.3), we find

$$\sigma_y \doteq 0.144\bar{\varepsilon} = 0.000\,0415$$

which is close in magnitude to the true error $\Delta y = 0.000\,01787$ for 4-digit arithmetic. Compare this with the error bound $0.000\,10478$ furnished by (1.3.17).

We denote by $M(x)$ the set of all quantities which, directly or indirectly, have entered the calculation of the quantity x. If $M(x) \cap M(y) \neq \emptyset$ for the algorithm in question, then the random variables x and y are in general dependent.

The statistical roundoff error analysis of an algorithm becomes extremely complicated if dependent random variables are present. It becomes quite easy, however, under the following simplifying assumptions:

(1.5.4)

(a) *The operands of each arithmetic operation are independent random variables.*

(b) *In calculating variances all terms of an order higher than the smallest one are neglected.*

(c) *All variances are so small that for elementary operations τ in first-order approximation, $E(x \tau y) \doteq E(x) \tau E(y) = \mu_x \tau \mu_y$.*

If in addition the expected values μ_x are replaced by the estimated values x, and relative variances $\varepsilon_x^2 := \sigma_x^2/\mu_x^2 \approx \sigma_x^2/x^2$ are introduced, then from (1.5.2) and (1.5.3) [compare (1.2.6), (1.3.5)],

$$z = \mathrm{fl}(x \pm y): \quad \varepsilon_z^2 \doteq \left(\frac{x}{z}\right)^2 \varepsilon_x^2 + \left(\frac{y}{z}\right)^2 \varepsilon_y^2 + \bar{\varepsilon}^2,$$

$$z = \mathrm{fl}(x \times y): \quad \varepsilon_z^2 \doteq \varepsilon_x^2 + \varepsilon_y^2 + \bar{\varepsilon}^2,$$

$$z = \mathrm{fl}(x \,/\, y): \quad \varepsilon_z^2 \doteq \varepsilon_x^2 + \varepsilon_y^2 + \bar{\varepsilon}^2.$$

It should be kept in mind, however, that these results are valid only if the hypotheses (1.5.4), in particular (1.5.4a), are met.

It is possible to evaluate above formulas in the course of a numerical computation and thereby to obtain an estimate of the error of the final results. As in the case of interval arithmetic, this leads to an arithmetic of paired quantities (x, ε_x^2) for which elementary operations are defined with the help of the above or similar formulas. Error bounds for the final results r are then obtained from the relative variance ε_r^2, assuming that the final error distribution is normal. This assumption is justified inasmuch as the distributions of propagated errors alone tend to become normal if subjected to many elementary operations. At each such operation the nonnormal roundoff error distribution is superimposed on the distribution of previous errors. However, after many operations, the propagated errors are large compared to the newly created roundoff errors, so that the latter do not appreciably affect the normality of the total error distribution. Assuming the final error distribution to be normal, the actual relative error of the final result r is bounded with probability 0.9 by $2\varepsilon_r$.

Exercises for Chapter 1

1. Show that with floating-point arithmetic of t decimal places

$$\operatorname{rd}(a) = \frac{a}{1+\varepsilon} \quad \text{with} \quad |\varepsilon| \leqslant 5 \times 10^{-t}$$

holds in analogy to (1.2.2). [In parallel with (1.2.6), as a consequence, $\operatorname{fl}(a * b) = (a * b)/(1 + \varepsilon)$ with $|\varepsilon| \leqslant 5 \times 10^{-t}$ for all arithmetic operations $* = +, -, \times, /$.]

2. Let a, b, c be fixed-point numbers with N decimal places after the decimal point, and suppose $0 < a, b, c < 1$. A substitute product $a * b$ is defined as follows: Add $10^{-N}/2$ to the exact product $a \cdot b$, and then delete the $(N + 1)$-st and subsequent digits.

   (a) Give a bound for $|(a * b) * c - abc|$.
   (b) By how many units of the Nth place can $(a * b) * c$ and $a * (b * c)$ differ?

3. Evaluating $\sum_{i=1}^{n} a_i$ in floating-point arithmetic may lead to an arbitrarily large relative error. If, however, all summands a_i are of the same sign, then this relative error is bounded. Derive a crude bound for this error, disregarding terms of higher order.

4. Show how to evaluate the following expressions in a numerically stable fashion:

$$\frac{1}{1+2x} - \frac{1-x}{1+x} \quad \text{for } |x| \ll 1,$$

$$\sqrt{x + \frac{1}{x}} - \sqrt{x - \frac{1}{x}} \quad \text{for } x \gg 1,$$

$$\frac{1 - \cos x}{x} \quad \text{for } x \neq 0,\ |x| \ll 1.$$

5. Suppose a computer program is available which yields values for arcsin y in floating-point representation with t decimal mantissa places and for $|y| \leqslant 1$ subject to a relative error ε with $|\varepsilon| \leqslant 5 \times 10^{-t}$. In view of the relation

$$\arctan x = \arcsin \frac{x}{\sqrt{1+x^2}},$$

this program could also be used to evaluate arctan x. Determine for which values x this procedure is numerically stable by estimating the relative error.

6. For given z, the function $\tan(z/2)$ can be computed according to the formula

$$\tan \frac{z}{2} = \pm \left(\frac{1 - \cos z}{1 + \cos z} \right)^{1/2}.$$

Is this method of evaluation numerically stable for $z \approx 0$, $z \approx \pi/2$? If necessary, give numerically stable alternatives.

7. The function

$$f(\varphi, k_c) := \frac{1}{\sqrt{\cos^2\varphi + k_c^2 \sin^2\varphi}}$$

is to be evaluated for $0 \leqslant \varphi \leqslant \pi/2$, $0 < k_c \leqslant 1$.
The method

$$k^2 := 1 - k_c^2,$$

$$f(\varphi, k_c) = \frac{1}{\sqrt{1 - k^2 \sin^2\varphi}}$$

avoids the calculation of $\cos\varphi$ and is therefore faster. Compare this with the direct evaluation of the original expression for $f(\varphi, k_c)$ with respect to numerical stability.

8. For the linear function $f(x) := a + bx$, where $a \neq 0$, $b \neq 0$, compute the first derivative $D_h f(0) = f'(0) = b$ by the formula

$$D_h f(0) = \frac{f(h) - f(-h)}{2h}$$

in binary floating-point arithmetic. Suppose that a and b are binary machine numbers, and h a power of 2. Multiplication by h and division by $2h$ can be therefore carried out exactly. Give a bound for the relative error of $D_h f(0)$. What is the behavior of this bound as $h \to 0$?

9. The square root $\pm(u + iv)$ of a complex number $x + iy$ with $y \neq 0$ may be calculated from the formulas

$$u = \pm\sqrt{\frac{x + \sqrt{x^2 + y^2}}{2}},$$

$$v = \frac{y}{2u}.$$

Compare the cases $x \geqslant 0$ and $x < 0$ with respect to their numerical stability. Modify the formulas if necessary to ensure numerical stability.

10. The variance S^2 of a set of observations $x_1, \ldots, x_n$ is to be determined. Which of the formulas

$$S^2 = \frac{1}{n-1}\left(\sum_{i=1}^{n} x_i^2 - n\bar{x}^2\right),$$

$$S^2 = \frac{1}{n-1}\sum_{i=1}^{n}(x_i - \bar{x})^2 \quad \text{with } \bar{x} := \frac{1}{n}\sum_{i=1}^{n} x_i$$

is numerically more trustworthy?

11. The coefficients $a_r, b_r (r = 0, \ldots, n)$ are, for fixed x, connected recursively:

$$b_n := a_n;$$

$$(*) \qquad \text{for } r = n-1, n-2, \ldots, 0: \quad b_r := x b_{r+1} + a_r.$$

(a) Show that the polynomials

$$A(z) := \sum_{r=0}^{n} a_r z^r, \qquad B(z) := \sum_{r=1}^{n} b_r z^{r-1}$$

satisfy

$$A(z) = (z - x)B(z) + b_0 .$$

(b) Suppose $A(x) = b_0$ is to be calculated by the recursion (*) for fixed x in floating-point arithmetic, the result being b'_0. Show, using the formulas (compare Exercise 1)

$$\mathrm{fl}(u + v) = \frac{u + v}{1 + \sigma}, \qquad |\sigma| \leqslant \mathrm{eps},$$

$$\mathrm{fl}(u \cdot v) = \frac{u \cdot v}{1 + \pi}, \qquad |\pi| \leqslant \mathrm{eps},$$

the inequality

$$|A(x) - b'_0| \leqslant \frac{\mathrm{eps}}{1 - \mathrm{eps}}(2e_0 - |b'_0|),$$

where e_0 is defined by the following recursion:

$$e_n := |a_n|/2;$$

$$\text{for } r = n-1, n-2, \ldots, 0; \quad e_r := |x| e_{r+1} + |b'_r|.$$

Hint: From

$$\left.\begin{aligned} b'_n &:= a_n, \\ p_r &:= \mathrm{fl}(xb'_{r+1}) = \frac{xb'_{r+1}}{1 + \pi_{r+1}} \\ b'_r &:= \mathrm{fl}(p_r + a_r) = \frac{p_r + a_r}{1 + \sigma_r} = xb'_{r+1} + a_r + \delta_r \end{aligned}\right\} r = n-1, \ldots, 0,$$

derive

$$\delta_r = -xb'_{r+1}\frac{\pi_{r+1}}{1 + \pi_{r+1}} - \sigma_r b'_r \qquad (r = n-1, \ldots, 0);$$

then show $b'_0 = \sum_{k=0}^{n} (a_k + \delta_k)x^k$, $\delta_n := 0$, and estimate $\sum_0^n |\delta_k| \, |x|^k$.

12. Assuming Earth to be spherical, two points on its surface can be expressed in Cartesian coordinates

$$p_i = [x_i, y_i, z_i] = [r \cos\alpha_i \cos\beta_i, r \sin\alpha_i \cos\beta_i, r \sin\beta_i], \qquad i = 1, 2,$$

where r is the earth radius and α_i, β_i are the longitudes and latitudes of the two points p_i, respectively. If

$$\cos\delta = \frac{p_1^T p_2}{r^2} = \cos(\alpha_1 - \alpha_2)\cos\beta_1 \cos\beta_2 + \sin\beta_1 \sin\beta_2,$$

then $r\delta$ is the *great-circle distance* between the two points.

(a) Show that using the arccos function to determine δ from the above expression is not numerically stable.
(b) Derive a numerically stable expression for δ.

References for Chapter 1

Ashenhurst, R. L., Metropolis, N.: Unnormalized floating-point arithmetic. *J. Assoc. Comput. Mach.* **6**, 415–428, 1959.

Bauer, F. L.: Computational graphs and rounding error. *SIAM J. Numer. Anal.* **11**, 87–96, 1974.

———, Heinhold, J., Samelson, K., Sauer, R.: *Moderne Rechenanlagen*. Stuttgart: Teubner 1965.

Henrici, P.: *Error Propagation for Difference Methods*. New York: Wiley 1963.

Knuth, D. E.: *The Art of Computer Programming. Vol. 2. Seminumerical Algorithms*. Reading, Mass.: Addison-Wesley 1969.

Kulisch, U.: Grundzüge der Intervallrechnung. In: D. Laugwitz: *Überblicke Mathematik* **2**, 51–98. Mannheim: Bibliographisches Institut 1969.

Moore, R. E.: *Interval Analysis*. Englewood Cliffs, N.J.: Prentice-Hall 1966.

Neumann, J. von, Goldstein, H. H.: Numerical inverting of matrices. *Bull. Amer. Math. Soc.* **53**, 1021–1099, 1947.

Rademacher, H. A.: On the accumulation of errors in processes of integration on high-speed calculating machines. Proceedings of a symposium on large-scale digital calculating machinery. *Annals Comput. Labor. Harvard Univ.* **16**, 176–185, 1948.

Scarborough, J. B.: *Numerical Mathematical Analysis*. Baltimore: Johns Hopkins Press 1930, 2nd edition 1950.

Sterbenz, P. H.: *Floating Point Computation*. Englewood Cliffs, N.J.: Prentice-Hall 1974.

Wilkinson, J. H.: Error analysis of floating-point computation. *Numer. Math.* **2**, 219–340, 1960.

———: *Rounding Errors in Algebraic Processes*. New York: Wiley 1963.

———: *The Algebraic Eigenvalue Problem*. Oxford: Clarendon Press 1965.

Interpolation 2

Consider a family of functions of a single variable x,

$$\Phi(x; a_0, \ldots, a_n),$$

having $n+1$ parameters $a_0, \ldots, a_n$, whose values characterize the individual functions in this family. The *interpolation problem* for Φ consists of determining these parameters a_i so that for $n+1$ given real or complex pairs of numbers (x_i, f_i), $i = 0, \ldots, n$, with $x_i \neq x_k$ for $i \neq k$,

$$\Phi(x_i; a_0, \ldots, a_n) = f_i, \qquad i = 0, \ldots, n,$$

holds. We will call the pairs (x_i, f_i) *support points*, the locations x_i *support abscissas*, and the values f_i *support ordinates*. Occasionally, the values of derivatives of Φ are also prescribed.

The above is a *linear interpolation problem* if Φ depends linearly on the parameters a_i:

$$\Phi(x; a_0, \ldots, a_n) \equiv a_0 \Phi_0(x) + a_1 \Phi_1(x) + \cdots + a_n \Phi_n(x).$$

This class of problems includes the classical one of *polynomial interpolation* (Section 2.1),

$$\Phi(x; a_0, \ldots, a_n) \equiv a_0 + a_1 x + a_2 x^2 + \cdots + a_n x^n,$$

as well as *trigonometric interpolation* (Section 2.3),

$$\Phi(x; a_0, \ldots, a_n) \equiv a_0 + a_1 e^{xi} + a_2 e^{2xi} + \cdots + a_n e^{nxi} \qquad (i^2 = -1).$$

In the past, polynomial interpolation was frequently used to interpolate function values gathered from tables. The availability of modern computing machinery has reduced the need for extensive table lookups. However, polynomial interpolation is also important as the basis of several types of numerical integration formulas in general use. In a more modern development, polynomial and rational interpolation (see below) are employed in the construction of "extrapolation methods" for integration, differential equations, and related problems (see for instance Sections 3.3 and 3.4).

Trigonometric interpolation is used extensively for the numerical Fourier analysis of time series and cyclic phenomena in general. In this context, the so-called "fast Fourier transforms" are particularly important and successful (Section 2.3.2).

The class of linear interpolation problems also contains *spline interpolation* (Section 2.4). In the special case of *cubic splines*, the functions Φ are assumed to be twice continuously differentiable for $x \in [x_0, x_n]$ and to coincide with some cubic polynomial on every subinterval $[x_i, x_{i+1}]$ of a given partition $x_0 < x_1 < \cdots < x_n$.

Spline interpolation is a fairly new development of growing importance. It provides a valuable tool for representing empirical curves and for approximating complicated mathematical functions. It is increasingly used when dealing with ordinary or partial differential equations.

Two nonlinear interpolation schemes are of importance: *rational interpolation*,

$$\Phi(x; a_0, \ldots, a_n, b_0, \ldots, b_m) \equiv \frac{a_0 + a_1 x + \cdots + a_n x^n}{b_0 + b_1 x + \cdots + b_m x^m},$$

and *exponential interpolation*,

$$\Phi(x; a_0, \ldots, a_n, \lambda_0, \ldots, \lambda_n) \equiv a_0 e^{\lambda_0 x} + a_1 e^{\lambda_1 x} + \cdots + a_n e^{\lambda_n x}.$$

Rational interpolation (Section 2.2) plays a role in the process of best approximating a given function by one which is readily evaluated on a digital computer. Exponential interpolation is used, for instance, in the analysis of radioactive decay.

Interpolation is a basic tool for the approximation of given functions. For a comprehensive discussion of these and related topics consult Davis (1965).

2.1 Interpolation by Polynomials

2.1.1 Theoretical Foundation: The Interpolation Formula of Lagrange

In what follows, we denote by Π_n the set of all real or complex polynomials P whose degrees do not exceed n:

$$P(x) = a_0 + a_1 x + \cdots + a_n x^n.$$

(2.1.1.1) Theorem *For $n + 1$ arbitrary support points*

$$(x_i, f_i), \qquad i = 0, \ldots, n, \quad x_i \neq x_k \textit{ for } i \neq k,$$

there exists a unique polynomial $P \in \Pi_n$ *with*

$$P(x_i) = f_i, \qquad i = 0, 1, \ldots, n.$$

PROOF. *Uniqueness:* For any two polynomials $P_1, P_2 \in \Pi_n$ with

$$P_1(x_i) = P_2(x_i) = f_i, \qquad i = 0, 1, \ldots, n,$$

the polynomial $P := P_1 - P_2 \in \Pi_n$ has degree at most n, and it has at least $n + 1$ different zeros, namely x_i, $i = 0, \ldots, n$. P must therefore vanish identically, and $P_1 = P_2$.

Existence: We will construct the interpolating polynomial P explicitly with the help of polynomials $L_i \in \Pi_n$, $i = 0, \ldots, n$, for which

(2.1.1.2)
$$L_i(x_k) = \delta_{ik} = \begin{cases} 1 & \text{if } i = k, \\ 0 & \text{if } i \neq k. \end{cases}$$

The following *Lagrange polynomials* satisfy the above conditions:

(2.1.1.3)
$$\begin{aligned} L_i(x) :&\equiv \frac{(x - x_0) \ldots (x - x_{i-1})(x - x_{i+1}) \ldots (x - x_n)}{(x_i - x_0) \ldots (x_i - x_{i-1})(x_i - x_{i+1}) \ldots (x_i - x_n)} \\ &\equiv \frac{\omega(x)}{(x - x_i)\omega'(x_i)} \quad \text{with } \omega(x) :\equiv \prod_{i=0}^{n} (x - x_i). \end{aligned}$$

Note that our proof so far shows that the Lagrange polynomials are uniquely determined by (2.1.1.2).

The solution P of the interpolation problem can now be expressed directly in terms of the polynomials L_i, leading to the *Lagrange interpolation formula*:

(2.1.1.4)
$$P(x) \equiv \sum_{i=0}^{n} f_i L_i(x) \equiv \sum_{i=0}^{n} f_i \prod_{\substack{k \neq i \\ k=0}}^{n} \frac{x - x_k}{x_i - x_k}. \qquad \square$$

The above interpolation formula shows that the coefficients of P depend linearly on the support ordinates f_i. While theoretically important, Lagrange's formula is, in general, not as suitable for actual calculations as some other methods to be described below, particularly for large numbers n of support points. Lagrange's formula may, however, be useful in some situations in which many interpolation problems are to be solved for the same support abscissae x_i, $i = 0, \ldots, n$, but different sets of support ordinates f_i, $i = 0, \ldots, n$.

EXAMPLE. Given for $n = 2$:

x_i	0	1	3
f_i	1	3	2

Wanted: $P(2)$, where $P \in \Pi_2$, $P(x_i) = f_i$ for $i = 0, 1, 2$.

Solution:

$$L_0(x) \equiv \frac{(x-1)(x-3)}{(0-1)(0-3)}, \qquad L_1(x) \equiv \frac{(x-0)(x-3)}{(1-0)(1-3)}, \qquad L_2(x) \equiv \frac{(x-0)(x-1)}{(3-0)(3-1)},$$

$$P(2) = 1 \cdot L_0(2) + 3 \cdot L_1(2) + 2 \cdot L_2(2) = 1 \cdot \frac{-1}{3} + 3 \cdot 1 + 2 \cdot \frac{1}{3} = \frac{10}{3}.$$

2.1.2 Neville's Algorithm

Instead of solving the interpolation problem all at once, one might consider solving the problem for smaller sets of support points first and then updating these solutions to obtain the solution to the full interpolation problem. This idea will be explored in the following two sections.

For a given set of support points (x_i, f_i), $i = 0, 1, \ldots, n$, we denote by

$$P_{i_0 i_1 \ldots i_k} \in \Pi_k$$

that polynomial in Π_k for which

$$P_{i_0 i_1 \ldots i_k}(x_{i_j}) = f_{i_j}, \qquad j = 0, 1, \ldots, k.$$

These polynomials are linked by the following recursion:

(2.1.2.1a) $\qquad P_i(x) \equiv f_i,$

(2.1.2.1b) $\quad P_{i_0 i_1 \ldots i_k}(x) \equiv \dfrac{(x - x_{i_0})P_{i_1 i_2 \ldots i_k}(x) - (x - x_{i_k})P_{i_0 i_1 \ldots i_{k-1}}(x)}{x_{i_k} - x_{i_0}}.$

Proof. (2.1.2.1a) is trivial. To prove (2.1.2.1b), we denote its right-hand side by $R(x)$, and go on to show that R has the characteristic properties of $P_{i_0 i_1 \ldots i_k}$. The degree of R is clearly no greater than k. By the definitions of $P_{i_0 \ldots i_{k-1}}$ and $P_{i_1 \ldots i_k}$,

$$R(x_{i_0}) = P_{i_0 \ldots i_{k-1}}(x_{i_0}) = f_{i_0},$$
$$R(x_{i_k}) = P_{i_1 \ldots i_k}(x_{i_k}) \quad = f_{i_k},$$

and

$$R(x_{i_j}) = \frac{(x_{i_j} - x_{i_0})f_{i_j} - (x_{i_j} - x_{i_k})f_{i_j}}{x_{i_k} - x_{i_0}} = f_{i_j}$$

for $j = 1, 2, \ldots, k-1$. Thus $R = P_{i_0 i_1 \ldots i_k}$, in view of the uniqueness of polynomial interpolation [Theorem (2.1.1.1)]. □

Neville's algorithm aims at determining the value of the interpolating polynomial P for a single value of x. It is less suited for determining the interpolating polynomial itself. Algorithms that are more efficient for the latter task, and also more efficient if values of P are sought for several arguments x simultaneously, will be described in Section 2.1.3.

Based on the recursion (2.1.2.1), Neville's algorithm constructs a symmetric tableau of the values of some of the partially interpolating polynomials $P_{i_0 i_1 \ldots i_k}$ for fixed x:

(2.1.2.2)

	$k=0$	1	2	3
x_0	$f_0 = P_0(x)$			
		$P_{01}(x)$		
x_1	$f_1 = P_1(x)$		$P_{012}(x)$	
		$P_{12}(x)$		$P_{0123}(x)$
x_2	$f_2 = P_2(x)$		$P_{123}(x)$	
		$P_{23}(x)$		
x_3	$f_3 = P_3(x)$			

The first column of the tableau contains the prescribed support ordinates f_i. Subsequent columns are filled by calculating each entry recursively from its two "neighbors" in the previous column according to (2.1.2.1b). The entry $P_{123}(x)$, for instance, is given by

$$P_{123}(x) = \frac{(x - x_1)P_{23}(x) - (x - x_3)P_{12}(x)}{x_3 - x_1}.$$

Example. Determine $P_{012}(2)$ for the same support points as in section 2.1.1.

	$k=0$	1	2
$x_0 = 0$	$f_0 = P_0(2) = 1$		
		$P_{01}(2) = 5$	
$x_1 = 1$	$f_1 = P_1(2) = 3$		$P_{012}(2) = \frac{10}{3}$
		$P_{12}(2) = \frac{5}{2}$	
$x_2 = 3$	$f_2 = P_2(2) = 2$		

$$P_{01}(2) = \frac{(2-0)\cdot 3 - (2-1)\cdot 1}{1-0} = 5,$$

$$P_{12}(2) = \frac{(2-1)\cdot 2 - (2-3)\cdot 3}{3-1} = \frac{5}{2},$$

$$P_{012}(2) = \frac{(2-0)\cdot 5/2 - (2-3)\cdot 5}{3-0} = \frac{10}{3}.$$

We will now discuss slight variants of Neville's algorithm, employing a frequently used abbreviation,

$$T_{i+k,k} := P_{i,i+1,\ldots,i+k}. \tag{2.1.2.3}$$

The tableau (2.1.2.2) becomes

$$
(2.1.2.4)\qquad
\begin{array}{llllll}
x_0 & f_0 = T_{00} & & & \\
 & & T_{11} & & \\
x_1 & f_1 = T_{10} & & T_{22} & \\
 & & T_{21} & & T_{33} \\
x_2 & f_2 = T_{20} & & T_{32} & \\
 & & T_{31} & & \\
x_3 & f_3 = T_{30} & & &
\end{array}
$$

The arrows indicate how the additional upward diagonal $T_{i0}, T_{i1}, \ldots, T_{ii}$ can be constructed if one more support point (x_i, f_i) is added.

The recursion (2.1.2.1) may be modified for more efficient evaluation:

$$(2.1.2.5a)\qquad T_{i0} := f_i$$

$$
(2.1.2.5b)\qquad
\begin{aligned}
T_{ik} &:= \frac{(x - x_{i-k})T_{i,k-1} - (x - x_i)T_{i-1,k-1}}{x_i - x_{i-k}} \\
&= T_{i,k-1} + \frac{T_{i,k-1} - T_{i-1,k-1}}{\dfrac{x - x_{i-k}}{x - x_i} - 1}, \qquad 1 \leqslant k \leqslant i, \quad i = 0, 1, \ldots, n.
\end{aligned}
$$

The following ALGOL algorithm is based on this modified recursion:

```
for i := 0 step 1 until n do
begin  t[i] := f[i];
       for j := i - 1 step - 1 until 0 do
       t[j] := t[j + 1] + (t[j + 1] - t[j]) × (z - x[i])/(x[i] - x[j])
end;
```

After the inner loop has terminated, $t[j] = T_{i,i-j}$, $0 \leqslant j \leqslant i$. The desired value $T_{nn} = P_{01\ldots n}$ of the interpolating polynomial can be found in $t[0]$.

Still another modification of Neville's algorithm serves to improve somewhat the accuracy of the interpolated polynomial value. For $i = 0, 1, \ldots, n$, let the quantities Q_{ik}, D_{ik} be defined by

$$Q_{i0} := D_{i0} := f_i,$$

$$
\left.
\begin{aligned}
Q_{ik} &:= T_{ik} - T_{i,k-1} \\
D_{ik} &:= T_{ik} - T_{i-1,k-1}
\end{aligned}
\right\} \qquad 1 \leqslant k \leqslant i.
$$

The recursion (2.1.2.5) then translates into

$$
(2.1.2.6)\qquad
\left.
\begin{aligned}
Q_{ik} &:= (D_{i,k-1} - Q_{i-1,k-1})\frac{x_i - x}{x_{i-k} - x_i} \\
D_{ik} &:= (D_{i,k-1} - Q_{i-1,k-1})\frac{x_{i-k} - x}{x_{i-k} - x_i}
\end{aligned}
\right\} \qquad 1 \leqslant k \leqslant i, \quad i = 0, 1, \ldots, n.
$$

Starting with $Q_{i0} := D_{i0} := f_i$, one calculates Q_{ik}, D_{ik} from the above recursion. Finally

$$T_{nn} := f_n + \sum_{k=1}^{n} Q_{nk}.$$

If the values $f_0, \ldots, f_n$ are close to each other, the quantities Q_{ik} will be small compared to f_i. This suggests forming the sum of the "corrections" $Q_{n1}, \ldots, Q_{nn}$ first [contrary to (2.1.2.5)] and then adding it to f_n, thereby avoiding unnecessary roundoff errors.

Note finally that for $x = 0$ the recursion (2.1.2.5) takes a particularly simple form

(2.1.2.7a) $$T_{i0} := f_i$$

(2.1.2.7b) $$T_{ik} := T_{i,k-1} + \frac{T_{i,k-1} - T_{i-1,k-1}}{\dfrac{x_{i-k}}{x_i} - 1}, \qquad 1 \leqslant k \leqslant i.$$

—as does its analog (2.1.2.6). These forms are encountered when applying extrapolation methods.

For historical reasons mainly, we mention *Aitken's algorithm*. It is also based on (2.1.2.1), but uses different intermediate polynomials. Its tableau is of the form

x_0	$f_0 = P_0(x)$				
x_1	$f_1 = P_1(x)$	$P_{01}(x)$			
x_2	$f_2 = P_2(x)$	$P_{02}(x)$	$P_{012}(x)$		
x_3	$f_3 = P_3(x)$	$P_{03}(x)$	$P_{013}(x)$	$P_{0123}(x)$	
x_4	$f_4 = P_4(x)$	$P_{04}(x)$	$P_{014}(x)$	$P_{0124}(x)$	$P_{01234}(x)$
$\vdots$	$\vdots$ $\vdots$	$\vdots$	$\vdots$	$\vdots$	$\vdots$

The first column again contains the prescribed values f_i. Each subsequent entry derives from the previous entry in the same row and the top entry in the previous column according to (2.1.2.1b).

2.1.3 Newton's Interpolation Formula: Divided Differences

Neville's algorithm is geared towards determining interpolating values rather than polynomials. If the interpolating polynomial itself is needed, or if one wants to find interpolating values for several arguments ξ_j simultaneously, then Newton's interpolation formula is to be preferred. Here we write the interpolating polynomial $P \in \Pi_n$, $P(x_i) = f_i$, $i = 0, 1, \ldots, n$, in the form

$$(2.1.3.1)\qquad \begin{aligned} P(x) &\equiv P_{01\ldots n}(x) \\ &\equiv a_0 + a_1(x - x_0) + a_2(x - x_0)(x - x_1) + \cdots \\ &\qquad + a_n(x - x_0)\ldots(x - x_{n-1}). \end{aligned}$$

Note that the evaluation of (2.1.3.1) for $x = \xi$ may be done recursively as indicated by the following expression:

$$P(\xi) = (\ldots(a_n(\xi - x_{n-1}) + a_{n-1})(\xi - x_{n-2}) + \cdots + a_1)(\xi - x_0) + a_0.$$

This requires fewer operations than evaluating (2.1.3.1) term by term. It corresponds to the so-called *Horner scheme* for evaluating polynomials which are given in the usual form, i.e. in terms of powers of x, and it shows that the representation (2.1.3.1) is well suited for evaluation.

It remains to determine the coefficients a_i in (2.1.3.1). In principle, they can be calculated successively from

$$\begin{aligned} f_0 &= P(x_0) = a_0, \\ f_1 &= P(x_1) = a_0 + a_1(x_1 - x_0), \\ f_2 &= P(x_2) = a_0 + a_1(x_2 - x_0) + a_2(x_2 - x_0)(x_2 - x_1), \\ &\ \ \vdots \end{aligned}$$

This can be done with n divisions and $n(n-1)$ multiplications. There is, however, a better way, which requires only $n(n+1)/2$ divisions and which produces useful intermediate results.

Observe that the two polynomials $P_{i_0 i_1 \ldots i_k}(x)$ and $P_{i_0 i_1 \ldots i_{k-1}}(x)$ differ by a polynomial of degree k with k zeros $x_{i_0}, x_{i_1}, \ldots, x_{i_{k-1}}$, since both polynomials interpolate the corresponding support points. Therefore there exists a unique coefficient

$$(2.1.3.2)\qquad f_{i_0 i_1 \ldots i_k}, \qquad k = 0, 1, \ldots, n,$$

such that

(2.1.3.3)

$$P_{i_0 i_1 \ldots i_k}(x) \equiv P_{i_0 i_1 \ldots i_{k-1}}(x) + f_{i_0 i_1 \ldots i_k}(x - x_{i_0})(x - x_{i_1})\ldots(x - x_{i_{k-1}}).$$

From this and from the identity $P_{i_0}(x) \equiv f_{i_0}$ it follows immediately that

$$(2.1.3.4)\qquad \begin{aligned} P_{i_0 i_1 \ldots i_k}(x) \equiv f_{i_0} &+ f_{i_0 i_1}(x - x_{i_0}) + \cdots \\ &+ f_{i_0 i_1 \ldots i_k}(x - x_{i_0})(x - x_{i_1})\ldots(x - x_{i_{k-1}}) \end{aligned}$$

is a Newton representation of the partially interpolating polynomial $P_{i_0 i_1 \ldots i_k}$. The coefficients (2.1.3.2) are called *k th divided differences.*

The recursion (2.1.2.1) for the partially interpolating polynomials translates into the recursion

(2.1.3.5) $$f_{i_0 i_1 \dots i_k} = \frac{f_{i_1 \dots i_k} - f_{i_0 \dots i_{k-1}}}{x_{i_k} - x_{i_0}}$$

for the divided differences, since by (2.1.3.3), $f_{i_1 \dots i_k}$ and $f_{i_0 \dots i_{k-1}}$ are the coefficients of the highest terms of the polynomials $P_{i_1 i_2 \dots i_k}$ and $P_{i_0 i_1 \dots i_{k-1}}$, respectively. The above recursion starts for $k = 0$ with the given support ordinates f_i, $i = 0, \dots, n$. It can be used in various ways for calculating divided differences $f_{i_0}, f_{i_0 i_1}, \dots, f_{i_0 i_1 \dots i_n}$, which then characterize the desired interpolating polynomial $P = P_{i_0 i_1 \dots i_n}$.

Because the polynomial $P_{i_0 i_1 \dots i_k}$ is uniquely determined by the support points it interpolates [Theorem (2.1.1.1)], the polynomial is invariant to any permutation of the indices $i_0, i_1, \dots, i_k$, and so is its coefficient $f_{i_0 i_1 \dots i_k}$ of x^k. Thus:

(2.1.3.6). *The divided differences $f_{i_0 i_1 \dots i_k}$ are invariant to permutations of the indices $i_0, i_1, \dots, i_k$: If*

$$(j_0, j_1, \dots, j_k) = (i_{s_0}, i_{s_1}, \dots, i_{s_k})$$

is a permutation of the indices $i_0, i_1, \dots, i_k$, then

$$f_{j_0 j_1 \dots j_k} = f_{i_0 i_1 \dots i_k}.$$

If we choose to calculate the divided differences in analogy to Neville's method—instead of, say, Aitken's method—then we are led to the following tableau, called the *divided-difference scheme*:

(2.1.3.7)

	$k=0$	$k=1$	$k=2$		$k=n$
x_0	f_0				
		f_{01}			
x_1	f_1		f_{012}		
		f_{12}		$\ddots$	
x_2	f_2		$\vdots$		$f_{012\dots n}$
		$\vdots$		$⋰$	
$\vdots$	$\vdots$		$f_{n-2,n-1,n}$		
		$f_{n-1,n}$			
x_n	f_n				

The entries in the second column are of the form

$$f_{01} = \frac{f_1 - f_0}{x_1 - x_0}, \qquad f_{12} = \frac{f_2 - f_1}{x_2 - x_1}, \qquad \dots,$$

those in the third column,

$$f_{012} = \frac{f_{12} - f_{01}}{x_2 - x_0}, \qquad f_{123} = \frac{f_{23} - f_{12}}{x_3 - x_1}, \qquad \ldots.$$

Clearly,

$$\begin{aligned} P(x) &\equiv P_{01\ldots n}(x) \\ &\equiv f_0 + f_{01}(x - x_0) + \cdots + f_{01\ldots n}(x - x_0)(x - x_1)\ldots(x - x_{n-1}) \end{aligned}$$

is the desired solution to the interpolation problem at hand. The coefficients of the above expansion are found in the top descending diagonal of the divided-difference scheme (2.1.3.7).

EXAMPLE. With the numbers of the example in sections 2.1.1 and 2.1.2, we have:

$$\begin{array}{llll} x_0 = 0 & \underline{f_0 = 1} & & \\ & & \underline{f_{01} = 2} & \\ x_1 = 1 & f_1 = 3 & & \underline{f_{012} = -\frac{5}{6}} \\ & & f_{12} = -\frac{1}{2} & \\ x_2 - 3 & f_2 = 2 & & \end{array}$$

$$\begin{aligned} P_{012}(x) &= 1 + 2(x - 0) - \tfrac{5}{6}(x - 0)(x - 1), \\ P_{012}(2) &= (-\tfrac{5}{6}(2 - 1) + 2)(2 - 0) + 1 = \tfrac{10}{3}. \end{aligned}$$

Instead of building the divided-difference scheme column by column, one might want to start with the upper left corner and add successive ascending diagonal rows. This amounts to adding new support points one at a time after having interpolated the previous ones. In the following ALGOL procedure, the entries in an ascending diagonal of (2.1.3.7) are found, after each increase of i, in the top portion of array t, and the first i coefficients $f_{01\ldots i}$ are found in array a.

```
for i := 0 step 1 until n do
begin  t[i] := f[i];
       for j := i - 1 step - 1 until 0 do
              t[j] := (t[j + 1] - t[j])/((x[i] - x[j]);
       a[i] := t[0]
end;
```

Afterwards, the interpolating polynomial (2.1.3.1) may be evaluated for any desired argument z:

```
p := a[n];
for i := n - 1 step - 1 until 0 do
        p := p × (z - x[i]) + a[i];
```

Some Newton representations of the same polynomial are numerically more trustworthy to evaluate than others. Choosing the permutation so that

$$|\xi - x_{i_k}| \geqslant |\xi - x_{i_{k-1}}|, \qquad k = 0, 1, \ldots, n-1,$$

dampens the error (see Section 1.3) during the Horner evaluation of

(2.1.3.8)

$$P(x) \equiv P_{i_0 \ldots i_n}(x) \equiv f_{i_0} + f_{i_0 i_1}(x - x_{i_0}) + \cdots + f_{i_0 i_1 \ldots i_n}(x - x_{i_0}) \ldots (x - x_{i_{n-1}}).$$

All Newton representations of the above kind can be found in the single divided-difference scheme which arises if the support arguments x_i, $i = 0, \ldots, n$, are ordered by size: $x_i < x_{i+1}$ for $i = 0, \ldots, n-1$. Then the preferred sequence of indices $i_0, i_1, \ldots, i_k$ is such that each index i_k is "adjacent" to some previous index. More precisely, either $i_k = \min\{i_l \mid 0 \leqslant l < k\} - 1$ or $i_k = \max\{i_l \mid 0 \leqslant l < k\} + 1$. Therefore the coefficients of (2.1.3.8) are found along a zigzag path—instead of the upper descending diagonal—of the divided-difference scheme. Starting with f_{i_0}, the path proceeds to the upper right neighbor if $i_k < i_{k-1}$, or to te lower right neighbor if $i_k > i_{k-1}$.

EXAMPLE. In the previous example, a preferred sequence for $\xi = 2$ is

$$i_0 = 1,\ i_1 = 2,\ i_2 = 0.$$

The corresponding path in the divided difference scheme is indicated below:

$$\begin{array}{llll}
x_0 = 0 & f_0 = 1 & & \\
 & & f_{01} = 2 & \\
x_1 = 1 & \underline{f_1 = 3} & & \underline{f_{012} = -\frac{5}{6}} \\
 & & \underline{f_{12} = -\frac{1}{2}} & \\
x_2 = 3 & f_2 = 2 & &
\end{array}$$

The desired Newton representation is:

$$P_{120}(x) \equiv 3 - \tfrac{1}{2}(x-1) - \tfrac{5}{6}(x-1)(x-3),$$

$$P_{120}(2) = (-\tfrac{5}{6}(2-3) - \tfrac{1}{2})(2-1) + 3 = \tfrac{10}{3}.$$

Frequently, the support ordinates f_i are the values $f(x_i) = f_i$ of a given function $f(x)$, which one wants to approximate by interpolation. In this case, the divided differences may be considered as multivariate functions of the support arguments x_i, and are historically written as

$$f[x_{i_0}, \ldots, x_{i_k}].$$

These functions satisfy (2.1.3.5). For instance,

$$
\begin{aligned}
f[x_0] &\equiv f(x_0),\\
f[x_0, x_1] &\equiv \frac{f[x_1]-f[x_0]}{x_1 - x_0} \equiv \frac{f(x_1)-f(x_0)}{x_1 - x_0},\\
f[x_0, x_1, x_2] &\equiv \frac{f[x_1, x_2]-f[x_0, x_1]}{x_2 - x_0}\\
&\equiv \frac{f(x_2)(x_1 - x_0) - f(x_1)(x_2 - x_0) + f(x_0)(x_2 - x_1)}{(x_1 - x_0)(x_2 - x_0)(x_2 - x_1)},\\
&\;\vdots\\
f[x_0, x_1, \ldots, x_k] &\equiv \frac{f[x_1, \ldots, x_k] - f[x_0, \ldots, x_{k-1}]}{x_k - x_0},\\
&\;\vdots
\end{aligned}
$$

Also, (2.1.3.6) gives immediately:

(2.1.3.9) Theorem. *The divided differences $f[x_{i_0}, \ldots, x_{i_k}]$ are symmetric functions of their arguments, i.e., they are invariant to permutations of the $x_{i_0}, \ldots, x_{i_k}$.*

If the function $f(x)$ is itself a polynomial, then we have the

(2.1.3.10) Theorem. *If $f(x)$ is a polynomial of degree N, then*

$$f[x_0, \ldots, x_k] = 0$$

for $k > N$.

PROOF. Because of the unique solvability of the interpolation problem (Theorem 2.1.1.1), $P_{0, \ldots, k}(x) \equiv f(x)$ for $k \geqslant N$. The coefficient of x^k in $P_{0, \ldots, k}(x)$ must therefore vanish for $k > N$. This coefficient, however, is given by $f[x_0, \ldots, x_k]$ according to (2.1.3.3). □

EXAMPLE. $f(x) = x^2$.

x_i	$k = 0$	1	2	3	4
0	0				
		1			
1	1		1		
		3		0	
2	4		1		0
		5		0	
3	9		1		
		7			
4	16				

If the function $f(x)$ is sufficiently often differentiable, then its divided differences $f[x_0, \ldots, x_k]$ can also be defined if some of the arguments x_i coincide. For instance, if $f(x)$ has a derivative at x_0, then it makes sense for certain purposes to define

$$f[x_0, x_0] := f'(x_0).$$

For a corresponding modification of the divided-difference scheme (2.1.3.7) see Section 2.1.5 on Hermite interpolation.

2.1.4 The Error in Polynomial Interpolation

Once again we consider a given function $f(x)$ and certain of its values

$$f_i = f(x_i), \qquad i = 0, 1, \ldots, n,$$

which are to be interpolated. We wish to ask how well the interpolating polynomial $P(x) \equiv P_{0 \ldots n}(x)$ with

$$P(x_i) = f_i, \qquad i = 0, 1, \ldots, n$$

reproduces $f(x)$ for arguments different from the support arguments x_i. The error

$$f(x) - P(x),$$

where $x \neq x_i$, $i = 0, 1$, can clearly become arbitrarily large for suitable functions f unless some restrictions are imposed on f. Under certain conditions, however, it is possible to bound the error. We have, for instance:

(2.1.4.1) Theorem. *If the function f has an $(n + 1)$st derivative, then for every argument $\bar{x}$ there exists a number ξ in the smallest interval $I[x_0, \ldots, x_n, \bar{x}]$ which contains $\bar{x}$ and all support abscissas x_i, satisfying*

$$f(\bar{x}) - P_{01 \ldots n}(\bar{x}) = \frac{\omega(\bar{x}) f^{(n+1)}(\xi)}{(n + 1)!},$$

where

$$\omega(x) \equiv (x - x_0)(x - x_1) \ldots (x - x_n).$$

PROOF. Let $P(x) :\equiv P_{01 \ldots n}(x)$ be the polynomial which interpolates the function at x_i, $i = 0, 1, \ldots, n$, and suppose $\bar{x} \neq x_i$ (for $\bar{x} = x_i$ there is nothing to show). We can find a constant K such that the function

$$F(x) :\equiv f(x) - P(x) - K\omega(x)$$

vanishes for $x = \bar{x}$:

$$F(\bar{x}) = 0.$$

Consequently, $F(x)$ has at least the $n+2$ zeros

$$x_0, \ldots, x_n, \bar{x}$$

in the interval $I[x_0, \ldots, x_n, \bar{x}]$. By Rolle's theorem, applied repeatedly, $F'(x)$ has at least $n+1$ zeros in the above interval, $F''(x)$ at least n zeros, and finally $F^{(n+1)}(x)$ at least one zero $\xi \in I[x_0, \ldots, x_n, \bar{x}]$.

Since $P^{(n+1)}(x) \equiv 0$,

$$F^{(n+1)}(\xi) = f^{(n+1)}(\xi) - K(n+1)! = 0$$

or

$$K = \frac{f^{(n+1)}(\xi)}{(n+1)!}.$$

This proves the proposition

$$f(\bar{x}) - P(\bar{x}) = K\omega(\bar{x}) = \frac{\omega(\bar{x})}{(n+1)!} f^{(n+1)}(\xi). \qquad \square$$

A different error term can be derived from Newton's interpolation formula (see Section 2.1.3):

$$P(x) \equiv P_{01\ldots n}(x) \equiv f[x_0] + f[x_0, x_1](x - x_0) + \cdots + f[x_0, x_1, \ldots, x_n](x - x_0) \ldots (x - x_{n-1}).$$

Here $f[x_0, x_1, \ldots, x_k]$ are the divided differences of the given function f. If in addition to the $n+1$ support points

$$(x_i, f_i): \quad f_i = f(x_i), \qquad i = 0, 1, \ldots, n,$$

we introduce an $(n+2)$nd support point

$$(x_{n+1}, f_{n+1}): \quad x_{n+1} := \bar{x}, \quad f_{n+1} := f(\bar{x}),$$

where

$$\bar{x} \neq x_i, \qquad i = 0, \ldots, n,$$

then by Newton's formula

$$f(\bar{x}) = P_{0\ldots n+1}(\bar{x}) = P_{0\ldots n}(\bar{x}) + f[x_0, \ldots, x_n, \bar{x}]\omega(\bar{x}),$$

or

(2.1.4.2) $$f(\bar{x}) - P_{0\ldots n}(\bar{x}) = \omega(\bar{x}) f[x_0, \ldots, x_n, \bar{x}].$$

The difference on the left-hand side appears in Theorem (2.1.4.1), and since $\omega(\bar{x}) \neq 0$, we must have

$$f[x_0, \ldots, x_n, \bar{x}] = \frac{f^{(n+1)}(\xi)}{(n+1)!} \quad \text{for some } \xi \in I[x_0, \ldots, x_n, \bar{x}].$$

This also yields

$$\text{(2.1.4.3)} \qquad f[x_0, \ldots, x_n] = \frac{f^{(n)}(\xi)}{n!} \quad \text{for some } \xi \in I[x_0, \ldots, x_n],$$

which relates derivatives and divided differences.

EXAMPLE. $f(x) = \sin x$:

$$x_i = \frac{\pi i}{10}, \quad i = 0, 1, 2, 3, 4, 5, \qquad n = 5,$$

$$\sin x - P(x) \equiv (x - x_0)(x - x_1) \ldots (x - x_5) \frac{-\sin \xi}{720},$$

$$|\sin x - P(x)| \leqslant \tfrac{1}{720} |(x - x_0)(x - x_1) \ldots (x - x_5)| = \frac{|\omega(x)|}{720}.$$

We end this section with two brief warnings, one against trusting the interpolating polynomial outside of $I[x_0, \ldots, x_n]$ and one against expecting too much of polynomial interpolation inside $I[x_0, \ldots, x_n]$.

In the exterior of the interval $I[x_0, \ldots, x_n]$, the value of $|\omega(x)|$ in Theorem (2.1.4.1) grows very fast. The use of the interpolation polynomial P for approximating f at some location outside the interval $I[x_0, \ldots, x_n]$—called *extrapolation*—should be avoided if possible.

Within $I[x_0, \ldots, x_n]$ on the other hand, it should not be assumed that finer and finer samplings of the function f will lead to better and better approximations through interpolation.

Consider a real function f which is infinitely often differentiable in a given interval $[a, b]$. To every interval partition $\Delta = \{a = x_0 < x_1 < \cdots < x_n = b\}$ there exists an interpolating polynomial $P_\Delta \in \Pi_n$ with $P_\Delta(x_i) = f_i$ for $x_i \in \Delta$. A sequence of interval partitions

$$\Delta_m = \{a = x_0^{(m)} < x_1^{(m)} < \cdots < x_{n_m}^{(m)} = b\}$$

gives rise to a sequence of interpolating polynomials P_{Δ_m}. One might expect the polynomials P_{Δ_m} to converge toward f if the fineness

$$\|\Delta_m\| := \max_i \; |x_{i+1}^{(m)} - x_i^{(m)}|$$

of the partitions tends to 0 as $m \to \infty$. In general this is not true. For example, it has been shown for the functions

$$f(x) \equiv \frac{1}{1 + x^2}, \quad [a, b] = [-5, 5], \quad \text{or} \quad f(x) \equiv \sqrt{x}, \quad [a, b] = [0, 1],$$

that the polynomials P_{Δ_m} do not converge pointwise to f for arbitrarily fine uniform partitions Δ_m, $x_i^{(m)} = a + i(b - a)/m$, $i = 0, \ldots, m$.

2.1.5 Hermite Interpolation

Consider the real numbers ξ_i, $y_i^{(k)}$, $k = 0, 1, \ldots, n_i - 1$, $i = 0, 1, \ldots, m$, with

$$\xi_0 < \xi_1 < \cdots < \xi_m.$$

The *Hermite interpolation problem* for these data consists of determining a polynomial P whose degree does not exceed n, where

$$n + 1 := \sum_{i=0}^{m} n_i,$$

and which satisfies the following interpolation conditions:

$$\text{(2.1.5.1)} \quad P^{(k)}(\xi_i) = y_i^{(k)}, \qquad k = 0, 1, \ldots, n_i - 1, \quad i = 0, 1, \ldots, m.$$

This problem differs from the usual interpolation problem for polynomials in that it prescribes at each support abscissa ξ_i not only the value but also the first $n_i - 1$ derivatives of the desired polynomial. The polynomial interpolation of Section 2.1.1 is the special case $n_i = 1$, $i = 0, 1, \ldots, m$.

There are exactly $\sum n_i = n + 1$ conditions (2.1.5.1) for the $n + 1$ coefficients of the interpolating polynomial, leading us to expect that the Hermite interpolation problem can be solved uniquely:

(2.1.5.2) Theorem. *For arbitrary numbers $\xi_0 < \xi_1 < \cdots < \xi_m$, $y_i^{(k)}$, $k = 0, 1, \ldots, n_i - 1$, $i = 0, 1, \ldots, m$, there exists precisely one polynomial*

$$P \in \Pi_n, \qquad n + 1 := \sum_{i=0}^{m} n_i,$$

which satisfies (2.1.5.1).

PROOF. We first show uniqueness. Consider the difference polynomial $Q(x) := P_1(x) - P_2(x)$ of two polynomials P_1, $P_2 \in \Pi_n$ for which (2.1.5.1) holds. Since

$$Q^{(k)}(\xi_i) = 0, \qquad k = 0, 1, \ldots, n_i - 1, \quad i = 0, 1, \ldots, m,$$

ξ_i is at least an n_i-fold root of Q, so that Q has altogether $\sum n_i = n + 1$ roots, each counted according to its multiplicity. Thus Q must vanish identically, since its degree is less than $n + 1$.

Existence is a consequence of uniqueness: For (2.1.5.1) is a system of n linear equations for n unknown coefficients c_j of $P(x) = c_0 + c_1 x + \cdots + c_n x^n$. The matrix of this system is not singular, because of the uniqueness of its solutions. Hence the linear system (2.1.5.1) has a unique solution for arbitrary right-hand sides $y_i^{(k)}$. □

Hermite interpolating polynomials can be given explicitly in a form analogous to the interpolation formula of Lagrange (2.1.1.4). The polynomial

$P \in \Pi_n$ given by

$$P(x) = \sum_{i=0}^{m} \sum_{k=0}^{n_i - 1} y_i^{(k)} L_{ik}(x) \tag{2.1.5.3}$$

satisfies (2.1.5.1). The polynomials $L_{ik} \in \Pi_n$ are generalized Lagrange polynomials. They are defined as follows: Starting with the auxiliary polynomials

$$l_{ik}(x) := \frac{(x - \xi_i)^k}{k!} \prod_{\substack{j=0 \\ j \neq i}}^{m} \left(\frac{x - \xi_j}{\xi_i - \xi_j} \right)^{n_j}, \qquad 0 \leqslant i \leqslant m, \quad 0 \leqslant k < n_i$$

[compare (2.1.1.3)], put

$$L_{i, n_i - 1}(x) := l_{i, n_i - 1}(x), \qquad i = 0, 1, \ldots, m,$$

and recursively for $k = n_i - 2, n_i - 3, \ldots, 0$,

$$L_{ik}(x) := l_{ik}(x) - \sum_{\nu = k+1}^{n_i - 1} l_{ik}^{(\nu)}(\xi_i) L_{i\nu}(x).$$

By induction

$$L_{ik}^{(\sigma)}(\xi_j) = \begin{cases} 1 & \text{if } i = j \text{ and } k = \sigma, \\ 0 & \text{otherwise} \end{cases}$$

Thus P in (2.1.5.3) is indeed the desired Hermite interpolating polynomial.

In order to describe alternative methods for determining P, it will be useful to represent the data ξ_i, $y_i^{(k)}$, $i = 0, 1, \ldots, m$, $k = 0, 1, \ldots, n_i - 1$, in a somewhat different form as a sequence $\mathscr{F}_n = \{(x_i, f_i)\}_{i=0, \ldots, n}$ of $n + 1$ pairs of numbers. The pairs

$$(x_0, f_0), (x_1, f_1), \ldots, (x_{n_0 - 1}, f_{n_0 - 1}), (x_{n_0}, f_{n_0}), \ldots, (x_n, f_n)$$

of $\mathscr{F}_n$ denote consecutively the pairs

$$(\xi_0, y_0^{(0)}), (\xi_0, y_0^{(1)}), \ldots, (\xi_0, y_0^{(n_1 - 1)}), (\xi_1, y_1^{(0)}), \ldots, (\xi_m, y_m^{(n_m - 1)}).$$

Note that $x_0 \leqslant x_1 \leqslant \cdots \leqslant x_n$ and that the number ξ_i occurs exactly n_i times in the sequence $\{x_i\}_{i=0, \ldots, n}$.

EXAMPLE 1. Suppose $m = 2$, $n_0 = 2$, $n_1 = 3$ and

$$\xi_0 = 0, \qquad y_0^{(0)} = -1, \qquad y_0^{(1)} = -2;$$

$$\xi_1 = 1, \qquad y_1^{(0)} = 0, \qquad y_1^{(1)} = 10, \qquad y_1^{(2)} = 40.$$

This problem is described by the sequence $\mathscr{F}_4 = \{(x_i, f_i)\}_{i=0, \ldots, 4}$:

$$(x_0, f_0) = (0, -1), \qquad (x_1, f_1) = (0, -2), \qquad (x_2, f_2) = (1, 0),$$

$$(x_3, f_3) = (1, 10), \qquad (x_4, f_4) = (1, 40).$$

Given any Hermite interpolation problem, it uniquely determines a sequence $\mathscr{F}_n$, as above. Conversely, every sequence $\mathscr{F}_n = \{(x_i, f_i)\}_{i=0,\ldots,n}$ of $n+1$ pairs of numbers with $x_0 \leqslant x_1 \leqslant \cdots \leqslant x_n$ determines a Hermite interpolation problem, which will be referred to simply as $\mathscr{F}_n$. It also will be convenient to denote by

$$[\![x - x_0]\!]^j$$

the polynomials

$$(2.1.5.4) \qquad \begin{aligned} &[\![x - x_0]\!]^0 := 1, \\ &[\![x - x_0]\!]^j := (x - x_0)(x - x_1) \ldots (x - x_{j-1}) \end{aligned}$$

of degree j.

Our next goal is to represent the polynomial P which interpolates $\mathscr{F}_n$ in Newton form [compare (2.1.3.1)]:

$$(2.1.5.5) \quad P(x) = a_0 + a_1[\![x - x_0]\!] + a_2[\![x - x_0]\!]^2 + \cdots + a_n[\![x - x_0]\!]^n$$

and to determine the coefficients a_i with the help again of divided differences

$$(2.1.5.6) \qquad a_k = f[x_0, x_1, \ldots, x_k], \qquad k = 0, 1, \ldots, n.$$

However, the recursive definition (2.1.3.5) of the divided differences has to be modified because there may be repetitions among the support abscissae $x_0 \leqslant x_1 \leqslant \cdots \leqslant x_n$. For instance, if $x_0 = x_1$, then the divided difference $f[x_0, x_1]$ can no longer be defined as $(f[x_0] - f[x_1])/(x_1 - x_0)$.

The extension of the definition of divided differences to the case of repeated arguments involves transition to a limit. To this end, let

$$\zeta_0 < \zeta_1 < \cdots < \zeta_n$$

be mutually distinct support abscissas, and consider the divided differences $f[\zeta_i, \ldots, \zeta_{i+k}]$ which belong to the function $f(x) := P(x)$, where the polynomial P is the solution of the Hermite interpolation problem $\mathscr{F}_n$. These divided differences are now well defined by the recursion (2.1.3.5), if we let $f_i := P(\zeta_i)$ initially. Therefore, and by (2.1.3.5),

$$(2.1.5.7a) \qquad P(x) = \sum_{j=0}^{n} a_j[\![x - \zeta_0]\!]^j, \qquad a_j := f[\zeta_0, \zeta_1, \ldots, \zeta_j],$$

$$(2.1.5.7b) \qquad f[\zeta_i] = P(\zeta_i),$$

(2.1.5.7c)

$$f[\zeta_i, \zeta_{i+1}, \ldots, \zeta_{i+k}] = \frac{f[\zeta_{i+1}, \zeta_{i+2}, \ldots, \zeta_{i+k}] - f[\zeta_i, \zeta_{i+1}, \ldots, \zeta_{i+k-1}]}{\zeta_{i+k} - \zeta_i},$$

for $i = 0, 1, \ldots, n$, $k = 1, \ldots, n - i$. Since $x_0 \leqslant x_1 \leqslant \cdots \leqslant x_n$, all limits

$$f[x_i, x_{i+1}, \ldots, x_{i+k}] := \lim_{\substack{\zeta_j \to x_j \\ i \le j \le i+k}} f[\zeta_i, \zeta_{i+1}, \ldots, \zeta_{i+k}]$$

exist provided they exist for indices i, k with $x_i = x_{i+1} = \cdots = x_{i+k}$. The latter follows from (2.1.4.3), which yields

(2.1.5.8) $$\lim_{\substack{\zeta_j \to x_j \\ i \le j \le i+k}} f[\zeta_i, \zeta_{i+1}, \ldots, \zeta_{i+k}] = \frac{1}{k!} P^{(k)}(x_i)$$

$$\text{if } x_i = x_{i+1} = \cdots = x_{i+k}.$$

We now denote by $r = r(i) \geqslant 0$ the smallest index such that

$$x_r = x_{r+1} = \cdots = x_i.$$

Then due to the interpolation properties of P with respect to $\mathscr{F}_n$,

$$P^{(k)}(x_i) = P^{(k)}(x_r) = f_{r+k},$$

so that by (2.1.5.8)

$$f[x_i, x_{i+1}, \ldots, x_{i+k}] = \frac{f_{r(i)+k}}{k!} \quad \text{if } x_i = x_{i+1} = \cdots = x_{i+k}.$$

In the limit $\zeta_j \to x_j$, (2.1.5.7) becomes

(2.1.5.9a) $$P(x) = \sum_{j=0}^{n} a_j [\![x - x_j]\!]^j, \quad a_j := f_j[x_0, x_1, \ldots, x_j]$$

(2.1.5.9b) $$f[x_i, x_{i+1}, \ldots, x_{i+k}] := \frac{f_{r(i)+k}}{k!} \qquad \text{if } x_i = x_{i+k}$$

(2.1.5.9c) $$f[x_i, x_{i+1}, \ldots, x_{i+k}] := \frac{f[x_{i+1}, x_{i+2}, \ldots, x_{i+k}] - f[x_i, x_{i+1}, \ldots, x_{i+k-1}]}{x_{i+k} - x_i},$$

otherwise.

(Note that $x_0 \leqslant x_1 \leqslant \cdots \leqslant x_n$ has been assumed.) These formulas now permit a recursive calculation of the divided differences and thereby the coefficients a_j of the interpolating polynomial P in Newton form.

EXAMPLE 2. We illustrate the calculation of the divided differences with the data of Example 1 $(m = 2, n_0 = 2, n_1 = 3)$:

$$\mathscr{F}_4 = \{(0, -1), (0, -2), (1, 0), (1, 10), (1, 40)\}.$$

The following difference scheme results:

$$\begin{array}{llllll}
x_0 = 0 & -1^* = f[x_0] & & & & \\
 & & -2^* = f[x_0, x_1] & & & \\
x_1 = 0 & -1^* = f[x_1] & & 3 = f[x_0, x_1, x_2] & & \\
 & & 1 = f[x_1, x_2] & & 6 = f[x_0, \ldots, x_3] & \\
x_2 = 1 & 0^* = f[x_2] & & 9 = f[x_1, x_2, x_3] & & 5 = f[x_0, \ldots, x_4] \\
 & & 10^* = f[x_2, x_3] & & 11 = f[x_1, \ldots, x_4] & \\
x_3 = 1 & 0^* = f[x_3] & & 20^* = f[x_2, x_3, x_4] & & \\
 & & 10^* = f[x_3, x_4] & & & \\
x_4 = 1 & 0^* = f[x_4] & & & &
\end{array}$$

The entries marked * have been calculated using (2.1.5.9b) rather than (2.1.5.9c). The coefficients of the Hermite interpolating polynomial can be found in the upper diagonal of the difference scheme:

$$\begin{aligned} P(x) &= -1 - 2[\![x - x_0]\!] + 3[\![x - x_0]\!]^2 + 6[\![x - x_0]\!]^3 + 5[\![x - x_0]\!]^4 \\ &= -1 - 2x + 3x^2 + 6x^2(x-1) + 5x^2(x-1)^2. \end{aligned}$$

The interpolation error which is incurred by Hermite interpolation can be estimated in the same fashion as for the usual interpolation by polynomials. In particular, the proof of the following theorem is entirely analogous to the proof of Theorem (2.1.4.1):

(2.1.5.10) Theorem. *Let the real function f be $n+1$ times differentiable on the interval $[a, b]$, and consider $m+1$ support abscissae $\xi_i \in [a, b]$,*

$$\xi_0 < \xi_1 < \cdots < \xi_m.$$

If the polynomial $P(x)$ is of degree at most n,

$$\sum_{i=0}^{m} n_i = n + 1,$$

and satisfies the interpolation conditions

$$P^{(k)}(\xi_i) = f^{(k)}(\xi_i), \qquad k = 0, 1, \ldots, n_i - 1, \quad i = 0, 1, \ldots, m,$$

then to every $\bar{x} \in [a, b]$ there exists $\bar{\xi} \in I[\xi_0, \ldots, \xi_m, \bar{x}]$ such that

$$f(\bar{x}) - P(\bar{x}) = \frac{\omega(\bar{x}) f^{(n+1)}(\bar{\xi})}{(n+1)!},$$

where

$$\omega(x) := (x - \xi_0)^{n_0}(x - \xi_1)^{n_1} \ldots (x - \xi_m)^{n_m}.$$

Hermite interpolation is frequently used to approximate a given real function f by a piecewise polynomial function φ. Given a partition

$$\Delta a = \xi_0 < \xi_1 < \cdots < \xi_m = b$$

of an interval $[a, b]$, the corresponding *Hermite function space* $H_\Delta^{(\nu)}$ is defined as consisting of all functions $\varphi\colon [a, b] \to \mathbb{R}$ with the following properties:

(2.1.5.11).

(*a*) $\varphi \in C^{\nu-1}[a, b]$: *The $(\nu - 1)$st derivative of φ exists and is continuous on $[a, b]$.*

(*b*) $\varphi | I_i \in \Pi_{2\nu-1}$: *On each subinterval $I_i := [\xi_i, \xi_{i+1}]$, $i = 0, 1, \ldots, m-1$, φ agrees with a polynomial of degree at most $2\nu - 1$.*

Thus the function φ consists of polynomial pieces of degree $2\nu - 1$ or less which are $\nu - 1$ times differentiable at the "knots" ξ_i. In order to approximate a given real function $f \in C^{\nu-1}[a, b]$ by a function $\varphi \in H_\Delta^{(\nu)}$, we choose the component polynomials $P_i = \varphi \mid I_i$ of φ so that $P_i \in \Pi_{2\nu-1}$ and so that the Hermite interpolation conditions

$$P_i^{(k)}(\xi_i) = f^{(k)}(\xi_i), \quad P_i^{(k)}(\xi_{i+1}) = f^{(k)}(\xi_{i+1}), \qquad k = 0, 1, \ldots, \nu - 1,$$

are satisfied.

Under the more stringent condition $f \in C^{2\nu}[a, b]$, Theorem (2.1.5.10) provides a bound to the interpolation error for $x \in I_i$ which arises if the component polynomial P_i replaces f:

$$\text{(2.1.5.12)} \qquad \begin{aligned} |f(x) - P_i(x)| &\leqslant \frac{|(x - \xi_i)(x - \xi_{i+1})|^\nu}{(2\nu)!} \max_{\xi \in I_i} |f^{(2\nu)}(\xi)| \\ &\leqslant \frac{|\xi_{i+1} - \xi_i|^{2\nu}}{2^{2\nu} \cdot (2\nu)!} \max_{\xi \in I_i} |f^{(2\nu)}(\xi)|. \end{aligned}$$

Combining these results for $i = 0, 1, \ldots, m$ gives for the function $\varphi \in H_\Delta^{(\nu)}$, which was defined earlier,

$$\text{(2.1.5.13)} \qquad \|f - \varphi\|_\infty := \max_{x \in [a, b]} |f(x) - \varphi(x)| \leqslant \frac{1}{2^{2\nu}(2\nu)!} \|f^{(2\nu)}\|_\infty \|\Delta\|^{2\nu},$$

where

$$\|\Delta\| = \max_{0 \le i \le m-1} |\xi_{i+1} - \xi_i|$$

is the "fineness" of the partition Δ.

The approximation error goes to zero with the 2νth power of the fineness $\|\Delta_j\|$ if we consider a sequence of partitions Δ_j of the interval $[a, b]$ with $\|\Delta_j\| \to 0$. Contrast this with the case of ordinary polynomial interpolation, where the approximation error does not necessarily go to zero as $\|\Delta_j\| \to 0$ (Section 2.1.4).

Ciarlet, Schultz, and Varga (1967) were able to show that also the first ν derivatives of φ are good approximations to the corresponding derivatives of f:

(2.1.5.14)

$$|f^{(k)}(x) - P_i^{(k)}(x)| \leqslant \frac{|(x - \xi_i)(x - \xi_{i+1})|^{\nu-k}}{k!\,(2\nu - 2k)!} (\xi_{i+1} - \xi_i)^k \max_{\xi \in I_i} |f^{(2\nu)}(\xi)|$$

for all $x \in I_i$, $k = 0, 1, \ldots, \nu$, $i = 0, 1, \ldots, m - 1$, and therefore

$$\text{(2.1.5.15)} \qquad \|f^{(k)} - \varphi^{(k)}\|_\infty \leqslant \frac{\|\Delta\|^{2\nu-k}}{2^{2\nu-2k}k!\,(2\nu - 2k)!} \|f^{(2\nu)}\|_\infty$$

for $k = 0, 1, \ldots, \nu$.

2.2 Interpolation by Rational Functions

2.2.1 General Properties of Rational Interpolation

Consider again a given set of support points (x_i, f_i), $i = 0, 1, 2, \ldots$. We will now examine the use of rational functions

$$\Phi^{\mu,\nu}(x) \equiv \frac{P^{\mu,\nu}(x)}{Q^{\mu,\nu}(x)} \equiv \frac{a_0 + a_1 x + \cdots + a_\mu x^\mu}{b_0 + b_1 x + \cdots + b_\nu x^\nu}$$

for interpolating these support points. Here the integers μ and ν denote the maximum degrees of the polynomials in the numerator and denominator, respectively. We call the pair of integers (μ, ν) the *degree type* of the rational interpolation problem.

The rational function $\Phi^{\mu,\nu}$ is determined by its $\mu + \nu + 2$ coefficients

$$a_0, a_1, \ldots, a_\mu, b_0, b_1, \ldots, b_\nu.$$

On the other hand, $\Phi^{\mu,\nu}$ determines these coefficients only up to a common factor $\rho \neq 0$. This suggests that $\Phi^{\mu,\nu}$ is fully determined by the $\mu + \nu + 1$ interpolation conditions

$$(2.2.1.1) \qquad \Phi^{\mu,\nu}(x_i) = f_i, \qquad i = 0, 1, \ldots, \mu + \nu.$$

We denote by $A^{\mu,\nu}$ the problem of calculating the rational function $\Phi^{\mu,\nu}$ from (2.2.1.1).

It is clearly necessary that the coefficients a_r, b_s of $\Phi^{\mu,\nu}$ solve the homogeneous system of linear equations

$$(2.2.1.2) \qquad P^{\mu,\nu}(x_i) - f_i Q^{\mu,\nu}(x_i) = 0, \qquad i = 0, 1, \ldots, \mu + \nu,$$

or written out in full,

$$a_0 + a_1 x_i + \cdots + a_\mu x_i^\mu - f_i(b_0 + b_1 x_i + \cdots + b_\nu x_i^\nu) = 0.$$

We denote the above system by $S^{\mu,\nu}$.

At first glance, substituting $S^{\mu,\nu}$ for $A^{\mu,\nu}$ does not seem to present a problem. The next example will show, however, that this is not the case, and that rational interpolation is inherently more complicated than polynomial interpolation.

Example. For support points

x_i	0	1	2
f_i	1	2	2

and $\mu = \nu = 1$:

$$\begin{aligned} a_0 \qquad\quad - 1 \cdot b_0 \qquad\quad &= 0, \\ a_0 + a_1 - 2(b_0 + b_1) \; &= 0, \\ a_0 + 2a_1 - 2(b_0 + 2b_1) &= 0. \end{aligned}$$

Up to a common nonzero factor, solving the above system $S^{1,1}$ yields the coefficients

$$a_0 = 0, \qquad b_0 = 0, \qquad a_1 = 2, \qquad b_1 = 1,$$

and therefore the rational expression

$$\Phi^{1,1}(x) \equiv \frac{2x}{x},$$

which for $x = 0$ leads to the indeterminate expression 0/0. After canceling the factor x, we arrive at the rational expression

$$\tilde{\Phi}^{1,1}(x) \equiv 2.$$

Both expressions $\Phi^{1,1}$ and $\tilde{\Phi}^{1,1}$ represent the same rational function, namely the constant function of value 2. This function misses the first support point $(x_0, f_0) = (0, 1)$. Therefore it does not solve $A^{1,1}$. Since solving $S^{1,1}$ is necessary for any solution of $A^{1,1}$, we conclude that no such solution exists.

The above example shows that the rational interpolation problem $A^{\mu,\nu}$ need not be solvable. Indeed, if $S^{\mu,\nu}$ has a solution which leads to a rational function that does not solve $A^{\mu,\nu}$—as was the case in the example—then the rational interpolation problem is not solvable. In order to examine this situation more closely, we have to distinguish between different representations of the same rational function $\Phi^{\mu,\nu}$, which arise from each other by canceling or by introducing a common polynomial factor in numerator and denominator. We say that two rational expressions,

$$\Phi_1(x) :\equiv \frac{P_1(x)}{Q_1(x)}, \quad \Phi_2(x) :\equiv \frac{P_2(x)}{Q_2(x)}, \qquad Q_1(x) \not\equiv 0, \quad Q_2(x) \not\equiv 0,$$

are *equivalent*, and write

$$\Phi_1 \sim \Phi_2,$$

if

$$P_1(x)Q_2(x) \equiv P_2(x)Q_1(x).$$

This is precisely when the two rational expressions represent the same rational function.

A rational expression is called *relatively prime* if its numerator and denominator are relatively prime, i.e., not both divisible by the same polynomial of positive degree. If a rational expression is not relatively prime, then canceling all common polynomial factors leads to an equivalent rational expression which is.

Finally we say that a rational expression $\Phi^{\mu,\nu}$ is a solution of $S^{\mu,\nu}$ if its coefficients solve $S^{\mu,\nu}$. As noted before, $\Phi^{\mu,\nu}$ solves $S^{\mu,\nu}$ if it solves $A^{\mu,\nu}$. Rational interpolation is complicated by the fact that the converse need not hold.

(2.2.1.3) Theorem. *The homogeneous linear system of equations $S^{\mu,\nu}$ always has nontrivial solutions. For each such solution*

$$\Phi^{\mu,\nu} \equiv \frac{P^{\mu,\nu}(x)}{Q^{\mu,\nu}(x)},$$

$Q^{\mu,\nu}(x) \not\equiv 0$ *holds, i.e., all nontrivial solutions define rational expressions.*

PROOF. The homogeneous linear system $S^{\mu,\nu}$ has $\mu + \nu + 1$ equations for $\mu + \nu + 2$ unknowns. As a homogeneous linear system with more unknowns than equations, $S^{\mu,\nu}$ has nontrivial solutions

$$(a_0, a_1, \ldots, a_\mu, b_0, \ldots, b_\nu) \neq (0, \ldots, 0, 0, \ldots, 0).$$

For any such solution, $Q^{\mu,\nu}(x) \not\equiv 0$, since

$$Q^{\mu,\nu}(x) \equiv b_0 + b_1 x + \cdots + b_\nu x^\nu \equiv 0$$

would imply that the polynomial $P^{\mu,\nu}(x) \equiv a_0 + a_1 x + \cdots + a_\mu x^\mu$ has the zeros

$$P^{\mu,\nu}(x_i) = 0, \qquad i = 0, 1, \ldots, \mu + \nu.$$

It would follow that $P^{\mu,\nu}(x) \equiv 0$, since the polynomial $P^{\mu,\nu}$ has at most degree μ, and vanishes at $\mu + \nu + 1 \geqslant \mu + 1$ different locations, contradicting

$$(a_0, a_1, \ldots, a_\mu, b_0, \ldots, b_\nu) \neq (0, \ldots, 0). \qquad \square$$

The following theorem shows that the rational interpolation problem has a unique solution if it has a solution at all.

(2.2.1.4) Theorem. *If Φ_1 and Φ_2 are both (nontrivial) solutions of the homogeneous linear system $S^{\mu,\nu}$, then they are equivalent ($\Phi_1 \sim \Phi_2$), that is, they determine the same rational function.*

PROOF. If both $\Phi_1(x) \equiv P_1(x)/Q_1(x)$ and $\Phi_2(x) \equiv P_2(x)/Q_2(x)$ solve $S^{\mu,\nu}$, then the polynomial

$$P(x) :\equiv P_1(x)Q_2(x) - P_2(x)Q_1(x)$$

has $\mu + \nu + 1$ different zeros

$$\begin{aligned} P(x_i) &= P_1(x_i)Q_2(x_i) - P_2(x_i)Q_1(x_i) \\ &= f_i Q_1(x_i)Q_2(x_i) - f_i Q_2(x_i)Q_1(x_i) \\ &= 0, \qquad i = 0, 1, \ldots, \mu + \nu. \end{aligned}$$

Since the degree of polynomial P does not exceed $\mu + \nu$, it must vanish identically, and it follows that $\Phi_1(x) \sim \Phi_2(x)$. $\square$

Note that the converse of the above theorem does not hold: a rational expression Φ_1 may well solve $S^{\mu,\nu}$ whereas some equivalent rational expression Φ_2 does not. The previously considered example furnishes a case in point. In fact, we will see that this situation is typical for unsolvable interpolation problems.

Combining Theorems (2.2.1.3) and (2.2.1.4), we find that there exists for each rational interpolation problem $A^{\mu,\nu}$ a unique rational function, which is represented by any rational expression $\Phi^{\mu,\nu}$ that solves the corresponding linear system $S^{\mu,\nu}$. Either this rational function satisfies (2.2.1.1), thereby solving $A^{\mu,\nu}$, or $A^{\mu,\nu}$ is not solvable at all. In the latter case, there must be some support point (x_i, f_i) which is "missed" by the rational function. Such a support point is called *inaccessible*. Thus $A^{\mu,\nu}$ is solvable if there are no inaccessible points.

Suppose $\Phi^{\mu,\nu}(x) \equiv P^{\mu,\nu}(x)/Q^{\mu,\nu}(x)$ is a solution to $S^{\mu,\nu}$. For any $i \in \{0, 1, \ldots, \mu + \nu\}$ we distinguish the two cases:

(1) $Q^{\mu,\nu}(x_i) \neq 0$,

(2) $Q^{\mu,\nu}(x_i) = 0$.

In the first case, clearly, $\Phi^{\mu,\nu}(x_i) = f_i$. In the second case, however, the support point (x_i, f_i) may be inaccessible. Here

$$P^{\mu,\nu}(x_i) = 0$$

must hold by (2.2.1.2). Therefore, both $P^{\mu,\nu}$ and $Q^{\mu,\nu}$ contain the factor $x - x_i$ and are consequently not relatively prime. Thus:

(2.2.1.5). *If $S^{\mu,\nu}$ has a solution $\Phi^{\mu,\nu}$ which is relatively prime, then there are no inaccessible points: $A^{\mu,\nu}$ is solvable.*

Given $\Phi^{\mu,\nu}$, let $\tilde{\Phi}^{\mu,\nu}$ be an equivalent rational expression which is relatively prime. We then have the general result:

(2.2.1.6) Theorem. *Suppose $\Phi^{\mu,\nu}$ solves $S^{\mu,\nu}$. Then $A^{\mu,\nu}$ is solvable—and $\Phi^{\mu,\nu}$ represents the solution—if and only if $\tilde{\Phi}^{\mu,\nu}$ solves $S^{\mu,\nu}$.*

PROOF. If $\tilde{\Phi}^{\mu,\nu}$ solves $S^{\mu,\nu}$, then $A^{\mu,\nu}$ is solvable by (2.2.1.5). If $\tilde{\Phi}^{\mu,\nu}$ does not solve $S^{\mu,\nu}$, its corresponding rational function does not solve $A^{\mu,\nu}$. □

Even if the linear system $S^{\mu,\nu}$ has full rank $\mu + \nu + 1$, the rational interpolation problem $A^{\mu,\nu}$ may not be solvable. However, since the solutions of $S^{\mu,\nu}$ are, in this case, uniquely determined up to a common constant factor $\rho \neq 0$, we have:

(2.2.1.7) Corollary to (2.2.1.6). *If $S^{\mu,\nu}$ has full rank, then $A^{\mu,\nu}$ is solvable if and only if the solution $\Phi^{\mu,\nu}$ of $S^{\mu,\nu}$ is relatively prime.*

We say that the support points (x_i, f_i), $i = 0, 1, \ldots, \sigma$ are in *special position* if they are interpolated by a rational expression of degree type (κ, λ)

with $\kappa + \lambda < \sigma$. In other words, the interpolation problem is solvable for a smaller combined degree of numerator and denominator than suggested by the number of support points. We observe that

(2.2.1.8). *The accessible support points of a nonsolvable interpolation problem $A^{\mu,\nu}$ are in special position.*

PROOF. Let $i_1, \ldots, i_\alpha$ be the subscripts of the inaccessible points, and let $\Phi^{\mu,\nu}$ be a solution of $S^{\mu,\nu}$. The numerator and the denominator of $\Phi^{\mu,\nu}$ were seen above to have the common factors $x - x_{i_1}, \ldots, x - x_{i_\alpha}$, whose cancellation leads to an equivalent rational expression $\Phi^{\kappa,\lambda}$ with $\kappa = \mu - \alpha$, $\lambda = \nu - \alpha$. $\Phi^{\kappa,\lambda}$ solves the interpolation problem $A^{\kappa,\lambda}$ which just consists of the $\mu + \nu + 1 - \alpha$ accessible points. As

$$\kappa + \lambda + 1 = \mu + \nu + 1 - 2\alpha < \mu + \nu + 1 - \alpha,$$

the accessible points of $A^{\mu,\nu}$ are clearly in special position. □

The observation (2.2.1.8) makes it clear that nonsolvability of the rational interpolation problem is a degeneracy phenomenon: solvability can be restored by arbitrarily small perturbations of the support points. In what follows, we will therefore restrict our attention to *fully nondegenerate* problems that is, problems for which no subset of the support points is in special position. Not only is $A^{\mu,\nu}$ solvable in this case, but so are all problems $A^{\kappa,\lambda}$ of $\kappa + \lambda + 1$ of the original support points where $\kappa + \lambda \leqslant \mu + \nu$. For further details see Milne (1950) and Maehly and Witzgall (1960).

Most of the following discussion will be of recursive procedures for solving rational interpolation problems $A^{m,n}$. With each step of such recursions there will be associated a rational expression $\Phi^{\mu,\nu}$ of degree type (μ, ν) with $\mu \leqslant m$ and $\nu \leqslant n$, and either the numerator or the denominator of $\Phi^{\mu,\nu}$ will be increased by 1. Because of the availability of this choice, the recursion methods for rational interpolation are more varied than those for polynomial interpolation. It will be helpful to plot the sequence of degree types (μ, ν) which are encountered in a particular recursion as paths in a diagram:

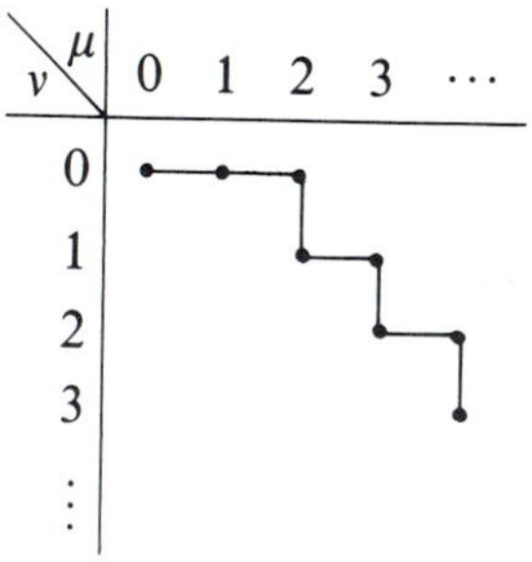

We will distinguish two kinds of algorithms. The first kind is analogous to Newton's method of interpolation: A tableau of quantities analogous to divided differences is generated from which coefficients are gathered for an interpolating rational expression. The second kind corresponds to the Neville–Aitken approach of generating a tableau of values of intermediate rational functions $\Phi^{\mu,\nu}$. These values relate to each other directly.

2.2.2 Inverse and Reciprocal Differences. Thiele's Continued Fraction

The algorithms to be described in this section calculate rational expressions along the main diagonal of the (μ, ν)-plane:

(2.2.2.1)

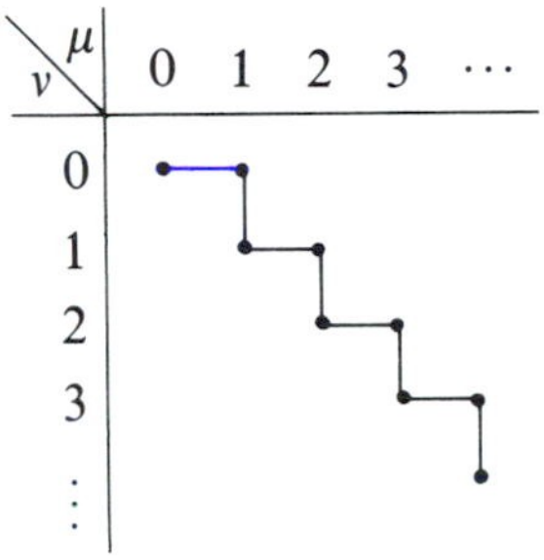

Starting from the support points (x_i, f_i), $i = 0, 1, \ldots$, we build the following tableau of *inverse differences*:

i	x_i	f_1			
0	x_0	f_0			
1	x_1	f_1	$\varphi(x_0, x_1)$		
2	x_2	f_2	$\varphi(x_0, x_2)$	$\varphi(x_0, x_1, x_2)$	
3	x_3	f_3	$\varphi(x_0, x_3)$	$\varphi(x_0, x_1, x_3)$	$\varphi(x_0, x_1, x_2, x_3)$
$\vdots$	$\vdots$	$\vdots$	$\vdots$	$\vdots$	$\vdots$

The inverse differences are defined recursively as follows:

(2.2.2.2)

$$\varphi(x_i, x_j) = \frac{x_i - x_j}{f_i - f_j},$$

$$\varphi(x_i, x_j, x_k) = \frac{x_j - x_k}{\varphi(x_i, x_j) - \varphi(x_i, x_k)},$$

$$\vdots$$

$$\varphi(x_i, \ldots, x_l, x_m, x_n) = \frac{x_m - x_n}{\varphi(x_i, \ldots, x_l, x_m) - \varphi(x_i, \ldots, x_l, x_n)}.$$

On occasion, certain inverse differences become ∞ because the denominators in (2.2.2.2) vanish.

Note that the inverse differences are, in general, *not* symmetric functions of their arguments.

Let P^μ, Q^ν be polynomials whose degree is bounded by μ and ν, respectively. We will now try to use inverse differences in order to find a rational expression

$$\Phi^{n,n}(x) = \frac{P^n(x)}{Q^n(x)}$$

with

$$\Phi^{n,n}(x_i) = f_i \quad \text{for } i = 0, 1, \ldots, 2n.$$

We must therefore have

$$\begin{aligned}\frac{P^n(x)}{Q^n(x)} &= f_0 + \frac{P^n(x)}{Q^n(x)} - \frac{P^n(x_0)}{Q^n(x_0)} \\ &= f_0 + (x - x_0)\frac{P^{n-1}(x)}{Q^n(x)} = f_0 + \frac{x - x_0}{Q^n(x)/P^{n-1}(x)}.\end{aligned}$$

The rational expression $Q^n(x)/P^{n-1}(x)$ satisfies

$$\frac{Q^n(x_i)}{P^{n-1}(x_i)} = \frac{x_i - x_0}{f_i - f_0} = \varphi(x_0, x_i)$$

for $i = 1, 2, \ldots, 2n$. It follows that

$$\begin{aligned}\frac{Q^n(x)}{P^{n-1}(x)} &= \varphi(x_0, x_1) + \frac{Q^n(x)}{P^{n-1}(x)} - \frac{Q^n(x_1)}{P^{n-1}(x_1)} \\ &= \varphi(x_0, x_1) + (x - x_1)\frac{Q^{n-1}(x)}{P^{n-1}(x)} \\ &= \varphi(x_0, x_1) + \frac{x - x_1}{P^{n-1}(x)/Q^{n-1}(x)},\end{aligned}$$

and therefore

$$\frac{P^{n-1}(x_i)}{Q^{n-1}(x_i)} = \frac{x_i - x_1}{\varphi(x_0, x_i) - \varphi(x_0, x_1)} = \varphi(x_0, x_1, x_i), \qquad i = 2, 3, \ldots, 2n.$$

Continuing in this fashion, we arrive at the following expression for $\Phi^{n,n}(x)$:

$$\Phi^{n,n}(x) = \frac{P^n(x)}{Q^n(x)} = f_0 + \frac{x - x_0}{Q^n(x)/P^{n-1}(x)}$$

$$= f_0 + \cfrac{x - x_0}{\varphi(x_0, x_1) + \cfrac{x - x_1}{P^{n-1}(x)/Q^{n-1}(x)}} = \cdots$$

$$= f_0 + \cfrac{x - x_0}{\varphi(x_0, x_1) + \cfrac{x - x_1}{\varphi(x_0, x_1, x_2) + \cfrac{x - x_2}{\varphi(x_0, x_1, x_2, x_3) + \begin{matrix} \vdots \\ + \cfrac{x - x_{2n-1}}{\varphi(x_0, \ldots, x_{2n})} \end{matrix}}}}$$

$\Phi^{n,n}(x)$ is thus represented by a *continued fraction*:

$$\begin{aligned} \Phi^{n,n}(x) = f_0 &+ \frac{x - x_0}{|\varphi(x_0, x_1)} + \frac{x - x_1}{|\varphi(x_0, x_1, x_2)} \\ &+ \frac{x - x_2}{|\varphi(x_0, x_1, x_2, x_3)} + \cdots \\ &+ \frac{x - x_{2n-1}}{|\varphi(x_0, x_1, \ldots, x_{2n})}. \end{aligned} \tag{2.2.2.3}$$

It is readily seen that the partial fractions of this continued fraction are nothing but the rational expressions $\Phi^{\mu,\mu}(x)$ and $\Phi^{\mu+1,\mu}(x)$, $\mu = 0, 1, \ldots, n-1$, which satisfy (2.2.1.1) and which are indicated in the diagram (2.2.2.1):

$$\Phi^{0,0}(x) = f_0,$$

$$\Phi^{1,0}(x) = f_0 + \frac{x - x_0}{|\varphi(x_0, x_1)},$$

$$\Phi^{1,1}(x) = f_0 + \frac{x - x_0}{|\varphi(x_0, x_1)} + \frac{x - x_1}{|\varphi(x_0, x_1, x_2)},$$

$$\vdots$$

Example

i	x_i	f_i	$\varphi(x_0, x_i)$	$\varphi(x_0, x_1, x_i)$	$\varphi(x_0, x_1, x_2, x_i)$
0	0	0			
1	1	-1	-1		
2	2	$-\frac{2}{3}$	-3	$-\frac{1}{2}$	
3	3	9	$\frac{1}{3}$	$\frac{3}{2}$	$\frac{1}{2}$

$$\Phi^{2,1}(x) = 0 + \frac{x}{|-1} + \frac{x - 1}{|-1/2} + \frac{x - 2}{|1/2} = (4x^2 - 9x)/(-2x + 7).$$

Because the inverse differences lack symmetry, the so-called *reciprocal differences*

$$\rho(x_i, x_{i+1}, \ldots, x_{i+k})$$

are often preferred. They are defined by the recursions

$$\begin{aligned}
&\rho(x_i) := f_i,\\
(2.2.2.4)\qquad &\rho(x_i, x_k) := \frac{x_i - x_k}{f_i - f_k},\\
&\vdots\\
\rho(x_i, x_{i+1}, \ldots, x_{i+k}) &:= \frac{x_i - x_{i+k}}{\rho(x_i, \ldots, x_{i+k-1}) - \rho(x_{i+1}, \ldots, x_{i+k})}\\
&\qquad + \rho(x_{i+1}, \ldots, x_{i+k-1}).
\end{aligned}$$

For a proof that the reciprocal differences are indeed symmetrical, see Milne–Thompson (1951).

The reciprocal differences are closely related to the inverse differences.

(2.2.2.5) Theorem. *For* $p = 1, 2, \ldots$ [*letting* $\rho(x_0, \ldots, x_{p-2}) = 0$ *for* $p = 1$],

$$\varphi(x_0, x_1, \ldots, x_p) = \rho(x_0, \ldots, x_p) - \rho(x_0, \ldots, x_{p-2}).$$

PROOF. The proposition is correct for $p = 1$. Assuming it true for p, we conclude from

$$\varphi(x_0, x_1, \ldots, x_{p+1}) = \frac{x_p - x_{p+1}}{\varphi(x_0, \ldots, x_p) - \varphi(x_0, \ldots, x_{p-1}, x_{p+1})}$$

that

$$\varphi(x_0, x_1, \ldots, x_{p+1}) = \frac{x_p - x_{p+1}}{\rho(x_0, \ldots, x_p) - \rho(x_0, \ldots, x_{p-1}, x_{p+1})}.$$

By (2.2.2.4),

$$\rho(x_{p+1}, x_0, \ldots, x_p) - \rho(x_0, \ldots, x_{p-1}) = \frac{x_{p+1} - x_p}{\rho(x_0, \ldots, x_p) - \rho(x_{p+1}, x_0, \ldots, x_p)}.$$

Since the $\rho(\ldots)$ are symmetric,

$$\varphi(x_0, x_1, \ldots, x_{p+1}) = \rho(x_0, \ldots, x_{p+1}) - \rho(x_0, \ldots, x_{p-1}),$$

whence (2.2.2.5) has been established for $p + 1$. □

The reciprocal differences can be arranged in the tableau

$$
\begin{array}{c|ccccc}
x_0 & f_0 & & & & \\
 & & \underline{\rho(x_0, x_1)} & & & \\
x_1 & f_1 & & \underline{\rho(x_0, x_1, x_2)} & & \\
 & & \rho(x_1, x_2) & & \underline{\rho(x_0, x_1, x_2, x_3)} & \\
x_2 & f_2 & & \rho(x_1, x_2, x_3) & \vdots & \\
 & & \rho(x_2, x_3) & \vdots & & \\
x_3 & f_3 & \vdots & & & \\
\vdots & \vdots & & & &
\end{array}
\tag{2.2.2.6}
$$

Using (2.2.2.5) to substitute reciprocal differences for inverse differences in (2.2.2.3) yields *Thiele's continued fraction*:

$$
\begin{aligned}
\Phi^{n,n}(x) = f_0 &+ \frac{x - x_0|}{|\rho(x_0, x_1)} + \frac{x - x_1|}{|\rho(x_0, x_1, x_2) - \rho(x_0)} \\
&+ \cdots + \frac{x - x_{2n-1}|}{|\rho(x_0, \ldots, x_{2n}) - \rho(x_0, \ldots, x_{2n-2})}.
\end{aligned}
\tag{2.2.2.7}
$$

2.2.3 Algorithms of the Neville Type

We proceed to derive an algorithm for rational interpolation which is analogous to Neville's algorithm for polynomial interpolation.

A quick reminder that, after discussing possible degeneracy effects in rational interpolation problems (Section 2.2.1), we have assumed that such effects are absent in the problems whose solution we are discussing. Indeed, such degeneracies are not likely to occur in numerical problems.

We use

$$\Phi_s^{\mu,\nu}(x) \equiv \frac{P_s^{\mu,\nu}(x)}{Q_s^{\mu,\nu}(x)}$$

to denote the rational expression with

$$\Phi_s^{\mu,\nu}(x_i) = f_i \quad \text{for } i = s, s+1, \ldots, s+\mu+\nu,$$

$P_s^{\mu,\nu}$, $Q_s^{\mu,\nu}$ being polynomials of degrees not exceeding μ and ν, respectively. Let $p_s^{\mu,\nu}$ and $q_s^{\mu,\nu}$ be the leading coefficients of these polynomials:

$$P_s^{\mu,\nu}(x) = p_s^{\mu,\nu} x^\mu + \cdots, \qquad Q_s^{\mu,\nu}(x) = q_s^{\mu,\nu} x^\nu + \cdots.$$

For brevity we put

$$\alpha_i := x - x_i \quad \text{and} \quad T_s^{\mu,\nu}(x, y) := P_s^{\mu,\nu}(x) - y Q_s^{\mu,\nu}(x),$$

noting that

$$T_s^{\mu,\nu}(x_i, f_i) = 0, \qquad i = s, s+1, \ldots, s+\mu+\nu.$$

(2.2.3.1) Theorem. *Starting with*

$$P_s^{0,0}(x) = f_s, \qquad Q_s^{0,0}(x) = 1,$$

the following recursions hold:

(a) *Transition* $(\mu - 1, \nu) \to (\mu, \nu)$:

$$P_s^{\mu,\nu}(x) = \alpha_s q_s^{\mu-1,\nu} P_{s+1}^{\mu-1,\nu}(x) - \alpha_{s+\mu+\nu} q_{s+1}^{\mu-1,\nu} P_s^{\mu-1,\nu}(x),$$
$$Q_s^{\mu,\nu}(x) = \alpha_s q_s^{\mu-1,\nu} Q_{s+1}^{\mu-1,\nu}(x) - \alpha_{s+\mu+\nu} q_{s+1}^{\mu-1,\nu} Q_s^{\mu-1,\nu}(x).$$

(b) *Transition* $(\mu, \nu - 1) \to (\mu, \nu)$:

$$P_s^{\mu,\nu}(x) = \alpha_s p_s^{\mu,\nu-1} P_{s+1}^{\mu,\nu-1}(x) - \alpha_{s+\mu+\nu} p_{s+1}^{\mu,\nu-1} P_s^{\mu,\nu-1}(x),$$
$$Q_s^{\mu,\nu}(x) = \alpha_s p_s^{\mu,\nu-1} Q_{s+1}^{\mu,\nu-1}(x) - \alpha_{s+\mu+\nu} p_{s+1}^{\mu,\nu-1} Q_s^{\mu,\nu-1}(x).$$

PROOF. We show only (a), the proof of (b) being analogous. Suppose the rational expressions $\Phi_s^{\mu-1,\nu}$ and $\Phi_{s+1}^{\mu-1,\nu}$ meet the interpolation requirements

$$\text{(2.2.3.2)} \qquad \begin{aligned} T_s^{\mu-1,\nu}(x_i, f_i) &= 0 \quad \text{for } i = s, \ldots, s + \mu + \nu - 1, \\ T_{s+1}^{\mu-1,\nu}(x_i, f_i) &= 0 \quad \text{for } i = s + 1, \ldots, s + \mu + \nu. \end{aligned}$$

If we define $P_s^{\mu,\nu}(x)$, $Q_s^{\mu,\nu}(x)$ by (a), then the degree of $P_s^{\mu,\nu}$ clearly does not exceed μ. The polynomial expression for $Q_s^{\mu,\nu}$ contains formally a term with $x^{\nu+1}$, whose coefficient, however, vanishes. The polynomial $Q_s^{\mu,\nu}$ is therefore of degree at most ν. Finally,

$$T_s^{\mu,\nu}(x, y) = \alpha_s q_s^{\mu-1,\nu} T_{s+1}^{\mu-1,\nu}(x, y) - \alpha_{s+\mu+\nu} q_{s+1}^{\mu-1,\nu} T_s^{\mu-1,\nu}(x, y).$$

From this and (2.2.3.2),

$$T_s^{\mu,\nu}(x_i, f_i) = 0 \quad \text{for } i = s, \ldots, s + \mu + \nu.$$

Under the general hypothesis that no combination (μ, ν, s) has inaccessible points, the above result shows that (a) indeed defines the numerator and denominator of $\Phi_s^{\mu,\nu}$. □

Unfortunately, the recursions (2.2.3.1) still contain the coefficients $p_s^{\mu,\nu-1}$, $q_s^{\mu-1,\nu}$. The formulas are therefore not yet suitable for the calculation of $\Phi_s^{m,n}(x)$ for a prescribed value of x. However, we can eliminate these coefficients on the basis of the following theorem.

(2.2.3.3) Theorem.

(a). $$\Phi_s^{\mu-1,\nu}(x) - \Phi_{s+1}^{\mu-1,\nu-1}(x) = k_1 \frac{(x - x_{s+1}) \ldots (x - x_{s+\mu+\nu-1})}{Q_s^{\mu-1,\nu}(x) Q_{s+1}^{\mu-1,\nu-1}(x)}$$

with $k_1 = -p_{s+1}^{\mu-1,\nu-1} q_s^{\mu-1,\nu}$,

(b). $$\Phi_{s+1}^{\mu-1,\nu}(x) - \Phi_{s+1}^{\mu-1,\nu-1}(x) = k_2 \frac{(x - x_{s+1}) \ldots (x - x_{s+\mu+\nu-1})}{Q_{s+1}^{\mu-1,\nu}(x) Q_{s+1}^{\mu-1,\nu-1}(x)}$$

with $k_2 = -p_{s+1}^{\mu-1,\nu-1} q_{s+1}^{\mu-1,\nu}$.

Proof. The numerator polynomial of the rational expression

$$\Phi_s^{\mu-1,\nu}(x) - \Phi_{s+1}^{\mu-1,\nu-1}(x) = \frac{P_s^{\mu-1,\nu}(x)Q_{s+1}^{\mu-1,\nu-1}(x) - P_{s+1}^{\mu-1,\nu-1}(x)Q_s^{\mu-1,\nu}(x)}{Q_s^{\mu-1,\nu}(x)Q_{s+1}^{\mu-1,\nu-1}(x)}$$

is at most of degree $\mu - 1 + \nu$ and has $\mu + \nu - 1$ different zeros

$$x_i, \qquad i = s+1, s+2, \ldots, s+\mu+\nu-1$$

by definition of $\Phi_s^{\mu-1,\nu}$ and $\Phi_{s+1}^{\mu-1,\nu-1}$. It must therefore be of the form

$$k_1 \cdot (x - x_{s+1}) \ldots (x - x_{s+\mu+\nu-1}) \quad \text{with } k_1 = -p_{s+1}^{\mu-1,\nu-1} q_s^{\mu-1,\nu}$$

This proves (a). (b) is shown analogously. □

(2.2.3.4) Theorem. *For* $\mu \geqslant 1$, $\nu \geqslant 1$,

$$\text{(a).}\quad \Phi_s^{\mu,\nu}(x) = \Phi_{s+1}^{\mu-1,\nu}(x) + \frac{\Phi_{s+1}^{\mu-1,\nu}(x) - \Phi_s^{\mu-1,\nu}(x)}{\dfrac{\alpha_s}{\alpha_{s+\mu+\nu}}\underbrace{\left[1 - \dfrac{\Phi_{s+1}^{\mu-1,\nu}(x) - \Phi_s^{\mu-1,\nu}(x)}{\Phi_{s+1}^{\mu-1,\nu}(x) - \Phi_{s+1}^{\mu-1,\nu-1}(x)}\right]}_{*} - 1},$$

$$\text{(b).}\quad \Phi_s^{\mu,\nu}(x) = \Phi_{s+1}^{\mu,\nu-1}(x) + \frac{\Phi_{s+1}^{\mu,\nu-1}(x) - \Phi_s^{\mu,\nu-1}(x)}{\dfrac{\alpha_s}{\alpha_{s+\mu+\nu}}\left[1 - \dfrac{\Phi_{s+1}^{\mu,\nu-1}(x) - \Phi_s^{\mu,\nu-1}(x)}{\Phi_{s+1}^{\mu,\nu-1}(x) - \Phi_{s+1}^{\mu-1,\nu-1}(x)}\right] - 1}.$$

Proof. By Theorem (2.2.3.1a),

$$\Phi_s^{\mu,\nu}(x) = \frac{\alpha_s q_s^{\mu-1,\nu} P_{s+1}^{\mu-1,\nu}(x) - \alpha_{s+\mu+\nu} q_{s+1}^{\mu-1,\nu} P_s^{\mu-1,\nu}(x)}{\alpha_s q_s^{\mu-1,\nu} Q_{s+1}^{\mu-1,\nu}(x) - \alpha_{s+\mu+\nu} q_{s+1}^{\mu-1,\nu} Q_s^{\mu-1,\nu}(x)}.$$

We now assume that $p_{s+1}^{\mu-1,\nu-1} \neq 0$, and multiply numerator and denominator of the above fraction by

$$\frac{-p_{s+1}^{\mu-1,\nu-1}(x - x_{s+1})(x - x_{s+2}) \ldots (x - x_{s+\mu+\nu-1})}{Q_{s+1}^{\mu-1,\nu}(x)Q_s^{\mu-1,\nu}(x)Q_{s+1}^{\mu-1,\nu-1}(x)}.$$

Taking Theorem (2.2.3.3) into account, we arrive at

$$\text{(2.2.3.5)} \qquad \Phi_s^{\mu,\nu}(x) = \frac{\alpha_s \Phi_{s+1}^{\mu-1,\nu}(x)[\;\;]_1 - \alpha_{s+\mu+\nu}\Phi_s^{\mu-1,\nu}(x)[\;\;]_2}{\alpha_s[\;\;]_1 - \alpha_{s+\mu+\nu}[\;\;]_2},$$

where

$$[\;\;]_1 = \Phi_s^{\mu-1,\nu}(x) - \Phi_{s+1}^{\mu-1,\nu-1}(x),$$

$$[\;\;]_2 = \Phi_{s+1}^{\mu-1,\nu}(x) - \Phi_{s+1}^{\mu-1,\nu-1}(x).$$

(a) follows by a straightforward transformation. (b) is derived analogously. □

The formulas in Theorem (2.2.3.4) can now be used to calculate the values of rational expressions for prescribed x successively, alternately increasing the degrees of numerators and denominators. This corresponds to a zigzag path in the (μ, ν)-diagram:

(2.2.3.6)

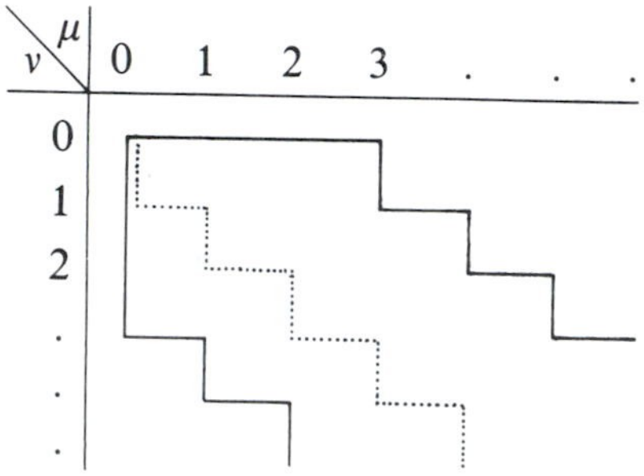

Special recursive rules are still needed for initial straight portions—vertically and horizontally—of such paths.

As long as $\nu = 0$ and only μ is being increased, one has a case of pure polynomial interpolation. One uses Neville's formulas [see (2.1.2.1)]

$$\Phi_s^{0,0}(x) := f_s,$$

$$\Phi_s^{\mu,0}(x) := \frac{\alpha_s \Phi_{s+1}^{\mu-1,0}(x) - \alpha_{s+\mu}\Phi_s^{\mu-1,0}(x)}{\alpha_s - \alpha_{s+\mu}}, \qquad \mu = 1, 2, \ldots.$$

Actually these are specializations of Theorem (2.2.3.4a) for $\nu = 0$, provided the convention $\Phi_{s+1}^{\mu-1,-1} := \infty$ is adopted, which causes the quotient marked * (on page 69) to vanish.

If $\mu = 0$ and only ν is being increased, then this case relates to polynomial interpolation with the support points $(x_i, 1/f_i)$, and one can use the formulas

$$\Phi_s^{0,0}(x) := f_s,$$

(2.2.3.7) $$\Phi_s^{0,\nu}(x) := \frac{\alpha_s - \alpha_{s+\nu}}{\dfrac{\alpha_s}{\Phi_{s+1}^{0,\nu-1}(x)} - \dfrac{\alpha_{s+\nu}}{\Phi_s^{0,\nu-1}(x)}}, \qquad \nu = 1, 2, \ldots,$$

which arise from Theorem (2.2.3.4) if one defines $\Phi_{s+1}^{-1,\nu-1}(x) := 0$.

Experience has shown that the (μ, ν)-sequence

$$(0, 0) \to (0, 1) \to (1, 1) \to (1, 2) \to (2, 2) \to \cdots$$

—indicated by the dotted line in the diagram (2.2.3.6)—holds particular advantages, especially in the important application area of extrapolation methods (Sections 3.4 and 3.5), where interest focuses on the values $\Phi_s^{\mu,\nu}(x)$ for $x = 0$. If we refer to this particular sequence, then it suffices to indicate $\mu + \nu$, instead of both μ and ν, and this permits the shorter notation

$$T_{ik} := \Phi_s^{\mu,\nu}(x) \quad \text{with } i = s + \mu + \nu,\ k = \mu + \nu.$$

The formulas (2.2.3.4) combine with (2.2.3.7) to yield the algorithm

$$T_{i0} := f_i, \qquad T_{i,-1} := 0,$$

$$\text{(2.2.3.8)} \qquad T_{ik} := T_{i,k-1} + \frac{T_{i,k-1} - T_{i-1,k-1}}{\dfrac{x - x_{i-k}}{x - x_i}\left[1 - \dfrac{T_{i,k-1} - T_{i-1,k-1}}{T_{i,k-1} - T_{i-1,k-2}}\right] - 1}$$

for $1 \leqslant k \leqslant i$, $i = 0, 1, \ldots$. Note that this recursion formula differs from the corresponding polynomial formula (2.1.2.5) only by the expression in brackets $[\ldots]$, which assumes the value 1 in the polynomial case.

If we display the values T_{ik} in the tableau below, letting i count the ascending diagonals and k the columns, then each instance of the recursion formula (2.2.3.8) interrelates the four corners of a rhombus:

$(\mu, \nu) =$	$(0, 0)$	$(0, 1)$	$(1, 1)$	$(1, 2) \ldots$
	$f_0 = T_{00}$			
$0 = T_{0,-1}$		T_{11}		
	$f_1 = T_{10}$		T_{22}	
$0 = T_{1,-1}$		T_{21}	$\longrightarrow$	T_{33}
	$f_2 = T_{20}$		T_{32}	$\vdots \ \ddots$
$0 = T_{2,-1}$		T_{31}	$\vdots$	
$\vdots$	$f_3 = T_{30}$	$\vdots$		
	$\vdots$			

If one is interested in the rational function itself, i.e. its coefficients, then the methods of Section 2.2.2, involving inverse or reciprocal differences, are suitable. However, if one desires the value of the interpolating function for just one single argument, then algorithms of the Neville type based on the formulas of Theorem (2.2.3.4) and (2.2.3.8) are to be preferred. The formula (2.2.3.8) is particularly useful in the context of extrapolation methods (see Sections 3.4, 3.5, 7.2.3, 7.2.14).

2.2.4 Comparing Rational and Polynomial Interpolations

Interpolation, as mentioned before, is frequently used for the purpose of approximating a given function $f(x)$. In many such instances, interpolation by polynomials is entirely satisfactory. The situation is different if the location x for which one desires an approximate value of $f(x)$ lies in the proximity of a pole or some other singularity of $f(x)$—like the value of $\tan x$ for x close to $\pi/2$. In such cases, polynomial interpolation does not give satisfactory results, whereas rational interpolation does, because rational functions themselves may have poles.

EXAMPLE [taken from Bulirsch and Rutishauser (1968)]. For the function $f(x) = \cot x$ the values cot 1°, cot 2°, ... have been tabulated. The problem is to determine an approximate value for cot 2°30′.

Polynomial interpolation of order 4, using the formulas (2.1.2.4), yields the tableau

x_i	$f_i = \cot(x_i)$				
1°	57.28996163				
		14.30939911			
2°	28.63625328		21.47137102		
		23.85869499		22.36661762	
3°	19.08113669		23.26186421		$22.\underline{63519158}$
		21.47137190		23.08281486	
4°	14.30066626		22.18756808		
		18.60658719			
5°	11.43005230				

Rational interpolation with $(\mu, \nu) = (2, 2)$ using the formulas (2.2.3.8) in contrast gives

1°	57.28996163				
		22.90760673			
2°	28.63625328		22.90341624		
		22.90201805		22.90369573	
3°	19.08113669		22.90411487		$22.9037655\underline{2}$
		22.91041916		22.90384141	
4°	14.30066626		22.90201975		
		22.94418151			
5°	11.43005230				

The exact value is cot 2°30′ = 22.903 765 5484...; incorrect digits are underlined.

A similar situation is encountered in extrapolation methods (see Sections 3.4, 3.5, 7.2.3, 7.2.14). Here a function $T(h)$ of the step length h is interpolated at small positive values of h.

2.3 Trigonometric Interpolation

2.3.1 Basic Facts

Trigonometric interpolation uses combinations of the trigonometric functions $\cos hx$ and $\sin hx$ for integer h. We will confine ourselves to linear interpolation, that is, interpolation by one of the trigonometric expressions

$$\Psi(x) = \frac{A_0}{2} + \sum_{h=1}^{M} (A_h \cos hx + B_h \sin hx) \tag{2.3.1.1a}$$

(2.3.1.1b) $$\Psi(x) = \frac{A_0}{2} + \sum_{h=1}^{M-1} (A_h \cos hx + B_h \sin hx) + \frac{A_M}{2} \cos Mx$$

of, respectively, $N = 2M + 1$ or $N = 2M$ support points (x_k, f_k), $k = 0, \ldots, N - 1$. Interpolation by such expressions is suitable for data which are periodic of known period. Indeed, the expressions $\Psi(x)$ in (2.3.1.1) represent periodic functions of x with the period 2π.[1]

Considerable conceptual and algebraic simplifications are achieved by using complex numbers and invoking De Moivre's formula

$$e^{kix} = \cos kx + i \sin kx.$$

Here and in what follows, i denotes the imaginary unit. If $c = a + ib$, a, b real, then $\bar{c} = a - ib$ is its complex conjugate, a is the real part of c, b its imaginary part, and $|c| = \sqrt{c\bar{c}} = \sqrt{a^2 + b^2}$ its absolute value.

Particularly important are uniform partitions of the interval $[0, 2\pi]$

$$x_k = 2\pi k/N, \qquad k = 0, \ldots, N - 1$$

to which we now restrict our attention. For such partitions, the trigonometric interpolation problem can be transformed into the problem of finding a *phase polynomial*

(2.3.1.2) $$p(x) = \beta_0 + \beta_1 e^{ix} + \cdots + \beta_{N-1} e^{(N-1)ix}$$

with complex coefficients β_j such that

$$p(x_k) = f_k, \qquad k = 0, \ldots, N - 1.$$

Indeed

$$e^{-hix_k} = e^{-2\pi ihk/N} = e^{2\pi i(N-h)k/N} = e^{(N-h)ix_k},$$

and therefore

(2.3.1.3) $$\cos hx_k = \frac{e^{hix_k} + e^{(N-h)ix_k}}{2}, \qquad \sin hx_k = \frac{e^{hix_k} - e^{(N-h)ix_k}}{2i}.$$

Making these substitutions in expressions (2.3.1.1) for $\Psi(x)$ and then collecting the powers of e^{ix_k} produces a phase polynomial $p(x)$, (2.3.1.2), with

[1] If $\sin u$ and $\cos u$ have to be both evaluated for the same argument u, then it may be advantageous to evaluate $t = \tan(u/2)$ and to express $\sin u$ and $\cos u$ in terms of t:

$$\sin u = \frac{2t}{1 + t^2}, \qquad \cos u = \frac{1 - t^2}{1 + t^2}.$$

This procedure is numerically stable for $0 \leqslant u \leqslant \pi/4$, and the problem can always be transformed so that the argument falls into that range.

coefficients $\beta_j, j = 0, \ldots, N-1$ which are related to the coefficients A_h, B_h of $\Psi(x)$ as follows:

(2.3.1.4)

(a) *If N is odd, then* $N = 2M + 1$ *and*

$$\beta_0 = \frac{A_0}{2}, \qquad \beta_j = \tfrac{1}{2}(A_j - iB_j),\ \beta_{N-j} = \tfrac{1}{2}(A_j + iB_j), \qquad j = 1, \ldots, M;$$

$$A_0 = 2\beta_0, \qquad A_h = \beta_h + \beta_{N-h}, \qquad B_h = i(\beta_h - \beta_{N-h}), \quad h = 1, \ldots, M.$$

(b) *If N is even, then* $N = 2M$ *and*

$$\beta_0 = \frac{A_0}{2}, \quad \beta_j = \tfrac{1}{2}(A_j - iB_j), \quad \beta_{N-j} = \tfrac{1}{2}(A_j + iB_j), \quad j = 1, \ldots, M-1,$$

$$\beta_M = \frac{A_M}{2};$$

$$A_0 = 2\beta_0, \qquad A_h = \beta_h + \beta_{N-h}, \quad B_h = i(\beta_h - \beta_{N-h}), \qquad h = 1, \ldots, M-1,$$

$$A_M = 2\beta_M.$$

The trigonometric expression $\Psi(x)$ and its corresponding phase polynomial $p(x)$ agree for all support arguments $x_k = 2\pi k/N$ of an equidistant partition of the interval $[0, 2\pi]$:

$$f_k = \Psi(x_k) = p(x_k), \qquad k = 0, 1, \ldots, N-1.$$

However, $\Psi(x) = p(x)$ need not hold at intermediate points $x \neq x_k$. The two interpolation problems are equivalent only insofar as a solution to one problem will produce a solution to the other via the coefficient relations (2.3.1.4).

The phase polynomials $p(x)$ in (2.3.1.2) are structurally simpler than the trigonometric expressions $\Psi(x)$ in (2.3.1.1). Upon abbreviating

$$\omega := e^{ix},$$

$$\omega_k := e^{ix_k},$$

$$P(\omega) := \beta_0 + \beta_1 \omega + \cdots + \beta_{N-1}\omega^{N-1},$$

and since $\omega_j \neq \omega_k$ for $j \neq k$, $0 \leqslant j, k \leqslant N-1$, it becomes clear that we are faced with just a standard polynomial interpolation problem in disguise: find the (complex) algebraic polynomial P of degree less than N with

$$P(\omega_k) = f_k, \qquad k = 0, \ldots, N-1.$$

The uniqueness of polynomial interpolation immediately gives the following

(2.3.1.5) Theorem. *For any support points* (x_k, f_k), $k = 0, \ldots, N-1$, *with* f_k *complex and* $x_k = 2\pi k/N$, *there exists a unique phase polynomial*

$$p(x) = \beta_0 + \beta_1 e^{ix} + \cdots + \beta_{N-1} e^{(N-1)ix}$$

with

$$p(x_k) = f_k$$

for $k = 0, 1, \ldots, N-1$.

The coefficients β_j of the interpolating phase polynomial can be expressed in closed form. To this end, we note that, for $0 \leqslant j, h \leqslant N-1$

(2.3.1.6) $$\omega_k^j = \omega_j^k \quad \text{and} \quad \omega_k^{-j} = \overline{\omega_k^j}.$$

More importantly, however, we have for $0 \leqslant j, h \leqslant N-1$

(2.3.1.7) $$\sum_{k=0}^{N-1} \omega_k^j \omega_k^{-h} = \begin{cases} N & \text{for } j = h, \\ 0 & \text{for } j \neq h. \end{cases}$$

PROOF. ω_{j-h} is a root of the polynomial

$$\omega^N - 1 = (\omega - 1) \sum_{k=0}^{N-1} \omega^k,$$

from which either $\omega_{j-h} = 1$, and therefore $j = h$, or

$$\sum_{k=0}^{N-1} \omega_k^j \omega_k^{-h} = \sum_{k=0}^{N-1} \omega_k^{j-h} = \sum_{k=0}^{N-1} \omega_{j-h}^k = 0. \qquad \square$$

Introducing the N-vectors

$$w^{(h)} = (1, \omega_1^h, \ldots, \omega_{N-1}^h)^T, \qquad h = 0, \ldots, N-1,$$

we see that the sums in (2.3.1.7) are the complex scalar products of the vectors $w^{(j)}$ and $w^{(h)}$:

(2.3.1.8) $$\sum_{k=0}^{N-1} \omega_k^j \omega_k^{-h} = w^{(j)T} \overline{w^{(h)}} = [w^{(j)}, w^{(h)}].$$

This definition of the scalar product of two complex vectors is standard; it implies that $[w, w] = \sum_{h=0}^{N-1} |w_h|^2 \geqslant 0$ for each complex vector w. Thus the vectors $w^{(h)}$ are seen to form an orthogonal basis of the complex space $\mathbb{C}^N$. Note that the vectors are of length $\sqrt{[w^{(h)}, w^{(h)}]} = \sqrt{N}$ instead of length 1, however.

From the orthogonality of the vectors $w^{(h)}$ follows:

(2.3.1.9) Theorem. *The phase polynomial* $p(x) = \sum_{j=0}^{N-1} \beta_j e^{jix}$ *satisfies*

$$p(x_k) = f_k, \qquad k = 0, 1, \ldots, N-1,$$

for f_k complex and $x_k = 2\pi k/N$, if and only if

$$\beta_j = \frac{1}{N}\sum_{k=0}^{N-1} f_k\omega_k^{-j} = \frac{1}{N}\sum_{k=0}^{N-1} f_k e^{-2\pi ijk/N}, \qquad j = 0, 1, \ldots, N-1.$$

PROOF. With the vector notation $f = (f_0, f_1, \ldots, f_{N-1})^T$,

$$\frac{1}{N}\sum_{k=0}^{N-1} f_k\omega_k^{-j} = \frac{1}{N}[f, w^{(j)}] = \frac{1}{N}[\beta_0 w^{(0)} + \cdots + \beta_{N-1}w^{(N-1)}, w^{(j)}] = \beta_j. \quad \square$$

For phase polynomials $q(x)$ of degree at most s, $s \leqslant N-1$ given, it is in general not possible to make all *residuals*

$$f_k - q(x_k), \qquad k = 0, \ldots, N-1,$$

vanish, as they would for the interpolating phase polynomial. In this context, the s-segments

$$p_s(x) = \beta_0 + \beta_1 e^{ix} + \cdots + \beta_s e^{six}$$

of the interpolating polynomial $p(x)$ have an interesting best-approximation property:

(2.3.1.10) Theorem. *The s-segment $p_s(x)$, $0 \leqslant s < N$, of the interpolating phase polynomial $p(x)$ minimizes the sum*

$$S(q) = \sum_{k=0}^{N-1} |f_k - q(x_k)|^2$$

[*note that* $S(p) = 0$] *of the squared absolute values of the residuals over all phase polynomials*

$$q(x) = \gamma_0 + \gamma_1 e^{ix} + \cdots + \gamma_s e^{six}.$$

The phase polynomial $p_s(x)$ is uniquely determined by this minimum property.

PROOF. We introduce the vectors

$$p_s = (p_s(x_0), \ldots, p_s(x_{N-1}))^T, \qquad q = (q(x_0), \ldots, q(x_{N-1}))^T$$

and use the scalar product (2.3.1.8) to write

$$S(q) = [f - q, f - q].$$

By Theorem (2.3.1.9), $\beta_j = (1/N)[f, w^{(j)}]$ for $j = 0, \ldots, N-1$. For $j \leqslant s$,

$$\frac{1}{N}[f - p_s, w^{(j)}] = \frac{1}{N}\left[f - \sum_{h=0}^{s}\beta_h w^{(h)}, w^{(j)}\right] = \beta_j - \beta_j = 0,$$

and

$$[f - p_s, p_s - q] = \sum_{j=0}^{s}[f - p_s, (\beta_j - \gamma_j)w^{(j)}] = 0.$$

But then we have

$$\begin{aligned}S(q) &= [f - q, f - q] \\ &= [f - p_s + p_s - q, f - p_s + p_s - q] \\ &= [f - p_s, f - p_s] + [p_s - q, p_s - q] \\ &\geqslant [f - p_s, f - p_s] \\ &= S(p_s).\end{aligned}$$

Equality holds only if $[p_s - q, p_s - q] = 0$, i.e., if the vectors p_s and q are equal. Then the phase polynomials $p_s(x)$ and $q(x)$ are identical by the uniqueness theorem (2.3.1.5). □

Returning to the original trigonometric expressions (2.3.1.1), we note that Theorems (2.3.1.5) and (2.3.1.9) translate into the following:

(2.3.1.11) Theorem. *The trigonometric expressions*

$$\Psi(x) = \frac{A_0}{2} + \sum_{h=0}^{M} (A_h \cos hx + B_h \sin hx)$$

$$\Psi(x) = \frac{A_0}{2} + \sum_{h=0}^{M-1} (A_h \cos hx + B_h \sin hx) + \frac{A_M}{2} \cos Mx,$$

where $N = 2M + 1$ and $N = 2M$, respectively, satisfy

$$\Psi(x_k) = f_k, \; k = 0, 1, \ldots, N - 1,$$

for $x_k = 2\pi k/N$ if and only if the coefficients of $\Psi(x)$ are given by

$$A_h = \frac{2}{N} \sum_{k=0}^{N-1} f_k \cos hx_k = \frac{2}{N} \sum_{k=0}^{N-1} f_k \cos \frac{2\pi hk}{N},$$

$$B_h = \frac{2}{N} \sum_{k=0}^{N-1} f_k \sin hx_k = \frac{2}{N} \sum_{k=0}^{N-1} f_k \sin \frac{2\pi hk}{N}.$$

PROOF. Only the expressions for A_h, B_h remain to be verified. For by (2.3.1.4)

$$A_h = \beta_h + \beta_{N-h} = \frac{1}{N} \sum_{k=0}^{N-1} f_k(e^{-hix_k} + e^{-(N-h)ix_k}),$$

$$B_h = i(\beta_h - \beta_{N-h}) = \frac{i}{N} \sum_{k=0}^{N-1} f_k(e^{-hix_k} - e^{-(N-h)ix_k}),$$

and the substitutions (2.3.1.3) yield the desired expressions. □

Note that if the support ordinates f_k are real, then so are the coefficients A_h, B_h in (2.3.1.11).

2.3.2 Fast Fourier Transforms

The interpolation of equidistant support points (x_k, f_k), $x_k = 2\pi k/N$, $k = 0, \ldots, N-1$, by a phase polynomial $p(x) = \sum_{j=0}^{N-1} \beta_j e^{jix}$ leads to expressions of the form [Theorem (2.3.1.9)]

$$(2.3.2.1) \qquad \beta_j = \frac{1}{N} \sum_{k=0}^{N-1} f_k e^{-2\pi ijk/N}, \qquad j = 0, \ldots, N-1.$$

The evaluation of such expressions is of prime importance in Fourier analysis. The expressions occur also as discrete approximations—for N equidistant arguments s—to the *Fourier transform*

$$H(s) := \int_{-\infty}^{+\infty} f(t) e^{-2\pi ist}\, dt,$$

which pervades many areas of applied mathematics. However, the numerical evaluation of expressions (2.3.2.1) had long appeared to require on the order of N^2 multiplications, putting it out of reach for even high-speed electronic computers for those large values of N necessary for a sufficiently accurate discrete representation of the above integrals. The discovery [Cooley and Tukey (1965)] of a method for rapidly evaluating (on the order of $N \log N$ multiplications) all expressions (2.3.2.1) for large special values of N has therefore opened up vast new areas of applications. This method and its variations are called *fast Fourier transforms*. For a detailed treatment see Brigham (1974) and Bloomfield (1976).

There are two main approaches, the original *Cooley–Tukey method* and one described by Gentleman and Sande (1966), commonly called the *Sande–Tukey method*. Both approaches rely on an integer factorization of N and decompose the problem accordingly into subproblems of lower degree. These decompositions are then carried out recursively. This works best when

$$N = 2^n, \qquad n > 0 \text{ integer}.$$

We restrict our presentation to this most important and most straightforward case, although analogous techniques will clearly work for the more general case $N = N_1 N_2 \ldots N_n$, N_m integer.

The Cooley–Tukey approach is best understood in terms of the interpolation problem described in the previous section (2.3.1). Suppose $N = 2M$ and consider the two interpolating phase polynomials $q(x)$ and $r(x)$ with

$$q(x_{2h}) = f_{2h}, \quad r(x_{2h}) = f_{2h+1}, \qquad h = 0, \ldots, M-1.$$

The phase polynomial $q(x)$ interpolates all support points of even index, whereas the phase polynomial $\hat{r}(x) = r(x - 2\pi/N) = r(x - \pi/M)$ interpolates all those of odd index. Since

$$e^{Mix_k} = e^{2\pi iMk/N} = e^{\pi ik} = \begin{cases} +1, & k \text{ even}, \\ -1, & k \text{ odd}, \end{cases}$$

the complete interpolating phase polynomial $p(x)$ is now readily expressed in terms of the two lower-degree phase polynomials $q(x)$ and $r(x)$:

$$p(x) = q(x)\left(\frac{1 + e^{Mix}}{2}\right) + r(x - \pi/M)\left(\frac{1 - e^{Mix}}{2}\right). \tag{2.3.2.2}$$

This suggests the following n step recursive scheme. For $m \leqslant n$, let

$$M = 2^{m-1} \quad \text{and} \quad R = 2^{n-m}.$$

Step m then consists of determining R phase polynomials

$$p_r^{(m)} = \beta_{r0}^{(m)} + \beta_{r1}^{(m)} e^{ix} + \cdots + \beta_{r,2M-1}^{(m)} e^{(2M-1)ix}, \qquad r = 0, \ldots, R-1,$$

from $2R$ phase polynomials $p_r^{(m-1)}(x)$, $r = 0, \ldots, 2R-1$, using the recursion (2.3.2.2):

$$2p_r^{(m)}(x) = p_r^{(m-1)}(x)(1 + e^{Mix}) + p_{R+r}^{(m-1)}(x - \pi/M)(1 - e^{Mix}).$$

This relation gives rise to the following recursive relationship between the coefficients of the above phase polynomials:

$$\left.\begin{aligned} 2\beta_{rj}^{(m)} &= \beta_{rj}^{(m-1)} + \beta_{R+r,j}^{(m-1)}\varepsilon_m^j \\ 2\beta_{r,M+j}^{(m)} &= \beta_{rj}^{(m-1)} - \beta_{R+r,j}^{(m-1)}\varepsilon_m^j \end{aligned}\right\} \qquad \begin{aligned} r &= 0, \ldots, R-1, \\ j &= 0, \ldots, M-1, \end{aligned} \tag{2.3.2.3}$$

where

$$\varepsilon_m := e^{-2\pi i/2^m}, \qquad m = 0, \ldots, n.$$

The recursion is initiated by putting

$$\beta_{k0}^{(0)} := f_k, \qquad k = 0, \ldots, N-1,$$

and terminates with

$$\beta_j := \beta_{0j}^{(n)}, \qquad j = 0, \ldots, N-1.$$

This recursion typifies the Cooley–Tukey method.

The Sande–Tukey approach chooses a clever sequence of additions in the sums $\sum_{k=0}^{N-1} f_k e^{-jix_k}$. Again with $M = N/2$, we assign to each term $f_k e^{-jix_k}$ an *opposite term* $f_{k+M} e^{-jix_{k+M}}$. Summing respective opposite terms in (2.3.2.1) produces N sums of $M = N/2$ terms each. Splitting those N sums into two sets, one for even indices $j = 2h$ and one for odd indices $j = 2h + 1$, will lead to two problems of evaluating expressions of the form (2.3.2.1), each problem being of reduced degree $M = N/2$.

Using the abbreviation

$$\varepsilon_m = e^{-2\pi i/2^m}$$

again, we can write the expressions (2.3.2.1) in the form

$$\beta_j = \frac{1}{N} \sum_{k=0}^{N-1} f_k \varepsilon_n^{jk}, \qquad j = 0, \ldots, N-1.$$

Here n is such that $N = 2^n$. Distinguishing between even and odd values of j and combining opposite terms gives

$$\beta_{2h} = \frac{1}{N}\sum_{k=0}^{N-1} f_k \varepsilon_n^{2hk} = \frac{1}{N}\sum_{k=0}^{M-1} (f_k + f_{k+M})\varepsilon_{n-1}^{hk} = \frac{1}{N}\sum_{k=0}^{M-1} f'_k \varepsilon_{n-1}^{hk}$$

$$\beta_{2h+1} = \frac{1}{N}\sum_{k=0}^{N-1} f_k \varepsilon_n^{(2h+1)k} = \frac{1}{N}\sum_{k=0}^{M-1} ((f_k - f_{k+M})\varepsilon_n^k)\varepsilon_{n-1}^{hk} = \frac{1}{N}\sum_{k=0}^{M-1} f''_k \varepsilon_{n-1}^{hk}$$

for $h = 0 \ldots, M - 1$ and $M := N/2$, since $\varepsilon_n^2 = \varepsilon_{n-1}$, $\varepsilon_n^M = -1$. Here

$$\left.\begin{aligned} f'_k &= f_k + f_{k+M} \\ f''_k &= (f_k - f_{k+M})\varepsilon_n^k \end{aligned}\right\} \quad k = 0, \ldots, M - 1.$$

In order to iterate this process for $m = n, n - 1, \ldots, 0$, we let $M := 2^{m-1}$, $R := 2^{n-m}$ and introduce the notation

$$f_{rk}^{(m)}, \qquad r = 0, \ldots, R - 1, \quad k = 0, \ldots, 2M - 1,$$

with $f_{0k}^{(n)} = f_k$, $k = 0, \ldots, N - 1$. $f_{0k}^{(n-1)}$ and $f_{1k}^{(n-1)}$ represent the quantities f'_k and f''_k, respectively, which were introduced above. In general we have, with $M = 2^{m-1}$ and $R = 2^{n-m}$,

$$\text{(2.3.2.4)} \quad \beta_{jR+r} = \frac{1}{N}\sum_{k=0}^{2M-1} f_{rk}^{(m)}\varepsilon_m^{jk}, \qquad r = 0, \ldots, R - 1, \quad j = 0, \ldots, 2M - 1,$$

with the quantities $f_{rk}^{(m)}$ satisfying the recursions:

$$\text{(2.3.2.5)} \quad \left.\begin{aligned} f_{rk}^{(m-1)} &= f_{rk}^{(m)} + f_{r,k+M}^{(m)} \\ f_{r+R,k}^{(m-1)} &= (f_{rk}^{(m)} - f_{r,k+M}^{(m)})\varepsilon_m^k \end{aligned}\right\} \quad \left\{\begin{aligned} m &= n, \ldots, 1, \\ r &= 0, \ldots, R - 1, \\ k &= 0, \ldots, M - 1. \end{aligned}\right.$$

Proof. Suppose (2.3.2.4) is correct for some $m \leqslant n$, and let $M' := M/2 = 2^{m-2}$, $R' := 2R = 2^{n-m+1}$. For $j = 2h$ and $j = 2h + 1$, respectively, we find by combining opposite terms

$$\beta_{hR'+r} = \beta_{jR+r} = \frac{1}{N}\sum_{k=0}^{M-1} (f_{r,k}^{(m)} + f_{r,k+M}^{(m)})\varepsilon_m^{jk} = \frac{1}{N}\sum_{k=0}^{2M'-1} f_{r,k}^{(m-1)}\varepsilon_{m-1}^{hk},$$

$$\begin{aligned} \beta_{hR'+r+R} = \beta_{jR+r} &= \frac{1}{N}\sum_{k=0}^{M-1} (f_{r,k}^{(m)} - f_{r,k+M}^{(m)})\varepsilon_m^{jk} \\ &= \frac{1}{N}\sum_{k=0}^{M-1} (f_{r,k}^{(m)} - f_{r,k+M}^{(m)})\varepsilon_m^k \varepsilon_{m-1}^{hk} = \frac{1}{N}\sum_{k=0}^{2M'-1} f_{r+R,k}^{(m-1)}\varepsilon_{m-1}^{hk}, \end{aligned}$$

where $r = 0, \ldots, R - 1, j = 0, \ldots, 2M - 1$. □

The recursion (2.3.2.5) typefies the Sande–Tukey method. It is initiated by putting

$$f_{0k}^{(n)} := f_k, \qquad k = 0, \ldots, N - 1,$$

and terminates with

$$\beta_r := \frac{1}{N} f_{r0}^{(0)}, \qquad r = 0, \ldots, N-1.$$

Returning to the Cooley–Tukey method for a more detailed algorithmic formulation, we are faced with the problem of arranging the quantities $\beta_{rj}^{(m)}$ in an array:

$$\tilde{\beta}[\kappa] := \beta_{rj}^{(m)}, \qquad \kappa = 0, \ldots, N-1.$$

Among suitable maps $\kappa = \kappa(m, r, j)$, the following is the most straightforward:

$$\kappa = 2^m r + j, \qquad m = 0, \ldots, n, \quad r = 0, \ldots, 2^{n-m} - 1, \quad j = 0, \ldots, 2^m - 1.$$

It has the advantage, that the final results are automatically in the correct order. However, two arrays $\tilde{\beta}'[\ ]$, $\tilde{\beta}[\ ]$ are necessary to accommodate the left- and right-hand sides of the recursion (2.3.2.3).

We can make do with only one array $\tilde{\beta}[\ ]$ if we execute the transformations "in place," that is, if we let each pair of quantities $\beta_{rj}^{(m)}$, $\beta_{r, M+j}^{(m)}$ occupy the same positions in $\tilde{\beta}[\ ]$ as the pair of quantities $\beta_{rj}^{(m-1)}$, $\beta_{R+r, j}^{(m-1)}$, from which the former are computed. In this case, however, the entries in the array $\tilde{\beta}[\ ]$ are being permuted, and the maps which assign the positions in $\tilde{\beta}[\ ]$ as a function of the integers m, r, j become more complicated. Let

$$\tau = \tau(m, r, j)$$

be a map with the above mentioned replacement properties, namely $\tilde{\beta}[\tau] = \beta_{rj}^{(m)}$ with

(2.3.2.6)
$$\left.\begin{aligned} \tau(m, r, j) &= \tau(m-1, r, j) \\ \tau(m, r, j + 2^{m-1}) &= \tau(m-1, r + 2^{n-m}, j) \end{aligned}\right\} \quad \left\{\begin{aligned} m &= 1, \ldots, n, \\ r &= 0, \ldots, 2^{n-m} - 1, \\ j &= 0, \ldots, 2^{m-1} - 1, \end{aligned}\right.$$

and

(2.3.2.7)
$$\tau(n, 0, j) = j, \qquad j = 0, \ldots, N-1.$$

The last condition means that the final result β_j will be found in position j in the array $\tilde{\beta}[\ ]$: $\beta_j = \tilde{\beta}[j]$.

The conditions (2.3.2.6) and (2.3.2.7) define the map τ recursively. It remains to determine it explicitly. To this end, let

$$t = \alpha_0 + \alpha_1 \cdot 2 + \cdots + \alpha_{n-1} \cdot 2^{n-1}, \qquad \alpha_p = 0, 1 \quad \text{for } p = 0, \ldots, n-1,$$

be a binary representation of an integer t, $0 \leqslant t < 2^n$. Then putting

(2.3.2.8)
$$\rho(t) := \alpha_{n-1} + \alpha_{n-2} \cdot 2 + \cdots + \alpha_0 \cdot 2^{n-1}$$

defines a permutation of the integers $t = 0, \ldots, 2^n - 1$ called *bit reversal*. The bit-reversal permutation is symmetric, i.e. $\rho(\rho(t)) = t$.

With the help of the bit-reversal permutation ρ, we can express $\tau(m, r, j)$ explicitly:

(2.3.2.9) $$\tau(m, r, j) = \rho(r) + j,$$
$$m = 0, \ldots, n, \quad r = 0, \ldots, 2^{n-m} - 1, \quad j = 0, \ldots, 2^m - 1.$$

PROOF. If again

$$t = \alpha_0 + \alpha_1 \cdot 2 + \cdots + \alpha_{n-1} \cdot 2^{n-1}, \qquad \alpha_p = 0, 1 \quad \text{for } p = 0, \ldots, n-1,$$

then by (2.3.2.6) and (2.3.2.7)

$$t = \tau(n, 0, t) = \begin{cases} \tau(n-1, 0, t) & \text{if } \alpha_{n-1} = 0 \\ \tau(n-1, 1, t - 2^{n-1}) & \text{if } \alpha_{n-1} = 1. \end{cases}$$

Thus

$$t = \tau(n, 0, t) = \tau(n-1, \alpha_{n-1}, \alpha_0 + \cdots + \alpha_{n-2} \cdot 2^{n-2}),$$

and, more generally,

$$t = \tau(n, 0, t) = \tau(m, \alpha_{n-1} + \cdots + \alpha_m \cdot 2^{n-m-1}, \alpha_0 + \cdots + \alpha_{m-1} \cdot 2^{m-1})$$

for $m = 0, \ldots, n-1$. For $r = \alpha_{n-1} + \cdots + \alpha_m \cdot 2^{n-m-1}$, we find

$$\rho(r) = \alpha_m \cdot 2^m + \cdots + \alpha_{n-1} \cdot 2^{n-1}$$

and $t = \rho(r) + j$. □

By the symmetry of bit reversal,

$$\tau(m, \rho(\bar{r}), j) = \bar{r} + j,$$

where $\bar{r}$ is a multiple of 2^m, $0 \leqslant \bar{r} < 2^n$, and $0 \leqslant j < 2^m$. Observe that if $0 \leqslant j < 2^{m-1}$, then

$$t = \tau(m, \rho(\bar{r}), j) = \tau(m-1, \rho(\bar{r}), j) = \bar{r} + j,$$
$$\bar{t} = \tau(m, \rho(\bar{r}), j + 2^{m-1}) = \tau(m-1, \rho(\bar{r}) + 2^{n-m}, j) = \bar{r} + j + 2^{m-1}$$

mark a pair of positions in $\tilde{\beta}[\ ]$ which contain quantities connected by the Cooley–Tukey recursions (2.3.2.3).

In the following pseudo-ALGOL formulation of the classical Cooley–Tukey method, we assume that the array $\tilde{\beta}[\ ]$ is initialized by putting

$$\tilde{\beta}[\rho(k)] := f_k, \qquad k = 0, \ldots, N-1,$$

where ρ is the bit-reversal permutation (2.3.2.8). This "scrambling" of the initial values can also be carried out "in place," because the bit-reversal permutation is symmetric and consists, therefore, of a sequence of pairwise interchanges or "transpositions." In addition, we have deleted the factor 2 which is carried along in the formulas (2.3.2.3), so that finally

$$\beta_j := \frac{1}{N} \tilde{\beta}[j], \qquad j = 0, \ldots, N-1.$$

The algorithm then takes the form

```
for m := 1 step1 until n do
begin for j := 0 step 1 until 2^(m-1) - 1 do
      begin e := ε_m^j;
            for r̄ := 0 step 2^m until 2^n - 1 do
            begin u := β̃[r̄ + j]; v := β̃[r̄ + j + 2^(m-1)] × e;
                  β̃[r̄ + j] := u + v; β̃[r̄ + j + 2^(m-1)] := u - v
            end
      end
end;
```

If the Sande–Tukey recursions (2.3.2.5) are used, there is again no problem if two arrays of length N are available for new and old values, respectively. However, if the recursions are to be carried out "in place" in a single array $\tilde{f}[\ ]$, then we must again map index triples m, r, j into single indices τ. This index map has to satisfy the relations

$$\tau(m-1, r, k) = \tau(m, r, k),$$

$$\tau(m-1, r+2^{n-m}, k) = \tau(m, r, k+2^{m-1})$$

for $m = n, n-1, \dots, 1$, $r = 0, 1, \dots, 2^{n-m}-1$, $k = 0, 1, \dots, 2^{m-1}-1$. If we assume

$$\tau(n, 0, k) = k \quad \text{for } k = 0, \dots, N-1,$$

that is, if we start out with the natural order, then these conditions are precisely the conditions (2.3.2.6) and (2.3.2.7) written in reverse. Thus $\tau = \tau(m, r, k)$ is identical to the index map τ considered for the Cooley–Tukey method.

In the following pseudo-ALGOL formulation of the Sande–Tukey method, we assume that the array $\tilde{f}[\ ]$ has been initialized directly with the values f_k:

$$\tilde{f}[k] := f_k, \quad k = 0, 1, \dots, N-1.$$

However, the final results have to be "unscrambled" using bit reversal,

$$\beta_j := \frac{1}{N}[\tilde{f}[\rho(j)], \quad j = 0, \dots, N-1:$$

```
for m := n step -1 until 1 do
begin for k := 0 step 1 until 2^(m-1) - 1 do
      begin e := ε_m^k;
            for r̄ := 0 step 2^m until 2^n - 1 do
            begin u := f̃[r̄ + k]; v := f̃[r̄ + k + 2^(m-1)];
                  f̃[r̄ + k] := u + v; f̃[r̄ + k + 2^(m-1)] := (u - v) × e
            end
      end
end;
```

If all values f_k, $k = 0, \ldots, N-1$ are real and $N = 2M$ is even, then the problem of evaluating the expressions (2.3.2.1) can be reduced in size by putting

$$g_h = f_{2h} + if_{2h+1}, \qquad h = 0, \ldots, M-1,$$

and evaluating the expressions

$$\gamma_j = \frac{1}{M} \sum_{h=0}^{M-1} g_h e^{-2\pi ijh/M}, \qquad j = 0, \ldots, M-1.$$

The desired values β_j, $j = 0, \ldots, N-1$, can be expressed in terms of the values γ_j, $j = 0, \ldots, M-1$. Indeed, one has with $\gamma_M := \gamma_0$

(2.3.2.10)
$$\beta_j = \frac{1}{4}(\gamma_j + \bar{\gamma}_{M-j}) + \frac{1}{4i}(\gamma_j - \bar{\gamma}_{M-j})e^{-2\pi ij/N}, \qquad j = 0, \ldots, M,$$
$$\beta_{N-j} = \bar{\beta}_j, \qquad j = 1, \ldots, M-1.$$

PROOF. It is readily verified that

$$\frac{1}{4}(\gamma_j + \bar{\gamma}_{M-j}) = \frac{1}{N} \sum_{h=0}^{M-1} f_{2h} e^{-2\pi ij2h/N},$$

$$\frac{1}{4i}(\gamma_j - \bar{\gamma}_{M-j}) = \frac{1}{N} \sum_{h=0}^{M-1} f_{2h+1} e^{-2\pi ij(2h+1)/N + 2\pi ij/N}. \qquad \square$$

In many cases, particularly if all values f_k are real, one is actually interested in the expressions

$$A_j = \frac{2}{N} \sum_{k=0}^{N-1} f_k \cos \frac{2\pi jk}{N}, \qquad B_j = \frac{2}{N} \sum_{k=0}^{N-1} f_k \sin \frac{2\pi jk}{N},$$

which occur, for instance, in Theorem (2.3.1.11). The values A_j, B_j are connected with the corresponding values for β_j via the relations (2.3.1.4).

2.3.3 The Algorithms of Goertzel and Reinsch

The problem of evaluating phase polynomials $p(x)$ from (2.3.1.2) or trigonometric expressions $\Psi(x)$ from (2.3.1.1) for some arbitrary argument $x = \xi$ is called *Fourier synthesis*. For phase polynomials, there are Horner-type evaluation schemes, as there are for expressions (2.3.1.1a) when written in the form $\Psi(x) = \sum_{j=-M}^{M} \beta_j e^{jix}$. The numerical behavior of such evaluation schemes, however, should be examined carefully.

For example, Goertzel (1958) proposed an algorithm for a problem closely related to Fourier synthesis, namely, for simultaneously evaluating the two sums

$$\sum_{k=0}^{N-1} y_k \cos k\xi, \quad \sum_{k=1}^{N-1} y_k \sin k\xi$$

for a given argument ξ and given values y_k, $k = 0, \ldots, N-1$. This algorithm is not numerically stable unless it is suitably modified. The algorithm is based on the following:

(2.3.3.1) Theorem. *For $\xi \neq r\pi$, $r = 0, \pm 1, \pm 2, \ldots$, define the quantities*

$$U_j := \frac{1}{\sin \xi} \sum_{k=j}^{N-1} y_k \sin(k - j + 1)\xi, \qquad j = 0, 1, \ldots, N-1,$$

$$U_N := U_{N+1} := 0.$$

These quantities satisfy the recursions

(2.3.3.1a) $$U_j = y + 2U_{j+1} \cos \xi - U_{j+2}, \qquad j = N-1, N-2, \ldots, 0.$$

In particular

(2.3.3.1b) $$\sum_{k=1}^{N-1} y_k \sin k\xi = U_1 \sin \xi,$$

(2.3.3.1c) $$\sum_{k=0}^{N-1} y_k \cos k\xi = y_0 + U_1 \cos \xi - U_2.$$

PROOF. For $0 \leqslant j \leqslant N-1$, let

$$A := y_j + 2U_{j+1} \cos \xi - U_{j+2}.$$

By the definition of U_{j+1}, U_{j+2},

$$\begin{aligned} A &= y_j + \frac{1}{\sin \xi}\left\{2(\cos \xi) \sum_{k=j+1}^{N-1} y_k \sin(k-j)\xi - \sum_{k=j+2}^{N-1} y_k \sin(k-j-1)\xi\right\} \\ &= y_j + \frac{1}{\sin \xi} \cdot \sum_{k=j+1}^{N-1} y_k[2 \cos \xi \sin(k-j)\xi - \sin(k-j-1)\xi]. \end{aligned}$$

Now

$$2 \cos \xi \sin(k-j)\xi = \sin(k-j+1)\xi + \sin(k-j-1)\xi.$$

Hence

$$A = \frac{1}{\sin \xi}\left[y_j \sin \xi + \sum_{k=j+1}^{N-1} y_k \sin(k-j+1)\xi\right] = U_j.$$

This proves (2.3.3.1a). (2.3.3.1b) restates the definition of U_1. To verify (2.3.3.1c), note that

$$U_2 = \frac{1}{\sin \xi} \cdot \sum_{k=2}^{N-1} y_k \sin(k-1)\xi = \frac{1}{\sin \xi} \cdot \sum_{k=1}^{N-1} y_k \sin(k-1)\xi,$$

and

$$\sin(k-1)\xi = \cos \xi \sin k\xi - \sin \xi \cos k\xi. \qquad \square$$

Goertzel's algorithm applies the recursions (2.3.3.1) directly:

```
U[N] := U[N + 1] := 0; c := cos(ξ); cc := 2 × c;
for j := N − 1 step −1 until 1 do
      U[j] := y[j] + cc × U[j + 1] − U[j + 2];
s1 := y[0] + c × U[1] − U[2];
s2 := U[1] × sin(ξ);
```

to find the desired results $s1 = \sum_{k=0}^{N-1} y_k \cos k\xi$, $s2 = \sum_{k=1}^{N-1} y_k \sin k\xi$.

This algorithm is unfortunately not numerically stable for small absolute values of ξ, $|\xi| \ll 1$. Indeed, having calculated $c = \cos \xi$, the quantity $s1 = \sum_{k=0}^{N-1} y_k \cos k\xi$ will depend solely on c and the values y_k. We can write $s1 = \varphi(c, y_0, \ldots, y_{N-1})$, where

$$\varphi(c, y_0, \ldots, y_{N-1}) = \sum_{k=0}^{N-1} y_k \cos k(\arccos c).$$

As in Section 1.2, we denote by eps the machine precision. The roundoff error $\Delta c = \varepsilon_c c$, $|\varepsilon_c| \leqslant \text{eps}$, which occurs during the calculation of c, causes an absolute error $\Delta_c s1$ in $s1$, which in first-order approximation amounts to

$$\begin{aligned}\Delta_c s1 &\doteq \frac{\partial \varphi}{\partial c} \Delta c = \frac{\varepsilon_c \cos \xi}{\sin \xi} \sum_{k=0}^{N-1} k y_k \sin k\xi \\ &= \varepsilon_c (\cot \xi) \sum_{k=0}^{N-1} k y_k \sin k\xi.\end{aligned}$$

An error $\Delta\xi = \varepsilon_\xi \xi$, $|\varepsilon_\xi| \leqslant \text{eps}$ in ξ, on the other hand, causes only the error

$$\begin{aligned}\Delta_\xi s1 &\doteq \frac{\partial}{\partial \xi} \left\{ \sum_{k=0}^{N-1} y_k \cos k\xi \right\} \cdot \Delta\xi \\ &= -\varepsilon_\xi \xi \sum_{k=0}^{N-1} k y_k \sin k\xi\end{aligned}$$

in $s1$. Now $\cot \xi \approx 1/\xi$ for small $|\xi|$. The influence of the roundoff error in c is consequently an order of magnitude more serious than that of a corresponding error in ξ. In other words, the algorithm is not numerically stable.

In order to overcome these numerical difficulties, Reinsch has modified Goertzel's algorithm [see Bulirsch and Stoer (1968)]. He distinguishes the two cases $\cos \xi > 0$ and $\cos \xi \leqslant 0$.

Case (a): $\cos \xi > 0$. The recursion (2.3.3.1a) yields for the difference

$$\delta U_j := U_j - U_{j+1}$$

the relation

$$\begin{aligned}\delta U_j = U_j - U_{j+1} &= y_j + (2 \cos \xi - 2) U_{j+1} + U_{j+1} - U_{j+2} \\ &= y_j + \lambda U_{j+1} + \delta U_{j+1},\end{aligned}$$

where

$$\lambda := 2(\cos \xi - 1) = -4 \sin^2(\xi/2).$$

This suggests the algorithm

```
λ := −4 sin²(ξ/2);
U[N + 1] := δU[N] := 0;
for j := N − 1 step −1 until 0 do
begin  U[j + 1] := δU[j + 1] + U[j + 2];
       δU[j] := λ × U[j + 1] + δU[j + 1] + y[j]
end;
s1 := δU[0] − λ/2 × U[1];
s2 :=  U[1] × sin(ξ);
```

This algorithm is well behaved as far as the propagation of the error $\Delta\lambda = \varepsilon_\lambda \lambda$, $|\varepsilon_\lambda| \leqslant \text{eps}$ in λ is concerned. The latter causes only the following error $\Delta_\lambda s1$ in $s1$:

$$\begin{aligned}\Delta_\lambda s1 &\doteq \frac{\partial s1}{\partial \lambda}\Delta\lambda = \varepsilon_\lambda \lambda \cdot \frac{\partial s1}{\partial \xi}\Big/\frac{\partial \lambda}{\partial \xi} \\ &= -\varepsilon_\lambda \frac{\sin^2(\xi/2)}{\sin(\xi/2)\cos(\xi/2)} \cdot \sum_{k=0}^{N-1} k y_k \sin k\xi \\ &= -\varepsilon_\lambda \left(\tan \frac{\xi}{2}\right) \sum_{k=0}^{N-1} k y_k \sin k\xi.\end{aligned}$$

and $\tan(\xi/2)$ is small for small $|\xi|$. Besides, $|\tan(\xi/2)| < 1$ for $\cos \xi > 0$.

Case (b): $\cos \xi \leqslant 0$. Here we put

$$\delta U_j := U_j + U_{j+1}$$

and find

$$\begin{aligned}\delta U_j = U_j + U_{j+1} &= y_j + (2\cos\xi + 2)U_{j+1} - U_{j+1} - U_{j+2} \\ &= y_j + \lambda U_{j+1} - \delta U_{j+1},\end{aligned}$$

where now

$$\lambda := 2(\cos \xi + 1) = 4 \cos^2(\xi/2).$$

This leads to the following algorithm:

```
λ := 4 cos²(ξ/2);
U[N + 1] := δU[N] := 0;
for j := N − 1 step −1 until 0 do
begin  U[j + 1] := δU[j + 1] − U[j + 2];
       δU[j] := λ × U[j + 1] − δU[j + 1] + y[j]
end;
s1 := δU[0] − U[1] × λ/2;
s2 :=  U[1] × sin(ξ);
```

It is readily confirmed that a roundoff error $\Delta\lambda = \varepsilon_\lambda \lambda$, $|\varepsilon_\lambda| \leqslant \text{eps}$, in λ causes an error of at most

$$\Delta_\lambda s1 \doteq \varepsilon_\lambda \left(\cot \frac{\xi}{2}\right) \sum_{k=0}^{N-1} k y_k \sin k\xi$$

in $s1$, and $|\cot(\xi/2)| \leqslant 1$ for $\cos \xi \leqslant 0$. The algorithm is therefore well behaved as far as the propagation of the error $\Delta\lambda$ is concerned.

2.3.4 The Calculation of Fourier Coefficients. Attenuation Factors

Let $\mathscr{F}$ be the set of all absolutely continuous[2] real functions $f: \mathbb{R} \to \mathbb{R}$ which are periodic with period 2π. It is well known [see for instance Achieser (1956)] that every function $f \in \mathscr{F}$ can be expanded into a Fourier series

(2.3.4.1) $$f(x) = \sum_{j=-\infty}^{\infty} c_j e^{jix},$$

which converges towards $f(x)$ for every $x \in \mathbb{R}$. The coefficients $c_j = c_j(f)$ of this series are given by

(2.3.4.2) $$c_j = c_j(f) := \frac{1}{2\pi} \int_0^{2\pi} f(x) e^{-jix}\, dx, \qquad j = 0, \pm 1, \pm 2, \ldots.$$

In practice, frequently all one knows of a function f are its values $f_k := f(x_k)$ at equidistant arguments $x_k := 2\pi k/N$, where N is a given fixed positive integer. The problem then is to find, under these circumstances, reasonable approximate values for the Fourier coefficients $c_j(f)$. We will show how the methods of trigonometric interpolation can be applied to this problem.

By Theorem (2.3.1.9), the coefficients β_j of the interpolating phase polynomial

$$p(x) = \beta_0 + \beta_1 e^{ix} + \cdots + \beta_{N-1} e^{(N-1)ix},$$

with

$$p(x_k) = f_k$$

[2] A real function $f: [a, b] \to \mathbb{R}$ is *absolutely continuous* on the interval $[a, b]$ if for every $\varepsilon > 0$ there exists $\delta > 0$ such that $\sum_i |f(b_i) - f(a_i)| < \varepsilon$ for every finite set of intervals $[a_i, b_i]$ with $a \leqslant a_1 < b_1 < \cdots < a_n < b_n \leqslant b$ and $\sum_i |b_i - a_i| < \delta$. If the function f is differentiable everywhere on the closed interval $[a, b]$ or, more generally, if it satisfies a "Lipschitz condition" $|f(x_1) - f(x_2)| \leqslant \theta |x_1 - x_2|$ on $[a, b]$, then f is absolutely continuous, but not conversely: there are absolutely continuous functions with unbounded derivatives. If the function is absolutely continuous, then it is continuous and its derivative f' exists almost everywhere. Moreover, $f(x) = f(a) + \int_a^x f'(t)\, dt$ for $x \in [a, b]$. The absolute continuity of the functions f, g in an integral of the form $\int_a^b f(t) g'(t)\, dt$ also ensures that integration by parts can be carried out safely.

for $k = 0, \pm 1, \pm 2, \ldots$, are given by

$$\beta_j = \frac{1}{N} \sum_{k=0}^{N-1} f_k e^{-jix_k}, \qquad j = 0, 1, \ldots, N-1.$$

Since $f_0 = f_N$, the quantities β_j can be thought of as a "trapezoidal sum" [compare (3.1.7)]

$$\beta_j = \frac{1}{N}\left[\frac{f_0}{2} + f_1 e^{-jix_1} + \cdots + f_{N-1} e^{-jix_{N-1}} + \frac{f_N}{2} e^{-jix_N}\right]$$

approximating the integral (2.3.4.2), so that one might think of using the sums

$$\beta_j(f) = \beta_j := \frac{1}{N} \sum_{k=0}^{N-1} f_k e^{-jix_k} \tag{2.3.4.3}$$

for all integers $j = 0, \pm 1, \pm 2, \ldots$ as approximate values to the desired Fourier coefficients $c_j(f)$. This approach appears attractive, since fast Fourier transforms can be utilized to calculate the quantities $\beta_j(f)$ efficiently. However, for large indices j the value $\beta_j(f)$ is a very poor approximation to $c_j(f)$. Indeed, $\beta_{j+kN} = \beta_j$ holds for all integers k, j, while on the other hand $\lim_{|j| \to \infty} c_j = 0$. [This follows immediately from the convergence of the Fourier series (2.3.4.1) for the argument $x = 0$.] A closer look also reveals that the asymptotic behavior of the Fourier coefficients $c_j(f)$ depends on the degree of differentiability of f:

(2.3.4.4) Theorem. *If the 2π-periodic function f has an absolutely continuous* r th *derivative $f^{(r)}$, then*

$$|c_j| = O\left(\frac{1}{|j|^{r+1}}\right).$$

Proof. Successive integration by parts yields

$$\begin{aligned} c_j &= \frac{1}{2\pi} \int_0^{2\pi} f(x) e^{-jix}\, dx \\ &= \frac{1}{2\pi ji} \int_0^{2\pi} f'(x) e^{-jix}\, dx \\ &= \cdots \\ &= \frac{1}{2\pi (ji)^r} \int_0^{2\pi} f^{(r)}(x) e^{-jix}\, dx \\ &= \frac{1}{2\pi (ji)^{r+1}} \int_0^{2\pi} e^{-jix}\, df^{(r)}(x). \end{aligned}$$

in view of the periodicity of f. This proves the proposition. □

To approximate the Fourier coefficients $c_j(f)$ by values which display the right asymptotic behavior, the following approach suggests itself: Determine for given values f_k, $k = 0, \pm 1, \pm 2, \ldots$, as simple a function $g \in \mathscr{F}$ as possible which approximates f in some sense (e.g., interpolates f for x_k) and share with f some degree of differentiability. The Fourier coefficients $c_j(g)$ of g are then chosen to approximate the Fourier coefficients $c_j(f)$ of the given function f. In pursuing this idea, it comes as a pleasant surprise that even for quite general methods of approximating the function f by a suitable function g, the Fourier coefficients $c_j(g)$ of g can be calculated in a straightforward manner from the coefficients $\beta_j(f)$ in (2.3.4.3). More precisely, there are so-called attenuation factors τ_j, j integer, which depend only on the choice of the approximation method and not on the particular function values f_k, $k = 0, \pm 1, \ldots$, and for which

$$c_j(\varphi) = \tau_j \beta_j(f), \qquad j = 0, \pm 1, \ldots.$$

To clarify what we mean by an "approximation method," we consider—besides the set $\mathscr{F}$ of all absolutely continuous 2π-periodic functions $f: \mathbb{R} \to \mathbb{R}$—the set

$$\mathbb{F} = \{(f_k)_{k \in \mathbb{Z}} \mid f_k \in \mathbb{R}, f_{k+N} = f_k \text{ for all } k \in \mathbb{Z}\},$$
$$\mathbb{Z} := \{k \mid k \text{ integer}\},$$

of all N-periodic sequences of real numbers

$$f = (\ldots, f_{-1}, f_0, f_1, \ldots).$$

For convenience, we denote by f both the function $f \in \mathscr{F}$ and its the corresponding sequence $(f_k)_{k \in \mathbb{Z}}$ with $f_k = f(x_k)$. The meaning of f will follow from the context.

Any method of approximation assigns to each sequence $f \in \mathbb{F}$ a function $g = P(f)$ in $\mathscr{F}$; it can therefore be described by a map

$$P: \mathbb{F} \to \mathscr{F}.$$

$\mathscr{F}$ and $\mathbb{F}$ are real vector spaces with the addition of elements and the multiplication by scalars defined in the usual straightforward fashion. It therefore makes sense to distinguish *linear* approximation methods P. The vector space $\mathbb{F}$ is of finite dimension N, a basis being formed by the sequences

$$\text{(2.3.4.5)} \qquad e^{(k)} = (e_j^{(k)})_{j \in \mathbb{Z}}, \qquad k = 0, 1, \ldots, N-1,$$

where

$$e_j^{(k)} := \begin{cases} 1 & \text{if } k \equiv j \bmod N, \\ 0 & \text{otherwise.} \end{cases}$$

In both $\mathbb{F}$ and $\mathscr{F}$ we now introduce *translation operators* $E: \mathbb{F} \to \mathbb{F}$ and $E: \mathscr{F} \to \mathscr{F}$, respectively, by

$$(Ef)_k = f_{k-1} \quad \text{for all } k \in \mathbb{Z} \quad \text{if } f \in \mathbb{F},$$
$$(Eg)(x) = g(x-h) \quad \text{for all } x \in \mathbb{R} \quad \text{if } g \in \mathscr{F}, \qquad h := 2\pi/N = x_1.$$

(For convenience, we use the same symbol for both kinds of translation operators.) We call an approximation method $P\colon \mathbb{F} \to \mathscr{F}$ *translation invariant* if

$$P(E(f)) = E(P(f))$$

for all $f \in \mathbb{F}$, that is, a "shifted" sequence is approximated by a "shifted" function. $P(E(f)) = E(P(f))$ yields $P(E^k(f)) = E^k(P(f))$, where $E^2 = E \circ E$, $E^3 = E \circ E \circ E$, etc. We can now prove the following theorem by Gautschi and Reinsch [for further details see W. Gautschi (1972)]:

(2.3.4.6) Theorem. *For each approximation method* $P\colon \mathbb{F} \to \mathscr{F}$ *there exist attenuation factors* τ_j, $j \in \mathbb{Z}$, *for which*

(2.3.4.7) $\qquad c_j(Pf) = \tau_j \beta_j(f) \quad$ *for all* $j \in \mathbb{Z}$ *and arbitrary* $f \in \mathbb{F}$

if and only if the approximation method P is linear and translation invariant.

PROOF. Suppose that P is linear and translation invariant. Every $f \in \mathbb{F}$ can be expressed in terms of the basis (2.3.4.5):

$$f = \sum_{k=0}^{N-1} f_k e^{(k)} = \sum_{k=0}^{N-1} f_k E^k e^{(0)}.$$

Therefore

$$g := Pf = \sum_{k=0}^{N-1} f_k E^k P e^{(0)},$$

by the linearity and the translation invariance of P. Equivalently,

$$g(x) = \sum_{k=0}^{N-1} f_k \eta_0(x - x_k),$$

where $\eta_0 := Pe^{(0)}$ is the function which approximates the sequence $e^{(0)}$. The periodicity of g yields

$$\begin{aligned} c_j(Pf) = c_j(g) &= \sum_{k=0}^{N-1} \frac{f_k}{2\pi} \int_0^{2\pi} \eta_0(x - x_k) e^{-jix}\, dx \\ &= \sum_{k=0}^{N-1} \frac{f_k}{2\pi} e^{-jix_k} \int_0^{2\pi} \eta_0(x) e^{-jix}\, dx \\ &= \tau_j \beta_j(f), \end{aligned}$$

where

(2.3.4.8) $$\tau_j := N c_j(\eta_0).$$

We have thus found expressions for the attenuation factors τ_j which depend only on the approximation method P and the number N of given function values f_k for arguments x_k, $0 \le x_k < 2\pi$. This proves the "if" direction of the theorem.

Suppose now that (2.3.4.7) holds for arbitrary $f \in \mathbb{F}$. Since all functions in $\mathscr{F}$ can be represented by their Fourier series, and in particular $Pf \in \mathscr{F}$, (2.3.4.7) implies

$$(2.3.4.9) \qquad (Pf)(x) = \sum_{j=-\infty}^{\infty} c_j(Pf)e^{jix} = \sum_{j=-\infty}^{\infty} \tau_j \beta_j(f)e^{jix}.$$

By the definition (2.3.4.3) of $\beta_j(f)$, β_j is a linear operator on $\mathscr{F}$ and, in addition,

$$\begin{aligned}\beta_j(Ef) &= \frac{1}{N}\sum_{k=0}^{N-1} f_{k-1}e^{-jix_k}\\ &= \frac{1}{N}e^{-jih}\sum_{k=0}^{N-1} f_k e^{-jix_k}\\ &= e^{-jih}\beta_j(f).\end{aligned}$$

Thus (2.3.4.9) yields the linearity and the translation invariance of P:

$$(P(E(f)))(x) = \sum_{j=-\infty}^{\infty} \tau_j\beta_j(f)e^{ji(x-h)} = (Pf)(x-h) = (E(P(f)))(x). \qquad \square$$

As a by-product of the above proof, we obtained an explicit formula (2.3.4.8) for the attenuation factors. An alternative way of determining the attenuation factors τ_j for a given approximation method P is to evaluate the formula

$$(2.3.4.10) \qquad \tau_j = \frac{c_j(Pf)}{\beta_j(f)}$$

for a suitable $f \in \mathbb{F}$.

EXAMPLE 1. For a given sequence $f \in \mathbb{F}$, let $g := Pf$ be the piecewise linear interpolation of f, that is, g is continuous and linear on each subinterval $[x_k, x_{k+1}]$, and satisfies $g(x_k) = f_k$ for $k = 0, \pm 1, \ldots$. This function $g = Pf$ is clearly absolutely continuous and has period 2π. It is also clear that the approximation method P is linear and translation invariant. Hence Theorem (2.3.4.6) ensures the existence of attenuation factors. In order to calculate them, we note that for the special sequence $f = e^{(0)}$ of (2.3.4.5)

$$\beta_j(f) = \frac{1}{N},$$

$$Pf(x) = \begin{cases} 1 - \dfrac{1}{h}|x - x_{kN}| & \text{if } |x - x_{kN}| \leqslant h, \qquad k = 0, \pm 1, \ldots, \\ 0 & \text{otherwise,}\end{cases}$$

$$c_j(Pf) = \frac{1}{2\pi}\int_0^{2\pi} Pf(x)e^{-jix}\,dx = \frac{1}{2\pi}\int_{-h}^{h}\left(1 - \frac{|x|}{h}\right)e^{-jix}\,dx.$$

Utilizing the symmetry properties of the above integrand, we find

$$c_j(Pf) = \frac{1}{\pi}\int_0^h \left(1 - \frac{x}{h}\right) \cos jx \, dx$$
$$= \frac{2}{j^2 \pi h} \sin^2\left(\frac{jh}{2}\right).$$

With $h = 2\pi/N$, the formula (2.3.4.10) gives

$$\tau_j = \left(\frac{N}{\pi j} \sin \frac{\pi j}{N}\right)^2, \qquad j = 0, \pm 1, \ldots.$$

EXAMPLE 2. Let $g := Pf$ be the periodic cubic spline function (see Section 2.4) with $g(x_k) = f_k$, $k = 0, \pm 1, \ldots$. Again, P is linear and translation invariant. Using the same technique as in the previous example, we find the following attenuation factors:

$$\tau_j = \left(\frac{\sin z}{z}\right)^4 \frac{3}{1 + 2\cos^2 z}, \quad \text{where } z := \frac{\pi j}{N}.$$

2.4 Interpolation by Spline Functions

Spline functions yield smooth interpolating curves which are less likely to exhibit the large oscillations characteristic of high-degree polynomials. They are finding applications in graphics and, increasingly, in numerical methods. For instance, spline functions may be used as trial functions in connection with the Rayleigh–Ritz–Galerkin method for solving boundary-value problems of ordinary and partial differential equations. Introductions are for instance Greville (1969), Schultz (1973), Böhmer (1974), and de Boor (1978).

2.4.1 Theoretical Foundations

Let $\Delta := \{a = x_0 < x_1 < \cdots < x_n = b\}$ be a partition of the interval $[a, b]$.

(2.4.1.1) Definition. *A cubic spline (function) S_Δ on Δ is a real function $S_\Delta : [a, b] \to \mathbb{R}$ with the properties:*

(a) *$S_\Delta \in C^2[a, b]$, that is, S_Δ is twice continuously differentiable on $[a, b]$.*
(b) *S_Δ coincides on every subinterval $[x_i, x_{i+1}]$, $i = 0, 1, \ldots, n - 1$, with a polynomial of degree three.*

Thus a cubic spline consists of cubic polynomials pieced together in such a fashion that their values and those of their first two derivatives coincide at the *knots* x_i, $i = 1, \ldots, n - 1$.

Consider a set $Y := \{y_0, y_1, \ldots, y_n\}$ of $n + 1$ real numbers. We denote by

$$S_\Delta(Y; \cdot)$$

an *interpolating spline function* S_Δ with $S_\Delta(Y; x_i) = y_i$ for $i = 0, 1, \ldots, n$.

Such an interpolating spline function $S_\Delta(Y; \cdot)$ is not uniquely determined by the set Y of support ordinates. Roughly speaking, there are still two degrees of freedom left, calling for suitable additional requirements. The following three additional requirements are most commonly considered:

(2.4.1.2)

(a) $S''_\Delta(Y; a) = S''_\Delta(Y; b) = 0$,

(b) $S^{(k)}_\Delta(Y; a) = S^{(k)}_\Delta(Y; b)$ *for* $k = 0, 1, 2$: $S_\Delta(Y; \cdot)$ *is periodic*,

(c) $S'_\Delta(Y; a) = y'_0$, $S'_\Delta(Y; b) = y'_n$ *for given numbers* y'_0, y'_n.

We will confirm that each of these three conditions by itself ensures uniqueness of the interpolating spline function $S_\Delta(Y; \cdot)$. A prerequisite of the condition (2.4.1.2b) is, of course, that $y_n = y_0$.

For this purpose, and to establish a characteristic minimum property of spline functions, we consider the sets

(2.4.1.3) $$\mathscr{K}^m[a, b],$$

$m > 0$ integer, of real functions $f: [a, b] \to \mathbb{R}$ for which $f^{(m-1)}$ is absolutely continuous[3] on $[a, b]$ and $f^{(m)} \in L^2[a, b]$.[4] By

$$\mathscr{K}^m_p[a, b]$$

we denote the set of all functions in $\mathscr{K}^m[a, b]$ with $f^{(k)}(a) = f^{(k)}(b)$ for $k = 0, 1, \ldots, m-1$. We call such functions *periodic*, because they arise as restrictions to $[a, b]$ of functions which are periodic with period $b - a$.

Note that $S_\Delta \in \mathscr{K}^3[a, b]$, and that $S_\Delta(Y; \cdot) \in \mathscr{K}^3_p[a, b]$ if (2.4.1.2b) holds.

If $f \in \mathscr{K}^2[a, b]$, then we can define

$$\|f\|^2 := \int_a^b |f''(x)|^2 \, dx.$$

Note that $\|f\| \geq 0$. However, $\|f\| = 0$ may hold for functions which are not identically zero, for instance, for all linear functions $f(x) \equiv cx + d$.

We proceed to show a fundamental identity due to Holladay [see for instance Ahlberg, Nilson, and Walsh (1967)].

(2.4.1.4) Theorem. *If* $f \in \mathscr{K}^2(a, b)$, *if* $\Delta = \{a = x_0 < x_1 < \cdots < x_n = b\}$ *is a partition of the interval* $[a, b]$, *and if* S_Δ *is a spline function with knots* $x_i \in \Delta$, *then*

$$\|f - S_\Delta\|^2 = \|f\|^2 - \|S_\Delta\|^2$$
$$- 2\left[(f'(x) - S'_\Delta(x))S''_\Delta(x)\Big|_a^b - \sum_{i=1}^{n} (f(x) - S_\Delta(x))S'''_\Delta(x)\Big|_{x_{i-1}^+}^{x_i^-}\right].$$

[3] See footnote 2 in Section 2.3.4.

[4] The set $L^2[a, b]$ denotes the set of all real functions whose squares are integrable on the interval $[a, b]$, i.e., $\int_a^b |f(t)|^2 \, dt$ exists and is finite.

Here $g(x)|_v^u$ stands for $g(u) - g(v)$, as it is commonly understood in the calculus of integrals. It should be realized, however, that $S_\Delta'''(x)$ is piecewise constant with possible discontinuities at the knots $x_1, \ldots, x_{n-1}$. Hence we have to use the left and right limits of $S_\Delta'''(x)$ at the locations x_i and x_{i-1}, respectively, in the above formula. This is indicated by the notation x_i^-, x_{i-1}^+.

PROOF. By the definition of $\|\cdot\|$,

$$\|f - S_\Delta\|^2 = \int_a^b |f''(x) - S_\Delta''(x)|^2\,dx$$

$$= \|f\|^2 - 2\int_a^b f''(x)S_\Delta''(x)\,dx + \|S_\Delta\|^2$$

$$= \|f\|^2 - 2\int_a^b (f''(x) - S_\Delta''(x))S_\Delta''(x)\,dx - \|S_\Delta\|^2.$$

Integration by parts gives for $i = 1, 2, \ldots, n$

$$\int_{x_{i-1}}^{x_i} (f''(x) - S_\Delta''(x))S_\Delta''(x)\,dx = (f'(x) - S_\Delta'(x))S_\Delta''(x)\Big|_{x_{i-1}}^{x_i}$$

$$- \int_{x_{i-1}}^{x_i} (f'(x) - S_\Delta'(x))S_\Delta'''(x)\,dx$$

$$= (f'(x) - S_\Delta'(x))S_\Delta''(x)\Big|_{x_{i-1}}^{x_i} - (f(x) - S_\Delta(x))S_\Delta'''(x)\Big|_{x_{i-1}^+}^{x_i^-}$$

$$+ \int_{x_{i-1}}^{x_i} (f(x) - S_\Delta(x))S_\Delta^{(4)}(x)\,dx.$$

Now $S^{(4)}(x) \equiv 0$ on the subintervals (x_{i-1}, x_i), and f', S_Δ', S_Δ'' are continuous on $[a, b]$. Adding these formulas for $i = 1, 2, \ldots, n$ yields the proposition of the theorem, since

$$\sum_{i=1}^{n} (f'(x) - S_\Delta'(x))S_\Delta''(x)\Big|_{x_{i-1}}^{x_i} = (f'(x) - S_\Delta'(x))S_\Delta''(x)\Big|_a^b. \qquad \square$$

With the help of this theorem we will prove the important *minimum-norm* property of spline functions.

(2.4.1.5) Theorem. *Given a partition* $\Delta := \{a = x_0 < x_1 < \cdots < x_n = b\}$ *of the interval* $[a, b]$, *values* $Y := \{y_0, \ldots, y_n\}$ *and a function* $f \in \mathscr{K}^2[a, b]$ *with* $f(x_i) = y_i$, *for* $i = 0, 1, \ldots, n$, *then* $\|f\|^2 \geqslant \|S_\Delta(Y; \cdot)\|^2$, *and more precisely*

$$\|f - S_\Delta(Y; \cdot)\|^2 = \|f\|^2 - \|S_\Delta(Y; \cdot)\|^2 \geqslant 0$$

holds for every spline function $S_\Delta(Y; \cdot)$, *provided one of the conditions* [*compare* (2.4.1.2)]

(a) $S_\Delta''(Y; a) = S_\Delta''(Y; b) = 0$,

(b) $f \in \mathscr{K}_p^2(a, b)$, $S_\Delta(Y; \cdot)$ *periodic*,

(c) $f'(a) = S_\Delta'(Y; a)$, $f'(b) = S_\Delta'(Y; b)$,

is met. In each of these cases, the spline function $S_\Delta(Y; \cdot)$ *is uniquely determined.*

The existence of such spline functions will be shown in Section 2.4.2.

PROOF. In each of the above three cases (2.4.1.5a, b, c), the expression

$$(f'(x) - S'_\Delta(x))S''_\Delta(x)\Big|_a^b - \sum_{i=1}^{n} (f(x) - S_\Delta(x))S'''_\Delta(x)\Big|_{x_{i-1}^+}^{x_i^-} = 0$$

vanishes in the Holladay identity (2.4.1.4) if $S_\Delta \equiv S_\Delta(Y; \cdot)$. This proves the minimum property of the spline function $S_\Delta(Y; \cdot)$. Its uniqueness can be seen as follows: suppose $\bar{S}_\Delta(Y; \cdot)$ is another spline function having the same properties as $S_\Delta(Y; \cdot)$. Letting $\bar{S}_\Delta(Y; \cdot)$ play the role of the function $f \in \mathscr{K}^2[a, b]$ in the theorem, the minimum property of $S_\Delta(Y; \cdot)$ requires that

$$\|\bar{S}_\Delta(Y; \cdot) - S_\Delta(Y; \cdot)\|^2 = \|\bar{S}_\Delta(Y; \cdot)\|^2 - \|S_\Delta(Y; \cdot)\|^2 \geqslant 0,$$

and since $S_\Delta(Y; \cdot)$ and $\bar{S}_\Delta(Y; \cdot)$ may switch roles,

$$\|\bar{S}_\Delta(Y; \cdot) - S_\Delta(Y; \cdot)\|^2 = \int_a^b (\bar{S}''_\Delta(Y; x) - S''_\Delta(Y; x))^2 \, dx = 0.$$

Since $S''_\Delta(Y; \cdot)$ and $\bar{S}''_\Delta(Y; \cdot)$ are both continuous,

$$\bar{S}''_\Delta(Y; x) \equiv S''_\Delta(Y; x),$$

from which

$$\bar{S}_\Delta(Y; x) \equiv S_\Delta(Y; x) + cx + d$$

follows by integration. But $\bar{S}_\Delta(Y; x) = S_\Delta(Y; x)$ holds for $x = a, b$, and this implies $c = d = 0$. □

The minimum-norm property of the spline function expressed in Theorem (2.4.1.5) implies in case (2.4.1.2a) that, among all functions f in $\mathscr{K}^2[a, b]$ with $f(x_i) = y_i$, $i = 0, 1, \ldots, n$, it is precisely the spline function $S_\Delta(Y; \cdot)$ with $S''_\Delta(Y; x) = 0$ for $x = a, b$ that minimizes the integral

$$\|f\|^2 = \int_a^b |f''(x)|^2 \, dx.$$

The spline function of case (2.4.1.2a) is therefore often referred to as the *natural* spline function. (In cases (2.4.1.2b) and (2.4.1.2c), the corresponding spline functions $S_\Delta(Y; \cdot)$ minimize $\|f\|$ over the more restricted sets $\mathscr{K}_p^2[a, b]$ and $\{f \in \mathscr{K}^2[a, b] \mid f'(a) = y'_0,\ f'(b) - y'_n\} \cap \{f \mid f(x_i) = y_i$ for $i = 0, 1, \ldots, n\}$, respectively.

The expression $f''(x)(1 + f'(k)^2)^{-3/2}$

indicates the curvature of the function $f(x)$ at $x \in [a, b]$. If $f'(x)$ is small compared to 1, then the curvature is approximately equal to $f''(x)$. The value $\|f\|$ provides us therefore with an approximate measure of the total curvature of the function f in the interval $[a, b]$. In this sense, the natural spline function is the "smoothest" function to interpolate given support points (x_i, y_i), $i = 0, 1, \ldots, n$.

Spline functions have been generalized in many ways. For instance, polynomials of degree k are used to define spline functions $S_\Delta \in C^{k-1}[a, b]$ of degree k as piecewise polynomial functions with continuous $(k-1)$-th derivatives. All these functions share many properties [see Greville (1969), de Boor (1972)] with the cubic splines considered in this and the next two sections.

2.4.2 Determining Interpolating Cubic Spline Functions

In this section, we will describe computational methods for determining cubic spline functions which assume prescribed values at their knots and satisfy one of the side conditions (2.4.1.2). In the course of this, we will have also proved the existence of such spline functions; their uniqueness has already been established by Theorem (2.4.1.5).

In what follows, $\Delta = \{x_i \mid i = 0, 1, \ldots, n\}$ will be a fixed partition of the interval $[a, b]$ by knots $a = x_0 < x_1 < \cdots < x_n = b$, and $Y = \{y_i \mid i = 0, 1, \ldots, n\}$ will be a set of $n + 1$ prescribed real numbers. In addition let

$$h_{j+1} := x_{j+1} - x_j, \qquad j = 0, 1, \ldots, n-1.$$

We refer to the values of the second derivatives at knots $x_j \in \Delta$,

$$\text{(2.4.2.1)} \qquad M_j := S''_\Delta(Y; x_j), \qquad j = 0, 1, \ldots, n.$$

of the desired spline function $S_\Delta(Y; \cdot)$ as the *moments* M_j of $S_\Delta(Y; \cdot)$. We will show that spline functions are readily characterized by their moments, and that the moments of the interpolating spline function can be calculated as the solution of a system of linear equations.

Note that the second derivative $S''_\Delta(Y; \cdot)$ of the spline function coincides with a linear function in each interval $[x_j, x_{j+1}]$, $j = 0, \ldots, n-1$, and that these linear functions can be described in terms of the moments M_i of $S_\Delta(Y; \cdot)$:

$$S''_\Delta(Y; x) = M_j \frac{x_{j+1} - x}{h_{j+1}} + M_{j+1} \frac{x - x_j}{h_{j+1}} \quad \text{for } x \in [x_j, x_{j+1}].$$

By integration,

$$\text{(2.4.2.2)} \quad S'_\Delta(Y; x) = -M_j \frac{(x_{j+1} - x)^2}{2h_{j+1}} + M_{j+1} \frac{(x - x_j)^2}{2h_{j+1}} + A_j,$$

$$S_\Delta(Y; x) = M_j \frac{(x_{j+1} - x)^3}{6h_{j+1}} + M_{j+1} \frac{(x - x_j)^3}{6h_{j+1}} + A_j(x - x_j) + B_j$$

for $x \in [x_j, x_{j+1}]$, $j = 0, 1, \ldots, n-1$, where A_j, B_j are constants of integration. From $S_\Delta(Y; x_j) = y_j$, $S_\Delta(Y; x_{j+1}) = y_{j+1}$, we obtain the following equations for these constants A_j and B_j:

$$M_j \frac{h_{j+1}^2}{6} \qquad\qquad\qquad + B_j = y_j,$$

$$M_{j+1} \frac{h_{j+1}^2}{6} + A_j h_{j+1} + B_j = y_{j+1}.$$

Consequently,

$$B_j = y_j - M_j \frac{h_{j+1}^2}{6}, \tag{2.4.2.3}$$

$$A_j = \frac{y_{j+1} - y_j}{h_{j+1}} - \frac{h_{j+1}}{6}(M_{j+1} - M_j).$$

This yields the following representation of the spline function in terms of its moments:

(2.4.2.4)

$$S_\Delta(Y; x) = \alpha_j + \beta_j(x - x_j) + \gamma_j(x - x_j)^2 + \delta_j(x - x_j)^3 \quad \text{for } x \in [x_j, x_{j+1}],$$

where

$$\alpha_j := y_j,$$

$$\gamma_j := \frac{M_j}{2},$$

$$\beta_j := S'_\Delta(Y; x_j) = -\frac{M_j h_{j+1}}{2} + A_j$$

$$= \frac{y_{j+1} - y_j}{h_{j+1}} - \frac{2M_j + M_{j+1}}{6} h_{j+1},$$

$$\delta_j := \frac{S'''_\Delta(Y; x_j^+)}{6} = \frac{M_{j+1} - M_j}{6h_{j+1}}.$$

Thus $S_\Delta(Y; \cdot)$ has been characterized by its moments M_j. The task of calculating these moments will now be addressed.

The continuity of $S'_\Delta(Y; \cdot)$ at the knots $x = x_j$, $j = 1, 2, \ldots, n-1$ [namely, the relations $S'_\Delta(Y; x_j^-) = S'_\Delta(Y; x_j^+)$] yields $n-1$ equations for the moments M_j. Substituting the values (2.4.2.3) for A_j and B_j in (2.4.2.2) gives

$$S'_\Delta(Y; x) = -M_j \frac{(x_{j+1} - x)^2}{2h_{j+1}} + M_{j+1} \frac{(x - x_j)^2}{2h_{j+1}}$$

$$+ \frac{y_{j+1} - y_j}{h_{j+1}} - \frac{h_{j+1}}{6}(M_{j+1} - M_j).$$

For $j = 1, 2, \ldots, n-1$, we have therefore

$$S'_\Delta(Y; x_j^-) = \frac{y_j - y_{j-1}}{h_j} + \frac{h_j}{3} M_j + \frac{h_j}{6} M_{j-1},$$

$$S'_\Delta(Y; x_j^+) = \frac{y_{j+1} - y_j}{h_{j+1}} - \frac{h_{j+1}}{3} M_j - \frac{h_{j+1}}{6} M_{j+1},$$

and since $S'_\Delta(Y; x_j^+) = S'_\Delta(Y; x_j^-)$,

(2.4.2.5) $$\frac{h_j}{6} M_{j-1} + \frac{h_j + h_{j+1}}{3} M_j + \frac{h_{j+1}}{6} M_{j+1} = \frac{y_{j+1} - y_j}{h_{j+1}} - \frac{y_j - y_{j-1}}{h_j}$$

for $j = 1, 2, \ldots, n-1$. These are $n-1$ equations for the $n+1$ unknown moments. Two further equations can be gained separately from each of the side conditions (a), (b), and (c) listed in (2.4.1.2).

Case (a): $S''_\Delta(Y; a) = M_0 = 0 = M_n = S''_\Delta(Y; b)$.

Case (b): $S''_\Delta(Y; a) = S''_\Delta(Y; b) \Rightarrow M_0 = M_n$,

$$S'_\Delta(Y; a) = S'_\Delta(Y; b) \Rightarrow \frac{h_n}{6} M_{n-1} + \frac{h_n + h_1}{3} M_n + \frac{h_1}{6} M_1$$

$$= \frac{y_1 - y_n}{h_1} - \frac{y_n - y_{n-1}}{h_n}.$$

The latter condition is identical with (2.4.2.5) for $j = n$ if we put

$$h_{n+1} := h_1, \qquad M_{n+1} := M_1, \qquad y_{n+1} := y_1.$$

Recall that (2.4.1.2b) requires $y_n = y_0$.

Case (c): $S'_\Delta(Y; a) = y'_0 \Rightarrow \dfrac{h_1}{3} M_0 + \dfrac{h_1}{6} M_1 = \dfrac{y_1 - y_0}{h_1} - y'_0$,

$$S'_\Delta(Y; b) = y'_n \Rightarrow \frac{h_n}{6} M_{n-1} + \frac{h_n}{3} M_n = y'_n - \frac{y_n - y_{n-1}}{h_n}.$$

The last two equations, as well as those in (2.4.2.5), can be written in a common format:

$$\mu_j M_{j-1} + 2M_j + \lambda_j M_{j+1} = d_j, \qquad j = 1, 2, \ldots, n-1,$$

upon introducing the abbreviations

(2.4.2.6) $$\left.\begin{aligned} \lambda_j &:= \frac{h_{j+1}}{h_j + h_{j+1}}, \qquad \mu_j := 1 - \lambda_j = \frac{h_j}{h_j + h_{j+1}} \\ d_j &:= \frac{6}{h_j + h_{j+1}} \left\{ \frac{y_{j+1} - y_j}{h_{j+1}} - \frac{y_j - y_{j-1}}{h_j} \right\} \end{aligned}\right\} \quad j = 1, 2, \ldots, n-1.$$

In case (a), we define in addition

(2.4.2.7) $$\lambda_0 := 0, \qquad d_0 := 0, \qquad \mu_n := 0, \qquad d_n := 0,$$

and in case (c)

(2.4.2.8) $$\lambda_0 := 1, \qquad d_0 := \frac{6}{h_1}\left(\frac{y_1 - y_0}{h_1} - y'_0\right),$$
$$\mu_n := 1, \qquad d_n := \frac{6}{h_n}\left(y'_n - \frac{y_n - y_{n-1}}{h_n}\right).$$

This leads in cases (a) and (c) to the following system of linear equations for the moments M_i:

$$\begin{aligned}
2M_0 + \lambda_0 M_1 \qquad\qquad\qquad\qquad\qquad &= d_0,\\
\mu_1 M_0 + 2M_1 + \lambda_1 M_2 \qquad\qquad\qquad &= d_1,\\
\cdots \qquad\qquad\qquad\qquad &\quad \cdot\\
\mu_{n-1} M_{n-2} + 2M_{n-1} + \lambda_{n-1} M_n &= d_{n-1},\\
\mu_n M_{n-1} + 2M_n &= d_n.
\end{aligned}$$

In matrix notation, we have

(2.4.2.9) $$\begin{bmatrix} 2 & \lambda_0 & & & 0 \\ \mu_1 & 2 & \lambda_1 & & \\ & \mu_2 & \cdot & \cdot & \\ & & \cdot & \cdot & \cdot \\ & & \cdot & \cdot\ 2 & \lambda_{n-1} \\ 0 & & & \mu_n & 2 \end{bmatrix} \begin{bmatrix} M_0 \\ M_1 \\ \cdot \\ \cdot \\ \cdot \\ M_n \end{bmatrix} = \begin{bmatrix} d_0 \\ d_1 \\ \cdot \\ \cdot \\ \cdot \\ d_n \end{bmatrix}.$$

The periodic case (b) also requires further definitions,

(2.4.2.10) $$\lambda_n := \frac{h_1}{h_n + h_1}, \qquad \mu_n := 1 - \lambda_n = \frac{h_n}{h_n + h_1},$$
$$d_n := \frac{6}{h_n + h_1}\left\{\frac{y_1 - y_n}{h_1} - \frac{y_n - y_{n-1}}{h_n}\right\},$$

which then lead to the following linear system of equations for the moments $M_1, M_2, \ldots, M_n (= M_0)$:

(2.4.2.11) $$\begin{bmatrix} 2 & \lambda_1 & & & \mu_1 \\ \mu_2 & 2 & \lambda_2 & & \\ & \mu_3 & \cdot & \cdot & \\ & & \cdot & \cdot & \cdot \\ & & & \cdot\ 2 & \lambda_{n-1} \\ \lambda_n & & & \mu_n & 2 \end{bmatrix} \begin{bmatrix} M_1 \\ M_2 \\ \cdot \\ \cdot \\ \cdot \\ M_n \end{bmatrix} = \begin{bmatrix} d_1 \\ d_2 \\ \cdot \\ \cdot \\ \cdot \\ d_n \end{bmatrix}.$$

The coefficients λ_i, μ_i, d_i in (2.4.2.9) and (2.4.2.11) are well defined by (2.4.2.6) and the additional definitions (2.4.2.7), (2.4.2.8), and (2.4.2.10), respectively. Note in particular that in (2.4.2.9) and (2.4.2.11)

$$\lambda_i \geqslant 0, \qquad \mu_i \geqslant 0, \qquad \lambda_i + \mu_i = 1 \tag{2.4.2.12}$$

for all coefficients λ_i, μ_i, and that these coefficients depend only on the location of the knots $x_j \in \Delta$ and not on the prescribed values $y_i \in Y$ nor on y_0', y_n' in case (c). We will use this observation when proving the following:

(2.4.2.13) Theorem. *The systems* (2.4.2.9) *and* (2.4.2.11) *of linear equations are nonsingular for any partition* Δ *of* $[a, b]$.

This means that the above systems of linear equations have unique solutions for arbitrary right-hand sides, and that consequently the problem of interpolation by cubic splines has a unique solution in each of the three cases (a), (b), (c) of (2.4.1.2).

PROOF. Consider the $(n + 1) \times (n + 1)$ matrix

$$A = \begin{bmatrix} 2 & \lambda_0 & & & & 0 \\ \mu_1 & 2 & \lambda_1 & & & \\ & \mu_2 & \cdot & \cdot & & \\ & & \cdot & \cdot & \cdot & \\ & & & \cdot & 2 & \lambda_{n-1} \\ 0 & & & & \mu_n & 2 \end{bmatrix}$$

of the linear system (2.4.2.9). This matrix has the following property:

$$Az = w \quad \Rightarrow \quad \max_i |z_i| \leqslant \max_i |w_i| \tag{2.4.2.14}$$

for every pair of vectors $z = (z_0, \ldots, z_n)^T$, $w = (w_0, \ldots, w_n)^T$, $z, w \in \mathbb{R}^{n+1}$. Indeed, let r be such that $|z_r| = \max_i |z_i|$. From $Az = w$,

$$\mu_r z_{r-1} + 2z_r + \lambda_r z_{r+1} = w_r \qquad (\mu_0 := 0, \quad \lambda_n := 0)$$

By the definition of r and because $\mu_r + \lambda_r = 1$,

$$\begin{aligned} \max_i |w_i| \geqslant |w_r| &\geqslant 2|z_r| - \mu_r|z_{r-1}| - \lambda_r|z_{r+1}| \\ &\geqslant 2|z_r| - \mu_r|z_r| - \lambda_r|z_r| \\ &= (2 - \mu_r - \lambda_r)|z_r| \\ &= |z_r| = \max_i |z_i| \end{aligned}$$

Suppose the matrix A were singular. Then there would exist a solution $z \neq 0$ of $Az = 0$, and (2.4.2.14) would lead to the contradiction

$$0 < \max_i |z_i| \leqslant 0.$$

The nonsingularity of the matrix in (2.4.2.11) is shown similarly. □

To solve the equations (2.4.2.9), we may proceed as follows: subtract $\mu_1/2$ times the first equation from the second, thereby annihilating μ_1, and then a suitable multiple of the second equation from the third to annihilate μ_2, and so on. This leads to a "triangular" system of equations which can be solved in a straightforward fashion [note that this method is the Gaussian elimination algorithm applied to (2.4.2.9); compare Section 4.1]:

(2.4.2.15)

$q_0 := -\lambda_0/2;\ u_0 := d_0/2;\quad \lambda_n := 0;$
for $k := 1, 2, \ldots, n$ **do**
begin $p_k := \mu_k q_{k-1} + 2;$
$\quad q_k := -\lambda_k/p_k;$
$\quad u_k := (d_k - \mu_k u_{k-1})/p_k$ **end**;
$M_n := u_n;$
for $k := n-1, n-2, \ldots, 0$ **do**
$\quad M_k := q_k M_{k+1} + u_k;$

[It can be shown that $p_k > 0$, so that (2.4.2.15) is well defined; see Exercise 25.] The linear system (2.4.2.11) can be solved in a similar, but not as straightforward, fashion. An ALGOL program by C. Reinsch can be found in Bulirsch and Rutishauser (1968).

The reader can find more details in Greville (1969) and de Boor (1972), ALGOL programs in Herriot and Reinsch (1971), and FORTRAN programs in de Boor (1978). These references also contain information and algorithms for the spline functions of degree $k \geq 3$ and B-splines, which are treated here in Sections 2.4.4 and 2.4.5.

2.4.3 Convergence Properties of Cubic Spline Functions

Interpolating polynomials may not converge to a function f whose values they interpolate, even if the partitions Δ are chosen arbitrarily fine (see Section 2.1.4). In contrast, we will show in this section that, under mild conditions on the function f and the partitions Δ, the interpolating spline functions do converge towards f as the fineness of the underlying partitions approaches zero.

We will show first that the moments (2.4.2.1) of the interpolating spline function converge to the second derivatives of the given function. More precisely, consider a fixed partition $\Delta = \{a = x_0 < x_1 < \cdots < x_n = b\}$ of $[a, b]$, and let

$$M = \begin{bmatrix} M_0 \\ \vdots \\ M_n \end{bmatrix}$$

be the vector of moments M_j of the spline function $S_\Delta(Y; \cdot)$ with $f(x_j) = y_j$ for $j = 1, \ldots, n-1$, as well as

$$S_\Delta(Y; a) = f'(a), \qquad S_\Delta(Y; b) = f'(b).$$

We are thus dealing with case (c) of (2.4.1.2). The vector M of moments satisfies the equation

$$AM = d,$$

which expresses the linear system of equations (2.4.2.9) in matrix form. The components d_j of d are given by (2.4.2.6) and (2.4.2.8). Let F and r be the vectors

$$F := \begin{bmatrix} f''(x_0) \\ f''(x_1) \\ \vdots \\ f''(x_n) \end{bmatrix}, \qquad r := d - AF = A(M - F).$$

Writing $\|z\| := \max_i |z_i|$ for vectors z, and $\|\Delta\|$ for the *fineness*

(2.4.3.1) $$\|\Delta\| := \max_j |x_{j+1} - x_j|$$

of the partition Δ, we have:

(2.4.3.2) *If $f \in C^4[a, b]$ and $|f^{(4)}(x)| \leqslant L$ for $x \in [a, b]$, then*

$$\|M - F\| \leqslant \|r\| \leqslant \tfrac{3}{4} L \|\Delta\|^2.$$

PROOF. By definition, $r_0 = d_0 - 2f''(x_0) - f''(x_1)$, and by (2.4.2.8),

$$r_0 = \frac{6}{h_1}\left(\frac{y_1 - y_0}{h_1} - y_0'\right) - 2f''(x_0) - f''(x_1).$$

Using Taylor's theorem to express $y_1 = f(x_1)$ and $f''(x_1)$ in terms of the value and the derivatives of the function f at x_0 yields

$$\begin{aligned} r_0 = \frac{6}{h_1}&\left[f'(x_0) + \frac{h_1}{2} f''(x_0) + \frac{h_1^2}{6} f'''(x_0) + \frac{h_1^3}{24} f^{(4)}(\tau_1) - f'(x_0)\right] \\ &- 2f''(x_0) - \left[f''(x_0) + h_1 f'''(x_0) + \frac{h_1^2}{2} f^{(4)}(\tau_2)\right] \\ = \frac{h_1^2}{4}& f^{(4)}(\tau_1) - \frac{h_1^2}{2} f^{(4)}(\tau_2) \end{aligned}$$

with $\tau_1, \tau_2 \in [x_0, x_1]$. Therefore

$$|r_0| \leqslant \tfrac{3}{4}L\|\Delta\|^2.$$

Analogously, we find for

$$r_n = d_n - f''(x_{n-1}) - 2f''(x_n)$$

that

$$|r_n| \leqslant \tfrac{3}{4}L\|\Delta\|^2.$$

For the remaining components of $r = d - AF$, we find similarly

$$\begin{aligned} r_j &= d_j - \mu_j f''(x_{j-1}) - 2f''(x_j) - \lambda_j f''(x_{j+1}) \\ &= \frac{6}{h_j + h_{j+1}}\left[\frac{y_{j+1} - y_j}{h_{j+1}} - \frac{y_j - y_{j-1}}{h_j}\right] \\ &\quad - \frac{h_j}{h_j + h_{j+1}} f''(x_{j-1}) - 2f''(x_j) - \frac{h_{j+1}}{h_j + h_{j+1}} f''(x_{j+1}). \end{aligned}$$

Taylor's formula at x_j then gives

$$\begin{aligned} r_j = \frac{1}{h_j + h_{j+1}}\Bigg\{6\bigg[& f'(x_j) + \frac{h_{j+1}}{2} f''(x_j) + \frac{h_{j+1}^2}{6} f'''(x_j) + \frac{h_{j+1}^3}{24} f^{(4)}(\tau_1) \\ & - f'(x_j) + \frac{h_j}{2} f''(x_j) - \frac{h_j^2}{6} f'''(x_j) + \frac{h_j^3}{24} f^{(4)}(\tau_2)\bigg] \\ & - h_j\left[f''(x_j) - h_j f'''(x_j) + \frac{h_j^2}{2} f^{(4)}(\tau_3)\right] \\ & - 2f''(x_j)(h_j + h_{j+1}) \\ & - h_{j+1}\left[f''(x_j) + h_{j+1} f'''(x_j) + \frac{h_{j+1}^2}{2} f^{(4)}(\tau_4)\right]\Bigg\} \\ = \frac{1}{h_j + h_{j+1}}\bigg[& \frac{h_{j+1}^3}{4} f^{(4)}(\tau_1) + \frac{h_j^3}{4} f^{(4)}(\tau_2) - \frac{h_j^3}{2} f^{(4)}(\tau_3) - \frac{h_{j+1}^3}{2} f^{(4)}(\tau_4)\bigg]. \end{aligned}$$

Here $\tau_1, \ldots, \tau_4 \in [x_{j-1}, x_{j+1}]$. Therefore

$$|r_j| \leqslant \tfrac{3}{4}L\frac{1}{h_j + h_{j+1}}[h_{j+1}^3 + h_j^3] \leqslant \tfrac{3}{4}L\|\Delta\|^2.$$

for $j = 1, 2, \ldots, n-1$. In sum,

$$\|r\| \leqslant \tfrac{3}{4}L\|\Delta\|^2$$

and since $r = A(M - F)$, (2.4.2.14) implies $\|M - F\| \leqslant \|r\|$. □

(2.4.3.3) Theorem. *Suppose $f \in C^4[a, b]$ and $|f^{(4)}(x)| \leqslant L$ for $x \in [a, b]$. Let Δ be a partition $\Delta = \{a = x_0 < \cdots < x_n = b\}$ of the interval $[a, b]$, and K a constant such that*

$$\frac{\|\Delta\|}{|x_{j+1} - x_j|} \leqslant K \quad \textit{for } j = 0, \ldots, n-1.$$

If S_Δ is the spline function which interpolates the values of the function f at the knots $x_0, \ldots, x_n \in \Delta$ and satisfies $S'_\Delta(x) = f'(x)$ for $x = a, b$, then there exist constants $C_k \leqslant 2$, which do not depend on the partition Δ, such that for $x \in [a, b]$,

$$|f^{(k)}(x) - S_\Delta^{(k)}(x)| \leqslant C_k L K \|\Delta\|^{4-k}, \qquad k = 0, 1, 2, 3.$$

Note that the constant $K \geqslant 1$ bounds the deviation of the partition Δ from uniformity.

PROOF. We prove the proposition first for $k = 3$. For $x \in [x_{j-1}, x_j]$,

$$\begin{aligned} S'''_\Delta(x) - f'''(x) &= \frac{M_j - M_{j-1}}{h_j} - f'''(x) \\ &= \frac{M_j - f''(x_j)}{h_j} - \frac{M_{j-1} - f''(x_{j-1})}{h_j} \\ &\quad + \frac{f''(x_j) - f''(x) - [f''(x_{j-1}) - f''(x)]}{h_j} - f'''(x). \end{aligned}$$

Using (2.4.3.2) and Taylor's theorem at x, we conclude that

$$\begin{aligned} |S'''_\Delta(x) - f'''(x)| &\leqslant \tfrac{3}{2} L \frac{\|\Delta\|^2}{h_j} + \frac{1}{h_j} \bigg| (x_j - x) f'''(x) + \frac{(x_j - x)^2}{2} f^{(4)}(\eta_1) \\ &\qquad - (x_{j-1} - x) f'''(x) - \frac{(x_{j-1} - x)^2}{2} f^{(4)}(\eta_2) - h_j f'''(x) \bigg| \\ &\leqslant \tfrac{3}{2} L \frac{\|\Delta\|^2}{h_j} + \frac{L}{2} \frac{\|\Delta\|^2}{h_j}, \qquad \eta_1, \eta_2 \in [x_{j-1}, x_j] \end{aligned}$$

By hypothesis, $\|\Delta\|/h_j \leqslant K$ for every j. Thus $|f'''(x) - S'''_\Delta(x)| \leqslant 2LK\|\Delta\|$.

To prove the proposition for $k = 2$, we observe: for each $x \in (a, b)$ there exists a closest knot $x_j = x_j(x)$, for which $|x_j(x) - x| \leqslant \frac{1}{2}\|\Delta\|$. From

$$f''(x) - S''_\Delta(x) = f''(x_j(x)) - S''_\Delta(x_j(x)) + \int_{x_j(x)}^{x} (f'''(t) - S'''_\Delta(t))\, dt,$$

and since $K \geqslant 1$,

$$\begin{aligned} |f''(x) - S''_\Delta(x)| &\leqslant \tfrac{3}{4} L \|\Delta\|^2 + \tfrac{1}{2}\|\Delta\| \cdot 2LK\|\Delta\| \\ &\leqslant \tfrac{7}{4} LK \|\Delta\|^2, \qquad x \in [a, b]. \end{aligned}$$

We consider $k = 1$ next. In addition to the boundary points $\xi_0 := a$, $\xi_{n+1} := b$, there exist, by Rolle's theorem, n further points $\xi_j \in (x_{j-1}, x_j)$, $j = 1, \ldots, n$, with

$$f'(\xi_j) = S'_\Delta(\xi_j), \qquad j = 0, 1, \ldots, n+1.$$

For any $x \in [a, b]$ there exists a closest one of the above points $\xi_j = \xi_j(x)$, for which consequently

$$|\xi_j(x) - x| < \|\Delta\|.$$

Thus

$$f'(x) - S'_\Delta(x) = \int_{\xi_j(x)}^{x} (f''(t) - S''_\Delta(t))\, dt,$$

and

$$|f'(x) - S'_\Delta(x)| \leqslant \tfrac{7}{4}LK\|\Delta\|^2 \cdot \|\Delta\| = \tfrac{7}{4}LK\|\Delta\|^3, \qquad x \in [a, b].$$

The case $k = 0$ remains. Since

$$f(x) - S_\Delta(x) = \int_{x_j(x)}^{x} (f'(t) - S'_\Delta(t))\, dt,$$

it follows from the above result for $k = 1$ that

$$|f(x) - S_\Delta(x)| \leqslant \tfrac{7}{4}LK\|\Delta\|^3 \cdot \tfrac{1}{2}\|\Delta\| = \tfrac{7}{8}LK\|\Delta\|^4, \qquad x \in [a, b].$$ [5] □

Clearly, (2.4.3.3) implies that for sequences

$$\Delta_m = \{a = x_0^{(m)} < x_1^{(m)} < \cdots < x_{n_m}^{(m)} = b\}, \qquad m = 0, 1, \ldots,$$

of partitions with $\Delta_m \to 0$ and

$$\sup_{m, j} \frac{\|\Delta_m\|}{x_{j+1}^{(m)} - x_j^{(m)}} \leqslant K < +\infty,$$

the corresponding spline functions S_{Δ_m} and their first three derivatives converge to f and its corresponding derivatives uniformly on $[a, b]$. Note that even the third derivative f''' is uniformly approximated by S'''_{Δ_m}, a usually discontinuous sequence of step functions.

[5] The estimates of Theorem (2.4.3.3) have been improved by Hall and Meyer (1976): $|f^{(k)}(x) - S_\Delta^{(k)}(x)| \leqslant c_k L\|\Delta\|^{4-k}$, $k = 0, 1, 2, 3$, with $c_0 := 5/384$, $c_1 := 1/24$, $c_2 := 3/8$, $c_3 := (K + K^{-1})/2$. Here c_0 and c_1 are optimal.

2.4.4 B-Splines

Spline functions are instances of piecewise polynomial functions associated with a partition

$$\Delta = \{a = x_0 < x_1 < \cdots < x_n = b\}$$

of an interval $[a, b]$. In general, a real function $f\colon [a, b] \to \mathbb{R}$ is called a *piecewise polynomial function* of order r or degree $r - 1$ if, for each $i = 0, \ldots, n - 1$, the restriction of f to the subinterval (x_i, x_{i+1}) agrees with a polynomial $p_i(x)$ of degree $\leq r - 1$. In order to get a 1-1 correspondence between f and the sequence $(p_0(x), p_1(x), \ldots, p_{n-1}(x))$, we define f at the knots x_i, $i = 0, \ldots, n - 1$, so that it becomes continuous from the right, $f(x_i) := f(x_i + 0)$, $0 \leq i \leq n - 1$ and $f(x_n) = f(b) := f(x_n - 0)$.

Thus, the spline functions S_Δ of degree k introduced earlier are polynomial functions of degree k that are $(k - 1)$-times differentiable at the interior knots x_i, $1 \leq i \leq n - 1$ of Δ. By $S_{\Delta,k}$ we denote the set of all spline functions S_Δ of degree k, which is easily seen to be a real vector space of dimension $n + k$: In fact, the polynomial $S_\Delta|[x_0, x_1]$ is uniquely determined by its $k + 1$ coefficients; this in turn already fixes the first $k - 1$ derivatives ($=k$ conditions) of the next polynomial $S_\Delta|[x_1, x_2]$ at x_1, so that only one degree of freedom is left for choosing $S_\Delta|[x_1, x_2]$. As the same holds for all further polynomials $S_\Delta|[x_i, x_{i+1}]$, $i = 2, \ldots, n - 1$, one finds $\dim S_{\Delta,k} = k + 1 + (n - 1)\cdot 1 = k + n$.

B-splines are special piecewise polynomial functions with remarkable properties: they are nonnegative and vanish everywhere except on a few contiguous intervals $[x_i, x_{i+1}]$. Moreover, the function space $S_{\Delta,k}$ has a basis consisting of B-splines. As a consequence, B-splines provide the framework for an efficient and numerically stable calculation of splines.

In order to define B-splines, we introduce the function $f_x\colon \mathbb{R} \to \mathbb{R}$

$$f_x(t) := (t - x)_+ := \max(t - x, 0) = \begin{cases} t - x & \text{for } t \geq x \\ 0 & \text{for } t < x \end{cases}$$

and its powers f_x^r, $f_x^r(t) := (t - x)_+^r$, $r \geq 0$. Note that f_x depends on a real parameter x. The function $f_x^r(\cdot)$ is composed of two polynomials of degree $\leq r$: the 0-polynomial $P_0(t) := 0$ for $t < x$ and the polynomial $P_1(t) := (t - x)^r$ for $t > x$. Clearly, $f_x^r \in C^{r-1}$ for $r \geq 1$. Further, we recall that under certain conditions, the divided difference $f[t_i, t_{i+1}, \ldots, t_{i+r}]$ of a real function $f(t)$, $f\colon \mathbb{R} \to \mathbb{R}$ is well defined for all $t_i \leq t_{i+1} \leq \cdots \leq t_{i+r}$, even if the t_j are not mutually distinct: The only requirement is that f be $n_j - 1$-times differentiable at $t = t_j$, $j = i, i + 1, \ldots, i + r$, if t_j occurs n_j times among the $t_i, t_{i+1}, \ldots, t_{i+r}$. In this case, by (2.1.5.9)

(2.4.4.1)
$$f[t_i, \ldots, t_{i+r}] := \frac{f^{(r)}(t_i)}{r!}, \qquad \text{if } t_i = t_{i+1} = \cdots = t_{i+r}$$

$$f[t_i, \ldots, t_{i+r}] := \frac{f[t_{i+1}, \ldots, t_{i+r}] - f[t_i, \ldots, t_{i+r-1}]}{t_{i+r} - t_i}, \qquad \text{otherwise.}$$

It follows by induction over r (see Example 2 of Section 2.1.5) that the divided differences of the function f are linear combinations of its derivatives at the points t_j:

(2.4.4.2) $$f[t_i, t_{i+1}, \ldots, t_{i+r}] = \sum_{j=i}^{i+r} \alpha_j \sum_{s=0}^{n_j-1} f^{(s)}(t_j),$$

where $f^{(0)}(t_j) := f(t_j)$.

Let

$$\mathbf{t} = \{t_j\}_{\underline{m} \le j \le \overline{m}}, \qquad -\infty \le \underline{m} < \overline{m} \le \infty,$$

be any finite or infinite nondecreasing sequence of reals. Then for any integer $r \ge 1$ and i with $\underline{m} \le i < i + r \le \overline{m}$, the ith B-spline of order r associated with $\mathbf{t}$ is defined as the following function in x:

(2.4.4.3) $$B_{i,r,\mathbf{t}}(x) := (t_{i+r} - t_i) f_x^{r-1}[t_i, t_{i+1}, \ldots, t_{i+r}],$$

for which we also write B_i or $B_{i,r}$. Clearly, $B_{i,r,\mathbf{t}}(x)$ is well defined for all $x \ne t_i, t_{i+1}, \ldots, t_{i+r}$, and by (2.4.4.2), is a linear combination of the functions (in x),

$$\left.\frac{d^s}{dt^s} f_x^{r-1}(t)\right|_{t=t_j}, \qquad s = 0, 1, \ldots, n_j - 1, \quad i \le j \le i + r,$$

if t_j occurs n_j times among the $t_i, t_{i+1}, \ldots, t_{i+r}$. Thus $B_{i,r,\mathbf{t}}$ is a linear combination of

(2.4.4.4) $$(t_j - x)_+^s, \qquad \text{where } \max\{r - n_j, 0\} \le s \le r - 1, \quad i \le j \le i + r.$$

Hence the function $B_{i,r,\mathbf{t}}$ coincides with a polynomial of degree at most $r - 1$ on each open interval (t_j, t_{j+1}) with $i \le j < i + r$ and $t_j < t_{j+1}$. That is, $B_{i,r,\mathbf{t}}$ is a piecewise polynomial function of order r with respect to a certain partition of the real axis given by t_j with $i \le j < i + r$, $t_j < t_{j+1}$. At the knots $x = t_j$, $t_j \in \mathbf{t}$, the function $B_{i,r,\mathbf{t}}(x)$ may have jump discontinuities. In that case, we follow our previously stated convention and define $B_{i,r,\mathbf{t}}(t_j) := B_{i,r,\mathbf{t}}(t_j + 0)$. Thus $B_{i,r,\mathbf{t}}$ is a piecewise polynomial function that is continuous from the right. Also by (2.4.4.4) for given $\mathbf{t} = \{t_j\}$, the B-spline $B_{i,r}(x) \equiv B_{i,r,\mathbf{t}}(x)$ is $(r - n_j - 1)$ times differentiable at $x = t_j$, if t_j occurs n_j times within t_i, $t_{i+1}, \ldots, t_{i+r}$. Hence the order of differentiability of the $B_{i,r}(x)$ at $x = t_j$ is governed by the number of repetitions of t_j in $\mathbf{t}$.

EXAMPLE. For the finite sequence $\mathbf{t}$: $t_0 < t_1 = t_2 = t_3 < t_4 = t_5 < t_6$, the B-spline $B_{1,5}$ of order 5 is a piecewise polynomial function of degree 4 with respect to the partition $t_3 < t_5 < t_6$ of $\mathbb{R}$. For $x = t_3, t_5, t_6$ it has continuous derivatives up to the orders 1, 2, and 3, respectively, as $n_3 = 3$, $n_5 = 2$, and $n_6 = 1$.

We note some important properties of B-Splines:

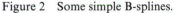

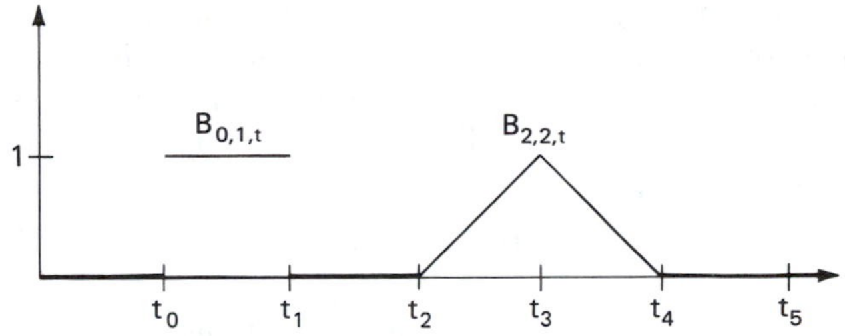

Figure 2 Some simple B-splines.

(2.4.4.5) Theorem. (a) $B_{i,r,t}(x) = 0$ *for* $x \notin [t_i, t_{i+r}]$

(b) $B_{i,r,t}(x) > 0$ *for* $t_i < x < t_{i+r}$.

(c) *For all* x *with* $\inf t_j < x < \sup t_j$

$$\sum_i B_{i,r,t}(x) = 1,$$

and the sum contains only finitely many nonzero terms.

By that theorem, the functions $B_{i,r} \equiv B_{i,r,t}$ are nonnegative weight functions with sum 1 and support $[t_i, t_{i+r}]$, if $t_i < t_{i+r}$: they form a "partition of unity."

PROOF. (a) For $x < t_i \le t \le t_{i+r}$, $f_x^{r-1}(t) = (t-x)^{r-1}$ is a polynomial of degree $(r-1)$ in t, which has a vanishing rth divided difference

$$f_x^{r-1}[t_i, t_{i+1}, \ldots, t_{i+r}] = 0 \quad \Rightarrow \quad B_{i,r}(x) = 0$$

by Theorem (2.1.3.10). On the other hand, if $t_i \le t \le t_{i+1} < x$, then $f_x^{r-1}(t) := (t-x)_+^{r-1} = 0$ is trivially true, so that again $B_{i,r}(x) = 0$.

(b) For $r = 1$ and $t_i < x < t_{i+1}$, the assertion follows from the definition

$$B_{i,1}(x) = [(t_{i+1} - x)_+^0 - (t_i - x)_+] = 1 - 0 = 1,$$

and for $r > 1$, from recursion (2.4.5.2) for B-splines $B_{i,r}$, which will be derived later on.

(c) Assume first $t_j < x < t_{j+1}$. Then by (a), $B_{i,r}(x) = 0$ for all i, r with $i + r \le j$ and all $i \ge j+1$, so that

$$\sum_i B_{i,r}(x) = \sum_{i=j-r+1}^{j} B_{i,r}(x).$$

Now, (2.4.4.1) implies

$$B_{i,r}(x) = f_x^{r-1}[t_{i+1}, t_{i+2}, \ldots, t_{i+r}] - f_x^{r-1}[t_i, t_{i+r}, \ldots, t_{i+r-1}].$$

Therefore,

$$\sum_i B_{i,r}(x) = f_x^{r-1}[t_{j+1}, \ldots, t_{j+r}] - f_x^{r-1}[t_{j-r+1}, \ldots, t_j] = 1 - 0.$$

Here, the last equality holds because the function $f_x^{r-1}(t) = (t - x)^{r-1}$ is a polynomial of degree $(r - 1)$ in t for $t_j < x < t_{j+1} \le t \le t_{j+r}$, for which $f_x^{r-1}[t_{j+1}, \ldots, t_{j+r}] = 1$ by (2.1.4.3), and the function $f_x^{(r-1)}(t) = (t - x)_+^{r-1}$ vanishes for $t_{j-r+1} \le t \le t_j < x < t_{j+1}$. For arbitrary $x = t_j < \sup t_i$, the assertion follows because

$$B_{j,r}(t_j) = \lim_{x \downarrow t_j} B_{i,r}(x). \qquad \square$$

We now return to the space $S_{\Delta,k}$ of spline functions of degree k. We want to construct a sequence $\mathbf{t} = \{t_j\}$ such that the corresponding B-Splines $B_{i,r,\mathbf{t}}(x)$ form a basis of $S_{\Delta,k}$. For this purpose, we associate with the partition Δ the particular finite sequence $\mathbf{t} = \{t_j\}_{-k \le j \le n+k}$ defined by

$$\text{(2.4.4.6)} \quad t_{-k} = \cdots = t_0 := x_0 < t_1 := x_1 < \cdots < t_n = \cdots = t_{n+k} := x_n.$$

Then the $n + k$ B-splines of order $k + 1$

$$\text{(2.4.4.7)} \quad B_{i,k+1,\mathbf{t}}(x) \equiv (t_{i+k+1} - t_i) f_x^k[t_i, \ldots, t_{i+k+1}], \qquad -k \le i \le n - 1,$$

will form a basis of $S_{\Delta,k}$: In order to show that $B_{i,k+1,\mathbf{t}} \in S_{\Delta,k}$ we note first that $B_{i,k+1,\mathbf{t}}$ agrees with a polynomial of degree $\le k$ on each subinterval $[x_i, x_{i+1}]$ of $[a, b]$, and that $B_{i,k+1,\mathbf{t}}$ has continuous derivatives up to order $(k + 1) - n_i - 1 = k - 1$ at the interior knots x_i, $i = 1, \ldots, n - 1$, of Δ, since any such $t_i = x_i$ occurs only once in $\mathbf{t}$ (2.4.4.6), $n_i = 1$.

On the other hand, by a result of Curry and Schoenberg (1968), which we quote without proof, the $n + k$ functions $B_{i,k+1,\mathbf{t}}(x)$, $-k \le i \le n - 1$, are linearly independent. Since the dimension of $S_{\Delta,k}$ is equal to $n + k$, these functions must form a basis of $S_{\Delta,k}$.

2.4.5 The Computation of B-Splines

B-splines can be computed recursively. The recursions are based on a remarkable generalization of the Leibniz formula for the derivatives of the product of two functions. Indeed, in terms of divided differences, we find the following.

(2.4.5.1) Product Rule for Divided Differences. *Suppose $t_i \le t_{i+1} \le \cdots \le t_{i+k}$. Assume further that the function $f(t) = g(t)h(t)$ is a product of two functions that are sufficiently often differentiable at $t = t_j$, $j = i, \ldots, i + k$ so that $g[t_i, t_{i+1}, \ldots, t_{i+k}]$ and $h[t_i, t_{i+1}, \ldots, t_{i+k}]$ are defined by* (2.4.4.1). *Then*

$$f[t_i, t_{i+1}, \ldots, t_{i+k}] = \sum_{r=i}^{i+k} g[t_i, t_{i+1}, \ldots, t_r] h[t_r, t_{r+1}, \ldots, t_{i+k}].$$

Proof. From (2.1.5.5) and (2.1.5.6), the polynomials

$$\sum_{r=i}^{i+k} g[t_i, \ldots, t_r](t - t_i) \ldots (t - t_{r-1})$$

and

$$\sum_{s=i}^{i+k} h[t_s, \ldots, t_{i+k}](t - t_{s+1}) \ldots (t - t_{i+k})$$

interpolate the functions g and h, respectively, at the points $t = t_i$, $t_{i+1}, \ldots, t_{i+k}$ (in the sense of Hermite interpolation, see (2.1.5), if the t_j are not mutually distinct). Therefore, the product polynomial

$$F(t) := \sum_{r=i}^{i+k} g[t_i, \ldots, t_r](t - x_i) \ldots (t - t_{i-1})$$
$$\cdot \sum_{s=i}^{i+k} h[t_s, \ldots, t_{i+k}](t - t_{s+1}) \ldots (t - t_{i+k})$$

also interpolates the function $f(t)$ at $t = t_i, \ldots, t_{i+k}$. This product can be written as the sum of two polynomials

$$F(t) = \sum_{r,s=i}^{i+k} \cdots = \sum_{r \le s} \cdots + \sum_{r > s} \cdots = P_1(t) + P_2(t).$$

Since each term of the second sum $\sum_{r>s}$ is a multiple of the polynomial $\prod_{j=i}^{i+k}(t - t_j)$, the polynomial $P_2(t)$ interpolates the 0-function at $t = t_i, \ldots, t_{i+k}$. Therefore, the polynomial $P_1(t)$, which is of degree $\le k$, interpolates $f(t)$ at $t = t_i, \ldots, t_{i+k}$. Hence, $P_1(t)$ is the unique (Hermite-) interpolant of $f(t)$ of degree $\le k$.

By (2.1.5.6) the highest coefficient of $P_1(t)$ is $f[t_i, \ldots, t_{i+k}]$. A comparison of the coefficients of t^k on both sides of the sum representation $P_1(t) = \sum_{r \le s} \ldots$ of P_1 proves the desired formula

$$f[t_i, \ldots, t_{i+k}] = \sum_{r=i}^{i+k} g[t_i, \ldots, r_r]h[t_r, \ldots, t_{i+k}]. \qquad \square$$

Now we use (2.4.5.1) to derive a recursion for the B-splines $B_{i,r}(x) \equiv B_{i,r,t}(x)$ (2.4.4.3). To do this, it will be convenient to renormalize the functions $B_{i,r}(x)$, letting

$$N_{i,r}(x) := \frac{B_{i,r}(x)}{t_{i+r} - t_i} \equiv f_x^{r-1}[t_i, t_{i+1}, \ldots, t_{i+r}],$$

for which the following simple recursion holds:

For $r \ge 2$ and $t_i < t_{i+r}$

(2.4.5.2) $$N_{i,r}(x) = \frac{x - t_i}{t_{i+r} - t_i} N_{i,r-1}(x) + \frac{t_{i+r} - x}{t_{i+r} - t_i} N_{i+1,r-1}(x).$$

PROOF. Suppose first $x \ne t_j$ for all j. We apply rule (2.4.5.1) to the product

$$f_x^{r-1}(t) \equiv (t - x)_+^{r-1} \equiv (t - x)(t - x)_+^{r-2} =: g(t) f_x^{r-2}(t).$$

Noting that $g(t)$ is a linear polynomial in t for which by (2.1.4.3)

$$g[t_i] = t_i - x,\ g[t_i, t_{i+1}] = 1,\ g[t_i, \ldots, t_j] = 0 \quad \text{for } j > i + 1,$$

we obtain

$$\begin{aligned} f_x^{r-1}[t_i, \ldots, t_{i+r}] &= (t_i - x) f_x^{r-2}[t_i, \ldots, t_{i+r}] + 1 \cdot f_x^{r-2}[t_{i+1}, \ldots, t_{i+r}] \\ &= \frac{(t_i - x)}{t_{i+r} - t_i}(f_x^{r-2}[t_{i+1}, \ldots, t_{i+r}] - f_x^{r-2}[t_i, \ldots, t_{i+r-1}]) \\ &\quad + 1 \cdot f_x^{r-2}[t_{i+1}, \ldots, t_{i+r}] \\ &= \frac{x - t_i}{t_{i+r} - t_i} f_x^{r-2}[t_i, \ldots, t_{i+r-1}] + \frac{t_{i+r} - x}{t_{i+r} - t_i} f_x^{r-2}[t_{i+1}, \ldots, t_{i+r}], \end{aligned}$$

and this proves (2.4.5.2) for $x \neq t_i, \ldots, t_{i+r}$. The result is furthermore true for all x since all $B_{i,r}(x)$ are continuous from the right and $t_i < t_{i+r}$.

The proof of (2.4.4.5), (c) can now be completed: By (2.4.5.2), the value $N_{i,r}(x)$ is a convex linear combination of $N_{i,r-1}(x)$ and $N_{i+1,r-1}(x)$ for $t_i < x < t_{i+r}$ with positive weights $\lambda_i(x) = (x - t_i)/(t_{i+r} - t_i) > 0$, $1 - \lambda_i(x) > 0$. Also $N_{i,r}(x)$ and $B_{i,r}(x)$ have the same sign, and we already know that $B_{i,1}(x) = 0$ for $x \notin [t_i, t_{i+1}]$ and $B_{i,1}(x) > 0$ for $t_i < x < t_{i+1}$. Induction over r using (2.4.5.2) shows that $B_{i,r}(x) > 0$ for $t_i < x < t_{i+r}$.

The formula

(2.4.5.3) $$B_{i,r}(x) = \frac{x - t_i}{t_{i+r-1} - t_i} B_{i,r-1}(x) + \frac{t_{i+r} - x}{t_{i+r} - t_{i+1}} B_{i+1,r-1}(x)$$

is equivalent to (2.4.5.2), and represents $B_{i,r}(x)$ directly as a positive linear combination of $B_{i,r-1}(x)$ and $B_{i+1,r-1}(x)$. It can be used to compute the values of all B-splines $B_{i,r}(x) = B_{i,r,\mathbf{t}}(x)$ for a given fixed value of x.

To show this, we assume that there is a $t_j \in \mathbf{t}$ with $t_j \le x < t_{j+1}$ (otherwise $B_{i,r}(x) = 0$ for all i, r for which $B_{i,r} = B_{i,r,\mathbf{t}}$ is defined, and the problem is trivial). By (2.4.4.5)a) we know $B_{i,r}(x) = 0$ for all i, r with $x \notin [t_i, t_{i+r}]$, i.e., for $i \le j - r$ and for $i \ge j + 1$. Therefore, in the following tableau of $B_{i,r} := B_{i,r}(x)$, the $B_{i,r}$ vanish at the positions denoted by 0:

(2.4.5.4) $$\begin{matrix} 0 & 0 & 0 & 0 & \cdots \\ 0 & 0 & 0 & B_{j-3,4} & \cdots \\ 0 & 0 & B_{j-2,3} & B_{j-2,4} & \cdots \\ 0 & B_{j-1,2} & B_{j-1,3} & B_{j-1,4} & \cdots \\ B_{j,1} & B_{j,2} & B_{j,3} & B_{j,4} & \cdots \\ 0 & 0 & 0 & 0 & \cdots \\ \vdots & \vdots & \vdots & \vdots & \end{matrix}$$

By definition, $B_{j,1} = B_{j,1}(x) = 1$ for $t_j \le x < t_{j+1}$, which determines the first column of (2.4.5.4). The remaining columns can be computed consecutively using recursion (2.4.5.3): Each element $B_{i,r}$ can be derived from its two left

neighbors $B_{i,r-1}$ and $B_{i+1,r-1}$

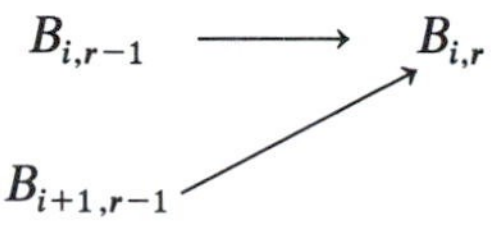

This method is numerically very stable because only nonnegative multiples of nonnegative numbers are added together.

EXAMPLE. For $t_i = i$, $i = 0, 1, \dots$ and $x = 3.5 \in [t_3, t_4)$ the following tableau of values $B_{i,r} = B_{i,r}(x)$ is obtained.

$r =$	1	2	3	4
$i = 0$	0	0	0	1/48
$i = 1$	0	0	1/8	23/48
$i = 2$	0	1/2	6/8	23/48
$i = 3$	1	1/2	1/8	1/48
$i = 4$	0	0	0	0

For instance, $B_{2,4}$ is obtained from

$$B_{2,4} = B_{2,4}(3.5) = \frac{3.5-2}{5-2}\cdot\frac{6}{8} + \frac{6-3.5}{6-3}\cdot\frac{1}{8} = \frac{23}{48}.$$

We now consider the interpolation problem for spline functions, namely, the problem of finding a spline $S_\Delta \in S_{\Delta,k}$ that assumes prescribed values at given locations. Since the vector space $S_{\Delta,k}$ has a basis of B-splines, see (2.4.7.7), we may proceed as follows. Assume that $r \geq 1$ is an integer and $\mathbf{t} = \{t_i\}_{1 \leq i \leq N+r}$ a finite sequence of real numbers satisfying

$$t_1 \leq t_2 \leq \cdots \leq t_{N+r}$$

and $t_i < t_{i+r}$ for $i = 1, 2, \dots N$. Denote by $B_i(x) \equiv B_{i,r,\mathbf{t}}(x)$, $i = 1, \dots, N$, the associated B-splines, and by

$$\mathscr{S}_{r,\mathbf{t}} := \left\{ \sum_{i=1}^{N} \alpha_i B_i(x) \,|\, \alpha_i \in \mathbb{R} \right\},$$

the vector space spanned by the B_i, $i = 1, \dots, N$. Further, assume that we are given N pairs (ξ_j, f_j), $j = 1, \dots, N$, of interpolation points with

$$\xi_1 < \xi_2 < \cdots < \xi_N.$$

These are the data for the interpolation problem of finding a function $S \in \mathscr{S}_{r,\mathbf{t}}$ satisfying

(2.4.5.5) $$S(\xi_j) = f_j, \qquad j = 1, \dots, N.$$

Since any $S \in \mathscr{S}_{r,\mathbf{t}}$ can be written as a linear combination of the B_i, $i =$

$1, \ldots, N$, this is equivalent to the problem of solving the linear equations

$$\sum_{i=1}^{N} \alpha_i B_i(\xi_j) = f_j, \qquad j = 1, \ldots, N. \tag{2.4.5.6}$$

The matrix of this system

$$A = \begin{bmatrix} B_1(\xi_1) & \cdots & B_N(\xi_1) \\ \vdots & \vdots & \vdots \\ B_1(\xi_N) & \cdots & B_N(\xi_N) \end{bmatrix}$$

has a special structure: A is a band matrix, because by (2.4.4.5) the functions $B_i(x) = B_{i,r,t}(x)$ have support $[t_i, t_{i+r}]$, so that within the jth row of A all elements $B_i(\xi_j)$ with $t_{i+r} < \xi_j$ or $t_i > \xi_j$ are zero. Therefore, each row of A contains at most r elements different from 0, and these elements are in consecutive positions. The components $B_i(\xi_j)$ of A can be computed by the recursion described previously. The system (2.4.5.6), and thereby the interpolation problem, is uniquely solvable if A is nonsingular. The nonsingularity of A can be checked by means of the following simple criterion due to Schoenberg and Whitney (1953), which is quoted without proof.

(2.4.5.7) Theorem. *The matrix $A = (B_i(\xi_j))$ of (2.4.5.6) is nonsingular if and only if all its diagonal elements $B_i(\xi_i) \neq 0$ are nonzero.*

It is possible to show [see Karlin (1968)] that the matrix A is *totally positive* in the following sense: all $r \times r$ submatrices B of A of the form

$$B = (a_{i_p, j_q})_{p,q=1}^{r} \quad \text{with } r \geq 1, \quad i_1 < i_2 < \cdots < i_r, \quad j_1 < j_2 < \cdots < j_r$$

have a nonnegative determinant, $\det(B) \geq 0$. As a consequence, solving (2.4.5.6) for nonsingular A by Gaussian elimination *without pivoting* is numerically stable [see de Boor and Pinkus (1977)]. Also the band structure of A can be exploited for additional savings.

For further properties of B-Splines, their applications, and algorithms the reader is referred to the literature, in particular to de Boor (1978), where one can also find numerous FORTRAN programs.

Exercises for Chapter 2

1. Let $L_i(x)$ be the Lagrange polynomials (2.1.1.3) for pairwise different support abscissas $x_0, \ldots, x_n$, and let $c_i := L_i(0)$. Show that

(a) $$\sum_{i=0}^{n} c_i x_i^j = \begin{cases} 1 & \text{for } j = 0, \\ 0 & \text{for } j = 1, 2, \ldots, n, \\ (-1)^n x_0 x_1 \ldots x_n & \text{for } j = n+1; \end{cases}$$

(b) $$\sum_{i=0}^{n} L_i(x) \equiv 1.$$

2. Interpolate the function $\ln x$ by a quadratic polynomial at $x = 10, 11, 12$.
   (a) Estimate the error committed for $x = 11.1$ when approximating $\ln x$ by the interpolating polynomial.
   (b) How does the sign of the error depend on x?

3. Consider a function f which is twice continuously differentiable on the interval $I = [-1, 1]$. Interpolate the function by a linear polynomial through the support points $(x_i, f(x_i))$, $i = 0, 1$, $x_0, x_1 \in I$. Verify that

$$\alpha = \tfrac{1}{2} \max_{\xi \in I} |f''(\xi)| \max_{x \in I} |(x - x_0)(x - x_1)|$$

is an upper bound for the maximal absolute interpolation error on the interval I. Which values x_0, x_1 minimize α? What is the connection between $(x - x_0) \times (x - x_1)$ and $\cos(2 \arccos x)$?

4. Suppose a function $f(x)$ is interpolated on the interval $[a, b]$ by a polynomial $P_n(x)$ whose degree does not exceed n. Suppose further that f is arbitrarily often differentiable on $[a, b]$ and that there exists M such that $|f^{(i)}(x)| \leqslant M$ for $i = 0, 1, 2, \ldots$ and any $x \in [a, b]$. Can it be shown, without additional hypotheses about the location of the support abscissas $x_i \in [a, b]$, that $P_n(x)$ converges uniformly on $[a, b]$ to $f(x)$ as $n \to \infty$?

5. (a) The Bessel function of order zero,

$$J_0(x) = \frac{1}{\pi} \int_0^\pi \cos(x \sin t)\, dt,$$

   is to be tabulated at equidistant arguments $x_i = x_0 + ih$, $i = 0, 1, 2, \ldots$. How small must the increment h be chosen so that the interpolation error remains below 10^{-6} if linear interpolation is used?

   (b) What is the behavior of the maximal interpolation error

$$\max_{0 \le x \le 1} |P_n(x) - J_0(x)|$$

   as $n \to \infty$, if $P_n(x)$ interpolates $J_0(x)$ at $x = x_i^{(n)} := i/n$, $i = 0, 1, \ldots, n$? *Hint:* It suffices to show that $|J_0^{(k)}(x)| \leqslant 1$ for $k = 0, 1, \ldots$.

   (c) Compare the above result with the behavior of the error

$$\max_{0 \le x \le 1} |S_{\Delta_n}(x) - J_0(x)|$$

   as $n \to \infty$, where S_{Δ_n} is the interpolating spline function with knot set $\Delta_n = \{x_i^{(n)}\}$ and $S'_{\Delta_n}(x) = J'_0(x)$ for $x = 0, 1$.

6. *Interpolation on product spaces:* Suppose every linear interpolation problem stated in terms of functions $\varphi_0, \varphi_1, \ldots, \varphi_n$ has a unique solution

$$\Phi(x) = \sum_{i=0}^{n} \alpha_i \varphi_i(x)$$

with $\Phi(x_k) = f_k$, $k = 0, \ldots, n$, for prescribed support arguments $x_0, \ldots x_n$ with $x_i \neq x_j$, $i \neq j$. Show the following: If $\psi_0, \ldots, \psi_m$ is also a set of functions for which every linear interpolation problem has a unique solution, then for every choice of abscissas

$$x_0, x_1, \ldots, x_n, \qquad x_i \neq x_j, \quad i \neq j,$$

$$y_0, y_1, \ldots, y_m, \qquad y_i \neq y_j, \quad i \neq j,$$

and support ordinates

$$f_{ik}, \qquad i = 0, \ldots, n, \quad k = 0, \ldots, m,$$

there exists a unique function of the form

$$\Phi(x, y) = \sum_{\nu=0}^{n} \sum_{\mu=0}^{m} \alpha_{\nu\mu} \varphi_\nu(x) \psi_\mu(y)$$

with $\Phi(x_i, y_k) = f_{ik}$, $i = 0, 1, \ldots, n$, $k = 0, \ldots, m$.

7. Specialize the general result of Exercise 6 to interpolation by polynomials. Give the explicit form of the function $\Phi(x, y)$ in this case.

8. Given the abscissas

$$y_0, y_1, \ldots, y_m, \qquad y_i \neq y_j, \quad i \neq j,$$

and, for each $k = 0, \ldots, m$, the values

$$x_0^{(k)}, x_1^{(k)}, \ldots, x_{n_k}^{(k)}, \qquad x_i^{(k)} \neq x_j^{(k)}, \quad i \neq j,$$

and support ordinates

$$f_{ik}, \qquad i = 0, \ldots, n_k, \quad k = 0, \ldots, m,$$

suppose without loss of generality that the y_k are numbered in such a fashion that

$$n_0 \geqslant n_1 \geqslant \cdots \geqslant n_m.$$

Prove by induction over m that exactly one polynomial

$$P(x, y) \equiv \sum_{\mu=0}^{m} \sum_{\nu=0}^{n_\mu} \alpha_{\nu\mu} x^\nu y^\mu$$

exists with

$$P(x_i^{(k)}, y_k) = f_{ik}, \qquad i = 0, \ldots, n_k, \quad k = 0, \ldots, m.$$

9. Is it possible to solve the interpolation problem of Exercise 8 by other polynomials

$$P(x, y) = \sum_{\mu=0}^{M} \sum_{\nu=0}^{N_\mu} \alpha_{\nu\mu} x^\nu y^\mu,$$

requiring only that the number of parameters $\alpha_{\nu\mu}$ agree with the number of support points, that is,

$$\sum_{\mu=0}^{m} (n_\mu + 1) = \sum_{\mu=0}^{M} (N_\mu + 1)?$$

Hint: Study a few simple examples.

10. Calculate the inverse and reciprocal differences for the support points

x_i	0	1	-1	2	-2
f_i	1	3	$\frac{3}{5}$	3	$\frac{3}{5}$

and use them to determine the rational expression $\Phi^{2,2}(x)$ whose numerator and denominator are quadratic polynomials and for which $\Phi^{2,2}(x_i) = f_i$, first in continued-fraction form and then as the ratio of polynomials.

11. Let $\Phi^{m,n}$ be the rational function which solves the system $S^{m,n}$ for given support points (x_k, f_k), $k = 0, 1, \ldots, m+n$:

$$(a_0 + a_1 x_k + \cdots + a_m x_k^m) - f_k(b_0 + b_1 x_k + \cdots + b_n x_k^n) = 0,$$
$$k = 0, 1, \ldots, m+n.$$

Show that $\Phi^{m,n}(x)$ can be represented as follows by determinants:

$$\Phi^{m,n}(x) = \frac{|f_k, x_k - x, \ldots, (x_k - x)^m, (x_k - x)f_k, \ldots, (x_k - x)^n f_k|_{k=0}^{m+n}}{|1, x_k - x, \ldots, (x_k - x)^m, (x_k - x)f_k, \ldots, (x_k - x)^n f_k|_{k=0}^{m+n}}.$$

Here the following abbreviation has been used:

$$|\alpha_k, \ldots, \zeta_k|_{k=0}^{m+n} := \det \begin{bmatrix} \alpha_0 & \cdots & \zeta_0 \\ \alpha_1 & \cdots & \zeta_1 \\ \vdots & & \vdots \\ \alpha_{m+n} & \cdots & \zeta_{m+n} \end{bmatrix}.$$

12. Generalize Theorem (2.3.1.11):

(a) For $2n + 1$ support abscissas x_k with

$$a \leqslant x_0 < x_1 < \cdots < x_{2n} < a + 2\pi$$

and support ordinates $y_0, \ldots, y_{2n}$, there exists a unique trigonometric polynomial

$$T(x) = \tfrac{1}{2}a_0 + \sum_{j=1}^{n} (a_j \cos jx + b_j \sin jx)$$

with

$$T(x_k) = y_k \quad \text{for } k = 0, 1, \ldots, 2n.$$

(b) If $y_0, \ldots, y_{2n}$ are real numbers, then so are the coefficients a_j, b_j.
Hint: Reduce the interpolation by trigonometric polynomials in (a) to (complex) interpolation by polynomials using the transformation $T(x) = \sum_{j=-n}^{n} c_j e^{ijx}$. Then show $c_{-j} = \overline{c_j}$ to establish (b).

13. (a) Show that, for real $x_1, \ldots, x_{2n}$, the function

$$t(x) = \prod_{k=1}^{2n} \sin \frac{x - x_k}{2}$$

is a trigonometric polynomial

$$\tfrac{1}{2}a_0 + \sum_{j=1}^{n} (a_j \cos jx + b_j \sin jx)$$

with real coefficients a_j, b_j.
Hint: Substitute $\sin \varphi = (1/2i)(e^{i\varphi} - e^{-i\varphi})$.

(b) Prove, using (a) that the interpolating trigonometric polynomial with support abscissas x_k,

$$a \leqslant x_0 < x_1 \cdots < x_{2n} < a + 2\pi,$$

and support ordinates $y_0, \ldots, y_{2n}$ is identical with

$$T(x) = \sum_{j=0}^{2n} y_j t_j(x),$$

where

$$t_j(x) := \prod_{\substack{k=0 \\ k \neq j}}^{2n} \sin \frac{x - x_k}{2} \Big/ \prod_{\substack{k=0 \\ k \neq j}}^{2n} \sin \frac{x_j - x_k}{2}.$$

14. Show that for $n + 1$ support abscissas x_k with

$$0 \leqslant x_0 < x_1 < \cdots < x_n < \pi$$

and support ordinates $y_0, \ldots, y_n$, a unique "cosine polynomial"

$$C(x) = \sum_{j=0}^{n} a_j \cos jx$$

exists with $C(x_k) = y_k$, $k = 0, 1, \ldots, n$.
Hint: See Exercise 12.

15. (a) Show that for any integer j

$$\sum_{k=0}^{2m} \cos jx_k = (2m + 1)h(j),$$

$$\sum_{k=0}^{2m} \sin jx_k = 0,$$

with

$$x_k := \frac{2\pi k}{2m + 1}, \qquad k = 0, 1, \ldots, 2m,$$

and

$$h(j) := \begin{cases} 1 & \text{for } j = 0 \bmod 2m + 1 \\ 0 & \text{otherwise.} \end{cases}$$

(b) Use (a) to derive for integers j, k the following orthogonality relations:

$$\sum_{i=0}^{2m} \sin jx_i \sin kx_i = \frac{2m + 1}{2} [h(j - k) - h(j + k)],$$

$$\sum_{i=0}^{2m} \cos jx_i \cos kx_i = \frac{2m + 1}{2} [h(j - k) + h(j + k)],$$

$$\sum_{i=0}^{2m} \cos jx_i \sin kx_i = 0.$$

16. Suppose the 2π-periodic function f: $\mathbb{R} \to \mathbb{R}$ has an absolutely convergent Fourier series

$$f(x) = \tfrac{1}{2}a_0 + \sum_{j=1}^{\infty} (a_j \cos jx + b_j \sin jx).$$

Let

$$\Psi(x) = \tfrac{1}{2}A_0 + \sum_{j=1}^{m} (A_j \cos jx + B_j \sin jx)$$

be trigonometric polynomials with

$$\Psi(x_k) = f(x_k), \qquad x_k = \frac{2\pi k}{2m+1},$$

for $k = 0, 1, \ldots, 2m$.
Show that

$$A_k = a_k + \sum_{p=1}^{\infty} [a_{p(2m+1)+k} + a_{p(2m+1)-k}], \qquad 0 \leqslant k \leqslant m,$$

$$B_k = b_k + \sum_{p=1}^{\infty} [b_{p(2m+1)+k} - b_{p(2m+1)-k}], \qquad 1 \leqslant k \leqslant m.$$

17. Formulate a Cooley–Tukey method in which the array $\tilde{\beta}[\ ]$ is initialized directly ($\tilde{\beta}[j] := f_j$) rather than in bit-reversed fashion.
Hint: Define and determine explicitly a map $\sigma = \sigma(m, r, j)$ with the same replacement properties as (2.3.2.6) but $\sigma(0, r, 0) := r$.

18. Let $N := 2^n$. Consider the N-vectors $f := [f_0, \ldots, f_{N-1}]^T$, $\beta := [\beta_0, \ldots, \beta_{N-1}]^T$. (2.3.2.1) expresses a linear transformation between these two vectors, $\beta = (1/N)Tf$, where $T = [t_{jk}]$ with $t_{jk} = e^{-2\pi i jk/N}$.

(a) Show that T can be factored as follows:

$$T = QSP(D_{n-1}SP) \ldots (D_1 SP),$$

where S is the $N \times N$ matrix

$$S := \begin{bmatrix} 1 & 1 & & & \\ 1 & -1 & & & \\ & & \ddots & & \\ & & & 1 & 1 \\ & & & 1 & -1 \end{bmatrix}.$$

The matrices $D_l = \operatorname{diag}(1, \delta_1^{(l)}, 1, \delta_3^{(l)}, \ldots, 1, \delta_{N-1}^{(l)})$, $l = 1, \ldots, n-1$, are diagonal matrices with

$$\delta_r^{(l)} = \exp(-2\pi i \bar{r}/2^{n-l-1}), \qquad \bar{r} = \left[\frac{r}{2^l}\right], \quad r \text{ odd}.$$

Q is the matrix of the bit-reversal permutation (2.3.2.8), and P is the matrix of the following *bit-cycling* permutation ζ:

$$\zeta(\alpha_0 + \alpha_1 2 + \cdots + \alpha_{n-1}2^{n-1}) := \alpha_{n-1} + \alpha_0 2 + \cdots + \alpha_{n-2}2^{n-1}.$$

(b) Show that the Sande–Tukey method for fast Fourier transforms corresponds to multiplying the vector f from the left by the (sparse) matrices in the above factorization.

(c) Which factorization of T corresponds to the Cooley–Tukey method?
Hint: T^H differs from T by a permutation.

19. Investigate the numerical stability of the methods for fast Fourier transforms described in Section 2.3.2.
Hint: The matrices in the factorization of Exercise 18 are almost orthogonal.

20. Given a set of knots $\Delta = \{x_0 < x_1 < \cdots < x_n\}$ and values $Y := \{y_0, \ldots, y_n\}$, prove independently of Theorem (2.4.1.5) that the spline function $S_\Delta(Y; \cdot)$ with $S''_\Delta(Y; x_0) = S''_\Delta(Y; x_n) = 0$ is unique.
Hint: Examine the number of zeros of the difference $S''_\Delta - \tilde{S}''_\Delta$ of two such spline functions. Note that this number is incompatible with $S_\Delta - \tilde{S}_\Delta \not\equiv 0$.

21. The existence of a spline function $S_\Delta(Y; \cdot)$ in cases (a), (b), and (c) of (2.4.1.2) can be established without explicitly calculating it, as was done in Section 2.4.2.

(a) The representation of $S_\Delta(Y; \cdot)$ requires $4n$ parameters $\alpha_j, \beta_j, \gamma_j, \delta_j$. Show that in each of the cases (a), (b), (c) a linear system of $4n$ equations results. ($n + 1$ = number of knots.)

(b) Use the uniqueness of $S_\Delta(Y; \cdot)$ (Exercise 20) to show that the system of linear equations is not singular, which ensures the existence of a solution.

22. Show that the quantities d_j of (2.4.2.6) and (2.4.2.8) satisfy

$$d_j = 3f''(x_j) + O(\|\Delta\|), \qquad j = 0, 1, \ldots, n,$$

and even

$$d_j = 3f''(x_j) + O(\|\Delta\|^2), \qquad j = 1, \ldots, n - 1,$$

in the case of $n + 1$ equidistant knots $x_j \in \Delta$.

23. Show that Theorem (2.4.1.5) implies: If the set of knots $\Delta' \subset [a, b]$ contains the set of knots Δ, $\Delta' \supset \Delta$, then in each of the cases (a), (b), and (c),

$$\|f\| \geqslant \|S_{\Delta'}(Y'; \cdot)\| \geqslant \|S_\Delta(Y; \cdot)\|.$$

24. Suppose $S_\Delta(x)$ is a spline function with the set of knots

$$\Delta = \{a = x_0 < x_1 < \cdots < x_n = b\}$$

interpolating $f \in \mathscr{K}^4(a, b)$. Show that

$$\|f - S_\Delta\|^2 = \int_a^b (f(x) - S_\Delta(x)) f^{(4)}(x)\, dx$$

if any one of the following additional conditions is met:

(a) $f'(x) = S'_\Delta(x)$ for $x = a, b$.

(b) $f''(x) = S''_\Delta(x)$ for $x = a, b$.

(c) S_Δ is periodic and $f \in \mathscr{K}^4_p(a, b)$.

25. Prove that $p_k > 1$ holds for the quantities p_k, $k = 1, \ldots, n$, encountered in solution method (2.4.2.15) of the system of linear equations (2.4.2.9). All the divisions required by this method can therefore be carried out.

26. Define the spline functions S_j for equidistant knots $x_i = a + ih$, $h > 0$, $i = 0, \ldots, n$, by

$$S_j(x_k) = \delta_{jk}, \qquad j, k = 0, \ldots, n, \quad \text{and} \quad S_j''(x_0) = S_j''(x_n) = 0.$$

Verify that the moments $M_1, \ldots, M_{n-1}$ of S_j are as follows:

$$M_i = -\frac{1}{\rho_i} M_{i+1}, \qquad i = 1, \ldots, j-2,$$

$$M_i = -\frac{1}{\rho_{n-i}} M_{i-1}, \qquad i = j+2, \ldots, n-1,$$

$$\left.\begin{aligned} M_j &= \frac{-6}{h^2} \frac{2 + 1/\rho_{j-1} + 1/\rho_{n-j-1}}{4 - 1/\rho_{j-1} - 1/\rho_{n-j-1}} \\ M_{j-1} &= \frac{1}{\rho_{j-1}}(6h^{-2} - M_j) \\ M_{j+1} &= \frac{1}{\rho_{n-j-1}}(6h^{-2} - M_j) \end{aligned}\right\} \quad \text{for } j \neq 0, 1, n-1, n,$$

where the numbers ρ_i are recursively defined by $\rho_1 := 4$ and

$$\rho_i := 4 - 1/\rho_{i-1}, \qquad i = 2, 3, \ldots.$$

It is readily seen that they satisfy the inequalities

$$4 = \rho_1 > \rho_2 > \cdots > \rho_i > \rho_{i+1} > 2 + \sqrt{3} > 3.7, \qquad 0.25 < 1/\rho_i < 0.3.$$

27. Show for the functions S_j defined in Exercise 26 that for $j = 2, 3, \ldots, n-2$ and either $x \in [x_i, x_{i+1}]$, $j + 1 \leqslant i \leqslant n - 1$ or $x \in [x_{i-1}, x_i]$, $1 \leqslant i \leqslant j - 1$,

$$|S_j(x)| \leqslant \frac{h^2}{8} |M_i|.$$

28. Let $S_{\Delta; f}$ denote the spline function which interpolates the function f at prescribed knots $x \in \Delta$ and for which

$$S''_{\Delta; f}(x_0) = S''_{\Delta; f}(x_n) = 0.$$

The map $f \to S_{\Delta; f}$ is linear, that is,

$$S_{\Delta; f+g} = S_{\Delta; f} + S_{\Delta; g}, \qquad S_{\Delta; \alpha f} = \alpha S_{\Delta; f}.$$

The effect on $S_{\Delta; f}$ of changing a single function value $f(x_j)$ is therefore that of adding a corresponding multiple of the function S_j which was defined in Exercise 26. Show, for equidistant knots and using the results of Exercise 27, that a perturbation of a function value subsides quickly as one moves away from the location of the perturbation. Consider the analogously defined Lagrange polynomials (2.1.1.2) in order to compare the behavior of interpolating polynomials.

29. Let $\Delta = \{x_0 < x_1 < \cdots < x_n\}$.

(a) Show that a spline function S_Δ with the boundary conditions

(*) $$S_\Delta^{(k)}(x_0) = S_\Delta^{(k)}(x_n) = 0, \qquad k = 0, 1, 2,$$

vanishes identically for $n < 4$.

(b) For $n = 4$, the spline function with (*) is uniquely determined for any value c and the normative condition

$$S_\Delta(x_2) = c.$$

Hint: Prove the uniqueness of S_Δ for $c = 0$ by determining the zeros of S''_Δ in the open interval (x_0, x_4). Deduce from this the existence of S_Δ by the reasoning employed in exercise 21.

(c) Calculate S_Δ explicitly in the following special case of (b):

$$x_i = -2, \ -1, 0, 1, 2, \qquad c = 1.$$

30. Let $\mathscr{S}$ be the linear space of all spline functions S_Δ with knot set $\Delta = \{x_0 < \cdots < x_n\}$ and $S''_\Delta(x_0) = S''_\Delta(x_n) = 0$. The spline functions $S_0, \ldots, S_n$ are the ones defined in Exercise 26. Show that for $Y := \{y_0, \ldots, y_n\}$,

$$S_\Delta(Y; x) \equiv \sum_{j=0}^{n} y_j S_j(x).$$

What is the dimension of $\mathscr{S}$?

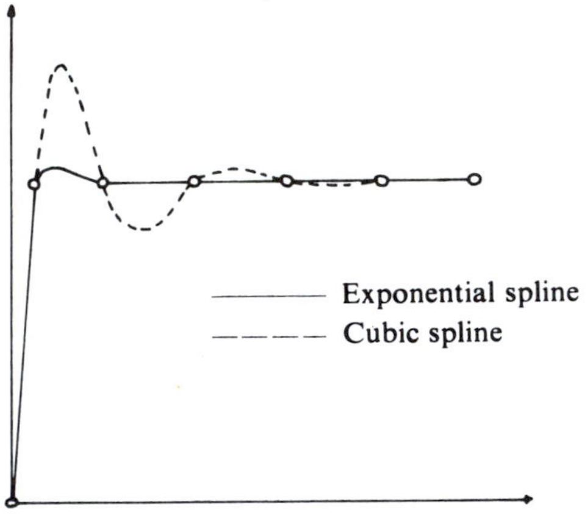

Figure 3 Comparison of spline functions.

31. Let $E_{\Delta, f}(x)$ denote the spline-like function which, for given λ_i, minimizes the functional

$$E[y] = \sum_{i=0}^{N-1} \int_{x_i}^{x_{i+1}} [(y''(x))^2 + \lambda_i^2 (y'(x))^2] \, dx$$

over $\mathscr{K}^2(a, b)$. [Compare Theorem (2.4.1.5).]

(a) Show that $E_{\Delta, f}$ is between knots of the form

$$E_{\Delta, f}(x) = \alpha_i + \beta_i(x - x_i) + \gamma_i \psi_i(x - x_i) + \delta_i \varphi_i(x - x_i), \quad x_i \leqslant x \leqslant x_{i+1},$$
$$i = 0, \ldots, N - 1,$$

$$\psi_i(x) = \frac{2}{\lambda_i^2} [\cosh(\lambda_i x) - 1], \qquad \varphi_i(x) = \frac{6}{\lambda_i^3} [\sinh(\lambda_i x) - \lambda_i x],$$

with constants $\alpha_i, \beta_i, \gamma_i, \delta_i$. $E_{\Delta, f}$ is called *exponential spline function.*

(b) Examine the limit as $\lambda_i \to 0$.
(c) Figure 3 illustrates the qualitative behavior of cubic and exponential spline functions interpolating the same set of support points.

References for Chapter 2

Achieser, N. I.: *Theory of Approximations.* Translated from the Russian by C. Hyman. New York: Frederick Ungar (1956).

Ahlberg, J., Nilson, E., Walsh, J.: *The Theory of Splines and Their Applications.* New York: Academic Press (1967).

Bloomfield, P.: *Fourier Analysis of Time Series.* New York: Wiley (1976).

Böhmer, E. O.: *Spline-Funktionen.* Stuttgart: Teubner (1974).

Brigham, E. O.: *The Fast Fourier Transform.* Englewood Cliffs, N.J.: Prentice-Hall (1974).

Bulirsch, R., Rutishauser, H.: Interpolation and genäherte Quadratur. In: Sauer, Szabó (1968).

———, Stoer, J.: Darstellung von Funktionen in Rechenautomaten. In: Sauer, Szabó (1968).

Ciarlet, P. G., Schultz, M. H., Varga, R. S.: Numerical methods of high-order accuracy for nonlinear boundary value problems I. One dimensional problems. *Numer. Math.* **9**, 294–430 (1967).

Cooley, J. W., Tukey, J. W.: An algorithm for the machine calculation of complex Fourier series. *Math. Comput.* **19**, 297–301 (1965).

Curry, H. B., Schoenberg, I. J.: On Polya frequency functions, IV: The fundamental spline functions and their limits. *J. d'Analyse Math.* **17**, 73–82 (1966).

Davis, P. J.: *Interpolation and Approximation.* New York: Blaisdell (1963), 2d printing (1965).

de Boor, C.: On calculating with B-splines. *J. Approximation Theory* **6**, 50–62 (1972).

———, *A practical Guide to Splines.* New York: Springer-Verlag (1978).

———, Pinkus, A.: Backward error analysis for totally positive linear systems. *Numer. Math.* **27**, 485–490 (1977).

Gautschi, W.: Attenuation factors in practical Fourier analysis. *Numer. Math.* **18**, 373–400 (1972).

Gentleman, W. M., Sande, G.: Fast Fourier transforms—For fun and profit. In: *Proc. AFIPS 1966 Fall Joint Computer Conference*, **29**, 503–578. Washington, D.C.: Spartan Books (1966).

Goertzel, G.: An algorithm for the evaluation of finite trigonometric series. *Amer. Math. Monthly* **65**, 34–35 (1958).

Greville, T. N. E.: Introduction to spline functions. In: *Theory and Applications of Spline Functions.* Edited by T. N. E. Greville. New York: Academic Press (1969).

Hall, C. A., Meyer, W. W.: Optimal error bounds for cubic spine interpolation. *J. Approximation Theory* **16**, 105–122 (1976).

Herriot, J. G., Reinsch, C.: ALGOL 60 procedures for the calculation of interpolating natural spline functions. Technical Report STAN-CS-71-200, Computer Science Department, Stanford University, CA (1971).

Karlin, S.: *Total Positivity*, Vol. 1. Stanford: Stanford University Press (1968).

Kuntzmann, J.: *Méthodes Numeriques, Interpolation—Dérivées.* Paris: Dunod (1959).

Maehly, H., Witzgall, Ch.: Tschebyscheff-approximation in kleinen intervallen II. *Numer. Math.* **2**, 293–307 (1960).

Milne, E. W.: *Numerical Calculus.* Princeton, N.J.: Princeton University Press (1949), 2d printing (1950).

Milne-Thomson, L. M.: *The Calculus of Finite Differences.* London: Macmillan (1933), reprinted (1951).

Reinsch, C.: Unpublished manuscript.

Sauer, R., Szabó, I. (Eds.): *Mathematische Hilfsmittel des Ingenieurs*, Part III. Berlin, Heidelberg, New York: Springer (1968).

Schoenberg, I. J., Whitney, A.: On Polya frequency functions, III: The positivity of translation determinants with an application to the interpolation problem by spline curves. *Trans. Amer. Math. Soc.* **74**, 246–259 (1953).

Schultz, M. H.: *Spline analysis.* Englewood Cliffs, N.J.: Prentice-Hall (1973).

Singleton, R. C.: On computing the fast Fourier transform. *Comm. ACM* **10**, 647–654 (1967).

———, Algorithm 338: ALGOL procedures for the fast Fourier transform. Algorithm 339: An ALGOL procedure for the fast Fourier transform with arbitrary factors. *Comm. ACM* **11**, 773–779 (1968).

Topics in Integration 3

Calculating the definite integral of a given real function $f(x)$,

$$\int_a^b f(x)\,dx,$$

is a classic problem. For some simple integrands $f(x)$, the indefinite integral

$$\int_a^x f(x)\,dx = F(x), \qquad F'(x) = f(x),$$

can be obtained in closed form as an algebraic expression in x and well-known transcendental functions of x. Then

$$\int_a^b f(x)\,dx = F(b) - F(a).$$

See Gröbner and Hofreiter (1961) for a comprehensive collection of formulas describing such indefinite integrals and many important definite integrals.

As a rule, however, definite integrals are computed using discretization methods which approximate the integral by finite sums corresponding to some partition of the interval of integration $[a, b]$ ("numerical quadrature"). A typical representative of this class of methods is Simpson's rule, which is still the best-known and most widely used integration method. It is described in Section 3.1, together with some other elementary integration methods. Peano's elegant and systematic representation of the error terms of integration rules is described in Section 3.2.

A closer investigation of the trapezoidal sum in Section 3.3 reveals that its deviation from the true value of the integral admits an asymptotic expansion in terms of powers of the discretization step h. This expansion is the classical summation formula of Euler and Maclaurin. Asymptotic expansions of this form are exploited in so-called "extrapolation methods," which increase the accuracy of a large class of discretization methods. An application of extrapolation methods to integration ("Romberg integration") is studied in Section 3.4. The general scheme is described in Section 3.5.

A description of Gaussian integration rules follows in Section 3.6. The chapter closes with remarks on the integration of functions with singularities. For a comprehensive treatment of integration, the reader is referred to Davis and Rabinowitz (1975).

3.1 The Integration Formulas of Newton and Cotes

The integration formulas of Newton and Cotes are obtained if the integrand is replaced by a suitable interpolating polynomial $P(x)$ and if then $\int_a^b P(x)\,dx$ is taken as an approximate value for $\int_a^b f(x)\,dx$. Consider a uniform partition of the closed interval $[a, b]$ given by

$$x_i = a + ih, \qquad i = 0, \ldots, n,$$

of step length $h := (b - a)/n$, $n > 0$ integer, and let P_n be the interpolating polynomial of degree n or less with

$$P_n(x_i) = f_i := f(x_i) \quad \text{for } i = 0, 1, \ldots, n.$$

By Lagrange's interpolation formula (2.1.1.4),

$$P_n(x) \equiv \sum_{i=0}^{n} f_i L_i(x), \qquad L_i(x) = \prod_{\substack{k=0\\ k\neq i}}^{n} \frac{x - x_k}{x_i - x_k},$$

or, introducing the new variable t such that $x = a + ht$,

$$L_i(x) = \varphi_i(t) := \prod_{\substack{k=0\\ k\neq i}}^{n} \frac{t - k}{i - k}.$$

Integration gives

$$\begin{aligned}\int_a^b P_n(x)\,dx &= \sum_{i=0}^{n} f_i \int_a^b L_i(x)\,dx \\ &= h \sum_{i=0}^{n} f_i \int_0^n \varphi_i(t)\,dt \\ &= h \sum_{i=0}^{n} f_i \alpha_i.\end{aligned}$$

Note that the coefficients or *weights*

$$\alpha_i := \int_0^n \varphi_i(t)\,dt$$

depend solely on n; in particular, they do not depend on the function f to be integrated, nor on the boundaries a, b of the integral.

If $n = 2$ for instance, then

$$\alpha_0 = \int_0^2 \frac{t-1}{0-1}\frac{t-2}{0-2}\,dt = \frac{1}{2}\int_0^2 (t^2 - 3t + 2)\,dt = \frac{1}{2}\left(\frac{8}{3} - \frac{12}{2} + 4\right) = \frac{1}{3},$$

$$\alpha_1 = \int_0^2 \frac{t-0}{1-0}\frac{t-2}{1-2}\,dt = -\int_0^2 (t^2 - 2t)\,dt = -\left(\frac{8}{3} - 4\right) = \frac{4}{3},$$

$$\alpha_2 = \int_0^2 \frac{t-0}{2-0}\frac{t-1}{2-1}\,dt = \frac{1}{2}\int_0^2 (t^2 - t)\,dt = \frac{1}{2}\left(\frac{8}{3} - \frac{4}{2}\right) = \frac{1}{3},$$

and we obtain the following approximate value:

$$\int_a^b P_2(x)\,dx = \frac{h}{3}(f_0 + 4f_1 + f_2)$$

for the integral $\int_a^b f(x)\,dx$. This is *Simpson's rule.*

For any natural number n, the *Newton–Cotes formulas*

$$\text{(3.1.1)} \qquad \int_a^b P_n(x)\,dx = h\sum_{i=0}^{n} f_i\alpha_i, \quad f_i = f(a + ih), \qquad h := \frac{b-a}{n},$$

provide approximate values for $\int_a^b f(x)\,dx$. The weights α_i, $i = 0, 1, \ldots, n$, have been tabulated. They are rational numbers with the property

$$\text{(3.1.2)} \qquad \sum_{i=0}^{n} \alpha_i = n.$$

This follows from (3.1.1) when applied to $f(x) :\equiv 1$, for which $P_n(x) \equiv 1$.

If s is a common denominator for the fractional weights α_i, that is, if the numbers

$$\sigma_i := s\alpha_i, \qquad i = 0, 1, \ldots, n,$$

are integers, then (3.1.1) becomes

$$\text{(3.1.3)} \qquad \int_a^b P_n(x)\,dx = h\sum_{i=0}^{n} f_i\alpha_i = \frac{b-a}{ns}\sum_{i=0}^{n} \sigma_i f_i.$$

It can be shown [see Steffensen (1950)] that the approximation error may be expressed as follows:

$$\text{(3.1.4)} \qquad \int_a^b P_n(x)\,dx - \int_a^b f(x)\,dx = h^{p+1}\cdot K \cdot f^{(p)}(\xi), \qquad \xi \in (a, b).$$

Here (a, b) denotes the open interval from a to b. The values of p and K depend only on n but not on the integrand f.

For $n = 1, 2, \ldots, 6$ we find the Newton–Cotes formulas given in the following table. For larger n, some of the values σ_i become negative and the corresponding formulas are unsuitable for numerical purposes, as cancellations tend to occur in computing the sum (3.1.3).

n	σ_i	ns	Error	Name
1	1 1	2	$h^3 \frac{1}{12} f^{(2)}(\xi)$	Trapezoidal rule
2	1 4 1	6	$h^5 \frac{1}{90} f^{(4)}(\xi)$	Simpson's rule
3	1 3 3 1	8	$h^5 \frac{3}{80} f^{(4)}(\xi)$	3/8-rule
4	7 32 12 32 7	90	$h^7 \frac{8}{945} f^{(6)}(\xi)$	Milne's rule
5	19 75 50 50 75 19	288	$h^7 \frac{275}{12096} f^{(6)}(\xi)$	———
6	41 216 27 272 27 216 41	840	$h^9 \frac{9}{1400} f^{(8)}(\xi)$	Weddle's rule

Additional integration rules may be found by Hermite interpolation (see Section 2.1.5) of the integrand f by a polynomial $P \in \Pi_n$ of degree n or less. In the simplest case, a polynomial $P \in \Pi_3$ with

$$P(a) = f(a), \qquad P'(a) = f'(a),$$
$$P(b) = f(b), \qquad P'(b) = f'(b)$$

is substituted for the integrand f. The generalized Lagrange formula (2.1.5.3) yields for P in the special case $a = 0$, $b = 1$,

$$\begin{aligned} P(t) = f(0)[(t-1)^2 + 2t(t-1)^2] + f(1)[t^2 - 2t^2(t-1)] \\ + f'(0)t(t-1)^2 + f'(1)t^2(t-1), \end{aligned}$$

integration of which gives

$$\int_0^1 P(t)\,dt = \tfrac{1}{2}(f(0) + f(1)) + \tfrac{1}{12}(f'(0) - f'(1)).$$

From this, we obtain by a simple variable transformation the following integration rule for general $a < b$ $(h := b - a)$:

(3.1.5) $$\int_a^b f(x)\,dx \approx M(h) := \frac{h}{2}(f(a) + f(b)) + \frac{h^2}{12}(f'(a) - f'(b)),$$

If $f \in C^4[a, b]$ then—using methods to be described in Section 3.2—the approximation error of the above rule can be expressed as follows:

(3.1.6) $$M(h) - \int_a^b f(x)\,dx = -\frac{h^5}{720} f^{(4)}(\xi), \qquad \xi \in (a, b), \quad h := (b - a).$$

If the support abscissas x_i, $i = 0, 1, \ldots, n$, $x_0 = a$, $x_n = b$ are not equally spaced, then interpolating the integrand $f(x)$ will lead to different integration rules, among them the ones given by Gauss. These will be described in Section 3.6.

The Newton–Cotes and related formulas are usually not applied to the entire interval of integration $[a, b]$, but are instead used in each one of a collection of subintervals into which the interval $[a, b]$ has been divided. The full integral is then approximated by the sum of the approximations to the subintegrals. The locally used integration rule is said to have been *extended*, giving rise to a corresponding *composite rule*. We proceed to examine some composite rules of this kind.

The trapezoidal rule ($n = 1$) provides the approximate value

$$I_i := \frac{h}{2}[f(x_i) + f(x_{i+1})]$$

in the subinterval $[x_i, x_{i+1}]$ of the partition $x_i = a + ih$, $i = 0, 1, \ldots, N$, $h := (b - a)/N$. For the entire interval $[a, b]$, we obtain the approximation

(3.1.7)

$$T(h) := \sum_{i=0}^{N-1} I_i = h\left[\frac{f(a)}{2} + f(a+h) + f(a+2h) + \cdots + f(b-h) + \frac{f(b)}{2}\right],$$

which is the *trapezoidal sum* for step length h. In each subinterval $[x_i, x_{i+1}]$ the error

$$I_i - \int_{x_i}^{x_{i+1}} f(x)\,dx = \frac{h^3}{12} f^{(2)}(\xi_i), \qquad \xi_i \in (x_i, x_{i+1}),$$

is incurred, assuming $f \in C^2[a, b]$. Summing these individual error terms gives

$$T(h) - \int_a^b f(x)\,dx = \frac{h^3}{12}\sum_{i=0}^{N-1} f^{(2)}(\xi_i) = \frac{h^2}{12}(b-a)\frac{1}{N}\sum_{i=0}^{N-1} f^{(2)}(\xi_i).$$

Since

$$\min_i f^{(2)}(\xi_i) \leqslant \frac{1}{N}\sum_{i=0}^{N-1} f^{(2)}(\xi_i) \leqslant \max_i f^{(2)}(\xi_i)$$

and $f^{(2)}(x)$ is continuous, there exists $\xi \in [\min_i \xi_i, \max_i \xi_i] \subset (a, b)$ with

$$f^{(2)}(\xi) = \frac{1}{N}\sum_{i=0}^{N-1} f^{(2)}(\xi_i).$$

Thus

$$T(h) - \int_a^b f(x)\,dx = \frac{b-a}{12} h^2 f^{(2)}(\xi), \qquad \xi \in (a, b).$$

Upon reduction of the step length h (increase of N) the approximation error approaches zero as fast as h^2, so we have a method of *order 2*.

If N is even, then Simpson's rule may be applied to each subinterval $[x_{2i}, x_{2i+1}, x_{2i+2}]$, $i = 0, 1, \ldots, N/2 - 1$, individually, yielding the approximation $(h/3)(f(x_{2i}) + 4f(x_{2i+1}) + f(x_{2i+2}))$. Summing these $N/2$ approximations results in the composite version of Simpson's rule,

$$S(h) = \frac{h}{3}[f(a) + 4f(a+h) + 2f(a+2h) + 4f(a+3h) + \cdots$$
$$+ 2f(b-2h) + 4f(b-h) + f(b)],$$

for the entire interval. The error of $S(h)$ is the sum of all $N/2$ individual errors

$$S(h) - \int_a^b f(x)\,dx = \frac{h^5}{90}\sum_{i=0}^{(N/2)-1} f^{(4)}(\xi_i) = \frac{h^4}{90}\frac{b-a}{2}\frac{2}{N}\sum_{i=0}^{(N/2)-1} f^{(4)}(\xi_i),$$

and we conclude, just as we did for the trapezoidal sum, that

$$S(h) - \int_a^b f(x)\,dx = \frac{b-a}{180}h^4 f^{(4)}(\xi), \qquad \xi \in (a, b),$$

provided $f \in C^4[a, b]$. The method is therefore of *order 4*.

Extending the rule of integration $M(h)$ in (3.1.5) has a remarkable effect: when the approximations to the individual subintegrals

$$\int_{x_i}^{x_{i+1}} f(x)\,dx \quad \text{for } i = 0, 1, \ldots, N-1$$

are added up, all the "interior" derivatives $f'(x_i)$, $0 < i < N$, cancel. The following approximation to the entire integral is obtained:

$$U(h) := h\left[\frac{f(a)}{2} + f(a+h) + \cdots + f(b-h) + \frac{f(b)}{2}\right] + \frac{h^2}{12}[f'(a) - f'(b)]$$
$$= T(h) + \frac{h^2}{12}[f'(a) - f'(b)].$$

This formula can be considered as a correction to the trapezoidal sum $T(h)$. It relates closely to the Euler–Maclaurin summation formula, which will be discussed in Section 3.3 [see also Schoenberg (1969)]. The error formula (3.1.6) for $M(h)$ can be extended to an error formula for the composite rule $U(h)$ in the same fashion as before. Thus

$$\text{(3.1.8)} \qquad U(h) - \int_a^b f(x)\,dx = -\frac{b-a}{720}h^4 f^{(4)}(\xi), \qquad \xi \in (a, b),$$

provided $f \in C^4[a, b]$. Comparing this error with that of the trapezoidal sum, we note that the order of the method has been improved by 2 with a minimum of additional effort, namely the computation of $f'(a)$ and $f'(b)$. If these two boundary derivatives are known to agree, e.g. for periodic functions, then the trapezoidal sum itself provides a method of order at least 4.

Replacing $f'(a), f'(b)$ by difference quotients with an approximation error of sufficiently high order, we obtain simple modifications ["end corrections": see Henrici (1964)] of the trapezoidal sum which do not involve derivatives but still lead to methods of orders higher than 2. The following variant of the trapezoidal sum is already a method of order 3:

$$\hat{T}(h) = h(\tfrac{5}{12}f(a) + \tfrac{13}{12}f(a+h) + f(a+2h) + \cdots + f(b-2h) + \tfrac{13}{12}f(b-h) + \tfrac{5}{12}f(b)).$$

For many additional integration methods and their systematic examination see, for instance Davis and Rabinowitz (1975).

3.2 Peano's Error Representation

All integration rules considered so far are of the form

$$(3.2.1)\quad I(f) := \sum_{k=0}^{m_0} a_{k0} f(x_{k0}) + \sum_{k=0}^{m_1} a_{k1} f'(x_{k1}) + \cdots + \sum_{k=0}^{m_n} a_{kn} f^{(n)}(x_{kn}).$$

The integration error

$$(3.2.2)\qquad R(f) := I(f) - \int_a^b f(x)\,dx$$

is a linear operator

$$R(\alpha f + \beta g) = \alpha R(f) + \beta R(g) \quad \text{for } f, g \in V,\ \alpha, \beta \in \mathbb{R}$$

on some suitable linear function space V. Examples are $V = C^n[a, b]$, the space of functions with continuous nth derivatives on the interval $[a, b]$, or $V = \Pi_n$, the space of all polynomials of degree no greater than n. The following elegant integral representation of the error $R(f)$ is a classical result due to Peano:

(3.2.3) Theorem. *Suppose $R(P) = 0$ holds for all polynomials $P \in \Pi_n$, that is, every polynomial whose degree does not exceed n is integrated exactly. Then for all functions $f \in C^{n+1}[a, b]$,*

$$R(f) = \int_a^b f^{(n+1)}(t) K(t)\,dt,$$

where

$$K(t) := \frac{1}{n!} R_x[(x-t)_+^n], \quad (x-t)_+^n := \begin{cases} (x-t)^n & \text{if } x \geqslant t, \\ 0 & \text{if } x < t, \end{cases}$$

and

$$R_x[(x-t)_+^n]$$

denotes the error of $(x-t)_+^n$ when the latter is considered as a function in x.

The function $K(t)$ is called the *Peano kernel* of the operator R.

Before proving the theorem, we will discuss its application in the case of Simpson's rule

$$R(f) = \tfrac{1}{3}f(-1) + \tfrac{4}{3}f(0) + \tfrac{1}{3}f(+1) - \int_{-1}^{1} f(x)\,dx.$$

We note that any polynomial $P \in \Pi_3$ is integrated exactly. Indeed, let $Q \in \Pi_2$ be the polynomial with $P(-1) = Q(-1)$, $P(0) = Q(0)$, $P(+1) = Q(+1)$. Putting $S(x) := P(x) - Q(x)$, we have $R(P) = R(S)$. Since the degree of $S(x)$ is no greater then 3, and since $S(x)$ has the three roots $-1, 0, +1$, it must be of the form $S(x) = a(x^2 - 1)x$, and

$$R(P) = R(S) = -a \int_{-1}^{1} x(x^2 - 1)\,dx = 0.$$

Thus Theorem (3.2.3) can be applied with $n = 3$. The Peano kernel becomes

$$\begin{aligned} K(t) &= \tfrac{1}{6}R_x[(x-t)_+^3] \\ &= \frac{1}{6}\left[\tfrac{1}{3}(-1-t)_+^3 + \tfrac{4}{3}(0-t)_+^3 + \tfrac{1}{3}(1-t)_+^3 - \int_{-1}^{1}(x-t)_+^3\,dx\right]. \end{aligned}$$

By definition of $(x - t)_+^n$, we find that for $t \in [-1, 1]$

$$\int_{-1}^{1}(x-t)_+^3\,dx = \int_t^1 (x-t)^3\,dx = \frac{(1-t)^4}{4},$$

$$(-1-t)_+^3 = 0, \qquad (1-t)_+^3 = (1-t)^3,$$

$$(-t)_+^3 = \begin{cases} 0 & \text{if } t \geqslant 0, \\ -t^3 & \text{if } t < 0. \end{cases}$$

The Peano kernel for Simpson's rule in the interval $[-1, +1]$ is then

$$K(t) = \begin{cases} \tfrac{1}{72}(1-t)^3(1+3t) & \text{if } 0 \leqslant t \leqslant 1 \\ K(-t) & \text{if } -1 \leqslant t \leqslant 0. \end{cases} \tag{3.2.4}$$

PROOF OF THEOREM (3.2.3). Consider the Taylor expansion of $f(x)$ at $x = a$:

$$f(x) = f(a) + f'(a)(x-a) + \cdots + \frac{f^{(n)}(a)}{n!}(x-a)^n + r_n(x). \tag{3.2.5}$$

Its remainder term can be expressed in the form

$$r_n(x) = \frac{1}{n!}\int_a^x f^{(n+1)}(t)(x-t)^n\,dt = \frac{1}{n!}\int_a^b f^{(n+1)}(t)(x-t)_+^n\,dt.$$

Applying the linear operator R to (3.2.5) gives

$$R(f) = R(r_n) = \frac{1}{n!}R_x\left(\int_a^b f^{(n+1)}(t)(x-t)_+^n\,dt\right), \tag{3.2.6}$$

since $R(P) = 0$ for $P \in \Pi_n$.

In order to transform this representation of $R(f)$ into the desired one, we have to interchange the R_x operator with the integration. To prove that this interchange is legal, we show first that

$$(3.2.7)\quad \frac{d^k}{dx^k}\left[\int_a^b f^{(n+1)}(t)(x-t)_+^n\,dt\right] = \int_a^b f^{(n+1)}(t)\frac{d^k}{dx^k}[(x-t)_+^n]\,dt$$

for $1 \leqslant k \leqslant n$. For $k < n$ this follows immediately from the fact that $(x-t)_+^n$ is $n-1$ times continuously differentiable. For $k = n-1$ we have in particular

$$\frac{d^{n-1}}{dx^{n-1}}\left[\int_a^b f^{(n+1)}(t)(x-t)_+^n\,dt\right] = \int_a^b f^{(n+1)}(t)\frac{d^{n-1}}{dx^{n-1}}[(x-t)_+^n]\,dt$$

and therefore

$$\begin{aligned}\frac{d^{n-1}}{dx^{n-1}}\left[\int_a^b f^{(n+1)}(t)(x-t)_+^n\,dt\right] &= n!\int_a^b f^{(n+1)}(t)(x-t)_+\,dt\\ &= n!\int_a^x f^{(n+1)}(t)(x-t)\,dt.\end{aligned}$$

The latter integral is differentiable as a function of x, since the integrand is jointly continuous in x and t; hence

$$\begin{aligned}&\frac{d}{dx}\left[\frac{d^{n-1}}{dx^{n-1}}\left[\int_a^b f^{(n+1)}(t)(x-t)_+^n\,dt\right]\right]\\ &\quad = n!f^{(n+1)}(x)(x-x) + n!\int_a^x f^{(n+1)}(t)\,dt\\ &\quad = \int_a^b f^{(n+1)}(t)\frac{d^n}{dx^n}[(x-t)_+^n]\,dt.\end{aligned}$$

Thus (3.2.7) holds also for $k = n$.

By (3.2.7), the differential operators

$$\frac{d^k}{dx^k}, \qquad k = 1, \ldots, n,$$

commute with integration. Because $I(f) = I_x(f)$ is a linear combination of differential operators, it also commutes with integration.

Finally the continuity properties of the integrand $f^{(n+1)}(t)(x-t)_+^n$ are such that the following two integrations can be interchanged:

$$\int_a^b\left[\int_a^b f^{(n+1)}(t)(x-t)_+^n\,dt\right]dx = \int_a^b f^{(n+1)}(t)\left[\int_a^b (x-t)_+^n\,dx\right]dt.$$

This then shows that the entire operator R_x commutes with integration, and we obtain the desired result

$$R(f) = \frac{1}{n!}\int_a^b f^{(n+1)}(t)R_x((x-t)_+^n)\,dt. \qquad \square$$

Note that Peano's integral representation of the error is not restricted to operators of the form (3.2.1). It holds for all operators R for which R_x commutes with integration.

For a surprisingly large class of integration rules, the Peano kernel $K(t)$ has constant sign on $[a, b]$. In this case, the mean-value theorem of integral calculus gives

$$R(f) = f^{(n+1)}(\xi) \int_a^b K(t)\,dt \quad \text{for some } \xi \in (a, b). \tag{3.2.8}$$

The above integral of $K(t)$ does not depend on f, and can therefore be determined by applying R, for instance, to the polynomial $f(x) := x^{n+1}$. This gives

$$R(f) = \frac{R(x^{n+1})}{(n+1)!} f^{(n+1)}(\xi) \quad \text{for some } \xi \in (a, b). \tag{3.2.9}$$

In the case of Simpson's rule, $K(t) \geqslant 0$ for $-1 \leqslant t \leqslant 1$ by (3.2.4). In addition

$$\frac{R(x^4)}{4!} = \frac{1}{24}\left(\tfrac{1}{3} \cdot 1 + \tfrac{4}{3} \cdot 0 + \tfrac{1}{3} \cdot 1 - \int_{-1}^{1} x^4\,dx\right) = \frac{1}{90},$$

so that we obtain for the error of Simpson's formula

$$\tfrac{1}{3}f(-1) + \tfrac{4}{3}f(0) + \tfrac{1}{3}f(1) - \int_{-1}^{1} f(t)\,dt = \tfrac{1}{90} f^{(4)}(\xi), \qquad \xi \in (a, b).$$

In general, the Newton–Cotes formulas of degree n integrate without error polynomials $P \in \Pi_n$ if n is odd, and $P \in \Pi_{n+1}$ if n is even (see Exercise 2). The Peano kernels for the Newton–Cotes formulas are of constant sign [see for instance Steffensen (1950)], and (3.2.9) confirms the error estimates in Section 3.1:

$$R_n(f) = \begin{cases} \dfrac{R_n(x^{n+1})}{(n+1)!} f^{(n+1)}(\xi) & \text{if } n \text{ is odd}, \\[2ex] \dfrac{R_n(x^{n+2})}{(n+2)!} f^{(n+2)}(\xi) & \text{if } n \text{ is even}. \end{cases} \qquad \xi \in (a, b),$$

Finally, we use (3.2.9) to prove the representation (3.1.6) of the error of integration rule (3.1.5), which was based on cubic Hermite interpolation. Here

$$R(f) = \frac{h}{2}(f(a) + f(b)) + \frac{h^2}{12}(f'(a) - f'(b)) - \int_a^b f(x)\,dx, \qquad h := b - a,$$

which vanishes for all polynomials $P \in \Pi_3$. For $n = 3$, we obtain the following Peano kernel:

$$K(t) = \tfrac{1}{6}R_x((x-t)_+^3)$$
$$= \frac{1}{6}\left[\frac{h}{2}((a-t)_+^3 + (b-t)_+^3) + \frac{h^2}{4}((a-t)_+^2 - (b-t)_+^2) - \int_a^b (x-t)_+^3\,dx\right]$$
$$= \frac{1}{6}\left[\frac{h}{2}(b-t)^3 - \frac{h^2}{4}(b-t)^2 - \tfrac{1}{4}(b-t)^4\right]$$
$$= -\tfrac{1}{24}(b-t)^2(a-t)^2.$$

Since $K(t) \leqslant 0$ in the interval of integration $[a, b]$, (3.2.9) is applicable. We find for $a = 0$, $b = 1$ that

$$\frac{R(x^4)}{4!} = \tfrac{1}{24}(\tfrac{1}{2}\cdot 1 + \tfrac{1}{12}\cdot(-4) - \tfrac{1}{5}) = -\tfrac{1}{720}.$$

Thus

$$R(f) = \frac{-1}{24}\int_a^b f^{(4)}(t)(b-t)^2(a-t)^2\,dt = -\frac{(b-a)^5}{720}f^{(4)}(\xi), \qquad \xi \in (a, b),$$

for $f \in C^4[a, b]$, which was to be shown.

3.3 The Euler–Maclaurin Summation Formula

The error formulas (3.1.6) and (3.1.8) are low-order instances of the famous Euler–Maclaurin summation formula, which in its simplest form reads (for $g \in C^{2m+2}[0, 1]$)

$$(3.3.1)\qquad \int_0^1 g(t)\,dt = \frac{g(0)}{2} + \frac{g(1)}{2} + \sum_{l=1}^{m}\frac{B_{2l}}{(2l)!}(g^{(2l-1)}(0) - g^{(2l-1)}(1)) - \frac{B_{2m+2}}{(2m+2)!}g^{(2m+2)}(\xi), \qquad 0 < \xi < 1.$$

Here B_k are the classical *Bernoulli numbers*

$$(3.3.2)\qquad B_2 = \tfrac{1}{6},\quad B_4 = -\tfrac{1}{30},\quad B_6 = \tfrac{1}{42},\quad B_8 = -\tfrac{1}{30},\ \ldots,$$

whose general definition will be given below. Extending (3.3.1) to its composite form in the same way as (3.1.6) was extended to (3.1.8), we obtain for $g \in C^{2m+2}[0, N]$

$$\int_0^N g(t)\,dt = \frac{g(0)}{2} + g(1) + \cdots + g(N-1) + \frac{g(N)}{2} + \sum_{l=1}^{m}\frac{B_{2l}}{(2l)!}(g^{(2l-1)}(0) - g^{(2l-1)}(N)) - \frac{B_{2m+2}}{(2m+2)!}Ng^{(2m+2)}(\xi), \qquad 0 < \xi < N.$$

Rearranging the terms of the above formula leads to the most frequently presented form of the Euler–Maclaurin summation formula:

$$(3.3.3)\qquad \frac{g(0)}{2} + g(1) + \cdots + g(N-1) + \frac{g(N)}{2}$$

$$= \int_0^N g(t)\,dt + \sum_{l=1}^{m} \frac{B_{2l}}{(2l)!}\left(g^{(2l-1)}(N) - g^{(2l-1)}(0)\right)$$

$$+ \frac{B_{2m+2}}{(2m+2)!} N g^{(2m+2)}(\xi), \qquad 0 < \xi < N.$$

For a general uniform partition $x_i = a + ih$, $i = 0, \ldots, N$, $x_N = b$, of the interval $[a, b]$, (3.3.3) becomes

$$(3.3.4)\quad T(h) = \int_a^b f(t)\,dt + \sum_{l=1}^{m} h^{2l} \frac{B_{2l}}{(2l)!}\left(f^{(2l-1)}(b) - f^{(2l-1)}(a)\right)$$

$$+ h^{2m+2} \frac{B_{2m+2}}{(2m+2)!}(b-a) f^{(2m+2)}(\xi),$$

$$a < \xi < b,$$

where $T(h)$ denotes the trapezoidal sum (3.1.7)

$$T(h) = h\left(\frac{f(a)}{2} + f(a+h) + \cdots + f(b-h) + \frac{f(b)}{2}\right).$$

In this form, the Euler–Maclaurin summation formula expands the trapezoidal sum $T(h)$ in terms of the step length $h = (b-a)/N$, and herein lies its importance for our purposes: the existence of such an expansion puts at one's disposal a wide arsenal of powerful general "extrapolation methods," which will be discussed in subsequent sections.

Proof of (3.3.1). We will use integration by parts and successively determine polynomials $B_k(x)$, starting with $B_1(x) \equiv x - \frac{1}{2}$, such that

$$\int_0^1 g(t)\,dt = B_1(t)g(t)\Big|_0^1 - \int_0^1 B_1(t)g'(t)\,dt,$$

$$(3.3.5)\qquad \int_0^1 B_1(t)g'(t)\,dt = \frac{1}{2}B_2(t)g'(t)\Big|_0^1 - \frac{1}{2}\int_0^1 B_2(t)g''(t)\,dt,$$

$$\vdots$$

$$\int_0^1 B_{k-1}(t)g^{(k-1)}(t)\,dt = \frac{1}{k}B_k(t)g^{(k-1)}(t)\Big|_0^1 - \frac{1}{k}\int_0^1 B_k(t)g^{(k)}(t)\,dt,$$

where

$$(3.3.6)\qquad B_{k+1}'(x) = (k+1)B_k(x), \qquad k = 1, 2, \ldots.$$

It is clear from (3.3.6) that each polynomial $B_k(x)$ is of degree k and that its highest-order term has coefficient 1. Given $B_k(x)$, the relation (3.3.6) determines $B_{k+1}(x)$ up to an arbitrary additive constant. We now select these constants so as to satisfy the additional conditions

$$B_{2l+1}(0) = B_{2l+1}(1) = 0 \quad \text{for } l > 0, \tag{3.3.7}$$

which determine the polynomials $B_k(x)$ uniquely. Indeed, if

$$B_{2l-1}(x) = x^{2l-1} + c_{2l-2}x^{2l-2} + \cdots + c_1 x + c_0,$$

then, with integration constants c and d,

$$B_{2l+1}(x) = x^{2l+1} + \frac{(2l+1)2l}{2l(2l-1)} c_{2l-2}x^{2l} + \cdots + (2l+1)cx + d.$$

$B_{2l+1}(0) = 0$ requires $d = 0$, and $B_{2l+1}(1) = 0$ determines c.

The polynomials

$$B_0(x) \equiv 1, \qquad B_1(x) \equiv x - \tfrac{1}{2}, \qquad B_2(x) \equiv x^2 - x + \tfrac{1}{6},$$
$$B_3(x) \equiv x^3 - \tfrac{3}{2}x^2 + \tfrac{1}{2}x, \qquad B_4(x) \equiv x^4 - 2x^3 + x^2 - \tfrac{1}{30}, \ldots$$

are known as *Bernoulli polynomials*. Their constant terms $B_k = B_k(0)$ are the Bernoulli numbers (3.3.2). All Bernoulli numbers of odd index $k > 1$ vanish because of (3.3.7).

The Bernoulli polynomials satisfy

$$(-1)^k B_k(1 - x) = B_k(x). \tag{3.3.8}$$

This follows from the fact that the polynomials $(-1)^k B_k(1 - x)$ satisfy the same recursion, namely (3.3.6) and (3.3.7), as the Bernoulli polynomials $B_k(x)$. Since they also start out the same, they must coincide.

The following relation—true for odd indices $k > 1$ by (3.3.7)—can now be made general, since (3.3.8) establishes it for even k:

$$B_k(0) = B_k(1) = B_k \quad \text{for } k > 1. \tag{3.3.9}$$

This gives

$$\int_0^1 B_k(t)\,dt = \frac{1}{k+1}(B_{k+1}(1) - B_{k+1}(0)) = 0. \tag{3.3.10}$$

We are now able to complete the expansion of

$$\int_0^1 g(t)\,dt.$$

Combining the first $2m + 1$ relations (3.3.5), observing

$$\frac{1}{k} B_k(t) g^{(k-1)}(t) \Big|_0^1 = -\frac{B_k}{k}(g^{(k-1)}(0) - g^{(k-1)}(1))$$

for $k > 1$ by (3.3.9), and accounting for $B_{2l+1} = 0$, we get

$$\text{(3.3.11)} \quad \int_0^1 g(t)\,dt = \frac{g(0)}{2} + \frac{g(1)}{2} + \sum_{l=1}^{m} \frac{B_{2l}}{(2l)!}\left(g^{(2l-1)}(0) - g^{(2l-1)}(1)\right) + r_{m+1},$$

where the error term r_{m+1} is given by the integral

$$\text{(3.3.12)} \qquad r_{m+1} = \frac{-1}{(2m+1)!}\int_0^1 B_{2m+1}(t)g^{(2m+1)}(t)\,dt.$$

We use integration by parts once more to transform the error term:

$$\int_0^1 B_{2m+1}(t)g^{(2m+1)}(t)\,dt = \frac{1}{2m+2}(B_{2m+2}(t) - B_{2m+2})g^{(2m+1)}(t)\Big|_0^1 - \frac{1}{2m+2}\int_0^1 (B_{2m+2}(t) - B_{2m+2})g^{(2m+2)}(t)\,dt.$$

The first term on the right-hand side vanishes again by (3.3.9). Thus

$$\text{(3.3.13)} \qquad r_{m+1} = \frac{1}{(2m+2)!}\int_0^1 (B_{2m+2}(t) - B_{2m+2})g^{(2m+2)}(t)\,dt.$$

In order to complete the proof of (3.3.1), we need to show that $B_{2m+2}(t) - B_{2m+2}$ does not change its sign between 0 and 1. We will show by induction that

$$\text{(3.3.14a)} \qquad (-1)^m B_{2m-1}(x) > 0 \quad \text{for } 0 < x < \tfrac{1}{2},$$

$$\text{(3.3.14b)} \qquad (-1)^m (B_{2m}(x) - B_{2m}) > 0 \quad \text{for } 0 < x < 1,$$

$$\text{(3.3.14c)} \qquad (-1)^{m+1} B_{2m} > 0.$$

Indeed, (3.3.14a) holds for $m = 1$. Suppose it holds for some $m \geqslant 1$. Then for $0 < x \le \frac{1}{2}$,

$$\frac{(-1)^m}{2m}(B_{2m}(x) - B_{2m}) = (-1)^m \int_0^x B_{2m-1}(t)\,dt > 0.$$

By (3.3.8), this extends to the other half of the unit interval, $\frac{1}{2} \leqslant x < 1$, proving (3.3.14b) for this value of m. In view of (3.3.10), we have

$$(-1)^{m+1} B_{2m} = (-1)^m \int_0^1 (B_{2m}(t) - B_{2m})\,dt > 0,$$

which takes care of (3.3.14c). We must now prove (3.3.14a) for $m + 1$. Since $B_{2m+1}(x)$ vanishes for $x = 0$ by (3.3.7), and for $x = \frac{1}{2}$ by (3.3.8), it cannot change its sign without having an inflection point $\bar{x}$ between 0 and $\frac{1}{2}$. But then $B_{2m-1}(\bar{x}) = 0$, in violation of the induction hypothesis. The sign of $B_{2m+1}(x)$ in $0 < x < \frac{1}{2}$ is equal to the sign of its first derivative at zero, whose value is $(2m + 1)B_{2m}(0) = (2m + 1)B_{2m}$. The sign of the latter is $(-1)^{m+1}$ by (3.3.14c).

Now for the final simplification of the error term (3.3.13). Since $B_{2m+2}(x) - B_{2m+2}$ does not change its sign in the interval of integration, there exists ξ, $0 < \xi < 1$, such that

$$r_{m+1} = \frac{1}{(2m+2)!}\int_0^1 (B_{2m+2}(t) - B_{2m+2})\,dt \cdot g^{(2m+2)}(\xi).$$

From (3.3.10),

$$r_{m+1} = -\frac{B_{2m+2}}{(2m+2)!}\,g^{(2m+2)}(\xi),$$

which completes the proof of (3.3.1). □

3.4 Integrating by Extrapolation

Let $f \in C^{2m+2}[a, b]$ be a real function to be integrated over the interval $[a, b]$. Consider the expansion (3.3.4) of the trapezoidal sum $T(h)$ of f in terms of the step length $h = (b - a)/n$. It is of the form

(3.4.1) $$T(h) = \tau_0 + \tau_1 h^2 + \tau_2 h^4 + \cdots + \tau_m h^{2m} + \alpha_{m+1}(h)h^{2m+2}.$$

Here

$$\tau_0 = \int_a^b f(t)\,dt$$

is the integral to be calculated,

$$\tau_k = \frac{B_{2k}}{(2k)!}\left(f^{(2k-1)}(b) - f^{(2k-1)}(a)\right), \qquad k = 1, 2, \ldots, m,$$

and

$$\alpha_{m+1}(h) = \frac{B_{2m+2}}{(2m+2)!}(b - a)f^{(2m+2)}(\xi(h)), \qquad a < \xi = \xi(h) < b,$$

is the error coefficient. Since $f^{(2m+2)}$ is continuous by hypothesis in the closed finite interval $[a, b]$, there exists a bound L such that $|f^{(2m+2)}(x)| \leqslant L$ for all $x \in [a, b]$. Therefore:

(3.4.2). *There exists a constant M_{m+1} such that*

$$|\alpha_{m+1}(h)| \leqslant M_{m+1}$$

for all $h = (b - a)/n$, $n = 1, 2, \ldots$.

Expansions of the form (3.4.1) are called *asymptotic expansions* in h if the coefficients τ_k, $k \leqslant m$, do not depend on h, and $\alpha_{m+1}(h)$ satisfies (3.4.2). The summation formula of Euler and Maclaurin is an example of an asymptotic expansion. If all derivatives of f exist in $[a, b]$, then by letting $m \to \infty$, the right-hand side of (3.4.1) becomes an infinite series:

$$\tau_0 + \tau_1 h^2 + \tau_2 h^4 + \cdots.$$

This power series may diverge for any $h \neq 0$. Nevertheless, because of (3.4.2), asymptotic expansions are capable of yielding for small h results which are often sufficiently accurate for practical purposes [see, for instance, Erdélyi (1956), Olver (1974)].

The above result (3.4.2) shows that the error term of the asymptotic expansion (3.4.1) becomes small relative to the other terms of (3.4.1) as h decreases. The expansion then behaves like a polynomial in h^2 which yields the value τ_0 of the integral for $h = 0$. This suggests the following method for finding τ_0: For each step length h_i in a sequence

$$h_0 = b - a, \quad h_1 = \frac{h_0}{n_1}, \quad h_2 = \frac{h_0}{n_2}, \ldots, \quad h_m = \frac{h_0}{n_m},$$

where $n_1, n_2, \ldots, n_m$ are strictly increasing positive integers, determine the corresponding trapezoidal sums

$$T_{i0} := T(h_i), \qquad i = 0, 1, \ldots, m.$$

Let

$$\tilde{T}_{mm}(h) := a_0 + a_1 h^2 + \cdots + a_m h^{2m}$$

be the interpolating polynomial in h^2 with

$$\tilde{T}_{mm}(h_i) = T(h_i), \qquad i = 0, 1, \ldots, m,$$

and take the "extrapolated" value $\tilde{T}_{mm}(0)$ as the approximation to the desired integral τ_0. This method of integration is known as *Romberg integration*, having been introduced by Romberg (1955) for the special sequence $h_i = (b - a)/2^i$. It has been closely examined by Bauer, Rutishauser, and Stiefel (1963).

Neville's interpolation algorithm is particularly well suited for calculating $\tilde{T}_{mm}(0)$. For indices i, k with $1 \leqslant k \leqslant i \leqslant m$, let $\tilde{T}_{ik}(h)$ be the polynomial of degree at most k in h^2 for which

$$\tilde{T}_{ik}(h_j) = T(h_j), \qquad j = i - k, i - k + 1, \ldots, i,$$

and let

$$T_{ik} := \tilde{T}_{ik}(0).$$

The recursion formula (2.1.2.7) becomes for $x_i = h_i^2$:

$$T_{ik} = T_{i,k-1} + \frac{T_{i,k-1} - T_{i-1,k-1}}{\left[\dfrac{h_{i-k}}{h_i}\right]^2 - 1}. \tag{3.4.3}$$

It will be advantageous to arrange the intermediate values T_{ik} in the triangular tableau (2.1.2.4), where each element derives from its two left neighbors:

$$
\text{(3.4.4)}\qquad
\begin{array}{l|lccc}
h_0^2 & T(h_0) = T_{00} & & & \\
 & & T_{11} & & \\
h_1^2 & T(h_1) = T_{10} & & T_{22} & \\
 & & T_{21} & & T_{33} \\
h_2^2 & T(h_2) = T_{20} & & T_{32} & \vdots \\
 & & T_{31} & \vdots & \\
h_3^2 & T(h_3) = T_{30} & \vdots & & \\
\vdots & \quad\vdots \qquad\quad \vdots & & &
\end{array}
$$

Example. Calculating the integral

$$\int_0^1 t^5 \, dt$$

by extrapolation over the step lengths $h_0 = 1$, $h_1 = \frac{1}{2}$, $h_2 = \frac{1}{4}$, we arrive at the following tableau using (3.4.3) and 6-digit arithmetic (the fractions in parentheses indicate the true values)

$h_0^2 = 1$	$T_{00} = 0.500\,000\ (= \frac{1}{2})$		
		$T_{11} = 0.187\,500\ (= \frac{3}{16})$	
$h_1^2 = \frac{1}{4}$	$T_{10} = 0.265\,625\ (= \frac{17}{64})$		$T_{22} = 0.166\,667\ (= \frac{1}{6})$
		$T_{21} = 0.167\,969\ (= \frac{43}{256})$	
$h_2^2 = \frac{1}{16}$	$T_{20} = 0.192\,383\ (= \frac{197}{1024})$		

Each entry T_{ik} of the polynomial extrapolation tableau (3.4.4) represents in fact a linear integration rule for step length $h_i = (b - a)/n_i$:

$$T_{ik} = \alpha_0 f(a) + \alpha_1 f(a + h_i) + \cdots + \alpha_{n_i - 1} f(b - h_i) + \alpha_{n_i} f(b).$$

Some of the rules, but not all, with $i = k$ turn out to be of the Newton–Cotes type (see Section 3.1).

For instance, if $h_1 = h_0/2 = (b - a)/2$, then T_{11} is Simpson's rule. Indeed,

$$T_{00} = (b - a)(\tfrac{1}{2} f(a) + \tfrac{1}{2} f(b)),$$

$$T_{10} = \tfrac{1}{2}(b - a)\left(\tfrac{1}{2} f(a) + f\left(\frac{a + b}{2}\right) + \tfrac{1}{2} f(b)\right).$$

By (3.4.3),

$$T_{11} = T_{10} + \frac{T_{10} - T_{00}}{3} = \tfrac{4}{3} T_{10} - \tfrac{1}{3} T_{00},$$

and therefore

$$T_{11} = \tfrac{1}{2}(b - a)\left(\tfrac{1}{3} f(a) + \tfrac{4}{3} f\left(\frac{a + b}{2}\right) + \tfrac{1}{3} f(b)\right).$$

If we go one step further and put $h_2 = h_1/2$, then T_{22} becomes Milne's rule. However, this pattern breaks down for $h_3 = h_2/2$, since T_{33} is no longer a Newton–Cotes formula (see Exercise 10).

In the above cases, T_{21} and T_{31} are composite Simpson rules:

$$T_{21} = \frac{b-a}{4}\left(\tfrac{1}{3}f(a) + \tfrac{4}{3}f(a+h_2) + \tfrac{2}{3}f(a+2h_2) + \tfrac{4}{3}f(a+3h_2) + \tfrac{1}{3}f(b)\right)$$

$$T_{31} = \frac{b-a}{8}\left(\tfrac{1}{3}f(a) + \tfrac{4}{3}f(a+h_3) + \tfrac{2}{3}f(a+2h_3) + \cdots + \tfrac{1}{3}f(b)\right).$$

Very roughly speaking, proceeding downward in tableau (3.4.4) corresponds to extending integration rules, whereas proceeding to the right increases their order.

The following sequences of step lengths are usually chosen for extrapolation methods:

(3.4.5).

(a) $$h_0 = b - a, \quad h_1 = \frac{h_0}{2}, \ldots, \quad h_i = \frac{h_{i-1}}{2}, \qquad i = 2, 3, \ldots.$$

(b) $$h_0 = b - a, \quad h_1 = \frac{h_0}{2}, \quad h_2 = \frac{h_0}{3}, \ldots, \quad h_i = \frac{h_{i-2}}{2}, \qquad i = 3, 4, \ldots.$$

The first sequence is characteristic of Romberg's method [Romberg (1955)]. The second has been proposed by Bulirsch (1964). It has the advantage that the effort for computing $T(h_i)$ does not increase quite as rapidly as for the Romberg sequence.

For the sequence (3.4.5a), half of the function values needed for calculating the trapezoidal sum $T(h_{i+1})$ have been previously encounterered in the calculation of $T(h_i)$, and their recalculation can be avoided. Clearly

$$T(h_{i+1}) = \tfrac{1}{2}T(h_i) + h_{i+1}(f(a+h_{i+1}) + f(a+3h_{i+1}) + \cdots + f(b-h_{i+1})).$$

Similar savings can be realized for the sequence (3.4.5b).

An ALGOL procedure which calculates the tableau (3.4.4) for given m and the interval $[a, b]$ using the Romberg sequence (3.4.5a) is given below. To save memory space, the tableau is built up by adding upward diagonals to the bottom of the tableau. Only the lowest elements in each column need to be stored for this purpose in the linear array $t[0:m]$.

```
procedure romberg (a, b, f, m);
value a, b, m;
integer m;
real a, b;
real procedure f;
begin real h, s;
```

```
        integer i, k, n, q;
        array t[0 : m];
        h := b − a; n := 1;
        t[0] := 0.5 × h × (f(a) + f(b));
        for k := 1 step 1 until m do
        begin s := 0; h := 0.5 × h; n := 2 × n; q := 1;
                for i := 1 step 2 until n − 1 do
                s := s + f(a + i × h);
                t[k] := 0.5 × t[k − 1] + s × h;
                print (t[k]);
                for i := k − 1 step − 1 until 0 do
                begin q := q × 4;
                        t[i] := t[i + 1] + (t[i + 1] − t[i])/(q − 1);
                        print (t[i])
                end
        end
end;
```

We emphasize that the above algorithm serves mainly as an illustration of integration by extrapolation methods. As it stands, it is not well suited for practical calculations. For one thing, one does not usually know ahead of time how big the parameter m should be chosen in order to obtain the desired accuracy. In practice, one calculates only a few (say seven) columns of (3.4.4), and stops the calculation as soon as $|T_{i,6} - T_{i+1,6}| \leqslant \varepsilon s$, where ε is a specified tolerance and s is a rough approximation to the integral

$$\int_a^b |f(t)|\, dt.$$

Such an approximation s can be obtained concurrently with calculating one of the trapezoidal sums $T(h_i)$. A more general stopping rule will be described, together with a numerical example, in Section 3.5. Furthermore, the sequence (3.4.5b) of step lengths is to be preferred over (3.4.5a). Finally, rational interpolation has been found to yield in most applications considerably better results than polynomial interpolation. A program with all these improvements can be found in Bulirsch and Stoer (1967).

When we apply rational interpolation (see Section 2.2), then the recursion (2.2.3.8) replaces (2.1.2.7):

(3.4.6)

$$T_{ik} = T_{i,k-1} + \frac{T_{i,k-1} - T_{i-1,k-1}}{\left[\dfrac{h_{i-k}}{h_i}\right]^2 \left[1 - \dfrac{T_{i,k-1} - T_{i-1,k-1}}{T_{i,k-1} - T_{i-1,k-2}}\right] - 1}, \qquad 1 \leqslant k \leqslant i \leqslant m.$$

The same triangular tableau arrangement is used as for polynomial extrapolation: k is the column index, and the recursion (3.4.6) relates each tableau element to its left-hand neighbors. The meaning of T_{ik}, however, is now as follows: The functions $\tilde{T}_{ik}(h)$ are rational functions in h^2,

$$\tilde{T}_{ik}(h) := \frac{p_0 + p_1 h^2 + \cdots + p_\mu h^{2\mu}}{q_0 + q_1 h^2 + \cdots + q_\nu h^{2\nu}},$$

$$\mu + \nu = k, \quad \mu = \nu \text{ or } \mu = \nu - 1,$$

with the interpolation property

$$\tilde{T}_{ik}(h_j) = T(h_j), \qquad j = i - k, i - k + 1, \ldots, i.$$

We then define

$$T_{ik} := \tilde{T}_{ik}(0),$$

and initiate the recursion (3.4.6) by putting $T_{i0} := T(h_i)$ for $i = 0, 1, \ldots, m$, and $T_{i,-1} := 0$ for $i = 0, 1, \ldots, m - 1$. The observed superiority of rational extrapolation methods reflects the more flexible approximation properties of rational functions (see Section 2.2.4).

In Section 3.5, we will illustrate how error estimates for extrapolation methods can be obtained from asymptotic expansions like (3.4.1). Under mild restrictions on the sequence of step lengths, it will follow that, for polynomial extrapolation methods based on even asymptotic expansions, the errors of T_{i0} behave like h_i^2, those of T_{i1} like $h_{i-1}^2 h_i^2$, and, in general, those of T_{ik} like $h_{i-k}^2 h_{i-k+1}^2 \ldots h_i^2$ as $i \to \infty$. For fixed k, consequently, the sequence T_{ik}, $i = k, k + 1, \ldots$, approximates the integral like a method of order $2k + 2$. For the sequence (3.4.5a) a stronger result has been found:

$$\text{(3.4.7)} \quad T_{ik} - \int_a^b f(x)\,dx = (b - a)h_{i-k}^2 h_{i-k+1}^2 \ldots h_i^2 \frac{(-1)^k B_{2k+2}}{(2k + 2)!} f^{(2k+2)}(\xi)$$

for a suitable $\xi \in (a, b)$ and $f \in C^{2k+2}[a, b]$ [see Bauer, Rutishauser, and Stiefel (1963), Bulirsch (1964)].

3.5 About Extrapolation Methods

Some of the numerical integration methods discussed in this chapter (as, for instance, the methods based on the formulas of Newton and Cotes) had a common feature: they utilized function information only on a discrete set of points whose distance—and consequently the coarseness of the sample—was governed by a "step length." To each such step length $h \neq 0$ corresponded an approximate result $T(h)$, which furthermore admitted an asymptotic expansion in powers of h. Analogous *discretization methods* are available for many other problems, of which the numerical integration of functions is but one instance. In all these cases, the asymptotic expansion of the result $T(h)$ is of the form

$$(3.5.1)\qquad T(h) = \tau_0 + \tau_1 h^{\gamma_1} + \tau_2 h^{\gamma_2} + \cdots + \tau_m h^{\gamma_m} + h^{\gamma_{m+1}}\alpha_{m+1}(h),$$
$$0 < \gamma_1 < \gamma_2 < \cdots < \gamma_{m+1},$$

where the exponents γ_i need not be integers. The coefficients τ_i are independent of h, the function $\alpha_{m+1}(h)$ is bounded for $h \to 0$, and $\tau_0 = \lim_{h\to 0} T(h)$ is the exact solution of the problem at hand.

Consider, for example, numerical differentiation. For $h \neq 0$, the *central difference quotient*

$$T(h) = \frac{f(x+h) - f(x-h)}{2h}$$

is an approximation to $f'(x)$. For functions $f \in C^{2m+3}[x-a, x+a]$ and $|h| \leqslant |a|$, Taylor's theorem gives

$$T(h) = \frac{1}{2h}\left\{ f(x) + hf'(x) + \frac{h^2}{2!}f''(x) + \cdots + \frac{h^{2m+3}}{(2m+3)!}[f^{(2m+3)}(x) + o(1)] \right.$$
$$\left. - f(x) + hf'(x) - \frac{h^2}{2!}f''(x) + \cdots + \frac{h^{2m+3}}{(2m+3)!}[f^{(2m+3)}(x) + o(1)] \right\}$$
$$= \tau_0 + \tau_1 h^2 + \cdots + \tau_m h^{2m} + h^{2m+2}\alpha_{m+1}(h),$$

where $\tau_0 = f'(x)$, $\tau_k = f^{(2k+1)}(x)/(2k+1)!$ for $k = 1, 2, \ldots, m+1$, and $\alpha_{m+1}(h) = \tau_{m+1} + o(1)$.

Using the one-sided difference quotient

$$T(h) := \frac{f(x+h) - f(x)}{h}$$

leads to the asymptotic expansion

$$T(h) = \tau_0 + \tau_1 h + \tau_2 h^2 + \cdots + \tau_m h^m + h^{m+1}(\tau_{m+1} + o(1))$$

with

$$\tau_k = \frac{f^{(k+1)}(x)}{(k+1)!}, \qquad k = 0, 1, 2, \ldots, m+1.$$

We will see later that the central difference quotient is a better approximation to base an extrapolation method on, as far as convergence is concerned, because its asymptotic expansion contains only even powers of the step length h. Other important examples of discretization methods which lead to such asymptotic expansions are those for the solution of ordinary differential equations (see Sections 7.2.3 and 7.2.12).

In order to derive an extrapolation method for a given discretization method, we select a sequence of step lengths

$$F = \{h_0, h_1, h_2, \ldots\}, \qquad h_0 > h_1 > h_2 > \cdots > 0,$$

and calculate the corresponding approximate solutions $T(h_i)$, $i = 0, 1, 2, \ldots$. For $i \geqslant k$, we introduce the "polynomials"

$$\tilde{T}_{ik}(h) = b_0 + b_1 h^{\gamma_1} + \cdots + b_k h^{\gamma_k},$$

for which

$$\tilde{T}_{ik}(h_j) = T(h_j), \qquad j = i-k, i-k+1, \ldots, i,$$

and we consider the values

$$T_{ik} := \tilde{T}_{ik}(0)$$

as approximations to the desired value τ_0. Rational functions $T_{ik}(h)$ are frequently preferred over polynomials. Also the exponents γ_k need not be integer [see Bulirsch and Stoer (1964)].

For the following discussion of the discretization errors, we will assume that the $\tilde{T}_{ik}(h)$ are polynomials with exponents of the form $\gamma_k = k\gamma$. Romberg integration (see Section 3.4) is a special case with $\gamma = 2$. We will use the abbreviations

$$z := h^\gamma, \qquad z_j := h_j^\gamma, \quad j = 0, 1, \ldots, m.$$

Applying Lagrange's interpolation formula (2.1.1.3) to the polynomial

$$\tilde{T}_{ik}(h) = P_{ik}(z) = b_0 + b_1 z + b_2 z^2 + \cdots + b_k z^k$$

yields for $z = 0$

$$T_{ik} = P_{ik}(0) = \sum_{j=i-k}^{i} c_{kj}^{(i)} P_{ik}(z_j) = \sum_{j=i-k}^{i} c_{kj}^{(i)} T(h_j)$$

with

$$c_{kj}^{(i)} := \prod_{\substack{\sigma \neq j \\ \sigma = i-k}}^{i} \frac{z_\sigma}{z_\sigma - z_j}.$$

Then

$$(3.5.2) \qquad \sum_{j=i-k}^{i} c_{kj}^{(i)} z_j^\tau = \begin{cases} 1 & \text{if } \tau = 0, \\ 0 & \text{if } \tau = 1, 2, \ldots, k, \\ (-1)^k z_{i-k} z_{i-k+1} \cdots z_i & \text{if } \tau = k+1. \end{cases}$$

PROOF. The Lagrange coefficients $c_{kj}^{(i)}$ depend only on the support abscissas z_j and not on the functions to be interpolated. Selecting the polynomials z^l, $l = 0, 1, \ldots, k$, Lagrange's interpolation formula gives therefore

$$z^l = \sum_{j=i-k}^{i} z_j^l \prod_{\substack{\sigma \neq j \\ \sigma = i-k}}^{i} \frac{z - z_\sigma}{z_j - z_\sigma}, \qquad l = 0, \ldots, k.$$

For $z = 0$, all but the last one of the relations (3.5.2) follow.

To prove the last of the relations (3.5.2), we note that

$$(3.5.3) \qquad z^{k+1} \equiv \sum_{j=i-k}^{i} z_j^{k+1} \prod_{\substack{\sigma \neq j \\ \sigma=i-k}}^{i} \left(\frac{z - z_k}{z_\sigma - z_k}\right) + \prod_{\sigma=i-k}^{i} (z - z_\sigma).$$

Indeed, since the coefficients of z^{k+1} are the same on both sides, the difference polynomial has degree at most k. Since it vanishes at the $k + 1$ points z_σ, $\sigma = i - k, \ldots, i$, it vanishes identically, and (3.5.3) holds. Letting $z = 0$ in (3.5.3) completes the proof of (3.5.2). □

(3.5.2) can be sharpened for sequences h_j for which there exists a constant b such that

$$\frac{h_{j+1}}{h_j} \leqslant b < 1 \quad \text{for all } j.$$

In this case, there exists a constant C_k which depends only on b and for which

$$(3.5.4) \qquad \sum_{j=i-k}^{i} |c_{kj}^{(i)}| z_j^{k+1} \leqslant C_k z_{i-k} z_{i-k+1} \cdots z_i.$$

We prove (3.5.4) only for the special case of geometric sequences $\{h_j\}$ with

$$h_j = h_0 b^j, \qquad 0 < b < 1, \quad j = 0, 1, \ldots.$$

For the general case see Bulirsch and Stoer (1964). It suffices to prove (3.5.4) for $i = k$. With the abbreviation $\theta := b^\gamma$ we have

$$z_j^\tau = (h_0 b^j)^{\gamma\tau} = z_0^\tau \theta^{j\tau}.$$

In view of (3.5.2), the polynomial

$$P_k(z) := \sum_{j=0}^{k} c_{kj}^{(k)} z^j$$

satisfies

$$P_k(\theta^\tau) = \sum_{j=0}^{k} c_{kj}^{(k)} \theta^{j\tau} = z_0^{-\tau} \sum_{j=0}^{k} c_{kj}^{(k)} z_j^\tau = \begin{cases} 1 & \text{for } \tau = 0 \\ 0 & \text{for } \tau = 1, 2, \ldots, k, \end{cases}$$

so that $P_k(z)$ has the k different roots θ^τ, $\tau = 1, \ldots, k$. Since $P_k(1) = 1$, the polynomial P_k must have the form

$$P_k(z) = \prod_{l=1}^{k} \frac{z - \theta^l}{1 - \theta^l}.$$

The coefficients of P_k alternate in sign, so that

$$\begin{aligned}\sum_{j=0}^{k} |c_{kj}^{(k)}| z_j^{k+1} &= z_0^{k+1} \sum_{j=0}^{k} |c_{kj}^{(k)}| (\theta^{k+1})^j = z_0^{k+1} |P_k(-\theta^{k+1})| \\ &= z_0^{k+1} \prod_{l=1}^{k} \frac{\theta^{k+1} + \theta^l}{1 - \theta^l} \\ &= z_0^{k+1} \theta^{1+2+\cdots+k} \prod_{l=1}^{k} \frac{1 + \theta^l}{1 - \theta^l} \\ &= C_k(\theta) z_0 z_1 \ldots z_k\end{aligned}$$

with

$$C_k = C_k(\theta) := \prod_{j=1}^{k} \frac{1 + \theta^l}{1 - \theta^l}. \tag{3.5.5}$$

This proves (3.5.4) for the special case of geometrically increasing step lengths h_j.

We are now able to make use of the asymptotic expansion (3.5.1) which gives for $k < m$

$$\begin{aligned}T_{ik} &= \sum_{j=i-k}^{i} c_{kj}^{(i)} T(h_j) \\ &= \sum_{j=i-k}^{i} c_{kj}^{(i)} [\tau_0 + \tau_1 z_j + \tau_2 z_j^2 + \cdots + \tau_k z_j^k + z_j^{k+1}(\tau_{k+1} + O(h_j))],\end{aligned}$$

and for $k = m$

$$T_{im} = \sum_{j=i-m}^{i} c_{mj}^{(i)} [\tau_0 + \tau_1 z_j + \tau_2 z_j^2 + \cdots + \tau_m z_j^m + z_j^{m+1} \alpha_{m+1}(h_j)].$$

By (3.5.2) and (3.5.4),

$$T_{ik} = \tau_0 + (-1)^k z_{i-k} z_{i-k+1} \ldots z_i (\tau_{k+1} + O(h_{i-k})) \quad \text{for } k < m, \tag{3.5.6}$$

and

$$|T_{im} - \tau_0| \leqslant M_{m+1} C_m z_{i-m} z_{i-m+1} \ldots z_i$$

if $|\alpha_{m+1}(h_j)| \leqslant M_{m+1}$ for $j \geqslant 0$ [see (3.4.2)]. Consequently, for fixed k and $i \to \infty$,

$$T_{ik} - \tau_0 = O(z_{i-k}^{k+1}) = O(h_{i-k}^{(k+1)\gamma}).$$

In other words, the elements T_{ik} of the $(k + 1)$st column of the tableau (3.4.4) converge to τ_0 like a method of order $(k + 1)\gamma$. Note that the increase of the order of convergence from column to column which can be achieved by extrapolation methods is equal to γ: $\gamma = 2$ is twice as good as $\gamma = 1$. This explains the preference for discretization methods whose corresponding asymptotic expansions contain only even powers of h, e.g., the asymptotic expansion of the trapezoidal sum (3.4.1) or the central difference quotient discussed in this section.

The formula (3.5.6) shows furthermore that the sign of the error remains constant for fixed $k < m$ and sufficiently large i provided $\tau_{k+1} \neq 0$. Advantage can be taken of this fact in the many cases in which

$$(3.5.7) \qquad 0 \leqslant \frac{|T_{i+1,k} - \tau_0|}{|T_{ik} - \tau_0|} \approx \frac{h_{i+1}^{\gamma}}{h_{i-k}^{\gamma}} \leqslant b^{\gamma(k+1)} < \frac{1}{2}.$$

If we put

$$U_{ik} := 2T_{i+1,k} - T_{ik},$$

then

$$U_{ik} - \tau_0 = 2(T_{i+1,k} - \tau_0) - (T_{ik} - \tau_0).$$

For $s := \operatorname{sign}(T_{i+1,k} - \tau_0) = \operatorname{sign}(T_{i,k} - \tau_0)$, we have

$$s(U_{ik} - \tau_0) = 2|T_{i,k+1} - \tau_0| - |T_{ik} - \tau_0| \approx -|T_{ik} - \tau_0| < 0.$$

Thus U_{ik} converges monotonically to τ_0 for $i \to \infty$ at roughly the same rate as T_{ik} but from the opposite direction, so that eventually T_{ik} and U_{ik} will include the limit τ_0 between them. This observation yields a convenient stopping criterion.

EXAMPLE. The exact value of the integral

$$\int_0^{\pi/2} 5(e^{\pi} - 2)^{-1} e^{2x} \cos x \, dx$$

is 1. Using the polynomial extrapolation method of Romberg, and carrying 12 digits, we obtain for T_{ik}, U_{ik}, $0 \leqslant i \leqslant 6$, $0 \leqslant k \leqslant 3$, the values given in the following table.

i	T_{i0}	T_{i1}	T_{i2}	T_{i3}
0	0.185 755 068 924			
1	0.724 727 335 089	0.904 384 757 145		
2	0.925 565 035 158	0.992 510 935 182	0.998 386 013 717	
3	0.981 021 630 069	0.999 507 161 706	0.999 973 576 808	0.999 998 776 222
4	0.995 232 017 388	0.999 968 813 161	0.999 999 589 925	1.000 000 002 83
5	0.998 806 537 974	0.999 998 044 836	0.999 999 993 614	1.000 000 000 02
6	0.999 701 542 775	0.999 999 877 709	0.999 999 999 901	1.000 000 000 00

i	U_{i0}	U_{i1}	U_{i2}	U_{i3}
0	1.263 699 601 26			
1	1.126 402 735 23	1.080 637 113 22		
2	1.036 478 224 98	1.006 503 388 23	1.001 561 139 90	
3	1.009 442 404 71	1.000 430 464 62	1.000 025 603 04	1.000 001 229 44
4	1.002 381 058 56	1.000 027 276 51	1.000 000 397 30	0.999 999 997 211
5	1.000 596 547 58	1.000 001 710 58	1.000 000 006 19	0.999 999 999 978
6	1.000 149 217 14	1.000 000 107 00	1.000 000 000 09	1.000 000 000 00

3.6 Gaussian Integration Methods

In this section, we broaden the scope of our examination by considering integrals of the form

$$I(f) := \int_a^b \omega(x) f(x)\, dx,$$

where $\omega(x)$ is a given nonnegative *weight function* on the interval $[a, b]$. Also, the interval $[a, b]$ may be infinite, e.g., $[0, +\infty]$ or $[-\infty, +\infty]$. The weight function must meet the following requirements:

(3.6.1).

(a) *$\omega(x) \geqslant 0$ is measurable on the finite or infinite interval $[a, b]$.*
(b) *All moments $\mu_k := \int_a^b x^k \omega(x)\, dx$, $k = 0, 1, \ldots$, exist and are finite.*
(c) *For polynomials $s(x)$ which are nonnegative on $[a, b]$, $\int_a^b \omega(x) s(x)\, dx = 0$ implies $s(x) \equiv 0$.*

The conditions (3.6.1) are met, for instance, if $\omega(x)$ is positive and continuous on a finite interval $[a, b]$. Condition (3.6.1c) is equivalent to $\int_a^b \omega(x)\, dx > 0$ (see Exercise 14).

We will again examine integration rules of the type

$$\tilde{I}(f) := \sum_{i=1}^{n} w_i f(x_i). \tag{3.6.2}$$

The Newton–Cotes formulas (see Section 3.1) are of this form, but the abscissas x_i were required to form a uniform partition of the interval $[a, b]$. In this section, we relax this restriction and try to choose the x_i as well as the w_i so as to maximize the order of the integration method, that is, to maximize the degree for which all polynomials are exactly integrated by (3.6.2). We will see that this is possible and leads to a class of well-defined so-called *Gaussian integration rules* or *Gaussian quadrature formulas* [see for instance Stroud and Secrest (1966)]. These Gaussian integration rules will be shown to be unique and of order $2n - 1$. Also $w_i > 0$ and $a < x_i < b$ for $i = 1, \ldots, n$. In order to establish these results and to determine the exact form of the Gaussian integration rules, we need some basic facts about orthogonal polynomials.

We introduce the notation

$$\bar{\Pi}_j := \{p \mid p(x) = x^j + a_1 x^{j-1} + \cdots + a_j\}$$

for the set of normed real polynomials of degree j, and, as before, we denote by

$$\Pi_j := \{p \mid \text{degree}(p) \leqslant j\}$$

the linear space of all polynomials whose degree does not exceed j. In addition, we define the scalar product

$$(f, g) := \int_a^b \omega(x) f(x) g(x)\, dx$$

on the linear space $L^2[a, b]$ of all functions for which the integral

$$(f, f) = \int_a^b \omega(x)(f(x))^2\, dx$$

is well defined and finite. The functions $f, g \in L^2[a, b]$ are called orthogonal if $(f, g) = 0$. The following theorem establishes the existence of a sequence of mutually orthogonal polynomials, the system of *orthogonal polynomials* associated with the weight function $\omega(x)$.

(3.6.3) Theorem. *There exist polynomials* $p_j \in \bar{\Pi}_j$, $j = 0, 1, 2, \ldots,$ *such that*

(3.6.4) $$(p_i, p_k) = 0 \quad \textit{for } i \neq k.$$

These polynomials are uniquely defined by the recursions

(3.6.5a) $$p_0(x) \equiv 1,$$

(3.6.5b) $$p_{i+1}(x) \equiv (x - \delta_{i+1})p_i(x) - \gamma_{i+1}^2 p_{i-1}(x) \quad \textit{for } i \geqslant 0,$$

where $p_{-1}(x) \equiv 0$ *and*[1]

(3.6.6a) $$\delta_{i+1} := (xp_i, p_i)/(p_i, p_i) \quad \textit{for } i \geqslant 0,$$

(3.6.6b) $$\gamma_{i+1}^2 := \begin{cases} 0 & \textit{for } i = 0, \\ (p_i, p_i)/(p_{i-1}, p_{i-1}) & \textit{for } i \geqslant 1. \end{cases}$$

PROOF. The polynomials can be constructed recursively by a technique known as *Gram–Schmidt orthogonalization*. Clearly $p_0(x) \equiv 1$. Suppose then, as an induction hypothesis, that all orthogonal polynomials with the above properties have been constructed for $j \leqslant i$ and have been shown to be unique. We proceed to show that there exists a unique polynomial $p_{i+1} \in \bar{\Pi}_{i+1}$ with

(3.6.7) $$(p_{i+1}, p_j) = 0 \quad \text{for } j \leqslant i,$$

and that this polynomial satisfies (3.6.5b). Any polynomial $p_{i+1} \in \bar{\Pi}_{i+1}$ can be written uniquely in the form

$$p_{i+1}(x) \equiv (x - \delta_{i+1})p_i(x) + c_{i-1}p_{i-1}(x) + c_{i-2}p_{i-2}(x) + \cdots + c_0 p_0(x),$$

[1] xp_i denotes the polynomial with values $xp_i(x)$ for all x.

because its leading coefficient and those of the polynomials $p_j, j \leqslant i$, have value 1. Since $(p_j, p_k) = 0$ for all $j, k \leqslant i$ with $j \neq k$, (3.6.7) holds if and only if

(3.6.8a) $\quad (p_{i+1}, p_i) = (xp_i, p_i) - \delta_{i+1}(p_i, p_i) = 0,$

(3.6.8b) $\quad (p_{i+1}, p_{j-1}) = (xp_{j-1}, p_i) + c_{j-1}(p_{j-1}, p_{j-1}) = 0 \quad \text{for } j \leqslant i.$

The condition (3.6.1c)—with p_i^2 and p_{j-1}^2, respectively, in the role of the nonnegative polynomial s—rules out $(p_i, p_i) = 0$ and $(p_{j-1}, p_{j-1}) = 0$ for $1 \leqslant j \leqslant i$. Therefore, the equations (3.6.8) can be solved uniquely. (3.6.8a) gives (3.6.6a). By the induction hypothesis,

$$p_j(x) \equiv (x - \delta_j)p_{j-1}(x) - \gamma_j^2 p_{j-2}(x)$$

for $j \leqslant i$. From this, by solving for $xp_{j-1}(x)$, we have $(xp_{j-1}, p_i) = (p_j, p_i)$ for $j \leqslant i$, so that

$$c_{j-1} = -\frac{(p_j, p_i)}{(p_{j-1}, p_{j-1})} = \begin{cases} -\gamma_{i+1}^2 & \text{for } j = i, \\ 0 & \text{for } j < i, \end{cases}$$

in view of (3.6.8). Thus (3.6.5b) has been established for $i + 1$. □

Every polynomial $p \in \Pi_k$ is clearly representable as a linear combination of the orthogonal polynomials p_i, $i \leqslant k$. We thus have:

(3.6.9) Corollary. $(p, p_n) = 0$ *for all* $p \in \Pi_{n-1}$.

(3.6.10) Theorem. *The roots* x_i, $i = 1, \ldots, n$, *of* p_n *are real and simple. They all lie in the open interval* (a, b).

PROOF. Consider those roots of p_n which lie in (a, b) and which are of odd multiplicity, that is, at which p_n changes sign:

$$a < x_1 < \cdots < x_l < b.$$

The polynomial

$$q(x) := \prod_{j=1}^{l} (x - x_j) \in \bar{\Pi}_l$$

is such that the polynomial $p_n(x)q(x)$ does not change sign in $[a, b]$, so that

$$(p_n, q) = \int_a^b \omega(x)p_n(x)q(x)\,dx \neq 0$$

by (3.6.1c). Thus degree$(q) = l = n$ must hold, as otherwise $(p_n, q) = 0$ by Corollary (3.6.9). □

Next we have the

(3.6.11) Theorem. *The $n \times n$ matrix*

$$A := \begin{bmatrix} p_0(t_1) & \cdots & p_0(t_n) \\ \vdots & & \vdots \\ p_{n-1}(t_1) & \cdots & p_{n-1}(t_n) \end{bmatrix}$$

is nonsingular for mutually distinct arguments t_i, $i = 1, \ldots, n$.

Proof. Assume A is singular. Then there is a row vector $c^T = (c_0, \ldots, c_{n-1}) \neq 0$ with $c^T A = 0$. The polynomial

$$q(x) :\equiv \sum_{i=0}^{n-1} c_i p_i(x),$$

with degree$(p) < n$, has the n distinct roots $t_1, \ldots, t_n$ and must vanish identically. Let l be the largest index with $c_l \neq 0$. Then

$$p_l(x) = -\frac{1}{c_l} \sum_{i=0}^{l-1} c_i p_i(x).$$

This is a contradiction, since the polynomial to the right has a lower degree than $p_l \in \Pi_l$. □

Theorem (3.6.11) shows that the interpolation problem of finding a function of the form

$$p(x) \equiv \sum_{i=0}^{n-1} c_i p_i(x)$$

with $p(t_i) = f_i$, $i = 1, \ldots, n$ is always uniquely solvable. The condition of the theorem is known as the *Haar condition.* Any sequence of functions $p_0, p_1, \ldots$ which satisfy the Haar condition is said to form a *Chebyshev system.* Theorem (3.6.11) states that sequences of orthogonal polynomials are Chebyshev systems.

Now we arrive at the main result of this section.

(3.6.12) Theorem.

(a) *Let $x_1, \ldots, x_n$ be the roots of the* nth *orthogonal polynomial $p_n(x)$, and let $w_1, \ldots, w_n$ be the solution of the (nonsingular) system of equations*

$$\sum_{i=1}^{n} p_k(x_i) w_i = \begin{cases} (p_0, p_0) & \text{if } k = 0, \\ 0 & \text{if } k = 1, 2, \ldots, n-1. \end{cases} \tag{3.6.13}$$

Then $w_i > 0$ for $i = 1, 2, \ldots, n$, and

$$\int_a^b \omega(x) p(x)\, dx = \sum_{i=1}^{n} w_i p(x_i) \tag{3.6.14}$$

holds for all polynomials $p \in \Pi_{2n-1}$. *The positive numbers* w_i *are called "weights."*

(b) *Conversely, if the numbers* $w_i, x_i, i = 1, \ldots, n$, *are such that* (3.6.14) *holds for all* $p \in \Pi_{2n-1}$, *then the* x_i *are the roots of* p_n *and the weights* w_i *satisfy* (3.6.13).

(c) *It is not possible to find numbers* $x_i, w_i, i = 1, \ldots, n$, *such that* (3.6.14) *holds for all polynomials* $p \in \Pi_{2n}$.

PROOF. By Theorem (3.6.10), the roots x_i, $i = 1, \ldots, n$, of p_n are real and mutually distinct numbers in the open interval (a, b). The matrix

$$A = \begin{bmatrix} p_0(x_1) & \cdots & p_0(x_n) \\ \vdots & & \vdots \\ p_{n-1}(x_1) & \cdots & p_{n-1}(x_n) \end{bmatrix} \tag{3.6.15}$$

is nonsingular by Theorem (3.6.11), so that the system of equations (3.6.13) has a unique solution.

Consider an arbitrary polynomial $p \in \Pi_{2n-1}$. It can be written in the form

$$p(x) \equiv p_n(x)q(x) + r(x), \tag{3.6.16}$$

where q, r are polynomials in Π_{n-1}, which we can express as linear combinations of orthogonal polynomials

$$q(x) \equiv \sum_{k=0}^{n-1} \alpha_k p_k(x), \qquad r(x) \equiv \sum_{k=0}^{n-1} \beta_k p_k(x).$$

Since $p_0(x) \equiv 1$, it follows from (3.6.16) and Corollary (3.6.9) that

$$\int_a^b \omega(x)p(x)\,dx = (p_n, q) + (r, p_0) = \beta_0(p_0, p_0).$$

On the other hand, by (3.6.16) [since $p_n(x_i) = 0$] and by (3.6.13),

$$\sum_{i=1}^n w_i p(x_i) = \sum_{i=1}^n w_i r(x_i) = \sum_{k=0}^{n-1} \beta_k \left(\sum_{i=1}^n w_i p_k(x_i) \right) = \beta_0(p_0, p_0).$$

Thus (3.6.14) is satisfied.

We observe that

(3.6.17). *If* $w_i, x_i, i = 1, \ldots, n$, *are such that* (3.6.14) *holds for all polynomials* $p \in \Pi_{2n-1}$, *then* $w_i > 0$ *for* $i = 1, \ldots, n$.

This is readily verified by applying (3.6.14) to the polynomials

$$\bar{p}_j(x) :\equiv \prod_{\substack{h=1 \\ h \neq j}}^{n} (x - x_h)^2 \in \Pi_{2n-2}, \qquad j = 1, \ldots, n,$$

and noting that

$$0 < \int_a^b \omega(x)\bar{p}_j(x)\,dx = \sum_{i=1}^{n} w_i \bar{p}_j(x_i) = w_j \prod_{\substack{h=1 \\ h \neq j}}^{n} (x_j - x_h)^2$$

by (3.6.1c). This completes the proof of (3.6.12a).

Assume that w_i, x_i, $i = 1, \ldots, n$, are such that (3.6.14) even holds for all polynomials $p \in \Pi_{2n}$. Then

$$\bar{p}(x) :\equiv \prod_{j=1}^{n} (x - x_j)^2 \in \Pi_{2n}$$

contradicts this claim, since by (3.6.1c)

$$0 < \int_a^b \omega(x)\bar{p}(x)\,dx = \sum_{i=1}^{n} w_i \bar{p}(x_i) = 0.$$

This proves (3.6.12c)

To prove (3.6.12b), suppose that w_i, x_i, $i = 1, \ldots, n$ are such that (3.6.14) holds for all $p \in \Pi_{2n-1}$. Note that the abscissas x_i must be mutually distinct, since otherwise we could formulate the same integration rule using only $n - 1$ of the abscissas x_i, contradicting (3.6.12c).

Applying (3.6.14) to the orthogonal polynomials $p = p_k$, $k = 0, \ldots, n - 1$, themselves, we find

$$\sum_{i=1}^{n} w_i p_k(x_i) = \int_a^b \omega(x) p_k(x)\,dx = (p_k, p_0) = \begin{cases} (p_0, p_0) & \text{if } k = 0 \\ 0 & \text{if } k = 1, \ldots, n-1. \end{cases}$$

In other words, the weights w_i must satisfy (3.6.13).

Applying (3.6.14) to $p(x) :\equiv p_k(x)p_n(x)$, $k = 0, \ldots, n - 1$, gives by (3.6.9)

$$0 = (p_k, p_n) = \sum_{i=1}^{n} w_i p_n(x_i) p_k(x_i), \qquad k = 0, \ldots, n-1.$$

In other words, the vector $c := (w_1 p_n(x_1), \ldots, w_n p_n(x_n))^T$ solves the homogeneous system of equations $Ac = 0$ with A the matrix (3.6.15). Since the abscissas x_i, $i = 1, \ldots, n$, are mutually distinct, the matrix A is nonsingular by Theorem (3.6.11). Therefore $c = 0$ and $w_i p_n(x_i) = 0$ for $i = 1, \ldots, n$. Since $w_i > 0$ by (3.6.17), we have $p_n(x_i) = 0$, $i = 1, \ldots, n$. This completes the proof of (3.6.12b). □

For the most common weight function $\omega(x) \equiv 1$ and the interval $[-1, 1]$, the results of Theorem (3.6.12) are due to Gauss. The corresponding orthogonal polynomials are (see Exercise 16)

(3.6.18) $$p_k(x) :\equiv \frac{k!}{(2k)!} \frac{d^k}{dx^k} (x^2 - 1)^k, \qquad k = 0, 1, \ldots,$$

Indeed, $p_k \in \bar{\Pi}_k$ and integration by parts establishes $(p_i, p_k) = 0$ for $i \neq k$. Up to a factor, the polynomials (3.6.18) are the *Legendre polynomials*. In the following table we give some values for w_i, x_i in this important special case. For further values see the National Bureau of Standard's *Handbook of Mathematical Functions* [Abramowitz and Stegun (1964)].

n	w_i	x_i
1	$w_1 = 2$	$x_1 = 0$
2	$w_1 = w_2 = 1$	$x_2 = -x_1 = 0.577\ 350\ 2692\ldots$
3	$w_1 = w_3 = \frac{5}{9}$ $w_2 = \frac{8}{9}$	$x_3 = -x_1 = 0.774\ 596\ 6692\ldots$ $x_2 = 0$
4	$w_1 = w_4 = 0.347\ 854\ 8451\ldots$ $w_2 = w_3 = 0.652\ 145\ 1549\ldots$	$x_4 = -x_1 = 0.861\ 136\ 3116\ldots$ $x_3 = -x_2 = 0.339\ 981\ 0436\ldots$
5	$w_1 = w_5 = 0.236\ 926\ 8851\ldots$ $w_2 = w_4 = 0.478\ 628\ 6705\ldots$ $w_3 = \frac{128}{225} = 0.568\ 888\ 8889\ldots$	$x_5 = -x_1 = 0.906\ 179\ 8459\ldots$ $x_4 = -x_2 = 0.538\ 469\ 3101\ldots$ $x_3 = 0$

Other important cases which lead to Gaussian integration rules are listed in the following table:

$[a, b]$	$\omega(x)$	Orthogonal polynomials
$[-1, 1]$	$(1 - x^2)^{-1/2}$	$T_n(x)$, Chebyshev polynomials
$[0, \infty)$	e^{-x}	$L_n(x)$, Laguerre polynomials
$(-\infty, \infty]$	e^{-x^2}	$H_n(x)$, Hermite polynomials

We have characterized the quantities w_i, x_i which enter the Gaussian integration rules for given weight functions, but we have yet to discuss methods for their actual calculation. We will examine this problem under the assumption that the coefficients δ_i, γ_i of the recursion (3.6.5) are given. Golub and Welsch (1969) and Gautschi (1968, 1970) discuss the much harder problem of finding the coefficients δ_i, γ_i.

The theory of orthogonal polynomials ties in with the theory of real tridiagonal matrices

$$(3.6.19) \qquad J_n = \begin{bmatrix} \delta_1 & \gamma_2 & & & \\ \gamma_2 & \cdot & \cdot & & \\ & \cdot & \cdot & \cdot & \\ & & \cdot & \cdot & \cdot \\ & & & \cdot & \cdot & \gamma_n \\ & & & & \gamma_n & \delta_n \end{bmatrix}$$

and their principal submatrices

$$J_j = \begin{bmatrix} \delta_1 & \gamma_2 & & & \\ \gamma_2 & \cdot & \cdot & & \\ & \cdot & \cdot & \cdot & \\ & & \cdot & \cdot & \gamma_j \\ & & & \gamma_j & \delta_j \end{bmatrix}.$$

Such matrices will be studied in Sections 5.5, 5.6 and 6.6.1. In section 5.5 it will be seen that the characteristic polynomials p_j of J_j satisfy the recursions (3.6.5) with the matrix elements δ_i, γ_i as the coefficients. Therefore, p_n is the characteristic polynomial of the tridiagonal matrix J_n. Consequently we have

(3.6.20) Theorem. *The roots x_i, $i = 1, \ldots, n$, of the nth orthogonal polynomial p_n are the eigenvalues of the tridiagonal matrix J_n in* (3.6.19).

The bisection method of Section 5.6, the *QR* method of section 6.6.6, and others are available to calculate the eigenvalues of these tridiagonal systems. With respect to the weights w_i, we have [Szegö (1959), Golub and Welsch (1969)].

(3.6.21) Theorem. *Let $v^{(i)} := (v_1^{(i)}, \ldots, v_n^{(i)})^T$ be an eigenvector of J_n* (3.6.19) *for the eigenvalue x_i, $J_n v^{(i)} = x_i v^{(i)}$. Suppose $v^{(i)}$ is scaled in such a way that*

$$v^{(i)\,T} v^{(i)} = (p_0, p_0) = \int_a^b \omega(x)\, dx.$$

Then the weights are given by

$$w_i = (v_1^{(i)})^2, \qquad i = 1, \ldots, n.$$

Proof. We verify that the vector

$$\tilde{v}^{(i)} = (\rho_0 p_0(x_i), \rho_1 p_1(x_i), \ldots, \rho_{n-1} p_{n-1}(x_i))^T$$

where—note that $\gamma_i \neq 0$ by (3.6.6b)—

$$\rho_j := \begin{cases} 1 & \text{for } j = 0, \\ \dfrac{1}{\gamma_2 \cdots \gamma_{j+1}} & \text{for } j = 1, \ldots, n-1 \end{cases}$$

is an eigenvector of J_n for the eigenvalue x_i: $J_n \tilde{v}^{(i)} = x_i \tilde{v}^{(i)}$. By (3.6.5), for any x,

$$\delta_1 \rho_0 p_0(x) + \gamma_2 \rho_1 p_1(x) = \delta_1 p_0(x) + p_1(x) = x p_0(x) = x \rho_0 p_0(x).$$

For $j = 2, \ldots, n-1$, similarly,

$$\begin{aligned}\gamma_j \rho_{j-2} p_{j-2}(x) + \delta_j \rho_{j-1} p_{j-1}(x) + \gamma_{j+1} \rho_j p_j(x) \\ = \rho_{j-1}[\gamma_j^2 p_{j-2}(x) + \delta_j p_{j-1}(x) + p_j(x)] \\ = x \rho_{j-1} p_{j-1}(x),\end{aligned}$$

and finally,

$$\rho_{n-1}[\gamma_n^2 p_{n-2} + \delta_n p_{n-1}(x)] = x\rho_{n-1} p_{n-1}(x) - \rho_{n-1} p_n(x),$$

so that

$$\gamma_n \rho_{n-2} p_{n-2}(x) + \delta_n \rho_{n-1} p_{n-1}(x) = x_i \rho_{n-1} p_{n-1}(x_i)$$

holds, provided $p_n(x_i) = 0$.

Since $\rho_j \neq 0$, $j = 0, \ldots, n-1$, the system of equations (3.6.13) for w_i is equivalent to

(3.6.22) $$(\tilde{v}^{(1)}, \ldots, \tilde{v}^{(n)})w = (p_0, p_0)e_1,$$

with $$w = (w_1, \ldots, w_n)^T, \quad e_1 = (1, 0, \ldots, 0)^T.$$

Eigenvectors of symmetric matrices for distinct eigenvalues are orthogonal. Therefore, multiplying (3.6.22) by $v^{(i)T}$ from the left yields

$$(\tilde{v}^{(i)T}\tilde{v}^{(i)})w_i = (p_0, p_0)\tilde{v}_1^{(i)}.$$

Since $\rho_0 = 1$ and $p_0(x) \equiv 1$, we have $\tilde{v}_1^{(i)} = 1$. Thus

(3.6.23) $$(\tilde{v}^{(i)T}\tilde{v}^{(i)})w_i = (p_0, p_0).$$

Using again the fact that $\tilde{v}_1^{(i)} = 1$, we find $v_1^{(i)}\tilde{v}^{(i)} = v^{(i)}$, and multiplying (3.6.23) by $(v_1^{(i)})^2$ gives

$$(v^{(i)T}v^{(i)})w_i = (v_1^{(i)})^2(p_0, p_0).$$

Since $v^{(i)T}v^{(i)} = (p_0, p_0)$ by hypothesis, we obtain $w_i = (v_1^{(i)})^2$. □

If the QR-method is employed for determining the eigenvalues of J_n, then the calculation of the first components $v_1^{(i)}$ of the eigenvectors $v^{(i)}$ is readily included in that algorithm: calculating the abscissas x_i and the weights w_i can be done concurrently [Golub and Welsch (1969)].

Finally, we will estimate the error of Gaussian integration:

(3.6.24) Theorem. *If $f \in C^{2n}[a, b]$, then*

$$\int_a^b \omega(x) f(x)\, dx - \sum_{i=1}^n w_i f(x_i) = \frac{f^{(2n)}(\xi)}{(2n)!}(p_n, p_n)$$

for some $\xi \in (a, b)$.

PROOF. Consider the solution $h \in \Pi_{2n-1}$ of the Hermite interpolation problem (see Section 2.1.5)

$$h(x_i) = f(x_i), \quad h'(x_i) = f'(x_i), \qquad i = 1, \ldots, n.$$

Since degree$(h) < 2n$,

$$\int_a^b \omega(x)h(x)\,dx = \sum_{i=1}^n w_i h(x_i) = \sum_{i=1}^n w_i f(x_i)$$

by Theorem (3.6.12). Therefore the error term has the integral representation

$$\int_a^b \omega(x)f(x)\,dx - \sum_{i=1}^n w_i f(x_i) = \int_a^b \omega(x)(f(x) - h(x))\,dx.$$

By Theorem (2.1.5.10), and since the x_i are the roots of $p_n(x) \in \tilde{\Pi}_n$,

$$f(x) - h(x) = \frac{f^{(2n)}(\zeta)}{(2n)!}(x - x_1)^2 \ldots (x - x_n)^2 = \frac{f^{(2n)}(\zeta)}{(2n)!}p_n^2(x)$$

for some $\zeta = \zeta(x)$ in the interval $I(x_1, \ldots, x_n, x)$ spanned by $x_1, \ldots, x_n, x$. Next,

$$\frac{f^{(2n)}(\zeta(x))}{(2n)!} = \frac{f(x) - h(x)}{p_n^2(x)}$$

is continuous on $[a, b]$ so that the mean-value theorem of integral calculus applies:

$$\int_a^b \omega(x)(f(x) - h(x))\,dx = \int_a^b \omega(x)f^{(2n)}(\zeta(x))p_n^2(x)\,dx = \frac{f^{(2n)}(\xi)}{(2n)!}(p_n, p_n)$$

for some $\xi \in (a, b)$. □

Comparing the various integration rules (Newton–Cotes formulas, extrapolation methods, Gaussian integration), we find that, computational efforts being equal, Gaussian integration yields the most accurate results. If only one knew ahead of time how to chose n so as to achieve specified accuracy for any given integral, then Gaussian integration would be clearly superior to other methods. Unfortunately, it is frequently not possible to use the error formula (3.6.24) for this purpose, because the $2n$th derivative is difficult to estimate. For these reasons, one will usually apply Gaussian integration for increasing values of n until successive approximate values agree within the specified accuracy. Since the function values which had been calculated for n cannot be used for $n + 1$ (at least not in the classical case $\omega(x) \equiv 1$), the apparent advantages of Gauss integration as compared with extrapolation methods are soon lost. There have been attempts to remedy this situation [e.g. Kronrod (1965)]. A collection of Fortran programs is given in Piessens et al. (1983).

3.7 Integrals with Singularities

Examining some frequently used integration methods in this chapter, we found that their application to a given integral

$$\int_a^b f(x)\,dx, \qquad a,\, b \text{ finite}$$

was justified provided the integrand $f(x)$ was sufficiently often differentiable in $[a, b]$. For many practical problems, however, the function $f(x)$ turns out to be not differentiable at the end points of $[a, b]$, or at some isolated points in its interior. In what follows, we suggest several ways of dealing with this and related situations.

(1) $f(x)$ is sufficiently often differentiable on the closed subintervals of a partition $a = a_1 < a_2 < \cdots < a_{m+1} = b$. Putting $f_i(x) :\equiv f(x)$ on $[a_i, a_{i+1}]$, and defining the derivatives of $f_i(x)$ at a_i as the one-sided right derivative and at a_{i+1} as the one-sided left derivative, we find that standard methods can be applied to integrate the functions $f_i(x)$ separately. Finally,

$$\int_a^b f(x)\,dx = \sum_{i=1}^{m} \int_{a_i}^{a_{i+1}} f_i(x)\,dx.$$

(2) Suppose there is a point $\bar{x} \in [a, b]$ for which not even one-sided derivatives of $f(x)$ exist. For instance, the function $f(x) = \sqrt{x}\sin x$ is such that $f'(x)$ will not be continuous for any choice of the value $f'(0)$. Nevertheless, the variable transformation $t := \sqrt{x}$ yields

$$\int_0^b \sqrt{x}\sin x\,dx = \int_0^{\sqrt{b}} 2t^2 \sin t^2\,dt$$

and leads to an integral with an integrand which is now arbitrarily often differentiable in $[0, \sqrt{b}]$.

(3) Another way to deal with the previously discussed difficulty is to split the integral:

$$\int_0^b \sqrt{x}\sin x\,dx = \int_0^\varepsilon \sqrt{x}\sin x\,dx + \int_\varepsilon^b \sqrt{x}\sin x\,dx, \qquad \varepsilon > 0.$$

The second integrand is arbitrarily often differentiable. The first integrand can be developed into a uniformly convergent series on $[0, \varepsilon]$ so that integration and summation can be interchanged:

$$\int_0^\varepsilon \sqrt{x}\sin x\,dx = \int_0^\varepsilon \sqrt{x}\left(x - \frac{x^3}{3!} \pm \cdots\right) dx = \sum_{v=0}^{\infty} (-1)^v \frac{\varepsilon^{2v+5/2}}{(2v+1)!\,(2v+5/2)}.$$

For sufficiently small ε, only few of the series need be considered. The difficulty lies in the choice of ε: if ε is selected too small, then the proximity of the singularity at $x = 0$ causes the speed of convergence to deteriorate when we calculate the remaining integral.

(4) Sometimes it is possible to subtract from the integrand $f(x)$ a function whose indefinite integral is known, and which has the same singularities as $f(x)$. For the above example, $x\sqrt{x}$ is such a function:

$$\int_0^b \sqrt{x}\sin x\,dx$$

$$= \int_0^b \sqrt{x}\,(\sin x - x)\,dx + \int_0^b x\sqrt{x}\,dx = \int_0^b \sqrt{x}(\sin x - x)\,dx + \tfrac{2}{5}b^{5/2}.$$

The new integrand has a continuous third derivative and is therefore better amenable to standard integration methods. In order to avoid cancellation when calculating the difference $\sin x - x$ for small x, it is recommended to evaluate the power series

$$\sin x - x = -x^3\left(\frac{1}{3!} - \frac{1}{5!}x^2 \pm \cdots\right) = -x^3 \sum_{\nu=0}^{\infty} \frac{(-1)^\nu}{(2\nu+3)!}x^{2\nu}.$$

(5) For certain types of singularities, as in the case of

$$I = \int_0^b x^\alpha f(x)\,dx, \qquad 0 < \alpha < 1,$$

with $f(x)$ sufficiently often differentiable on $[0, b]$, the trapezoidal sum $T(h)$ does not have an asymptotic expansion of the form (3.4.1), but rather of the more general form (3.5.1):

$$T(h) \sim \tau_0 + \tau_1 h^{\gamma_1} + \tau_2 h^{\gamma_2} + \cdots,$$

where

$$\{\gamma_i\} = \{1 + \alpha, 2, 2 + \alpha, 4, 4 + \alpha, 6, 6 + \alpha, \ldots\}$$

[see Bulirsch (1964)]. Suitable step-length sequences for extrapolation methods in this case are discussed in Bulirsch and Stoer (1964).

(6) Often the following scheme works surprisingly well: if the integrand of

$$I = \int_a^b f(x)\,dx$$

is not, or not sufficiently often, differentiable for $x = a$, put

$$a_j := a + \frac{b-a}{j}, \qquad j = 1, 2, \ldots,$$

in effect partitioning the half-open interval $(a, b]$ into infinitely many subintervals over which to integrate separately:

$$I_j := \int_{a_{j+1}}^{a_j} f(x)\,dx,$$

using standard methods. Then

$$I = I_1 + I_2 + I_3 + \cdots.$$

The convergence of this sequence can often be accelerated using, for instance, Aitken's Δ^2 method (see Section 5.10). Obviously, this scheme can be adapted to calculating *improper* integrals

$$\int_a^\infty f(x)\,dx.$$

(7) The range of improper integrals can be made finite by suitable variable transformations. For $x = 1/t$ we have, for instance,

$$\int_1^\infty f(x)\,dx = \int_0^1 \frac{1}{t^2} f\left(\frac{1}{t}\right) dt.$$

If the new integrand is singular at 0, then one of the above approaches may be tried. Note that the Gaussian integration rules based on Laguerre and Hermite polynomials (Section 3.6) apply directly to improper integrals of the forms

$$\int_0^\infty f(x)\,dx, \quad \int_{-\infty}^{+\infty} f(x)\,dx,$$

respectively.

Exercises for Chapter 3

1. Let $a \leqslant x_0 < x_1 < \cdots < x_n \leqslant b$ be an arbitrary fixed partition of the interval $[a, b]$. Show that there exist unique numbers $\gamma_0, \gamma_1, \ldots, \gamma_n$ with

$$\sum_{i=0}^n \gamma_i P(x_i) = \int_a^b P(x)\,dx$$

for all polynomials P with degree$(P) \leqslant n$. *Hint*: $P(x) = 1, x, \ldots, x^n$. Compare the resulting system of linear equations with that representing the polynomial interpolation problem with support abscissas x_i, $i = 0, \ldots, n$.

2. By construction, the nth Newton–Cotes formula yields the exact value of the integral for integrands which are polynomials of degree at most n. Show that for even values of n, polynomials of degree $n + 1$ are also integrated exactly. *Hint*: Consider the integrand x^{n+1} in the interval $[-k, +k]$, $n = 2k + 1$.

3. If $f \in C^2[a, b]$ then there exists an $\bar{x} \in (a, b)$ such that the error of the trapezoidal rule is expressed as follows:

$$\int_a^b f(x)\,dx - \tfrac{1}{2}(b-a)(f(a) + f(b)) = \tfrac{1}{12}(b-a)^3 f''(\bar{x}).$$

Derive this result from the error formula in (2.1.4.1) by showing that $f''(\xi(x))$ is continuous in x.

4. Derive the error formula (3.1.6) using Theorem (2.1.5.10). *Hint*: See Exercise 3.

5. Let $f \in C^6[-1, +1]$, and let $P \in \Pi_5$ be the Hermite interpolation polynomial with $P(x_i) = f(x_i)$, $P'(x_i) = f'(x_i)$, $x_i = -1, 0, +1$.

(a) Show that

$$\int_{-1}^{+1} P(t)\,dt = \tfrac{7}{15}f(-1) + \tfrac{16}{15}f(0) + \tfrac{7}{15}f(+1) + \tfrac{1}{15}f'(-1) - \tfrac{1}{15}f'(+1).$$

(b) By construction, the above formula represents an integration rule which is exact for all polynomials of degree 5 or less. Show that it need not be exact for polynomials of degree 6.

(c) Use Theorem (2.1.5.10) to derive an error formula for the integration rule in (a).

(d) Given a uniform partition $x_i = a + ih$, $i = 0, \ldots, 2n$, $h = (b - a)/2n$ of the interval $[a, b]$, what composite integration rule can be based on the integration rule in (a)?

6. Consider an arbitrary partition $\Delta := \{a = x_0 < \cdots < x_n = b\}$ of a given interval $[a, b]$. In order to approximate

$$\int_a^b f(t)\,dt$$

using the function values $f(x_i)$, $i = 0, \ldots, n$, spline interpolation (see Section 2.4) may be considered. Derive an integration rule in terms of $f(x_i)$ and the moments (2.4.2.1) of the "natural" spline (2.4.1.2a).

7. Determine the Peano kernel for Simpson's rule and $n = 2$ instead of $n = 3$ in $[-1, +1]$. Does it change sign in the interval of integration?

8. Consider the integration rule of Exercise 5.

(a) Show that its Peano kernel does not change its sign in $[-1, +1]$.

(b) Use (3.2.8) to derive an error term.

9. Prove

$$\sum_{k=0}^{n} k^3 = \left(\frac{n(n+1)}{2}\right)^2$$

using the Euler–Maclaurin summation formula.

10. Integration over the interval [0, 1] by Romberg's method using Neville's algorithm leads to the tableau

$$\begin{array}{l|lllll}
h_0^2 = 1 & T_{00} = T(h_0) & & & \\
 & & T_{11} & & \\
h_1^2 = \frac{1}{4} & T_{10} = T(h_1) & & T_{22} & \\
 & & T_{21} & & T_{33} \\
h_2^2 = \frac{1}{16} & T_{20} = T(h_2) & & T_{32} & \\
 & & T_{31} & & \\
h_3^2 = \frac{1}{64} & T_{30} = T(h_3) & & &
\end{array}$$

In Section 3.4, it is shown that T_{11} is Simpson's rule.

(a) Show that T_{22} is Milne's rule.

(b) Show that T_{33} is not the Newton–Cotes formula for $n = 8$.

11. Let $h_0 := b - a$, $h_1 := h_0/3$. Show that extrapolating $T(h_0)$ and $T(h_1)$ linearly to $h = 0$ gives the 3/8-rule.

12. One wishes to approximate the number e by an extrapolation method.
    (a) Show that $T(h) = (1 + h)^{1/h}$, $h \neq 0$, $|h| < 1$, has an expansion of the form
    $$T(h) = e + \sum_{i=1}^{\infty} \tau_i h^i$$
    which converges if $|h| < 1$.
    (b) Modify $T(h)$ in such a way that extrapolation to $h = 0$ yields, for a fixed value x, an approximation to e^x.

13. Consider integration by a polynomial extrapolation method based on a geometric step-size sequence $h_j = h_0 b^j$, $j = 0, 1, \ldots, 0 < b < 1$. Show that small errors ΔT_j in the computation of the trapezoidal sums $T(h_j)$, $j = 0, 1, \ldots, m$, will cause an error T_{mm} in the extrapolated value T_{mm} satisfying
    $$|\Delta T_{mm}| \leqslant C_m(b^2) \max_{0 \leqslant j \leqslant m} |\Delta T_j|,$$
    where $C_m(\theta)$ is the constant given in (3.5.5). Note that $C_m(\theta) \to \infty$ as $\theta \to 1$, so that the stability of the extrapolation method deteriorates sharply as b approaches 1.

14. Consider a weight function $\omega(x) \geqslant 0$ which satisfies (3.6.1a) and (b). Show that (3.6.1c) is equivalent to
    $$\int_a^b \omega(x)\,dx > 0.$$
    Hint: The mean-value theorem of integral calculus applied to suitable subintervals of $[a, b]$.

15. The integral
    $$(f, g) := \int_{-1}^{+1} f(x)g(x)\,dx$$
    defines a scalar product for functions $f, g \in C[-1, +1]$. Show that if f and g are polynomials of degree less than n, if x_i, $i = 1, 2, \ldots, n$, are the roots of the nth Legendre polynomial (3.6.18), and if
    $$\gamma_i := \int_{-1}^{+1} L_i(x)\,dx$$
    with
    $$L_i(x) := \prod_{\substack{k \neq i \\ k=1}}^{n} \frac{x - x_k}{x_i - x_k}, \qquad i = 1, 2, \ldots, n,$$
    then
    $$(f, g) = \sum_{i=1}^{n} \gamma_i f(x_i)g(x_i).$$

16. Consider the Legendre polynomials $p_j(x)$ in (3.6.18).
    (a) Show that the leading coefficient of $p_j(x)$ has value 1.
    (b) Verify the orthogonality of these polynomials: $(p_i, p_j) = 0$ if $i < j$.

Hint: Integration by parts, noting that

$$\frac{d^{2i+1}}{dx^{2i+1}}(x^2-1)^i \equiv 0$$

and that the polynomial

$$\frac{d^l}{dx^l}(x^2-1)^k$$

is divisible by $x^2 - 1$ if $l < k$.

17. Consider Gaussian integration, $[a, b] = [-1, +1]$, $\omega(x) \equiv 1$.
    (a) Show that $\delta_i = 0$ for $i > 0$ in the recursion (3.6.5) for the corresponding orthogonal polynomials $p_j(x)$ (3.6.18). *Hint*: $p_j(x) \equiv (-1)^{j+1}p_j(-x)$.
    (b) Verify

    $$\int_{-1}^{+1}(x^2-1)^j\,dx = \frac{(-1)^j 2^{2j+1}}{\binom{2j}{j}(2j+1)}.$$

    Hint: Repeated integration by parts of the integrand $(x^2-1)^j \equiv (x+1)^j(x-1)^j$.
    (c) Calculate (p_j, p_j) using integration by parts (see Exercise 16) and the result (b) of this exercise. Show that

    $$\gamma_i^2 = \frac{i^2}{(2i+1)(2i-1)}$$

    for $i > 0$ in the recursion (3.6.5).

18. Consider Gaussian integration in the interval $[-1, +1]$ with the weight function

    $$\omega(x) \equiv \frac{1}{\sqrt{1-x^2}}.$$

    In this case, the orthogonal polynomials $p_j(x)$ are the classical Chebychev polynomials, $T_0(x) \equiv 1$, $T_1(x) \equiv x$, $T_2(x) \equiv 2x^2 - 1$, $T_3(x) \equiv 4x^3 - 3x, \ldots$, $T_{j+1}(x) \equiv 2xT_j(x) - T_{j-1}(x)$, up to scalar factors.
    (a) Prove that $p_j(x) \equiv (1/2^{j-1})T_j(x)$ for $j \geqslant 1$. What is the form of the tridiagonal matrix (3.6.19) in this case?
    (b) For $n = 3$, determine the equation system (3.6.13). Verify that $w_1 = w_2 = w_3 = \pi/3$. (In the Chebychev case, the weights w_i are equal for general n.)

19. Denote by $T(f; h)$ the trapezoidal sum of step length h for the integral

    $$\int_0^1 f(x)\,dx.$$

    For $\alpha > 1$, $T(x^\alpha; h)$ has the asymptotic expansion

    $$T(x^\alpha; h) \sim \int_0^1 x^\alpha\,dx + a_1 h^{1+\alpha} + a_2 h^2 + a_4 h^4 + a_6 h^6 + \cdots.$$

Show that, as a consequence, every function $f(x)$ which is analytic on a disk $|z| \leqslant r$ in the complex plane with $r > 1$ admits an asymptotic expansion of the form

$$T(x^\alpha f(x); h) \sim \int_0^1 x^\alpha f(x)\,dx + b_1 h^{1+\alpha} + b_2 h^{2+\alpha} + b_3 h^{3+\alpha} + \cdots$$
$$+ c_2 h^2 + c_4 h^4 + c_6 h^6 + \cdots.$$

Hint: Expand $f(x)$ into a power series and apply $T(\varphi + \psi; h) = T(\varphi; h) + T(\psi; h)$.

References for Chapter 3

Abramowitz, M., Stegun, I. A.: *Handbook of Mathematical Functions*. National Bureau of Standards, Applied Mathematics Series 55, Washington, D.C.: U.S. Government Printing Office 1964, 6th printing 1967.

Bauer, F. L., Rutishauser, H., Stiefel, E.: New aspects in numerical quadrature. *Proc. of Symposia in Applied Mathematics* **15**, 199–218, Amer. Math. Soc. 1963.

Bulirsch, R.: Bemerkungen zur Romberg-Integration. *Numer. Math.* **6**, 6–16 (1964).

———, Stoer, J.: Fehlerabschätzungen und Extrapolation mit rationalen Funktionen bei Verfahren vom Richardson-Typus. *Numer. Math.* **6**, 413–427 (1964).

———, ———: Numerical quadrature by extrapolation. *Numer. Math.* 271–278 (1967).

Davis, P. J.: *Interpolation and Approximation*. New York: Blaisdell 1963, 2nd printing 1965.

———, Rabinowitz, P.: *Methods of Numerical Integration*. New York: Academic Press 1975.

Erdelyi, A.: *Asymptotic Expansions*. New York: Dover 1956.

Gautschi, W.: Construction of Gauss–Christoffel quadrature formulas. *Math. Comp.* **22**, 251–270 (1968).

———: On the construction of Gaussian quadrature rules from modified moments. *Math. Comput.* **24**, 245–260 (1970).

Golub, G. H., Welsch, J. H.: Calculation of Gauss quadrature rules. *Math. Comput.* **23**, 221–230 (1969).

Gröbner, W., Hofreiter, N.: *Integraltafel*, 2 vols. Berlin: Springer Verlag 1961.

Kronrod, A. S.: *Nodes and Weights of Quadrature Formulas*. Authorized translation from the Russian. New York: Consultants Bureau 1965.

Henrici, P.: *Elements of Numerical Analysis*. New York: Wiley 1964.

Olver, F. W. J.: *Asymptotics and Special Functions*. New York: Academic Press 1974.

Piessens, R., de Doncker, E., Überhuber, C. W., Kahaner, D. K.: *Quadpack, A subroutine package for automatic integration*. Berlin, Heidelberg, New York: Springer-Verlag 1983.

Romberg, W.: Vereinfachte numerische Integration. *Det. Kong. Norske Videnskabers Selskab Forhandlinger* **28**, Nr. 7, Trondheim 1955.

Schoenberg, I. J.: Monosplines and quadrature formulae. In: *Theory and Applications of Spline Functions*. Edited by T. N. E. Greville. 157–207. New York: Academic Press 1969.

Steffensen, J. F.: *Interpolation* (1927) 2nd edition. New York: Chelsea 1950.

Stroud, A. H., Secrest, D.: *Gaussian Quadrature Formulas*. Englewood Cliffs, N.J.: Prentice-Hall 1966.

Szegö, G.: *Orthogonal Polynomials*. New York: Amer. Math. Soc. 1959.

4 Systems of Linear Equations

In this chapter direct methods for solving systems of linear equations

$$Ax = b, \qquad A = \begin{bmatrix} a_{11} & \dots & a_{1n} \\ \vdots & & \vdots \\ a_{n1} & \dots & a_{nn} \end{bmatrix}, \quad b = \begin{bmatrix} b_1 \\ \vdots \\ b_n \end{bmatrix}$$

will be presented. Here A is a given $n \times n$ matrix, and b is a given vector. We assume in addition that A and b are real, although this restriction is inessential in most of the methods. In contrast to the iterative methods (Chapter 8), the direct methods discussed here produce the solution in finitely many steps, assuming computations without roundoff errors.

This problem is closely related to that of computing the inverse A^{-1} of the matrix A provided this inverse exists. For if A^{-1} is known, the solution x of $Ax = b$ can be obtained by matrix vector multiplication, $x = A^{-1}b$. Conversely, the ith column $\bar{a}_i$ of $A^{-1} = (\bar{a}_1, \dots, \bar{a}_n)$ is the solution of the linear system $Ax = e_i$, where $e_i = (0, \dots, 0, 1, 0, \dots, 0)^T$ is the ith unit vector.

A general introduction to numerical linear algebra is given in Golub and van Loan (1983) and Stewart (1973). ALGOL programs are found in Wilkinson and Reinsch (1971), FORTRAN programs in Dongarra, Bunch, Moler, and Stewart (1979).

4.1 Gaussian Elimination. The Triangular Decomposition of a Matrix

In the method of Gaussian elimination for solving a system of linear equations

(4.1.1) $$Ax = b,$$

where A is an $n \times n$ matrix and $b \in \mathbb{R}^n$, the given system (4.1.1) is transformed in steps by appropriate rearrangements and linear combinations of equations into a system of the form

$$Rx = c, \qquad R = \begin{bmatrix} r_{11} & \cdots & r_{1n} \\ & \ddots & \vdots \\ 0 & & r_{nn} \end{bmatrix}$$

which has the same solution as (4.1.1). R is an upper triangular matrix, so that $Rx = c$ can easily be solved by "back substitution" (so long as $r_{ii} \neq 0$, $i = 1, \ldots, n$):

$$x_i := \frac{c_i - \sum_{k=i+1}^{n} r_{ik} x_k}{r_{ii}} \quad \text{for } i = n, n-1, \ldots, 1.$$

In the first step of the algorithm an appropriate multiple of the first equation is subtracted from all of the other equations in such a way that the coefficients of x_1 vanish in these equations; hence, x_1 remains only in the first equation. This is possible only if $a_{11} \neq 0$, of course, which can be achieved by rearranging the equations if necessary, as long as at least one $a_{i1} \neq 0$. Instead of working with the equations themselves, the operations are carried out on the matrix

$$(A, b) = \begin{bmatrix} a_{11} & \cdots & a_{1n} & b_1 \\ \vdots & & \vdots & \vdots \\ a_{n1} & \cdots & a_{nn} & b_n \end{bmatrix}$$

which corresponds to the full system given in (4.1.1). The first step of the Gaussian elimination process leads to a matrix (A', b') of the form

$$\text{(4.1.2)} \qquad (A', b') = \begin{bmatrix} a'_{11} & a'_{12} & \cdots & a'_{1n} & b'_1 \\ 0 & a'_{22} & \cdots & a'_{2n} & b'_2 \\ \vdots & \vdots & & \vdots & \vdots \\ 0 & a'_{n2} & \cdots & a'_{nn} & b'_n \end{bmatrix},$$

and this step can be described formally as follows:

(4.1.3)

(a) *Determine an element $a_{r1} \neq 0$ and proceed with* (b); *if no such r exists, A is singular; set $(A', b') := (A, b)$; stop.*

(b) *Interchange rows r and 1 of (A, b). The result is the matrix $(\bar{A}, \bar{b})$.*

(c) *For $i = 2, 3, \ldots, n$, subtract the multiple*

$$l_{i1} := \bar{a}_{i1} / \bar{a}_{11}$$

of row 1 from row i of the matrix $(\bar{A}, \bar{b})$. The desired matrix (A', b') is obtained as the result.

The transition $(A, b) \to (\bar{A}, \bar{b}) \to (A', b')$ can be described by using matrix multiplications:

(4.1.4) $$(\bar{A}, \bar{b}) = P_1(A, b), \qquad (A', b') = G_1(\bar{A}, \bar{b}) = G_1 P_1(A, b),$$

where P_1 is a permutation matrix, and G_1 is a lower triangular matrix:

(4.1.5)

$$P_1 := \begin{bmatrix} 0 & & & 1 & & & 0 \\ & 1 & & & & & \\ & & \ddots & & & & \\ & & & 1 & & & \\ 1 & & & 0 & & & \\ & & & & 1 & & \\ & & & & & 1 & \\ 0 & & & & & & 1 \end{bmatrix} \leftarrow r, \qquad G_1 := \begin{bmatrix} 1 & & & 0 \\ -l_{21} & 1 & & \\ \vdots & & \ddots & \\ -l_{n1} & 0 & & 1 \end{bmatrix}$$

Matrices such as G_1, which differ in at most one column from an identity matrix, are called *Frobenius matrices*. Both matrices P_1 and G_1 are nonsingular; in fact

$$P_1^{-1} = P_1, \qquad G_1^{-1} = \begin{bmatrix} 1 & & & 0 \\ l_{21} & 1 & & \\ \vdots & & \ddots & \\ l_{n1} & & 0 & 1 \end{bmatrix}.$$

For this reason, the equation systems $Ax = b$ and $A'x = b'$ have the same solution: $Ax = b$ implies $G_1 P_1 Ax = A'x = b' = G_1 P_1 b$, and $A'x = b'$ implies $P_1^{-1}G_1^{-1}A'x = Ax = b = P_1^{-1}G_1^{-1}b'$.

The element $a_{r1} = \bar{a}_{11}$ which is determined in (a) is called the *pivot element* (or simply the *pivot*), and step (a) itself is called *pivot selection* (or *pivoting*). In the pivot selection one can, in theory, choose any $a_{r1} \neq 0$ as the pivot element. For reasons of numerical stability (see Section 4.5) it is not recommended that an arbitrary $a_{r1} \neq 0$ be chosen. Usually the choice

$$|a_{r1}| = \max_i |a_{i1}|$$

is made; that is, among all candidate elements the one of largest absolute value is selected. (It is assumed in making this choice however—see Section 4.5—that the matrix A is "equilibrated", that is, that the orders of magnitudes of the elements of A are "roughly equal".) This sort of pivot selection is called *partial pivot selection* (or *partial pivoting*), in contrast to *complete pivot selection* (or *complete pivoting*), in which the search for a pivot is not restricted to the first column; that is, (a) and (b) in (4.1.3) are replaced by (a′) and (b′):

(a′) *Determine r and s so that*

$$|a_{rs}| = \max_{i,j} |a_{ij}|$$

and continue with (b′) *if* $a_{rs} \neq 0$. *Otherwise* A *is singular; set* $(A', b') := (A, b)$*; stop.*

(b′) *Interchange rows* 1 *and* r *of* (A, b), *as well as columns* 1 *and* s. *Let the resulting matrix be* $(\bar{A}, \bar{b})$.

After the first elimination step, the resulting matrix has the form (4.1.2):

$$(A', b') = \left[\begin{array}{c|c|c} a'_{11} & a'^T & b'_1 \\ \hline 0 & \tilde{A} & \tilde{b} \end{array}\right]$$

with an $(n-1)$-row matrix $\tilde{A}$. The next elimination step consists simply of applying the process described in (4.1.3) for (A, b) to the smaller matrix $(\tilde{A}, \tilde{b})$. Carrying on in this fashion, a sequence of matrices

$$(A, b) =: (A^{(0)}, b^{(0)}) \to (A^{(1)}, b^{(1)}) \to \cdots \to (A^{(n-1)}, b^{(n-1)}) =: (R, c)$$

is obtained which begins with the given matrix (A, b) (4.1.1) and ends with the desired matrix (R, c). In this sequence the jth intermediate matrix $(A^{(j)}, b^{(j)})$ has the form

$$\text{(4.1.6)} \qquad (A^{(j)}, b^{(j)}) = \left[\begin{array}{cccc|ccc} * & \cdots & * & * & * & \cdots & * \; * \\ 0 & \ddots & & * & \vdots & & \vdots \; \vdots \\ \vdots & \ddots & \ddots & \vdots & \vdots & & \vdots \; \vdots \\ 0 & \cdots & 0 & * & * & \cdots & * \; * \\ \hline 0 & \cdots & & 0 & * & \cdots & * \; * \\ \vdots & & & \vdots & \vdots & & \vdots \; \vdots \\ 0 & \cdots & & 0 & * & \cdots & * \; * \end{array}\right] = \left[\begin{array}{c|c|c} A_{11}^{(j)} & A_{12}^{(j)} & b_1^{(j)} \\ \hline 0 & A_{22}^{(j)} & b_2^{(j)} \end{array}\right]$$

with a j-row upper triangular matrix $A_{11}^{(j)}$. The transition $(A^{(j)}, b^{(j)}) \to (A^{(j+1)}, b^{(j+1)})$ consists of the application of (4.1.3) on the $(n-j) \times (n-j+1)$ matrix $(A_{22}^{(j)}, b_2^{(j)})$. The elements of $A_{11}^{(j)}$, $A_{12}^{(j)}$, $b_1^{(j)}$ do not change from this step on; hence they agree with the corresponding elements of (R, c). As in the first step, (4.1.4) and (4.1.5), the ensuing steps can be described using matrix multiplication. As can be readily seen

$$\text{(4.1.7)} \qquad \begin{aligned} (A^{(j)}, b^{(j)}) &= G_j P_j (A^{(j-1)}, b^{(j-1)}), \\ (R, c) &= G_{n-1} P_{n-1} G_{n-2} P_{n-2} \cdots G_1 P_1 (A, b), \end{aligned}$$

with permutation matrices P_j and nonsingular Frobenius matrices G_j of the form

$$\text{(4.1.8)} \qquad G_j = \begin{bmatrix} 1 & & & & & 0 \\ & \ddots & & & & \\ & & 1 & & & \\ & & -l_{j+1,j} & 1 & & \\ & & \vdots & & \ddots & \\ 0 & & -l_{n,j} & 0 & & 1 \end{bmatrix}.$$

In the jth elimination step $(A^{(j-1)}, b^{(j-1)}) \to (A^{(j)}, b^{(j)})$ the elements below the diagonal in the jth column are anihilated. In the implementation of this algorithm on a computer, the locations which were occupied by these elements can now be used for the storage of the important quantities l_{ij}, $i \geqslant j+1$, of G_j; that is, we work with a matrix of the form

$$T^{(j)} = \left[\begin{array}{cccc|cccc} r_{11} & r_{12} & \cdots & r_{1j} & r_{1,j+1} & \cdots & r_{1n} & c_1 \\ \lambda_{21} & r_{22} & \cdots & r_{2j} & \vdots & & \vdots & \vdots \\ \lambda_{31} & \lambda_{32} & & \vdots & & & & \\ \vdots & \vdots & & r_{jj} & r_{j,j+1} & \cdots & r_{j,n} & c_j \\ \hline \vdots & \vdots & & \lambda_{j+1,j} & a^{(j)}_{j+1,j+1} & \cdots & a^{(j)}_{j+1,n} & b^{(j)}_{j+1} \\ & & & \vdots & \vdots & & \vdots & \vdots \\ \lambda_{n1} & \lambda_{n2} & \cdots & \lambda_{nj} & a^{(j)}_{n,j+1} & \cdots & a^{(j)}_{n,n} & b^{(j)}_n \end{array}\right]$$

Here the subdiagonal elements $\lambda_{k+1,k}, \lambda_{k+2,k}, \ldots, \lambda_{nk}$ of the kth column are a certain permutation of the elements $l_{k+1,k}, \ldots, l_{n,k}$ of G_k in (4.1.8).

Based on this arrangement, the jth step $T^{(j-1)} \to T^{(j)}$, $j = 1, 2, \ldots, n-1$, can be described as follows (for simplicity the elements of $T^{(j-1)}$ are denoted by t_{ik}, and those of $T^{(j)}$ by t'_{ik}, $1 \leqslant i \leqslant n$, $1 \leqslant k \leqslant n+1$):

(a) *Partial pivot selection: Determine r so that*

$$|t_{rj}| = \max_{i \geqslant j} |t_{ij}|.$$

If $t_{rj} = 0$, *set* $T^{(j)} := T^{(j-1)}$; *A is singular; stop. Otherwise carry on with* (b).

(b) *Interchange rows r and j of* $T^{(j-1)}$, *and denote the result by* $\bar{T} = (\bar{t}_{ik})$.

(c) *Replace*

$$t'_{ij} := l_{ij} := \bar{t}_{ij}/\bar{t}_{jj} \quad \textit{for } i = j+1, j+2, \ldots, n,$$

$$t'_{ik} := \bar{t}_{ik} - l_{ij}\bar{t}_{jk} \quad \textit{for } i = j+1, \ldots, n \textit{ and } k = j+1, \ldots, n+1,$$

$$t'_{ik} := \bar{t}_{ik} \textit{ otherwise.}$$

We note that in (c) the important elements $l_{j+1,j}, \ldots, l_{nj}$ of G_j are stored in their natural order as $t'_{j+1,j}, \ldots, t'_{nj}$. This order may, however, be changed in the subsequent elimination steps $T^{(k)} \to T^{(k+1)}$, $k \geqslant j$, because in (b) the rows of the entire matrix $T^{(k)}$ are rearranged. This has the following effect: The lower triangular matrix L and the upper triangular matrix R,

$$L = \begin{bmatrix} 1 & & & 0 \\ t_{21} & \ddots & & \\ \vdots & & \ddots & \\ t_{n1} & \cdots & t_{n,n-1} & 1 \end{bmatrix}, \qquad R = \begin{bmatrix} t_{11} & \cdots & t_{1n} \\ & \ddots & \vdots \\ 0 & & t_{nn} \end{bmatrix},$$

which are contained in the final matrix $T^{(n-1)} = (t_{ik})$, provide a triangular decomposition of the matrix PA:

$$LR = PA. \tag{4.1.9}$$

In this decomposition P is the product of all of the permutations appearing in (4.1.7):

$$P = P_{n-1} P_{n-2} \dots P_1.$$

We will only show here that a triangular decomposition is produced if no row interchanges are necessary during the course of the elimination process, i.e., if $P_1 = \dots = P_{n-1} = P = I$. In this case,

$$L = \begin{bmatrix} 1 & & & 0 \\ l_{21} & \ddots & & \\ \vdots & & \ddots & \\ l_{n1} & \dots & l_{n,n-1} & 1 \end{bmatrix},$$

since in all of the minor steps (b) nothing is interchanged. Now, because of (4.1.7),

$$R = G_{n-1} \dots G_1 A;$$

therefore

$$G_1^{-1} \dots G_{n-1}^{-1} R = A. \tag{4.1.10}$$

Since

$$G_j^{-1} = \begin{bmatrix} 1 & & & & 0 \\ & \ddots & & & \\ & & 1 & & \\ & & l_{j+1,j} & \ddots & \\ & & \vdots & & \ddots \\ & & l_{nj} & & & 1 \end{bmatrix},$$

it is easily verified that

$$G_1^{-1} \dots G_{n-1}^{-1} = \begin{bmatrix} 1 & & & 0 \\ l_{21} & \ddots & & \\ \vdots & & \ddots & \\ l_{n1} & \dots & l_{n,n-1} & 1 \end{bmatrix} = L.$$

Then the assertion follows from (4.1.10).

EXAMPLE.

$$\begin{bmatrix} 3 & 1 & 6 \\ 2 & 1 & 3 \\ 1 & 1 & 1 \end{bmatrix} \begin{bmatrix} x_1 \\ x_2 \\ x_3 \end{bmatrix} = \begin{bmatrix} 2 \\ 7 \\ 4 \end{bmatrix},$$

$$\left[\begin{array}{ccc|c} \textcircled{3} & 1 & 6 & 2 \\ 2 & 1 & 3 & 7 \\ 1 & 1 & 1 & 4 \end{array}\right] \rightarrow \left[\begin{array}{ccc|c} 3 & 1 & 6 & 2 \\ \frac{2}{3} & \frac{1}{3} & -1 & \frac{17}{3} \\ \frac{1}{3} & \textcircled{\tfrac{2}{3}} & -1 & \frac{10}{3} \end{array}\right] \rightarrow \left[\begin{array}{ccc|c} 3 & 1 & 6 & 2 \\ \frac{1}{3} & \frac{2}{3} & -1 & \frac{10}{3} \\ \frac{2}{3} & \frac{1}{2} & \textcircled{-\tfrac{1}{2}} & 4 \end{array}\right].$$

The pivot elements are marked. The triangular equation system is

$$\begin{bmatrix} 3 & 1 & 6 \\ 0 & \frac{2}{3} & -1 \\ 0 & 0 & -\frac{1}{2} \end{bmatrix} \begin{bmatrix} x_1 \\ x_2 \\ x_3 \end{bmatrix} = \begin{bmatrix} 2 \\ \frac{10}{3} \\ 4 \end{bmatrix}.$$

Its solution is

$$\begin{aligned} x_3 &= -8, \\ x_2 &= \tfrac{3}{2}(\tfrac{10}{3} + x_3) = -7, \\ x_1 &= \tfrac{1}{3}(2 - x_2 - 6x_3) = 19. \end{aligned}$$

Further

$$P = \begin{bmatrix} 1 & 0 & 0 \\ 0 & 0 & 1 \\ 0 & 1 & 0 \end{bmatrix}, \qquad PA = \begin{bmatrix} 3 & 1 & 6 \\ 1 & 1 & 1 \\ 2 & 1 & 3 \end{bmatrix},$$

and the matrix PA has the triangular decomposition $PA = LR$ with

$$L = \begin{bmatrix} 1 & 0 & 0 \\ \frac{1}{3} & 1 & 0 \\ \frac{2}{3} & \frac{1}{2} & 1 \end{bmatrix}, \qquad R = \begin{bmatrix} 3 & 1 & 6 \\ 0 & \frac{2}{3} & -1 \\ 0 & 0 & -\frac{1}{2} \end{bmatrix}.$$

Triangular decompositions (4.1.9) are of great practical importance in solving systems of linear equations. If the decomposition (4.1.9) is known for a matrix A (that is, the matrices L, R, P are known), then the equation system

$$Ax = b$$

can be solved immediately with any right-hand side b; for it follows that

$$PAx = LRx = Pb,$$

from which x can be found by solving both of the triangular systems

$$Lu = Pb, \qquad Rx = u$$

(provided all $r_{ii} \neq 0$).

Thus, with the help of the Gaussian elimination algorithm, it can be shown constructively that each square nonsingular matrix A has a triangular decomposition of the form (4.1.9). However, not every such matrix A has a triangular decomposition in the more narrow sense $A = LR$, as the example

$$A = \begin{bmatrix} 0 & 1 \\ 1 & 0 \end{bmatrix}$$

shows. In general, the rows of A must be permuted appropriately at the outset.

The triangular decomposition (4.1.9) can be obtained directly without forming the intermediate matrices $T^{(j)}$. For simplicity, we will show this under the assumption that the rows of A do not have to be permuted in order for a triangular decomposition $A = LR$ to exist. The equations $A = LR$ are regarded as n^2 defining equations for the n^2 unknown quantities

$$\begin{aligned} &r_{ik}, \quad i \leqslant k, \\ &l_{ik}, \quad i \geqslant k \qquad (l_{ii} = 1); \end{aligned}$$

that is,

$$\text{(4.1.11)} \qquad a_{ik} = \sum_{j=1}^{\min(i,k)} l_{ij} r_{jk} \qquad (l_{ii} = 1).$$

The order in which the l_{ij}, r_{jk} are to be computed remains open. The following versions are common:

In the Crout method the $n \times n$ matrix $A = LR$ is partitioned as follows:

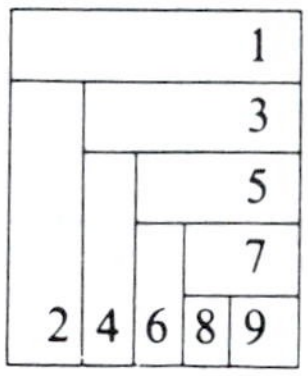

and the equations $A = LR$ are solved for L and R in an order indicated by this partitioning:

$$(1)\ a_{1i} = \sum_{j=1}^{1} l_{1j} r_{ji}, \quad r_{1i} := a_{1i}, \qquad i = 1, 2, \ldots, n$$

$$(2)\ a_{i1} = \sum_{j=1}^{1} l_{ij} r_{j1}, \quad l_{i1} := a_{i1} := a_{i1}/r_{11}, \quad i = 2, 3, \ldots, n$$

$$(3)\ a_{2i} = \sum_{j=1}^{2} l_{2j} r_{ji}, \quad r_{2i} := a_{2i} - l_{21} r_{1i}, \quad i = 2, 3, \ldots, n, \text{ etc.}$$

And in general, for $i = 1, 2, \ldots, n$,

$$(4.1.12)\qquad \begin{aligned} r_{ik} &:= a_{ik} - \sum_{j=1}^{i-1} l_{ij} r_{jk} && k = i, i+1, \ldots, n, \\ l_{ki} &:= \frac{a_{ki} - \sum_{j=1}^{i-1} l_{kj} r_{ji}}{r_{ii}} && k = i+1, i+2, \ldots, n. \end{aligned}$$

In all of the steps above $l_{ii} = 1$ for $i = 1, 2, \ldots, n$.

In the Banachiewicz method, the partitioning

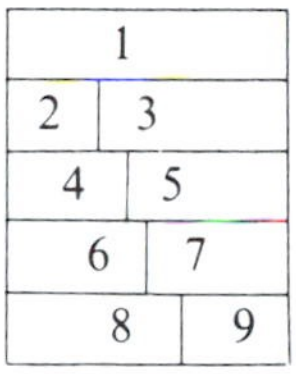

is used; that is, L and R are computed by rows.

The formulas above are valid only if no pivot selection is carried out. Triangular decomposition by the methods of Crout or Banachiewicz with pivot selection leads to more complicated algorithms; see Wilkinson (1965).

Gaussian elimination and direct triangular decomposition differ only in the ordering of operations. Both algorithms are, theoretically and numerically, entirely equivalent. Indeed, the jth partial sums

$$(4.1.13)\qquad a_{ik}^{(j)} := a_{ik} - \sum_{s=1}^{j} l_{is} r_{sk}$$

of (4.1.12) produce precisely the elements of the matrix $A^{(j)}$ in (4.1.6), as can easily be verified. In Gaussian elimination, therefore, the scalar products (4.1.12) are formed only in pieces, with temporary storing of the intermediate results; direct triangular decomposition, on the other hand, forms each scalar product as a whole. For these organizational reasons, direct triangular decomposition must be preferred if one chooses to accumulate the scalar products in double-precision arithmetic in order to reduce roundoff errors (without storing double-precision intermediate results). Further, these methods of triangular decomposition require about $n^3/3$ operations (1 operation = 1 multiplication + 1 addition). Thus, they also offer a simple way of evaluating the determinant of a matrix A: From (4.1.9) it follows, since $\det(P) = \pm 1$, $\det(L) = 1$, that

$$\det(PA) = \pm \det(A) = \det(R) = r_{11} r_{22} \ldots r_{nn}.$$

Up to its sign, $\det(A)$ is exactly the product of the pivot elements. (It should be noted that the direct evaluation of the formula

$$\det(A) = \sum_{\substack{\mu_1, \ldots, \mu_n = 1 \\ \mu_i \neq \mu_k \text{ for } i \neq k}}^{n} \operatorname{sign}(\mu_1, \ldots, \mu_n)\, a_{1\mu_1} a_{2\mu_2} \ldots a_{n\mu_n}$$

requires $n! \gg n^3/3$ operations.)

In the case that $P = I$, the pivot elements r_{ii} are representable as quotients of the determinants of the principal minors of A. If, in the representation $LR = A$, the matrices are partitioned as follows:

$$\begin{bmatrix} L_{11} & 0 \\ L_{21} & L_{22} \end{bmatrix} \begin{bmatrix} R_{11} & R_{12} \\ 0 & R_{22} \end{bmatrix} = \begin{bmatrix} A_{11} & A_{21} \\ A_{12} & A_{22} \end{bmatrix},$$

it is found that $L_{11} R_{11} = A_{11}$; hence $\det(R_{11}) = \det(A_{11})$, or

$$r_{11} \ldots r_{ii} = \det(A_{11}),$$

where A_{11} is an $i \times i$ matrix. In general, if A_i denotes the ith principal minor of A, then

$$r_{ii} = \det(A_i)/\det(A_{i-1}), \qquad i \geqslant 2,$$
$$r_{11} = \det(A_1).$$

A further practical and important property of the method of triangular decomposition is that, for band matrices with bandwidth m,

$$A = \left[\begin{array}{ccccc} * & \cdots & * & & \\ \vdots & \ddots & & \ddots & \\ * & & \ddots & & * \\ & \ddots & & \ddots & \vdots \\ & & * & \cdots & * \end{array}\right] \begin{matrix} \\ \\ \\ \left.\right\} m \end{matrix}, \qquad a_{ij} = 0 \quad \text{for } |i - j| \geqslant m,$$

$$\underbrace{\qquad\qquad}_{m}$$

the matrices L and R of the decomposition $LR = PA$ of A are not full: R is a band matrix with bandwidth $2m - 1$,

$$R = \left[\begin{array}{cccc} * & \cdots & * & 0 \\ & \ddots & & \ddots \\ & & \ddots & * \\ & & & \vdots \\ 0 & & & * \end{array}\right] \left.\right\} 2m - 1,$$

and in each column of L there are at most m elements different from zero. In contrast, the inverses A^{-1} of band matrices are usually filled with nonzero entries.

Thus, if $m \ll n$, using the triangular decomposition of A to solve $Ax = b$ results in a considerable saving in computation and storage over using A^{-1}. Additional savings are possible by making use of the symmetry of A if A is a positive definite matrix (see Sections 4.3 and 4.A).

4.2 The Gauss–Jordan Algorithm

In practice, the inverse A^{-1} of a nonsingular $n \times n$ matrix A is not frequently needed. Should a particular situation call for an inverse, however, it may be readily calculated using the triangular decomposition described in Section 4.1 or using the Gauss–Jordan algorithm, which will be described below. Both methods require the same amount of work.

If the triangular decomposition $PA = LR$ of (4.1.9) is available, then the ith column $\bar{a}_i$ of A^{-1} is obtained as the solution of the system

$$LR\bar{a}_i = Pe_i, \tag{4.2.1}$$

where e_i is the ith coordinate vector. If the simple structure of the right-hand side of (4.2.1), Pe_i, is taken into account, then the n equation systems (4.2.1) $(i = 1, \ldots, n)$ can be solved in about $\frac{2}{3}n^3$ operations. Adding the cost of producing the decomposition gives a total of n^3 operations to determine A^{-1}. The Gauss–Jordan method requires this amount of work, too, and offers advantages only of an organizational nature. The Gauss–Jordan method is obtained if one attempts to invert the mapping $x \to Ax = y$, $x \in \mathbb{R}^n$, $y \in \mathbb{R}^n$ determined by A in a systematic manner. Consider the system $Ax = y$:

$$\begin{aligned} a_{11}x_1 + \cdots + a_{1n}x_n &= y_1, \\ &\;\vdots \\ a_{n1}x_1 + \cdots + a_{nn}x_n &= y_n. \end{aligned} \tag{4.2.2}$$

In the first step of the Gauss–Jordan method, the variable x_1 is exchanged for one of the variables y_r. To do this, an $a_{r1} \neq 0$ is found, for example (*partial pivot selection*)

$$|a_{r1}| = \max_i |a_{i1}|,$$

and equations r and 1 of (4.2.2) are interchanged. In this way, a system

$$\begin{aligned} \bar{a}_{11}x_1 + \cdots + \bar{a}_{1n}x_n &= \bar{y}_1, \\ &\;\vdots \\ \bar{a}_{n1}x_1 + \cdots + \bar{a}_{nn}x_n &= \bar{y}_n \end{aligned} \tag{4.2.3}$$

is obtained in which the variables $\bar{y}_1, \ldots, \bar{y}_n$ are a permutation of $y_1, \ldots, y_n$ and $\bar{a}_{11} = a_{r1}$, $\bar{y}_1 = y_r$ holds. Now, $\bar{a}_{11} \neq 0$, for otherwise we would have $a_{i1} = 0$ for all i, making A singular, contrary to assumption. By solving the first equation of (4.2.3) for x_1 and substituting the result into the remaining equations, the system

(4.2.4)
$$\begin{aligned} a'_{11}\bar{y}_1 + a'_{12}x_2 + \cdots + a'_{1n}x_n &= x_1,\\ a'_{21}\bar{y}_1 + a'_{22}x_2 + \cdots + a'_{2n}x_n &= \bar{y}_2,\\ &\vdots\\ a'_{n1}\bar{y}_1 + a'_{n2}x_2 + \cdots + a'_{nn}x_n &= \bar{y}_n \end{aligned}$$

is obtained with

(4.2.5)
$$\begin{aligned} a'_{11} &:= \frac{1}{\bar{a}_{11}}, \quad a'_{1k} := -\frac{\bar{a}_{1k}}{\bar{a}_{11}},\\ a'_{i1} &:= \frac{\bar{a}_{i1}}{\bar{a}_{11}}, \quad a'_{ik} := \bar{a}_{ik} - \frac{\bar{a}_{i1}\bar{a}_{1k}}{\bar{a}_{11}} \quad \text{for } i, k = 2, 3, \ldots, n. \end{aligned}$$

In the next step, the variable x_2 is exchanged for one of the variables $\bar{y}_2, \ldots, \bar{y}_n$; then x_3 is exchanged for one of the remaining y variables, and so on. If the successive equation systems are represented by their matrices, then starting from $A^{(0)} := A$, a sequence

$$A^{(0)} \to A^{(1)} \to \cdots \to A^{(n)}$$

is obtained. The matrix $A^{(j)} = (a_{ik}^{(j)})$ stands for a "mixed equation system" of the form

(4.2.6)
$$\begin{aligned} a_{11}^{(j)}\tilde{y}_1 + \cdots + a_{1j}^{(j)}\tilde{y}_j + a_{1,j+1}^{(j)}x_{j+1} + \cdots + a_{1n}^{(j)}x_n &= x_1,\\ &\vdots\\ a_{j1}^{(j)}\tilde{y}_1 + \cdots + a_{jj}^{(j)}\tilde{y}_j + a_{j,j+1}^{(j)}x_{j+1} + \cdots + a_{jn}^{(j)}x_n &= x_j,\\ a_{j+1,1}^{(j)}\tilde{y}_1 + \cdots + a_{j+1,j}^{(j)}\tilde{y}_j + a_{j+1,j+1}^{(j)}x_{j+1} + \cdots + a_{j+1,n}^{(j)}x_n &= \tilde{y}_{j+1},\\ &\vdots\\ a_{n1}^{(j)}\tilde{y}_1 + \cdots + a_{nj}^{(j)}\tilde{y}_j + a_{n,j+1}^{(j)}x_{j+1} + \cdots + a_{nn}^{(j)}x_n &= \tilde{y}_n. \end{aligned}$$

In this system $(\tilde{y}_1, \ldots, \tilde{y}_j, \tilde{y}_{j+1}, \ldots, \tilde{y}_n)$ is a certain permutation of the original variables $(y_1, \ldots, y_n)$. In the transition $A^{(j-1)} \to A^{(j)}$ the variable x_j is exchanged for $\tilde{y}_j$. Thus, $A^{(j)}$ is obtained from $A^{(j-1)}$ according to the rules given below. For simplicity, the elements of $A^{(j-1)}$ are denoted by a_{ik}, and those of $A^{(j)}$ are denoted by a'_{ik}.

(4.2.7)

(a) *Partial pivot selection: Determine r so that*

$$|a_{rj}| = \max_{i \geq j} |a_{ij}|.$$

If $a_{rj} = 0$, the matrix is singular. Stop.

(b) *Interchange rows r and j of* $A^{(j-1)}$, *and call the result* $\bar{A} = (\bar{a}_{ik})$.
(c) *Compute* $A^{(j)} = (a'_{ik})$ *according to the formulas* [*compare with* (4.2.5)]

$$a'_{jj} := 1/\bar{a}_{jj},$$

$$a'_{jk} = -\frac{\bar{a}_{jk}}{\bar{a}_{jj}}, \quad a'_{ij} = \frac{\bar{a}_{ij}}{\bar{a}_{jj}} \quad \textit{for} \quad i, k \neq j,$$

$$a'_{ik} = \bar{a}_{ik} - \frac{\bar{a}_{ij}\bar{a}_{jk}}{\bar{a}_{jj}}.$$

(4.2.6) implies that

(4.2.8) $$A^{(n)}\hat{y} = x, \qquad \hat{y} = (\hat{y}_1, \ldots, \hat{y}_n)^T,$$

where $\hat{y}_1, \ldots, \hat{y}_n$ is a certain permutation of the original variables $y_1, \ldots, y_n$, $\hat{y} = Py$ which, since it corresponds to the interchange step (4.2.7b), can easily be determined. From (4.2.8) it follows that

$$(A^{(n)}P)y = x,$$

and therefore, since $Ax = y$,

$$A^{-1} = A^{(n)}P.$$

EXAMPLE.

$$A = A^{(0)} := \begin{bmatrix} \textcircled{1} & 1 & 1 \\ 1 & 2 & 3 \\ 1 & 3 & 6 \end{bmatrix} \to A^{(1)} = \begin{bmatrix} 1 & -1 & -1 \\ 1 & \textcircled{1} & 2 \\ 1 & 2 & 5 \end{bmatrix} \to A^{(2)} = \begin{bmatrix} 2 & -1 & 1 \\ -1 & 1 & -2 \\ -1 & 2 & \textcircled{1} \end{bmatrix}$$

$$\to A^{(3)} = \begin{bmatrix} 3 & -3 & 1 \\ -3 & 5 & -2 \\ 1 & -2 & 1 \end{bmatrix} = A^{-1}.$$

The pivot elements are marked.

The following ALGOL program is a formulation of the Gauss–Jordan method with partial pivoting. The inverse of the $n \times n$ matrix A is stored back into A. The array $p[i]$ serves to store the information about the row permutations which take place.

```
for j := 1 step 1 until n do p[j] := j;
for j := 1 step 1 until n do
begin
pivotsearch:
    max := abs (a[j, j]); r := j;
     for i := j + 1 step 1 until n do
          if abs (a[i, j]) greater max then
               begin max := abs (a[i, j]);
                       r := i
               end;
```

```
        if max = 0 then goto singular;
        rowinterchange:
        if r > j then
              begin for k := 1 step 1 until n do
                        begin
                              hr := a[j, k]; a[j, k] := a[r, k];
                              a[r, k] := hr
                        end;
                        hi := p[j]; p[j] := p[r]; p[r] := hi
              end;
        transformation:
        hr := 1/a[j, j];
        for i := 1 step 1 until n do
              a[i, j] := hr × a[i, j];
        a[j, j] := hr;
        for k := 1 step 1 until j − 1, j + 1 step 1 until n do
        begin
              for i := 1 step 1 until j − 1, j + 1 step 1 until n do
              a[i, k] := a[i, k] − a[i, j] × a[j, k];
              a[j, k] := −hr × a[j, k]
        end k
  end j;
  columninterchange:
  for i := 1 step 1 until n do
  begin
        for k := 1 step 1 until n do hv[p[k]] := a[i, k];
        for k := 1 step 1 until n do a[i, k] := hv[k]
  end;
```

4.3 The Cholesky Decomposition

The methods discussed so far for solving equations can fail if no pivot selection is carried out, i.e. if we restrict ourselves to taking the diagonal elements in order as pivots. Even if no failure occurs, as we will show in the next sections, pivot selection is advisable in the interest of numerical stability. However, there is an important class of matrices for which no pivot selection is necessary in computing triangular factors: the choice of each diagonal element in order always yields a nonzero pivot element. Furthermore, it is numerically stable to use these pivots. We refer to the class of positive definite matrices.

(4.3.1) Definition. A (complex) $n \times n$ matrix A is said to be *positive definite* if it satisfies:

(a) $A = A^H$, i.e., A is a Hermitian matrix.
(b) $x^H A x > 0$ for all $x \in \mathbb{C}^n$, $x \neq 0$.

$A = A^H$ is called *positive semidefinite* if $x^H A x \geqslant 0$ holds for all $x \in \mathbb{C}^n$.

(4.3.2) Theorem. *For any positive definite matrix A the matrix A^{-1} exists and is also positive definite. All principal submatrices of a positive definite matrix are also positive definite, and all principal minors of a positive definite matrix are positive.*

PROOF. The inverse of a positive definite matrix A exists: If this were not the case, an $x \neq 0$ would exist with $Ax = 0$ and $x^H A x = 0$, in contradiction to the definiteness of A. A^{-1} is positive definite: We have $(A^{-1})^H = (A^H)^{-1} = A^{-1}$, and if $y \neq 0$ it follows that $x = A^{-1} y \neq 0$. Hence $y^H A^{-1} y = x^H A^H A^{-1} A x = x^H A x > 0$. Every principal submatrix

$$\tilde{A} = \begin{bmatrix} a_{i_1 i_1}, & \cdots & a_{i_1 i_k} \\ \vdots & & \vdots \\ a_{i_k i_1}, & \cdots & a_{i_k i_k} \end{bmatrix}$$

of a positive definite matrix A is also positive definite: Obviously $\tilde{A}^H = \tilde{A}$. Moreover, every

$$\tilde{x} = \begin{bmatrix} \tilde{x}_1 \\ \vdots \\ \tilde{x}_k \end{bmatrix} \in \mathbb{C}^k, \qquad \tilde{x} \neq 0,$$

can be expanded to

$$x = \begin{bmatrix} x_1 \\ \vdots \\ x_n \end{bmatrix} \in \mathbb{C}^n, \qquad x \neq 0, \quad x_\mu := \begin{cases} \tilde{x}_j & \text{for } \mu = i_j,\ j = 1, \ldots, k, \\ 0 & \text{otherwise,} \end{cases}$$

and it follows that

$$\tilde{x}^H \tilde{A} \tilde{x} = x^H A x > 0.$$

In order to complete the proof of (4.3.2), then, it suffices to show that $\det(A) > 0$ for positive definite A. This is shown by using induction on n.

For $n = 1$ this is true from (4.3.1b). Now assume that the theorem is true for positive definite matrices of order $n - 1$, and let A be a positive definite matrix of order n. According to the preceeding parts of the proof,

$$A^{-1} = \begin{bmatrix} \alpha_{11} & \cdots & \alpha_{1n} \\ \vdots & & \vdots \\ \alpha_{n1} & \cdots & \alpha_{nn} \end{bmatrix}$$

is positive definite, and consequently $\alpha_{11} > 0$. As is well known,

$$\alpha_{11} = \det\left(\begin{bmatrix} a_{22} & \cdots & a_{2n} \\ \vdots & & \vdots \\ a_{n2} & \cdots & a_{nn} \end{bmatrix}\right) \bigg/ \det(A).$$

By the induction assumption, however,

$$\det\left(\begin{bmatrix} a_{22} & \dots & a_{2n} \\ \vdots & & \vdots \\ a_{n2} & \dots & a_{nn} \end{bmatrix}\right) > 0,$$

and hence $\det(A) > 0$ follows from $\alpha_{11} > 0$. □

(4.3.3) Theorem. *For each $n \times n$ positive definite matrix A there is a unique $n \times n$ lower triangular matrix L ($l_{ik} = 0$ for $k > i$) with $l_{ii} > 0$, $i = 1, 2, \dots, n$, satisfying $A = LL^H$. If A is real, so is L.*

(Note that $l_{ii} = 1$ is *not* required.)

PROOF. The theorem is established by induction on n. For $n = 1$ the theorem is trivial: A positive definite 1×1 matrix $A = (\alpha)$ is a positive number $\alpha > 0$, which can be written uniquely in the form

$$\alpha = l_{11}\bar{l}_{11}, \qquad l_{11} = +\sqrt{\alpha}.$$

Assume that the theorem is true for positive definite matrices of order $n - 1$. An $n \times n$ positive definite matrix A can be partitioned into

$$A = \begin{bmatrix} A_{n-1} & b \\ b^H & a_{nn} \end{bmatrix},$$

where $b \in \mathbb{C}^{n-1}$ and A_{n-1} is a positive definite matrix of order $n - 1$ by (4.3.2). By the induction hypothesis, there is a unique matrix L_{n-1} of order $n - 1$ satisfying

$$A_{n-1} = L_{n-1}L_{n-1}^H, \qquad l_{ik} = 0 \qquad \text{for } k > i, \quad l_{ii} > 0.$$

We consider a matrix L of the form

$$L = \begin{bmatrix} L_{n-1} & 0 \\ c^H & \alpha \end{bmatrix}$$

and try to determine $c \in \mathbb{C}^{n-1}$, $\alpha > 0$ so that

$$\begin{bmatrix} L_{n-1} & 0 \\ c^H & \alpha \end{bmatrix}\begin{bmatrix} L_{n-1}^H & c \\ 0 & \alpha \end{bmatrix} = \begin{bmatrix} A_{n-1} & b \\ b^H & a_{nn} \end{bmatrix} = A. \tag{4.3.4}$$

This means that we must have

$$L_{n-1}c = b,$$

$$c^Hc + \alpha^2 = a_{nn}, \qquad \alpha > 0.$$

The first equation must have a unique solution $c = L_{n-1}^{-1}b$, since L_{n-1}, as a triangular matrix with positive diagonal entries, has $\det(L_{n-1}) > 0$. As for the second equation, if $c^Hc \geqslant a_{nn}$ (that is, $\alpha^2 \leqslant 0$), then from (4.3.1) we would have a contradiction with $\alpha^2 > 0$, which follows from

$$\det(A) = |\det(L_{n-1})|^2\alpha^2,$$

$\det(A) > 0$ (4.3.2), and $\det(L_{n-1}) > 0$. Therefore, from (4.3.4), there exists exactly one $\alpha > 0$ giving $LL^H = A$, namely

$$\alpha = +\sqrt{a_{nn} - c^H c}. \qquad \square$$

The decomposition $A = LL^H$ can be determined in a manner similar to the methods given in Section 4.1. If it is assumed that all l_{ij} are known for $j \leqslant k - 1$, then as defining equations for l_{kk} and l_{ik}, $i \geqslant k + 1$, we have

$$\text{(4.3.5)} \qquad \begin{aligned} a_{kk} &= |l_{k1}|^2 + |l_{k2}|^2 + \cdots + |l_{kk}|^2, \qquad l_{kk} > 0, \\ a_{ik} &= l_{i1}\bar{l}_{k1} + l_{i2}\bar{l}_{k2} + \cdots + l_{ik}\bar{l}_{kk}. \end{aligned}$$

from $A = LL^H$.

For a real A, the following algorithm results:

```
for i := 1 step 1 until n do
for j := i step 1 until n do
begin x := a[i, j];
      for k := i - 1 step - 1 until 1 do
            x := x - a[j, k] × a[i, k];
      if i = j then begin
            if x ⩽ 0 then goto fail;
            p[i] := 1/sqrt (x)
      end else
      a[j, i] := x × p[i]
end i, j;
```

Note that only the upper triangular portion of A is used. The lower triangular matrix L is stored in the lower triangular portion of A, with the exception of the diagonal elements of L, whose reciprocals are stored in p.

This method is due to Cholesky. During the course of the computation, n square roots must be taken. Theorem (4.3.3) assures us that the arguments of these square roots will be positive. About $n^3/6$ operations (multiplications and additions) are needed beyond the n square roots. Further substantial savings are possible for sparse matrices, see Section 4.A. Finally, note as an important implication of (4.3.5) that

$$\text{(4.3.6)} \qquad |l_{kj}| \leqslant \sqrt{a_{kk}}, \qquad j = 1, \ldots, k, \quad k = 1, \ldots, n.$$

That is, the elements of L cannot grow too large.

4.4 Error Bounds

If any one of the methods described in the previous sections is used to determine the solution of a linear equation system $Ax = b$, then in general only an approximation $\tilde{x}$ to the true solution x is obtained, and there arises the question of how the accuracy of $\tilde{x}$ is judged. In order to measure the error

$$x - \tilde{x}$$

we have to have the means of measuring the "size" of a vector. To do this, a

$$\text{(4.4.1)} \qquad \textit{norm}: \quad \|x\|$$

is introduced on $\mathbb{C}^n$; that is, a function

$$\|\cdot\| : \mathbb{C}^n \to \mathbb{R},$$

which assigns to each vector $x \in \mathbb{C}^n$ a real value $\|x\|$ serving as a measure for the "size" of x. The function must have the following properties:

(4.4.2)

(a) $\|x\| > 0$ for all $x \in \mathbb{C}^n$, $x \neq 0$ (*positivity*),
(b) $\|\alpha x\| = |\alpha| \, \|x\|$ for all $\alpha \in \mathbb{C}$, $x \in \mathbb{C}^n$ (*homogeneity*),
(c) $\|x + y\| \leq \|x\| + \|y\|$ for all $x, y \in \mathbb{C}^n$ (*triangle inequality*).

In the following we use only the norms

$$\text{(4.4.3)} \qquad \begin{aligned} \|x\|_2 &:= \sqrt{x^H x} = \sqrt{\sum_{i=1}^{n} |x_i|^2} \qquad &(\textit{Euclidian norm}),\\ \|x\|_\infty &:= \max_i |x_i| \qquad &(\textit{maximum norm}). \end{aligned}$$

The norm properties (a), (b), (c) are easily verified.

For each norm $\|\cdot\|$ the inequality

$$\text{(4.4.4)} \qquad \|x - y\| \geqslant \big| \|x\| - \|y\| \big| \quad \text{for all } x, y \in \mathbb{C}^n$$

holds. From (4.4.2c) it follows that

$$\|x\| = \|(x - y) + y\| \leqslant \|x - y\| + \|y\|,$$

and consequently $\|x - y\| \geqslant \|x\| - \|y\|$. By interchanging the roles of x and y and using (4.4.2b), it follows that

$$\|x - y\| = \|y - x\| \geqslant \|y\| - \|x\|,$$

and hence (4.4.4).

It is easy to establish the following:

(4.4.5) Theorem. *Each norm $\|\cdot\|$ on $\mathbb{R}^n$ (or $\mathbb{C}^n$) is a uniformly continuous function with respect to the metric $\rho(x, y) = \max_i |x_i - y_i|$ on $\mathbb{R}^n$ ($\mathbb{C}^n$).*

PROOF. From (4.4.4) it follows that

$$\big| \|x + h\| - \|x\| \big| \leqslant \|h\|.$$

Now $h = \sum_{i=1}^{n} h_i e_i$, where $h = (h_1, \ldots, h_n)^T$, and e_i are the usual coordinate (unit) vectors of $\mathbb{R}^n(\mathbb{C}^n)$. Therefore

$$\|h\| \leq \sum_{i=1}^{n} |h_i| \, \|e_i\| \leqslant \max_i |h_i| \sum_{j=1}^{n} \|e_j\| = M \max_i |h_i|$$

with $M := \sum_{j=1}^{n} \|e_j\|$. Hence, for each $\varepsilon > 0$ and all h satisfying $\max_i |h_i| \leqslant \varepsilon/M$, the inequality

$$\big| \|x + h\| - \|x\| \big| \leqslant \varepsilon$$

holds. That is, $\|\cdot\|$ is uniformly continuous. □

This result is used to show:

(4.4.6) Theorem. *All norms on $\mathbb{R}^n(\mathbb{C}^n)$ are equivalent in the following sense: For each pair of norms $p_1(x)$, $p_2(x)$ there are positive constants m and M satisfying*

$$m p_2(x) \leqslant p_1(x) \leqslant M p_2(x) \quad \textit{for all } x.$$

PROOF. We will prove this only in the case that $p_2(x) := \|x\| := \max_i |x_i|$. The general case follows easily from this special result. The set

$$S = \left\{ x \in \mathbb{C}^n \,\middle|\, \max_i |x_i| = 1 \right\}$$

is a compact set in $\mathbb{C}^n$. Since $p_1(x)$ is continuous by (4.4.5), $\max_{x \in S} p_1(x) = M > 0$ and $\min_{x \in S} p_1(x) = m > 0$ exist. Thus, for all $y \neq 0$, since $y/\|y\| \in S$, it follows that

$$m \leqslant p_1\left(\frac{y}{\|y\|}\right) = \frac{1}{\|y\|} p_1(y) \leqslant M,$$

and therefore $m\|y\| \leqslant p_1(y) \leqslant M\|y\|$. □

For matrices as well, $A \in M(m, n)$ of fixed dimensions, norms $\|A\|$ can be introduced. In analogy to (4.4.2), the properties

$$\|A\| > 0 \quad \text{for all } A \neq 0,\ A \in M(m, n),$$

$$\|\alpha A\| = |\alpha| \, \|A\|,$$

$$\|A + B\| \leqslant \|A\| + \|B\|$$

are required. The matrix norm $\|\cdot\|$ is said to be *consistent* with the vector norms $\|\cdot\|_a$ on $\mathbb{C}^n$ and $\|\cdot\|_b$ on $\mathbb{C}^m$ if

$$\|Ax\|_b \leqslant \|A\| \, \|x\|_a \quad \text{for all } x \in \mathbb{C}^n,\ A \in M(m, n).$$

A matrix norm $\|\cdot\|$ for square matrices $A \in M(n, n)$ is called *submultiplicative* if

$$\|AB\| \leqslant \|A\| \, \|B\| \quad \text{for all } A, B \in M(n, n).$$

Frequently used matrix norms are

(4.4.7a) $$\|A\| = \max_i \sum_{k=1}^{n} |a_{ik}| \qquad \text{(row-sum norm)},$$

(4.4.7b) $$\|A\| = \Bigg(\sum_{i,k=1}^{n} |a_{ik}|^2\Bigg)^{1/2} \qquad \text{(Schur norm)},$$

(4.4.7c) $$\|A\| = \max_{i,k} |a_{ik}|.$$

(a) and (b) are submultiplicative; (c) is not; (b) is consistent with the Euclidian vector norm. Given a vector norm $\|x\|$, a corresponding matrix norm for square matrices, the *subordinate matrix norm*, can be defined by

(4.4.8) $$\operatorname{lub}(A) := \max_{x \neq 0} \frac{\|Ax\|}{\|x\|}.$$

Such a matrix norm is consistent with the vector norm $\|\cdot\|$ used to define it:

(4.4.9) $$\|Ax\| \leqslant \operatorname{lub}(A)\, \|x\|.$$

Obviously $\operatorname{lub}(A)$ is the smallest of all of the matrix norms $\|A\|$ which are consistent with the vector norm $\|x\|$:

$$\|Ax\| \leqslant \|A\|\,\|x\| \quad \text{for all } x \quad \Rightarrow \quad \operatorname{lub}(A) \leqslant \|A\|.$$

Each subordinate norm $\operatorname{lub}(\cdot)$ is submultiplicative:

$$\begin{aligned}\operatorname{lub}(AB) = \max_{x \neq 0} \frac{\|ABx\|}{\|x\|} &= \max_{x \neq 0} \frac{\|A(Bx)\|}{\|Bx\|}\,\frac{\|Bx\|}{\|x\|} \\ &\leqslant \max_{y \neq 0} \frac{\|Ay\|}{\|y\|} \max_{x \neq 0} \frac{\|Bx\|}{\|x\|} = \operatorname{lub}(A)\operatorname{lub}(B),\end{aligned}$$

and furthermore $\operatorname{lub}(I) = \max_{x \neq 0} \|Ix\|/\|x\| = 1$.

(4.4.9) shows that $\operatorname{lub}(A)$ is the greatest magnification which a vector may attain under the mapping determined by A: It shows how much $\|Ax\|$, the norm of an image point, can exceed $\|x\|$, the norm of a source point.

EXAMPLE.

(a) For the maximum norm $\|x\|_\infty = \max_\nu |x_\nu|$ the subordinate matrix norm is the row-sum norm

$$\operatorname{lub}_\infty(A) = \max_{x \neq 0} \frac{\|Ax\|_\infty}{\|x\|_\infty} = \max_{x \neq 0} \left\{ \frac{\max_i \left| \sum_{k=1}^{n} a_{ik} x_k \right|}{\max_k |x_k|} \right\} = \max_i \sum_{k=1}^{n} |a_{ik}|.$$

(b) Associated with the Euclidian norm $\|x\|_2 = \sqrt{x^H x}$ we have the subordinate matrix norm

$$\operatorname{lub}_2(A) = \max_{x \neq 0} \sqrt{\frac{x^H A^H A x}{x^H x}} = \sqrt{\lambda_{\max}(A^H A)},$$

which is expressed in terms of the largest eigenvalue $\lambda_{\max}(A^H A)$ of the matrix $A^H A$. With regard to this matrix norm, we note that

$$\text{lub}(U) = 1 \tag{4.4.10}$$

for unitary matrices U, that is, for matrices defined by $U^H U = I$.

In the following we assume that $\|x\|$ is an arbitrary vector norm and $\|A\|$ is a consistent submultiplicative matrix norm. Specifically, we can always take the subordinate norm $\text{lub}(A)$ as $\|A\|$ if we want to obtain particularly good estimates in the results below. We shall show how norms can be used to bound the influence due to changes in A and b on the solution x to a linear equation system

$$Ax = b.$$

If the solution $x + \Delta x$ corresponds to the right-hand side $b + \Delta b$,

$$A(x + \Delta x) = b + \Delta b,$$

then the relation

$$\Delta x = A^{-1}\,\Delta b$$

follows from $A\,\Delta x = \Delta b$, as does the bound

$$\|\Delta x\| \leqslant \|A^{-1}\|\,\|\Delta b\|. \tag{4.4.11}$$

For the relative change $\|\Delta x\|/\|x\|$, the bound

$$\frac{\|\Delta x\|}{\|x\|} \leqslant \|A\|\,\|A^{-1}\|\frac{\|\Delta b\|}{\|b\|} = \text{cond}(A)\frac{\|\Delta b\|}{\|b\|} \tag{4.4.12}$$

follows from $\|b\| = \|Ax\| \leqslant \|A\|\,\|x\|$. In this estimate, $\text{cond}(A) := \|A\|\,\|A^{-1}\|$. For the special case that $\text{cond}(A) := \text{lub}(A)\,\text{lub}(A^{-1})$, this so-called *condition of* A is a measure of the sensitivity of the relative error in the solution to changes in the right-hand side b. Since $AA^{-1} = I$, $\text{cond}(A)$ satisfies

$$\text{lub}(I) = 1 \leqslant \text{lub}(A)\,\text{lub}(A^{-1}) \leqslant \|A\|\,\|A^{-1}\| = \text{cond}(A).$$

The relation (4.4.11) can be interpreted as follows: If $\tilde{x}$ is an approximate solution to $Ax = b$ with residual

$$r(\tilde{x}) = b - A\tilde{x} = A(x - \tilde{x}),$$

then $\tilde{x}$ is the exact solution of

$$A\tilde{x} = b - r(\tilde{x}),$$

and the estimate

$$\|\Delta x\| \leqslant \|A^{-1}\|\,\|r(\tilde{x})\| \tag{4.4.13}$$

must hold for the error $\Delta x = x - \tilde{x}$.

Next, in order to investigate the influence of changes in the matrix A upon the solution x of $Ax = b$, we establish the following

(4.4.14) Lemma. *If F is an $n \times n$ matrix with $\|F\| < 1$, then $(I + F)^{-1}$ exists and satisfies*

$$\|(I + F)^{-1}\| \leqslant \frac{1}{1 - \|F\|}.$$

PROOF. From (4.4.4) the inequality

$$\|(I + F)x\| = \|x + Fx\| \geqslant \|x\| - \|Fx\| \geqslant (1 - \|F\|)\|x\|$$

follows for all x. From $1 - \|F\| > 0$ it follows that $\|(I + F)x\| > 0$ if $x \neq 0$; that is, $(I + F)x = 0$ has only the trivial solution $x = 0$, and $I + F$ is nonsingular.

Using the abbreviation $C := (I + F)^{-1}$, it follows that

$$\begin{aligned} 1 = \|I\| = \|(I + F)C\| &= \|C + FC\| \\ &\geqslant \|C\| - \|C\| \, \|F\| \\ &= \|C\|(1 - \|F\|) > 0, \end{aligned}$$

from which we have the desired result

$$\|(I + F)^{-1}\| \leqslant \frac{1}{1 - \|F\|}. \qquad \square$$

We can now show:

(4.4.15) Theorem. *Let A be a nonsingular $n \times n$ matrix, $B = A(I + F)$, $\|F\| < 1$, and x and Δx be defined by $Ax = b$, $B(x + \Delta x) = b$. It follows that*

$$\frac{\|\Delta x\|}{\|x\|} \leqslant \frac{\|F\|}{1 - \|F\|},$$

as well as

$$\frac{\|\Delta x\|}{\|x\|} \leqslant \frac{\operatorname{cond}(A)}{1 - \operatorname{cond}(A)\dfrac{\|B - A\|}{\|A\|}} \frac{\|B - A\|}{\|A\|}$$

if $\operatorname{cond}(A) \cdot \|B - A\|/\|A\| < 1$.

PROOF. B^{-1} exists from (4.4.14), and

$$\Delta x = B^{-1}b - A^{-1}b = B^{-1}(A - B)A^{-1}b, \qquad x = A^{-1}b,$$

$$\begin{aligned} \frac{\|\Delta x\|}{\|x\|} &\leqslant \|B^{-1}(A - B)\| = \|-(I + F)^{-1}A^{-1}AF\| \\ &\leqslant \|(I + F)^{-1}\| \, \|F\| \leqslant \frac{\|F\|}{1 - \|F\|}. \end{aligned}$$

Since $F = A^{-1}(B - A)$ and $\|F\| \leqslant \|A^{-1}\| \; \|A\| \; \|B - A\|/\|A\|$, the rest of the theorem follows. □

According to Theorem (4.4.15), cond(A) also measures the sensitivity of the solution x of $Ax = b$ to changes in the matrix A.

If the relations

$$C = (I + F)^{-1} = B^{-1}A,$$
$$F = A^{-1}B - I$$

are taken into account, it follows from (4.4.14) that

$$\|B^{-1}A\| \leqslant \frac{1}{1 - \|I - A^{-1}B\|}.$$

By interchanging A and B, it follows immediately from $A^{-1} = A^{-1}BB^{-1}$ that

(4.4.16) $$\|A^{-1}\| \leqslant \|A^{-1}B\| \; \|B^{-1}\| \leqslant \frac{\|B^{-1}\|}{1 - \|I - B^{-1}A\|}.$$

In particular, the residual estimate (4.4.13) leads to the bound

(4.4.17) $$\|\tilde{x} - x\| \leqslant \frac{\|B^{-1}\|}{1 - \|I - B^{-1}A\|} \|r(\tilde{x})\|, \qquad r(\tilde{x}) = b - A\tilde{x},$$

where B^{-1} is an approximate inverse to A with $\|I - B^{-1}A\| < 1$.

The estimates obtained up to this point show the significance of the quantity cond(A) for determining the influence on the solution of changes in the given data. These estimates give bounds on the error $\tilde{x} - x$, but the evaluation of the bounds requires at least an approximate knowledge of the inverse A^{-1} to A. The estimates to be discussed next, due to Prager and Oettli (1964), are based upon another principle and do not require any knowledge of A^{-1}.

The results are obtained through the following considerations:

Usually the given data A_0, b_0 of an equation system $A_0 x = b_0$ are inexact, being tainted, for example, by measurement errors ΔA, Δb. Hence, it is reasonable to accept an approximate solution $\tilde{x}$ to the system $A_0 x = b_0$ as "correct" if $\tilde{x}$ is the exact solution to a "neighboring" equation system

$$A\tilde{x} = b$$

with

(4.4.18) $$A \in \mathfrak{A} := \{A \mid |A - A_0| \leqslant \Delta A\}$$
$$b \in \mathfrak{B} := \{b \mid |b - b_0| \leqslant \Delta b\}.$$

The notation used here is

$$|A| = (|\alpha_{ik}|), \qquad \text{where } A = (\alpha_{ik}),$$
$$|b| = (|\beta_1|, \ldots |\beta_n|)^T, \quad \text{where } b = (\beta_1, \ldots \beta_n)^T,$$

and the relation $\leqslant$ between vectors and matrices is to be understood as holding componentwise. Prager and Oettli prove:

(4.4.19) Theorem. *Let* $\Delta A \geqslant 0$, $\Delta b \geqslant 0$, *and let* $\mathfrak{A}$, $\mathfrak{B}$ *be defined by* (4.4.18). *Associated with any approximate solution* $\tilde{x}$ *of the system* $A_0 x = b_0$ *there is a matrix* $A \in \mathfrak{A}$ *and a vector* $b \in \mathfrak{B}$ *satisfying*

$$A\tilde{x} = b,$$

if and only if

$$|r(\tilde{x})| \leqslant \Delta A\,|\tilde{x}| + \Delta b,$$

where $r(\tilde{x}) := b_0 - A_0\tilde{x}$ *is the residual of* $\tilde{x}$.

PROOF.

(1) We assume first that

$$A\tilde{x} = b.$$

holds for some $A \in \mathfrak{A}$, $b \in \mathfrak{B}$. Then it follows from

$$\begin{aligned} A &= A_0 + \delta A, \quad \text{where } |\delta A| \leqslant \Delta A, \\ b &= b_0 + \delta b, \quad \text{where } |\delta b| \leqslant \Delta b, \end{aligned}$$

that

$$\begin{aligned} |r(\tilde{x})| = |b_0 - A_0\tilde{x}| &= |b - \delta b - (A - \delta A)\tilde{x}| \\ &= |-\delta b + (\delta A)\tilde{x}| \leqslant |\delta b| + |\delta A|\,|\tilde{x}| \\ &\leqslant \Delta b + \Delta A\,|\tilde{x}|. \end{aligned}$$

(2) On the other hand, if

$$|r(\tilde{x})| \leqslant \Delta b + \Delta A\,|\tilde{x}|, \tag{4.4.20}$$

and if r and s stand for the vectors

$$\begin{aligned} r &:= r(\tilde{x}) = (\rho_1, \ldots, \rho_n)^T, \\ s &:= \Delta b + \Delta A\,|\tilde{x}| \geqslant 0, \qquad s = (\sigma_1, \ldots, \sigma_n)^T, \end{aligned}$$

then set

$$\delta A = (\delta\alpha_{ij}), \qquad \delta b = \begin{bmatrix} \delta\beta_1 \\ \vdots \\ \delta\beta_n \end{bmatrix}, \qquad \tilde{x} = \begin{bmatrix} \tilde{\xi}_1 \\ \vdots \\ \tilde{\xi}_n \end{bmatrix},$$

$$\begin{aligned} \delta\alpha_{ij} &:= \rho_i\,\Delta\alpha_{ij}\,\operatorname{sign}(\tilde{\xi}_j)/\sigma_i, \\ \delta\beta_i &:= -\rho_i\,\Delta\beta_i/\sigma_i, \quad \text{where } \rho_i/\sigma_i := 0 \text{ if } \sigma_i = 0. \end{aligned}$$

From (4.4.20) it follows that $|\rho_i/\sigma_i| \leqslant 1$, and consequently

$$A = A_0 + \delta A \in \mathfrak{A}, \qquad b = b_0 + \delta b \in \mathfrak{B}$$

as well as the following for $i = 1, 2, \ldots, n$:

$$\rho_i = \beta_i - \sum_{j=1}^{n} \alpha_{ij}\xi_j = \left(\Delta\beta_i + \sum_{j=1}^{n} \Delta\alpha_{ij}|\xi_j|\right)\frac{\rho_i}{\sigma_i}$$

$$= -\delta\beta_i + \sum_{j=1}^{n} \delta\alpha_{ij}\xi_j,$$

or

$$\sum_{j=1}^{n} (\alpha_{ij} + \delta\alpha_{ij})\xi_j = \beta_i + \delta\beta_i,$$

that is,

$$A\tilde{x} = b,$$

which was to be shown. □

The criterion expressed in Theorem (4.4.19) permits us to draw conclusions about the fitness of a solution from the smallness of its residual. For example, if all components of A_0 and b_0 have the same relative accuracy ε,

$$\Delta A = \varepsilon|A_0|, \qquad \Delta b = \varepsilon|b_0|,$$

then (4.4.19) is satisfied if

$$|A_0\tilde{x} - b_0| \leqslant \varepsilon(|b_0| + |A_0|\,|\tilde{x}|).$$

From this inequality, the smallest ε can be computed for which a given $\tilde{x}$ can still be accepted as a useable solution.

4.5 Roundoff-Error Analysis for Gaussian Elimination

In the discussion of methods for solving linear equations, the pivot selection played only the following role: it guaranteed for any nonsingular matrix A that the algorithm would not terminate prematurely if some pivot element happened to vanish. We will now show that the numerical behavior of the equation-solving methods which have been covered depends upon the choice of pivots. To illustrate this, we consider the following simple example:

Solve the system

(4.5.1)
$$\begin{bmatrix} 0.005 & 1 \\ 1 & 1 \end{bmatrix}\begin{bmatrix} x \\ y \end{bmatrix} = \begin{bmatrix} 0.5 \\ 1 \end{bmatrix}$$

with the use of Gaussian elimination. The exact solution is $x = 5000/9950 = 0.503\ldots$, $y = 4950/9950 = 0.497\ldots$. If the element $a_{11} = 0.005$ is taken as the pivot in the first step, then we obtain

$$\begin{bmatrix} 0.005 & 1 \\ 0 & -200 \end{bmatrix}\begin{bmatrix} x \\ y \end{bmatrix} = \begin{bmatrix} 0.5 \\ -99 \end{bmatrix}, \qquad y = 0.5, \quad x = 0.$$

using 2-place floating-point arithmetic. If the element $a_{21} = 1$ is taken as the pivot, then 2-place floating-point arithmetic yields

$$\begin{bmatrix} 1 & 1 \\ 0 & 1 \end{bmatrix} \begin{bmatrix} x \\ y \end{bmatrix} = \begin{bmatrix} 1 \\ 0.5 \end{bmatrix}, \qquad y = 0.50, \quad x = 0.50.$$

In the second case, the accuracy of the result is considerably higher. This could lead to the impression that the largest element in magnitude should be chosen as a pivot from among the candidates in a column to get the best numerical results. However, a moment's thought shows that this cannot be unconditionally true. If the first row of the equation system (4.5.1) is multiplied by 200, for example, the result is the system

$$\text{(4.5.2)} \qquad \begin{bmatrix} 1 & 200 \\ 1 & 1 \end{bmatrix} \begin{bmatrix} x \\ y \end{bmatrix} = \begin{bmatrix} 100 \\ 1 \end{bmatrix},$$

which has the same solution as (4.5.1). The element $\tilde{a}_{11} = 1$ is now just as large as the element $\tilde{a}_{21} = 1$. However, the choice of $\tilde{a}_{11}$ as pivot element leads to the same inexact result as before. We have replaced the matrix A of (4.5.1) by $\tilde{A} = DA$, where D is the diagonal matrix

$$D = \begin{bmatrix} 200 & 0 \\ 0 & 1 \end{bmatrix}.$$

Obviously, we can also adjust the column norms of A—i.e., replace A by $\tilde{A} = AD$ (where D is a diagonal matrix)—without changing the solution x to $Ax = b$ in any essential way. If x is the solution to $Ax = b$, then $y = D^{-1}x$ is the solution of $\tilde{A}y = (AD)(D^{-1}x) = b$. In general, we refer to a *scaling* of a matrix A if A is replaced by $D_1 A D_2$, where D_1, D_2 are diagonal matrices. The example shows that it is not reasonable to propose a particular choice of pivots unless assumptions about the scaling of a matrix are made. Unfortunately, no one has yet determined satisfactorily how to carry out scaling so that partial pivot selection is numerically stable for any matrix A. Practical experience, however, suggests the following scaling for partial pivoting: Choose D_1 and D_2 so that

$$\sum_{k=1}^{n} |\tilde{a}_{ik}| \approx \sum_{j=1}^{n} |\tilde{a}_{jl}|$$

holds approximately for all $i, l = 1, 2, \ldots, n$ in the matrix $\tilde{A} = D_1 A D_2$. The sum of the absolute values of the elements in the rows (and the columns) of $\tilde{A}$ should all have about the same magnitude. Such matrices are said to be *equilibrated*. In general, it is quite difficult to determine D_1 and D_2 so that $D_1 A D_2$ is equilibrated. Usually we must get by with the following: Let $D_2 = I$, $D_1 = \operatorname{diag}(s_1, \ldots, s_n)$, where

$$s_i := \frac{1}{\sum_{k=1}^{n} |a_{ik}|}.$$

then for $\tilde{A} = D_1 A D_2$, it is true at least that

$$\sum_{k=1}^{n} |\tilde{a}_{ik}| = 1 \quad \text{for } i = 1, 2, \ldots, n.$$

Now, instead of replacing A by $\tilde{A}$, i.e., instead of actually carrying out the transformation, we replace the rule for pivot selection instead in order to avoid the explicit scaling of A. The pivot selection for the jth elimination step $A^{(j-1)} \to A^{(j)}$ is given by the following:

(4.5.3). *Determine $r \geq j$ so that*

$$|a_{rj}^{(j-1)}| s_r = \max_{i \geq j} |a_{ij}^{(j-1)}| s_i \neq 0,$$

and take $a_{rj}^{(j-1)}$ as the pivot.

The example above shows that it is not sufficient, in general, to scale A prior to carrying out partial pivot selection by making the largest element in absolute value within each row and each column have roughly the same magnitude:

$$\text{(4.5.4)} \qquad \max_k |a_{ik}| \approx \max_j |a_{jl}| \quad \text{for all } i, l = 1, 2, \ldots, n.$$

For, if the scaling matrices

$$D_1 = \begin{bmatrix} 200 & 0 \\ 0 & 1 \end{bmatrix}, \qquad D_2 = \begin{bmatrix} 1 & 0 \\ 0 & 0.005 \end{bmatrix}$$

are chosen, then the matrix A of our example (4.5.1) becomes

$$\tilde{A} = D_1 A D_2 = \begin{bmatrix} 1 & 1 \\ 1 & 0.005 \end{bmatrix}.$$

The condition (4.5.4) is satisfied, but inexact answers will be produced, as before, if $\tilde{a}_{11} = 1$ is used as a pivot.

We would like to make a detailed study of the effect of the rounding errors which occur in Gaussian elimination or direct triangular decomposition (see Section 4.1). We assume that the rows of the $n \times n$ matrix A are already so arranged that A has a triangular decomposition of the form $A = LR$. Hence, L and R can be determined using the formulas of (4.1.12). In fact, we only have to evaluate expressions of the form

$$b_n := \mathrm{fl}\left(\frac{c - a_1 b_1 - \cdots - a_{n-1} b_{n-1}}{a_n}\right),$$

which were analyzed in (1.4.4)–(1.4.11). Instead of the exact triangular decomposition $LR = A$, (4.1.12) shows that the use of floating-point arithmetic will result in matrices $\bar{L} = (\bar{l}_{ik})$, $\bar{R} = (\bar{r}_{ik})$ for which the residual $F := (f_{ik}) =$

$A - LR$ is not zero in general. Since, according to (4.1.13), the jth partial sums

$$\bar{a}_{ik}^{(j)} = \mathrm{fl}\left(a_{ik} - \sum_{s=1}^{j} \bar{l}_{is}\bar{r}_{sk}\right)$$

are exactly the elements of the matrix $\bar{A}^{(j)}$, which is produced instead of $A^{(j)}$ of (4.1.6) from the jth step of Gaussian elimination in floating-point arithmetic, the estimates (1.4.7) applied to (4.1.12) yield

(4.5.5)
$$|f_{ik}| = \left|a_{ik} - \sum_{j=1}^{i} \bar{l}_{ij}\bar{r}_{jk}\right| \leqslant \frac{\text{eps}}{1-\text{eps}} \sum_{j=1}^{i-1} (|\bar{a}_{ik}^{(j)}| + |\bar{l}_{ij}|\,|\bar{r}_{jk}|) \quad \text{for } k \geqslant i,$$
$$|f_{ki}| = \left|a_{ki} - \sum_{j=1}^{i} \bar{l}_{kj}\bar{r}_{ji}\right| \leqslant \frac{\text{eps}}{1-\text{eps}} \left[|\bar{a}_{ki}^{(i-1)}| + \sum_{j=1}^{i-1} (|\bar{a}_{ki}^{(j)}| + |\bar{l}_{kj}|\,|\bar{r}_{ji}|)\right] \quad \text{for } k > i.$$

Further,

(4.5.6)
$$\bar{r}_{ik} = \bar{a}_{ik}^{(i-1)} \quad \text{for } i \leqslant k,$$

since the first $j + 1$ rows of $\bar{A}^{(j)}$ [or $A^{(j)}$ of (4.1.6)] do not change in the subsequent elimination steps, and so they already agree with the corresponding rows of $\bar{R}$. We assume, in addition, that $|\bar{l}_{ik}| \leqslant 1$ for all i, k (which is satisfied, for example, when partial or complete pivoting is used). Setting

$$a_j := \max_{i,k} |\bar{a}_{ik}^{(j)}|, \qquad a := \max_{0<i<n-1} a_i,$$

it follows immediately from (4.5.5) and (4.5.6) that

(4.5.7)
$$|f_{ik}| \leqslant \frac{\text{eps}}{1-\text{eps}} (a_0 + 2a_1 + 2a_2 + \cdots + 2a_{i-2} + a_{i-1})$$
$$\leqslant 2(i-1)a \frac{\text{eps}}{1-\text{eps}} \quad \text{for } k \geqslant i,$$
$$|f_{ik}| \leqslant \frac{\text{eps}}{1-\text{eps}} (a_0 + 2a_1 + \cdots + 2a_{k-2} + 2a_{k-1})$$
$$\leqslant 2ka \frac{\text{eps}}{1-\text{eps}} \quad \text{for } k < i.$$

For the matrix F, then, the inequality

(4.5.8)
$$|F| \leqslant 2a \frac{\text{eps}}{1-\text{eps}} \begin{bmatrix} 0 & 0 & 0 & \ldots & 0 & 0 \\ 1 & 1 & 1 & \ldots & 1 & 1 \\ 1 & 2 & 2 & \ldots & 2 & 2 \\ 1 & 2 & 3 & \ldots & 3 & 3 \\ \vdots & \vdots & \vdots & & & \vdots \\ 1 & 2 & 3 & \ldots & n-1 & n-1 \end{bmatrix}$$

holds, where $|F| := (|f_{jk}|)$. If a has the same order of magnitude as a_0, that is, if the matrices $\bar{A}^{(j)}$ do not grow too much, then the computed matrices $\bar{L}\bar{R}$ form the exact triangular decomposition of the matrix $A - F$, which differs little from A. Gaussian elimination is stable in this case.

The value a can be estimated with the help of $a_0 = \max_{r,s} |a_{rs}|$. For partial pivot selection it can easily be shown that

$$a_{k-1} \leqslant 2^k a_0,$$

and hence that $a \leqslant 2^{n-1} a_0$. This bound is much too pessimistic in most cases; however, it can be attained, for example, in forming the triangular decomposition of the matrix

$$A = \begin{bmatrix} 1 & 0 & \cdots & 0 & 1 \\ -1 & 1 & \cdots & 0 & 1 \\ -1 & -1 & \cdots & 0 & 1 \\ \vdots & \vdots & & \vdots & \vdots \\ -1 & -1 & \cdots & 0 & 1 \\ -1 & -1 & \cdots & 1 & 1 \\ -1 & -1 & \cdots & -1 & 1 \end{bmatrix}.$$

Better estimates hold for special types of matrices. For example in the case of upper Hessenberg matrices, that is, matrices of the form

$$A = \begin{bmatrix} x & \cdots & \cdots & x \\ x & \ddots & & \vdots \\ & \ddots & \ddots & \vdots \\ 0 & & x & x \end{bmatrix},$$

the bound $a \leqslant (n-1)a_0$ can be shown. (Hessenberg matrices arise in eigenvalue problems.)

For tridiagonal matrices

$$A = \begin{bmatrix} \alpha_1 & \beta_2 & & & 0 \\ \gamma_2 & \ddots & \ddots & & \\ \vdots & \ddots & \ddots & \ddots & \\ \vdots & & \ddots & \ddots & \beta_n \\ 0 & & & \gamma_n & \alpha_n \end{bmatrix}$$

it can even be shown that

$$a = \max_k |a_k| \leqslant 2a_0$$

holds for partial pivot selection. Hence, Gaussian elimination is quite numerically stable in this case.

For complete pivot selection, Wilkinson (1965) has shown that

$$a_{k-1} \leqslant f(k) a_0$$

with the function

$$f(k) := k^{1/2} [2^1 \, 3^{1/2} \, 4^{1/3} \cdots k^{1/(k-1)}]^{1/2}.$$

This function grows relatively slowly with k:

k	10	20	50	100
$f(k)$	19	67	530	3300

Even this estimate is too pessimistic in practice. Up until now, no real matrix has been found which fails to satisfy

$$a_k \leqslant (k+1)a_0, \qquad k = 1, 2, \ldots, n-1,$$

when complete pivot selection is used. This indicates that Gaussian elimination with complete pivot selection is usually a stable process. Despite this, partial pivot selection is preferred in practice, for the most part, because:

(1) Complete pivot selection is more costly than partial pivot selection. (To compute $A^{(i)}$, the maximum from among $(n-i+1)^2$ elements must be determined instead of $n-i+1$ elements as in partial pivot selection.)
(2) Special structures in a matrix, i.e. the band structure of a tridiagonal matrix, are destroyed in complete pivot selection.

If the weaker estimates (1.4.11) are used instead of (1.4.7), then the following bounds replace those of (4.5.5) for the f_{ik}:

$$|f_{ik}| \leqslant \frac{\text{eps}}{1-n\cdot\text{eps}}\left[\sum_{j=1}^{i} j\,|\bar{l}_{ij}|\,|\bar{r}_{jk}| - |\bar{r}_{ik}|\right], \qquad k \geqslant i,$$

$$|f_{ki}| \leqslant \frac{\text{eps}}{1-n\cdot\text{eps}}\left[\sum_{j=1}^{i} j\,|\bar{l}_{kj}|\,|\bar{r}_{ji}|\right], \qquad k \geqslant i+1,$$

or

(4.5.9)
$$|F| \leqslant \frac{\text{eps}}{1-n\cdot\text{eps}}[|\bar{L}|D|\bar{R}| - |\bar{R}|],$$

where $D := \begin{bmatrix} 1 & & & 0 \\ & 2 & & \\ & & \ddots & \\ 0 & & & n \end{bmatrix}$.

4.6 Roundoff Errors in Solving Triangular Systems

As a result of applying Gaussian elimination in floating-point arithmetic to the matrix A, a lower triangular matrix $\bar{L}$ and an upper triangular matrix $\bar{R}$ are obtained whose product $\bar{L}\bar{R}$ approximately equals A. Solving the sytem $Ax = b$ is thereby reduced to solving the triangular systems

$$\bar{L}y = b, \qquad \bar{R}x = y.$$

In this section we investigate the influence of roundoff errors on the solution of such equation systems. If we use $\bar{y}$ to denote the solution obtained using t-digit floating-point arithmetic, then the definition of $\bar{y}$ gives

$$(4.6.1)\qquad \bar{y}_r = \text{fl}((-\bar{l}_{r1}\bar{y}_1 - \bar{l}_{r2}\bar{y}_2 - \cdots \bar{l}_{r,r-1}\bar{y}_{r-1} + b_r)/\bar{l}_{rr}).$$

From (1.4.10), (1.4.11) it follows immediately that

$$\left| b_r - \sum_{j=1}^{r} \bar{l}_{rj}\bar{y}_j \right| \leqslant \frac{\text{eps}}{1 - n \cdot \text{eps}} \left[\sum_{j=1}^{r} j|\bar{l}_{rj}|\,|\bar{y}_j| - |\bar{y}_r| \right],$$

or

$$(4.6.2)\quad |b - \bar{L}\bar{y}| \leqslant \frac{\text{eps}}{1 - n \cdot \text{eps}} (|\bar{L}|D - I)|\bar{y}|, \qquad D := \begin{bmatrix} 1 & & & 0 \\ & 2 & & \\ & & \ddots & \\ 0 & & & n \end{bmatrix}.$$

In other words, there exists a matrix $\Delta\bar{L}$ satisfying

$$(4.6.3)\qquad (\bar{L} + \Delta\bar{L})\bar{y} = b, \qquad |\Delta\bar{L}| \leqslant \frac{\text{eps}}{1 - n \cdot \text{eps}} (|\bar{L}|D - I).$$

Thus, the computed solution can be interpreted as the exact solution of a slightly changed problem, showing that the process of solving a triangular system is stable. Similarly, the computed solution $\bar{x}$ of $\bar{R}x = \bar{y}$ is found to satisfy the bound

$$(4.6.4)\qquad \begin{aligned} &|\bar{y} - \bar{R}\bar{x}| \leqslant \frac{\text{eps}}{1 - n \cdot \text{eps}} |\bar{R}|E|\bar{x}|, \qquad E := \begin{bmatrix} n & & & 0 \\ & \ddots & & \\ & & 2 & \\ 0 & & & 1 \end{bmatrix}, \\ &(\bar{R} + \Delta\bar{R})\bar{x} = \bar{y}, \qquad |\Delta\bar{R}| \leqslant \frac{\text{eps}}{1 - n \cdot \text{eps}} |\bar{R}|E. \end{aligned}$$

By combining the estimates (4.5.9), (4.6.3), and (4.6.4), we obtain the following result (due to Sautter (1971)) for the approximate solution $\bar{x}$ produced by floating-point arithmetic to the linear system $Ax = b$:

$$(4.6.5)\qquad |b - A\bar{x}| \leqslant \frac{2(n+1) \cdot \text{eps}}{1 - n \cdot \text{eps}} |\bar{L}|\,|\bar{R}|\,|\bar{x}| \quad \text{if } n \cdot \text{eps} \leqslant \tfrac{1}{2}.$$

PROOF. Using the abbreviation $\epsilon := \text{eps}/(1 - n \cdot \text{eps})$, it follows from (4.5.9), (4.6.3) and (4.6.4) that

$$\begin{aligned} |b - A\bar{x}| &= |b - (\bar{L}\bar{R} + F)\bar{x}| = |-F\bar{x} + b - \bar{L}(\bar{y} - \Delta\bar{R}\,\bar{x})| \\ &= |(-F + \Delta\bar{L}(\bar{R} + \Delta\bar{R}) + \bar{L}\,\Delta\bar{R})\bar{x}| \\ &\leqslant \epsilon[2(|\bar{L}|D - I)|\bar{R}| + |\bar{L}|\,|\bar{R}|E + \epsilon(|\bar{L}|D - I)|\bar{R}|E]|\bar{x}|. \end{aligned}$$

The (i, k) component of the matrix $[\ldots]$ appearing in the last line above has the form

$$\sum_{j=1}^{\min(i,k)} |\bar{l}_{ij}|(2j - 2\delta_{ij} + n + 1 - k + \epsilon(j - \delta_{ij} + n + 1 - k))|\bar{r}_{jk}|,$$

$$\delta_{ij} = \begin{cases} 1 & \text{for } i = j, \\ 0 & \text{for } i \neq j. \end{cases}$$

It is easily verified for all $j \leqslant \min(i, k)$, $1 \leqslant i, k \leqslant n$, that

$$\begin{aligned} 2j - 2\delta_{ij} + n + 1 - k + \epsilon(j - \delta_{ij} + n + 1 - k) \\ &\leqslant \begin{cases} 2n - 1 + \epsilon n & \text{if } j \leqslant i \leqslant k, \\ 2n + \epsilon(n+1) & \text{if } j \leqslant k < i, \end{cases} \\ &\leqslant 2n + 2, \end{aligned}$$

since $n \cdot \text{eps} \leqslant \frac{1}{2}$ implies $\epsilon n \leqslant 2n \cdot \text{eps} \leqslant 1$. This completes the proof of (4.6.5). □

A comparison of (4.6.5) with the result (4.4.19) due to Oettli and Prager (1964) shows, finally, that the computed solution $\bar{x}$ can be interpreted as the exact solution of a slightly changed equation system, provided that the matrix $n|\bar{L}||\bar{R}|$ has the same order of magnitude as $|A|$. In that case, computing the solution via Gaussian elimination is a numerically stable algorithm.

4.7 Orthogonalization Techniques of Householder and Gram–Schmidt

The methods discussed up to this point for solving a system of equations

$$Ax = b \tag{4.7.1}$$

consisted of multiplying (4.7.1) on the left by appropriate matrices P_j, $j = 1, \ldots, n$, so that the system obtained as the final outcome,

$$A^{(n)}x = b^{(n)},$$

could be solved directly. The sensitivity of the result x to changes in the arrays $A^{(j)}$, $b^{(j)}$ of the intermediate systems

$$A^{(j)}x = b^{(j)}, \qquad (A^{(j)}, b^{(j)}) = P_j(A^{(j-1)}, b^{(j-1)})$$

is given by

$$\text{cond}(A^{(j)}) = \text{lub}(A^{(j)})\, \text{lub}((A^{(j)})^{-1}).$$

If we denote the roundoff error incurred in the transition from $(A^{(j-1)}, b^{(j-1)})$ to $(A^{(j)}, b^{(j)})$ by $\varepsilon^{(j)}$, then these roundoff errors are amplified by the factors $\text{cond}(A^{(j)})$ in their effect on the final result x, and we have

$$\frac{\|\Delta x\|}{\|x\|} \leqslant \sum_{j=0}^{n-1} \varepsilon^{(j)} \text{cond}(A^{(j)}),$$

In the right-hand expression above, $\varepsilon^{(0)}$ stands for the error in the initial data A, b. If there is an $A^{(j)}$ with

$$\text{cond}(A^{(j)}) \gg \text{cond}(A^{(0)}),$$

then the sequence of computations is not numerically stable: The roundoff error $\varepsilon^{(j)}$ has a stronger influence on the final result than the initial error $\varepsilon^{(0)}$.

For this reason, it is important to choose the P_j so that the condition numbers $\text{cond}(A^{(j)})$ do not grow. For condition numbers derived from an arbitrary norm $\|x\|$, that is difficult to do. For the Euclidian norm

$$\|x\| = \sqrt{x^H x}$$

and its subordinate matrix norm

$$\text{lub}(A) = \max_{x \neq 0} \sqrt{\frac{x^H A^H A x}{x^H x}},$$

however, the choice of transformations is more easily made. For this reason, only the above norms will be used in the present section. If U is a unitary matrix, $U^H U = I$, then the above matrix norm satisfies

$$\begin{aligned}\text{lub}(A) = \text{lub}(U^H U A) &\leqslant \text{lub}(U^H)\,\text{lub}(UA) = \text{lub}(UA)\\ &\leqslant \text{lub}(U)\,\text{lub}(A) = \text{lub}(A);\end{aligned}$$

hence

$$\text{lub}(UA) = \text{lub}(A),$$

and similarly $\text{lub}(AU) = \text{lub}(A)$.

In particular, it follows that

$$\text{cond}(A) = \text{lub}(A)\,\text{lub}(A^{-1}) = \text{cond}(UA)$$

for unitary U. If the transformation matrices P_j are chosen to be unitary, then the condition numbers associated with the systems $A^{(j)}x = b^{(j)}$ do not change (so they certainly don't get worse). Furthermore, the matrices P_j should be chosen so that the matrices $A^{(j)}$ become simpler. As was suggested by Householder, this can be accomplished in the following manner:

The unitary matrix P is chosen to be

$$P = I - 2ww^H, \quad \text{where } w^H w = 1,\ w \in \mathbb{C}^n.$$

This matrix is Hermitian:

$$P^H = I^H - (2ww^H)^H = I - 2ww^H = P,$$

unitary:

$$\begin{aligned} P^H P = PP = P^2 &= (I - 2ww^H)(I - 2ww^H) \\ &= I - 2ww^H - 2ww^H + 4ww^H ww^H \\ &= I, \end{aligned}$$

and therefore involutory:

$$P^2 = I.$$

If two vectors x, y satisfy

$$y = Px = x - 2(w^H x)w,$$

then it follows that

$$y^H y = x^H P^H P x = x^H x, \tag{4.7.2}$$

$$x^H y = x^H P x = (x^H P x)^H \tag{4.7.3}$$

and $x^H y$ is real. We wish to determine a vector w, and thereby P, so that a given

$$x = (x_1, \ldots, x_n)^T$$

is transformed into a multiple of the first coordinate vector e_1:

$$ke_1 = Px.$$

From (4.7.2) it follows immediately that k satisfies

$$|k|^2 = \|x\|^2 = x^H x$$

and, since $kx^H e_1$ must be real according to (4.7.3),

$$k = \mp e^{i\alpha}\sigma, \qquad \sigma := \sqrt{x^H x},$$

if $x_1 = e^{i\alpha}|x_1|$. Hence, it follows from

$$Px = x - 2(w^H x)w = ke_1$$

and the requirement $w^H w = 1$ that

$$w = \frac{x - ke_1}{\|x - ke_1\|}.$$

Now, since $x_1 = e^{i\alpha}|x_1|$,

$$\begin{aligned} \|x - ke_1\| = \|x \pm \sigma e^{i\alpha} e_1\| &= \sqrt{|x_1 \pm \sigma e^{i\alpha}|^2 + |x_2|^2 + \cdots + |x_n^2|} \\ &= \sqrt{(|x_1| \pm \sigma)^2 + |x_2|^2 + \cdots + |x_n|^2}. \end{aligned}$$

In order that no cancellation may occur in the computation of $|x_1| \pm \sigma$, we choose the sign in the definition of k to be

$$k = -\sigma e^{i\alpha},$$

which gives

$$(4.7.4) \qquad |x_1 - k|^2 = |x_1 + \sigma e^{i\alpha}|^2 = |\sigma + |x_1||^2 = \sigma^2 + 2\sigma|x_1| + |x_1|^2.$$

It follows that

$$\|x - ke_1\|^2 = 2\sigma^2 + 2\sigma|x_1|, \qquad 2ww^H = 2\frac{(x - ke_1)(x - ke_1)^H}{\|x - ke_1\|^2}.$$

The matrix $P = I - 2ww^H$ can be written in the form

$$P = I - \beta uu^H$$

with

$$\sigma = \sqrt{\sum_{i=1}^{n} |x_i|^2}, \qquad x_1 = e^{i\alpha}|x_1|, \qquad k = -\sigma e^{i\alpha},$$

$$(4.7.5) \qquad u = x - ke_1 = \begin{bmatrix} e^{i\alpha}(|x_1| + \sigma) \\ x_2 \\ \vdots \\ x_n \end{bmatrix},$$

$$\beta = (\sigma(\sigma + |x_1|))^{-1}.$$

A matrix $A \equiv A^{(0)}$ can be reduced step by step using these unitary "Householder matrices" P_j,

$$A^{(j)} = P_j A^{(j-1)},$$

into an upper triangular matrix

$$A^{(n-1)} = R = \begin{bmatrix} r_{11} & \cdots & r_{1n} \\ & \ddots & \vdots \\ 0 & & r_{nn} \end{bmatrix}.$$

To do this, the $n \times n$ unitary matrix P_1 is determined according to (4.7.5) so that

$$P_1 a_1^{(0)} = ke_1,$$

where $a_1^{(0)}$ stands for the first column of $A^{(0)}$.

If the matrix $A^{(j-1)}$ obtained after $j - 1$ steps has the form

(4.7.6)

$$A^{(j-1)} = \left[\begin{array}{ccc|ccc} x & \cdots & x & x & \cdots & x \\ & \ddots & \vdots & \vdots & & \vdots \\ 0 & & x & x & \cdots & x \\ \hline & & & a_{jj}^{(j-1)} & \cdots & a_{jn}^{(j-1)} \\ & 0 & & \vdots & & \vdots \\ & & & a_{nj}^{(j-1)} & \cdots & a_{nn}^{(j-1)} \end{array}\right] \begin{array}{l} \left.\vphantom{\begin{array}{c} x \\ \vdots \\ x \end{array}}\right\} j-1 \\ \left.\vphantom{\begin{array}{c} x \\ \vdots \\ x \end{array}}\right\} n-j+1 \end{array} = \begin{bmatrix} D & B \\ 0 & \tilde{A}^{(j-1)} \end{bmatrix},$$

then we determine the $(n-j+1) \times (n-j+1)$ unitary matrix $\tilde{P}_j$ according to (4.7.5) so that

$$\tilde{P}_j \begin{bmatrix} a_{jj}^{(j-1)} \\ \vdots \\ a_{nj}^{(j-1)} \end{bmatrix} = k \begin{bmatrix} 1 \\ 0 \\ \vdots \\ 0 \end{bmatrix} \in \mathbb{C}^{n-j+1}.$$

Using $\tilde{P}_j$ the desired $n \times n$ unitary matrix is constructed as

$$P_j = \begin{bmatrix} I_{j-1} & 0 \\ 0 & \tilde{P}_j \end{bmatrix} \begin{matrix} \}\ j-1 \\ \}\ n-j+1 \end{matrix}.$$

After forming $A^{(j)} = P_j A^{(j-1)}$, the elements $a_{ij}^{(j)}$ for $i > j$ are annihilated, and the rows in (4.7.6) above the horizontal dashed line remain unchanged. In this way an upper triangular matrix

$$R := A^{(n-1)}$$

is obtained after $n-1$ steps.

It should be noticed, when applying Householder transformations in practice, that the locations which are set to zero beneath the diagonal of $A^{(j)}$ can be used to store u, which contains the important information about the transformation P. However, since the vector u which belongs to $\tilde{P}_j$ contains $n-j+1$ components, while only $n-j$ locations are set free in $A^{(j)}$, it is usual to store the diagonal elements of $A^{(j)}$ in a separate array d in order to obtain enough space for u.

The transformation of a matrix by

$$\tilde{P}_j = I - \beta_j u_j u_j^H$$

is carried out as follows:

$$\tilde{P}_j \tilde{A}^{(j-1)} = \tilde{A}^{(j-1)} - u_j y_j^H \quad \text{with } y_j^H = \beta_j u_j^H \tilde{A}^{(j-1)};$$

that is, the vector y_j is computed first, and then $\tilde{A}^{j-1}$ is modified as indicated.

The following ALGOL program contains the essentials of the reduction of a real matrix, stored in the array a, to triangular form using Householder transformations:

```
for j := 1 step 1 until n do
begin
        sigma := 0;
        for i := j step 1 until n do sigma := sigma + a[i, j]↑2;
        if sigma = 0 then goto singular;
        s := d[j] := if a[j, j] < 0 then sqrt (sigma) else −sqrt (sigma);
        beta := 1/(s × a[j, j] − sigma); a[j, j] := a[j, j] − s;
        for k := j + 1 step 1 until n do
```

```
        begin
                  sum := 0;
                  for i := j step 1 until n do
                            sum := sum + a[i, j] × a[i, k];
                  sum := beta × sum;
                  for i := j step 1 until n do
                            a[i, k] := a[i, k] + a[i, j] × sum
        end
end
```

The Householder reduction of a matrix to triangular form requires about $2n^3/3$ operations. In this process an $n \times n$ unitary matrix $P = P_n \dots P_1$ consisting of Householder matrices P_i and an $n \times n$ upper triangular matrix R are determined so that

$$PA = R$$

or

$$A = P^{-1}R = QR \tag{4.7.7}$$

holds.

An upper triangular matrix R, with the property (4.7.7) that $AR^{-1} = Q$ is a matrix with orthonormal columns, can also be produced directly by the application of Gram-Schmidt orthogonalization to the columns a_i of $A = (a_1, \dots, a_n)$. The equation $A = QR$ shows that the kth column a_k of A

$$a_k = \sum_{i=1}^{k} r_{ik} q_i, \qquad k = 1, \dots, n$$

is a linear combination of the vectors $q_1, q_2, \dots, q_k$, so that, conversely, q_k is a linear combination of the first k columns $a_1, \dots, a_k$ of A. The Gram-Schmidt process determines the columns of Q and R recursively as follows: Begin with

$$r_{11} := \|a_1\|, \qquad q_1 := a_1/r_{11}.$$

If the orthonormal vectors $q_1, \dots, q_{k-1}$ and the elements r_{ij} with $j \leqslant k - 1$ of R are known, then the numbers $r_{1k}, \dots, r_{k-1,k}$ are determined so that the vector

$$b_k := a_k - r_{1k} q_1 - \cdots - r_{k-1,k} q_{k-1} \tag{4.7.8}$$

is orthogonal to all q_i, $i = 1, \dots, k - 1$.

Because

$$q_i^H q_j = \begin{cases} 1 & \text{for } i = j \\ 0 & \text{otherwise} \end{cases} \qquad \text{for } 1 \leqslant i, j \leqslant k - 1,$$

the conditions $q_i^H b_k = 0$ lead immediately to

(4.7.9) $$r_{ik} := q_i^H a_k, \qquad i = 1, \ldots, k-1.$$

After r_{ik}, $i \leqslant k-1$, and thereby b_k have been determined, the quantities

(4.7.10) $$r_{kk} := \|b_k\|, \qquad q_k := b_k / r_{kk},$$

are computed, so that (4.7.8) is equivalent to

$$a_k = \sum_{i=1}^{k} r_{ik} q_i,$$

and moreover

$$q_i^H q_j = \begin{cases} 1 & \text{for } i = j, \\ 0 & \text{otherwise} \end{cases}$$

for all $1 \leqslant i, j \leqslant k$.

As given above, the algorithm has a serious disadvantage: it is not numerically stable if the columns of the matrix A are nearly linearly dependent. In this case the vector $\bar{b}_k$, which is obtained from the formulas (4.7.8), (4.7.9) instead of the vector b_k (due to the influence of roundoff errors), is no longer orthogonal to the vectors $q_1, \ldots, q_{k-1}$. The slightest roundoff error incurred in the determination of the r_{ik} in (4.7.9) destroys orthogonality to a greater or lesser extent.

We will discuss this effect in the special case corresponding to $k = 1$ in (4.7.8). Let two real vectors a and q be given, which we regard, for simplicity, as normalized: $\|a\| = \|q\| = 1$. This means that their scalar product $\rho_0 := q^T a$ satisfies $|\rho_0| \leq 1$. We assume that $|\rho_0| < 1$ holds, that is, a and q are linearly independent. The vector

$$b = b(\rho) := a - \rho q$$

is orthogonal to q for $\rho = \rho_0$. In general, the angle $\alpha(\rho)$ between $b(\rho)$ and q satisfies

$$f(\rho) := \cos \alpha(\rho) = \frac{q^T b(\rho)}{\|b(\rho)\|} = \frac{q^T a - \rho}{\|a - \rho q\|},$$

$$\alpha(\rho_0) = \pi/2, \qquad f(\rho_0) = 0.$$

Differentiating with respect to ρ, we find

$$f'(\rho_0) = \frac{-1}{\|a - \rho_0 q\|} = \frac{-1}{\sqrt{1 - \rho_0^2}},$$

$$\alpha'(\rho_0) = \frac{1}{\sqrt{1 - \rho_0^2}},$$

since

$$\|a - \rho_0 q\|^2 = a^T a - 2\rho_0 a^T q + \rho_0^2 q^T q = 1 - \rho_0^2.$$

Therefore, to a first approximation, we have

$$\alpha(\rho_0 + \Delta\rho_0) \doteq \frac{\pi}{2} + \frac{\Delta\rho_0}{\sqrt{1-\rho_0^2}}$$

for $\Delta\rho_0$ small. The closer $|\rho_0|$ lies to 1, i.e., the more linearly dependent a and q are, the more the orthogonality between q and $b(\rho)$ is destroyed by even tiny errors $\Delta\rho_0$, in particular by the roundoff errors which occur in computing $\rho_0 = q^T a$.

Since it is precisely the orthogonality of the vectors q_i that is essential to the process, the following trick is used in practice. The vectors $\bar{b}_k$ which are obtained instead of the exact b_k from the evaluation of (4.7.8), (4.7.9) are subjected to a "reorthogonalization". That is, scalars Δr_{ik} and a vector $\tilde{b}_k$ are computed from

$$\tilde{b}_k = \bar{b}_k - \Delta r_{1k} q_1 - \cdots - \Delta r_{k-1,k} q_{k-1},$$

where

$$\Delta r_{ik} := q_i^T \bar{b}_k, \qquad i = 1, \ldots, k-1,$$

so that in exact arithmetic $\tilde{b}_k^T q_i = 0$, $i = 1, \ldots, k-1$. Since $\bar{b}_k$ was at least approximately orthogonal to the q_i, the Δr_{ik} are small, and according to the theory just given, the roundoff errors which are made in the computation of the Δr_{ik} have only a minor influence on the orthogonality of $\tilde{b}_k$ and q_i. This means that the vector

$$q_k := \tilde{b}_k / \tilde{r}_{kk}, \qquad \tilde{r}_{kk} := \|\tilde{b}_k\|$$

will be orthogonal within machine precision to the already known vectors $q_1, \ldots, q_{k-1}$. The values $\bar{r}_{ik}$ which have been found by evaluating (4.7.9) are corrected appropriately:

$$r_{ik} := \bar{r}_{ik} + \Delta r_{ik}.$$

Clearly, reorthogonalization requires twice as much computing effort as the straightforward Gram–Schmidt process.

4.8 Data Fitting

In many scientific observations, one is concerned with determining the values of certain constants

$$x_1, x_2, \ldots, x_n.$$

Often, however, it is exceedingly difficult or impossible to measure the quantities x_i directly. In such cases, the following indirect method is used: instead of observing the x_i, another, more easily measurable quantity y is sampled, which depends in a known way on the x_i and on further controllable "experimental conditions", which we symbolize by z:

$$y = f(z; x_1, \ldots, x_n).$$

In order to determine the x_i, experiments are carried out under m different conditions $z_1, \ldots, z_m$, and the corresponding results

$$y_k = f(z_k; x_1, \ldots, x_n), \qquad k = 1, 2, \ldots, m, \tag{4.8.0.1}$$

are measured. Values for $x_1, \ldots, x_n$ are sought so that the equations (4.8.0.1) are satisfied. In general, of course, at least m experiments, $m \geq n$, must be carried out in order that the x_i may be uniquely determined. If $m > n$, however, the equations (4.8.0.1) form an overdetermined system for the unknown parameters $x_1, \ldots, x_n$, which does not usually have a solution because the observed quantities y_i are perturbed by measurement errors. Consequently, instead of finding an exact solution to (4.8.0.1), the problem becomes one of finding a "best possible solution". Such a solution to (4.8.0.1) is taken to mean a set of values for the unknown parameters for which the expression

$$\sum_{k=1}^{m} (y_k - f_k(x_1, \ldots, x_n))^2 \tag{4.8.0.2}$$

or the expression

$$\sum_{k=1}^{m} |y_k - f_k(x_1, \ldots, x_n)| \tag{4.8.0.3}$$

is minimized. Here we have denoted $f(z_k; x_1, \ldots, x_n)$ by $f_k(x_1, \ldots, x_n)$.

In the first case, the Euclidian norm of the residuals is minimized, and we are presented with a fitting problem of the special type studied by Gauss (in his "method of least squares"). In mathematical statistics [e.g. see Guest (1961), Seber (1977), or Grossmann (1969)] it is shown that the "least-squares solution" has particularly simple statistical properties. It is the most reasonable point to determine if the errors in the measurements are independent and normally distributed. More recently, however, points x which minimize the norm (4.8.0.3) of the residuals have been considered in cases where the measurement errors are subject to occasional anomalies. Since the computation of the best solution x is more difficult in this case than in the method of least squares, we will not pursue this matter.

If the functions $f_k(x_1, \ldots, x_n)$ have continuous partial derivatives in all of the variables x_i, then we can readily give a necessary condition for $x = (x_1, \ldots, x_n)^T$ to minimize (4.8.0.2):

$$\frac{\partial}{\partial x_i} \sum_{k=1}^{m} (y_k - f_k(x_1, \ldots, x_n))^2 = 0, \qquad i = 1, \ldots, n. \tag{4.8.0.4}$$

These are the *normal equations* for x.

An important special case, the *linear least-squares problem*, is obtained if the functions $f_k(x_1, \ldots, x_n)$ are linear in the x_i. In this case there is an $m \times n$ matrix A with

$$\begin{bmatrix} f_1(x_1, \ldots, x_n) \\ \vdots \\ f_m(x_1, \ldots, x_n) \end{bmatrix} = Ax.$$

In this case the normal equations reduce to a linear system

$$\operatorname{grad}_x((y - Ax)^T(y - Ax)) = 2A^TAx - 2A^Ty = 0,$$

or

$$A^TAx = A^Ty. \tag{4.8.0.5}$$

We will concern ourselves in the following sections (4.8.1–4.8.3), with methods for solving the linear problem, and in particular, we will show that more numerically stable methods exist for finding the solution than by means of the normal equations.

Least-squares problems are studied in more detail in Björck (1990) and the book by Lawson and Hanson (1974), which also contains FORTRAN programs; ALGOL programs are found in Wilkinson and Reinsch (1971).

4.8.1 Linear Least Squares. The Normal Equations

In the following sections $\|x\|$ will always denote the Euclidian norm $\sqrt{x^Tx}$. Let an $m \times n$ matrix A and a vector $y \in \mathbb{R}^m$ be given, and let

$$\|y - Ax\|^2 = (y - Ax)^T(y - Ax) \tag{4.8.1.1}$$

be minimized as a function of x. We want to show that $x \in \mathbb{R}^n$ is a solution of the normal equations

$$A^TAx = A^Ty \tag{4.8.1.2}$$

if and only if x is also a minimum point for (4.8.1.1). We have the following:

(4.8.1.3) Theorem. *The linear least-squares problem*

$$\min_{x \in \mathbb{R}^n} \|y - Ax\|$$

has at least one minimum point x_0. If x_1 is another minimum point, then $Ax_0 = Ax_1$. The residual $r := y - Ax_0$ is uniquely determined and satisfies the equation $A^Tr = 0$. Every minimum point x_0 is also a solution of the normal equations (4.8.1.2) *and conversely.*

PROOF. Let $L \subseteq \mathbb{R}^m$ be the linear subspace

$$L = \{Ax \mid x \in \mathbb{R}^n\}$$

which is spanned by the columns of A, and let $L^\perp$ be the orthogonal complement

$$L^\perp := \{r \mid r^T z = 0 \quad \text{for all } z \in L\}$$
$$= \{r \mid r^T A = 0\}.$$

Because $\mathbb{R}^m = L \oplus L^\perp$, the vector $y \in \mathbb{R}^m$ can be written uniquely in the form

(4.8.1.4) $$y = s + r, \qquad s \in L, \quad r \in L^\perp,$$

and there is at least one x_0 with $Ax_0 = s$. Because $A^T r = 0$, x_0 satisfies

$$A^T y = A^T s = A^T A x_0,$$

that is, x_0 is a solution of the normal equations. Conversely, each solution x_1 of the normal equations corresponds to a representation (4.8.1.4):

$$y = s + r, \qquad s := Ax_1, \quad r = y - Ax_1, \quad s \in L, \quad r \in L^\perp.$$

Because this representation is unique, it follows that $Ax_0 = Ax_1$ for all solutions x_0, x_1 of the normal equations. Further, each solution x_0 of the normal equations is a minimum point for the problem

$$\min_{x \in \mathbb{R}^n} \|y - Ax\|.$$

To see this, let x be arbitrary, and set

$$z := Ax - Ax_0, \qquad r := y - Ax_0.$$

Then, since $r^T z = 0$,

$$\|y - Ax\|^2 = \|r - z\|^2 = \|r\|^2 + \|z\|^2 \geqslant \|r\|^2 = \|y - Ax_0\|^2,$$

that is, x_0 is a minimum point. This establishes Theorem (4.8.1.3). □

If the columns of A are linearly independent, that is, if $x \neq 0$ implies $Ax \neq 0$, then the matrix $A^T A$ is nonsingular (and positive definite). If this were not the case, there would exist an $x \neq 0$ satisfying $A^T Ax = 0$, from which

$$0 = x^T A^T A x = \|Ax\|^2$$

would yield a contradiction, since $Ax \neq 0$. Therefore the normal equations

$$A^T A x = A^T y$$

have a unique solution $x = (A^T A)^{-1} A^T y$, which can be computed using the methods of Section 4.3 (that is, using the Cholesky factorization of $A^T A$). However, we will see in the following sections that there are more numerically stable ways of solving the linear least squares problem.

We digress here to go briefly into the statistical meaning of the matrix $(A^T A)^{-1}$. To do this, we assume that the components y_i, $i = 1, \ldots, m$, are independent, normally distributed random variables having μ_i as means and all having the same variance σ^2:

$$E[y_i] = \mu_i,$$

$$E[(y_i - \mu_i)(y_k - \mu_k)] = \begin{cases} \sigma^2 & \text{for } i = k, \\ 0 & \text{otherwise.} \end{cases}$$

If we set $\mu := (\mu_1, \ldots, \mu_m)^T$, then the above can be expressed as

(4.8.1.5) $$E[y] = \mu, \qquad E[(y - \mu)(y - \mu)^T] = \sigma^2 I.$$

The covariance matrix of the random vector y is $\sigma^2 I$. The optimum $x = (A^T A)^{-1} A^T y$ of the least-squares problem is also a random vector, having mean

$$\begin{aligned} E[x] &= E[(A^T A)^{-1} A^T y] \\ &= (A^T A)^{-1} A^T E[y] \\ &= (A^T A)^{-1} A^T \mu, \end{aligned}$$

and covariance matrix

$$\begin{aligned} E[(x - E(x))(x - E(x))^T] &= E[(A^T A)^{-1} A^T (y - \mu)(y - \mu)^T A (A^T A)^{-1}] \\ &= (A^T A)^{-1} A^T E[(y - \mu)(y - \mu)^T] A (A^T A)^{-1} \\ &= \sigma^2 (A^T A)^{-1}. \end{aligned}$$

4.8.2 The Use of Orthogonalization in Solving Linear Least-Squares Problems

The problem of determining an $x \in \mathbb{R}^n$ which minimizes

$$\|y - Ax\| \qquad (A \in M(m, n), \quad m \geqslant n)$$

can be solved using the orthogonalization techniques discussed in Section 4.7. Let the matrix $A \equiv A^{(0)}$ and the vector $y \equiv y^{(0)}$ be transformed by a sequence of Householder transformations P_i, $A^{(i)} = P_i A^{(i-1)}$, $y^{(i)} = P_i y^{(i-1)}$. The final matrix $A^{(n)}$ has the form

(4.8.2.1) $$A^{(n)} = \overbrace{\begin{bmatrix} R \\ \cdots \\ 0 \end{bmatrix}}^{n} \begin{matrix} \}\, n \\ \\ \}\, m-n \end{matrix} \qquad \text{with} \qquad R = \begin{bmatrix} r_{11} & \cdots & r_{1n} \\ & \ddots & \vdots \\ 0 & & r_{nn} \end{bmatrix},$$

since $m \geqslant n$. Let the vector $h := y^{(n)}$ be partitioned correspondingly:

(4.8.2.2) $$h = \begin{bmatrix} h_1 \\ h_2 \end{bmatrix}, \qquad h_1 \in \mathbb{R}^n, \quad h_2 \in \mathbb{R}^{m-n}.$$

The matrix $P = P_n \ldots P_1$, being the product of unitary matrices, is unitary itself:

$$P^H P = P_1^H \ldots P_n^H P_n \ldots P_1 = I,$$

and satisfies

$$A^{(n)} = PA, \qquad h = Py.$$

Unitary transformations leave the norm $\|u\|$ of a vector u invariant ($\|Pu\| = \sqrt{u^H P^H P u} = \sqrt{u^H u} = \|u\|$), so

$$\|y - Ax\| = \|P(y - Ax)\| = \|y^{(n)} - A^{(n)}x\|.$$

However, from (4.8.2.1) and (4.8.2.2), the vector $y^{(n)} - A^{(n)}x$ has the structure

$$y^{(n)} - A^{(n)}x = \begin{bmatrix} h_1 - Rx \\ h_2 \end{bmatrix} \begin{matrix} \}n \\ \}m-n \end{matrix}.$$

Hence $\|y - Ax\|$ is minimized if x is chosen so that

(4.8.2.3) $$h_1 = Rx.$$

The matrix R has an inverse R^{-1} if and only if the columns $a_1, \ldots, a_n$ of A are linearly independent. $Az = 0$ for $z \neq 0$ is equivalent to

$$PAz = 0$$

and therefore to

$$Rz = 0.$$

If we assume that the columns of A are linearly independent, then

$$h_1 = Rx,$$

which is a triangular system, can be solved uniquely for x. This x is, moreover, the unique minimum point for the given least-squares problem. (If the columns of A, and with them the columns of R, are linearly dependent, then, although the value of $\min_z \|y - Az\|$ is uniquely determined, there are many minimum points x).

The residual $\|y - Ax\|$ is seen to be

(4.8.2.4) $$\|y - Ax\| = \|h_2\|.$$

We conclude by mentioning that instead of using unitary transformations, the Gram–Schmidt technique with reorthogonalization can be used to obtain the solution, as should be evident.

4.8.3 The Condition of the Linear Least-Squares Problem

We begin this section by investigating how a minimum point x for the linear least-squares problem

(4.8.3.1) $$\min_x \|y - Ax\|$$

changes if the matrix A and the vector y are perturbed. We assume that the columns of A are linearly independent. If the matrix A is replaced by $A + B$, and y is replaced by $y + \Delta y$, then the solution $x = (A^TA)^{-1}A^Ty$ of (4.8.3.1) changes to

$$x + \Delta x = ((A + B)^T(A + B))^{-1}(A + B)^T(y + \Delta y).$$

If B is small relative to A, then $((A + B)^T(A + B))^{-1}$ exists and satisfies, to a first approximation,

$$\begin{aligned}((A + B)^T(A + B))^{-1} &\doteq (A^TA(I + (A^TA)^{-1}[A^TB + B^TA]))^{-1}\\ &\doteq (I - (A^TA)^{-1}[A^TB + B^TA])(A^TA)^{-1}.\end{aligned}$$

[To a first approximation, $(I + F)^{-1} \doteq I - F$ if the matrix F is "small" relative to I.] Thus it follows that

$$\text{(4.8.3.2)}\quad \begin{aligned}x + \Delta x \doteq\ & (A^TA)^{-1}A^Ty - (A^TA)^{-1}[A^TB + B^TA](A^TA)^{-1}A^Ty\\ &+ (A^TA)^{-1}B^Ty + (A^TA)^{-1}A^T\,\Delta y.\end{aligned}$$

Noting that

$$x = (A^TA)^{-1}A^Ty$$

and introducing the residual

$$r := y - Ax,$$

it follows immediately from (4.8.3.2) that

$$\Delta x \doteq -(A^TA)^{-1}A^TBx + (A^TA)^{-1}B^Tr + (A^TA)^{-1}A^T\,\Delta y.$$

Therefore, for the Euclidian norm $\|\cdot\|$ and the associated matrix norm lub,

$$\text{(4.8.3.3)}\quad \begin{aligned}\|\Delta x\| \mathrel{\dot{\leqslant}}\ & \operatorname{lub}((A^TA)^{-1}A^T)\operatorname{lub}(A)\frac{\operatorname{lub}(B)}{\operatorname{lub}(A)}\|x\|\\ &+ \operatorname{lub}((A^TA)^{-1})\operatorname{lub}(A^T)^2\frac{\operatorname{lub}(B^T)}{\operatorname{lub}(A^T)}\frac{\|r\|}{\operatorname{lub}(A^T)}\\ &+ \operatorname{lub}((A^TA)^{-1}A^T)\operatorname{lub}(A)\frac{\|y\|}{\operatorname{lub}(A)}\frac{\|\Delta y\|}{\|y\|}.\end{aligned}$$

[Observe that the definition given in (4.4.8) for lub makes sense even for nonsquare matrices.] This approximate bound can be simplified. According to the results of Section 4.8.2, a unitary matrix P and an upper triangular matrix R can be found such that

$$PA = \begin{bmatrix} R \\ 0 \end{bmatrix}, \qquad A = P^T\begin{bmatrix} R \\ 0 \end{bmatrix},$$

and it follows that

$$(4.8.3.4)\qquad \begin{aligned} A^TA &= R^TR,\\ (A^TA)^{-1} &= R^{-1}(R^T)^{-1},\\ (A^TA)^{-1}A^T &= (R^{-1}, 0)P. \end{aligned}$$

If it is observed that

$$\operatorname{lub}(C^T) = \operatorname{lub}(C),$$
$$\operatorname{lub}(PC) = \operatorname{lub}(CP) = \operatorname{lub}(C)$$

holds for the Euclidian norm, where P is unitary, then (4.8.3.3) and (4.8.3.4) imply

$$(4.8.3.5)\qquad \begin{aligned} \frac{\|\Delta x\|}{\|x\|} \leqslant{} & \operatorname{cond}(R)\frac{\operatorname{lub}(B)}{\operatorname{lub}(A)} + \operatorname{cond}(R)^2 \frac{\|r\|}{\operatorname{lub}(A)\|x\|}\frac{\operatorname{lub}(B)}{\operatorname{lub}(A)}\\ & + \operatorname{cond}(R)\frac{\|y\|}{\operatorname{lub}(A)\,\|x\|}\frac{\|\Delta y\|}{\|y\|}. \end{aligned}$$

The second term of this bound dominates the first term if

$$(4.8.3.6)\qquad \|x\| \leqslant \frac{\operatorname{cond}(R)\,\|r\|}{\operatorname{lub}(A)} = \operatorname{lub}(R^{-1})\,\|r\|.$$

But $x = (A^TA)^{-1}A^Ty$, and therefore, according to (4.8.3.4),

$$x = (R^{-1}, 0)Py.$$

If Py is partitioned accordingly,

$$Py = \begin{bmatrix} h_1 \\ h_2 \end{bmatrix},$$

it follows that

$$(4.8.3.7)\qquad x = R^{-1}h_1$$

and hence

$$(4.8.3.8)\qquad \|x\| \leqslant \operatorname{lub}(R^{-1})\,\|h_1\|.$$

Now, however, from (4.8.3.7),

$$Pr = P(y - Ax) = Py - \begin{bmatrix} R \\ 0 \end{bmatrix} x = \begin{bmatrix} 0 \\ h_2 \end{bmatrix}.$$

From $\|y\|^2 = \|Py\|^2 = \|h_1\|^2 + \|h_2\|^2 = \|h_1\|^2 + \|Pr\|^2 = \|h_1\|^2 + \|r\|^2$ and (4.8.3.8),

$$\|x\|^2 \leqslant \operatorname{lub}(R^{-1})^2\,(\|y\|^2 - \|r\|^2).$$

The condition (4.8.3.6) is satisfied, therefore, if the relationship

$$\operatorname{lub}(R^{-1})^2\,(\|y\|^2 - \|r\|^2) \leqslant \operatorname{lub}(R^{-1})^2\,\|r\|^2$$

holds, which is equivalent to the easily verified inequality

$$\tfrac{1}{2}\|y\|^2 \leqslant \|r\|^2. \tag{4.8.3.9}$$

If the above holds, then the second term in (4.8.3.5) dominates the first. Furthermore, if $\operatorname{cond}(R) \gg 1$, then the linear least-squares problem is very badly conditioned. Finally, the influence of $\|\Delta y\|/\|y\|$ will dominate the influence of $\operatorname{lub}(B)/\operatorname{lub}(A)$ in (4.8.3.5) if

$$\|y\| \gg \operatorname{lub}(A)\,\|x\| \tag{4.8.3.10}$$

holds.

In summary, the linear least-squares problem becomes more strongly ill conditioned as $\operatorname{cond}(R)$ increases, or as either of the inequalities (4.8.3.6) [alternatively (4.8.3.9)] or (4.8.3.10) is more nearly satisfied.

If the minimum point x for the linear least-squares problem is found using the orthogonalization method described in (4.8.2), then in exact arithmetic, a unitary matrix P (consisting of a product of Householder matrices), an upper triangular matrix R, a vector $h^T = (h_1^T, h_2^T)$, and the solution x all satisfying

$$PA = \begin{bmatrix} R \\ 0 \end{bmatrix}, \qquad h = Py, \qquad Rx = h_1 \tag{4.8.3.11}$$

are produced. Using floating-point arithmetic with relative machine precision eps, an exactly unitary matrix will not be obtained, nor will R, h, x exactly satisfy (4.8.3.11). Wilkinson (1965) has shown, however, that there exist a unitary matrix P', matrices ΔA and ΔR, and a vector Δy such that

$$\operatorname{lub}(P' - P) \leqslant f(m)\,\text{eps},$$

$$P'(A + \Delta A) = \begin{bmatrix} R \\ 0 \end{bmatrix}, \qquad \operatorname{lub}(\Delta A) \leqslant f(m)\,\text{eps}\,\operatorname{lub}(A),$$

$$P'(y + \Delta y) = h, \qquad \|\Delta y\| \leqslant f(m)\,\text{eps}\,\|y\|,$$

$$(R + \Delta R)x = h_1, \qquad \operatorname{lub}(\Delta R) \leqslant f(m)\,\text{eps}\,\operatorname{lub}(R).$$

In these relations $f(m)$ is a slowly growing function of m, roughly $f(m) \approx O(m)$.

Up to terms of higher order

$$\operatorname{lub}(R) \doteq \operatorname{lub}(A),$$

and therefore, since

$$F := \Delta A + P'^H \begin{bmatrix} \Delta R \\ 0 \end{bmatrix},$$

it also follows that

$$P'(A+F) = \begin{bmatrix} R+\Delta R \\ 0 \end{bmatrix}, \qquad \operatorname{lub}(F) \leqslant 2f(m) \text{ eps } \operatorname{lub}(A).$$

In other words, the computed solution x can be interpreted as an exact minimum point of the following linear least-squares problem:

$$\min_z \|(y+\Delta y) - (A+F)z\|, \tag{4.8.3.12}$$

wherein the matrix $A + F$ and the right-hand-side vector $y + \Delta y$ differ only slightly from A and y respectively. The technique of orthogonalization is, therefore, numerically stable.

If, on the other hand, one computes the solution x by way of the normal equations

$$A^T A x = A^T y,$$

then the situation is quite different. It is known from Wilkinson's (1965) estimates (see also Section 4.5) that a solution vector $\tilde{x}$ is obtained which satisfies

$$(A^T A + G)\tilde{x} = A^T y, \qquad \operatorname{lub}(G) \leqslant g(n) \text{ eps } \operatorname{lub}(A^T A), \tag{4.8.3.13}$$

when floating point computation with the relative machine precision eps is used (even under the assumption that $A^T y$ and $A^T A$ are computed exactly). If $x = (A^T A)^{-1} A^T y$ is the exact solution, then (4.4.15) shows that

$$\begin{aligned} \frac{\|\tilde{x} - x\|}{\|x\|} &\leqslant \operatorname{cond}(A^T A) \frac{\operatorname{lub}(G)}{\operatorname{lub}(A^T A)} \\ &= \operatorname{cond}(R)^2 \frac{\operatorname{lub}(G)}{\operatorname{lub}(A^T A)} \end{aligned} \tag{4.8.3.14}$$

to a first approximation. The roundoff errors, represented here as the matrix G, are amplified by the factor $\operatorname{cond}(R)^2$.

This shows that the use of the normal equations is not numerically stable if the first term dominates in (4.8.3.5). Another situation holds if the second term dominates. If $\|r\|/(\operatorname{lub}(A)\,\|x\|) \geqslant 1$, for example, then the use of the normal equations will be numerically stable and will yield results which are comparable to those obtained through the use of orthogonal transformations.

EXAMPLE 1 (*Läuchli*). For the 6×5 matrix

$$A = \begin{bmatrix} 1 & 1 & 1 & 1 & 1 \\ \varepsilon & & & & 0 \\ & \varepsilon & & & \\ & & \varepsilon & & \\ & & & \varepsilon & \\ 0 & & & & \varepsilon \end{bmatrix}$$

it follows that

$$A^T A = \begin{bmatrix} 1+\varepsilon^2 & 1 & 1 & 1 & 1 \\ 1 & 1+\varepsilon^2 & 1 & 1 & 1 \\ 1 & 1 & 1+\varepsilon^2 & 1 & 1 \\ 1 & 1 & 1 & 1+\varepsilon^2 & 1 \\ 1 & 1 & 1 & 1 & 1+\varepsilon^2 \end{bmatrix}.$$

If $\varepsilon = 0.5 \times 10^{-5}$, then 10-place decimal arithmetic yields

$$\mathrm{fl}(A^T A) = \begin{bmatrix} 1 & 1 & 1 & 1 & 1 \\ 1 & 1 & 1 & 1 & 1 \\ 1 & 1 & 1 & 1 & 1 \\ 1 & 1 & 1 & 1 & 1 \\ 1 & 1 & 1 & 1 & 1 \end{bmatrix},$$

since $\varepsilon^2 = 0.25 \times 10^{-10}$. This matrix has rank 1 and has no inverse. The normal equations cannot be solved, whereas the orthogonalization technique can be applied without difficulty. [Note for $A^T A$ that $\mathrm{cond}(A^T A) = \mathrm{cond}(R)^2 = (5 + \varepsilon^2)/\varepsilon^2$.]

EXAMPLE 2. The following example is intended to offer a computational comparison between the two methods which have been discussed for the linear least-squares problem.

(4.8.3.12) shows that, for the solution of a least-squares problem obtained using orthogonal transformations, the approximate bound given by (4.8.3.5) holds with $B = F$ and

$$\mathrm{lub}(B) \leqslant 2f(m) \text{ eps } \mathrm{lub}(A).$$

(4.8.3.14) shows that the solution obtained directly from the normal equations satisfies a somewhat different bound:

$$(A^T A + G)(x + \Delta x) = A^T y + \Delta b$$

where $\|\Delta b\| \leqslant \text{eps } \mathrm{lub}(A)\ \|y\|$, that is,

(4.8.3.15) $$\frac{\|\Delta x\|_{\text{normal}}}{\|x\|} \leqslant \mathrm{cond}(R)^2 \left(\frac{\mathrm{lub}(G)}{\mathrm{lub}(A^T A)} + \frac{\|\Delta b\|}{\|A^T y\|} \right),$$

which holds because of (4.4.12) and (4.4.15).

Let the fundamental relationship

(4.8.3.16) $$y(s) := x_1 \frac{1}{s} + x_2 \frac{1}{s^2} + x_2 \frac{1}{s^3} \quad \text{with } x_1 = x_2 = x_3 = 1$$

be given. A series of observed values [computed from (4.8.3.16)]

$$\{s_i,\ y(s_i)\}_{i=1,\ldots,10}$$

were produced on a machine having $\text{eps} = 10^{-11}$.

(a) Determine x_1, x_2, x_3 from the data $\{s_i, y_i\}$. If exact function values $y_i = y(s_i)$ are used, then the residual will satisfy

$$r(x) = 0.$$

The following table presents some computational results for this example. The size of the errors $\|\Delta x\|_{\text{orth}}$ from the orthogonalization method and $\|\Delta x\|_{\text{normal}}$ from the normal equations are shown together with a lower bound for the condition number of R. The example was solved using $s_i = s_0 + i$, $i = 1, \ldots, 10$ for a number of values of s_0:

s_0	cond(R)	$\|\Delta x\|_{\text{orth}}$	$\|\Delta x\|_{\text{normal}}$
10	6.6×10^3	8.0×10^{-10}	8.8×10^{-7}
50	1.3×10^6	6.4×10^{-7}	8.2×10^{-5}
100	1.7×10^7	3.3×10^{-6}	4.2×10^{-2}
150	8.0×10^7	1.8×10^{-5}	6.9×10^{-1}
200	2.5×10^8	1.8×10^{-3}	2.7×10^0

(b) We introduce a perturbation into the y_i, replacing y by $y + \lambda v$, where v is chosen so that $A^T v = 0$. This means that, in theory, the solution should remain unchanged:

$$(A^T A)x = A^T(y + \lambda v) = A^T y.$$

Now the residual satisfies

$$r(x) = y + \lambda v - Ax = \lambda v, \qquad \lambda \in \mathbb{R}.$$

The following table presents the errors Δx, as before, and the norm of the residual $\|r(x)\|$ together with the corresponding values of λ:

$$s_0 = 10,$$

$$v = (0.1331, -0.5184, 0.6591, -0.2744, 0, 0, 0, 0, 0, 0)^T,$$

$$\text{lub}(A) \approx 0.22, \qquad \text{eps} = 10^{-11}.$$

λ	$\|r(x)\|$	$\|\Delta x\|_{\text{orth}}$	$\|\Delta x\|_{\text{normal}}$
0	0	8×10^{-10}	8.8×10^{-7}
10^{-6}	9×10^{-7}	9.5×10^{-9}	8.8×10^{-7}
10^{-4}	9×10^{-5}	6.2×10^{-10}	4.6×10^{-7}
10^{-2}	9×10^{-3}	9.1×10^{-9}	1.3×10^{-6}
10^0	9×10^{-1}	6.1×10^{-7}	8.8×10^{-7}
10^{+2}	9×10^1	5.7×10^{-5}	9.1×10^{-6}

4.8.4 Nonlinear Least-Squares Problems

Nonlinear data-fitting problems can generally be solved only by iteration, as for instance the problem given in (4.8.0.2) of determining a point $x^* = (x_1^*, \dots, x_n^*)^T$ for given functions $f_k(x) \equiv f_k(x_1, \dots, x_n)$ and numbers y_k, $k = 1, \dots, m$,

$$y = \begin{bmatrix} y_1 \\ \vdots \\ y_m \end{bmatrix}, \qquad f(x) = \begin{bmatrix} f_1(x_1, \dots, x_n) \\ \vdots \\ f_m(x_1, \dots, x_n) \end{bmatrix}$$

which minimizes

$$\|y - f(x)\|^2 = \sum_{k=1}^{m} (y_k - f_k(x_1, \dots, x_n))^2.$$

For example, linearization techniques (see Section 5.1) can be used to reduce this problem to a sequence of linear least squares problems. If each f_k is continuously differentiable, and if

$$Df(\xi) = \begin{bmatrix} \dfrac{\partial f_1}{\partial x_1} & \cdots & \dfrac{\partial f_1}{\partial x_n} \\ \vdots & & \vdots \\ \dfrac{\partial f_m}{\partial x_1} & \cdots & \dfrac{\partial f_m}{\partial x_n} \end{bmatrix}_{x=\xi}$$

represents the Jacobian matrix at the point $x = \xi$, then

$$f(\bar{x}) = f(x) + Df(x)(\bar{x} - x) + h, \qquad \|h\| = o(\|\bar{x} - x\|).$$

holds. If x is close to the minimum point of the nonlinear least-squares problem, then the solution $\bar{x}$ of the linear least-squares problem

$$\text{(4.8.4.1)} \qquad \begin{aligned} &\min_{z \in \mathbb{R}^n} \|y - f(x) - Df(x)(z - x)\|^2 = \|r(x) - Df(x)(\bar{x} - x)\|^2, \\ &r(x) := y - f(x), \end{aligned}$$

will often be still closer than x; that is,

$$\|y - f(\bar{x})\|^2 < \|y - f(x)\|^2.$$

This relation is not always true in the form given. However, it can be shown that the direction

$$s = s(x) := \bar{x} - x$$

satisfies the following:

There is a $\lambda > 0$ such that the function

$$\varphi(\tau) := \|y - f(x + \tau s)\|^2$$

is strictly monotone decreasing for all $0 \leqslant \tau \leqslant \lambda$. *In particular*

$$\varphi(\lambda) = \|y - f(x + \lambda s)\|^2 < \varphi(0) = \|y - f(x)\|^2.$$

PROOF. φ is a continuously differentiable function of τ and satisfies

$$\begin{aligned}\varphi'(0) &= \frac{d}{d\tau}[(y - f(x + \tau s))^T(y - f(x + \tau s))]\Big|_{\tau=0} \\ &= -2(Df(x)s)^T(y - f(x)) = -2(Df(x)s)^T r(x).\end{aligned}$$

Now, by the definition of $\bar{x}$, $s = (\bar{x} - x)$ is a solution to the normal equations

$$\text{(4.8.4.2)} \qquad Df(x)^T Df(x)s = Df(x)^T r(x),$$

of the linear least-squares problem (4.8.4.1). It follows immediately from (4.8.4.2) that

$$\|Df(x)s\|^2 = s^T Df(x)^T Df(x)s = (Df(x)s)^T r(x)$$

and therefore

$$\varphi'(0) = -2\|Df(x)s\|^2 < 0,$$

so long as rank $Df(x) = n$ and $s \neq 0$. The existence of a $\lambda > 0$ satisfying $\varphi'(\tau) < 0$ for $0 \leqslant \tau \leqslant \lambda$ is a result of continuity, from which observation the assertion follows. □

This result suggests the following algorithm, called the *Gauss–Newton algorithm*, for the iterative solution of nonlinear least-squares problems. Beginning with an initial point $x^{(0)}$, determine successive approximations $x^{(i)}$, $i = 1, 2, \ldots$, as follows:

(1) For $x^{(i)}$ compute a minimum point $s^{(i)}$ for the linear least-squares problem

$$\min_{s \in \mathbb{R}^n} \|r(x^{(i)}) - Df(x^{(i)})s\|^2.$$

(2) Let $\varphi(\tau) := \|y - f(x^{(i)} + \tau s^{(i)})\|^2$, and further, let k be the smallest integer $k \geqslant 0$ with

$$\varphi(2^{-k}) < \varphi(0) = \|r(x^{(i)})\|^2.$$

(3) Define $x^{(i+1)} := x^{(i)} + 2^{-k}s^{(i)}$.

In Section 5.4 we will study the convergence of algorithms which are closely related to the process described here.

4.8.5 The Pseudoinverse of a Matrix

For any arbitrary (complex) $m \times n$ matrix A there is an $n \times m$ matrix A^+, the so-called *pseudoinverse* (or *Moore-Penrose inverse*). It is associated with A in a natural fashion and agrees with the inverse A^{-1} of A in case $m = n$ and A is nonsingular.

Consider the range space $R(A)$ and the null space $N(A)$ of A,

$$R(A) := \{Ax \in \mathbb{C}^m \mid x \in \mathbb{C}^n\},$$
$$N(A) := \{x \in \mathbb{C}^n \mid Ax = 0\},$$

together with their orthogonal complement spaces $R(A)^\perp \subset \mathbb{C}^m$, $N(A)^\perp \subset \mathbb{C}^n$. Further, let P be the $n \times n$ matrix which projects $\mathbb{C}^n$ onto $N(A)^\perp$, and let $\bar{P}$ be the $m \times m$ matrix which projects $\mathbb{C}^m$ onto $R(A)$:

$$P = P^H = P^2, \quad Px = 0 \quad \Leftrightarrow \quad x \in N(A),$$
$$\bar{P} = \bar{P}^H = \bar{P}^2, \quad \bar{P}y = y \quad \Leftrightarrow \quad y \in R(A).$$

For each $y \in R(A)$ there is a uniquely determined $x_1 \in N(A)^\perp$ satisfying $Ax_1 = y$; i.e., there is a well-defined mapping f: $R(A) \to \mathbb{C}^n$ with

$$Af(y) = y, \quad f(y) \in N(A)^\perp \quad \text{for all } y \in R(A).$$

For, given $y \in R(A)$, there is an x which satisfies $y = Ax$; hence $y = A(Px + (I - P)x) = APx = Ax_1$, where $x_1 = Px \in N(A)^\perp$, since $(I - P)x \in N(A)$. Further, if $x_1, x_2 \in N(A)^\perp$, $Ax_1 = Ax_2$, it follows that

$$x_1 - x_2 \in N(A) \cap N(A)^\perp = \{0\},$$

which implies that $x_1 = x_2$. f is obviously linear.

The composite mapping $f \circ \bar{P}$: $y \in \mathbb{C}^m \to f(\bar{P}y) \in \mathbb{C}^n$ is well defined and linear, since $\bar{P}y \in R(A)$; hence it is represented by an $n \times m$ matrix, which is precisely A^+, the pseudoinverse of A; $A^+y := f(\bar{P}(y))$ for all $y \in \mathbb{C}^m$. A^+ has the following properties:

(4.8.5.1) Theorem. *Let A be an $m \times n$ matrix. The pseudoinverse A^+ is an $n \times m$ matrix satisfying:*

(1) *$A^+A = P$ is the orthogonal projector P: $\mathbb{C}^n \to N(A)^\perp$ and $AA^+ = \bar{P}$ is the orthogonal projector $\bar{P}$: $\mathbb{C}^m \to R(A)$.*
(2) *The following formulas hold:*
   (a) $A^+A = (A^+A)^H$,
   (b) $AA^+ = (AA^+)^H$,
   (c) $AA^+A = A$,
   (d) $A^+AA^+ = A^+$.

PROOF. According to the definition of A^+,

$$A^+Ax = f(\bar{P}Ax) = f(Ax) = Px \quad \text{for all } x,$$

so that $A^+A = P$. Since $P^H = P$, (4.8.5.1.2a) is satisfied. Further, from the definition of f,

$$AA^+y = A(f(\bar{P}y)) = \bar{P}y$$

for all $y \in \mathbb{C}^m$; hence $AA^+ = \bar{P}$. Since $\bar{P}^H = \bar{P}$, (4.8.5.1.2b) follows too. Finally, for all $x \in \mathbb{C}^n$

$$(AA^+)Ax = \bar{P}Ax = Ax$$

according to the definition of $\bar{P}$, and for all $y \in \mathbb{C}^m$

$$A^+(AA^+)y = A^+\bar{P}y = f(\bar{P}^2y) = A^+y;$$

hence, (4.8.5.1.2c, 2d) hold. □

The properties (2a–d) of (4.8.5.1) uniquely characterize A^+:

(4.8.5.2) Theorem. *If Z is a matrix satisfying*

(a′) $ZA = (ZA)^H$,
(b′) $AZ = (AZ)^H$,
(c′) $AZA = A$,
(d′) $ZAZ = Z$,

then $Z = A^+$.

PROOF. From (a)–(d) and (a′)–(d′) we have the following chain of equalities:

$$\begin{aligned}
Z = ZAZ &= Z(AA^+A)A^+(AA^+A)Z && \text{from (c), (d), (d}'\text{)}\\
&= (A^H Z^H A^H A^{+H})A^+(A^{+H}A^H Z^H A^H) && \text{from (a), (a}'\text{), (b), (b}'\text{)}\\
&= (A^H A^{+H})A^+(A^{+H}A^H) && \text{from (c)}\\
&= (A^+A)A^+(AA^+) && \text{from (a), (b)}\\
&= A^+AA^+ = A^+ && \text{from (d).}
\end{aligned}$$

□

We note the following

(4.8.5.3) Corollary. *For all matrices A,*

$$A^{++} = A, \qquad (A^+)^H = (A^H)^+.$$

This holds because $Z := A$ [respectively $Z := (A^+)^H$] has the properties of $(A^+)^+$ [respectively $(A^H)^+$] in (4.8.5.2).

An elegant representation of the solution to the least-squares problem

$$\min_x \|Ax - y\|_2$$

can be given with the aid of the pseudoinverse A^+:

(4.8.5.4) Theorem. *The vector $\bar{x} := A^+y$ satisfies:*

(a) $\|Ax - y\|_2 \geqslant \|A\bar{x} - y\|_2$ *for all* $x \in \mathbb{C}^n$.
(b) $\|Ax - y\|_2 = \|A\bar{x} - y\|_2$ *and* $x \neq \bar{x}$ *imply* $\|x\|_2 > \|\bar{x}\|_2$.

In other words, $\bar{x} = A^+y$ is that minimum point of the least squares problem which has the smallest Euclidean norm, in the event that the problem does not have a unique minimum point.

PROOF. From (4.8.5.1), AA^+ is the orthogonal projector on $R(A)$; hence, for all $x \in \mathbb{C}^n$ it follows that

$$Ax - y = u - v,$$

$$u := A(x - A^+y) \in R(A), \quad v := (I - AA^+)y = y - A\bar{x} \in R(A)^\perp.$$

Consequently, for all $x \in \mathbb{C}^n$

$$\|Ax - y\|_2^2 = \|u\|_2^2 + \|v\|_2^2 \geqslant \|v\|_2^2 = \|A\bar{x} - y\|_2^2,$$

and $\|Ax - y\|_2 = \|A\bar{x} - y\|_2$ holds precisely if

$$Ax = AA^+y.$$

Now, A^+A is the projector on $N(A)^\perp$. Therefore, for all x such that $Ax = AA^+y$,

$$x = u_1 + v_1, \qquad u_1 := A^+Ax = A^+AA^+y = A^+y = \bar{x} \in N(A)^\perp,$$
$$v_1 := x - \bar{x} \in N(A),$$

from which it follows that $\|x\|_2^2 > \|\bar{x}\|_2^2$ for all $x \in \mathbb{C}^n$ satisfying $x - \bar{x} \neq 0$ and $\|Ax - y\|_2 = \|A\bar{x} - y\|_2$. □

4.9 Modification Techniques for Matrix Decompositions

Given any $n \times n$ matrix A, Gaussian elimination (see Section 4.1.7) produces an $n \times n$ upper triangular matrix R and a nonsingular $n \times n$ matrix $F = G_{n-1}P_{n-1} \dots G_1P_1$, a product of Frobenius matrices G_j and permutation matrices P_j, which satisfy

$$FA = R.$$

Alternatively, the orthogonalization algorithms of Section 4.7 produce $n \times n$ unitary matrices P, Q and an upper triangular matrix R (different from the one above) for which

$$PA = R \quad \text{or} \quad A = QR.$$

[compare with (4.7.7)]. These algorithms can also be applied to rectangular $m \times n$ matrices A, $m \geqslant n$. Then they produce nonsingular $m \times n$ matrices F (alternatively $m \times m$ unitary matrices P or $m \times n$ matrices Q with orthonormal columns) and $n \times n$ upper triangular matrices R satisfying

(4.9.1) $$FA = \begin{bmatrix} R \\ 0 \end{bmatrix} \begin{matrix} \}\, n \\ \}\, m-n \end{matrix},$$

or

(4.9.2a) $$PA = \begin{bmatrix} R \\ 0 \end{bmatrix}, \qquad P^H P = PP^H = I_m,$$

(4.9.2b) $$A = QR, \qquad Q^H Q = I_n.$$

For $m = n$, these decompositions were seen to be useful in that they reduced the problem of solving the equation systems

(4.9.3) $$Ax = y \quad \text{or} \quad A^T x = y$$

to a process of solving triangular systems and carrying out matrix multiplications. When $m > n$, the orthogonal decompositions mentioned in (4.9.2) permit the solutions $\bar{x}$ of linear least-squares problems to be obtained in a similar manner:

(4.9.4) $$\min_x \|Ax - y\| = \min_x \left\| \begin{bmatrix} R \\ 0 \end{bmatrix} x - Py \right\|, \qquad R\bar{x} = Q^H y.$$

Further, these techniques offer an efficient way of obtaining the solutions to linear-equation problems (4.9.3) or to least-squares problems (4.9.4) in which a single matrix A is given with a number of right-hand sides y.

Frequently, a problem involving a matrix A is given and solved, following which a "simple" change $A \to \bar{A}$ is made. Clearly it would be desirable to determine a decomposition (4.9.1) or (4.9.2) of $\bar{A}$, starting with the corresponding decomposition of A, in some less expensive way than by using the algorithms in Sections 4.1 or 4.7. For certain simple changes this is possible. We will consider:

(1) the change of a row or column of A,
(2) the deletion of a column of A,
(3) the addition of a column to A,
(4) the addition of a row to A,
(5) the deletion of a row of A,

where A is an $m \times n$ matrix, $m \geqslant n$. [See Gill, Golub, Murray, and Saunders (1974) and Daniel, Gragg, Kaufman, and Stewart (1976) for a more detailed description of modification techniques for matrix decompositions.]

Our principal tool for devising modification techniques will be certain simple *elimination matrices* E_{ij}. These are nonsingular matrices of order m which differ from the identity matrix only in columns i and j and have the following form:

$$E_{ij} = \begin{bmatrix} 1 & & & & & & & & 0 \\ & \ddots & & & & & & & \\ & & 1 & & & & & & \\ & & & a & & b & & & \\ & & & & 1 & & & & \\ & & & & & \ddots & & & \\ & & & & & & 1 & & \\ & & & c & & d & & & \\ & & & & & & & 1 & \\ & & & & & & & & \ddots \\ 0 & & & & & & & & 1 \end{bmatrix} \begin{matrix} \\ \\ \\ \leftarrow i \\ \\ \\ \\ \leftarrow j \\ \\ \\ \\ \end{matrix}$$

A matrix multiplication $y = E_{ij}x$ changes only components x_i and x_j of the vector $x = (x_1, \ldots, x_m)^T \in \mathbb{R}^m$:

$$\begin{aligned} y_i &= ax_i + bx_j, \\ y_j &= cx_i + dx_j, \\ y_k &= x_k \quad \text{for } k \neq i, j. \end{aligned}$$

Given a vector x and indices $i \neq j$, the 2×2 matrix

$$\hat{E} := \begin{bmatrix} a & b \\ c & d \end{bmatrix},$$

and thereby E_{ij}, can be chosen in several ways so that the jth component y_j of the result $y = E_{ij}x$ vanishes and so that E_{ij} is nonsingular:

$$\begin{bmatrix} y_i \\ y_j \end{bmatrix} \equiv \begin{bmatrix} y_i \\ 0 \end{bmatrix} = \hat{E} \begin{bmatrix} x_i \\ x_j \end{bmatrix}.$$

For numerical reasons $\hat{E}$ should also be chosen so that the condition number $\operatorname{cond}(E_{ij})$ is not too large. The simplest possibility, which finds application in decompositions of the type (4.9.1), is to construct $\hat{E}$ as a Gaussian elimination of a certain type [see (4.1.8)]:

$$\text{(4.9.5)} \qquad \hat{E} = \begin{cases} \begin{bmatrix} 1 & 0 \\ 0 & 1 \end{bmatrix} & \text{if } x_j = 0, \\[2ex] \begin{bmatrix} 1 & 0 \\ -x_j/x_i & 1 \end{bmatrix} & \text{if } |x_i| \geqslant |x_j| > 0, \\[2ex] \begin{bmatrix} 0 & 1 \\ 1 & -x_i/x_j \end{bmatrix} & \text{if } |x_i| < |x_j|. \end{cases}$$

In the case of orthogonal decompositions (4.9.2) we choose $\hat{E}$, and thereby E_{ij}, to be the unitary matrices known as *Givens matrices*. One possible form such a matrix may take is given by the following:

(4.9.6) $$\hat{E} = \begin{bmatrix} c & s \\ s & -c \end{bmatrix}, \qquad c := \cos \varphi, \quad s := \sin \varphi.$$

$\hat{E}$ and E_{ij} are Hermitian and unitary, and satisfy $\det(\hat{E}) = -1$. Since $\hat{E}$ is unitary, it follows immediately from the condition

$$\begin{bmatrix} c & s \\ s & -c \end{bmatrix} \begin{bmatrix} x_i \\ x_j \end{bmatrix} = \begin{bmatrix} y_i \\ 0 \end{bmatrix}$$

that $y_i = k = \pm\sqrt{x_i^2 + x_j^2}$, which can be satisfied by

$$c := 1, \quad s := 0 \quad \text{if } x_i = x_j = 0$$

or alternatively, if $\mu := \max\{|x_i|, |x_j|\} > 0$, by

$$c := x_i/k, \qquad s := x_j/k.$$

$|k|$ is to be computed as

$$|k| = \mu\sqrt{(x_i/\mu)^2 + (x_j/\mu)^2},$$

and the sign of k is, for the moment, arbitrary. The computation of $|k|$ in this form avoids problems which would arise from exponent overflow or underflow given extreme values for the components x_i and x_j. On numerical grounds the sign of k will be chosen according to

$$k = |k| \operatorname{sign}(x_i), \operatorname{sign}(x_i) := \begin{cases} 1 & \text{if } x_i \geq 0, \\ -1 & \text{if } x_i < 0, \end{cases}$$

so that the computation of the auxiliary value

$$v := s/(1 + c)$$

can take place without cancellation. Using v, the significant components z_i, z_j of the transformation $z := E_{ij}u$ of the vector $u \in \mathbb{R}^n$ can be computed somewhat more efficiently (in which a multiplication is replaced by an addition) as follows:

$$z_i := cu_i + su_j,$$

$$z_j := v(u_i + z_i) - u_j.$$

The type of matrix $\hat{E}$ shown in (4.9.6), together with its associated E_{ij}, is known as a *Givens reflection*. We can just as easily use a matrix $\hat{E}$ of the form

$$\hat{E} = \begin{bmatrix} c & s \\ -s & c \end{bmatrix}, \qquad c = \cos \varphi, \quad s = \sin \varphi,$$

which, together with its associated E_{ij}, is known as a *Givens rotation*. In this case $\hat{E}$ and E_{ij} are orthogonal matrices which describe a rotation of $\mathbb{R}^m$

about the origin through the angle φ in the i-j plane; $\det(\hat{E}) = 1$. We will present the following material, however, only in terms of Givens reflections.

Since modification techniques for the decomposition (4.9.1) differ from those for (4.9.2a) only in that, in place of Givens matrices (4.9.6), corresponding elimination matrices $\hat{E}$ of type (4.9.5) are used, we will only study the orthogonal factorization (4.9.2). The techniques for (4.9.2b) are similar to, but more complicated than, those for (4.9.2a). Consequently, for simplicity's sake, we will restrict our attention to (4.9.2a). The corresponding discussion for (4.9.2b) can be found in Daniel, Gragg, Kaufman, and Stewart (1976).

In the following let A be a real $m \times n$ matrix with $m \geqslant n$, and let

$$PA = \begin{bmatrix} R \\ 0 \end{bmatrix}$$

be a factorization of type (4.9.2a).

(1) *Change of a row or column*, or more generally, the *change of A to* $\bar{A} := A + vu^T$, where $v \in \mathbb{R}^m$, $u \in \mathbb{R}^n$ are given vectors. From (4.9.2a) we have

$$\text{(4.9.7)} \qquad P\bar{A} = \begin{bmatrix} R \\ 0 \end{bmatrix} + wu^T, \qquad w := Pv \in \mathbb{R}^m.$$

In the first step of the modification we annihilate the successive components $m, m-1, \ldots, 2$ of the vector w using appropriate Givens matrices $G_{m-1,m}$, $G_{m-2,m-1}, \ldots, G_{12}$, so that

$$\tilde{w} = ke_1 = G_{12}G_{23} \ldots G_{m-1,m}w = (k, 0, \ldots, 0)^T \in \mathbb{R}^m,$$
$$k = \pm\|w\| = \pm\|u\|.$$

The sketch below in the case $m = 4$ should clarify the effect of carrying out the successive transformations $G_{i,i+1}$. We denote by * here, and in the further sketches of this section, those elements which are changed by a transformation, other than the one set to zero:

$$w = \begin{bmatrix} x \\ x \\ x \\ x \end{bmatrix} \xrightarrow{G_{34}} \begin{bmatrix} x \\ x \\ * \\ 0 \end{bmatrix} \xrightarrow{G_{23}} \begin{bmatrix} x \\ * \\ 0 \\ 0 \end{bmatrix} \xrightarrow{G_{12}} \begin{bmatrix} * \\ 0 \\ 0 \\ 0 \end{bmatrix} = \tilde{w}.$$

If (4.9.7) is multiplied on the left as well by $G_{m-1,m}, \ldots, G_{12}$ in order, we obtain

$$\text{(4.9.8)} \qquad \tilde{P}\bar{A} = R' + ke_1u^T =: \tilde{R},$$

where

$$\tilde{P} := GP, \qquad R' := G\begin{bmatrix} R \\ 0 \end{bmatrix}, \qquad G := G_{12}G_{23} \ldots G_{m-1,m}.$$

Here $\tilde{P}$ is unitary, as are G and P; the upper triangular matrix $\left[\begin{smallmatrix} R \\ 0 \end{smallmatrix}\right]$ is changed step by step into an *upper Hessenberg* matrix

$$R' = G\begin{bmatrix} R \\ 0 \end{bmatrix},$$

that is a matrix with $(R')_{ik} = 0$ for $i > k + 1$:

$$(m = 4, \quad n = 3)$$

$$\begin{bmatrix} R \\ 0 \end{bmatrix} = \begin{bmatrix} x & x & x \\ 0 & x & x \\ 0 & 0 & x \\ 0 & 0 & 0 \end{bmatrix} \xrightarrow{G_{34}} \begin{bmatrix} x & x & x \\ 0 & x & x \\ 0 & 0 & * \\ 0 & 0 & * \end{bmatrix} \xrightarrow{G_{23}} \begin{bmatrix} x & x & x \\ 0 & * & * \\ 0 & * & * \\ 0 & 0 & x \end{bmatrix} \xrightarrow{G_{12}} \begin{bmatrix} * & * & * \\ * & * & * \\ 0 & x & x \\ 0 & 0 & x \end{bmatrix}$$

$$= G\begin{bmatrix} R \\ 0 \end{bmatrix} = R'.$$

$\tilde{R} = R' + ke_1 u^T$ in (4.9.8) is also an upper Hessenberg matrix, since the addition changes only the first row of R'. In the second step of the modification we annihilate the subdiagonal elements $(\tilde{R})_{i+1,i}$, $i = 1, 2, \ldots, \mu := \min(m - 1, n - 1)$ of $\tilde{R}$ by means of further Givens matrices H_{12}, $H_{23}, \ldots, H_{\mu,\mu+1}$, so that [see (4.9.8)]

$$H\tilde{P}\bar{A} = H\tilde{R} =: \begin{bmatrix} \bar{R} \\ 0 \end{bmatrix}, \qquad H := H_{\mu,\mu+1} \cdots H_{23} H_{12},$$

where $\bar{R}$ is, again, an $n \times n$ upper triangular matrix and $\bar{P} := H\tilde{P}$ is an $m \times m$ unitary matrix. A decomposition of $\bar{A}$ of the type (4.9.2a) is provided once again:

$$\bar{P}\bar{A} = \begin{bmatrix} \bar{R} \\ 0 \end{bmatrix}.$$

The sequence of transformations $\tilde{R} \to H_{12}\tilde{R} \to H_{23}(H_{12}\tilde{R}) \to \cdots \to H\tilde{R}$ will be sketched for $m = 4$, $n = 3$:

$$\tilde{R} = \begin{bmatrix} x & x & x \\ x & x & x \\ 0 & x & x \\ 0 & 0 & x \end{bmatrix} \xrightarrow{H_{12}} \begin{bmatrix} * & * & * \\ 0 & * & * \\ 0 & x & x \\ 0 & 0 & x \end{bmatrix} \xrightarrow{H_{23}} \begin{bmatrix} x & x & x \\ 0 & * & * \\ 0 & 0 & * \\ 0 & 0 & x \end{bmatrix} \xrightarrow{H_{34}} \begin{bmatrix} x & x & x \\ 0 & x & x \\ 0 & 0 & * \\ 0 & 0 & 0 \end{bmatrix}$$

$$= \begin{bmatrix} \bar{R} \\ 0 \end{bmatrix}.$$

(2) *Deletion of a column.* If $\bar{A}$ is obtained from A by the deletion of the kth column, then from (4.9.2a) the matrix $\tilde{R} := P\bar{A}$ is an upper Hessenberg matrix of the following form (sketched for $m = 4$, $n = 4$, $k = 2$):

$$\tilde{R} := P\bar{A} = \begin{bmatrix} x & x & x \\ & x & x \\ & x & x \\ & & x \end{bmatrix}.$$

The subdiagonal elements of $\tilde{R}$ are to be annihilated as described before, using Givens matrices

$$H_{k,k+1},\ H_{k+1,k+2},\ \ldots,\ H_{n-1,n}.$$

The decomposition of $\bar{A}$ is

$$\bar{P}\bar{A} = \begin{bmatrix} \bar{R} \\ 0 \end{bmatrix}, \qquad \bar{P} := HP,$$

$$\begin{bmatrix} \bar{R} \\ 0 \end{bmatrix} := H \begin{bmatrix} \tilde{R} \\ 0 \end{bmatrix}, \qquad H := H_{n-1,n} H_{n-2,n-1} \cdots H_{k,k+1}.$$

(3) *Addition of a column.* If $\bar{A} = (A, a)$, $a \in \mathbb{R}^m$, and A is an $m \times n$ matrix with $m > n$, then (4.9.2a) implies

$$P\bar{A} = \left[\begin{array}{c|c} R & Pa \\ 0 & \end{array}\right] =: \tilde{R} = \begin{bmatrix} x & \cdots & x & x \\ & \ddots & \vdots & \vdots \\ & & x & x \\ & & & x \\ & & & \vdots \\ 0 & & & x \end{bmatrix}.$$

The subdiagonal elements of $\tilde{R}$, which appear as a "spike" in the last column, can be annihilated by means of a single Householder transformation H from (4.7.5):

$$H\tilde{R} = \begin{bmatrix} x & \cdots & x & x \\ & \ddots & \vdots & \vdots \\ & & x & x \\ & & & * \\ & & & 0 \\ & & & \vdots \\ 0 & & & 0 \end{bmatrix} = \begin{bmatrix} \bar{R} \\ 0 \end{bmatrix}.$$

$\bar{P} := HP$ and $\bar{R}$ are components of the decomposition (4.9.2a) of $\bar{A}$:

$$\bar{P}\bar{A} = \begin{bmatrix} \bar{R} \\ 0 \end{bmatrix}.$$

(4) *Addition of a row.* If

$$\bar{A} = \begin{bmatrix} A \\ a^T \end{bmatrix}, \qquad a \in \mathbb{R}^n,$$

then there is a permutation matrix Π of order $m + 1$ which satisfies

$$\Pi\bar{A} = \begin{bmatrix} a^T \\ A \end{bmatrix}.$$

The unitary matrix of order $m + 1$,

$$\tilde{P} := \begin{bmatrix} 1 & 0 \\ 0 & P \end{bmatrix} \Pi,$$

satisfies, according to (4.9.2a),

$$\tilde{P}\bar{A} = \begin{bmatrix} 1 & 0 \\ 0 & P \end{bmatrix} \begin{bmatrix} a^T \\ A \end{bmatrix} = \begin{bmatrix} a^T \\ PA \end{bmatrix} = \begin{bmatrix} a^T \\ R \\ 0 \end{bmatrix} =: \tilde{R} = \begin{bmatrix} x & \dots & x \\ x & & x \\ & \ddots & \vdots \\ & & x \\ & & \vdots \\ 0 & & 0 \end{bmatrix}.$$

$\tilde{R}$ is an upper Hessenberg matrix whose subdiagonal elements can be annihilated using Givens matrices $H_{12}, \dots, H_{n,n+1}$:

$$\bar{P} = H\tilde{P}, \qquad \begin{bmatrix} \bar{R} \\ 0 \end{bmatrix} = H\tilde{R}, \qquad H = H_{n,n+1} \cdots H_{23} H_{12},$$

as described before, to produce the decomposition

$$\bar{P}\bar{A} = \begin{bmatrix} \bar{R} \\ 0 \end{bmatrix}.$$

(5) *Deletion of a row:* Let A be an $m \times n$ matrix with $m > n$. We assume without loss of generality (see the use of the permutation above) that the last row a^T of A is to be dropped:

$$A = \begin{bmatrix} \bar{A} \\ a^T \end{bmatrix}.$$

We partition the matrix

$$P = (\tilde{P}, p), \qquad p \in \mathbb{R}^m$$

accordingly and obtain, from (4.9.2a),

$$(4.9.9) \qquad (\tilde{P}, p)\begin{bmatrix} \bar{A} \\ a^T \end{bmatrix} = \begin{bmatrix} R \\ 0 \end{bmatrix}.$$

We choose Givens matrices $H_{m,m-1}, H_{m,m-2}, \dots, H_{m1}$ to annihilate successive components $m-1, m-2, \dots, 1$ of p: $H_{m1}H_{m2} \cdots H_{m,m-1}\, p = (0, \dots, 0, \pi)^T$. A sketch for $m = 4$ is

$$(4.9.10) \qquad p = \begin{bmatrix} x \\ x \\ x \\ x \end{bmatrix} \xrightarrow{H_{43}} \begin{bmatrix} x \\ x \\ 0 \\ * \end{bmatrix} \xrightarrow{H_{42}} \begin{bmatrix} x \\ 0 \\ 0 \\ * \end{bmatrix} \xrightarrow{H_{41}} \begin{bmatrix} 0 \\ 0 \\ 0 \\ * \end{bmatrix} = \begin{bmatrix} 0 \\ 0 \\ 0 \\ \pi \end{bmatrix}.$$

Now, P is unitary, so $\|p\| = |\pi| = 1$. Therefore, the transformed matrix HP, $H := H_{m1}H_{m2} \cdots H_{m,m-1}$ has the form

$$(4.9.11) \qquad HP = \begin{bmatrix} \bar{P} & 0 \\ q & \pi \end{bmatrix} = \begin{bmatrix} \bar{P} & 0 \\ 0 & \pi \end{bmatrix}, \qquad \pi = \pm 1,$$

since $|\pi| = 1$ implies $q = 0$ because of the unitarity of HP. Consequently $\bar{P}$ is a unitary matrix of order $m - 1$. On the other hand, the H_{mi} transform the upper triangular matrix $\binom{R}{0}$, with the exception of its last row, into an upper triangular matrix

$$H\begin{bmatrix} R \\ 0 \end{bmatrix} = H_{m1} H_{m2} \dots H_{m,\, m-1} \begin{bmatrix} R \\ 0 \end{bmatrix} = \begin{bmatrix} \bar{R} \\ 0 \\ z^T \end{bmatrix} \begin{matrix} \}\, n \\ \}\, m-n-1 \\ \}\, 1 \end{matrix}$$

Sketching this for $m = 4$, $n = 3$, we have

$$\begin{bmatrix} R \\ 0 \end{bmatrix} = \begin{bmatrix} x & x & x \\ 0 & x & x \\ 0 & 0 & x \\ 0 & 0 & 0 \end{bmatrix} \xrightarrow{H_{43}} \begin{bmatrix} x & x & x \\ 0 & x & x \\ 0 & 0 & * \\ 0 & 0 & * \end{bmatrix} \xrightarrow{H_{42}} \begin{bmatrix} x & x & x \\ 0 & * & * \\ 0 & 0 & x \\ 0 & * & * \end{bmatrix} \xrightarrow{H_{41}} \begin{bmatrix} * & * & * \\ 0 & x & x \\ 0 & 0 & x \\ * & * & * \end{bmatrix}$$

$$= \begin{bmatrix} \bar{R} \\ z^T \end{bmatrix}.$$

From (4.9.9), (4.9.11) it follows that

$$HP\begin{bmatrix} \bar{A} \\ a^T \end{bmatrix} = \begin{bmatrix} \bar{P} & 0 \\ 0 & \pi \end{bmatrix} \begin{bmatrix} \bar{A} \\ a^T \end{bmatrix} = \begin{bmatrix} \bar{R} \\ 0 \\ z^T \end{bmatrix},$$

and therefore

$$\bar{P}\bar{A} = \begin{bmatrix} \bar{R} \\ 0 \end{bmatrix}$$

for the $(m-1)$-order unitary matrix $\bar{P}$ and the upper triangular matrix $\bar{R}$ of order n. This has produced a decomposition for $\bar{A}$ of the form (4.9.2a). It can be shown that the techniques of this section are numerically stable in the following sense. Let P and R be given matrices with the following property: There exists an exact unitary matrix P' and a matrix A' such that

$$P'A' = \begin{bmatrix} R \\ 0 \end{bmatrix}$$

is a decomposition of A' in the sense of (4.9.2a) and the differences $\|P - P'\|$, $\|A - A'\|$ are "small." Then the methods of this section, applied to P, R in floating-point arithmetic with relative machine precision eps, produce matrices $\bar{P}$, $\bar{R}$ to which are associated an exact unitary matrix $\bar{P}'$ and a matrix $\bar{A}'$ satisfying:

(a) $\|\bar{P} - \bar{P}'\|$, $\|\bar{A} - \bar{A}'\|$ are "small," and
(b) $\bar{P}'\bar{A}' = \binom{\bar{R}}{0}$ is a decomposition of $\bar{A}'$ in the sense of (4.9.2a).

By "small" we mean that the differences above satisfy

$$\|\Delta P\| = O(m^\alpha \text{ eps}) \quad \text{or} \quad \|\Delta A\| = O(m^\alpha \text{ eps } \|A\|),$$

where α is small (for example $\alpha = \frac{3}{2}$).

4.10 The Simplex Method

Linear algebraic techniques can be applied advantageously, in the context of the simplex method, to solve *linear programming problems*. These problems arise frequently in practice, particularly in the areas of economic planning and management science. Linear programming is also the means by which a number of important discrete approximation problems (e.g. data fitting in the L_1 and L_∞ norms) can be solved. At this introductory level of treatment we can cover only the most basic aspects of linear programming; for a more thorough treatment, the reader is referred to the special literature on this subject [e.g. Dantzig (1963), Gass (1969), Hadley (1962), or Murty (1976)].

A general linear programming problem (or *linear program*) has the following form:

$$\text{(4.10.1)} \qquad \text{minimize} \quad c_1x_1 + c_2x_2 + \cdots + c_nx_n \equiv c^Tx$$

with respect to all $x \in \mathbb{R}^n$ which satisfy finitely many constraints of the form

$$\text{(4.10.2)} \qquad \begin{aligned} a_{i1}x_1 + a_{i2}x_2 + \cdots + a_{in}x_n &\leqslant b_i, \qquad i = 1, 2, \ldots, m_1, \\ a_{i1}x_1 + a_{i2}x_2 + \cdots + a_{in}x_n &= b_i, \qquad i = m_1 + 1, m_1 + 2, \ldots, m. \end{aligned}$$

The numbers c_k, a_{ik}, b_i are given real numbers. The function c^Tx to be minimized is called the *objective function*. Each $x \in \mathbb{R}^n$ which satisfies all of the conditions (4.10.2) is said to be a *feasible point* for the problem. By introducing additional variables and equations, the linear programming problem (4.10.1), (4.10.2) can be put in a form in which the only constraints which appear are equalities or *elementary inequalities* (inequalities, for example, of the form $x_i \geqslant 0$). It is useful, for various reasons, to require that the objective function c^Tx have the form $c^Tx \equiv -x_p$. In order to bring a linear programming problem to this form, we replace each nonelementary inequality (4.10.2)

$$a_{i1}x_1 + \cdots + a_{in}x_n \leqslant b_i$$

by an equality and an elementary inequality using a *slack variable* x_{n+i},

$$a_{i1}x_1 + \cdots + a_{in}x_n + x_{n+i} = b_i, \qquad x_{n+i} \geqslant 0.$$

If the objective function $c_1x_1 + \cdots + c_nx_n$ is not elementary, we introduce an additional variable x_{n+m_1+1} and include an additional equation

$$c_1x_1 + \cdots + c_nx_n + x_{n+m_1+1} = 0$$

among the constraints (4.10.2). The minimization of c^Tx is equivalent to the maximization of x_{n+m_1+1} under this extended system of constraints.

Hence we can assume, without loss of generality, that the linear programming problem is already given in the following *standard form*:

$$\text{(4.10.3)} \qquad \begin{aligned} \text{LP}(I, p)\text{:} \quad & \text{maximize} \quad x_p \\ & \text{over all} \quad x \in \mathbb{R}^n \quad \text{with } Ax = b, \\ & \qquad\qquad\qquad x_i \geqslant 0 \quad \text{for } i \in I. \end{aligned}$$

In this formulation, $I \subsetneq N := \{1, 2, \ldots, n\}$ is a (possibly empty) index set, p is a fixed index satisfying $p \in N \backslash I$, $A = (a_1, a_2, \ldots, a_n)$ is a real $m \times n$ matrix having columns a_i, and $b \in \mathbb{R}^m$ is a given vector. The variables x_i for which $i \in I$ are the *restricted* variables, while those for which $i \notin I$ are the *free* variables. By

$$P := \{x \in \mathbb{R}^n \,|\, Ax = b \;\&\; x_i \geqslant 0 \text{ for all } i \in I\}$$

we denote the set of all feasible points of LP(I, p), $x^* \in P$ is an *optimum point* of LP(I, p) if $x_p^* = \max\{x_p | x \in P\}$.

As an illustration:

$$\begin{aligned} \text{minimize} \quad & -x_1 - 2x_2 \\ \text{subject to} \quad & -x_1 + x_2 \leqslant 2, \\ & x_1 + x_2 \leqslant 4, \\ & x_1 \geqslant 0, \quad x_2 \geqslant 0. \end{aligned}$$

After the introduction of x_3, x_4 as slack variables and x_5 as an objective function variable, the following standard form is obtained, $I = \{1, 2, 3, 4\}$, $p = 5$:

$$\begin{aligned} \text{maximize} \quad & x_5 \\ \text{subject to} \quad & -x_1 + x_2 + x_3 = 2, \\ & x_1 + x_2 + x_4 = 4, \\ & -x_1 - 2x_2 + x_5 = 0, \\ & x_i \geqslant 0 \quad \text{for } i \leqslant 4. \end{aligned}$$

This can be shown graphically in $\mathbb{R}^2$. The set P (shaded in Figure 4) is a polygon. (In higher dimensions P would be a polyhedron.)

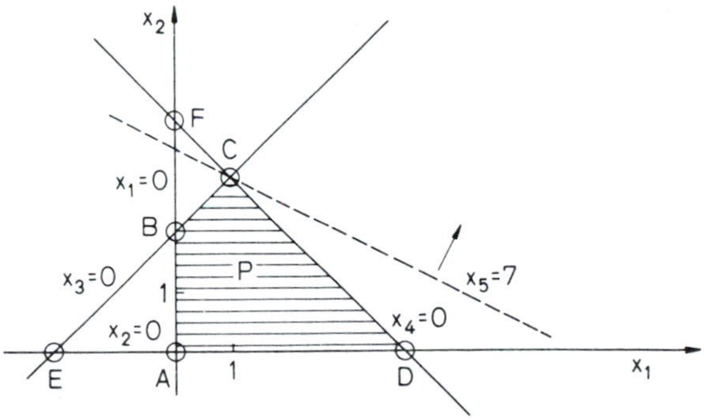

Figure 4 Feasible region and objective function.

We begin by considering the linear equation system $Ax = b$ of LP(I, p). For any index vector $J = (j_1, \ldots, j_r)$, $j_i \in N$, we let $A_J := (a_{j_1}, \ldots, a_{j_r})$ denote the submatrix of A having columns a_{j_i}; x_J denotes the vector $(x_{j_1}, \ldots, x_{j_r})^T$.

For simplicity we will denote the set

$$\{j_i \mid i = 1, 2, \ldots, r\}$$

of the components of J by J also, and we will write $p \in J$ if there exists an i with $p = j_i$.

(4.10.4) Definition. An index vector $J = (j_1, \ldots, j_m)$ of m distinct indices $j_i \in N$ is called a *basis* of $Ax = b$ [and of LP(I, p)] if A_J *is* nonsingular. A_J is also referred to as a basis; the variables x_i for $i \in J$ are referred to as *basic variables*, while the remaining variables x_k, $k \notin J$, are the *nonbasic variables*. If $K = (k_1, \ldots, k_{n-m})$ is an index vector containing the nonbasic indices, we will use the notation $J \oplus K = N$.

In the above example $J_A := (3, 4, 5)$ and $J_B := (4, 5, 2)$ are bases.

Associated with any basis J, $J \oplus K = N$, there is a unique solution $\bar{x} = \bar{x}(J)$ of $Ax = b$, called the *basic solution*, with $\bar{x}_K = 0$. Since

$$A\bar{x} = A_J\bar{x}_J + A_K\bar{x}_K = A_J\bar{x}_J = b,$$

$\bar{x}$ is given by

$$\bar{x}_J := \bar{b}, \quad \bar{x}_K := 0 \quad \text{with } \bar{b} := A_J^{-1}b. \tag{4.10.5}$$

Moreover, given a basis J, each solution x of $Ax = b$ is uniquely determined by its nonbasic segment x_K and the basic solution $\bar{x}$. This follows from the multiplication of $Ax = A_J x_J + A_K x_K = b$ by A_J^{-1} and from (4.10.5):

$$\begin{aligned} x_J &= \bar{b} - A_J^{-1}A_K x_K \\ &= \bar{x}_J - A_J^{-1}A_K x_K. \end{aligned} \tag{4.10.6}$$

If $x_K \in \mathbb{R}^{n-m}$ is chosen arbitrarily, and if x_J (and thereby x) is defined by (4.10.6), then x is a solution of $Ax = b$. Therefore, (4.10.6) provides a parametrization of the solution set $\{x \mid Ax = b\}$ through the components of $x_K \in \mathbb{R}^{n-m}$.

If the basic solution $\bar{x}$ of $Ax = b$ associated with the basis J is a feasible point of LP(I, p), $\bar{x} \in P$, that is, if

$$\bar{x}_i \geqslant 0 \quad \text{for all } i \in I \quad \Leftrightarrow \quad \bar{x}_i \geqslant 0 \quad \text{for all } i \in I \cap J, \tag{4.10.7}$$

then J is a *feasible basis* of LP(I, p) and $\bar{x}$ is a *basic feasible solution*. Finally, a feasible basis is said to be *nondegenerate* if

$$\bar{x}_i > 0 \quad \text{for all } i \in I \cap J. \tag{4.10.8}$$

The linear programming problem as a whole is nondegenerate if all of its feasible bases are nondegenerate.

Geometrically, the basic feasible solutions given by the various bases of LP(I, p) correspond to the vertices of the polyhedron P of feasible points, assuming that P does have vertices. In the above example (see Figure 3) the vertex $A \in P$ belongs to the feasible basis $J_A = (3, 4, 5)$, since A is determined by $x_1 = x_2 = 0$ and $\{1, 2\}$ is in the complementary set corresponding to J_A in $N = \{1, 2, 3, 4, 5\}$; B is associated with $J_B = (4, 5, 2)$, C is associated with $J_C = (1, 2, 5)$, etc. The basis $J_E = (1, 4, 5)$ is not feasible, since the associated basic solution E is not a feasible point ($E \notin P$).

The simplex method for solving linear programming problems, due to G. B. Dantzig, is a process which begins with a feasible basis J_0 of LP(I, p) satisfying $p \in J_0$ and recursively generates a sequence $\{J_i\}$ of feasible bases J_i of LP(I, p) with $p \in J_i$ for each i by means of *simplex steps*

$$J_i \rightarrow J_{i+1}.$$

These steps are designed to ensure that the objective function values $\bar{x}(J_i)_p$ corresponding to the bases J_i are nondecreasing:

$$\bar{x}(J_i)_p \leqslant \bar{x}(J_{i+1})_p \quad \text{for all } i \geqslant 0.$$

In fact, if all of the bases encountered are nondegenerate, this sequence is strictly increasing, and if LP(I, p) does have an optimum, then the sequence $\{J_i\}$ terminates after finitely many steps in a basis J_M whose basic solution $\bar{x}(J_M)$ is an optimum point for LP(I, p) and is such that $\bar{x}(J_i)_p < \bar{x}(J_{i+1})_p$ for all $0 \leqslant i \leqslant M - 1$. Furthermore, each two successive bases $J = (j_1, \ldots, j_m)$ and $\tilde{J} = (\tilde{j}_1, \ldots, \tilde{j}_m)$ are *neighboring*, that is, J and $\tilde{J}$ have exactly $m - 1$ components in common. This means that $\tilde{J}$ may be obtained from J through an index exchange: there are precisely two indices q, $s \in N$ satisfying $q \in J$, $s \notin J$ and $q \notin \tilde{J}$, $s \in \tilde{J}$: thus $\tilde{J} = (J \cup \{s\}) \setminus \{q\}$.

For nondegenerate problems, neighboring feasible bases correspond geometrically to neighboring vertices of P. In the example above (Figure 3), $J_A = (3, 4, 5)$ and $J_B = (4, 5, 2)$ are neighboring bases, and A and B are neighboring vertices.

The simplex method, or more precisely, "phase two" of the simplex method, assumes that a feasible basis J of LP(I, p) satisfying $p \in J$ is already available. Given that LP(I, p) does have feasible bases, one such basis can be found using a variant known as "phase one" of the simplex method. We will begin by describing a typical step of phase two of the simplex method, which leads from a feasible basis J to a neighboring basis $\tilde{J}$ of LP(I, p):

(4.10.9) Simplex Step. *Requirements: $J = (j_1, \ldots, j_m)$ is assumed to be a feasible basis of* LP(I, p) *having $p = j_t \in J$, $J \oplus K = N$.*
(1) *Compute the vector*

$$\bar{b} := A_J^{-1} b$$

which gives the basic feasible solution $\bar{x}$ corresponding to J, where

$$\bar{x}_J := \bar{b}, \qquad \bar{x}_K := 0.$$

(2) *Compute the row vector*

$$\pi := e_t^T A_J^{-1},$$

where $e_t = (0, \ldots, 1, \ldots, 0)^T \in \mathbb{R}^m$ *is the* t*th coordinate vector for* $\mathbb{R}^m$. *Determine the values*

$$c_k := \pi a_k, \qquad k \in K.$$

using π.

(3) *Determine whether*

(4.10.10)
$$c_k \geqslant 0 \quad \textit{for all } k \in K \cap I$$
$$\textit{and} \quad c_k = 0 \quad \textit{for all } k \in K \backslash I.$$

(a) *If so, then stop. The basic feasible solution* $\bar{x}$ *is optimal for* LP(I, p).

(b) *If not, determine an* $s \in K$ *such that*

$$c_s < 0, \quad s \in K \cap I \quad \textit{or} \quad c_s \neq 0, \qquad s \in K \backslash I,$$

and set $\sigma := -\operatorname{sign} c_s$.

(4) *Calculate the vector*

$$\bar{\alpha} := (\bar{\alpha}_1, \bar{\alpha}_2, \ldots, \bar{\alpha}_m)^T := A_J^{-1} a_s.$$

(5) *If*

(4.10.11)
$$\sigma \bar{\alpha}_i \leqslant 0 \quad \textit{for all } i \textit{ with } j_i \in I,$$

stop: LP(I, p) *has no finite optimum. Otherwise,*

(6) *Determine an index,* r *with* $j_r \in I$, $\sigma \bar{\alpha}_r > 0$, *and*

$$\frac{\bar{b}_r}{\sigma \bar{\alpha}_r} = \min\left\{\frac{\bar{b}_i}{\sigma \bar{\alpha}_i} \,\middle|\, i : j_i \in I \;\&\; \sigma \bar{\alpha}_i > 0\right\}.$$

(7) *Take as* $\tilde{J}$ *any suitable index vector with*

$$\tilde{J} := (J \cup \{s\}) \backslash \{j_r\},$$

for example

$$\tilde{J} := (j_1, \ldots, j_{r-1}, s, j_{r+1}, \ldots, j_m)$$

or

$$\tilde{J} := (j_1, \ldots, j_{r-1}, j_{r+1}, \ldots, j_m, s).$$

We wish to justify these rules. Let us assume that $J = (j_1, \ldots, j_m)$ is a feasible basis for LP(I, p) with $p = j_t \in J$, $J \oplus K = N$. Rule (1) of (4.10.9) produces the associated basic feasible solution $\bar{x} = \bar{x}(J)$, as is implied by (4.10.5). Since all solutions of $Ax = b$ can be represented in the form given by (4.10.6), the objective function can be written as

(4.10.12)
$$\begin{aligned} x_p = e_t^T x_J &= \bar{x}_p - e_t^T A_J^{-1} A_K x_K \\ &= \bar{x}_p - \pi A_K x_K \\ &= \bar{x}_p - c_K x_K. \end{aligned}$$

This uses the fact that $p = j_t$, and it uses the definitions of the vector π and of the components c_K, $c_K^T \in \mathbb{R}^{n-m}$, as given in rule (2). c_K is known as the vector of *reduced costs*. As a result of (4.10.12), c_k, $k \in K$, gives the amount by which the objective function x_p changes as x_k is changed by one unit. If the condition (4.10.10) is satisfied [see rule (3)], it follows from (4.10.12) for each *feasible* point x of LP(I, p), since $x_i \geq 0$ for $i \in I$, that

$$x_p = \bar{x}_p - \sum_{k \in K \cap I} c_k x_k \geqslant \bar{x}_p;$$

that is, the basic feasible solution $\bar{x}$ is an optimum point for LP(I, p). This motivates the test (4.10.10) and statement (a) of rule (3). If (4.10.10) is not satisfied, there is an index $s \in K$ for which either

$$(4.10.13) \qquad c_s < 0, \qquad s \in K \cap I$$

or else

$$(4.10.14) \qquad c_s \neq 0, \qquad s \in K \backslash I$$

holds. Let s be such an index. We set $\sigma := -\operatorname{sign} c_s$. Because (4.10.12) implies that an increase of σx_s leads to an increase of the objective function x_p, we consider the following family of vectors $x(\theta) \in \mathbb{R}^n$, $\theta \in \mathbb{R}$:

$$(4.10.15) \qquad \begin{aligned} x(\theta)_J &:= \bar{b} - \theta\sigma A_J^{-1} a_s = \bar{b} - \theta\sigma\bar{\alpha}, \\ x(\theta)_s &:= \theta\sigma, \\ x(\theta)_k &:= 0 \quad \text{for } k \in K,\ k \neq s. \end{aligned}$$

Here $\bar{\alpha} := A_J^{-1} a_s$ is defined as in rule (4) of (4.10.9).

In our example $I = \{1, 2, 3, 4\}$ and $J_0 = J_A = (3, 4, 5)$ is a feasible basis; $K_0 = (1, 2)$, $p = 5 \in J_0$, $t_0 = 3$. (4.10.9) produces, starting from J_0,

$$A_{J_0} = \begin{bmatrix} 1 & & \\ & 1 & \\ & & 1 \end{bmatrix}, \qquad \bar{b} = \begin{bmatrix} 2 \\ 4 \\ 0 \end{bmatrix},$$

$\bar{x}(J_0) = (0, 0, 2, 4, 0)^T$ (which is point A in Figure 3), and $\pi A_{J_0} = e_{t_0}^T \Leftrightarrow \pi = (0, 0, 1)$. The reduced costs are $c_1 = \pi a_1 = -1$, $c_2 = \pi a_2 = -2$. This implies that J_0 is not optimal. If we choose the index $s = 2$ for (4.10.9), rule (3)(b), then

$$\bar{\alpha} = A_{J_0}^{-1} a_2 = \begin{bmatrix} 1 \\ 1 \\ -2 \end{bmatrix}.$$

The family of vectors $x(\theta)$ is given by

$$x(\theta) = (0, \theta, 2 - \theta, 4 - \theta, 2\theta)^T.$$

Geometrically $x(\theta)$, $\theta \geqslant 0$ describes a semi-infinite ray in Figure 3 which extends along the edge of the polyhedron P from vertex $A(\theta = 0)$ in the direction of the neighboring vertex $B(\theta = 2)$.

From (4.10.6) it follows that $Ax(\theta) = b$ for all $\theta \in \mathbb{R}$. In particular (4.10.12) implies, from the fact that $\bar{x} = x(0)$ and as a result of the choice of σ, that

$$x(\theta)_p = \bar{x}_p - c_s x(\theta)_s = \bar{x}_p + \theta |c_s|, \tag{4.10.16}$$

so that the objective function is a strictly increasing function of θ on the family $x(\theta)$. It is reasonable to select the best *feasible* solution of $Ax = b$ from among the $x(\theta)$; that is, we wish to find the largest $\theta \geqslant 0$ satisfying

$$x(\theta)_i \geqslant 0 \quad \text{for all } i \in I.$$

From (4.10.15), this is equivalent to finding the largest $\theta \geqslant 0$ such that

$$x(\theta)_{j_i} \equiv \bar{b}_i - \theta\sigma\bar{\alpha}_i \geqslant 0 \quad \text{for all } i \text{ with } j_i \in I, \tag{4.10.17}$$

because $x(\theta)_k \geqslant 0$ is automatically satisfied, for all $k \in K \cap I$ and $\theta \geqslant 0$, because of (4.10.15). If $\sigma\bar{\alpha}_i \leqslant 0$ for all $j_i \in I$ [see rule (5) of (4.10.9)], then $x(\theta)$ is a feasible point for all $\theta \geqslant 0$, and because of (4.10.17) $\sup\{x(\theta)_p | \theta \geqslant 0\} = +\infty$: LP($I$, p) does not have a finite optimum. This justifies rule (5) of (4.10.9). Otherwise there is a largest $\theta = \bar{\theta}$ for which (4.10.17) holds:

$$\bar{\theta} = \frac{\bar{b}_r}{\sigma\bar{\alpha}_r} = \min\left\{\frac{\bar{b}_i}{\sigma\bar{\alpha}_i} \,\middle|\, i : j_i \in I \,\&\, \sigma\bar{\alpha}_i > 0\right\}.$$

This determines an index r with $j_r \in I$, $\sigma\bar{\alpha}_r > 0$ and

$$x(\bar{\theta})_{j_r} = \bar{b}_r - \bar{\theta}\sigma\bar{\alpha}_r = 0, \quad x(\bar{\theta}) \text{ feasible.} \tag{4.10.18}$$

In the example

$$\bar{\theta} = 2 = \frac{\bar{b}_1}{\bar{\alpha}_1} = \min\left\{\frac{\bar{b}_1}{\bar{\alpha}_1}, \frac{\bar{b}_2}{\bar{\alpha}_2}\right\}, \qquad r = 1,$$

$x(\bar{\theta}) = (0, 2, 0, 2, 4)^T$ corresponds to vertex B of Figure 3.

From the feasibility of J, $\bar{\theta} \geqslant 0$, and it follows from (4.10.16) that

$$x(\bar{\theta})_p \geqslant \bar{x}_p.$$

If J is nondegenerate, as defined in (4.10.8), then $\bar{\theta} > 0$ holds and further,

$$x(\bar{\theta})_p > \bar{x}_p.$$

From (4.10.6), (4.10.15), and (4.10.18), $x = x(\bar{\theta})$ is the uniquely determined solution of $Ax = b$ having the additional property

$$x_k = 0 \quad \text{for } k \in K, k \neq s$$

$$x_{j_r} = 0,$$

that is, $x_{\tilde{K}} = 0$, $\tilde{K} := (K \cup \{j_r\})\backslash\{s\}$. From the uniqueness of x it follows that $A_{\tilde{J}}$, $\tilde{J} := (J \cup \{s\})\backslash\{j_r\}$, is nonsingular; $x(\theta) = \bar{x}(\tilde{J})$ is, therefore, the basic solution associated with the neighboring feasible basis $\tilde{J}$, and we have

$$(4.10.19) \qquad \begin{array}{ll} \bar{x}(\tilde{J})_p > \bar{x}(J)_p & \text{if } J \text{ is nondegenerate,} \\ \bar{x}(\tilde{J})_p \geqslant \bar{x}(J)_p & \text{otherwise.} \end{array}$$

In our example we obtain the new basis $J_1 = (2, 4, 5) = J_B$, $K_1 = (1, 3)$, which corresponds to the vertex B of Figure 3. With respect to the objective function x_5, B is better than A: $\bar{x}(J_B)_5 = 4 > \bar{x}(J_A)_5 = 0$.

Since the definition of r implies that $j_r \in I$ always holds, it follows that

$$J\backslash I \subseteq \tilde{J}\backslash I;$$

that is, in the transition $J \to \tilde{J}$ at most one nonnegatively constrained variable x_{j_r}, $j_r \in I$, leaves the basis; as soon as a free variable x_s, $s \notin I$, becomes a basis variable, it remains in the basis throughout all subsequent simplex steps. In particular, $p \in \tilde{J}$, because $p \in J$ and $p \notin I$. Hence, the new basis $\tilde{J}$ also satisfies the requirements of (4.10.9), so that rules (1)–(7) can be applied to $\tilde{J}$ in turn. Thus, beginning with a first feasible basis J_0 of LP(I, p) with $p \in J_0$, we obtain a sequence

$$J_0 \to J_1 \to J_2 \to \cdots$$

of feasible bases J_i of LP(I, p) with $p \in J_i$, for which

$$\bar{x}(J_0)_p < \bar{x}(J_1)_p < \bar{x}(J_2)_p < \cdots$$

in case all of the J_i are nondegenerate. Since there are only finitely many index vectors J, and since no J_i can be repeated if the above chain of inequalities holds, the simplex method must terminate after finitely many steps. Thus, we have proven the following for the simplex method:

(4.10.20) Theorem. *Let J_0 be a feasible basis for* LP(I, p) *with $p \in J_0$. If* LP(I, p) *is nondegenerate, then the simplex method generates a finite sequence of feasible bases J_i for* LP(I, p) *with $p \in J_i$ which begin with J_0 and for which $\bar{x}(J_i)_p < \bar{x}(J_{i+1})_p$. Either the final basic solution is optimal for* LP(I, p), *or* LP(I, p) *has no finite optimum.*

We continue with our example: as a result of the first simplex step we have a new feasible basis $J_1 = (2, 4, 5) = J_B$, $K_1 = (1, 3)$, $t_1 = 3$, so that

$$A_{J_1} = \begin{bmatrix} 1 & 0 & 0 \\ 1 & 1 & 0 \\ -2 & 0 & 1 \end{bmatrix}, \qquad \bar{b} = \begin{bmatrix} 2 \\ 2 \\ 4 \end{bmatrix}, \qquad \bar{x}(J_1) = (0, 2, 0, 2, 4)^T (\equiv B),$$

$$\pi A_{J_1} = e_{t_1}^T \quad \Rightarrow \quad \pi = (2, 0, 1).$$

The reduced costs are $c_1 = \pi a_1 = -3$, $c_3 = \pi a_3 = 2$; hence J_1 is not optimal:

$$s = 1,\ \bar{\alpha} = A_{J_1}^{-1} a_1 \quad \Rightarrow \quad \bar{\alpha} = \begin{bmatrix} -1 \\ 2 \\ -3 \end{bmatrix} \quad \Rightarrow \quad r = 2.$$

Therefore

$$J_2 = (2, 1, 5) = J_C, \qquad K_2 = (3, 4), \qquad t_2 = 3,$$

$$A_{J_2} = \begin{bmatrix} 1 & -1 & 0 \\ 1 & 1 & 0 \\ -2 & -1 & 1 \end{bmatrix}, \qquad \bar{b} = \begin{bmatrix} 3 \\ 1 \\ 7 \end{bmatrix}.$$

$$\bar{x}(J_2) = (1, 3, 0, 0, 7)(\equiv C),$$

$$\pi A_{J_2} = e_{t_2}^T \quad \Rightarrow \quad \pi = (\tfrac{1}{2}, \tfrac{3}{2}, 1).$$

The reduced costs are $c_3 = \pi a_3 = \frac{1}{2} > 0$, $c_4 = \pi a_4 = \frac{3}{2} > 0$.

The optimality criterion is satisfied, so $\bar{x}(J_2)$ is optimal; i.e., $\bar{x}_1 = 1$, $\bar{x}_2 = 3$, $\bar{x}_3 = 0$, $\bar{x}_4 = 0$, $\bar{x}_5 = 7$. The optimal value of the objective function x_5 is $\bar{x}_5 = 7$.

We observe that the important quantities of each simplex step—$\bar{b}$, π, and $\bar{\alpha}$—are determined by the following systems of linear equations:

$$\begin{aligned} A_J \bar{b} &= b \ \Rightarrow \bar{b} \quad [(4.10.9), \text{rule } (1)], \\ \pi A_J &= e_t^T \Rightarrow \pi \quad [(4.10.9), \text{rule } (2)], \\ A_J \bar{a} &= a_s \ \Rightarrow \bar{a} \quad [(4.10.9), \text{rule } (4)]. \end{aligned} \tag{4.10.21}$$

The computational effort required to solve these systems for successive bases $J \to \tilde{J} \to \cdots$ can be significantly reduced if it is noted that the successive bases are neighboring: each new basis matrix $A_{\tilde{J}}$ is obtained from its predecessor A_J by the exchange of one column of A_J for one column of A_K. Suppose, for example, a decomposition of the basis matrix A_J of the form (4.9.1), $FA_J = R$, F nonsingular, R upper triangular, is used (see Section 4.9). On the one hand, such a decomposition can be used to solve the equation systems (4.10.21) in an efficient manner:

$$R\bar{b} = Fb \ \Rightarrow \bar{b},$$

$$R^T z = e_t \ \Rightarrow z \ \Rightarrow \pi = z^T F,$$

$$R\bar{\alpha} = Fa_s \ \Rightarrow \bar{\alpha}.$$

On the other hand, the techniques of Section 4.9 can be used on the decomposition $FA_J = R$ of A_J in each simplex step to obtain the corresponding decomposition $\tilde{F}A_{\tilde{J}} = \tilde{R}$ of the neighboring basis

$$\tilde{J} = (j_1, \ldots, j_{r-1}, j_{r+1}, \ldots, j_m, s)$$

[compare (4.10.9), rule (7)]. The matrix $FA_{\tilde{J}}$, with this choice of index vector $\tilde{J}$, is an upper Hessenberg matrix of the form depicted here for $m = 4$, $r = 2$:

$$FA_{\tilde{J}} = \begin{bmatrix} x & x & x & x \\ & x & x & x \\ & x & x & x \\ & & x & x \end{bmatrix} =: R'.$$

The subdiagonal elements can easily be eliminated using transformations of the form (4.9.5), $E_{r,r+1}, E_{r+1,r+2}, \ldots, E_{m-1,m}$, which will change R' into an upper triangular matrix $\tilde{R}$:

$$\tilde{F}A_{\tilde{J}} = \tilde{R}, \qquad \tilde{F} := EF, \qquad \tilde{R} := ER', \qquad E := E_{m-1,m}E_{m-2,m-1} \cdots E_{r,r+1}.$$

Thus it is easy to implement the simplex method in a practical fashion by taking a quadruple $\mathfrak{M} = \{J; t; F, R\}$ with the property that

$$p = j_t, \qquad FA_J = R$$

and changing it at each simplex step $J \to \tilde{J}$ into an analogous quadruple $\tilde{\mathfrak{M}} = \{\tilde{J}; \tilde{t}; \tilde{F}, \tilde{R}\}$. To begin this variant of the simplex method, it is necessary to find a factorization $F_0 A_{J_0} = R_0$ for A_{J_0} of the form given in (4.9.1) as well as finding a feasible basis J_0 of LP(I, p) with $p \in J_0$. ALGOL programs for such an implementation of the simplex method are to be found in Wilkinson and Reinsch (1971), and a roundoff investigation is to be found in Bartels (1971).

[In practice, particularly for large problems in which the solution of any one of the systems (4.10.21) takes a significant amount of time, the vector $\bar{x}_J = A_J^{-1}b = \bar{b}$ of one simplex step is often updated to the vector $\bar{x}_{\tilde{J}} = x(\bar{\theta})_{\tilde{J}} = A_{\tilde{J}}^{-1}b$ of the next step by using (4.10.15) with the value $\bar{\theta} = \bar{b}_r/\sigma\bar{\alpha}_r$ as given by (4.10.9), rule (6). The chance of incurring errors, particularly in the selection of the index r, should be borne in mind if this is done.]

The more usual implementations of the simplex method use other quantities than the decomposition $FA_J = R$ of (4.9.1) in order to solve the equation systems (4.10.21) in an efficient manner. The "basis inverse method" uses a quintuple of the form

$$\hat{\mathfrak{M}} = \{J; t; B, \bar{b}, \pi\}$$

with

$$p = j_t, \quad B := A_J^{-1}, \quad \bar{b} := A_J^{-1}b, \quad \pi := e_t^T A_J^{-1}.$$

Another variant uses the quintuple

$$\bar{\mathfrak{M}} = \{J; t; \bar{A}, \bar{b}, \pi\}$$

with

$$j_t = p, \quad \bar{A} := A_J^{-1}A_K, \quad \bar{b} := A_J^{-1}b, \quad \pi := e_t^T A_J^{-1}, \quad J \oplus K = N.$$

In the simplex step $J \to \tilde{J}$ efficiency can be gained by observing that the inverse $A_{\tilde{J}}^{-1}$ is obtainable from A_J^{-1} through multiplication by an appropriate Frobenius matrix G (see Exercise 4 of Chapter 4): $A_{\tilde{J}}^{-1} = GA_J^{-1}$. In this manner somewhat more computational effort can be saved than when the decomposition $FA_J = R$ is used. The disadvantage of updating the inverse of the basis matrix using Frobenius transformations, however, lies in numerical instability: it is possible for a Frobenius matrix which is used in this inverse updating process to be ill conditioned, particularly if the basis matrix itself is ill conditioned. In this case, errors can appear in $A_{J_i}^{-1}$, $A_{J_i}^{-1}A_{K_i}$ for $\hat{\mathfrak{M}}_i$, $\bar{\mathfrak{M}}_i$, and they will be propagated throughout all quintuples $\hat{\mathfrak{M}}_j$, $\bar{\mathfrak{M}}_j$, $j > i$. When the factorization $FA_J = R$ is used, however, it is always possible to update the decomposition with well-conditioned transformations, so that error propagation is not likely to occur.

The following practical example is typical of the gain in numerical stability which one obtains if the triangular factorization of the basis (4.9.1) is used to implement the simplex method rather than the basis inverse method. Consider a linear programming problem with constraints of the form

(4.10.22)
$$Ax = b, \qquad A = (\text{A1}, \text{A2}),$$
$$x \geqslant 0.$$

The 10×20 matrix A is composed of two 10×10 submatrices A1, A2:

$$\text{A1} = (\text{a1}_{ik}), \quad \text{a1}_{ik} = \frac{1}{i+k}, \qquad i, k = 1, \ldots, 10,$$

$$\text{A2} = I = \text{identity matrix of order 10.}$$

A1 is very ill conditioned, and A2 is very well conditioned. The vector

$$b_i := \sum_{k=1}^{10} \frac{1}{i+k},$$

that is,

$$b := \text{A1} \cdot e, \quad \text{where } e = (1, \ldots, 1)^T \in \mathbb{R}^{10},$$

is chosen as a right-hand side. Hence, the bases $J_1 := (1, 2, \ldots, 10)$ and $J_2 := (11, \ldots, 20)$ are both feasible for (4.10.22); the corresponding basic solutions are

(4.10.23)
$$\bar{x}(J_1) := \begin{pmatrix} \bar{b}_1 \\ 0 \end{pmatrix}, \qquad \bar{b}_1 := A_{J_1}^{-1} b = A1^{-1} b = e,$$
$$\bar{x}(J_2) := \begin{pmatrix} 0 \\ \bar{b}_2 \end{pmatrix}, \qquad \bar{b}_2 := A_{J_2}^{-1} b = A2^{-1} b = b.$$

We choose $J_2 = (11, 12, \ldots, 20)$ as a starting basis and transform it, using the basis inverse method on the one hand and triangular decompositions on the other, via a sequence of single column exchanges, into the basis J_1. From there another sequence of single column exchanges is used to return to J_2:

$$J_2 \to \cdots \to J_1 \cdots \to J_2.$$

The resulting basic solutions (4.10.23), produced on a machine having relative precision eps $\approx 10^{-11}$, are shown in the following table.

Basis	Exact basic solution		Basis inverse		Triangular decomposition
$J_2\bar{b}_2 =$	b	$=$	b	$=$	b
$J_1\bar{b}_1 =$	1	$\neq$	$9.995\ 251\ 6281_{10}-1$	$\neq$	$9.998\ 637\ 8918_{10}-1$
	1		$1.018\ 551\ 9965_{10}\ \ 0$		$1.005\ 686\ 0944_{10}\ \ 0$
	1		$7.596\ 751\ 9456_{10}-1$		$9.196\ 219\ 3606_{10}-1$
	1		$2.502\ 031\ 9231_{10}\ \ 0$		$1.555\ 895\ 6482_{10}\ \ 0$
	1		$-4.265\ 502\ 0405_{10}\ \ 0$		$-1.177\ 196\ 3556_{10}\ \ 0$
	1		$1.208\ 552\ 0735_{10}\ \ 1$		$6.157\ 172\ 0228_{10}\ \ 0$
	1		$-1.331\ 094\ 5894_{10}\ \ 1$		$-6.533\ 076\ 2136_{10}\ \ 0$
	1		$1.208\ 139\ 8560_{10}\ \ 1$		$7.633\ 553\ 7514_{10}\ \ 0$
	1		$-3.717\ 873\ 7849_{10}\ \ 0$		$-2.228\ 141\ 2212_{10}\ \ 0$
	1		$1.847\ 607\ 2006_{10}\ \ 0$		$1.666\ 626\ 9133_{10}\ \ 0$
$J_2\bar{b}_2 =$	$2.019\ 877\ 3448_{10}\ \ 0$	$\neq$	$-1.199\ 958\ 5030_{10}\ \ 0$	$\neq$	$2.019\ 877\ 3448_{10}\ \ 0$
	$1.603\ 210\ 6781_{10}\ \ 0$		$-1.114\ 741\ 3211_{10}\ \ 0$		$1.603\ 210\ 6780_{10}\ \ 0$
	$1.346\ 800\ 4217_{10}\ \ 0$		$-9.987\ 647\ 8557_{10}-1$		$1.346\ 800\ 4217_{10}\ \ 0$
	$1.168\ 228\ 9932_{10}\ \ 0$		$-8.910\ 752\ 3957_{10}-1$		$1.168\ 228\ 9933_{10}\ \ 0$
	$1.034\ 895\ 6598_{10}\ \ 0$		$-7.980\ 878\ 7783_{10}-1$		$1.034\ 895\ 6598_{10}\ \ 0$
	$9.307\ 289\ 9322_{10}-1$		$-7.191\ 967\ 3857_{10}-1$		$9.307\ 289\ 9617_{10}-1$
	$8.466\ 953\ 7977_{10}-1$		$-6.523\ 444\ 2973_{10}-1$		$8.466\ 953\ 7978_{10}-1$
	$7.772\ 509\ 3533_{10}-1$		$-5.954\ 308\ 9303_{10}-1$		$7.772\ 509\ 3560_{10}-1$
	$7.187\ 714\ 0316_{10}-1$		$-5.466\ 517\ 9294_{10}-1$		$7.187\ 714\ 0320_{10}-1$
	$6.687\ 714\ 0316_{10}-1$		$-5.045\ 374\ 4089_{10}-1$		$6.687\ 714\ 0328_{10}-1$

The following is to be observed: Since $A_{J_2} = I_{10}$, both computational methods yield the exact solution at the start. For the basis J_1, both methods give equally inexact results, which reflects the ill-conditioning of A_{J_1}. No computational method could produce better results than these without resorting to higher-precision arithmetic. After passing through this ill-conditioned basis A_{J_1}, however, the situation changes radically in favor of the triangular decomposition method. This method yields, once again, the basic solution corresponding to J_2 essentially to full machine accuracy. The basis inverse method, in contrast, produces a basic solution for J_2 with the same inaccuracy as it did for J_1. With the basis inverse method, the effect of ill-conditioning encountered while processing one basis matrix A_J is felt throughout all further bases; this is not the case using triangular factorization.

4.11 Phase One of the Simplex Method

In order to start phase two of the simplex method, we require a feasible basis J_0 of LP(I, p) with $p = j_{t_0} \in J_0$; alternatively, we must find a quadruple $\mathfrak{M}_0 = \{J_0; t_0; F_0, R_0\}$ in which a nonsingular matrix F_0 and a nonsingular triangular matrix R_0 form a decomposition $F_0 A_{J_0} = R_0$ as in (4.9.1) of the basis matrix A_{J_0}.

In some special cases, finding $J_0(\mathfrak{M}_0)$ presents no problem, e.g. if LP(I, p) results from a linear programming problem of the special form

$$\begin{aligned} \text{minimize} \quad & c_1 x_1 + \cdots + c_n x_n \\ \text{subject to} \quad & a_{i1} x_1 + \cdots + a_{in} x_n \leqslant b_i, \quad i = 1, 2, \ldots, m, \\ & x_i \geqslant 0 \quad \text{for } i \in I_1 \subset \{1, 2, \ldots, n\}, \end{aligned}$$

with the additional property

$$b_i \geqslant 0 \quad \text{for } i = 1, 2, \ldots, m.$$

Such problems may be transformed by the introduction of slack variables $x_{n+1}, \ldots, x_{n+m}$ into the form

$$\begin{aligned} & \text{maximize} \quad x_{n+m+1} \\ (4.11.1) \quad & \text{subject to} \quad a_{i1} x_1 + \cdots + a_{in} x_n + x_{n+i} = b_i, \quad i = 1, 2, \ldots, m, \\ & \qquad c_1 x_1 + \cdots + c_n x_n + x_{n+m+1} = 1, \\ & \qquad x_i \geqslant 0 \quad \text{for } i \in I := I_1 \cup \{n+1, \ldots, n+m\}. \end{aligned}$$

Note that $x_{n+m+1} = 1 - c_1 x_1 - \cdots - c_n x_n$. The extra positive constant (arbitrarily selected to be 1) prevents the initially chosen basis from being degenerate. (4.11.1) has the standard form LP(I, p) with

$$A = \begin{bmatrix} a_{11} & \ldots & a_{1n} & 1 & & \\ \vdots & & \vdots & & \ddots & \\ a_{m1} & \ldots & a_{mn} & & & 1 & \\ c_1 & \ldots & c_n & & & & 1 \end{bmatrix}, \quad b = \begin{bmatrix} b_1 \\ \vdots \\ b_m \\ 1 \end{bmatrix},$$

$$p = n + m + 1, \qquad I := I_1 \cup \{n+1, \ldots, n+m\}$$

and an initial basis J_0 with $p \in J_0$ and a corresponding $\mathfrak{M}_0 = (J_0; t_0; F_0, R_0)$ given by

$$J_0 := (n+1, n+2, \ldots, n+m+1) \qquad t_0 \simeq m+1$$

$$F_0 = R_0 := I_{m+1} = \begin{bmatrix} 1 & & 0 \\ & \ddots & \\ 0 & & 1 \end{bmatrix} \quad (\text{order } m+1).$$

Since $b_i \geqslant 0$ for $i = 1, \ldots, m$, the basis J_0 is feasible for (4.11.1).

"Phase one of the simplex method" is a name for a class of techniques which are applicable in general. Essentially, all of these techniques consist of applying phase two of the simplex method to some linear programming problem derived from the given problem. The optimal basis obtained from the derived problem provides a starting basis for the given problem. We will sketch one such method here in the briefest possible fashion. This sketch is included for the sake of completeness. For full details on starting techniques for the simplex method, the reader is referred to the extensive literature on linear programming, for example to Dantzig (1963), Gass (1969), Hadley (1962), and Murty (1976).

Consider a general linear programming problem which has been cast into the form

$$\text{(4.11.2)}\qquad \begin{aligned} &\text{minimize} && c_1 x_1 + \cdots + c_n x_n \\ &\text{subject to} && a_{j1} x_1 + \cdots + a_{jn} x_n = b_j, \quad j = 1, \ldots, m, \\ & && x_i \geqslant 0 \quad \text{for } i \in I, \end{aligned}$$

where $I \subseteq \{1, \ldots, n\}$. We may assume without loss of generality that $b_j \geqslant 0$ for $j = 1, \ldots, m$ (otherwise multiply the nonconforming equations through by -1).

We begin by extending the constraints of the problem by introducing *artificial variables* $x_{n+1}, \ldots, x_{n+m}$:

$$\text{(4.11.3)}\qquad \begin{aligned} a_{11} x_1 + \cdots + a_{1n} x_n + x_{n+1} \qquad\qquad &= b_1, \\ \vdots \qquad\qquad \ddots \qquad\qquad & \quad \vdots \\ a_{m1} x_1 + \cdots + a_{mn} x_n \qquad\qquad + x_{n+m} &= b_m, \\ x_i \geqslant 0 \quad \text{for } i \in I \cup \{n+1, \ldots, n+m\}. & \end{aligned}$$

Clearly there is a one-to-one correspondence between feasible points for (4.11.2) and those feasible points for (4.11.3) which satisfy

$$\text{(4.11.4)}\qquad x_{n+1} = x_{n+2} = \cdots = x_{n+m} = 0.$$

We will set up a derived problem with constraints (4.11.3) whose maximum should be taken on, if possible, by a point satisfying (4.11.4). Consider

LP($\hat{I}$, $\hat{p}$):

$$\begin{aligned} &\text{maximize} && x_{n+m+1} \\ &\text{subject to} && a_{11} x_1 + \cdots + a_{1n} x_n + x_{n+1} \qquad\qquad = b_1, \\ & && \vdots \qquad\qquad \ddots \qquad\qquad \vdots \\ & && a_{m1} x_1 + \cdots + a_{mn} x_n \qquad + x_{n+m} \qquad = b_m, \\ & && x_{n+1} + \cdots + x_{n+m} + x_{n+m+1} = 2\sum_{i=1}^{m} b_i, \\ & && x_i \geqslant 0 \text{ for all } i \in \hat{I} := I \cup \{n+1, \ldots, n+m\} \\ & && \text{with } \hat{p} = n + m + 1. \end{aligned}$$

We may take $\hat{J}_0 = \{n+1, \ldots, n+m+1\}$ as a starting basis with

$$\bar{b}_0 = \left(b_1, \ldots, b_m, \sum_{i=1}^{m} b_i\right)^T.$$

It is evident that $x_{n+m+1} \leqslant 2 \sum_{j=1}^{m} b_j$; hence $\mathrm{LP}(\hat{I}, \hat{p})$ is bounded. Furthermore

$$x_{n+m+1} = 2 \sum_{i=1}^{m} b_i$$

if and only if (4.11.4) holds.

Corresponding to $\hat{J}_0$ we have the quadruple $\mathfrak{M}_0 = (\hat{J}_0 ; \hat{t}_0 ; \hat{F}_0, \hat{R}_0)$ given by

$$\hat{t}_0 = m + 1,$$

$$\hat{F}_0 = \begin{bmatrix} 1 & & & 0 \\ & \ddots & & \\ 0 & & 1 & \\ -1 & \dots & -1 & 1 \end{bmatrix},$$

$$\hat{R}_0 = \begin{bmatrix} 1 & & 0 \\ & \ddots & \\ 0 & & 1 \end{bmatrix}.$$

Phase 2 can be applied immediately to $\mathrm{LP}(\hat{I}, \hat{p})$, and it will terminate with one of three possible outcomes:

(1) $x_{n+m+1} < 2 \sum_{i=1}^{m} b_i$ [i.e., (4.11.4) does not hold],
(2) $x_{n+m+1} = 2 \sum_{i=1}^{m} b_i$ and all artificial variables are nonbasic,
(3) $x_{n+m+1} = 2 \sum_{i=1}^{m} b_i$ and some artificial variables are basic.

In case (1), (4.11.2) is not a feasible problem [since any feasible point for (4.11.2) would correspond to a feasible point for $\mathrm{LP}(\hat{I}, \hat{p})$ with $x_{n+m+1} = 2 \sum_{i=1}^{m} b_i$]. In case (2), the optimal basis for $\mathrm{LP}(\hat{I}, \hat{p})$ clearly provides a feasible basis for (4.11.2). Phase 2 can be applied to the problem as originally given. In case 3, we are faced with a degenerate problem, since the artificial variables which are basic must have the value zero. We may assume without loss of generality (by renumbering equations and artificial variables as necessary) that $x_{n+1}, \dots, x_{n+k}$ are the artificial variables which are basic and have value zero. We may replace x_{n+m+1} by a new variable x_{n+k+1} and use the optimal basis for $\mathrm{LP}(\hat{I}, \hat{p})$ to provide an initial feasible basis for the problem

$$\begin{array}{llllll}
\text{maximize} & c_1 x_1 + \cdots\cdots + c_n x_n & & & & \\
\text{subject to} & a_{11} x_1 + \cdots\cdots + a_{1n} x_n + x_{n+1} & & & & = b_1, \\
& \vdots \qquad\qquad \vdots & & \ddots & & \vdots \\
& a_{k1} x_1 + \cdots\cdots + a_{kn} x_n & & + x_{n+k} & & = b_k, \\
& & x_{n+1} + \cdots & + x_{n+k} & + x_{n+k+1} & = 0, \\
& a_{k+1,1} x_1 + \cdots + a_{k+1,n} x_n & & & & = b_{k+1} \\
& \vdots \qquad\qquad \vdots & & & & \vdots \\
& a_{m1} x_1 + \cdots\cdots + a_{mn} x_n & & & & = b_m
\end{array}$$

$$x_i \geqslant 0 \quad \text{for } i \in I \cup \{n+1, \dots, n+k+1\}.$$

This problem is evidently equivalent to (4.11.2).

Appendix to Chapter 4

4.A Elimination Methods for Sparse Matrices

Many practical applications require solving systems of linear equations $Ax = b$ with a matrix A that is very large but *sparse*, i.e., only a small fraction of the elements a_{ik} of A are nonzero. Such applications include the solution of partial differential equations by means of discretization methods (see Sections 7.4, 7.5, 8.4), network problems, or structural design problems in engineering. The corresponding linear systems can be solved only if the sparsity of A is used in order to reduce memory requirements, and if solution methods are designed accordingly. Many sparse linear systems, in particular, those arising from partial differential equations, are solved by iterative methods (see Chapter 8). In this section, we consider only elimination methods, in particular, the Cholesky method (see 4.3) for solving linear systems with a positive-definite matrix A, and explain in this context some basic techniques for exploiting the sparsity of A. For further results, we refer to the literature, e.g., Reid (1971), Rose and Willoughby (1972), Tewarson (1973), and Barker (1974). A systematic exposition of these methods for positive-definite systems is found in George and Liu (1981), and for general sparse linear systems in Duff, Erisman, and Reid (1986).

We first illustrate some basic storage techniques for sparse matrices. Consider, for instance, the matrix

$$A = \begin{bmatrix} 1 & 0 & 0 & 0 & -2 \\ 3 & 0 & 2 & 0 & 1 \\ 0 & -4 & 0 & 7 & 0 \\ 0 & -5 & 0 & 0 & 0 \\ 0 & -6 & 0 & 0 & 6 \end{bmatrix}.$$

One possibility is to store such a matrix by rows, for instance, in three one-dimensional arrays, say a, ja, and ip. Here, $a[k]$, $k = 1, 2, \ldots$, are the values of the (potentially) nonzero elements of A, $ja[k]$ records the column index of the matrix component stored in $a[k]$. The array ip holds pointers: if $ip[i] = p$ and $ip[i+1] = q$ then the segment of nonzero elements in row i of A begins with $a[p]$ and ends with $a[q-1]$. In particular, if $ip[i] = ip[i+1]$ then row i of A is zero. So, the matrix could be stored in memory as follows

$i =$	1	2	3	4	5	6	7	8	9	10	11
$ip[i] =$	1	3	6	8	9	11					
$ja[i] =$	5	1	1	3	5	4	2	2	2	5	
$a[i] =$	-2	1	3	2	1	7	-4	-5	-6	6	

Of course, for symmetric matrices A, further savings are possible if one stores only the nonzero elements a_{ik} with $i \le k$.

When using this kind of data structure, it is difficult to insert additional nonzero elements into the rows of A, as might be necessary with elimination methods. This drawback is avoided if the nonzero elements of A are stored as a *linked list*. It requires only one additional array *next*: if $a[k]$ contains an element of row i of A, then the "next" nonzero element of row i is found in $a[next[k]]$, if $next[k] \neq 0$. If $next[k] = 0$ then $a[k]$ was the "last" nonzero component of row i.

Using linked lists the matrix can be stored as follows:

$i =$	1	2	3	4	5	6	7	8	9	10
$ip[i] =$	6	4	5	10	9					
$ja[i] =$	2	3	5	1	4	1	5	5	2	2
$a[i] =$	−4	2	−2	3	7	1	1	6	−6	−5
$next[i] =$	0	7	0	2	1	3	0	0	8	0

Now it is easy to insert a new element into a row of A: e.g., a new element a_{31} could be incorporated by extending the vectors a, ja, and *next* each by one component $a[11]$, $ja[11]$, and $next[11]$. The new element a_{31} is stored in $a[11]$; $ja[11] = 1$ records its column index; and the vectors *next* and *ip* are adjusted as follows: $next[11] := ip[3]\ (=5)$, $ip[3] := 11$. On the other hand, with this technique the vector *ip* no longer contains information on the number of nonzero elements in the rows of A as it did before.

Refined storage techniques are also necessary for the efficient implementation of iterative methods to solve large sparse systems (see Chapter 8). However, with elimination methods there are additional difficulties, if the storage (data structure) used for the matrix A is also to be used for storing the factors of the triangular decomposition generated by these methods (see Sections 4.1 and 4.3): These factors may contain many more nonzero elements than A. In particular, the number of these extra nonzeros (the "fill-in" of A) created during the elimination depends heavily on the choice of the pivot elements. Thus a bad choice of a pivot may not only lead to numerical instabilities (see Section 4.5), but also spoil the original sparsity pattern of the matrix A. It is therefore desirable to find a sequence of pivots that not only ensures numerical stability but also limits fill-in as much as possible. In the case of Cholesky's method for positive-definite matrices A, the situation is particularly propitious because, in that case, pivot selection is not crucial for numerical stability: Instead of choosing consecutive diagonal pivots, as described in Section 4.3, one may just as well select them in any other order without losing numerical stability (see (4.3.6)). One is therefore free to select the diagonal pivots in any order that tends to minimize the fill-in generated during the elimination. This amounts to finding a permutation P such that the matrix $PAP^T = LL^T$ has a Cholesky factor L that is as sparse as possible.

The following example shows that the choice of P may influence the sparsity of L drastically. Here, the diagonal elements are numbered so as to indicate their ordering under a permutation of A, and x denotes elements $\neq 0$: The Cholesky factor L of a positive-definite matrix A having the form

$$A = \begin{bmatrix} 1 & x & x & x & x \\ x & 2 & & & \\ x & & 3 & & \\ x & & & 4 & \\ x & & & & 5 \end{bmatrix} = LL^T, \qquad L = \begin{bmatrix} x & & & & \\ x & x & & & \\ x & x & x & & \\ x & x & x & x & \\ x & x & x & x & x \end{bmatrix}$$

is in general a "full" matrix, whereas the permuted matrix PAP^T obtained by interchanging the first and last rows and columns of A has a sparse Cholesky factor

$$PAP^T = \begin{bmatrix} 5 & & & & x \\ & 2 & & & x \\ & & 3 & & x \\ & & & 4 & x \\ x & x & x & x & 1 \end{bmatrix} = LL^T, \qquad L = \begin{bmatrix} x & & & & \\ & x & & & \\ & & x & & \\ & & & x & \\ x & x & x & x & x \end{bmatrix}.$$

Efficient elimination methods for sparse positive-definite systems therefore consist of three parts:

1. *Determine P so that the Cholesky factor L of $PAP^T = LL^T$ is as sparse as possible, and determine the sparsity structure of L.*
2. *Compute L numerically.*
3. *Compute the solution x of $Ax = b$, i.e., of $(PAP^T)Px = LL^TPx = Pb$ by solving the triangular systems $Lz = Pb$, $L^Tu = z$, and letting $x := P^Tu$.*

In step 1, only the sparsity pattern of A as given by index set

$$\text{Nonz}(A) := \{(i, j) \mid j < i \text{ und } a_{ij} \neq 0\},$$

is used to find Nonz(L), not the numerical values of the components of L: In this step only a "symbolic factorization" of PAP^T takes place; the "numerical factorization" is left until step 2.

It is convenient to describe the sparsity structure of a symmetric $n \times n$ matrix A, i.e., the set Nonz(A), by means of an *undirected graph* $G^A = (V^A, E^A)$ with a finite set of vertices (nodes) $V^A = \{v_1, v_2, \ldots, v_n\}$ and a finite set

$$E^A := \{\{v_i, v_j\} \mid (i, j) \in \text{Nonz}(A)\}$$

of "undirected" edges $\{v_i, v_j\}$ between the nodes v_i and $v_j \neq v_i$ (thus an edge is a subset of V^A containing two elements). The vertex v_i is associated with the diagonal element a_{ii} (i.e., also with row i and column i) of A, and the vertices $v_i \neq v_j$ are connected by an edge in G^A if and only if $a_{ij} \neq 0$.

EXAMPLE 1. The matrix

$$\begin{bmatrix} 1 & x & & x & & & \\ x & 2 & & & & & \\ & & 3 & & x & x & \\ x & & & 4 & x & & \\ & & x & x & 5 & & x \\ & & x & & & 6 & x \\ & & & & x & x & 7 \end{bmatrix}$$

is associated with the graph G^A

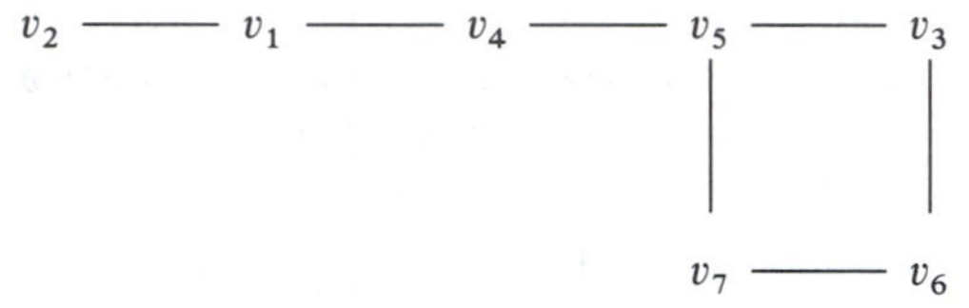

We need a few concepts from graph theory. If $G = (V, E)$ is an undirected graph and $S \subset V$ a subset of its vertices then $Adj_G(S)$, or briefly

$$Adj(S) := \{v \in V \backslash S \mid \{s, v\} \in E \text{ for some } s \in S\}$$

denotes the set of all vertices $v \in V \backslash S$ that are connected to (a vertex of) S by an edge of G. For example, $Adj(\{v\})$ or briefly $Adj(v)$ is the set of neighbors of the node v in G. The number of neighbors of the vertex $v \in V$ $\deg v := |Adj(v)|$ is called the *degree* of v. Finally, a subset $M \subset V$ of the vertices is called a *clique* in G if each vertex $x \in M$ of M is connected to any other vertex $y \in M$, $y \neq x$, by an edge of G.

We return to the discussion of elimination methods. First we try to find a permutation so that the number of nonzero elements of the Cholesky factor L of $PAP^T = LL^T$ becomes minimal. Unfortunately, an efficient method for computing an optimal P is not known, but there is a simple heuristic method, the *minimal degree algorithm* of Rose (1972), to find a P that nearly minimizes the sparsity of L. Its basic idea is to select the next pivot element in Cholesky's method as that diagonal element that is likely to destroy as few 0-elements as possible in the elimination step at hand.

To make this precise, we have to analyze only the first step of the Cholesky algorithm, since this step is already typical for the procedure in general. By partitioning the $n \times n$ matrices A and L in the equation $A = LL^T$

$$A = \begin{bmatrix} d & a^T \\ a & \tilde{A} \end{bmatrix} = \begin{bmatrix} \alpha & 0 \\ l & \bar{L} \end{bmatrix} \cdot \begin{bmatrix} \alpha & l^T \\ 0 & \bar{L}^T \end{bmatrix} = LL^T, \qquad a^T = (a_2, a_3, \ldots, a_n),$$

where $d = a_{11}$ and (d, a^T) is the first row of A, we find the following relations

$$\alpha = \sqrt{d}, \qquad l = a/\sqrt{d}, \qquad \bar{L}\bar{L}^T = \bar{A} := \tilde{A} - ll^T.$$

Hence, the first column L_1 of L and the first row L_1^T of L^T are given by

$$L_1 = \begin{bmatrix} \sqrt{d} \\ a/\sqrt{d} \end{bmatrix},$$

and the computation of the remaining columns of L, that is, of the columns of $\bar{L}$, is equivalent to the Cholesky factorization $\bar{A} = \bar{L}\bar{L}^T$ of the $(n-1) \times (n-1)$ matrix $\bar{A} = (\bar{a}_{ik})_{i,k=2}^n$:

$$\text{(4.A.1)} \qquad \begin{aligned} \bar{A} &= \tilde{A} - ll^T = \tilde{A} - \frac{aa^T}{d} \\ \bar{a}_{ik} &= a_{ik} - \frac{a_i a_k}{d} \quad \text{for all } i, k \geq 2. \end{aligned}$$

Disregarding the exceptional case in which the numerical values of $a_{ik} \neq 0$ and $a_i a_k = a_{1i} a_{1k} \neq 0$ are such that one obtains $\bar{a}_{ik} = 0$ by cancellation, we have

$$\text{(4.A.2)} \qquad \bar{a}_{ik} \neq 0 \Leftrightarrow a_{ik} \neq 0 \quad \text{or} \quad a_{1i} a_{1k} \neq 0.$$

Therefore, the elimination step with the pivot $d = a_{11}$ will generate a number of new nonzero elements to $\bar{A}$, which is roughly proportional to the number of nonzero components of the vector $a^T = (a_2, \ldots, a_n) = (a_{12}, \ldots, a_{1n})$.

The elimination step $A \to \bar{A}$ can be described in terms of the graphs $G = (V, E) := G^A$ and $\bar{G} = (\bar{V}, \bar{E}) := G^{\bar{A}}$ associated with A and $\bar{A}$: The vertices $1, 2, \ldots, n$ of $G = G^A$ (resp. $2, 3, \ldots, n$ of $\bar{G} = G^{\bar{A}}$) correspond to the diagonal elements of A (resp. $\bar{A}$); in particular, the pivot vertex 1 of G belongs to the pivot element a_{11}. By (4.A.2), the vertices $i \neq k$, $i, k \geq 2$, of $\bar{G}$ are connected by an edge in $\bar{G}$ if and only if they are either connected by an edge in G ($a_{ik} \neq 0$), or both vertices i, k are neighbors of pivot node 1 in G ($a_{1i} a_{1k} \neq 0$, $i, k \in Adj_G(1)$). The number of nonzero elements $a_{1i} \neq 0$ with $i \geq 2$ in the first row of A is equal to the degree $\deg_G(1)$ of pivot vertex 1 in G. Therefore, a_{11} is a favorable pivot, if vertex 1 is a vertex of minimum degree in G. Moreover, the set $Adj_G(1)$ describes exactly which nondiagonal elements of the first row of L^T (first column of L) are nonzero.

EXAMPLE 2. Choosing a_{11} as pivot in the following matrix A leads to fill-in at the positions denoted by $\otimes$ in the matrix $\bar{A}$:

$$A := \begin{bmatrix} 1 & x & & & x & \\ x & 2 & x & x & & x \\ & x & 3 & & & x \\ & x & & 4 & x & x \\ x & & & x & 5 & \\ & x & x & x & & 6 \end{bmatrix} \Rightarrow \bar{A} = \begin{bmatrix} 2 & x & x & \otimes & x \\ x & 3 & & & x \\ x & & 4 & x & x \\ \otimes & & x & 5 & \\ x & x & x & & 6 \end{bmatrix}.$$

The associated graphs are:

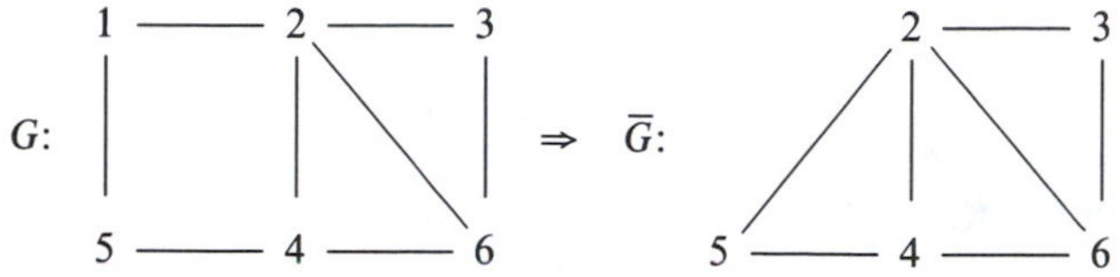

Generally, the choice of a diagonal pivot in A corresponds to the choice of a vertex $x \in V$ (pivot vertex) in the graph $G = (V, E) = G^A$, and the elimination step $A \to \bar{A}$ with this pivot corresponds to a transformation of the graph $G\ (= G^A)$ into the graph $\bar{G} = (\bar{V}, \bar{E})\ (= G^{\bar{A}})$ (which is also denoted by $G_x = \bar{G}$ to stress its dependence on x) according to the following rules

1) $\bar{V} := V \backslash \{x\}$.
2) *Connect the nodes $y \neq z$, $y, z \in \bar{V}$ by an undirected edge ($\{y, z\} \in \bar{E}$), if y and z are connected by an edge in G ($\{y, z\} \in E$), or if y and z are neighbors of x in G ($y, z \in Adj_G(x)$).*

For obvious reasons, we say that the graph G_x is obtained by "elimination of the vertex x" from G.

By definition, we have in G_x for $y \in \bar{V} = V \backslash \{x\}$

$$(4.A.3) \qquad Adj_{G_x}(y) = \begin{cases} Adj_G(y) & \text{if } y \notin Adj_G(x) \\ (Adj_G(x) \cup Adj_G(y)) \backslash \{x, y\} & \text{otherwise.} \end{cases}$$

Every pair of vertices in $Adj_G(x)$ is connected by an edge in G_x. Those vertices thus form a clique in G_x, the so-called *pivot clique*, associated with pivot vertex x chosen in G. Moreover, $Adj_G(x)$ describes the nonzero off-diagonal elements of the column of L (resp. the row of L^T) that correspond to pivot vertex x.

As we have seen, the fill-in newly generated during an elimination step is probably small if the degree of the vertex selected as pivot vertex for this step is small. This motivates the minimal degree algorithm of Rose (1972) for finding a suitable sequence of pivots:

(4.A.4) Minimal Degree Algorithm. *Let A be a positive definite $n \times n$ matrix and $G^0 = (V^0, E^0) := G^A$ be the graph associated to A.*

For $i = 1, 2, \ldots, n$:

1) *Determine a vertex $x_i \in V^{i-1}$ of minimal degree in G^{i-1}.*
2) *Set $G^i := G^{i-1}_{x_i}$.*

Remark. A minimal degree vertex x_i need not be uniquely determined.

EXAMPLE 3. Consider the matrix A of example 2. Here, the nodes 1, 3, and 5 have minimal degree in $G^0 := G^A$. The choice of vertex 1 as pivot vertex x_1 leads to the graph $\bar{G}$ of Example 2, $G^1 := G^0_1 = \bar{G}$. The pivot clique associated with $x_1 = 1$ is

$$Adj_{G^0}(1) = \{2, 5\}.$$

In the next step of (4.A.4) one may choose vertex 5 as pivot vertex x_2, since vertices 5 and 3 have minimal degree in $G^1 = \overline{G}$. The pivot clique associated with $x_2 = 5$ is $Adj_{G^1}(5) = \{2, 4\}$, and by "elimination" of x_2 from G^1 we obtain the next graph $G^2 := G^1_5$. All in all, a possible sequence of pivot elements given by (4.A.4) is (1, 5, 4, 2, 3, 6), which leads to the series of graphs ($G^0 := G$ and $G^1 := \overline{G}$ are as in example 2, G^6 is the empty graph):

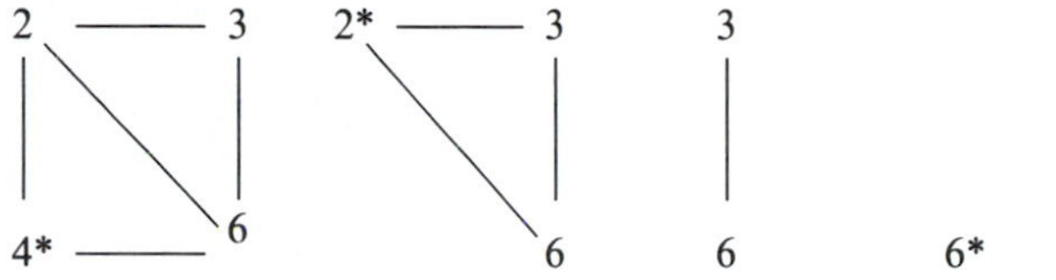

$G^2 = G^1_5$ $\qquad$ $G^3 = G^2_4$ $\qquad$ $G^3 = G^2_4$ $\qquad$ $G^5 = G^4_3$

The pivot vertices are marked. The associated pivot cliques are

Pivot	1	5	4	2	3	6
Clique	{2, 5}	{2, 4}	{2, 6}	{3, 6}	{6}	∅

The matrix $PAP^T = LL^T$ arising from a permutation of the rows and columns corresponding to the sequence of pivots selected and its Cholesky factor L have the following structure, which is determined by the pivot cliques (fill-in is denoted by $\otimes$):

$$PAP^T = \begin{bmatrix} 1 & x & & x & & \\ x & 5 & x & & & \\ & x & 4 & x & & x \\ x & & x & 2 & x & x \\ & & & x & 3 & x \\ & & x & x & x & 6 \end{bmatrix} = LL^T, \qquad L^T = \begin{bmatrix} 1 & x & & x & & \\ & 5 & x & \otimes & & \\ & & 4 & x & & x \\ & & & 2 & x & x \\ & & & & 3 & x \\ & & & & & 6 \end{bmatrix}.$$

For instance, the position of the nonzero off-diagonal elements of the third row of L^T, which belongs to the pivot vertex $x_3 = 4$, is given by the pivot clique $\{2, 6\} = Adj_{G^2}(4)$ of this vertex.

For large problems, any efficient implementation of algorithm (4.A.4) hinges on the efficient determination of the degree function deg_G of graphs $G^i = G^{i-1}_{x_i}$, $i = 1, 2, \ldots, n - 1$. One could use, for instance,

$$Adj_{G_x}(y) = Adj_G(y), \qquad deg_{G_x}(y) = deg_G(y),$$

since by (4.A.3) the degree function does not change in the transition $G \to G_x$ at all vertices $y \neq x$ with $y \notin Adj_G(x)$. Numerous other methods for efficiently implementing (4.A.4) have been proposed [see George and Liu (1989) for a review]. These proposals also involve an appropriate representation of graphs. For our purposes, it proved to be useful to represent the graph $G = (V, E)$ by a set of cliques $M = \{K_1, K_2, \ldots, K_q\}$ such that each edge is covered by at least one clique $K_i \in M$. Then the set E of edges can be recovered:

$$E = \{\{x, y\} \mid x \neq y \ \&\ \exists i\colon x, y \in K_i\}.$$

Under those conditions, M is called a *clique representation* of G. Such representations exist for any graph $G = (V, E)$. Indeed, $M := E$ is a clique representation of G since every edge $\{x, y\} \in E$ is a clique of G.

EXAMPLE 4. A clique representation of the graph $G = G^A$ of Example 2 is, for instance,

$$\{\{1, 5\}, \{4, 5\}, \{1, 2\}, \{1, 5\}, \{2, 3, 6\}, \{2, 4, 6\}\}.$$

The degree $deg_G(x)$ and the sets $Adj_G(x)$, $x \in V$, can be computed from a clique representation M by

$$Adj_G(x) = \bigcup_{i:\, x \in K_i} K_i, \qquad deg_G(x) = |Adj_G(x)|.$$

Let $x \in V$ be arbitrary. Then, because of (4.A.3), one can easily compute a clique representation M_x for the elimination graph G_x of $G = (V, E)$ from such a representation $M = \{K_1, \ldots, K_q\}$ of G: Denote by $\{K_{s_1}, \ldots, K_{s_t}\}$ the set of all cliques in M that contain x, and set $K := \bigcup_{i=1}^{t} K_{s_i} \backslash \{x\}$. Then

$$M_x = \{K_1, \ldots, K_q, K\} \backslash \{K_{s_1}, \ldots, K_{s_t}\}$$

gives a clique representation of G_x.

Assuming that cliques are represented as lists of their elements, then M_x takes even up less memory space than M because

$$|K| < \sum_{j=1}^{t} |K_{s_j}|.$$

Recall now the steps leading up to this point. A suitable pivot sequence was determined using algorithm (4.A.4). To this pivot sequence corresponds a permutation matrix P, and we are interested in the Cholesky factor L of the matrix PAP^T $(=LL^T)$ and, in particular, in a suitable data structure for storing L and associated information. Let Nonz(L) denote the sparsity pattern, that is, the set of locations of nonzero elements of L. The nonzero elements of L may, for instance, be stored by columns—this corresponds to storing the nonzero elements of L^T by rows—in three arrays *ip*, *ja*, *a* as described at the beginning of this section. Alternatively, a linked list *next* may be employed in addition to such arrays.

Because Nonz$(L) \supset$ Nonz(PAP^T), the data structure for L can also be used to store the nonzero elements of A.

Next follows the *numerical factorization* of $PAP^T = LL^T$: Here it is important to organize the computation of L^T by rows, that is, the array a is overwritten step by step with the corresponding data of the consecutive rows of L^T. Also the programs to solve the triangular systems

$$Lz = Pb, \qquad L^T u = z$$

for z and u, respectively, can be coded based on the data structure by accessing each row of the matrix L^T only once when computing z and again only once when computing u. Finally, the solution x of $Ax = b$ is given by $x = P^T u$.

The reader can find more details on methods for solving sparse linear systems in the literature, e.g., in Duff, Erisman, and Reid (1986) and George and Liu (1981), where numerous FORTRAN programs for solving positive-definite systems are also given. Large program packages exist, such as the Harwell package MA27 (see Duff and Reid (1982)), the Yale sparse matrix package YSMP (see Eisenstat et al. (1982)), and SPARSEPAK (see George, Liu, and Ng (1980)).

Exercises for Chapter 4

1. Consider the following vector norms defined on $\mathbb{R}^n$ (or $\mathbb{C}^n$):

$$\|x\|_\infty := \max_{1 \leqslant i \leqslant n} |x_i|,$$

$$\|x\|_2 := \sqrt{\sum_{i=1}^{n} |x_i|^2},$$

$$\|x\|_1 := \sum_{i=1}^{n} |x_i|.$$

Show:

(a) that the norm properties are satisfied by each;
(b) that $\|x\|_\infty \leqslant \|x\|_2 \leqslant \|x\|_1$;
(c) that $\|x\|_2 \leqslant \sqrt{n}\,\|x\|_\infty$, $\|x\|_1 \leqslant \sqrt{n}\,\|x\|_2$.

Can equality hold in (b), (c)?

(d) Determine what $\operatorname{lub}(A)$ is, in general, for the norm $\|\cdot\|_1$
(e) Starting from the definition

$$\operatorname{lub}(A) = \max_{x \neq 0} \frac{\|Ax\|}{\|x\|},$$

show that

$$\frac{1}{\operatorname{lub}(A^{-1})} = \min_{y \neq 0} \frac{\|Ay\|}{\|y\|}$$

for nonsingular A.

2. Consider the following class of vector norms defined on $\mathbb{C}^n$:

$$\|x\|_D := \|Dx\|,$$

where $\|\cdot\|$ is a fixed vector norm, and D is any member of the class of nonsingular matrices.

(a) Show that $\|\cdot\|_D$ is, indeed, a vector norm.
(b) Show that $m\|x\| \leqslant \|x\|_D \leqslant M\|x\|$ with

$$m = 1/\operatorname{lub}(D^{-1}), \qquad M = \operatorname{lub}(D),$$

where $\operatorname{lub}(D)$ is defined with respect to $\|\cdot\|$.
(c) Express $\operatorname{lub}_D(A)$ in terms of the lub norm defined with respect to $\|\cdot\|$.

(d) For a nonsingular matrix A, $\operatorname{cond}(A)$ depends upon the underlying vector norm used. To see an example of this, show that $\operatorname{cond}_D(A)$ as defined from $\|\cdot\|_D$ can be arbitrarily large, depending on the choice of D. Give an estimate of $\operatorname{cond}_D(A)$ using m, M.
(e) How different can $\operatorname{cond}(A)$ defined using $\|\cdot\|_\infty$ be from that defined using $\|\cdot\|_2$? [Use the results of Exercise 1(b, c) above.]

3. Show that, for an $n \times n$ nonsingular matrix A and vectors $u, v \in \mathbb{R}^n$,

(a)

$$(A + uv^T)^{-1} = A^{-1} - \frac{A^{-1}uv^TA^{-1}}{1 + v^TA^{-1}u}$$

if $v^TA^{-1}u \neq -1$.
(b) If $v^TA^{-1}u = -1$, then $A + uv^T$ is singular.

Hint: Find a vector $z \neq 0$ such that $(A + uv^T)z = 0$.

4. Let A be a nonsingular $n \times n$ matrix with columns a_i,

$$A = (a_1, \ldots, a_n).$$

(a) Let $\tilde{A} = (a_1, \ldots, a_{i-1}, b, a_{i+1}, \ldots, a_n)$, $b \in \mathbb{R}^n$ be the matrix obtained from A by replacing the ith column a_i by b. Determine, using the formula of Exercise 3(a), under what conditions $\tilde{A}^{-1}$ exists, and show that $\tilde{A}^{-1} = FA^{-1}$ for some Frobenius matrix F.
(b) Let A_α be the matrix obtained from A by changing a single element a_{ik} to $a_{ik} + \alpha$. For what α will A_α^{-1} exist?

5. Consider the following theorem [see Theorem (6.4.10)]: *If A is a real, nonsingular $n \times n$ matrix, then there exist two real orthogonal matrices U, V satisfying*

$$U^TAV = D,$$

where $D = \operatorname{diag}(\mu_1, \ldots, \mu_n)$ and

$$\mu_1 \geqslant \mu_2 \geqslant \cdots \geqslant \mu_n > 0.$$

Using this theorem, and taking the Euclidian norm as the underlying vector norm:

(a) Express $\operatorname{cond}(A)$ in terms of the quantities μ_i.
(b) Give an expression for the vectors b and Δb in terms of U for which the bounds (4.4.11), (4.4.12), and

$$\|b\| \geqslant \operatorname{lub}(A)\|x\|$$

are satisfied with equality.
(c) Is there a vector b such that for all Δb in (4.4.12),

$$\frac{\|\Delta x\|}{\|x\|} \leqslant \frac{\|\Delta b\|}{\|b\|}$$

holds? Determine such vectors b with the help of U.

Hint: Look at b satisfying $\operatorname{lub}(A^{-1})\|b\| = \|x\|$.

6. Let $Ax = b$ be given with

$$A = \begin{bmatrix} 0.780 & 0.563 \\ 0.913 & 0.659 \end{bmatrix} \quad \text{and} \quad b = \begin{bmatrix} 0.217 \\ 0.254 \end{bmatrix}.$$

The exact solution is $x^T = (1, -1)$. Further, let two approximate solutions

$$x_1^T = (0.999, -1.001),$$
$$x_2^T = (0.341, -0.087)$$

be given.

(a) Compute the residuals $r(x_1)$, $r(x_2)$. Does the more accurate solution have a smaller residual?

(b) Determine $\operatorname{cond}(A)$ with respect to the maximum norm, given

$$A^{-1} = \begin{bmatrix} 659\,000 & -563\,000 \\ -913\,000 & 780\,000 \end{bmatrix}.$$

(c) Express $\tilde{x} - x = \Delta x$ using $r(\tilde{x})$, the residual for $\tilde{x}$. Does this provide an explanation for the discrepancy observed in (a)? (Compare with Exercise 5.)

7. Show that:

(a) The largest elements in magnitude in a positive definite matrix appear on the diagonal and are positive.

(b) If all leading principal minors of an $n \times n$ Hermitian matrix $A = (a_{ik})$ are positive, i.e.,

$$\det\left(\begin{bmatrix} a_{11} & \cdots & a_{1i} \\ \vdots & & \vdots \\ a_{i1} & \cdots & a_{ii} \end{bmatrix}\right) > 0 \quad \text{for } i = 1, \ldots, n,$$

then A is positive definite.

Hint: Study the induction proof for Theorem (4.3.3).

8. Let A' be a given, $n \times n$, real, positive definite matrix partitioned as follows:

$$A' = \begin{bmatrix} A & B \\ B^T & C \end{bmatrix},$$

where A is an $m \times m$ matrix. First, show:

(a) $C - B^T A^{-1} B$ is positive definite.

[*Hint:* Partition x correspondingly:

$$x = \begin{bmatrix} x_1 \\ x_2 \end{bmatrix}, \qquad x_1 \in \mathbb{R}^m, \quad x_2 \in \mathbb{R}^{n-m},$$

and determine an x_1 for fixed x_2 such that

$$x^T A' x = x_2^T (C - B^T A^{-1} B) x_2$$

holds.]

According to Theorem (4.3.3), A' has a decomposition

$$A' = R^T R,$$

where R is an upper triangular matrix, which may be partitioned consistently with A':

$$R = \begin{bmatrix} R_{11} & R_{12} \\ 0 & R_{22} \end{bmatrix}.$$

Now show:

(b) Each matrix $M = N^T N$, where N is a nonsingular matrix, is positive definite.
(c) $R_{22}^T R_{22} = C - B^T A^{-1} B$.
(d) The result

$$r_{ii}^2 > 0, \qquad i = 1, \ldots, n,$$

follows from (a), where r_{ii} is any diagonal element of R.

(e)

$$r_{ii}^2 \geqslant \min_{x \neq 0} \frac{x^T A' x}{x^T x} = \frac{1}{\operatorname{lub}(R^{-1})^2}$$

for $i = 1, \ldots, n$, where $\operatorname{lub}(R^{-1})$ is defined using the Euclidian vector norm.

Hint: Exercise 1(e).

(f)

$$\operatorname{lub}(R)^2 = \max_{x \neq 0} \frac{x^T A' x}{x^T x} \geqslant r_{ii}^2, \qquad i = 1, \ldots, n,$$

for $\operatorname{lub}(R)$ defined with the Euclidian norm.
(g) $\operatorname{cond}(R)$ satisfies

$$\operatorname{cond}(R) \geqslant \max_{1 \leqslant i, k \leqslant n} \left| \frac{r_{ii}}{r_{kk}} \right|.$$

9. A sequence A_n of real or complex $r \times r$ matrices converges componentwise to a matrix A if and only if the A_n form a Cauchy sequence; that is, given any vector norm $\|\cdot\|$ and any $\varepsilon > 0$, $\operatorname{lub}(A_n - A_m) < \varepsilon$ for m and n sufficiently large. Using this, show that if $\operatorname{lub}(A) < 1$, then the sequence A^n and the series $\sum_{n=0}^{\infty} A^n$ converge. Further, show that $I - A$ is nonsingular, and

$$(I - A)^{-1} = \sum_{n=0}^{\infty} A^n.$$

Use this result to prove (4.4.14).

10. Suppose that we attempt to find the inverse of an $n \times n$ matrix A using the Gauss–Jordan method and partial pivot selection. Show that the columns of A are linearly dependent if no nonzero pivot element can be found by the partial pivot selection process at some step of the method.
[*Caution:* The converse of this result is *not* true if floating-point arithmetic is used. A can have linearly dependent columns, yet the Gauss–Jordan method may never encounter all zeros among the candidate pivots at any step, due to round-off errors. Similar statements can be made about any of the decomposition

methods we have studied. For a discussion of how the determination of rank (or singularity) is best made numerically, see Chapter 6, Section 6, of Stewart (1973).]

11. Let A be a positive definite $n \times n$ matrix. Let Gaussian elimination be carried out on A without pivot selection. After k steps of elimination, A will be reduced to the form

$$A^{(k)} = \begin{bmatrix} A_{11}^{(k)} & A_{12}^{(k)} \\ 0 & A_{22}^{(k)} \end{bmatrix},$$

where $A_{22}^{(k)}$ is an $(n-k) \times (n-k)$ matrix. Show by induction that

(a) $A_{22}^{(k)}$ is positive definite,
(b) $a_{ii}^{(k)} \leqslant a_{ii}^{(k-1)}$ for $k \leqslant i \leqslant n$, $k = 1, 2, \ldots, n-1$.

12. In the error analysis of Gaussian elimination (Section 4.5) we used certain estimates of the growth of the maximal elements of the matrices $A^{(i)}$. Let

$$a_i := \max_{r,s} |a_{rs}^{(i)}|, \qquad A^{(i)} := (a_{rs}^{(i)}).$$

Show that for partial pivot selection:

(a) $a_k \leqslant 2^k a_0$, $k = 1, \ldots, n-1$ for arbitrary A.
(b) $a_k \leqslant k a_0$, $k = 1, \ldots, n-1$ for Hessenberg matrices A.
(c) $a = \max_{1 \leqslant k \leqslant n-1} a_k \leqslant 2a_0$ for tridiagonal matrices A.

13. The following decomposition of a positive definite matrix A,

$$A = SDS^H,$$

where S is a lower triangular matrix with $s_{ii} = 1$ and D is a diagonal matrix $D = \text{diag}(d_i)$, gives rise to a variant of the Cholesky method. Show:

(a) that such a decomposition is possible [Theorem (4.3.3)];
(b) that $d_i = (l_{ii})^2$, where $A = LL^H$ and L is a lower triangular matrix;
(c) that this decomposition does not require the n square roots which are required by the Cholesky method.

14. We have the following mathematical model:

$$y = x_1 z + x_2,$$

which depends upon the two unknown parameters x_1, x_2. Moreover, let a collection of data be given:

$$\{y_l, z_l\}_{l=1,\ldots,m} \quad \text{with } z_l = l.$$

Try to determine the parameters x_1, x_2 from the data using least-square fitting.

(a) What are the normal equations?
(b) Carry out the Cholesky decomposition of the normal equation matrix $B = A^T A = LL^T$.
(c) Give an estimate for cond(L) based upon the Euclidian vector norm.

[*Hint:* Use the estimate for cond(L) from Exercise 8(g).]

(d) How does the condition vary as m, the number of data, is increased [Schwarz, Rutishauser, and Stiefel (1968)]?

15. The straight line

$$y(x) = \alpha + \beta x$$

is to be fitted to the data

x_i	-2	-1	0	1	2
y_i	0.5	0.5	2	3.5	3.5

so that

$$\sum_i [y(x_i) - y_i]^2$$

is minimized.
Determine the parameters α and β.

16. Determine α and β as in Exercise 15 under the condition that

$$\sum_i |y(x_i) - y_i|$$

is to be minimized.
[*Hint:* Let $\rho_i - \sigma_i = y(x_i) - y_i$ for $i = 1, \ldots, 5$, where $\rho_i \geqslant 0$ and $\sigma_i \geqslant 0$. Then $\sum_i |y(x_i) - y_i| = \sum_i (\rho_i + \sigma_i)$. Set up a linear programming problem in the variables α, β (unrestricted) and ρ_i, σ_i (nonnegative).]

References for Chapter 4

Barker, V. A. (Ed.): *Sparse Matrix Techniques.* Lecture Notes in Mathematics 572. Berlin, Heidelberg, New York: Springer-Verlag (1977).

Bartels, R. H.: A stabilization of the simplex method. *Numer. Math.* **16**, 414–434 (1971).

Bauer, F. L.: Genauigkeitsfragen bei der Lösung linearer Gleichungssysteme. *ZAMM* **46**, 409–421 (1966).

Björck, Å.: Least squares methods. In: Ciarlet, Lions, Eds. (1990), 465–647 (1990).

Blum, E., Oettli, W.: *Mathematische Optimierung.* Berlin, Heidelberg, New York: Springer (1975).

Ciarlet, P. G., Lions, J. L. (Eds): *Handbook of Numerical Analysis*, Vol. 1. Finite Difference Methods (Part 1), Solution of Equations in $\mathbb{R}^n$ (Part 1). Amsterdam: North Holland (1990).

Collatz, L.: *Functional Analysis and Numerical Mathematics.* New York: Academic Press (1966).

Daniel, J. W., Gragg, W. B., Kaufmann, L., Stewart, G. W.: Reorthogonalization and stable algorithms for updating the Gram-Schmidt *QR* factorization. *Math. Comp.* **30**, 772–795 (1976).

Dantzig, G. B.: *Linear Programming and Extensions.* Princeton, N.J.: Princeton University Press (1963).

Dongarra, J. J., Bunch, J. R., Moler, C. B., Stewart, G. W.: *LINPACK Users Guide.* Philadelphia: SIAM Publications (1979).

Duff, I. S., Erisman, A. M., Reid, J. K.: *Direct Methods for Sparse Matrices.* Oxford: Oxford University Press (1986).

———, Reid, J. K.: MA27: A set of FORTRAN subroutines for solving sparse symmetric sets of linear equations. Techn. Rep. AERE R 10533, Harwell, U.K. (1982).

Eisenstat, S. C., Gursky, M. C., Schultz, M. H., Sherman, A. H.: The Yale sparse matrix package I. The symmetric codes. *Internat. J. Numer. Methods Engrg.* **18**, 1145–1151 (1982).

Forsythe, G. E., Moler, C. B.: *Computer Solution of Linear Algebraic Systems.* Series in Automatic Computation. Englewood Cliffs, N.J.: Prentice-Hall (1967).

Gass, S. T.: *Linear Programming.* 3d edition. New York: McGraw-Hill (1969).

George, J. A., Liu, J. W.: *Computer Solution of Large Sparse Positive Definite Systems.* Englewood Cliffs, N.J.: Prentice-Hall (1981).

———, ———: The evolution of the minimum degree ordering algorithm. *SIAM Review* **31**, 1–19 (1989).

———, ———, Ng, E. G.: Users guide for SPARSEPAK: Waterloo sparse linear equations package. Tech. Rep. CS-78-30, Dept. of Computer Science, University of Waterloo, Waterloo (1980).

Gill, P. E., Golub, G. H., Murray, W., Saunders, M. A.: Methods for modifying matrix factorizations. *Math. Comp.* **28**, 505–535 (1974).

Golub, G. H., van Loan, C. F.: *Matrix Computations.* Baltimore: The John Hopkins University Press (1983).

Grossmann, W.: *Grundzüge der Ausgleichsrechnung.* 3. Aufl. Berlin, Heidelberg, New York: Springer (1969).

Guest, P. G.: *Numerical Methods of Curve Fitting.* Cambridge: University Press (1961).

Hadley, G.: *Linear Programming.* Reading, MA: Addison-Wesley (1962).

Householder, A. S.: *The Theory of Matrices in Numerical Analysis.* New York: Blaisdell (1964).

Lawson, C. L., Hanson, H. J.: *Solving Least Squares Problems.* Englewood Cliffs, N.J.: Prentice-Hall (1974).

Murty, K. G.: *Linear and Combinatorial Programming.* New York: Wiley (1976).

Prager, W., Oettli, W.: Compatibility of approximate solution of linear equations with given error bounds for coefficients and right hand sides. *Num. Math.* **6**, 405–409 (1964).

Reid, J. K. (Ed): *Large Sparse Sets of Linear Equations.* London, New York: Academic Press (1971).

Rose, D. J.: A graph-theoretic study of the numerical solution of sparse positive definite systems of linear equations, pp. 183–217. In: *Graph Theory and Computing*, R. C. Read, ed., New York: Academic Press (1972).

———, Willoughby, R. A. (Eds.): *Sparse Matrices and Their Applications.* New York: Plenum Press (1972).

Sautter, W.: Dissertation TU München (1971).

Schwarz, H. R., Rutishauser, H., Stiefel, E.: *Numerik symmetrischer Matrizen.* Leitfäden der angewandten Mathematik, Bd. 11. Stuttgart: Teubner 1968.

Seber, G. A. F.: *Linear Regression Analysis.* New York: Wiley (1977).

Stewart, G. W.: *Introduction to Matrix Computations.* New York: Academic Press (1973).

Tewarson, R. P.: *Sparse Matrices.* New York: Academic Press (1973).

Wilkinson, J. H.: *The Algebraic Eigenvalue Problem.* Monographs on Numerical Analysis, Oxford: Clarendon Press (1965).

———, Reinsch, Ch.: *Linear Algebra.* Handbook for Automatic Computation, Vol. II. Grundlehren der mathematischen Wissenschaften in Einzeldarstellungen, Bd. 186. Berlin, Heidelberg, New York: Springer (1971).

5 Finding Zeros and Minimum Points by Iterative Methods

Finding the *zeros* of a given function f, that is arguments ξ for which $f(\xi) = 0$, is a classical problem. In particular, determining the zeros of a polynomial (the zeros of a polynomial are also known as its *roots*)

$$p(x) = a_0 + a_1 x + \cdots + a_n x^n$$

has captured the attention of pure and applied mathematicians for centuries. However, much more general problems can be formulated in terms of finding zeros, depending upon the definition of the function $f: E \to F$, its domain E, and its range F.

For example, if $E = F = \mathbb{R}^n$, then a transformation $f: \mathbb{R}^n \to \mathbb{R}^n$ is described by n real functions $f_i(x^1, \ldots, x^n)$ of n real variables $x^1, \ldots, x^n$ (we will use superscripts in this chapter to denote the components of vectors $x \in \mathbb{R}^n$, $n > 1$, and subscripts to denote elements in a set or sequence of vectors x_i, $i = 1, 2, \ldots$):

$$f(x) = \begin{bmatrix} f_1(x^1, \ldots, x^n) \\ \vdots \\ f_n(x^1, \ldots, x^n) \end{bmatrix}, \qquad x^T = (x^1, \ldots, x^n).$$

The problem of solving $f(x) = 0$ becomes that of solving a system of (nonlinear) equations:

(5.0.1) $$f_i(x^1, \ldots, x^n) = 0, \qquad i = 1, \ldots, n.$$

Even more general problems result if E and F are linear vector spaces of infinite dimension, e.g. function spaces.

Problems of finding zeros are closely associated with problems of the form

$$\underset{x \in \mathbb{R}^n}{\text{minimize}}\ h(x)$$

for a real function $h\colon \mathbb{R}^n \to \mathbb{R}$ of n variables $h(x) = h(x^1, x^2, \ldots, x^n)$. For if h is differentiable and $g(x) := (\partial h/\partial x^1, \ldots, \partial h/\partial x^n)^T$ is the gradient of h, then each *minimum point* $\bar{x}$ of $h(x)$ is a zero of the gradient $g(\bar{x}) = 0$. Conversely, each zero of f (5.0.1) is also the minimum point of some function h, for example $h(x) := \| f(x)\|^2$.

The minimization problem described above is an *unconstrained* minimization problem. More generally, one encounters *constrained* problems such as the following:

$$\text{minimize } h_0(x)$$

subject to

$$h_i(x) \leqslant 0 \quad \text{for } i = 1, 2, \ldots, m_1,$$

$$h_i(x) = 0 \quad \text{for } i = m_1 + 1, m_1 + 2, \ldots, m.$$

Finding minimum points for functions subject to constraints is one of the most important problems in applied mathematics. In this chapter, however, we will consider only unconstrained minimization. The special case of constrained *linear* minimization, for which all $h_i\colon \mathbb{R}^n \to \mathbb{R}$ are linear (or, more exactly, affine) functions has been discussed in Sections 4.10 and 4.11. For a more thorough treatment of finding zeros and minimum points the reader is referred to the extensive literature on that subject [for example Ortega, Rheinboldt (1970), Luenberger (1973), Himmelblau (1972), Traub (1964)].

5.1 The Development of Iterative Methods

Usually it is not possible to determine a zero ξ of a function $f\colon E \to F$ explicitly within a finite number of steps, so we have to resort to approximation methods. These methods are usually iterative and have the following form: beginning with a starting value x_0, successive approximates x_i, $i = 1, 2, \ldots$, to ξ are computed with the aid of an *iteration function* $\Phi\colon E \to E$:

$$x_{i+1} := \Phi(x_i), \qquad i = 0, 1, 2, \ldots.$$

If ξ is a fixed point of Φ [$\Phi(\xi) = \xi$], if all fixed points of Φ are also zeros of f, and if Φ is continuous in a neighborhood of each of its fixed points, then each limit point of the sequence x_i, $i = 1, 2, \ldots$, is a fixed point of Φ, and hence a zero of f.

The following questions arise in this connection:

(1) How is a suitable iteration function Φ to be found?
(2) Under what conditions will the sequence x_i converge?
(3) How quickly will the sequence x_i converge?

Our discussion of these questions will be restricted to the finite-dimensional case $E = F = \mathbb{R}^n$.

Let us examine how iteration functions Φ might be constructed. Frequently such functions are suggested by the formulation of the problem. For example, if the equation $x - \cos x = 0$ is to be solved, then it is natural to try the iterative process

$$x_{i+1} = \cos x_i, \qquad i = 0, 1, 2, \ldots,$$

for which $\Phi(x) := \cos x$.

More systematically, iteration functions Φ can be obtained as follows: If ξ is the zero of a function $f\colon \mathbb{R} \to \mathbb{R}$, and if f is sufficiently differentiable in a neighborhood $\mathcal{N}(\xi)$ of this point, then the Taylor series expansion of f about $x_0 \in \mathcal{N}(\xi)$ is

$$f(\xi) = 0 = f(x_0) + (\xi - x_0)f'(x_0) + \frac{(\xi - x_0)^2}{2!} f''(x_0) + \cdots$$
$$+ \frac{(\xi - x_0)^k}{k!} f^{(k)}(x_0 + \vartheta(\xi - x_0)), \qquad 0 < \vartheta < 1.$$

If the higher powers $(\xi - x_0)^\nu$ are ignored, we arrive at equations which must express the point ξ approximately in terms of a given, nearby point x_0, e.g.

(5.1.1) $$0 = f(x_0) + (\bar{\xi} - x_0)f'(x_0)$$

or

(5.1.2) $$0 = f(x_0) + (\bar{\bar{\xi}} - x_0)f'(x_0) + \frac{(\bar{\bar{\xi}} - x_0)^2}{2!} f''(x_0).$$

These produce the approximations

$$\bar{\xi} = x_0 - \frac{f(x_0)}{f'(x_0)}$$

and

$$\bar{\bar{\xi}} = x_0 - \frac{f'(x_0) \pm \sqrt{(f'(x_0))^2 - 2f(x_0)f''(x_0)}}{f''(x_0)},$$

respectively. In general $\bar{\xi}$, $\bar{\bar{\xi}}$ are merely close to the desired zero: they must be corrected further, for instance by the scheme from which they were themselves derived. In this manner we arrive at the iteration methods

(5.1.3)
$$x_{i+1} := \Phi(x_i), \qquad \Phi(x) := x - \frac{f(x)}{f'(x)},$$
$$x_{i+1} := \Phi_\pm(x_i), \qquad \Phi_\pm(x) := x - \frac{f'(x) \pm \sqrt{(f'(x))^2 - 2f(x)f''(x)}}{f''(x)}.$$

The first is the classical *Newton–Raphson* method. The second is an obvious extension. In general such methods can be obtained by truncating the Taylor expansion after the $(\xi - x_0)^\nu$ term. Geometrically these methods amount to

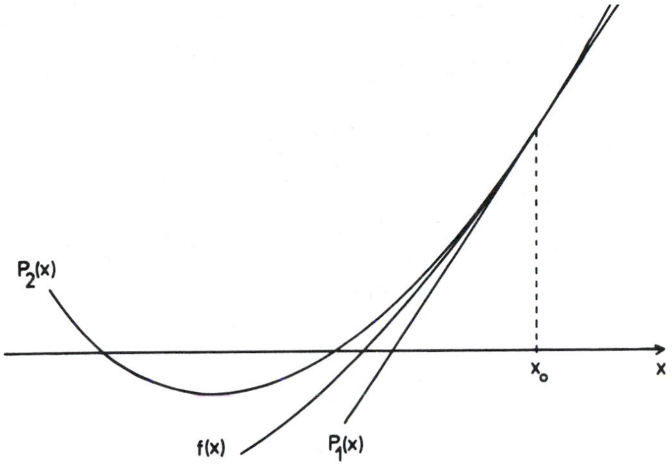

Figure 5 Newton–Raphson methods (5.1.3).

replacing the function f by a polynomial of degree ν, $P_\nu(x)$ $(\nu = 1, 2, \ldots)$, which has the same derivatives $f^{(k)}(x_0)$, $k = 0, 1, \ldots, \nu$, as f at the point x_0. One of the roots of the polynomial is taken as an approximation to the desired zero ξ of f. (See Figure 5.)

The classical Newton–Raphson method is obtained by linearizing f. Linearization is also a means of constructing iterative methods to solve equation systems of the form

(5.1.4)
$$f(x) = \begin{bmatrix} f_1(x^1, \ldots, x^n) \\ \vdots \\ f_n(x^1, \ldots, x^n) \end{bmatrix} = 0.$$

If we assume that $x = \xi$ is a zero for f, that x_0 is an approximation to ξ, and that f is differentiable for $x = x_0$ then to a first approximation

$$0 = f(\xi) \approx f(x_0) + Df(x_0)(\xi - x_0),$$

where

(5.1.5)
$$Df(x_0) = \begin{bmatrix} \dfrac{\partial f_1}{\partial x^1} & \cdots & \dfrac{\partial f_1}{\partial x^n} \\ \vdots & & \vdots \\ \dfrac{\partial f_n}{\partial x^1} & \cdots & \dfrac{\partial f_n}{\partial x^n} \end{bmatrix}_{x = x_0}, \qquad \xi - x_0 = \begin{bmatrix} \xi^1 - x_0^1 \\ \vdots \\ \xi^n - x_0^n \end{bmatrix}.$$

If the Jacobian $Df(x_0)$ is nonsingular, then the equation

$$f(x_0) + Df(x_0)(x_1 - x_0) = 0$$

can be solved for x_1:

$$x_1 = x_0 - (Df(x_0))^{-1} f(x_0)$$

and x_1 may be taken as a closer approximation to the zero ξ. The generalized Newton method for solving systems of equations (5.1.4) is given by

$$(5.1.6) \qquad x_{i+1} = x_i - (Df(x_i))^{-1} f(x_i), \qquad i = 0, 1, 2, \ldots.$$

In addition to Newton's method for such equation systems there are, for example, generalized secant methods [see (5.9.7)] for functions of many variables, and there are generalizations to nonlinear systems of the iteration methods given in Chapter 8 for systems of linear equations. A good survey can be found in Ortega and Rheinboldt (1970).

5.2 General Convergence Theorems

In this section we will study the convergence behavior of a sequence x_i which has been generated by an iteration function Φ:

$$x_{i+1} := \Phi(x_i), \qquad i = 0, 1, 2, \ldots,$$

in the neighborhood of a fixed point ξ of Φ. We will concentrate on the case $E = \mathbb{R}^n$ rather than considering general normed linear vector spaces. Using a norm $\|\cdot\|$ on $\mathbb{R}^n$ we can measure the difference between two vectors $x, y \in \mathbb{R}^n$ by $\|x - y\|$. A sequence of vectors $x_i \in \mathbb{R}^n$ converges to a vector x if for each $\varepsilon > 0$ there is an integer $N(\varepsilon)$ such that

$$\|x_l - x\| < \varepsilon \quad \text{for all } l \geqslant N(\varepsilon).$$

It can be shown that this definition of the convergence of vectors in $\mathbb{R}^n$ is independent of the chosen norm [see Theorem (4.4.6)]. Finally, it is known that the space $\mathbb{R}^n$ is *complete* in the sense that the Cauchy convergence criterion is satisfied:

A sequence $x_i \in \mathbb{R}^n$ is convergent if and only if for each $\varepsilon > 0$ there exists an $N(\varepsilon)$ such that $\|x_l - x_m\| < \varepsilon$ for all $l, m \geqslant N(\varepsilon)$.

Let Φ be an iteration function on $\mathbb{R}^n$. Let ξ be a fixed point of Φ. For all initial vectors x_0 taken from a neighborhood $\mathcal{N}(\xi)$ and for the generated sequence $x_{i+1} = \Phi(x_i)$, $i = 0, 1, \ldots$, let an inequality of the form

$$\|x_{i+1} - \xi\| \leqslant C\|x_i - \xi\|^p$$

hold for all $i \geqslant 0$, where $C < 1$ if $p = 1$. Then the iteration method defined by Φ is said to be a *method of at least pth order* for determining ξ. The following is easily shown:

(5.2.1) Theorem. *Each method of at least pth order for determining a fixed point ξ is locally convergent, in the sense that there is a neighborhood $\mathcal{N}(\xi)$ of ξ with the property that for all initial $x_0 \in \mathcal{N}(\xi)$, the sequence x_i generated by Φ converges to ξ.*

[If $\mathcal{N}(\xi)$ can be taken as $\mathbb{R}^n$, then the method is said to be *globally convergent*.]

In the one-dimensional case, $E = \mathbb{R}$, the order of a method defined by Φ can often be determined if Φ is sufficiently often differentiable in the neighborhood $\mathcal{N}(\xi)$. If $x_i \in \mathcal{N}(\xi)$ and if $\Phi^{(k)}(\xi) = 0$ for $k = 1, 2, \ldots, p-1$, but $\Phi^{(p)}(\xi) \neq 0$, it follows that

$$x_{i+1} = \Phi(x_i) = \Phi(\xi) + \frac{(x_i - \xi)^p}{p!}\Phi^{(p)}(\xi) + O(\|x_i - \xi\|^{p+1}),$$

$$\lim_{i\to\infty} \frac{x_{i+1} - \xi}{(x_i - \xi)^p} = \frac{\Phi^{(p)}(\xi)}{p!}.$$

For $p = 2, 3, \ldots$ the method is of (precisely) pth order. A method is of first order if, besides $p = 1$, it is true that $|\Phi'(\xi)| < 1$.

EXAMPLE 1. $E = \mathbb{R}$, Φ is differentiable in a neighborhood $\mathcal{N}(\xi)$. If $0 < \Phi'(\xi) < 1$, then convergence will be linear (first order). In fact, the x_i will converge monotonically to ξ. (See Figure 6.)

If $-1 < \Phi'(\xi) < 0$, then the x_i will alternate about ξ during convergence (see Figure 7).

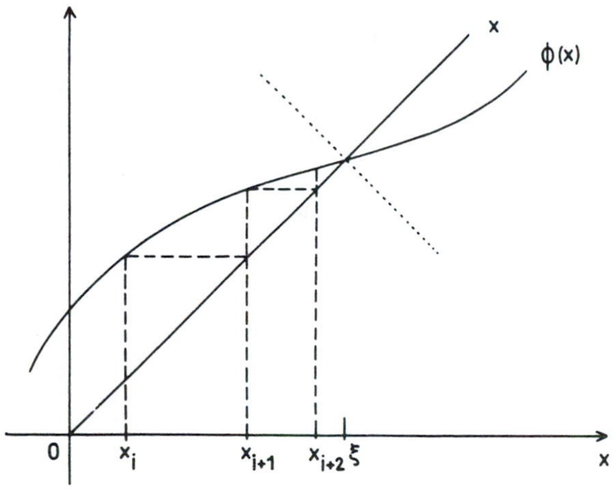

Figure 6 Monotone convergence.

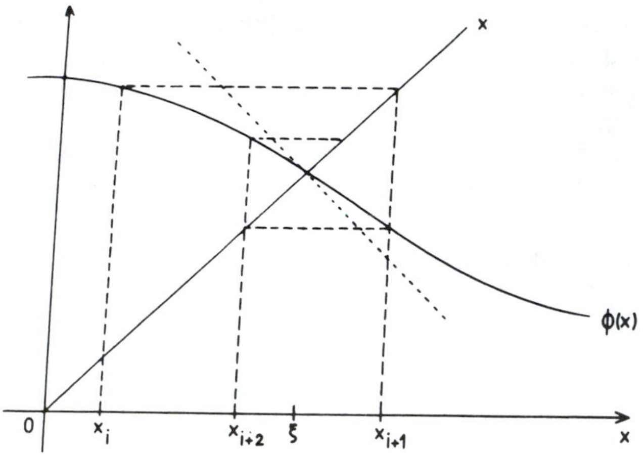

Figure 7 Alternating convergence.

EXAMPLE 2. $E = \mathbb{R}$, $\Phi(x) = x - f(x)/f'(x)$ (Newton's method). Assume that f has a sufficient number of continuous derivatives in a neighborhood of the simple zero ξ of f [i.e., $f'(\xi) \neq 0$]. It follows that

$$\Phi(\xi) = \xi,$$

$$\Phi'(\xi) = \left.\frac{f(x)f''(x)}{(f'(x))^2}\right|_{x=\xi} = 0,$$

$$\Phi''(\xi) = \frac{f''(\xi)}{f'(\xi)}.$$

Newton's method is at least locally quadratically (second order) convergent.

EXAMPLE 3. In the more general case that ξ is an m-fold zero of f, i.e.,

$$f^{(\nu)}(\xi) = 0 \quad \text{for } \nu = 0, 1, \ldots, m - 1,$$

then f and f' have a representation of the form

$$f(x) = (x - \xi)^m g(x), \qquad g(\xi) \neq 0,$$

$$f'(x) = m(x - \xi)^{m-1} g(x) + (x - \xi)^m g'(x),$$

with a differentiable function g. It follows that

$$\Phi(x) \equiv x - \frac{f(x)}{f'(x)} \equiv x - \frac{(x - \xi)g(x)}{mg(x) + (x - \xi)g'(x)},$$

and therefore

$$\Phi'(\xi) = 1 - \frac{1}{m}.$$

Thus for $m > 1$—that is, for multiple zeros of f—Newton's method is only linearly convergent.

The following general convergence theorems show that a sequence x_i generated by $\Phi\colon E \to E$ will converge to a fixed point ξ of Φ if Φ is a contractive mapping. As usual, $\|\cdot\|$ represents some norm on $E = \mathbb{R}^n$.

(5.2.2) Theorem. *Let the function $\Phi\colon E \to E$, $E = \mathbb{R}^n$ have a fixed point $\xi\colon \Phi(\xi) = \xi$. Further let $S_r(\xi) := \{z \mid \|z - \xi\| < r\}$ be a neighborhood of ξ such that Φ is a contractive mapping in $S_r(\xi)$, that is,*

$$\|\Phi(x) - \Phi(y)\| \leqslant K\|x - y\|, \qquad 0 \leqslant K < 1,$$

for all $x, y \in S_r(\xi)$. Then for any $x_0 \in S_r(\xi)$ the generated sequence $x_{i+1} = \Phi(x_i)$, $i = 0, 1, 2, \ldots$, has the properties

(a) $x_i \in S_r(\xi)$ *for all* $i = 0, 1, \ldots,$
(b) $\|x_{i+1} - \xi\| \leqslant K\|x_i - \xi\| \leqslant K^{i+1}\|x_0 - \xi\|,$

i.e., $\{x_i\}$ *converges at least linearly to* ξ.

PROOF. The proof follows immediately from the contraction property. Properties (a) and (b) are true for $i = 0$. If we assume that they are true for $j = 0, 1, \ldots, i$, then it follows immediately that

$$\|x_{i+1} - \xi\| = \|\Phi(x_i) - \Phi(\xi)\| \leqslant K\|x_i - \xi\| \leqslant K^{i+1}\|x_0 - \xi\| < r. \qquad \square$$

The following theorem is more precise. Note that the existence of a fixed point is no longer assumed *a priori*.

(5.2.3) Theorem. *Let $\Phi\colon E \to E$, $E = \mathbb{R}^n$ be an iteration function, $x_0 \in E$ be a starting point, and $x_{i+1} = \Phi(x_i)$, $i = 0, 1, \ldots$. Further, let a neighborhood $S_r(x_0) := \{x \mid \|x - x_0\| < r\}$ of x_0 and a constant K, $0 < K < 1$, exist such that*

(a) $\|\Phi(x) - \Phi(y)\| \leqslant K\|x - y\|$ *for all* $x, y \in \overline{S_r(x_0)} := \{x \mid \|x - x_0\| \leqslant r\}$,
(b) $\|x_1 - x_0\| = \|\Phi(x_0) - x_0\| \leqslant (1 - K)r < r$.

Then it follows that

(1) $x_i \in S_r(x_0)$ *for all* $i = 0, 1, \ldots,$
(2) Φ *has exactly one fixed point* ξ, $\Phi(\xi) = \xi$, in $\overline{S_r(x_0)}$, *and*

$$\lim_{i\to\infty} x_i = \xi, \qquad \|x_{i+1} - \xi\| \leqslant K\|x_i - \xi\|,$$

as well as

$$\|x_i - \xi\| \leqslant \frac{K^i}{1 - K}\|x_1 - x_0\|.$$

PROOF. The proof is by induction.

(1): From (b) it follows that $x_1 \in S_r(x_0)$. If it is true that $x_j \in S_r(x_0)$ for $j = 0, 1, \ldots, i$ and $i \geqslant 1$, then (a) implies

(5.2.4) $$\|x_{i+1} - x_i\| = \|\Phi(x_i) - \Phi(x_{i-1})\| \leqslant K\|x_i - x_{i-1}\| \leqslant K^i\|x_1 - x_0\|,$$

and therefore, from the triangle inequality and from (b),

$$\begin{aligned}\|x_{i+1} - x_0\| &\leqslant \|x_{i+1} - x_i\| + \|x_i - x_{i-1}\| + \cdots + \|x_1 - x_0\| \\ &\leqslant (K^i + K^{i-1} + \cdots + 1)\|x_1 - x_0\| \\ &\leqslant (1 + K + \cdots + K^i)(1 - K)r = (1 - K^{i+1})r < r.\end{aligned}$$

(2): First we show that $\{x_i\}$ is a Cauchy sequence. From (5.2.4) and from (b) it follows for $m > l$ that

$$\begin{aligned}(5.2.5)\quad \|x_m - x_l\| &\leqslant \|x_m - x_{m-1}\| + \|x_{m-1} - x_{m-2}\| + \cdots + \|x_{l+1} - x_l\| \\ &\leqslant K^l(1 + K + \cdots + K^{m-l-1})\|x_1 - x_0\| \\ &< \frac{K^l}{1 - K}\|x_1 - x_0\| < K^l r.\end{aligned}$$

Because $0 < K < 1$, we have $K^l r < \varepsilon$ for sufficiently large $l \geqslant N(\varepsilon)$. Hence $\{x_i\}$ is a Cauchy sequence. Since $E = \mathbb{R}^n$ is complete, there exists a limit

$$\lim_{i\to\infty} x_i = \xi.$$

Because $x_i \in S_r(x_0)$ for all i, ξ must lie in the closure $\overline{S_r(x_0)}$. Furthermore ξ is a fixed point of Φ, because for all $i \geqslant 0$

$$\begin{aligned}\|\Phi(\xi) - \xi\| &\leqslant \|\Phi(\xi) - \Phi(x_i)\| + \|\Phi(x_i) - \xi\| \\ &\leqslant K\|\xi - x_i\| + \|x_{i+1} - \xi\|.\end{aligned}$$

Since $\lim_{i\to\infty} \|x_i - \xi\| = 0$, it follows at once that $\|\Phi(\xi) - \xi\| = 0$, and hence $\Phi(\xi) = \xi$.

If $\hat{\xi} \in \overline{S_r(x_0)}$ were another fixed point of Φ, then

$$\|\xi - \hat{\xi}\| = \|\Phi(\xi) - \Phi(\hat{\xi})\| \leqslant K\|\xi - \hat{\xi}\|,$$

$0 < K < 1$, which implies that $\|\xi - \hat{\xi}\| = 0$.

Finally, (5.2.5) implies

$$\|\xi - x_l\| = \lim_{m\to\infty} \|x_m - x_l\| \leqslant \frac{K^l}{1 - K}\|x_1 - x_0\|$$

and

$$\|x_{i+1} - \xi\| = \|\Phi(x_i) - \Phi(\xi)\| \leqslant K\|x_i - \xi\|,$$

which concludes the proof. □

5.3 The Convergence of Newton's Method in Several Variables

Consider the system $f(x) = 0$ given by the function $f\colon \mathbb{R}^n \to \mathbb{R}^n$. Such a function is said to be *differentiable* at a point $x_0 \in \mathbb{R}^n$ if an $n \times n$ matrix A exists for which

$$\lim_{x \to x_0} \frac{\|f(x) - f(x_0) - A(x - x_0)\|}{\|x - x_0\|} = 0.$$

In this case A agrees with the Jacobian matrix $Df(x_0)$ [see (5.1.5)].

We first note the following

(5.3.1) Lemma. *If $Df(x)$ exists for all x in a convex region $C_0 \subseteq \mathbb{R}^n$, and if a constant γ exists with*

$$\|Df(x) - Df(y)\| \leqslant \gamma \|x - y\| \quad \textit{for all } x, y \in C_0,$$

then for all $x, y \in C_0$ the estimate

$$\|f(x) - f(y) - Df(y)(x - y)\| \leqslant \frac{\gamma}{2} \|x - y\|^2$$

holds.

(Recall that a set $M \subseteq \mathbb{R}^n$ is convex if $x, y \in M$ implies that the line segment $[x, y] := \{z = \lambda x + (1 - \lambda)y \mid 0 \leqslant \lambda \leqslant 1\}$ is contained within M.)

PROOF. The function $\varphi\colon [0, 1] \to \mathbb{R}^n$ given by

$$\varphi(t) := f(y + t(x - y))$$

is differentiable for all $0 \leqslant t \leqslant 1$, where $x, y \in C_0$ are arbitrary. This follows from the chain rule:

$$\varphi'(t) = Df(y + t(x - y))(x - y).$$

Hence it follows for $0 \leqslant t \leqslant 1$ that

$$\begin{aligned}\|\varphi'(t) - \varphi'(0)\| &= \|(Df(y + t(x - y)) - Df(y))(x - y)\| \\ &\leqslant \|Df(y + t(x - y)) - Df(y)\| \, \|x - y\| \\ &\leqslant \gamma t \|x - y\|^2.\end{aligned}$$

On the other hand,

$$\begin{aligned}\Delta := f(x) - f(y) - Df(y)(x - y) &= \varphi(1) - \varphi(0) - \varphi'(0) \\ &= \int_0^1 (\varphi'(t) - \varphi'(0))\, dt,\end{aligned}$$

so the above inequality yields

$$\|\Delta\| \leqslant \int_0^1 \|\varphi'(t) - \varphi'(0)\|\, dt \leqslant \gamma\|x - y\|^2 \int_0^1 t\, dt$$
$$= \frac{\gamma}{2}\|x - y\|^2.$$

This completes the proof. □

We can now show that Newton's method is quadratically convergent:

(5.3.2) Theorem. *Let $C \subseteq \mathbb{R}^n$ be a given open set. Further, let C_0 be a convex set with $\bar{C}_0 \subseteq C$, and let $f\colon C \to \mathbb{R}^n$ be a function which is differentiable for all $x \in C_0$ and continuous for all $x \in C$.*

For $x_0 \in C_0$ let positive constants $r, \alpha, \beta, \gamma, h$ be given with the following properties:

$$S_r(x_0) := \{x \mid \|x - x_0\| < r\} \subseteq C_0,$$
$$h := \alpha\beta\gamma/2 < 1,$$
$$r := \alpha/(1 - h),$$

and let $f(x)$ have the properties

(a) $\|Df(x) - Df(y)\| \leqslant \gamma\|x - y\|$ *for all* $x, y \in C_0$
(b) $Df(x)^{-1}$ *exists and satisfies* $\|Df(x)^{-1}\| \leqslant \beta$ *for all* $x \in C_0$
(c) $\|Df(x_0)^{-1}f(x_0)\| \leqslant \alpha$.

Then

(1) *Beginning at x_0, each point*

$$x_{k+1} := x_k - Df(x_k)^{-1}f(x_k), \qquad k = 0, 1, \ldots,$$

is well defined and satisfies $x_k \in S_r(x_0)$ for all $k \geqslant 0$.
(2) $\lim_{k\to\infty} x_k = \xi$ *exists and satisfies $\xi \in \overline{S_r(x_0)}$ and $f(\xi) = 0$.*
(3) *For all $k \geqslant 0$*

$$\|x_k - \xi\| \leqslant \alpha \frac{h^{2^k - 1}}{1 - h^{2^k}}.$$

Since $0 < h < 1$, Newton's method is at least quadratically convergent.

Proof. (1): Since $Df(x)^{-1}$ exists for $x \in C_0$, x_{k+1} is well defined for all k if $x_k \in S_r(x_0)$ for all $k \geqslant 0$. This is valid for $k = 0$ and $k = 1$ by assumption (c). Now, if $x_j \in S_r(x_0)$ for $j = 0, 1, \ldots, k$, then from assumption (b)

$$\|x_{k+1} - x_k\| = \|-Df(x_k)^{-1}f(x_k)\| \leqslant \beta\|f(x_k)\|$$
$$= \beta\|f(x_k) - f(x_{k-1}) - Df(x_{k-1})(x_k - x_{k-1})\|,$$

since the definition of x_k implies

$$f(x_{k-1}) + Df(x_{k-1})(x_k - x_{k-1}) = 0.$$

But, according to Lemma (5.3.1),

$$\|x_{k+1} - x_k\| \leqslant \frac{\beta\gamma}{2} \|x_k - x_{k-1}\|^2, \tag{5.3.3}$$

and therefore

$$\|x_{k+1} - x_k\| \leqslant \alpha h^{2^k - 1}. \tag{5.3.4}$$

This last inequality is correct for $k = 0$ because of (c). If it is correct for $k \geqslant 0$, then it is correct for $k + 1$, since (5.3.3) implies

$$\|x_{k+1} - x_k\| \leqslant \frac{\beta\gamma}{2} \|x_k - x_{k-1}\|^2 \leqslant \frac{\beta\gamma}{2} \alpha^2 h^{2^k - 2} = \alpha h^{2^k - 1}.$$

Furthermore, (5.3.4) implies

$$\begin{aligned}\|x_{k+1} - x_0\| &\leqslant \|x_{k+1} - x_k\| + \|x_k - x_{k-1}\| + \cdots + \|x_1 - x_0\| \\ &\leqslant \alpha(1 + h + h^3 + h^7 + \cdots + h^{2^k - 1}) < \alpha/(1 - h) = r,\end{aligned}$$

and consequently $x_{k+1} \in S_r(x_0)$.

(2): From (5.3.4) it is easily determined that $\{x_k\}$ is a Cauchy sequence, since for $m \geqslant n$ we have

$$\begin{aligned}\|x_{m+1} - x_n\| &\leqslant \|x_{m+1} - x_m\| + \|x_m - x_{m-1}\| + \cdots + \|x_{n+1} - x_n\| \\ &\leqslant \alpha h^{2^n - 1}(1 + h^{2^n} + (h^{2^n})^2 + \cdots) \\ &< \frac{\alpha h^{2^n - 1}}{1 - h^{2^n}} < \varepsilon\end{aligned} \tag{5.3.5}$$

for sufficiently large $n \geqslant N(\varepsilon)$, because $0 < h < 1$.

Consequently, there is a limit

$$\lim_{k \to \infty} x_k = \xi \in \overline{S_r(x_0)},$$

whose inclusion in the closure follows from the fact that $x_k \in S_r(x_0)$ for all $k \geqslant 0$.

By passing to the limit $m \to \infty$ in (5.3.5) we obtain (3) as a side result:

$$\lim_{m \to \infty} \|x_m - x_n\| = \|\xi - x_n\| \leqslant \frac{\alpha h^{2^n - 1}}{1 - h^{2^n}}.$$

We must still show that ξ is a zero of f in $\overline{S_r(x_0)}$.

Because of (a), and because $x_k \in S_r(x_0)$ for all $k \geqslant 0$,

$$\|Df(x_k) - Df(x_0)\| \leqslant \gamma\|x_k - x_0\| < \gamma r,$$

and therefore

$$\|Df(x_k)\| \leqslant \gamma r + \|Df(x_0)\| =: K.$$

The inequality

$$\|f(x_k)\| \leqslant K\|x_{k+1} - x_k\|.$$

follows from the equation

$$f(x_k) = -Df(x_k)(x_{k+1} - x_k).$$

Hence

$$\lim_{k\to\infty} \|f(x_k)\| = 0,$$

and, since f is continuous at ξ,

$$\lim_{k\to\infty} \|f(x_k)\| = \|f(\xi)\| = 0,$$

i.e., ξ is a zero of f. □

Under somewhat stronger assumptions it can be shown that ξ is the only zero of f in $S_r(x_0)$:

(5.3.6) Theorem (Newton-Kantorovich). *Given the function $f\colon C \subseteq \mathbb{R}^n \to \mathbb{R}^n$ and the convex set $C_0 \subseteq C$, let f be continuously differentiable on C_0 and satisfy the conditions*
(a) $\|Df(x) - Df(y)\| \leqslant \gamma\|x - y\|$ *for all* $x, y \in C_0$,
(b) $\|Df(x_0)^{-1}f(x_0)\| \leqslant \alpha$,
(c) $\|Df(x_0)^{-1}\| \leqslant \beta$,
for some $x_0 \in C_0$. Consider the quantities

$$h := \alpha\beta\gamma,$$

$$r_{1,2} := \frac{1 \mp \sqrt{1 - 2h}}{h}\alpha.$$

If $h \leqslant \frac{1}{2}$ and $\overline{S_{r_1}(x_0)} \subset C_0$, then the sequence $\{x_k\}$ defined by

$$x_{k+1} := x_k - Df(x_k)^{-1}f(x_k) \quad \textit{for } k = 0, 1, \ldots$$

remains in $S_{r_1}(x_0)$ and converges to the unique zero of $f(x)$ in $C_0 \cap S_{r_2}(x_0)$.

For the proof see Ortega and Rheinboldt (1970) or Collatz (1968).

5.4 A Modified Newton Method

Theorem (5.3.2) guarantees the convergence of Newton's method only if the starting point x_0 of the iteration is chosen "sufficiently close" to the desired solution ξ of

$$f(x) = 0, \qquad f\colon \mathbb{R}^n \to \mathbb{R}^n.$$

The following example shows that Newton's method may diverge otherwise.

EXAMPLE. Let $f: \mathbb{R} \to \mathbb{R}$ be given by $f(x) = \arctan x$. Then $\xi = 0$ is a solution of $f(x) = 0$.

The Newton iteration is defined by

$$x_{k+1} := x_k - (1 + x_k^2) \arctan x_k.$$

If we choose x_0 so that

$$\arctan |x_0| > \frac{2|x_0|}{1 + x_0^2},$$

then the sequence $\{|x_k|\}$ diverges: $\lim_{k \to \infty} |x_k| = \infty$.

We describe a modification of Newton's method for which global convergence can be proven for a large class of functions f. The modification involves the introduction of an extra parameter λ and a *search direction* s to define the sequence

(5.4.0.1) $$x_{k+1} := x_k - \lambda_k s_k,$$

where typically $s_k := d_k \equiv [Df(x_k)]^{-1} f(x_k)$, and the λ_k are chosen so that the sequence $\{h(x_k)\}$, $h(x) = f(x)^T f(x)$, is strictly monotone decreasing and the x_k converge to a minimum point of $h(x)$. (Compare this with the problem of nonlinear least-squares data fitting mentioned in Section 4.8.)

Since $h(x) \geqslant 0$ for all x,

$$h(\bar{x}) = 0 \quad \Leftrightarrow \quad f(\bar{x}) = 0.$$

Every local minimum point $\bar{x}$ of h which satisfies $h(x) = 0$ is also a global minimum point $\bar{x}$ of h as well as a zero of f.

In the following section we will consider first a few general results about the convergence of a class of minimization methods for arbitrary functionals $h(x)$. These results will then be used in Section 5.4.2 to investigate the convergence of the modified Newton method.

5.4.1 On the Convergence of Minimization Methods

Let $\|\cdot\|$ be the Euclidean vector norm and $0 < \gamma \leq 1$. We consider the set

(5.4.1.1) $$D(\gamma, x) := \{s \in \mathbb{R}^n \mid \|s\| = 1 \text{ with } Dh(x)s \geq \gamma \|Dh(x)\|\}$$

of all directions s forming a not-too-large acute angle with the gradient

$$Dh(x) = \left(\frac{\partial h(x)}{\partial x^1}, \ldots, \frac{\partial h(x)}{\partial x^n}\right), \quad \text{where } x = (x^1, \ldots, x^n)^T.$$

The following lemma shows, given an x, under what conditions a scalar λ and an $s \in \mathbb{R}^n$ exist such that $h(x - \lambda s) < h(x)$.

(5.4.1.2) Lemma. *Let $h: \mathbb{R}^n \to \mathbb{R}$ be a function whose gradient is defined and continuous for all $x \in V(\bar{x})$ in a neighborhood $V(\bar{x})$ of $\bar{x}$. Suppose further that $Dh(\bar{x}) \neq 0$, and let $1 \geqslant \gamma > 0$. Then there is a neighborhood $U(\bar{x}) \subseteq V(\bar{x})$ of $\bar{x}$ and a number $\lambda > 0$ such that $h(x - \mu s) \leqslant h(x) - (\mu\gamma/4)\|Dh(\bar{x})\|$ for all $x \in U(\bar{x})$, $s \in D(\gamma, x)$, and $0 \leqslant \mu \leqslant \lambda$.*

PROOF. The set

$$U^1(\bar{x}) := \left\{ x \in V(\bar{x}) \,\middle|\, \|Dh(x) - Dh(\bar{x})\| \leqslant \frac{\gamma}{4} \|Dh(\bar{x})\| \right\}$$

is nonempty and a neighborhood of $\bar{x}$, since $Dh(\bar{x}) \neq 0$ and since $Dh(x)$ is continuous on $V(\bar{x})$. Similarly

$$U^2(\bar{x}) := \left\{ x \in V(\bar{x}) \,\middle|\, D(\gamma, x) \subseteq D\left(\frac{\gamma}{2}, \bar{x}\right) \right\}$$

is nonempty and a neighborhood of $\bar{x}$. Choose a $\lambda > 0$ so that

$$\overline{S_{2\lambda}(\bar{x})} = \{x \mid \|x - \bar{x}\| \leqslant 2\lambda\} \subseteq U^1(\bar{x}) \cap U^2(\bar{x}),$$

and let

$$U(\bar{x}) := \overline{S_\lambda(\bar{x})} = \{x \mid \|x - \bar{x}\| \leqslant \lambda\}.$$

Then if $x \in U(\bar{x})$, $0 \leqslant \mu \leqslant \lambda$, $s \in D(\gamma, x)$, there exist θ, $0 < \theta < 1$ such that

$$\begin{aligned} h(x) - h(x - \mu s) &= \mu Dh(x - \theta\mu s)s \\ &= \mu[(Dh(x - \theta\mu s) - Dh(\bar{x}))s + Dh(\bar{x})s]. \end{aligned}$$

Since $x \in U(\bar{x})$ implies x, $x - \mu s$, $x - \theta\mu s \in U^1 \cap U^2$, it follows that

$$\begin{aligned} h(x) - h(x - \mu s) &\geqslant -\frac{\mu\gamma}{4}\|Dh(\bar{x})\| + \mu Dh(\bar{x})s \\ &\geqslant -\frac{\mu\gamma}{4}\|Dh(\bar{x})\| + \mu\frac{\gamma}{2}\|Dh(\bar{x})\| \\ &= \frac{\mu\gamma}{4}\|Dh(\bar{x})\|. \end{aligned}$$

□

We consider the following method for minimizing a differentiable function $h: \mathbb{R}^n \to \mathbb{R}$.

(5.4.1.3).

(a) *Choose numbers γ_k, σ_k, $k = 0, 1, \ldots,$ with*

$$\sup_k \gamma_k \leqslant 1, \qquad \inf_k \gamma_k > 0, \qquad \inf_k \sigma_k > 0,$$

and choose a starting point $x_0 \in \mathbb{R}^n$.

(b) *For all* $k = 0, 1, \ldots,$ *choose an* $s_k \in D(\gamma_k, x_k)$ *and set*

$$x_{k+1} := x_k - \lambda_k s_k$$

where $\lambda_k \in [0, \sigma_k \|Dh(x_k)\|]$ *is such that*

$$h(x_{k+1}) = \min_{\mu} \{h(x_k - \mu s_k) \mid 0 \leqslant \mu \leqslant \sigma_k \|Dh(x_k)\|\}.$$

The convergence properties of this method are given by the following

(5.4.1.4) Theorem. *Let* $h: \mathbb{R}^n \to \mathbb{R}$ *be a function, and let* $x_0 \in \mathbb{R}^n$ *be chosen so that*

(a) $K := \{x \mid h(x) \leqslant h(x_0)\}$ *is compact, and*
(b) h *is continuously differentiable in some open set containing* K.

Then for any sequence $\{x_k\}$ *defined by a method of the type* (5.4.1.3):

(1) $x_k \in K$ *for all* $k = 0, 1, \ldots.$ $\{x_k\}$ *has at least one accumulation point* $\bar{x}$ *in* K.
(2) *Each accumulation point of* $\{x_k\}$ *is a stationary point of* h:

$$Dh(\bar{x}) = 0.$$

Proof. (1): From the definition of the sequence $\{x_k\}$ it follows immediately that the sequence $\{h(x_k)\}$ is monotone: $h(x_0) \geqslant h(x_1) \geqslant \cdots$. Hence $x_k \in K$ for all k. K is compact; therefore $\{x_k\}$ has at least one accumulation point $\bar{x} \in K$.

(2): Assume that $\bar{x}$ is an accumulation point of $\{x_k\}$ but is not a stationary point of h:

$$Dh(\bar{x}) \neq 0. \tag{5.4.1.5}$$

Without loss of generality, let $\lim_{k\to\infty} x_k = \bar{x}$. Let $\gamma := \inf_k \gamma_k > 0$, $\sigma := \inf_k \sigma_k > 0$.

According to Lemma (5.4.1.2) there is a neighborhood $U(\bar{x})$ of $\bar{x}$ and a number $\lambda > 0$ satisfying

$$h(x - \mu s) \leqslant h(x) - \mu \frac{\gamma}{4} \|Dh(\bar{x})\| \tag{5.4.1.6}$$

for all $x \in U(\bar{x})$, $s \in D(\gamma, x)$, and $0 \leqslant \mu \leqslant \lambda$.

Since $\lim_{k\to\infty} x_k = \bar{x}$, the continuity of $Dh(x)$, together with (5.4.1.5), implies the existence of a k_0 such that for all $k \geqslant k_0$

(a) $x_k \in U(\bar{x})$,
(b) $\|Dh(x_k)\| \geqslant \frac{1}{2}\|Dh(\bar{x})\|$.

Let $\Lambda := \min\{\lambda, \frac{1}{2}\sigma\|Dh(\bar{x})\|\}$, $\varepsilon := \Lambda(\gamma/4)\|Dh(\bar{x})\| > 0$. Since $\sigma_k \geqslant \sigma$, it follows that $[0, \Lambda] \subseteq [0, \sigma_k \|Dh(x_k)\|]$ for all $k \geqslant k_0$. Therefore, from the definition of x_{k+1},

$$h(x_{k+1}) \leqslant \min_{\mu}\{h(x_k - \mu s_k) \mid 0 \leqslant \mu \leqslant \Lambda\}.$$

Since $\Lambda \leqslant \lambda$, $x_k \in U(\bar{x})$, $s_k \in D(\gamma_k, x_k) \subseteq D(\gamma, x_k)$, (5.4.1.6) implies that

$$h(x_{k+1}) \leqslant h(x_k) - \frac{\Lambda\gamma}{4} \|Dh(\bar{x})\| = h(x_k) - \varepsilon$$

for all $k \geqslant k_0$. This means that $\lim_{k\to\infty} h(x_k) = -\infty$, which contradicts $h(x_k) \geqslant h(x_{k+1}) \geqslant \cdots \geqslant h(\bar{x})$. Hence, $\bar{x}$ is a stationary point of h. □

Step (b) of (5.4.1.3) is known as the *line search*. Even though the method given by (5.4.1.3) is quite general, its practical application is limited by the fact that the line search must be *exact*, i.e., it requires that the exact minimum point of the function

$$\varphi(\mu) := h(x_k - \mu s_k)$$

be found on the interval $[0, \sigma_k \|Dh(x_k)\|]$ in order to determine x_{k+1}. Generally a great deal of effort is required to obtain even an approximate minimum point. The following variant of (5.4.1.3) has the virtue that in step (b) the exact minimization is replaced by an *inexact line search*, in particular by a finite search process:

(5.4.1.7).

(a) *Choose numbers* γ_k, σ_k, $k = 0, 1, \ldots,$ *so that*

$$\sup_k \gamma_k \leqslant 1, \qquad \inf_k \gamma_k > 0, \qquad \inf_k \sigma_k > 0.$$

Choose a starting point $x_0 \in \mathbb{R}^n$.

(b) *For each* $k = 0, 1, \ldots$ *obtain* x_{k+1} *from* x_k *as follows:*

(α) *Select*

$$s_k \in D(\gamma_k, x_k),$$

define

$$\rho_k := \sigma_k \|Dh(x_k)\|, \qquad h_k(\mu) := h(x_k - \mu s_k),$$

and determine the smallest integer $j \geqslant 0$ *such that*

$$h_k(\rho_k 2^{-j}) \leqslant h_k(0) - \rho_k 2^{-j} \frac{\gamma_k}{4} \|Dh(x_k)\|.$$

(β) *Determine* $\bar{i} \in \{0, 1, \ldots, j\}$ *such that* $h_k(\rho_k 2^{-\bar{i}})$ *is minimum, and let* $x_{k+1} = x_k - \lambda_k s_k$, *where* $\lambda_k := \rho_k 2^{-\bar{i}}$.

[Note that $h(x_{k+1}) = \min_{0 \leqslant i \leqslant j} h_k(\rho_k 2^{-i})$.]

It is easily seen that an integer $j \geqslant 0$ exists with the properties (5.4.1.7bα): If x_k is a stationary point, then $j = 0$. If x_k is not stationary, then the existence of j follows immediately from Lemma (5.4.1.2) applied to $\bar{x} := x_k$. In any case j (and λ_k) can be found after a finite number of steps.

The modified process (5.4.1.7) satisfies an analog to (5.4.1.4):

(5.4.1.8) Theorem. *Under the hypotheses of Theorem* (5.4.1.4) *each sequence* $\{x_k\}$ *produced by a method of the type* (5.4.1.7) *satisfies the conclusions of Theorem* (5.4.1.4).

PROOF. We assume as before that $\bar{x}$ is an accumulation point of a sequence $\{x_k\}$ defined by (5.4.1.7), but not a stationary point, i.e.,

$$Dh(\bar{x}) \neq 0.$$

Again, without loss of generality, let $\lim x_k = \bar{x}$. Also let $\sigma := \inf_k \sigma_k > 0$, $\gamma := \inf_k \gamma_k > 0$. According to Lemma (5.4.1.2) there is a neighborhood $U(\bar{x})$ and a number $\lambda > 0$ such that

$$h(x - \mu s) \leqslant h(x) - \mu \frac{\gamma}{4} \|Dh(\bar{x})\| \tag{5.4.1.9}$$

for all $x \in U(\bar{x})$, $s \in D(\gamma, x)$, $0 \leqslant \mu \leqslant \lambda$. Again, the fact that $\lim_k x_k = \bar{x}$, that $Dh(x)$ is continuous, and that $Dh(\bar{x}) \neq 0$ imply the existence of a k_0 such that

$$x_k \in U(\bar{x}), \tag{5.4.1.10a}$$

$$\|Dh(x_k)\| \geqslant \tfrac{1}{2}\|Dh(\bar{x})\|, \tag{5.4.1.10b}$$

for all $k \geqslant k_0$.

We need to show that there is an $\varepsilon > 0$ for which

$$h(x_{k+1}) \leqslant h(x_k) - \varepsilon \text{ for all } k \geqslant k_0.$$

Note first that (5.4.1.10) and $\gamma_k \geqslant \gamma$ imply

$$\gamma_k \|Dh(x_k)\| \geqslant \frac{\gamma}{2} \|Dh(\bar{x})\| \text{ for all } k \geqslant k_0.$$

Consequently, according to the definition of x_{k+1} and j

$$\begin{aligned} h(x_{k+1}) \leqslant h_k(\rho_k 2^{-j}) &\leqslant h(x_k) - \rho_k 2^{-j} \frac{\gamma_k}{4} \|Dh(x_k)\| \\ &\leqslant h(x_k) - \rho_k 2^{-j} \frac{\gamma}{8} \|Dh(\bar{x})\|. \end{aligned} \tag{5.4.1.11}$$

Now let $\bar{j} \geqslant 0$ be the smallest integer satisfying

$$h_k(\rho_k 2^{-\bar{j}}) \leqslant h(x_k) - \rho_k 2^{-\bar{j}} \frac{\gamma}{8} \|Dh(\bar{x})\|. \tag{5.4.1.12}$$

According to (5.1.1.11), $\bar{j} \leqslant j$, and the definition of x_{k+1} we have

$$h(x_{k+1}) \leqslant h_k(\rho_k 2^{-\bar{j}}). \tag{5.4.1.13}$$

There are two cases:

Case 1, $\bar{j}=0$. Let $\rho_k := \sigma_k \|Dh(x_k)\|$, and note that $\rho_k \geqslant \sigma/2\|Dh(\bar{x})\|$. Then (5.4.1.12) and (5.4.1.13) imply

$$h(x_{k+1}) \leqslant h(x_k) - \rho_k \frac{\gamma}{8} \|Dh(\bar{x})\|$$

$$\leqslant h(x_k) - \frac{\sigma\gamma}{16} \|Dh(\bar{x})\|^2 = h(x_k) - \varepsilon_1$$

with $\varepsilon_1 > 0$ independent of x_k.

Case 2, $\bar{j} > 0$. From the minimality of $\bar{j}$ we have

$$h_k(\rho_k 2^{-(\bar{j}-1)}) > h(x_k) - \rho_k 2^{-(\bar{j}-1)} \frac{\gamma}{8} \|Dh(\bar{x})\|$$

$$\geqslant h(x_k) - \rho_k 2^{-(\bar{j}-1)} \frac{\gamma}{4} \|Dh(\bar{x})\|.$$

Because $x_k \in U(\bar{x})$ and $s_k \in D(\gamma_k, x_k) \subseteq D(\gamma, x_k)$, it follows immediately from (5.4.1.9) that

$$\rho_k 2^{-(\bar{j}-1)} > \lambda.$$

Combining this with (5.4.1.12) and (5.4.1.13) yields

$$h(x_{k+1}) \leqslant h_k(\rho_k 2^{-\bar{j}}) \leqslant h(x_k) - \frac{\lambda\gamma}{16} \|Dh(\bar{x})\| = h(x_k) - \varepsilon_2$$

with $\varepsilon_2 > 0$ independent of x_k.

Hence, for $\varepsilon = \min(\varepsilon_1, \varepsilon_2)$

$$h(x_{k+1}) \leqslant h(x_k) - \varepsilon$$

for all $k \geqslant k_0$, contradicting the fact that $h(x_k) \geqslant h(\bar{x})$ for all k. Therefore $\bar{x}$ is a stationary point of h. □

5.4.2 Application of the Convergence Criteria to the Modified Newton Method

In order to solve the equation $f(x) = 0$, we let $h(x) = f(x)^T f(x)$ and apply one of the methods (5.4.1.3) or (5.4.1.7) to minimize $h(x)$. We use the Newton direction

$$d_k := Df(x_k)^{-1} f(x_k), \qquad s_k = \frac{d}{\|d_k\|},$$

as the search direction s_k to be taken from the point x_k. This direction will be defined if $Df(x_k)^{-1}$ exists and $f(x_k) \neq 0$. ($\|\cdot\|$ denotes the Euclidian norm.)

To apply the theorems of the last section requires a little preparation. We show first that, for every x such that

$$d = d(x) := Df(x)^{-1}f(x) \quad \text{and} \quad s = s(x) = \frac{d}{\|d\|}$$

exist [i.e., $Df(x)^{-1}$ exists and $d \neq 0$], we have

$$\text{(5.4.2.1)} \quad s \in D(\gamma, x) \quad \text{for all } 0 < \gamma \leqslant \bar{\gamma}(x), \quad \bar{\gamma}(x) := \frac{1}{\operatorname{cond}(Df(x))}.$$

In the above,

$$\|Df(x)\| := \operatorname{lub}(Df(x))$$

and

$$\operatorname{cond}(Df(x)) := \|Df(x)^{-1}\| \; \|Df(x)\|$$

are to be defined with respect to the Euclidian norm.

PROOF. Since $h(x) = f(x)^T f(x)$, we have

$$\text{(5.4.2.2)} \qquad Dh(x) = 2f^T(x)Df(x).$$

The inequalities

$$\|f^T(x)Df(x)\| \leqslant \|Df(x)\| \; \|f(x)\|,$$

$$\|Df(x)^{-1}f(x)\| \leqslant \|Df(x)^{-1}\| \; \|f(x)\|$$

clearly hold, and consequently

$$\frac{Dh(x)s}{\|Dh(x)\|} = \frac{f(x)^T Df(x) Df(x)^{-1} f(x)}{\|Df(x)^{-1}f(x)\| \; \|f^T(x)Df(x)\|} \geqslant \frac{1}{\operatorname{cond}(Df(x))} > 0.$$

Now, for all γ with $0 < \gamma \leqslant 1/\operatorname{cond}(Df(x))$, it follows that $s \in D(\gamma, x)$ according to the definition of $D(\gamma, x)$ given in (5.4.1.1). □

As a consequence of (5.4.2.2) we observe: If $Df(x)^{-1}$ exists, then

$$\text{(5.4.2.3)} \qquad Dh(x) = 0 \iff f(x) = 0,$$

i.e., x is a stationary point of h if and only if x is a zero of f.

Consider the following modified Newton method [compare with (5.4.1.7)]:

(5.4.2.4).

(a) *Select a starting point* $x_0 \in \mathbb{R}^n$.
(b) *For each* $k = 0, 1, \ldots$ *define* x_{k+1} *from* x_k *as follows:*
 (α) *Set*

$$d_k := Df(x_k)^{-1} f(x_k),$$

$$\gamma_k := \frac{1}{\operatorname{cond}(Df(x_k))},$$

and let $h_k(\tau) := h(x_k - \tau d_k)$, *where* $h(x) := f(x)^T f(x)$. *Determine the smallest integer* $j \geqslant 0$ *satisfying*

$$h_k(2^{-j}) \leqslant h_k(0) - 2^{-j}\frac{\gamma_k}{4}\|d_k\|\,\|Dh(x_k)\|.$$

(β) *Determine* λ_k *so that* $h(x_{k+1}) = \min_{0 \leqslant i \leqslant j} h_k(2^{-i})$, *and let*

$$x_{k+1} := x_k - \lambda_k d_k.$$

As an analog to Theorem (5.4.1.8) we have

(5.4.2.5) Theorem. *Let* $f\colon \mathbb{R}^n \to \mathbb{R}^n$ *be a given function, and let* $x_0 \in \mathbb{R}^n$ *be a point with the following properties:*

(a) *The set* $K := \{x \mid h(x) \leqslant h(x_0)\}$, *where* $h(x) := f(x)^T f(x)$, *is compact;*
(b) f *is continuously differentiable on some open set containing* K;
(c) $Df(x)^{-1}$ *exists for all* $x \in K$.

Then the sequence $\{x_k\}$ *defined by* (5.4.2.4) *is well defined and satisfies the following:*

(1) $x_k \in K$ *for all* $k = 0, 1, \ldots,$ *and* $\{x_k\}$ *has at least one accumulation point* $\bar{x} \in K$.
(2) *Each accumulation point* $\bar{x}$ *of* $\{x_k\}$ *is a zero of* f, $f(\bar{x}) = 0$.

Proof. By construction, $\{h(x_k)\}$ is monotone:

$$h(x_0) \geqslant h(x_1) \geqslant \cdots.$$

Hence $x_k \in K$, $k = 0, 1, \ldots.$ Because of assumption (c), d_k and γ_k are well defined if x_k is defined. From (5.4.2.1)

$$s_k \in D(\gamma_k, x_k),$$

where $s_k := d_k/\|d_k\|$.

As was the case for (5.4.1.7), there is a $j \geqslant 0$ with the properties given in (5.4.2.4). Hence x_{k+1} is defined for each x_k.

Now (5.4.2.4) becomes formally identical to the process given by (5.4.1.7) if σ_k is defined by

$$\sigma_k := \frac{\|d_k\|}{\|Dh(x_k)\|}.$$

The remainder of the theorem follows from (5.4.1.8) as soon as we establish that

$$\inf_k \gamma_k > 0, \qquad \inf_k \sigma_k > 0.$$

According to assumptions (b) and (c), $Df(x)^{-1}$ is continuous on the compact set K. Therefore $\operatorname{cond}(Df(x))$ is continuous, and

$$\gamma := \frac{1}{\max\limits_{x \in K} \operatorname{cond}(Df(x))} > 0$$

exists.

Without loss of generality, we may assume that x_k is not a stationary point of h; which means that it is no zero of f, because of (5.4.2.3) and assumption (c). [If $f(x_k) = 0$, then it follows immediately that $x_k = x_{k+1} = x_{k+2} = \cdots$, and there is nothing left to show.] Thus, since $x_k \in K$ for $k = 0, 1, \ldots,$

$$\inf_k \gamma_k \geqslant \gamma > 0.$$

On the other hand, from the fact that $f(x_k) \neq 0$, from (5.4.2.2), and from the inequalities

$$\|d_k\| = \|Df(x_k)^{-1} f(x_k)\| \geqslant \frac{1}{\|Df(x_k)\|} \|f(x_k)\|,$$

$$\|Dh(x_k)\| \leqslant 2\|Df(x_k)\| \, \|f(x_k)\|,$$

it follows immediately that

$$\sigma_k \geqslant \frac{1}{2\|Df(x_k)\|^2} \geqslant \sigma > 0$$

[from the continuity of $Df(x)$ in the set K, which is compact]. Thus, all the results of Theorem (5.4.1.8) [or (5.4.1.4)] apply to the sequence $\{x_k\}$. Since assumption (c) and (5.4.2.3) together imply that each stationary point of h is also a zero of f, the proof is complete. □

The method (5.4.2.4) requires that $\|Dh(x_k)\|$ and $\gamma_k := 1/\operatorname{cond}(Df(x_k))$ be computed at each iteration step. The proof of (5.4.1.8), however, shows that it would be sufficient to replace all γ_k by a lower bound $\gamma > 0$, $\gamma_k \geqslant \gamma$. In accord with this, λ_k is usually determined in practice so that

$$h_k(2^{-j}) < h_k(0).$$

However, since this only requires that $\gamma_k > 0$, the methods used for the above proofs are not sufficiently strong to guarantee the convergence of this variant.

A further remark about the behavior of (5.4.2.4): In a sufficiently small neighborhood of a zero the method chooses $\lambda_k = 1$ automatically. This means that the method conforms to the ordinary Newton method and converges quadratically. We can see this as follows:

Since $\lim_{k \to \infty} x_k = \bar{x}$ and $f(\bar{x}) = 0$, there is a neighborhood $V_1(\bar{x})$ of $\bar{x}$ in which every iteration step $z_k \to z_{k+1}$ which would be carried out by the ordinary Newton method would satisfy the condition

(5.4.2.6) $$\|z_{k+1} - \bar{x}\| \leqslant a\|z_k - \bar{x}\|^2$$

and

$$(5.4.2.7) \qquad 32a^2c^2\|z_k - \bar{x}\|^2 \leqslant 1, \qquad c := \operatorname{cond}(Df(\bar{x})).$$

Taylor's expansion of f about $\bar{x}$ gives

$$f(x) = Df(\bar{x})(x - \bar{x}) + o(\|x - \bar{x}\|).$$

Since $\lim_{x \to \bar{x}} o(\|x - \bar{x}\|)/\|x - \bar{x}\| = 0$, there is another neighborhood $V_2(\bar{x})$ of $\bar{x}$ such that

$$\tfrac{1}{2}\|Df(\bar{x})^{-1}\|^{-1}\|x - \bar{x}\| \leqslant f(x) = \sqrt{h(x)} \leqslant 2\|Df(\bar{x})\| \, \|x - \bar{x}\|$$

for all $x \in V_2(\bar{x})$.

Choose a neighborhood

$$U(\bar{x}) \subset V_1(\bar{x}) \cap V_2(\bar{x})$$

and let k_0 be such that

$$x_k \in U(\bar{x})$$

for $k \geqslant k_0$. This is possible because $\lim_{k \to \infty} x_k = \bar{x}$. Considering

$$x_{k+1} := x_k - Df(x_k)^{-1}f(x_k), \qquad \text{i.e.,} \quad \lambda_k = 1$$

in (5.4.2.4), and using (5.4.2.6), (5.4.2.7), we are led to

$$h(x_{k+1}) \leqslant 4\|Df(\bar{x})\|^2 \, \|x_{k+1} - \bar{x}\|^2 \leqslant 16a^2c^2\|x_k - \bar{x}\|^2 h(x_k) \leqslant h(x_k)(1 - \tfrac{1}{2}).$$

From (5.4.2.4bα),

$$\gamma_k\|d_k\| \, \|Dh(x_k)\| \leqslant 2\gamma_k\|Df(x_k)^{-1}\| \, \|Df(x_k)\| h(x_k) = 2h(x_k).$$

This implies

$$h(x_{k+1}) \leqslant h(x_k)(1 - \tfrac{1}{2}) \leqslant h(x_k) - \frac{\gamma_k}{4}\|d_k\| \, \|Dh(x_k)\|.$$

That is, there exists a k_0 such that for all $k \geqslant k_0$ the choice $j = 0$ and $\lambda_k = 1$ will be made in the process given by (5.4.2.4). Thus (5.4.2.4) is identical to the ordinary Newton method in a sufficiently small neighborhood of $\bar{x}$, which means that it is locally quadratically convergent.

Assumption (a)–(c) in Theorem (5.4.2.5) characterize the class of functions for which the algorithm (5.4.2.4) is applicable. In one of the assigned problems for this chapter, two examples will be given of function classes which do not satisfy (a)–(c).

5.4.3 Suggestions for a Practical Implementation of the Modified Newton Method. A Rank-One Method Due to Broyden

Newton's method for solving the system $f(x) = 0$, where $f: \mathbb{R}^n \to \mathbb{R}^n$, is quite expensive even in its modified form (5.4.2.4), since the Jacobian $Df(x_k)$ and the solution to the linear system $Df(x_k)d = f(x_k)$ must be computed at each iteration. The evaluation of explicit formulas for the components of $Df(x)$ is frequently complicated and costly; indeed, explicit formulas may not even be available. In such cases it is reasonable to replace

$$Df(x) = \left(\frac{\partial f(x)}{\partial x^1}, \ldots, \frac{\partial f(x)}{\partial x^n}\right)$$

at each $x = x_k$ by a matrix

$$\Delta f(x) = (\Delta_1 f, \ldots, \Delta_n f), \tag{5.4.3.1}$$

where

$$\Delta_i f(x) := \frac{f(x^1, \ldots, x^i + h_i, \ldots, x^n) - f(x^1, \ldots, x^i, \ldots, x^n)}{h_i}$$
$$= \frac{f(x + h_i e_i) - f(x)}{h_i},$$

that is, we may replace the partial derivatives $\partial f/\partial x^i$ by suitable difference quotients $\Delta_i f$. Note that the matrix $\Delta f(x)$ can be computed with only n additional evaluations of the function f (beyond that required at the point $x = x_k$). However it can be difficult to choose the stepsizes h_i. If any h_i is too large, then $\Delta f(x)$ can be a bad approximation to $Df(x)$, so that the iteration

$$x_{k+1} = x_k - \lambda_k \, \Delta f(x_k)^{-1} f(x_k) \tag{5.4.3.2}$$

converges, if it converges at all, much more slowly than (5.4.2.4). On the other hand, if any h_i is too small, then $f(x + h_i e_i) \approx f(x)$, and cancellations can occur which materially reduce the accuracy of the difference quotients. The following compromise seems to work the best: if we assume that all components of $f(x)$ can be computed with a relative error of the same order of magnitude as the machine precision eps, then choose h_i so that $f(x)$ and $f(x + h_i e_i)$ have roughly the first $t/2$ digits in common, given that t-digit accuracy is being maintained inside the computer. That is,

$$|h_i| \, \|\Delta_i f(x)\| \approx \sqrt{\text{eps}} \, \|f(x)\|.$$

In this case the influence of cancellations is usually not too bad.

If the function f is very complicated, however, even the n additional evaluations of f needed to produce $\Delta f(x)$ can be too expensive to bear at each iteration. In this case we try replacing $Df(x_k)$ by some matrix B_k which is even simpler than $\Delta f(x_k)$. Suitable matrices can be obtained using the following result due to Broyden (1965).

(5.4.3.3) Theorem. *Let A and B be arbitrary $n \times n$ matrices; let $b \in \mathbb{R}^n$, and let $F\colon \mathbb{R}^n \to \mathbb{R}^n$ be the affine mapping $F(u) := Au + b$. Suppose $x, x' \in \mathbb{R}^n$ are distinct vectors, and define p, q by*

$$p := x' - x, \qquad q := F(x') - F(x) = Ap.$$

Then the $n \times n$ matrix B' given by

$$B' := B + \frac{1}{p^T p}(q - Bp)p^T$$

satisfies

$$\operatorname{lub}_2(B' - A) \leqslant \operatorname{lub}_2(B - A)$$

with respect to the Euclidian norm, and it also satisfies the equation

$$B'p = Ap = q.$$

PROOF. The equality $(B' - A)p = 0$ is immediate from the definition of B'. Each vector $u \in \mathbb{R}^n$ satisfying $\|u\|_2 = 1$ has an orthogonal decomposition of the form

$$u = \alpha p + v, \qquad v^T p = 0, \quad \|v\|_2 \leqslant 1, \quad \alpha \in \mathbb{R}.$$

Thus it follows from the definition of B' that, for $\|u\|_2 = 1$,

$$\begin{aligned}\|(B' - A)u\|_2 = \|(B' - A)v\|_2 &= \|(B - A)v\|_2 \\ &\leqslant \operatorname{lub}_2(B - A)\|v\|_2 \leqslant \operatorname{lub}_2(B - A).\end{aligned}$$

Hence

$$\operatorname{lub}_2(B' - A) = \sup_{\|u\|_2 = 1} \|(B' - A)u\|_2 \leqslant \operatorname{lub}_2(B - A). \qquad \square$$

This result shows that the Jacobian $DF(x) \equiv A$ of an affine function F is approximated by B' at least as well as it is approximated by B, and furthermore B' and $DF(x)$ will both map p into the same vector. Since a differentiable nonlinear function $f\colon \mathbb{R}^n \to \mathbb{R}^n$ can be approximated to first order in the neighborhood of one of its zeros $\bar{x}$ by an affine function, this suggests using the above construction of B' from B even in the nonlinear case. Doing so yields an iteration of the form

(5.4.3.4)
$$\begin{aligned} d_k &:= B_k^{-1} f(x_k), \\ x_{k+1} &:= x_k - \lambda_k d_k, \\ p_k := x_{k+1} - x_k, &\qquad q_k := f(x_{k+1}) - f(x_k), \\ B_{k+1} &:= B_k + \frac{1}{p_k^T p_k}(q_k - B_k p_k)p_k^T. \end{aligned}$$

The formula for B_{k+1} was suggested by Broyden. Since $\operatorname{rank}(B_{k+1} - B_k) \leqslant 1$, it is called *Broyden's rank-one update*. The stepsizes λ_k may be determined from an approximate minimization of $\|f(x)\|^2$:

$$\|f(x_{k+1})\|^2 \approx \min_{\lambda \geqslant 0} \|f(x_k - \lambda d_k)\|^2,$$

using, for example, a finite search process

(5.4.3.5) $\lambda_k := 2^{-j}, \qquad j := \min\{i \geqslant 0 \mid \|f(x_k - 2^{-i}d_k)\| < \|f(x_k)\|\}$

as in (5.4.2.4).

A suitable starting matrix B_0 can be obtained using difference quotients: $B_0 = \Delta f(x_0)$. It does not make good sense, however, to compute all following matrices B_k, $k \geqslant 1$, from the updating formula. Various suggestions have been made about which iterations of (5.4.3.4) are to be modified by replacing $B_k + (1/p_k^T p_k)(q_k - B_k p_k)p_k^T$ with $\Delta f(x_k)$ ("reinitialization"). As one possibility we may obtain B_{k+1} from B_k using Broyden's update only on those iterations where the step produced by (5.4.3.5) lies in the interval $2^{-1} \leqslant \lambda \leqslant 1$. A justification for this is given by observing that the bisection method (5.4.3.5) automatically picks $\lambda_k = 1$ when $\|f(x_k - d_k)\| < \|f(x_k)\|$. The following result due to Broyden, Dennis, and Moré (1973) shows that this will be true for all x_k sufficiently close to $\bar{x}$ (in which case making an affine approximation to f is presumably justified):
Under the assumption that

(a) $Df(x)$ exists and is continuous in a neighborhood $U(\bar{x})$ of a zero point $\bar{x}$,
(b) $\|Df(x) - Df(\bar{x})\| \leqslant \Lambda\|x - \bar{x}\|$ for some $\Lambda > 0$ and all $x \in U(\bar{x})$,
(c) $[Df(\bar{x})]^{-1}$ exists,

then the iteration (5.4.3.4) is well defined using $\lambda_k = 1$ for all $k \geqslant 0$ (i.e., all B_k are nonsingular) provided x_0 and B_0 are "sufficiently close" to $\bar{x}$ and $Df(\bar{x})$. Moreover, the iteration generates a sequence $\{x_k\}$ which converges superlinearly to $\bar{x}$,

$$\lim_{k\to\infty} \frac{\|x_{k+1} - \bar{x}\|}{\|x_k - \bar{x}\|} = 0,$$

if $x_k \neq \bar{x}$ for all $k \geqslant 0$.

The direction $d_k = B_k^{-1} f(x_k)$ appearing in (5.4.3.4) is best obtained by solving the linear system $B_k d = f(x_k)$ using a decomposition $F_k B_k = R_k$ of the kind given in (4.9.1). Observe that the factors F_{k+1}, R_{k+1} of B_{k+1} can be found from the factors F_k, R_k of B_k by employing the techniques of Section 4.9, since modification of B_k by a rank-one matrix is involved.

We remark that all of the foregoing also has application to function minimization as well as the location of zeros. Let $h: \mathbb{R}^n \to \mathbb{R}$ be a given function. The minimum points of h are among the zeros of $f(x) = \nabla h(x)$. Moreover, the Jacobian $Df(x)$ is the Hessian matrix $\nabla^2 h(x)$ of h; hence it can be expected to be positive definite near a strong local minimum point $\bar{x}$ of h. This suggests that the matrix B_k which is taken to approximate $\nabla^2 h(x_k)$ should be positive definite. Consequently, for function minimization, the Broyden rank-one updating formula used in (5.4.3.4) should be replaced by an updating formula which guarantees the positive definiteness of B_{k+1} given that B_k is positive definite. A number of such formulas have been suggested. The most successful are of rank two and can be expressed as two stage updates:

$$\left.\begin{aligned} B_{k+1/2} &:= B_k + \alpha u_k u_k^T \\ B_{k+1} &:= B_{k+1/2} - \beta v_k v_k^T \end{aligned}\right\} \qquad (\alpha, \beta > 0),$$

where $B_{k+1/2}$ is guaranteed to be positive definite as well as B_{k+1}. For such updates the Choleski decomposition is the most reasonable to use in solving the system $B_k d = f(x_k)$. More details on these topics may be found in the reference by Gill, Golub, Murray, and Saunders (1974). Rank-two updates to positive definite approximations H_k of the inverse $[\nabla^2 h(x_k)]^{-1}$ of the Hessian are described in Section 5.11.

5.5 Roots of Polynomials. Application of Newton's Method

Sections 5.5–5.8 deal with roots of polynomials and some typical methods for their determination. There are a host of methods available for this purpose which we will not be covering. See, for example, Bauer (1956), Jenkins and Traub (1970), Nickel (1966), and Henrici (1974), to mention just a few.

The importance of general methods for determining roots of general polynomials may sometimes be overrated. Polynomials found in practice are frequently given in some special form, such as characteristic polynomials of matrices. In the latter case, the roots are eigenvalues of matrices, and methods to be described in Chapter 6 are to be preferred.

We proceed to describe how the Newton method applies to finding the roots of a given polynomial $p(x)$. In order to evaluate the iteration function of Newton's method,

$$x_{k+1} := x_k - \frac{p(x_k)}{p'(x_k)},$$

we have to calculate the value of the polynomial p, as well as the value of its first derivative, at the point $x = x_k$. Assume the polynomial p is given in the form

$$p(x) = a_0 x^n + a_1 x^{n-1} + \cdots + a_n.$$

Then $p(x_k)$ and $p'(x_k)$ can be calculated as follows: For $x = \xi$,

$$p(\xi) = (\cdots((a_0\xi + a_1)\xi + a_2)\xi + \cdots)\xi + a_n.$$

The multipliers of ξ in this expression are recursively of the form

(5.5.1)
$$\begin{aligned} b_0 &:= a_0 \\ b_i &:= b_{i-1}\xi + a_i, \qquad i = 1, 2, \ldots, n. \end{aligned}$$

The value of the polynomial p at ξ is then given by

$$p(\xi) = b_n.$$

The algorithm for evaluating polynomials using the recursion (5.5.1) is known as *Horner's scheme*. The quantities b_i, thus obtained, are also the coefficients of the polynomial

$$p_1(x) := b_0 x^{n-1} + b_1 x^{n-2} + \cdots + b_{n-1}$$

which results if the polynomial $p(x)$ is divided by $x - \xi$:

$$p(x) = (x - \xi)p_1(x) + b_n, \tag{5.5.2}$$

This is readily verified by comparing the coefficients of the powers of x on both sides of (5.5.2). Furthermore, differentiating the relation (5.5.2) with respect to x and setting $x = \xi$ yields

$$p'(\xi) = p_1(\xi).$$

Therefore, the first derivative $p'(\xi)$ can be determined by repeating the Horner scheme, using the results b_i of the first as coefficients for the second:

$$p'(\xi) = (\cdots (b_0 \xi + b_1)\xi + \cdots)\xi + b_{n-1}.$$

Frequently, however, the polynomial $p(x)$ is given in some form other than

$$p(x) = a_0 x^n + \cdots + a_n.$$

Particularly important is the case in which $p(x)$ is the characteristic polynomial of a symmetric tridiagonal matrix

$$J = \begin{bmatrix} \alpha_1 & \beta_2 & & & 0 \\ \beta_2 & \cdot & \cdot & & \\ & \cdot & \cdot & \cdot & \\ & & \cdot & \cdot & \beta_n \\ 0 & & & \beta_n & \alpha_n \end{bmatrix}, \qquad \alpha_i, \beta_i \text{ real}.$$

Denoting by $p_i(x)$ the characteristic polynomial

$$p_i(x) := \det\left(\begin{bmatrix} \alpha_1 - x & \beta_2 & & & 0 \\ \beta_2 & \cdot & \cdot & & \\ & \cdot & \cdot & \cdot & \\ & & \cdot & \cdot & \beta_i \\ 0 & & & \beta_i & \alpha_i - x \end{bmatrix}\right)$$

of the principal minor formed by the first i rows and columns of the matrix J, we have the recursions

$$\begin{aligned} &p_0(x) := 1, \\ &p_1(x) := (\alpha_1 - x) \cdot 1, \\ (5.5.3) \quad &p_i(x) := (\alpha_i - x)p_{i-1}(x) - \beta_i^2 p_{i-2}(x), \qquad i = 2, 3, \ldots, n, \\ &p(x) := \det(J - xI) := p_n(x). \end{aligned}$$

These can be used to calculate $p(\xi)$ for any $x = \xi$ and any given matrix elements α_i, β_i. A similar recursion for calculating $p'(x)$ is obtained by differentiating (5.5.3):

$$\begin{aligned} &p_0'(x) := 0, \\ &p_1'(x) := -1, \\ (5.5.4) \quad &p_i'(x) := -p_{i-1}(x) + (\alpha_i - x)p_{i-1}'(x) - \beta_i^2 p_{i-2}'(x), \qquad i = 2, 3, \ldots, n, \\ &p'(x) := p_n'(x), \end{aligned}$$

The two recursions (5.5.3) and (5.5.4) can be evaluated concurrently.

During our general discussion of the Newton method in Section 5.3 it became clear that the convergence of a sequence x_k towards a zero ξ of a function is assured only if the starting point x_0 is sufficiently close to ξ. A bad initial choice x_0 may cause the sequence x_k to diverge even for polynomials. If the real polynomial $p(x)$ has no real roots [e.g., $p(x) = x^2 + 1$], then the Newton method must diverge for any initial value $x_0 \in \mathbb{R}$. There are no known fail-safe rules for selecting initial values in the case of arbitrary polynomials. However, such a rule exists in an important special case, namely, if all roots ξ_i, $i = 1, 2, \ldots, n$, are real:

$$\xi_1 \geqslant \xi_2 \geqslant \cdots \geqslant \xi_n.$$

In Section 5.6, Theorem (5.6.5), we will show that the polynomials defined by (5.5.3) have this property if the matrix elements α_i, β_i are real.

(5.5.5) Theorem. *Let $p(x)$ be a polynomial of degree $n \geqslant 2$ with real coefficients. If all roots ξ_i,*

$$\xi_1 \geqslant \xi_2 \geqslant \cdots \geqslant \xi_n,$$

of $p(x)$ are real, then Newton's method yields a convergent strictly decreasing sequence x_k for any initial value $x_0 > \xi_1$.

PROOF. Without loss of generality, we may assume that $p(x_0) > 0$.

Since $p(x)$ does not change sign for $x > \xi_1$, we have

$$p(x) = a_0 x^n + \cdots + a_n > 0$$

for $x > \xi_1$ and therefore $a_0 > 0$. The derivative p' has $n - 1$ real zeros α_i with

$$\xi_1 \geqslant \alpha_1 \geqslant \xi_2 \geqslant \alpha_2 \geqslant \cdots \geqslant \alpha_{n-1} \geqslant \xi_n$$

by Rolle's theorem. Since p' is of degree $n-1 \geqslant 1$, these are all its roots, and $p'(x) > 0$ for $x > \alpha_1$ because $a_0 > 0$. Applying Rolle's theorem again, and recalling that $n \geqslant 2$, we obtain

$$(5.5.6) \qquad \begin{aligned} p''(x) &> 0 \quad \text{for } x > \alpha_1, \\ p'''(x) &\geqslant 0 \quad \text{for } x \geqslant \alpha_1. \end{aligned}$$

Thus p and p' are convex functions for $x \geqslant \alpha_1$

Now $x_k > \xi_1$ implies that

$$x_{k+1} = x_k - \frac{p(x_k)}{p'(x_k)} < x_k,$$

since $p'(x_k) > 0$, $p(x_k) > 0$. It remains to be shown, that we do not "overshoot," i.e., that $x_{k+1} > \xi_1$. From (5.5.6), $x_k > \xi_1 \geqslant \alpha_1$, and Taylor's theorem we conclude that

$$\begin{aligned} 0 = p(\xi_1) &= p(x_k) + (\xi_1 - x_k)p'(x_k) + \tfrac{1}{2}(\xi_1 - x_k)^2 p''(\delta), \qquad \xi_1 < \delta < x_k \\ &> p(x_k) + (\xi_1 - x_k)p'(x_k). \end{aligned}$$

$p(x_k) = p'(x_k)(x_k - x_{k+1})$ holds by the definition of x_{k+1}. Thus

$$0 > p'(x_k)(x_k - x_{k+1} + \xi_1 - x_k) = p'(x_k)(\xi_1 - x_{k+1}),$$

and $x_{k+1} > \xi_1$ follows, since $p'(x_k) > 0$. □

For later use we note the following consequence of (5.5.6):

(5.5.7) Lemma. *Let $p(x) = a_0 x^n + \cdots + a_n$, $a_0 > 0$, be a real polynomial of degree $n \geqslant 2$ all roots of which are real. If α_1 is the largest root of p', then $p'''(x) \geqslant 0$ for $x \geqslant \alpha_1$, i.e., p' is a convex function for $x \geqslant \alpha_1$.*

We are still faced with the problem of finding a number $x_0 > \xi_1$, without knowing ξ_1 beforehand. The following inequalities are available for this purpose:

(5.5.8) Theorem. *For all roots ξ_i of an arbitrary polynomial $p(x) = a_0 x^n + a_1 x^{n-1} + \cdots + a_n$,*

$$|\xi_i| \leqslant \max\left\{\left|\frac{a_n}{a_0}\right|, 1 + \left|\frac{a_{n-1}}{a_0}\right|, \ldots, 1 + \left|\frac{a_1}{a_0}\right|\right\},$$

$$|\xi_i| \leqslant \max\left\{1, \sum_{j=1}^{n} \left|\frac{a_j}{a_0}\right|\right\},$$

$$|\xi_i| \leqslant \max\left\{\left|\frac{a_n}{a_{n-1}}\right|, 2\left|\frac{a_{n-1}}{a_{n-2}}\right|, \ldots, 2\left|\frac{a_1}{a_0}\right|\right\},$$

$$|\xi_i| \leqslant \sum_{j=0}^{n-1} \left|\frac{a_{j+1}}{a_j}\right|,$$

$$|\xi_i| \leqslant 2\max\left\{\left|\frac{a_1}{a_0}\right|, \sqrt{\left|\frac{a_2}{a_0}\right|}, \sqrt[3]{\left|\frac{a_3}{a_0}\right|}, \ldots, \sqrt[n]{\left|\frac{a_n}{a_0}\right|}\right\}.$$

Some of these inequalities will be proved in Section 6.9. Compare also Householder (1970). Additional inequalities can be found in Marden (1949).

Quadratic convergence does not necessarily mean fast convergence. If the initial value x_0 is far from a root, then the sequence x_k obtained by Newton's method may converge very slowly in the beginning. Indeed, if x_k is large, then

$$x_{k+1} = x_k - \frac{x_k^n + \cdots}{n x_k^{n-1} + \cdots} \approx x_k\left(1 - \frac{1}{n}\right),$$

so that there is little change between x_k and x_{k+1}. This observation has led to considering the following *double-step method*:

$$x_{k+1} = x_k - 2\frac{p(x_k)}{p'(x_k)}, \qquad k = 0, 1, 2, \ldots,$$

in lieu of the straightforward Newton method.

Of course, there is now the danger of "overshooting." In particular, in the case of polynomials with real roots only and an initial point $x_0 > \xi_1$, some x_{k+1} may overshoot ξ_1, negating the benefit of Theorem (5.5.5). However, this overshooting can be detected, and, due to some remarkable properties of polynomials, a good initial value y $(\xi_1 \geqslant y > \xi_2)$ with which to start a subsequent Newton procedure for the calculation of ξ_2 can be recovered. The latter is a consequence of the following theorem:

(5.5.9) Theorem. *Let $p(x)$ be a real polynomial of degree $n \geqslant 2$, all roots of which are real, $\xi_1 \geqslant \xi_2 \geqslant \cdots \geqslant \xi_n$. Let α_1 be the largest root of $p'(x)$:*

$$\xi_1 \geqslant \alpha_1 \geqslant \xi_2.$$

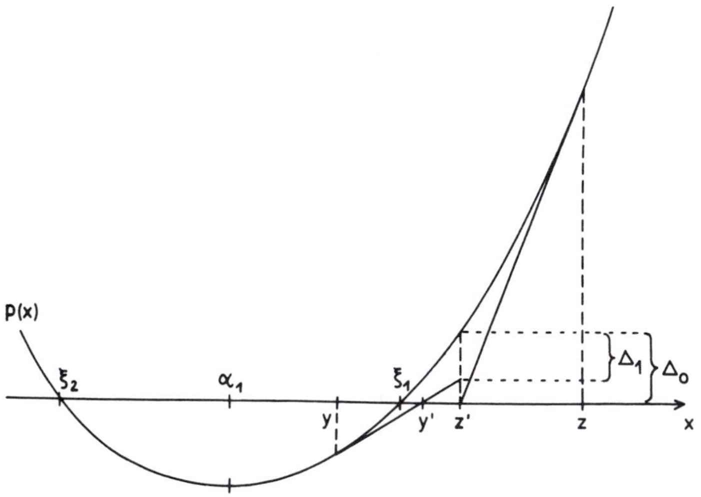

Figure 8 Geometric interpretation of the double-step method.

For $n = 2$, we require also that $\xi_1 > \xi_2$. Then for every $z > \xi_1$, the numbers

$$z' := z - \frac{p(z)}{p'(z)}, \qquad y := z - 2\frac{p(z)}{p'(z)}, \qquad y' := y - \frac{p(y)}{p'(y)}$$

(Figure 8) are well defined and satisfy

(5.5.10a) $$\alpha_1 < y,$$

(5.5.10b) $$\xi_1 \leqslant y' \leqslant z'.$$

It is readily verified that $n = 2$ and $\xi_1 = \xi_2$ imply $y = \xi_1$ for any $z > \xi_1$.

PROOF. Assume again that $p(z) > 0$ for $z > \xi_1$. For such values z, we consider the quantities Δ_0, Δ_1 (Figure 8), which are defined as follows:

$$\Delta_0 := p(z') = p(z') - p(z) - (z' - z)p'(z) = \int_z^{z'} [p'(t) - p'(z)]\, dt,$$

$$\Delta_1 := p(z') - p(y) - (z' - y)p'(y) = \int_y^{z'} [p'(t) - p'(y)]\, dt.$$

Δ_0 and Δ_1 can be interpreted as areas over and under the graph of $p'(x)$, respectively (Figure 9).

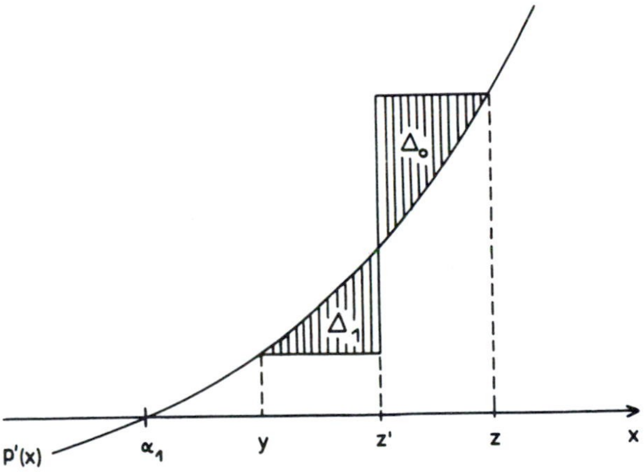

Figure 9 The quantities Δ_0 and Δ_1 interpreted as areas.

By Lemma (5.5.7), $p'(x)$ is a convex function for $x \geqslant \alpha_1$. Therefore, and because $z' - y = z - z' > 0$—the latter being positive by Theorem (5.5.5)—we have

(5.5.11) $$\Delta_1 \leqslant \Delta_0 \quad \text{if } y \geqslant \alpha_1,$$

with equality $\Delta_1 = \Delta_0$ holding if and only if p' is a linear function, that is, if p is a polynomial of degree 2. Now we distinguish the three cases $y > \xi_1$,

$y = \xi_1$, $y < \xi_1$. For $y > \xi_1$, the proposition of the theorem follows immediately from Theorem (5.5.5). For $y = \xi_1$, we show first that $\xi_2 < \alpha_1 < \xi_1$, that is, ξ_1 is a simple root of p. If $y = \xi_1 = \xi_2 = \alpha_1$ were a multiple root, then by hypothesis $n \geqslant 3$, and consequently $\Delta_1 < \Delta_0$ would hold in (5.5.11). This would lead to the contradiction

$$\Delta_1 = p(z') - p(\xi_1) - (z' - \xi_1)p'(\xi_1) = p(z') < \Delta_0 = p(z').$$

Thus ξ_1 must be a simple root; hence $\alpha_1 < \xi_1 = y' = y < z'$, and the proposition is seen to be correct in the second case, too.

The case $y < \xi_1$ remains. If $\alpha_1 < y$, then the validity of the proposition can be established as follows. Since $p(z) > 0$ and $\xi_2 < \alpha_1 < y < \xi_1$, we have $p(y) < 0$, $p'(y) > 0$. In particular, y' is well defined. Furthermore, since $p(y) = (y - y')p'(y)$ and $\Delta_1 \leqslant \Delta_0$, we have

$$\Delta_0 - \Delta_1 = p(y) + (z' - y)p'(y) = p'(y)(z' - y') \geqslant 0.$$

Therefore $z' \geqslant y'$. By Taylor's theorem, finally,

$$p(\xi_1) = 0 = p(y) + (\xi_1 - y)p'(y) + \tfrac{1}{2}(\xi_1 - y)^2 p''(\delta), \qquad y < \delta < \xi_1,$$

and since $p''(x) \geqslant 0$ for $x \geqslant \alpha_1$, $p(y) = (y - y')p'(y)$, and $p'(y) > 0$,

$$0 \geqslant p(y) + (\xi_1 - y)p'(y) = p'(y)(\xi_1 - y').$$

Therefore $\xi_1 \leqslant y'$.

To complete the proof, we proceed to show that

$$y = y(z) > \alpha_1 \tag{5.5.12}$$

for any $z > \xi_1$. Again we distinguish two cases, $\xi_1 > \alpha_1 > \xi_2$ and $\xi_1 = \alpha_1 = \xi_2$.

If $\xi_1 > \alpha_1 > \xi_2$, then (5.5.12) holds whenever

$$\xi_1 < z < \xi_1 + (\xi_1 - \alpha_1).$$

This is because Theorem (5.5.5) implies $z > z' \geqslant \xi_1$, and therefore

$$y = z' - (z - z') > \xi_1 - (\xi_1 - \alpha_1) = \alpha_1$$

holds by the definition of $y = y(z)$. Hence we can select a z_0 with $y(z_0) > \alpha_1$. Assume that there exists a $z_1 > \xi_1$ with $y(z_1) \leqslant \alpha_1$. By the intermediate-value theorem for continuous functions, there exists a $\bar{z} \in [z_0, z_1]$ with $\bar{y} = y(\bar{z}) = \alpha_1$. From (5.5.11) for $z = \bar{z}$,

$$\Delta_1 = p(\bar{z}') - p(\bar{y}) - (\bar{z}' - \bar{y})p'(\bar{y}) = p(\bar{z}') - p(\bar{y}) \leqslant \Delta_0 = p(\bar{z}'),$$

and therefore $p(\bar{y}) = p(\alpha_1) \geqslant 0$. On the other hand, $p(\alpha_1) < 0$, since ξ_1 is a simple root, in our case, causing $p(x)$ to change sign. This is a contradiction, and (5.5.12) must hold for all $z > \xi_1$.

If $\xi_1 = \alpha_1 = \xi_2$, then by hypothesis $n \geqslant 3$. Assume, without loss of generality, that

$$p(x) = x^n + a_1 x^{n-1} + \cdots + a_n.$$

Then

$$z' = z - \frac{p(z)}{p'(z)} = z - \frac{z}{n} \frac{1 + \dfrac{a_1}{z} + \cdots + \dfrac{a_n}{z^n}}{1 + \dfrac{n-1}{n}\dfrac{a_1}{z} + \cdots + \dfrac{a_{n-1}}{nz^{n-1}}}$$

$$= z - \frac{z}{n}\left(1 + O\left(\frac{1}{z}\right)\right).$$

Therefore

$$y = y(z) = z + 2(z' - z) = z - \frac{2z}{n}\left(1 + O\left(\frac{1}{z}\right)\right)$$

$$= z\left(1 - \frac{2}{n}\right) + O(1).$$

Since $n \geqslant 3$, the value of $y(z)$ increases indefinitely as $z \to \infty$, and we conclude again that there exists a $z_0 > \xi_1$ with $y_0 = y(z_0) > \alpha_1$. If (5.5.12) did not hold for all $z > \xi_1$, then we could conclude, just as before, that there exists $\bar{z} > \xi_1$ with $\bar{y} = y(\bar{z}) = \alpha_1$. However, the existence of such a value $\bar{y} = \alpha_1 = \xi_1 = \xi_2$ has been shown to be impossible earlier in the proof of this theorem. □

The practical significance of this theorem is as follows. If we have started with $x_0 > \xi_1$, then either the approximate values generated by the double-step method

$$x_{k+1} = x_k - 2\frac{p(x_k)}{p'(x_k)}$$

satisfy

$$x_0 \geqslant x_1 \geqslant \cdots \geqslant x_k \geqslant x_{k+1} \geqslant \cdots \geqslant \xi_1 \quad \text{and} \quad \lim_{k \to \infty} x_k = \xi_1,$$

or there exists a first $x_{k_0} := y$ such that

$$p(x_0)p(x_k) > 0 \ \text{ for } 0 \leqslant k < k_0, \quad \text{and} \quad p(x_0)p(x_{k_0}) < 0.$$

In the first case, all values $p(x_k)$ are of the same sign,

$$p(x_0)p(x_k) \geqslant 0 \quad \text{for all } k,$$

and the x_k converge monotonically (and faster than for the straightforward Newton method) towards the root ξ_1. In the second case,

$$x_0 > x_1 > \cdots > x_{k_0 - 1} > \xi_1 > y = x_{k_0} > \alpha_1 > \xi_2.$$

Using $y_0 := y$ as the starting point of a subsequent straightforward Newton procedure,

$$y_{k+1} = y_k - \frac{p(y_k)}{p'(y_k)}, \qquad k = 0, 1, \ldots,$$

will also provide monotonic convergence:

$$y_1 \geqslant y_2 \geqslant \cdots \geqslant \xi_1, \qquad \lim_{k\to\infty} y_k = \xi_1.$$

Having found the largest root ξ_1 of a polynomial p, there are still the other roots $\xi_2, \xi_3, \ldots, \xi_n$ to be found. The following idea suggests itself immediately: "divide off" the known root ξ_1, that is, form the polynomial

$$p_1(x) := \frac{p(x)}{x - \xi_1}$$

of degree $n - 1$. This process is called *deflation*. The largest root of $p_1(x)$ is ξ_2, and may be determined by the previously described procedures. Here ξ_1 or, even better, the value $y = x_{k_0}$ found by overshooting may serve as a starting point. In this fashion, all roots will be found eventually.

Deflation, in general, is not without hazard, because roundoff will preclude an exact determination of $p_1(x)$. The polynomial actually found in place of p_1 will have roots different from $\xi_2, \xi_3, \ldots, \xi_n$. These are then found by means of further approximations, with the result that the last roots may be quite inaccurate. However, deflation has been found to be numerically stable if done with care. In dividing off a root, the coefficients of the deflated polynomial

$$p_1(x) = a'_0 x^{n-1} + a'_1 x^{n-2} + \cdots + a'_{n-1}$$

may be computed in the order $a'_0, a'_1, \ldots, a'_{n-1}$ (*forward deflation*) or in the reverse order (*backward deflation*). The former is numerically stable if the root of smallest absolute value is divided off; the latter is numerically stable if the root of largest absolute value is divided off. A mixed process of determining the coefficients will be stable for roots of intermediate absolute value. See Peters and Wilkinson (1971) for details.

Deflation can be avoided altogether as suggested by Maehly (1954). He expresses the derivative of the deflated polynomial $p_1(x)$ as follows:

$$p'_1(x) = \frac{p'(x)}{x - \xi_1} - \frac{p(x)}{(x - \xi_1)^2};$$

and substitutes this expression into the Newton iteration function for p_1:

$$x_{k+1} = x_k - \frac{p_1(x_k)}{p'_1(x_k)} = x_k - \frac{p(x_k)}{p'(x_k) - \dfrac{p(x_k)}{x_k - \xi_1}}.$$

In general, we find for the polynomials

$$p_j(x) := \frac{p(x)}{(x - \xi_1) \ldots (x - \xi_j)}$$

that

$$p'_j(x) = \frac{p'(x)}{(x - \xi_1) \ldots (x - \xi_j)} - \frac{p(x)}{(x - \xi_1) \ldots (x - \xi_j)} \cdot \sum_{i=1}^{j} \frac{1}{x - \xi_i}.$$

The Maehly version of the (straightforward) Newton method for finding the root ξ_{j+1} is therefore as follows:

$$(5.5.13) \qquad x_{k+1} = \Phi_j(x_k) \quad \text{with } \Phi_j(x) := x - \frac{p(x)}{p'(x) - \sum_{i=1}^{j} \frac{p(x)}{x - \xi_i}}.$$

The advantage of this formula lies in the fact that the iteration given by Φ_j converges quadratically to ξ_{j+1} even if the numbers $\xi_1, \ldots, \xi_j$ in Φ_j are not roots of p (the convergence is only local in this case). Thus the calculation of ξ_{j+1} is not sensitive to the errors incurred in calculating the previous roots. This technique is an example of *zero suppression* as opposed to deflation [Peters and Wilkinson (1971)].

Note that $\Phi_j(x)$ is not defined if $x = \xi_k$, $k = 1, \ldots, j$, is a previous root of $p(x)$. Such roots cannot be selected as starting values. Instead one may use the values found by overshooting if the double-step method is employed.

The following pseudo-ALGOL program for finding all roots of a polynomial p having only real roots incorporates these features. Function procedures $p(z)$ and $p'(z)$ for the polynomial and its derivative are presumed available.

```
zo := starting point x0;
for j := 1 step 1 until n do
     begin m := 2; zs := zo;
          Iteration: z := zs; s := 0;
          for i := 1 step 1 until j - 1 do
               s := s + 1/(z - xi[i]);
               zs := p(z); zs := z - m × zs/(p'(z) - zs × s);
               if zs < z then goto Iteration;
               if m = 2 then
                  begin zs := z; m := 1; goto Iteration end;
     ξj := z
     end;
```

EXAMPLE. The following example illustrates the advantages of Maehly's method. The coefficients a_i of the polynomial

$$p(x) := \prod_{j=0}^{13} (x - 2^{-j}) = \sum_{i=0}^{14} a_i x^{14-i}$$

are calculated in general with a relative error of magnitude ε. Considerations in Section 5.8 will show the roots of $p(x)$ to be well conditioned. The following table shows that the Newton–Maehly method yields the roots of the above polynomial up to an absolute error of 40ε ($\varepsilon = 10^{-12}$ = machine precision). If forward deflation is employed but the roots are divided off as shown, i.e. in order of decreasing magnitude, then the fifth root is already completely wrong. The absolute errors below are understood as multiples of ε.

	(Absolute error) $\times 10^{12}$	
$\xi_j = 2^{-j}$	Newton–Maehly	Deflation
1.0	0	0
0.5	6.8	3.7×10^2
0.25	1.1	1.0×10^6
0.125	0.2	1.4×10^9
0.062 5	4.5	
0.031 25	4.0	
0.015 625	3.3	
0.007 812 5	39.8	$> 10^{12}$
0.003 906 25	10.0	
0.001 953 125	5.3	
0.000 976 562 5	0	
0.000 488 281 25	0	
0.000 244 140 625	0.4	
0.000 122 070 3125	0	

The polynomial $p_1(x)$,

$$p_1(x) = \prod_{j=1}^{13} (x - 2^{-j}) = \frac{p(x)}{(x - 1)}$$

has the roots 2^{-j}, $j = 1, 2, \ldots, 13$. If $\tilde{p}_1$ is produced by numerically dividing $p(x)$ by $x - 1$,

$$\tilde{p}_1(x) := \mathrm{fl}\left(\frac{p(x)}{(x - 1)}\right),$$

then we observe that the roots of $\tilde{p}_1(x)$ are already quite different from those of $p_1(x)$:

j	Roots of $\tilde{p}_1(x)$ computed by Newton–Maehly
1	0.499 999 996 335
2	0.250 001 00...
3	0.123 697...
4	0.092 4...
5	−0.098 4...
6	−0.056...
7	−0.64...
8	+1.83...
⋮	⋮

However, if forward deflation is employed and if the roots are divided off in the sequence of increasing absolute values, starting with the root of smallest absolute value, then this process will also yield the roots of $p(x)$ up to small multiples of machine precision [Wilkinson 1963, Peters and Wilkinson (1971)]:

	Sequence of dividing off roots of $p(x)$													
j	13	12	11	10	9	8	7	6	5	4	3	2	1	0
(Absolute error) $\times 10^{12}$	0.2	0.4	2	5	3	1	14	6	12	6	2	2	12	1

5.6 Sturm Sequences and Bisection Methods

Let $p(x)$ be a polynomial of degree n,

$$p(x) = a_0 x^n + a_1 x^{n-1} + \cdots + a_n, \qquad a_0 \neq 0.$$

It is possible [see Henrici (1974) for a thorough treatment of related results] to determine the number of real roots of $p(x)$ in a specified region by examining the number of *sign changes* $w(a)$ for certain points $x = a$ of a sequence of polynomials $p_i(x)$, $i = 0, 1, \ldots, m$, of descending degrees. Such a sign change happens whenever the sign of a polynomial value differs from that of its successor. Furthermore, if $p_i(a) = 0$, then this entry is to be removed from the sequence of polynomial values before the sign changes are counted. Suitable sequences of polynomials are the so-called *Sturm sequences.*

(5.6.1) Definition. The sequence

$$p(x) = p_0(x), p_1(x), \ldots, p_m(x)$$

of real polynomials is a *Sturm sequence* for the polynomial $p(x)$ if:

(a) All real roots of $p_0(x)$ are simple.
(b) sign $p_1(\xi) = -$sign $p_0'(\xi)$ if *ξ is a real root of $p_0(x)$.*
(c) For $i = 1, 2, \ldots, m-1$,

$$p_{i+1}(\xi)p_{i-1}(\xi) < 0$$

if *ξ is a real root of $p_i(x)$.*
(d) The last polynomial $p_m(x)$ *has no real roots.*

For such Sturm sequences we have the following

(5.6.2) Theorem. *The number of real roots of $p(x) \equiv p_0(x)$ in the interval $a \leqslant x < b$ equals $w(b) - w(a)$, where $w(x)$ is the number of sign changes of a Sturm sequence*

$$p_0(x), \ldots, p_m(x)$$

at location x.

Before proving this theorem, we show briefly how a simple recursion can be used to construct a Sturm sequence for the polynomial $p(x)$, provided all its real roots are simple. We define initially

$$p_0(x) := p(x), \qquad p_1(x) := -p_0'(x) = -p'(x),$$

and form the remaining polynomials $p_{i+1}(x)$ recursively, dividing $p_{i-1}(x)$ by $p_i(x)$,

$$\text{(5.6.3)} \qquad p_{i-1}(x) = q_i(x)p_i(x) - c_i p_{i+1}(x), \qquad i = 1, 2, \ldots,$$

where

$$[\text{degree of } p_i(x)] > [\text{degree of } p_{i+1}(x)]$$

and the constants $c_i > 0$ are positive but otherwise arbitrary. This recursion is the well-known *Euclidean algorithm*. Because the degree of the polynomials decreases, the algorithm must terminate after $m \leqslant n$ steps:

$$p_{m-1}(x) = q_m(x)p_m(x), \qquad p_m(x) \not\equiv 0.$$

The final polynomial $p_m(x)$ is a greatest common divisor of the two initial polynomials $p(x)$ and $p_1(x) = -p'(x)$. (This is the purpose of the Euclidean algorithm.) If all real roots of $p(x)$ are simple, then $p(x)$ and $p'(x)$ have no real roots in common. Thus $p_m(x)$ has no real roots and satisfies (5.6.1d). If $p_i(\xi) = 0$, then (5.6.3) gives $p_{i-1}(\xi) = -c_i p_{i+1}(\xi)$. Assume that $p_{i+1}(\xi) = 0$. Then (5.6.3) would imply $p_{i+1}(\xi) = \cdots = p_m(\xi) = 0$, contradicting $p_m(\xi) \neq 0$. Thus (5.6.1c) is satisfied. The two remaining conditions of (5.6.1) are immediate.

Proof of Theorem (5.6.2). We examine how a perturbation of the value a affects the number of sign changes $w(a)$ of the sequence

$$p_0(a), p_1(a), \ldots, p_m(a).$$

So long as a is not a root of any of the polynomials $p_i(x)$, $i = 0, 1, \ldots, m$, there is, of course, no change. If a is a root of $p_i(x)$, we consider the two cases $i > 0$ and $i = 0$.

In the first case, $i < m$ by (5.6.1d) and $p_{i+1}(a) \neq 0$, $p_{i-1}(a) \neq 0$ by (5.6.1c). If $p_i(x)$ changes sign at $x = a$, then for a sufficiently small perturbation $h > 0$, the signs of the polynomials $p_j(a)$, $j = i-1, i, i+1$, display the behavior illustrated in one of the following four tables:

	$a-h$	a	$a+h$
$i-1$	$-$	$-$	$-$
i	$-$	0	$+$
$i+1$	$+$	$+$	$+$

	$a-h$	a	$a+h$
$i-1$	$+$	$+$	$+$
i	$-$	0	$+$
$i+1$	$-$	$-$	$-$

	$a-h$	a	$a+h$
$i-1$	$-$	$-$	$-$
i	$+$	0	$-$
$i+1$	$+$	$+$	$+$

	$a-h$	a	$a+h$
$i-1$	$+$	$+$	$+$
i	$+$	0	$-$
$i+1$	$-$	$-$	$-$

In each instance, $w(a-h) = w(a) = w(a+h)$: the number of sign changes remains the same. This is also true if $p_i(x)$ does not change sign at $x = a$.

In the second case, we conclude from (5.6.1b) that the following sign patterns pertain:

i	$a-h$	a	$a+h$
0	$-$	0	$+$
1	$-$	$-$	$-$

i	$a-h$	a	$a+h$
0	$+$	0	$-$
1	$+$	$+$	$+$

In each instance, $w(a-h) = w(a) = w(a+h) - 1$: exactly one sign change is gained as we pass through a root of $p_0(x) \equiv p(x)$.

For $a < b$ and sufficiently small $h > 0$,

$$w(b) - w(a) = w(b-h) - w(a-h)$$

indicates the number of roots of $p(x)$ in the interval $a - h < x < b - h$. Since $h > 0$ can be chosen arbitrarily small, the above difference indicates the number of roots also in the interval $a \leqslant x < b$. □

An important use of Sturm sequences is in *bisection methods* for determining the eigenvalues of real symmetric matrices which are tridiagonal:

$$J = \begin{bmatrix} \alpha_1 & \beta_2 & & & & \\ \beta_2 & \cdot & \cdot & & 0 & \\ & \cdot & \cdot & \cdot & & \\ & & \cdot & \cdot & \cdot & \\ & 0 & & \cdot & \cdot & \beta_n \\ & & & & \beta_n & \alpha_n \end{bmatrix}.$$

Recall the characteristic polynomials $p_i(x)$ of the principal minor formed by the first i rows and columns of the matrix J, which were mentioned before as satisfying the recursion (5.5.3):

$$p_0(x) := 1,$$

$$p_1(x) := \alpha_1 - x,$$

$$p_i(x) := (\alpha_i - x)p_{i-1}(x) - \beta_i^2 p_{i-2}(x), \qquad i = 2, 3, \ldots, n.$$

The key observation is that the polynomials

(5.6.4) $$p_n(x),\ p_{n-1}(x),\ \ldots,\ p_0(x)$$

are a Sturm sequence for the characteristic polynomial $p_n(x) = \det(J - xI)$ [note that the polynomials (5.6.4) are indexed differently than in (5.6.1)] provided the off-diagonal elements β_i, $i = 2, \ldots, n$, of the tridiagonal matrix J are all nonzero. This is readily apparent from the following

(5.6.5) Theorem. *Let α_j, β_j be real numbers and $\beta_j \neq 0$ for $j = 2, \ldots, n$. Suppose the polynomials $p_i(x)$, $i = 0, \ldots, n$, are defined by the recursion* (5.5.3).

Then all roots $x_k^{(i)}$, $k = 1, \ldots, i$, *of* p_i, $i = 1, \ldots, n$, *are real and simple:*

$$x_1^{(i)} > x_2^{(i)} > \cdots > x_i^{(i)},$$

and the roots of p_{i-1} *and* p_i, *respectively, separate each other strictly:*

$$x_1^{(i)} > x_1^{(i-1)} > x_2^{(i)} > x_2^{(i-1)} > \cdots > x_{i-1}^{(i-1)} > x_i^{(i)}.$$

PROOF. We proceed by induction with respect to i. The theorem is plainly true for $i = 1$. Assume that it is true for some $i \geqslant 1$, that is, that the roots $x_k^{(i)}$ and $x_k^{(i-1)}$ of p_i and p_{i-1}, respectively, satisfy

$$x_1^{(i)} > x_1^{(i-1)} > x_2^{(i)} > x_2^{(i-1)} > \cdots > x_{i-1}^{(i-1)} > x_i^{(i)}. \tag{5.6.6}$$

By (5.5.3), p_k is of the form $p_k(x) = (-1)^k x^k + \cdots$. In particular, the degree of p_k equals k. Thus $p_{i-1}(x)$ does not change its sign for $x > x_1^{(i-1)}$, and since the roots $x_k^{(i-1)}$ are all simple, (5.6.6) gives immediately

$$\operatorname{sign} p_{i-1}(x_k^{(i)}) = (-1)^{i+k} \quad \text{for } k = 1, 2, \ldots, i. \tag{5.6.7}$$

Also by (5.5.3),

$$p_{i+1}(x_k^{(i)}) = -\beta_i^2 p_{i-1}(x_k^{(i)}), \qquad k = 1, 2, \ldots, i.$$

Since $\beta_i^2 > 0$,

$$\operatorname{sign} p_{i+1}(x_k^{(i)}) = (-1)^{i+k+1}, \qquad k = 1, 2, \ldots, i,$$
$$\operatorname{sign} p_{i+1}(+\infty) = (-1)^{i+1}, \qquad \operatorname{sign} p_{i+1}(-\infty) = 1$$

holds, and $p_{i+1}(x)$ changes sign in each of the intervals $[x_1^{(i)}, \infty)$, $(-\infty, x_i^{(i)}]$, $[x_{k+1}^{(i)}, x_k^{(i)}]$, $k = 1, \ldots, n-1$. The roots $x_k^{(i+1)}$ of p_{i+1} are therefore real and simple, and they separate the roots $x_k^{(i)}$ of p_i:

$$x_1^{(i+1)} > x_1^{(i)} > x_2^{(i+1)} > x_2^{(i)} > \cdots > x_i^{(i)} > x_{i+1}^{(i+1)}.$$ □

The polynomials of the above theorem,

$$p_n(x),\ p_{n-1}(x),\ \ldots,\ p_0(x),$$

indeed form a Sturm sequence: By (5.6.5), $p_n(x)$ has simple real roots $\xi_1 > \xi_2 > \cdots > \xi_n$, and by (5.6.7)

$$\operatorname{sign} p_{n-1}(\xi_k) = (-1)^{n+k},$$
$$\operatorname{sign} p_n'(\xi_k) = (-1)^{n+k+1} = -\operatorname{sign} p_{n-1}(\xi_k),$$

for $k = 1, 2, \ldots, n$.

For $x = -\infty$ the Sturm sequence (5.6.4) has the sign pattern

$$+,\ +,\ \ldots,\ +.$$

Thus $w(-\infty) = 0$. By Theorem (5.6.2), $w(\mu)$ indicates the number of roots ξ of $p_n(x)$ with $\xi < \mu$: $w(\mu) \geqslant n + 1 - i$ holds if and only if $\xi_i < \mu$.

The bisection method for determining the ith root ξ_i of $p_n(x)$ $(\xi_1 > \xi_2 > \cdots > \xi_n)$ now is as follows. Start with an interval

$$[a_0, b_0]$$

which is known to contain ξ_i; e.g., choose $b_0 > \xi_1$, $a_0 < \xi_n$. Then divide this interval with its midpoint and check by means of the Sturm sequence which of the two subintervals contains ξ_i. The subinterval which contains ξ_i is again divided, and so on. More precisely, we form for $j = 0, 1, 2, \ldots$

$$\mu_j := (a_j + b_j)/2,$$

$$a_{j+1} := \begin{cases} a_j & \text{if } w(\mu_j) \geqslant n + 1 - i, \\ \mu_j & \text{if } w(\mu_j) < n + 1 - i, \end{cases}$$

$$b_{j+1} := \begin{cases} \mu_j & \text{if } w(\mu_j) \geqslant n + 1 - i, \\ b_j & \text{if } w(\mu_j) < n + 1 - i. \end{cases}$$

Then

$$[a_{j+1}, b_{j+1}] \subseteq [a_j, b_j],$$
$$|a_{j+1} - b_{j+1}| = |a_j - b_j|/2,$$
$$\xi_i \in [a_{j+1}, b_{j+1}].$$

The quantities a_j increase, and the quantities b_j decrease, to the desired root ξ_i. The convergence process is linear with convergence rate 0.5. This method for determining the roots of a real polynomial all roots of which are real is relatively slow but very accurate. It has the additional advantage that each root can be determined independently of the others.

5.7 Bairstow's Method

If a real polynomial has any complex conjugate roots, they cannot be found using the ordinary Newton's method if it is carried out in real arithmetic and begun at a real starting point: complex starting points and complex arithmetic must be used. Bairstow's method avoids complex arithmetic. The method follows from the observation that the roots of a real quadratic polynomial

$$x^2 - rx - q$$

are roots of a given real polynomial

$$p(x) = a_0 x^n + \cdots + a_n, \qquad a_0 \neq 0,$$

if and only if $p(x)$ can be divided by $x^2 - rx - q$ without remainder. Now generally

$$p(x) = p_1(x)(x^2 - rx - q) + Ax + B, \tag{5.7.1}$$

where the degree of p_1 is $n-2$, and the remainder has been expressed as $Ax+B$. The coefficients of the remainder depend, of course, upon r and q, that is

$$A = A(r, q) \quad \text{and} \quad B = B(r, q),$$

and the remainder vanishes when r, q satisfy the system

$$A(r, q) = 0, \qquad B(r, q) = 0. \tag{5.7.2}$$

Bairstow's method is nothing more than Newton's method (5.3) applied to (5.7.2):

$$\begin{bmatrix} r_{i+1} \\ q_{i+1} \end{bmatrix} = \begin{bmatrix} r_i \\ q_i \end{bmatrix} - \begin{bmatrix} \dfrac{\partial A}{\partial r} & \dfrac{\partial A}{\partial q} \\ \dfrac{\partial B}{\partial r} & \dfrac{\partial B}{\partial q} \end{bmatrix}^{-1}_{\substack{r=r_i \\ q=q_i}} \cdot \begin{bmatrix} A(r_i, q_i) \\ B(r_i, q_i) \end{bmatrix}. \tag{5.7.3}$$

In order to carry out (5.7.3), we must first determine the partial derivatives

$$A_r = \frac{\partial A}{\partial r}, \qquad A_q = \frac{\partial A}{\partial q}, \qquad B_r = \frac{\partial B}{\partial r}, \qquad B_q = \frac{\partial B}{\partial q}.$$

Now (5.7.1) is an identity in r, q, and x. Hence, differentiating with respect to r and q,

$$\begin{aligned} \frac{\partial}{\partial r} p(x) &\equiv 0 = (x^2 - rx - q)\frac{\partial p_1(x)}{\partial r} - xp_1(x) + A_r x + B_r, \\ \frac{\partial}{\partial q} p(x) &\equiv 0 = (x^2 - rx - q)\frac{\partial p_1(x)}{\partial q} - p_1(x) + A_q x + B_q. \end{aligned} \tag{5.7.4}$$

After a further division of $p_1(x)$ by $x^2 - rx - q$ we obtain the representation

$$p_1(x) = p_2(x)(x^2 - rx - q) + A_1 x + B_1. \tag{5.7.5}$$

Assuming that $x^2 - rx - q = 0$ has two distinct roots x_0, x_1, it follows that

$$p_1(x_i) = A_1 x_i + B_1$$

for $x = x_i$, $i = 0, 1$. Therefore the equations

$$\left.\begin{aligned} -x_i(A_1 x_i + B_1) + A_r x_i + B_r &= 0 \\ -(A_1 x_i + B_1) + A_q x_i + B_q &= 0 \end{aligned}\right\} \qquad i = 0, 1,$$

follow from (5.7.4).

From the second of these equations we have

$$A_q = A_1, \qquad B_q = B_1, \tag{5.7.6}$$

since $x_0 \neq x_1$, and therefore the first equation yields

$$-x_i^2 A_q + x_i(A_r - B_q) + B_r = 0, \qquad i = 0, 1.$$

Since $x_i^2 = rx_i + q$ it follows that

$$x_i(A_r - B_q - A_q \cdot r) + B_r - A_q \cdot q = 0, \qquad i = 0, 1,$$

and therefore

$$A_r - B_q - A_q \cdot r = 0,$$
$$B_r - A_q \cdot q = 0,$$

since $x_0 \neq x_1$.

Putting this together with (5.7.6) yields

$$A_q = A_1, \qquad B_q = B_1,$$
$$A_r = rA_1 + B_1, \qquad B_r = qA_1.$$

The values A, B (or A_1, B_1) can be found by means of a Horner-type scheme. Using $p(x) = a_0 x^n + \cdots + a_n$, $p_1(x) = b_0 x^{n-2} + \cdots + b_{n-2}$ and comparing coefficients, we find the following recursion for A, B, b_i from (5.7.2):

$$b_0 := a_0,$$
$$b_1 := b_0 r + a_1,$$
$$b_i := b_{i-2} q + b_{i-1} r + a_i \quad \text{for } i = 2, 3, \ldots, n-2,$$
$$A := b_{n-3} q + b_{n-2} r + a_{n-1},$$
$$B := b_{n-2} q + a_n.$$

Similarly, (5.7.5) shows how the b_i can be used to find A_1 and B_1.

5.8 The Sensitivity of Polynomial Roots

We will consider the condition of a root ξ of a given polynomial $p(x)$. By this we mean the influence on ξ of a small perturbation of the coefficients of the polynomial $p(x)$:

$$p_\varepsilon(x) = p(x) + \varepsilon g(x),$$

where $g(x) \not\equiv 0$ is an arbitrary polynomial.

Later on it will be shown [Theorem (6.9.8)] that if ξ is a simple root of p, then for sufficiently small absolute values of ε there exists an analytic function $\xi(\varepsilon)$, with $\xi(0) = \xi$, such that $\xi(\varepsilon)$ is a (simple) root of the perturbed polynomial $p_\varepsilon(x)$:

$$p(\xi(\varepsilon)) + \varepsilon g(\xi(\varepsilon)) \equiv 0.$$

From this, by differentiation with respect to ε,

$$kp'(\xi(0)) + g(\xi(0)) = 0,$$

$$k = \frac{-g(\xi)}{p'(\xi)},$$

where

$$k := \left.\frac{d\xi(\varepsilon)}{d\varepsilon}\right|_{\varepsilon=0}.$$

Thus, to a first order of approximation (i.e., disregarding terms in powers of ε greater than 1), based on the Taylor expansion of $\xi(\varepsilon)$, we have

$$\xi(\varepsilon) \doteq \xi - \varepsilon \frac{g(\xi)}{p'(\xi)}. \tag{5.8.1}$$

In the case of a multiple root ξ, of order m, it can be shown that $p(x) + \varepsilon g(x)$ has a root of the form

$$\xi(\varepsilon) = \xi + h(\varepsilon^{1/m}),$$

where $h(t)$ is, for small $|t|$, an analytic function with $h(0) = 0$. Differentiating m times with respect to t and noting that $p(\xi) = p'(\xi) = \cdots = p^{(m-1)}(\xi) = 0$, $p^{(m)}(\xi) \neq 0$, we obtain from

$$0 \equiv p_\varepsilon(\xi(\varepsilon)) = p(\xi + h(t)) + t^m g(\xi + h(t)), \qquad t^m = \varepsilon,$$

the relations

$$p^{(m)}(\xi)k^m + m!\,g(\xi) = 0,$$

$$k = \left[-\frac{m!\,g(\xi)}{p^{(m)}(\xi)}\right]^{1/m},$$

where

$$k := \left.\frac{dh(t)}{dt}\right|_{t=0}.$$

Again to a first order of approximation,

$$\xi(\varepsilon) \doteq \xi + \varepsilon^{1/m}\left[-\frac{m!\,g(\xi)}{p^{(m)}(\xi)}\right]^{1/m}. \tag{5.8.2}$$

For $m = 1$ this formula reduces to (5.8.1) for single roots.

Let us assume that the polynomial $p(x)$ is given in the usual form

$$p(x) = a_0 x^n + \cdots + a_n,$$

by its coefficients a_i. For

$$g_i(x) := a_i x^{n-i}$$

the polynomial $p_\varepsilon(x)$ is the one which results if the coefficient a_i of $p(x)$ is replaced by $a_i(1+\varepsilon)$. The formula (5.8.2) then yields the following estimate of the effect on the root ξ of a relative error ε of a_i:

$$\xi(\varepsilon) - \xi \doteq \varepsilon^{1/m}\left[-\frac{m!\,a_i\xi^{n-i}}{p^{(m)}(\xi)}\right]^{1/m}. \tag{5.8.3}$$

It thus becomes apparent that in the case of multiple roots the changes in root values $\xi(\varepsilon) - \xi$ are proportional to $\varepsilon^{1/m}$, $m > 1$, whereas in the case of single roots they are proportional to just ε: multiple roots are always badly conditioned. But single roots may be badly conditioned too. This happens if the factor of ε in (5.8.3),

$$k(i, \xi) := \left|\frac{a_i\xi^{n-i}}{p'(\xi)}\right|,$$

is large compared to ξ, which may be the case for seemingly "harmless" polynomials.

EXAMPLE [Wilkinson (1959)].

(1) The roots $\xi_k = k$, $k = 1, 2, \ldots, 20$, of the polynomial

$$p(x) = (x-1)(x-2)\ldots(x-20) = \sum_{i=0}^{20} a_i x^{20-i}$$

are well separated. For $\xi_{20} = 20$, we find $p'(20) = 19!$, and replacing the coefficient $a_1 = -(1+2+\cdots+20) = -210$ by $a_1(1+\varepsilon)$ causes an estimated change of

$$\xi_{20}(\varepsilon) - \xi_{20} \doteq \varepsilon\frac{210 \times 20^{19}}{19!} \approx \varepsilon \cdot 0.9 \times 10^{10}.$$

The most drastic changes are caused in ξ_{16} by perturbations of a_5. Since $\xi_{16} = 16$ and $a_5 \approx -10^{10}$,

$$\xi_{16}(\varepsilon) - \xi_{16} \doteq -\varepsilon a_5\frac{16^{15}}{4!\,15!} \approx \varepsilon \cdot 3.7 \times 10^{14}.$$

This means that the roots of the polynomial p are so badly conditioned that even computing with 14-digit arithmetic will not guarantee any correct digit ξ_{16}.

(2) By contrast, the roots of the polynomial

$$p(x) = \sum_{i=0}^{20} a_i x^{20-i} := \prod_{j=1}^{20}(x - 2^{-j}), \qquad \xi_j = 2^{-j},$$

while not well separated and "accumulating" at zero, are all well conditioned. For instance, changing a_{20} to $a_{20}(1+\varepsilon)$ causes a variation of ξ_{20} which to a first order of approximation can be bounded as follows:

$$\left|\frac{\xi_{20}(\varepsilon) - \xi_{20}}{\xi_{20}}\right| \doteq \left|\varepsilon\frac{1}{(2^{-1}-1)(2^{-2}-1)\ldots(2^{-19}-1)}\right| \leqslant 4|\varepsilon|.$$

More generally, it can be shown for all roots ξ_j and changes $a_i \to a_i(1+\varepsilon)$ that

$$\left|\frac{\xi_j(\varepsilon)-\xi_j}{\xi_j}\right| \mathrel{\dot{\leqslant}} 64|\varepsilon|.$$

However, the roots are well conditioned only with respect to small relative changes of the coefficients a_i, and *not* for small absolute changes. If we replace $a_{20} = 2^{-210}$ by $\bar{a}_{20} = a_{20} + \Delta a_{20}$, $\Delta a_{20} = 2^{-48} (\approx 10^{-14})$—this can be considered a small absolute change—then the modified polynomial has roots $\bar{\xi}_i$ with

$$\bar{\xi}_1 \ldots \bar{\xi}_{20} = \bar{a}_{20} = 2^{-210} + 2^{-48} = (2^{162}+1)(\xi_1 \ldots \xi_{20}).$$

In other words, there exists at least one subscript r with $|\bar{\xi}_r/\xi_r| \geqslant (2^{162}+1)^{1/20} > 2^8 = 256$.

It should be emphasized that the formula (5.8.3) refers only to the sensitivity of the roots of a polynomial

$$p(x) := \sum_{i=0}^{n} a_i x^{n-i}$$

in its usual representation by coefficients. There are other ways of representing polynomials—for instance, as the characteristic polynomials of tridiagonal matrices by the elements of these matrices [see (5.5.3)]. The effect on the roots of a change in the parameters of such an alternative representation may differ by an order of magnitude from that described by the formula (5.8.3). The condition of roots is defined always with a particular type of representation in mind.

EXAMPLE. In Theorem (6.9.7) it will be shown that, for each real tridiagonal matrix

$$J = \begin{bmatrix} \alpha_1 & \beta_2 & & & \\ \beta_2 & \cdot & \cdot & & \\ & \cdot & \cdot & \cdot & \\ & & \cdot & \cdot & \beta_{20} \\ & & & \beta_{20} & \alpha_{20} \end{bmatrix}$$

whose characteristic polynomial is $p(x) \equiv (x-1)(x-2)\ldots(x-20)$, small relative changes in α_i and β_i cause only small relative changes of the roots $\xi_j = j$. With respect to this representation, all roots are well conditioned, although they are very badly conditioned with respect to the usual representation by coefficients, as was shown in the previous example. For detailed discussions of this topic see Peters and Wilkinson (1969) and Wilkinson (1965).

5.9 Interpolation Methods for Determining Roots

The interpolation methods to be discussed in this section are very useful for determining zeros of arbitrary real functions $f(x)$. Compared to Newton's method, they have the advantage that the derivatives of f need not be computed. Moreover, in a sense yet to be made precise, they converge even faster than Newton's method.

The simplest among these methods is known as the *method of false position* or *regula falsi*. It is similar to the bisection method in that two numbers x_i and a_i with

$$f(x_i)f(a_i) < 0 \tag{5.9.1}$$

are determined at each step. The interval $[x_i, a_i]$ contains therefore at least one zero of f, and the values x_i are determined so that they converge towards one of these zeros. In order to define x_{i+1}, a_{i+1}, let μ_i be the zero of the interpolating linear function

$$p(x) := f(x_i) + (x - x_i)\frac{f(x_i) - f(a_i)}{x_i - a_i},$$

where $p(x_i) = f(x_i)$, $p(a_i) = f(a_i)$, that is,

$$\mu_i = x_i - f(x_i)\frac{x_i - a_i}{f(x_i) - f(a_i)} = \frac{a_i f(x_i) - x_i f(a_i)}{f(x_i) - f(a_i)}. \tag{5.9.2}$$

Since $f(x_i)f(a_i) < 0$ implies $f(x_i) - f(a_i) \neq 0$, μ_i is always well defined and satisfies either $x_i < \mu_i < a_i$ or $a_i < \mu_i < x_i$. Unless $f(\mu_i) = 0$, put

$$\begin{aligned} &\left.\begin{aligned} x_{i+1} &:= \mu_i \\ a_{i+1} &:= a_i \end{aligned}\right\} \quad \text{if } f(\mu_i)f(x_i) > 0, \\ &\left.\begin{aligned} x_{i+1} &:= \mu_i \\ a_{i+1} &:= x_i \end{aligned}\right\} \quad \text{if } f(\mu_i)f(x_i) < 0. \end{aligned} \tag{5.9.3}$$

If $f(\mu_i) = 0$, then the method terminates with μ_i the zero. An improvement on (5.9.3) is due to Dekker and is described in Peters and Wilkinson (1969).

In order to discuss the convergence behavior of the *regula falsi*, we assume for simplicity that f'' exists and that for some i

$$x_i < a_i, \tag{5.9.4a}$$

$$f(x_i) < 0, \qquad f(a_i) > 0, \tag{5.9.4b}$$

$$f''(x) \geqslant 0 \quad \text{for all } x \in [x_i, a_i]. \tag{5.9.4c}$$

Under these assumptions, either $f(\mu_i) = 0$ or

$$f(\mu_i)f(x_i) > 0,$$

and consequently

$$x_i < x_{i+1} = \mu_i < a_{i+1} = a_i.$$

To see this, note that (5.9.4) and the definition of μ_i imply immediately that

$$x_i < \mu_i < a_i.$$

The remainder formula (2.1.4.1) for polynomial interpolation yields, for $x \in [x_i, a_i]$ and a suitable value $\xi \in [x_i, a_i]$, the representation

$$f(x) - p(x) = (x - x_i)(x - a_i)f''(\xi)/2.$$

By (5.9.4c), $f(x) - p(x) \leqslant 0$ in the interval $x \in [x_i, a_i]$. Thus $f(\mu_i) \leqslant 0$ since $p(\mu_i) = 0$, which was to be shown.

It is now easy to see that (5.9.4) holds for all $i \geqslant i_0$ provided it holds for some i_0. Therefore $a_i = a$ for $i \geqslant i_0$, the x_i form a monotone increasing sequence bounded by a, and $\lim_{i\to\infty} x_i = \xi$ exists. Since f is continuous, and because of (5.9.4) and (5.9.2),

$$f(\xi) \leqslant 0, \qquad f(a) > 0,$$

$$\xi = \frac{af(\xi) - \xi f(a)}{f(\xi) - f(a)}.$$

Thus $(\xi - a)f(\xi) = 0$. But $\xi \neq a$, since $f(a) > 0$ and $f(\xi) \leqslant 0$, and we conclude $f(\xi) = 0$. The values x_i converge to a zero of the function f.

Under the assumptions (5.9.4) the *regula falsi* method can be formulated with the help of an iteration function:

(5.9.5) $$x_{i+1} = \Phi(x_i), \qquad \Phi(x) := \frac{af(x) - xf(a)}{f(x) - f(a)}.$$

Since $f(\xi) = 0$,

$$\Phi'(\xi) = \frac{-(af'(\xi) - f(a))f(a) + \xi f(a)f'(\xi)}{f(a)^2} = 1 - f'(\xi)\frac{\xi - a}{f(\xi) - f(a)}.$$

By the mean-value theorem, there exist η_1, η_2 such that

(5.9.6) $$\frac{f(\xi) - f(a)}{\xi - a} = f'(\eta_1), \qquad \xi < \eta_1 < a,$$
$$\frac{f(x_i) - f(\xi)}{x_i - \xi} = f'(\eta_2), \qquad x_i < \eta_2 < \xi.$$

Because $f''(x) \geqslant 0$ holds for $x \in [x_i, a]$, $f'(x)$ increases monotonically in this interval. Thus (5.9.6), $x_i < \xi$, and $f(x_i) < 0$ imply immediately that

$$0 < f'(\eta_2) \leqslant f'(\xi) \leqslant f'(\eta_1),$$

and therefore

$$0 \leqslant \Phi'(\xi) < 1.$$

In other words, the *regula falsi* method converges linearly under the assumptions (5.9.4).

The previous discussion shows that, under the assumption (5.9.4), *regula falsi* will eventually utilize only the first two of the recursions (5.9.3). We will now describe an important variant of *regula falsi*, called the *secant method*, which is based exclusively on the second pair of recursions (5.9.3):

(5.9.7) $$x_{i+1} = \frac{x_{i-1} f(x_i) - x_i f(x_{i-1})}{f(x_i) - f(x_{i-1})}, \qquad i = 0, 1, \ldots.$$

In this case, the linear function always interpolates the two latest points of the interpolation. While the original method (5.9.3) is numerically stable because it enforces (5.9.1), this may not be the case for the secant method. Whenever $f(x_i) \approx f(x_{i-1})$, digits are lost due to cancellation. Moreover, x_{i+1} need not lie in the interval $[x_i, x_{i-1}]$, and it is only in a sufficiently small neighborhood of the zero ξ that the secant method is guaranteed to converge. We will examine the convergence behavior of the secant method (5.9.7) in such a neighborhood and show that the method has a superlinear order of convergence. To this end, we subtract ξ from both sides of (5.9.7), and obtain, using divided differences (2.1.3.5),

$$
\begin{aligned}
x_{i+1} - \xi &= (x_i - \xi) - f(x_i)\frac{x_i - x_{i-1}}{f(x_i) - f(x_{i-1})} \\
&= (x_i - \xi) - \frac{f(x_i)}{f[x_{i-1}, x_i]} = (x_i - \xi)\left(1 - \frac{f[x_i, \xi]}{f[x_{i-1}, x_i]}\right) \\
&= (x_i - \xi)(x_{i-1} - \xi)\frac{f[x_{i-1}, x_i, \xi]}{f[x_{i-1}, x_i]}.
\end{aligned} \tag{5.9.8}
$$

If f is twice differentiable, then (2.1.4.3) gives

$$
\begin{aligned}
f[x_{i-1}, x_i] &= f'(\eta_1), & \eta_1 &\in I[x_{i-1}, x_i], \\
f[x_{i-1}, x_i, \xi] &= \tfrac{1}{2} f''(\eta_2), & \eta_2 &\in I[x_{i-1}, x_i, \xi].
\end{aligned} \tag{5.9.9}
$$

For a simple zero ξ of f, $f'(\xi) \neq 0$, and there exist a bound M and an interval $J = \{x \mid |x - \xi| \leqslant \varepsilon\}$ such that

$$
\left| \frac{1}{2} \frac{f''(\eta_2)}{f'(\eta_1)} \right| \leqslant M \quad \text{for any } \eta_1, \eta_2 \in J. \tag{5.9.10}
$$

Let $e_i := M|x_i - \xi|$ and e_0, $e_1 < \min\{1, \varepsilon M\}$. Then, using (5.9.8) and (5.9.10), it can be easily shown by induction that

$$
e_{i+1} \leqslant e_i e_{i-1} \quad \text{for } i = 1, 2, \ldots \tag{5.9.11}
$$

and $|e_i| \leqslant \min\{1, \varepsilon M\}$. But note that

$$
e_i \leqslant K^{q^i} \quad \text{for } i = 0, 1, 2, \ldots, \tag{5.9.12}
$$

where $q = (1 + \sqrt{5})/2 = 1.618\ldots$ is the positive root of the equation $\mu^2 - \mu - 1 = 0$, and $K := \max\{e_0, \sqrt[q]{e_1}\} < 1$. The proof is inductive: This choice of K makes (5.9.12) valid for $i = 0$ and $i = 1$. If (5.9.12) holds for $i - 1$ and i, then (5.9.11) yields

$$
e_{i+1} \leqslant K^{q^i} K^{q^{i-1}} = K^{q^{i+1}},
$$

since $q^2 = q + 1$. Thus (5.9.12) holds also for $i + 1$, and must therefore hold in general.

According to (5.9.12), the secant method converges at least as well as a method of order $q = 1.618\ldots$. Since one step of the secant method requires only one additional function evaluation, two secant steps are at most as expensive as a single Newton step. Since $K^{q^{i+2}} = (K^{q^i})^{q^2} = (K^{q^i})^{q+1}$, two secant steps lead to a method of order $q^2 = q + 1 = 2.618\ldots$. With comparable effort, the secant method converges locally faster than the Newton method, which is of order 2.

The secant method suggests the following generalization. Suppose there are $r + 1$ different approximations $x_i, x_{i-1}, \ldots, x_{i-r}$ to a zero ξ of $f(x)$. Determine the interpolating polynomial $Q(x)$ of degree r with

$$Q(x_{i-j}) = f(x_{i-j}), \qquad j = 0, 1, \ldots, r,$$

and choose the root of $Q(x)$ closest to x_i as the new approximation x_{i+1}. For $r = 1$ this is the secant method. For $r = 2$ we obtain the method of Muller. The methods for $r \geqslant 3$ are rarely considered, because there are no practical formulas for the roots of the interpolating polynomial.

Muller's method has gained a reputation as an efficient and fairly reliable method for finding a zero of a function defined on the complex plane and, in particular, for finding a simple or multiple root of a polynomial. It will find real as well as complex roots. The approximating values x_i may be complex even if the coefficients and the roots of the polynomial as well as the starting values x_1, x_2, x_3 are all real. Our exposition employs divided differences (see Section 2.1.3), following Traub (1964).

By Newton's interpolation formula (2.1.3.8), the quadratic polynomial which interpolates a function f (in our case the given polynomial p) at x_{i-2}, x_{i-1}, x_i can be written as

$$Q_i(x) = f[x_i] + f[x_{i-1}, x_i](x - x_i) + f[x_{i-2}, x_{i-1}, x_i](x - x_{i-1})(x - x_i),$$

or

$$Q_i(x) = a_i(x - x_i)^2 + 2b_i(x - x_i) + c_i,$$

where

$$a_i := f[x_{i-2}, x_{i-1}, x_i],$$
$$b_i := \tfrac{1}{2}(f[x_{i-1}, x_i] + f[x_{i-2}, x_{i-1}, x_i](x_i - x_{i-1})),$$
$$c_i := f[x_i].$$

If h_i is the root of smallest absolute value of the quadratic equation $a_i h^2 + 2b_i h + c_i = 0$, then $x_{i+1} := x_i + h_i$ is the root of $Q_i(x)$ closest to x_i.

In order to express the smaller root of a quadratic equation in a numerically stable fashion, the reciprocal of the standard solution formula for quadratic equations should be used. Then Muller's iteration takes the form

$$x_{i+1} := x_i - \frac{c_i}{b_i \pm \sqrt{b_i^2 - a_i c_i}}, \tag{5.9.13}$$

where the sign of the square root is chosen so as to maximize the absolute value of the denominator. If $a_i = 0$, then a linear interpolation step as in the secant method results. If $a_i = b_i = 0$, then $f(x_{i-2}) = f(x_{i-1}) = f(x_i)$, and the iteration has to be restarted with different initial values. In solving the quadratic equation, complex arithmetic has to be used, even if the coefficients of the polynomial are all real, since $b_i^2 - a_i c_i$ may be negative.

Once a new approximate value x_{i+1} has been found, the function f is evaluated at x_{i+1} to find

$$\begin{aligned} f[x_{i+1}] &:= f(x_{i+1}), \\ f[x_i, x_{i+1}] &:= \frac{f[x_{i+1}] - f[x_i]}{x_{i+1} - x_i}, \\ f[x_{i-1}, x_i, x_{i+1}] &:= \frac{f[x_i, x_{i+1}] - f[x_{i-1}, x_i]}{x_{i+1} - x_{i-1}}. \end{aligned}$$

These quantities determine the next quadratic interpolating polynomial $Q_{i+1}(x)$.

It can be shown that the errors $\varepsilon_i = (x_i - \xi)$ of Muller's method in the proximity of a single zero ξ of $f(x) = 0$ satisfy

$$\text{(5.9.14)} \qquad \begin{aligned} \varepsilon_{i+1} &= \varepsilon_i \varepsilon_{i-1} \varepsilon_{i-2} \left(-\frac{f^{(3)}(\xi)}{6f'(\xi)} + O(\varepsilon) \right), \\ \varepsilon &= \max(|\varepsilon_i|, |\varepsilon_{i-1}|, |\varepsilon_{i-2}|) \end{aligned}$$

[compare (5.9.8)]. By an argument analogous to the one used for the secant method, it can be shown that Muller's method is at least of order $q = 1.84 \ldots$, where q is the largest root of the equation $\mu^3 - \mu^2 - \mu - 1 = 0$.

The secant method can be generalized in a different direction. Consider again $r + 1$ approximations $x_i, x_{i-1}, \ldots, x_{i-r}$ to the zero ξ of the function $f(x)$. If the inverse g of the function f exists in a neighborhood of ξ,

$$f(g(y)) = y, \qquad g(f(x)) = x, \qquad g(0) = \xi,$$

then determining ξ amounts to calculating $g(0)$. Since

$$g(f(x_j)) = x_j, \qquad j = i, i-1, \ldots, i-r,$$

the following suggests itself: determine the interpolating polynomial $Q(y)$ of degree r or less with $Q(f(x_j)) = x_j$, $j = i, i-1, \ldots, i-r$, and approximate $g(0)$ by $Q(0)$. Then select $x_{i+1} = Q(0)$ as the next approximate point to be included in the interpolation. This method is called determining zeros by *inverse interpolation*. Note that it does not require solving polynomial equations at each step even if $r \geqslant 3$. For $r = 1$ the secant method results. The interpolation formulas of Neville and Aitken (see Section 2.1.2) are particularly useful in implementing inverse interpolation methods. The methods are locally convergent of superlinear order. For details see Ostrowski (1966) and Brent (1973).

In Section 5.5, in connection with Newton's method, we considered two techniques for determining additional roots: deflation and zero suppression. Both are applicable to the root-finding methods discussed in this section. Here zero suppression simply amounts to evaluating the original polynomial $p(x)$ and numerically dividing it by the value of the product $(x - \xi_1) \ldots (x - \xi_k)$ to calculate the function whose zero is to be determined next. This process is safe if computation is restricted away from values x which fall in close neighborhoods of the previously determined roots $\xi_1, \ldots, \xi_k$. Contrary to deflation, zero suppression works also if the methods of this section are applied to finding the zeros of an arbitrary function $f(x)$.

5.10 The Δ^2-Method of Aitken

The Δ^2-method of Aitken belongs to the class of methods for accelerating the convergence of a given convergent sequence of values x_i,

$$\lim_{i \to \infty} x_i = \xi.$$

These methods transform the sequence $\{x_i\}$ into a sequence $\{x_i'\}$ which in general converges faster towards ξ than the original sequence of values x_i. Such methods apply to finding zeros inasmuch as they can be used to accelerate the convergence of sequences $\{x_i\}$ furnished by one of the methods previously discussed.

In order to illustrate the Δ^2-method, let us assume that the sequence $\{x_i\}$ converges towards ξ like a geometric sequence with factor k, $|k| < 1$:

$$x_{i+1} - \xi = k(x_i - \xi), \qquad i = 0, 1, \ldots.$$

Then k and ξ can be determined from x_i, x_{i+1}, x_{i+2} using the equations

$$\text{(5.10.1)} \qquad x_{i+1} - \xi = k(x_i - \xi), \qquad x_{i+2} - \xi = k(x_{i+1} - \xi).$$

By subtraction of these equations,

$$k = \frac{x_{i+2} - x_{i+1}}{x_{i+1} - x_i},$$

and by substitution into the first equation, since $k \neq 1$,

$$\xi = \frac{x_i x_{i+2} - x_{i+1}^2}{x_{i+2} - 2x_{i+1} + x_i}.$$

Using the difference operator $\Delta x_i := x_{i+1} - x_i$ and noting that $\Delta^2 x_i = \Delta x_{i+1} - \Delta x_i = x_{i+2} - 2x_{i+1} + x_i$, this can be written as

$$\text{(5.10.2)} \qquad \xi = x_i - \frac{(\Delta x_i)^2}{\Delta^2 x_i}.$$

The method is named after this formula. It is based on the expectation—to be confirmed below—that the value (5.10.2) provides at least an improved approximation to the limit of the sequence of values x_i even if the hypothesis that $\{x_i\}$ is a geometrically convergent sequence should not be valid.

The Δ^2-method of Aitken thus consists of generating from a given sequence $\{x_i\}$ the transformed sequence of values

$$x_i' := x_i - \frac{(x_{i+1} - x_i)^2}{x_{i+2} - 2x_{i+1} + x_i}. \tag{5.10.3}$$

The following theorem shows that the x_i' converge faster than the x_i to ξ as $i \to \infty$, provided $\{x_i\}$ behaves asymptotically as a geometric sequence:

(5.10.4) Theorem. *Suppose there exists k, $|k| < 1$, such that for the sequence $\{x_i\}$, $x_i \neq \xi$,*

$$x_{i+1} - \xi = (k + \delta_i)(x_i - \xi), \qquad \lim_{i\to\infty} \delta_i = 0,$$

holds, then the values x_i' defined by (5.10.3) *all exist for sufficiently large i, and*

$$\lim_{i\to\infty} \frac{x_i' - \xi}{x_i - \xi} = 0.$$

PROOF. By hypothesis, the errors $e_i := x_i - \xi$ satisfy $e_{i+1} = (k + \delta_i)e_i$. It follows that

$$\begin{aligned} x_{i+2} - 2x_{i+1} + x_i &= e_{i+2} - 2e_{i+1} + e_i \\ &= e_i((k + \delta_{i+1})(k + \delta_i) - 2(k + \delta_i) + 1) \\ &= e_i((k-1)^2 + \mu_i), \quad \text{where } \mu_i \to 0, \end{aligned} \tag{5.10.5}$$

$$x_{i+1} - x_i = e_{i+1} - e_i = e_i((k-1) + \delta_i).$$

Therefore

$$x_{i+2} - 2x_{i+1} + x_i \neq 0$$

for sufficiently large i, since $e_i \neq 0$, $k \neq 1$, and $\mu_i \to 0$. Hence (5.10.3) guarantees that x_i' is well defined. By (5.10.3) and (5.10.5),

$$x_i' - \xi = e_i - e_i \frac{((k-1) + \delta_i)^2}{(k-1)^2 + \mu_i},$$

for sufficiently large i, and consequently

$$\lim_{i\to\infty} \frac{x_i' - \xi}{x_i - \xi} = \lim_{i\to\infty} \left| 1 - \frac{((k-1) + \delta_i)^2}{((k-1)^2 + \mu_i)} \right| = 0. \qquad \square$$

Consider an iteration method with iteration function $\Phi(x)$,

$$x_{i+1} = \Phi(x_i), \qquad i = 0, 1, 2, \ldots, \tag{5.10.6}$$

for determining the zero ξ of a function $f(x) = 0$. The formula (5.10.3) can then be used to determine, from triples of successive elements x_i, x_{i+1}, x_{i+2} of the sequence $\{x_i\}$ generated by the above iteration function Φ, a new sequence $\{x_i'\}$, which hopefully converges faster. However, it appears to be advantageous to make use of the improved approximations immediately by putting

(5.10.7)
$$y_i := \Phi(x_i), \qquad z_i = \Phi(y_i) \qquad\qquad i = 0, 1, 2, \ldots.$$
$$x_{i+1} := x_i - \frac{(y_i - x_i)^2}{z_i - 2y_i + x_i}$$

This method is due to Steffensen. (5.10.7) leads to a new iteration function Ψ,

(5.10.8)
$$x_{i+1} = \Psi(x_i),$$
$$\Psi(x) := \frac{x\Phi(\Phi(x)) - \Phi(x)^2}{\Phi(\Phi(x)) - 2\Phi(x) + x}.$$

Both iteration functions Φ and Ψ have, in general, the same fixed points:

(5.10.9) Theorem. $\Psi(\xi) = \xi$ *implies* $\Phi(\xi) = \xi$. *Conversely, if* $\Phi(\xi) = \xi$ *and* $\Phi'(\xi) \neq 1$ *exists, then* $\Psi(\xi) = \xi$.

PROOF. By the definition (5.10.8) of Ψ,

$$(\xi - \Psi(\xi))(\Phi(\Phi(\xi)) - 2\Phi(\xi) + \xi) = (\xi - \Phi(\xi))^2.$$

Thus $\Psi(\xi) = \xi$ implies $\Phi(\xi) = \xi$. Next we assume that $\Phi(\xi) = \xi$, Φ is differentiable for $x = \xi$, and $\Phi'(\xi) \neq 1$. L'Hôpital's rule applied to the definition (5.10.8) gives

$$\Psi(\xi) = \frac{\Phi(\Phi(\xi)) + \xi\Phi'(\Phi(\xi))\Phi'(\xi) - 2\Phi(\xi)\Phi'(\xi)}{\Phi'(\Phi(\xi))\Phi'(\xi) - 2\Phi'(\xi) + 1} = \frac{\xi + \xi\Phi'(\xi)^2 - 2\xi\Phi'(\xi)}{1 + \Phi'(\xi)^2 - 2\Phi'(\xi)} = \xi.$$

□

In order to examine the convergence behavior of Ψ in the neighborhood of a fixed point ξ of Ψ (and Φ), we assume that Φ is $p + 1$ times differentiable in a neighborhood of $x = \xi$ and that it defines a method of order p, that is,

(5.10.10)
$$\Phi'(\xi) = \cdots = \Phi^{(p-1)}(\xi) = 0, \qquad \Phi^{(p)}(\xi) = p!\,A \neq 0$$

(see Section 5.2). For $p = 1$, we require that in addition

(5.10.11)
$$A = \Phi'(\xi) \neq 1,$$

and without loss of generality we assume $\xi = 0$. Then for small x

$$\Phi(x) = Ax^p + \frac{x^{p+1}}{(p+1)!}\Phi^{(p+1)}(\theta x), \qquad 0 < \theta < 1.$$

Thus

$$\begin{aligned}
\Phi(x) &= Ax^p + O(x^{p+1}),\\
\Phi(\Phi(x)) &= A(Ax^p + O(x^{p+1}))^p + O((Ax^p + O(x^{p+1}))^{p+1})\\
&= \begin{cases} O(x^{p^2}) & \text{if } p > 1,\\ A^2x + O(x^2) & \text{if } p = 1,\end{cases}\\
\Phi(x)^2 &= (Ax^p + O(x^{p+1}))^2 = A^2x^{2p} + O(x^{2p+1}).
\end{aligned}$$

For $p > 1$, because of (5.10.8),

$$\text{(5.10.12)}\quad \Psi(x) = \frac{O(x^{p^2+1}) - A^2x^{2p} + O(x^{2p+1})}{O(x^{p^2}) - 2Ax^p + O(x^{p+1}) + x} = -A^2x^{2p-1} + O(x^{2p}).$$

For $p = 1$, however, since $A \neq 1$,

$$\Psi(x) = \frac{A^2x^2 + O(x^3) - A^2x^2 + O(x^3)}{A^2x + O(x^2) - 2Ax + O(x^2) + x} = O(x^2).$$

This proves the following

(5.10.13) Theorem. *Let Φ be an iteration function defining a method of order p for computing its fixed point ξ. For $p > 1$ the corresponding iteration function Ψ of* (5.10.8) *determines a method of order $2p - 1$ for computing ξ. For $p = 1$ this method is at least of order 2 provided $\Phi'(\xi) \neq 1$.*

Note that Ψ yields a second-order method, that is, a locally quadratically convergent method, even if $|\Phi'(\xi)| > 1$ and the Φ-method diverges as a consequence (see Section 5.2)).

It is precisely in the case $p = 1$ that the method of Steffensen is important. For $p > 1$ the Ψ-method does generally not improve upon the Φ-method. This is readily seen as follows. Let $x_i - \xi = \varepsilon$ with ε sufficiently small. Neglecting terms of higher order, we find

$$\begin{aligned}
\Phi(x_i) - \xi &\doteq A\varepsilon^p,\\
\Phi(\Phi(x_i)) - \xi &\doteq A^{p+1}\varepsilon^{p^2},
\end{aligned}$$

whereas for $x_{i+1} - \xi$, $x_{i+1} = \Psi(x_i)$,

$$x_{i+1} - \xi = -A^2\varepsilon^{2p-1}$$

by (5.10.12). Now

$$|A^{p+1}\varepsilon^{p^2}| \ll |A^2\varepsilon^{2p-1}|$$

for $p > 1$ and sufficiently small ε, which establishes $\Phi(\Phi(x_i))$ as a much better approximation to ξ than $x_{i+1} = \Psi(x_i)$. For this reason, Steffensen's method is only recommended for $p = 1$.

EXAMPLE. The iteration function $\Phi(x) = x^2$ has fixed points $\xi_1 = 0$, $\xi_2 = 1$ with

$$\Phi'(\xi_1) = 0, \qquad \Phi''(\xi_1) = 2,$$
$$\Phi'(\xi_2) = 2.$$

The iteration $x_{i+1} = \Phi(x_i)$ converges quadratically to ξ_1 if $|x_0| < 1$. But for $|x_0| > 1$ the sequence $\{x_i\}$ diverges.

The transformation (5.10.8) yields

$$\Psi(x) = \frac{x^3}{x^2 + x - 1} = \frac{x^3}{(x - r_1)(x - r_2)} \quad \text{with } r_{1,2} = \frac{-1 \pm \sqrt{5}}{2}.$$

We proceed to show that the iteration $x_{i+1} = \Psi(x_i)$ reaches both fixed points for suitable choices of the starting point x_0.

For $|x| \leqslant 0.5$, $\Psi(x)$ is a contraction mapping. If $|x_0| \leqslant 0.5$ then $x_{i+1} = \Psi(x_i)$ converges towards $\xi_1 = 0$. In sufficient proximity of ξ_1 the iteration behaves as

$$x_{i+1} = \Psi(x_i) \approx x_i^3,$$

whereas the iteration $x_{i+1} = \Phi(\Phi(x_i))$ has 4th-order convergence for $|x_0| < 1$:

$$x_{i+1} = \Phi(\Phi(x_i)) = x_i^4.$$

For $|x_0| > r_1$, $x_{i+1} = \Psi(x_i)$ converges towards $\xi_2 = 1$. It is readily verified that

$$\Psi'(1) = 0, \qquad \Psi''(1) \neq 0.$$

and that therefore quadratic convergence holds (in spite of the fact that $\Phi(x)$ did not provide a convergent iteration).

5.11 Minimization Problems without Constraints

We will consider the following minimization problem for a real function $h: \mathbb{R}^n \to \mathbb{R}$ of n variables:

(5.11.1) $$\text{determine } \min_{x \in \mathbb{R}^n} h(x).$$

We assume that h has continuous second partial derivatives with respect to all of its variables, $h \in C^2(\mathbb{R}^n)$, and we denote the gradient of h by

$$g(x)^T = Dh(x) = \left(\frac{\partial h}{\partial x^1}, \ldots, \frac{\partial h}{\partial x^n}\right)$$

and the matrix of second derivatives of h by

$$H(x) = \left(\frac{\partial^2 h}{\partial x^i \, \partial x^k}\right), \qquad i, k = 1, \ldots, n.$$

Almost all minimization methods start with a point $x_0 \in \mathbb{R}^n$ and generate a sequence of points x_k, $k \geqslant 0$, which are supposed to approximate the desired minimum point $\bar{x}$. In each step of the iteration, $x_k \to x_{k+1}$, for which

$g_k = g(x_k) \neq 0$ [see (5.4.1.3)], a search direction s_k is determined by some computation which characterizes the method, and the next point

$$x_{k+1} = x_k - \lambda_k s_k$$

is obtained by a *line search*; i.e., the stepsize λ_k is determined so that

$$h(x_{k+1}) \approx \min_{\lambda} \varphi_k(\lambda), \qquad \varphi_k(\lambda) := h(x_k - \lambda s_k)$$

holds at least approximately for x_{k+1}. Usually the direction s_k is taken to be a *descent direction* for h, i.e.,

$$\varphi_k'(0) = -g_k^T s_k < 0. \tag{5.11.2}$$

This ensures that only positive values for λ need to be considered in the minimization of $\varphi_k(\lambda)$.

In Section 5.4.1 we established general convergence theorems [(5.4.1.4) and (5.4.1.8)] which apply to all methods of this type with only mild restrictions. In this section we wish to become acquainted with a few, special methods, which means primarily that we will discuss a few specific ways of selecting s_k which have become important in practice. A (local) minimum point $\bar{x}$ of h is a zero of $g(x)$; hence we can use any zero-finding method on the system $g(x) = 0$ as an approach to finding minimum points of h. Most important in this regard is Newton's method (5.1.6), at each step $x_k \to x_{k+1} = x_k - \lambda_k s_k$ of which the Newton direction

$$s_k := H(x_k)^{-1} g_k$$

is taken as a search direction. This, when used with the constant step length $\lambda_k = 1$, has the advantage of defining a locally quadratically convergent method [Theorem (5.3.2)]. But it has the disadvantage of requiring that the matrix $H(x_k)$ of all second partial derivatives of h be computed at each step. If n is large and if h is complicated, the computation of $H(x_k)$ can be very costly. Therefore methods have been devised wherein the matrices $H(x_k)^{-1}$ are replaced by suitable matrices H_k,

$$s_k := H_k g_k,$$

which are easy to compute. A method is said to be a *quasi-Newton method* if for each $k \geqslant 0$ the matrix H_{k+1} satisfies the so-called *quasi-Newton equation*

$$H_{k+1}(g_{k+1} - g_k) = x_{k+1} - x_k. \tag{5.11.3}$$

This equation causes H_{k+1} to behave—in the direction $x_{k+1} - x_k$—like the Newton matrix $H(x_{k+1})^{-1}$. For quadratic functions $h(x) = \frac{1}{2}x^T Ax + b^T x + c$, for example, where A is an $n \times n$ positive definite matrix, the Newton matrix $H(x_{k+1})^{-1} \equiv A^{-1}$ satisfies (5.11.3) because $g(x) \equiv Ax + b$. Further, it seems reasonable to insist that the matrix H_k be positive definite for each $k \geqslant 0$. This will guarantee that the direction $s_k = H_k g_k$ will be a descent direction for the function h if $g_k \neq 0$ [see (5.11.2)]:

$$g_k^T s_k = g_k^T H_k g_k > 0.$$

The above demands can, in fact, be met: Generalizing earlier work by Davidon (1959), Fletcher and Powell (1963), and Broyden (1965), Oren and Luenberger (1974) have found a two-parameter recursion for producing H_{k+1} from H_k with the desired properties. Using the notation

$$p_k := x_{k+1} - x_k,\; q_k := g_{k+1} - g_k$$

and the parameters

(5.11.4) $$\gamma_k > 0, \qquad \theta_k \geqslant 0,$$

this recursion has the form

$$H_{k+1} := \Psi(\gamma_k, \theta_k, H_k, p_k, q_k),$$

(5.11.5) $$\Psi(\gamma, \theta, H, p, q) := \gamma H + \left(1 + \gamma\theta \frac{q^T H q}{p^T q}\right) \frac{pp^T}{p^T q} - \gamma \frac{(1-\theta)}{q^T H q} Hq \cdot q^T H - \frac{\gamma\theta}{p^T q}(pq^T H + Hqp^T).$$

The "update function" Ψ is only defined if $p^T q \neq 0$, $q^T H q \neq 0$. Observe that H_{k+1} is obtained from H_k by adding a correction of rank $\leqslant 2$ to the matrix $\gamma_k H_k$:

$$\operatorname{rank}(H_{k+1} - \gamma_k H_k) \leqslant 2.$$

Hence (5.11.5) is said to define a *rank-two method.*

The following special cases are contained in (5.11.5):

(a) $\gamma_k \equiv 1$, $\theta_k \equiv 0$: the method of Davidon (1959) and Fletcher and Powell (1963) ("DFP method");
(b) $\gamma_k \equiv 1$, $\theta_k \equiv 1$: the rank two method of Broyden, Fletcher, Goldfarb, and Shanno ("BFGS method") [see, for example, Broyden (1970)];
(c) $\gamma_k \equiv 1$, $\theta_k = p_k^T q_k / (p_k^T q_k - q_k^T H_k q_k)$: the symmetric, rank-one method of Broyden.

The last method is only defined for $p_k^T q_k \neq q_k^T H_k q_k$. It is possible that $\theta_k < 0$, in which case H_{k+1} can be indefinite even when H_k is positive definite [see Theorem (5.11.9)]. If we substitute the value of θ_k into (5.11.5), we obtain:

$$H_{k+1} = h_k + \frac{z_k z_k^T}{\alpha_k}, \qquad z_k := p_k - H_k q_k, \quad \alpha_k := p_k^T q_k - q_k^T H_k q_k,$$

which explains why this is referred to as a rank-one method.

A minimization method of the Oren–Luenberger class has the following form:

(5.11.6).

(0) *Start: Choose a starting point $x_0 \in \mathbb{R}^n$ and an $n \times n$ positive definite matrix H_0 (e.g. $H_0 = I$), and set $g_0 := g(x_0)$.*

For $k = 0, 1, \ldots$ obtain x_{k+1}, H_{k+1} from x_k, H_k as follows:

(1) *If $g_k = 0$, stop: x_k is at least a stationary point for h. Otherwise*
(2) *compute $s_k := H_k g_k$.*
(3) *Determine $x_{k+1} = x_k - \lambda_k s_k$ by means of an (approximate) minimization*

$$h(x_{k+1}) \approx \min\{h(x_k - \lambda s_k) \mid \lambda \geqslant 0\},$$

and set

$$g_{k+1} := g(x_{k+1}), \qquad p_k := x_{k+1} - x_k, \qquad q_k := g_{k+1} - g_k.$$

(4) *Choose suitable parameter values $\gamma_k > 0$, $\theta_k \geqslant 0$, and compute $H_{k+1} = \Psi(\gamma_k, \theta_k, H_k, p_k, q_k)$ according to (5.11.5).*

The method is uniquely defined through the choice of the sequences $\{\gamma_k\}$, $\{\theta_k\}$ and through the line-search procedure in step (3). The characteristics of the line search can be described with the aid of the parameter μ_k defined by

$$g_{k+1}^T s_k = \mu_k g_k^T s_k = \mu_k g_k^T H_k g_k. \tag{5.11.7}$$

If s_k is a descent direction, $g_k^T s_k > 0$, then μ_k is uniquely determined by x_{k+1}. If the line search is exact, then $\mu_k = 0$, since $g_{k+1}^T s_k = -\varphi_k'(\lambda_k) = 0$, $\varphi_k(\lambda) := h(x_k - \lambda_k s_k)$. In what follows we assume that

$$\mu_k < 1. \tag{5.11.8}$$

If $g_k \neq 0$ and H_k is positive definite, it follows from (5.11.8) that $\lambda_k > 0$; therefore

$$q_k^T p_k = -\lambda_k(g_{k+1} - g_k)s_k = \lambda_k(1 - \mu_k)g_k^T s_k = \lambda_k(1 - \mu_k)g_k^T H_k g_k > 0,$$

and also $q_k \neq 0$, $q_k^T H_k q_k > 0$: the matrix H_{k+1} is well defined via (5.11.5). The condition (5.11.8) on the line search cannot be satisfied only if

$$\varphi_k'(\lambda) = -g(x_k - \lambda s_k)^T s_k \leqslant \varphi_k'(0) = -g_k^T s_k < 0 \quad \text{for all } \lambda \geqslant 0.$$

But then

$$h(x_k - \lambda s_k) - h(x_k) = \int_0^\lambda \varphi_k'(\tau)\, d\tau \leqslant -\lambda g_k^T s_k < 0 \quad \text{for all } \lambda > 0,$$

so that $h(x_k - \lambda s_k)$ is not bounded below as $\lambda \to \infty$. The condition (5.11.8) does not, therefore, pose any actual restriction.

At this point we have proved the first part of the following theorem, which states that the method (5.11.6) meets our requirements above:

(5.11.9) Theorem. *If there is a $k \geq 0$ in (5.11.6) for which H_k is positive definite, $g_k \neq 0$, and for which the line search satisfies $\mu_k < 1$, then for all $\gamma_k > 0$, $\theta_k \geqslant 0$ the matrix $H_{k+1} = \Psi(\gamma_k, \theta_k, H_k, p_k, q_k)$ is well defined and positive definite. The matrix H_{k+1} satisfies the quasi-Newton equation $H_{k+1} q_k = p_k$.*

PROOF. It is only necessary to show the following property of the function Ψ given by (5.11.5): Assuming that

$$H \text{ is positive definite}, \quad p^T q > 0, \quad q^T H q > 0, \quad \gamma > 0, \quad \theta \geqslant 0,$$

then $\bar{H} := \Psi(\gamma, \theta, H, p, q)$ is also positive definite.

Let $y \in \mathbb{R}^n$, $y \neq 0$, be an arbitrary vector, and let $H = LL^T$ be the Cholesky decomposition of H [Theorem (4.3.3)]. Using the vectors

$$u := L^T y, \qquad v := L^T q$$

and using (5.11.5), $y^T \bar{H} y$ can be written

$$\begin{aligned} y^T \bar{H} y &= \gamma u^T u + \left(1 + \gamma\theta \frac{v^T v}{p^T q}\right) \frac{(p^T y)^2}{p^T q} - \gamma \frac{(1-\theta)}{v^T v} (v^T u)^2 - \frac{2\gamma\theta}{p^T q} p^T y \cdot u^T v \\ &= \gamma\left(u^T u - \frac{(u^T v)^2}{v^T v}\right) + \frac{(p^T y)^2}{p^T q} + \theta\gamma \left[\sqrt{v^T v}\, \frac{p^T y}{p^T q} - \frac{v^T u}{\sqrt{v^T v}}\right]^2 \\ &\geqslant \gamma\left(u^T u - \frac{(u^T v)^2}{v^T v}\right) + \frac{(p^T y)^2}{p^T q}. \end{aligned}$$

The Cauchy–Schwarz inequality implies that $u^T u - (u^T v)^2 / v^T v \geqslant 0$, with equality if and only if $u = \alpha v$ for some $\alpha \neq 0$ (since $y \neq 0$). If $u \neq \alpha v$, then $y^T \bar{H} y > 0$. If $u = \alpha v$, it follows from the nonsingularity of H and L that $0 \neq y = \alpha q$, so that

$$y^T \bar{H} y \geqslant \frac{(p^T y)^2}{p^T q} = \alpha^2 p^T q > 0.$$

Because $0 \neq y \in \mathbb{R}^n$ was arbitrary, $\bar{H}$ must be positive definite. The quasi-Newton equation $\bar{H} q = p$ can be verified directly from (5.11.5). □

The following theorem establishes that the method (5.11.6) yields the minimum point of any quadratic function $h: \mathbb{R}^n \to \mathbb{R}$ after at most n steps, provided that the line searches are performed exactly. Since each sufficiently differentiable function h can be approximated arbitrarily closely in a small enough neighborhood of a local minimum point by a quadratic function, this property suggests that the method will be rapidly convergent even when applied to nonquadratic functions.

(5.11.10) Theorem. *Let $h(x) = \frac{1}{2} x^T A x + b^T x + c$ be a quadratic function, where A is an $n \times n$ positive definite matrix. Further, let $x_0 \in \mathbb{R}^n$, and let H_0 be an $n \times n$ positive definite matrix. If the method (5.11.6) is used to minimize h, starting with x_0, H_0 and carrying out exact line searches ($\mu_i = 0$ for all $i \geqslant 0$), then sequences $x_i, H_i, g_i, p_i := x_{i+1} - x_i, q_i := g_{i+1} - g_i$ are produced with the properties:*

(a) *There is a smallest index $m \leqslant n$ with $x_m = \bar{x} = -A^{-1} b$ such that $x_m = \bar{x}$ is the minimum of h, $g_m = 0$.*

(b) $p_i^T q_k = p_i^T A p_k = 0$ *for* $0 \leqslant i \neq k \leqslant m-1$, $p_i^T q_i > 0$ *for* $0 \leqslant i \leqslant m-1$. *That is, the vectors* p_i *are* A*-conjugate.*

(c) $p_i^T g_k = 0$ *for all* $0 \leqslant i < k \leqslant m$.

(d) $H_k q_i = \gamma_{i,k} p_i$ *for* $0 \leqslant i < k \leqslant m$, *with*

$$\gamma_{i,k} := \begin{cases} \gamma_{i+1}\gamma_{i+2}\cdots\gamma_{k-1} & \text{for } i < k-1, \\ 1 & \text{for } i = k-1. \end{cases}$$

(e) *If* $m = n$, *then additionally*

$$H_m = H_n = PDP^{-1}A^{-1},$$

where $D := \operatorname{diag}(\gamma_{0,n}, \gamma_{1,n}, \ldots, \gamma_{n-1,n})$, $P := (p_0, p_1, \ldots, p_{n-1})$. *And if* $\gamma_i = 1$, *it follows that* $H_n = A^{-1}$.

Proof. Consider the following conditions for an arbitrary index $l \geqslant 0$:

$$(\mathrm{A}_l) \qquad \begin{cases} p_i^T q_k = p_i^T A p_k = 0 & \text{for } 0 \leqslant i \neq k \leqslant l-1, \\ p_i^T q_i > 0 & \text{for } 0 \leqslant i \leqslant l-1, \\ H_l \text{ is positive definite;} \end{cases}$$

$$(\mathrm{B}_l) \qquad p_i^T g_k = 0 \quad \text{for all } 0 \leqslant i < k \leqslant l;$$

$$(\mathrm{C}_l) \qquad H_k q_i = \gamma_{i,k} p_i \quad \text{for } 0 \leqslant i < k \leqslant l.$$

If these conditions hold, and if in addition $g_l = g(x_l) \neq 0$, then we will show that $(\mathrm{A}_{l+1}, \mathrm{B}_{l+1}, \mathrm{C}_{l+1})$ hold.

Since H_l is positive definite by (A_l), $g_l^T H_l g_l > 0$ and $s_l := H_l g_l \neq 0$ follow immediately from $g_l \neq 0$. Because the line search is exact, λ_l is a zero point of

$$0 = g_{l+1}^T s_l = (g_l - \lambda_l A s_l)^T s_l,$$

$$\lambda_l = \frac{g_l^T H_l g_l}{s_l^T A s_l} > 0;$$

hence $p_l = -\lambda_l s_l \neq 0$ and

$$(5.11.11) \qquad \begin{aligned} p_l^T g_{l+1} &= -\lambda_l s_l^T g_{l+1} = 0, \\ p_l^T q_l &= -\lambda_l s_l^T (g_{l+1} - g_l) = \lambda_l s_l^T g_l = \lambda_l g_l^T H_l g_l > 0. \end{aligned}$$

According to Theorem (5.11.9), H_{l+1} is positive definite. Further,

$$p_i^T q_l = p_i^T A p_l = q_i^T p_l = -\lambda_l q_i^T H_l g_l = -\lambda_l \gamma_{i,l} p_i^T g_l = 0$$

for $i < l$, because $Ap_k = q_k$, (B_l) and (C_l) hold. This establishes (A_{l+1}).

To prove (B_{l+1}), we have to show that $p_i^T g_{l+1} = 0$ for all $i < l+1$. (5.11.11) takes care of the case $i = l$. For $i < l$, we can write

$$p_i^T g_{l+1} = p_i^T \left(g_{i+1} + \sum_{j=i+1}^{l} q_j \right)$$

since $q_j = g_{j+1} - g_j$ by (5.11.6). The above expression vanishes according to (B_l) and (A_{l+1}). Thus (B_{l+1}) holds.

Using (5.11.5), it is immediate that $H_{l+1}q_l = p_l$. Further (A_{l+1}), (C_l) imply $p_l^T q_i = 0$, $q_l^T H_l q_i = \gamma_{i,l} q_l^T p_i = 0$ for $i < l$, so that

$$H_{l+1}q_i = \gamma_l H_l q_i = \gamma_l \gamma_{i,l} p_i = \gamma_{i,l+1} p_i$$

follows for $i < l$ from (5.11.5). Thus (C_{l+1}) holds, too.

We note that (A_0, B_0, C_0) hold trivially. So long as x_l satisfies (b)–(d) and $g(x_l) \neq 0$, we can use the above results to generate, implicitly, a point x_{l+1} which also satisfies (b)–(d). The sequence $x_0, x_1, \ldots$ must terminate: (A_l) could only hold for $l \leqslant n$, since the l vectors $p_0, \ldots, p_{l-1}$ are linearly independent [if $\sum_{i \leqslant l-1} \alpha_i p_i = 0$, then multiplying by $p_k^T A$ for $k = 0, \ldots, l-1$, gives $\alpha_k p_k^T A p_k = 0 \Rightarrow \alpha_k = 0$ from (A_l)]. No more than n vectors in $\mathbb{R}^n$ can be linearly independent. When the sequence terminates, say at $l = m$, $0 \leqslant m \leqslant n$, it must do so because

$$g_m = 0, \qquad x_m = -A^{-1}b,$$

i.e. (a) holds. In case $m = n$, (d) implies

$$H_n Q = PD$$

for the matrices $P = (p_0, \ldots, p_{n-1})$, $Q = (q_0, \ldots, q_{n-1})$. Since $AP = Q$, the nonsingularity of P implies

$$H_n = PDP^{-1}A^{-1},$$

which proves (e) and, hence, the theorem. □

We will now discuss briefly how to choose the parameters γ_k, θ_k in order to obtain as good a method as possible. Theorem (5.11.10e) seems to imply that the choice $\gamma_k \equiv 1$ would be good, because it appears to ensure that $\lim_i H_i = A^{-1}$ (in general this is true for nonquadratic functions only under certain additional assumptions), which suggests that the quasi-Newton method will converge "in the same way as the Newton method." Practical experience indicates that the choice

$$\gamma_k \equiv 1, \quad \theta_k \equiv 1 \qquad \text{(BFGS method)}$$

is good [see Dixon (1971)]. Oren and Spedicato (1974) have been able to give an upper bound $\Phi(\gamma_k, \theta_k)$ on the quotient $\text{cond}(H_{k+1})/\text{cond}(H_k)$ of the condition numbers (with respect to the Euclidian norm) of H_{k+1} and H_k. Minimizing Φ with respect to γ_k and θ_k leads to the prescription

$$\text{if } \frac{\varepsilon}{\sigma} \leqslant 1 \qquad \text{then choose } \gamma_k := \frac{\varepsilon}{\sigma}, \ \theta_k := 0;$$

$$\text{if } \frac{\sigma}{\tau} \geqslant 1 \qquad \text{then choose } \gamma_k := \frac{\sigma}{\tau}, \ \theta_k := 1;$$

$$\text{if } \frac{\sigma}{\tau} \leqslant 1 \leqslant \frac{\varepsilon}{\sigma} \qquad \text{then choose } \gamma_k := 1, \ \theta_k := \frac{\sigma(\varepsilon - \sigma)}{\varepsilon\tau - \sigma^2}.$$

Here we have used

$$\varepsilon := p_k^T H_k^{-1} p_k, \qquad \sigma := p_k^T q_k, \qquad \tau := q_k^T H_k q_k.$$

Another promising suggestion has been made by Davidon (1975). He was able to show that the choice

$$\theta_k := \begin{cases} \dfrac{\sigma(\varepsilon - \sigma)}{\varepsilon\tau - \sigma^2} & \text{if } \sigma \leqslant \dfrac{2\varepsilon\tau}{\varepsilon + \tau}, \\[2ex] \dfrac{\sigma}{\sigma - \tau} & \text{otherwise}, \end{cases}$$

used with $\gamma_k \equiv 1$ in the method (5.11.5), minimizes the quotient $\lambda_{\max}/\lambda_{\min}$ of the largest to the smallest eigenvalues satisfying the generalized eigenproblem

$$\text{determine } \lambda \in \mathbb{C},\ y \neq 0 \quad \text{so that } H_{k+1} y = \lambda H_k y,\ \det(H_k^{-1} H_{k+1} - \lambda I) = 0.$$

Theorem (5.11.10) suggests that methods of the type (5.11.5) will converge quickly even on nonquadratic functions. This has been proved formally for some individual methods of the Oren–Luenberger class. These results rest, for the most part, on the local behavior in a sufficiently small neighborhood $U(\bar{x})$ of a local minimum $\bar{x}$ of h under the following assumptions:

(5.11.12a) $H(\bar{x})$ is positive definite,

(5.11.12b) $H(x)$ is Lipschitz continuous at $x = \bar{x}$, i.e., there is a Λ with $\|H(x) - H(\bar{x})\| \leqslant \Lambda \|x - \bar{x}\|$ for all $x \in U(\bar{x})$.

Further, certain mild restrictions have to be placed on the line search, for example:

(5.11.13) For given constants $0 < c_1 < c_2 < 1$, $c_1 \leqslant \frac{1}{2}$, $x_{k+1} = x_k - \lambda_k s_k$ is chosen so that

$$h(x_{k+1}) \leqslant h(x_k) - c_1 \lambda_k g_k^T s_k,$$

$$g_{k+1}^T s_k \leqslant c_2 g_k^T s_k,$$

or

(5.11.14) $\lambda_k = \min\{\lambda \geqslant 0 \mid g(x_k - \lambda s_k)^T s_k = \mu_k g_k^T s_k\}, \qquad |\mu_k| < 1.$

Under the conditions (5.11.12), (5.11.13) Powell (1975) was able to show for the BFGS method ($\gamma_k \equiv 1$, $\theta_k \equiv 1$): There is a neighborhood

$$V(\bar{x}) \subseteq U(\bar{x})$$

such that the method is *superlinearly convergent* for all positive definite initial matrices H_0 and all $x_0 \in V(\bar{x})$. That is,

$$\lim_{i \to \infty} \frac{\|x_{i+1} - \bar{x}\|}{\|x_i - \bar{x}\|} = 0,$$

as long as $x_i \neq \bar{x}$ for all $i \geqslant 0$. In his result the stepsize $\lambda_k \equiv 1$ satisfies (5.11.13) for all $k \geqslant 0$ sufficiently large.

Another convergence result has been established for the subclass of the Oren–Luenberger class (5.11.5) which additionally satisfies

$$(5.11.15) \qquad 0 \leqslant \theta_k \leqslant 1, \qquad \frac{p_k^T q_k}{q_k^T H_k q_k} \leqslant \gamma_k \leqslant \frac{p_k^T H_k^{-1} p_k}{p_k^T q_k}.$$

For this subclass, using (5.11.12), (5.11.14), and the additional demand that the line search be asymptotically exact, i.e.

$$|\mu_k| \leqslant c\|g_k\| \quad \text{for large enough } k,$$

it can be shown [Stoer (1977), Baptist and Stoer (1977)] that

$$\lim_k x_k = \bar{x}, \qquad \|x_{k+n} - \bar{x}\| \leqslant \gamma \|x_k - \bar{x}\|^2 \quad \text{for all } k \geqslant 0$$

for all positive definite initial matrices H_0 and for $\|x_0 - \bar{x}\|$ small enough. The proofs of all of the above convergence results are long and difficult.

The following simple example illustrates the typical behaviors of the DFP method, the BFGS method, and the steepest-descent method ($s_k := g_k$ in each iteration step).

We let

$$h(x, y) := 100(y^2(3 - x) - x^2(3 + x))^2 + \frac{(2 + x)^2}{1 + (2 + x)^2},$$

which has the minimum point $\bar{x} := -2$, $\bar{y} := 0.894\,271\,9099\ldots$, $h(\bar{x}, \bar{y}) = 0$. For each method we take

$$x_0 := 0.1, \qquad y_0 := 4.2$$

as the starting point and let

$$H_0 := \begin{bmatrix} 1 & 0 \\ 0 & 1 \end{bmatrix}$$

for the BFGS and DFP methods.

Using the same line-search procedure for each method, we obtained the following results on a machine with eps $= 10^{-11}$:

	BFGS	DFP	Steepest descent
N	54	47	201
F	374	568	1248
ε	$\leqslant 10^{-11}$	$\leqslant 10^{-11}$	0.7

N denotes the number of iteration steps $(x_i, y_i) \to (x_{i+1}, y_{i+1})$; F denotes the number of evaluations of h; and $\varepsilon := \|g(x_N, y_N)\|$ was the accuracy attained at termination.

The steepest descent method is clearly inferior to both the DFP and BFGS methods. The DFP method is slightly inferior to the BFGS method. (The line searches were not done in a particularly efficient manner. More than 6 function evaluations were needed, on the average, for each iteration step. It is possible to do better, but the same relative performance would have been evident among the methods even with a better line-search procedure.)

Exercises for Chapter 5

1. Let the continuously differentiable iteration function $\Phi: \mathbb{R}^n \to \mathbb{R}^n$ be given. If
$$\operatorname{lub}(D\Phi(x)) \leqslant K < 1 \quad \text{for all } x \in \mathbb{R}^n,$$
then the conditions for Theorem (5.2.2) are fulfilled for all $x, y \in \mathbb{R}^n$.

2. Show that the iteration
$$x_{k+1} = \cos x_k$$
converges to the fixed point $\xi = \cos \xi$ for all $x_0 \in \mathbb{R}$.

3. Give a locally convergent method for determining the fixed point $\xi = \sqrt[3]{2}$ of $\Phi(x) = x^3 + x - 2$. (Do not use the Aitken transformation.)

4. The polynomial $p(x) = x^3 - x^2 - x - 1$ has its only positive root near $\xi = 1.839 \ldots$. Without using $f'(x)$, construct an iteration function $\Phi(x)$ having the fixed point $\xi = \Phi(\xi)$ and having the property that the iteration converges for any starting point $x_0 > 0$.

5. Show that
$$\lim_{i \to \infty} x_i = 2,$$
where
$$x_0 := 0, \qquad x_{i+1} := \sqrt{2 + x_i}.$$

6. Let the function $f: \mathbb{R}^2 \to \mathbb{R}^2$
$$f(z) = \begin{bmatrix} \exp(x^2 + y^2) - 3 \\ x + y - \sin(3(x + y)) \end{bmatrix}, \qquad z = \begin{bmatrix} x \\ y \end{bmatrix},$$
be given. Compute the first derivative $Df(z)$. For which z is $Df(z)$ singular?

7. The polynomial $p_0(x) = x^4 - 8x^3 + 24x^2 - 32x + a_4$ has a quadruple root $x = 2$ for $a_4 = 16$. To first approximation, where are its roots if $a_4 = 16 \pm 10^{-4}$?

8. Consider the sequence $\{z_i\}$ with $z_{i+1} = \Phi(z_i)$, $\Phi: \mathbb{R} \to \mathbb{R}$. Any fixed point ξ of Φ is a zero of
$$F(z) = z - \Phi(z).$$
Show that if one step of *regula falsi* is applied to $F(z)$ with
$$a_i = z_i, \qquad x_i = z_{i+1},$$
then one obtains Steffensen's (or Aitken's) method for transforming the sequence $\{z_i\}$ into $\{\mu_i\}$.

9. Let $f: \mathbb{R} \to \mathbb{R}$ have a single, simple zero x. Show that, if $\Phi(x) = x - f(x)$ and the recursion (5.10.7) are used, the result is the quasi-Newton method
$$x_{n+1} := x_n - \frac{f(x_n)^2}{f(x_n) - f(x_n - f(x_n))}, \qquad n = 0, 1, \ldots.$$

Show that this iteration converges at least quadratically to simple zeros and linearly to multiple zeros.
Hint: (5.10.13).

10. Calculate $x = 1/a$ for any given $a \neq 0$ without using division. For which starting values x_0 will the method converge?

11. Give an iterative method for computing $\sqrt[n]{a}$, $a > 0$, which converges locally in second order. (The method may only use the four fundamental arithmetic operations.)

12. Let A be a nonsingular matrix and $\{X_k\}$, $k = 0, 1, \ldots$, be a sequence of matrices satisfying

$$X_{k+1} := X_k + X_k(I - AX_k)$$

(Schulz's method).

(a) Show that $\operatorname{lub}(I - AX_0) < 1$ is sufficient to ensure the convergence of $\{X_k\}$ to A^{-1}. Further, $E_k := I - AX_k$ satisfies

$$E_{k+1} = E_k E_k.$$

(b) Show that Schulz's method is locally quadratically convergent.
(c) If, in addition, $AX_0 = X_0 A$ then

$$AX_k = X_k A \quad \text{for all } k \geqslant 0.$$

13. Let the function $f\colon \mathbb{R} \to \mathbb{R}$ be twice continuously differentiable for all $x \in U(x) := \{x \mid |x - \xi| \leqslant r\}$ in a neighborhood of a simple zero $\xi, f(\xi) = 0$. Show that the iteration

$$y := x_n - f'(x_n)^{-1} f(x_n),$$
$$x_{n+1} := y - f'(x_n)^{-1} f(y), \qquad n = 0, 1, \ldots,$$

converges locally at least cubically to ξ.

14. Let the function $f\colon \mathbb{R} \to \mathbb{R}$ have the zero ξ. Let f be twice continuously differentiable and satisfy $f'(x) \neq 0$ for all $x \in I := \{x \mid |x - \xi| \leqslant r\}$. The method

$$x_{n+1} := x_n - \frac{f(x_n)^2}{f(x_n) - f(x_n + f(x_n))}, \qquad n = 0, 1, \ldots,$$

is a quasi-Newton method. Show:

(a) The method has the form

$$x_{n+1} := x_n - q(x_n) f(x_n)$$

Give $q(x_n)$, and show that there is a constant c such that

$$|q(x) - f'(x)^{-1}| \leqslant c\,|f(x)|.$$

(b) One can construct a majorizing sequence y_n sufficiently close to ξ,

$$|x_n - \xi| \leqslant y_n \quad \text{for all } n \geqslant 0.$$

Give conditions which ensure that y_n converges to zero. Using y_n, determine the local order of convergence.

15. Let the function $f\colon \mathbb{R}^n \to \mathbb{R}^n$ satisfy the assumptions

    (1) $f(x)$ is continuously differentiable for all $x \in \mathbb{R}^n$;
    (2) for all $x \in \mathbb{R}^n$, $Df(x)^{-1}$ exists;
    (3) $x^T f(x) \geqslant \gamma(\|x\|)\|x\|$ for all $x \in \mathbb{R}^n$, where $\gamma(\rho)$ is a continuous function for $\rho \geqslant 0$ and satisfies $\gamma(\rho) \to +\infty$ as $\rho \to \infty$;
    (4) for all $x, h \in \mathbb{R}^n$

    $$h^T Df(x)h \geqslant \mu(\|x\|)\|h\|^2$$

    with $\mu(\rho)$ monotone increasing in $\rho > 0$, $\mu(0) = 0$, and

    $$\int_0^\infty \mu(\rho)\, d\rho = +\infty.$$

    Then (1), (2), (3) or (1), (2), (4) are enough to ensure that conditions (a)–(c) of Theorem (5.4.2.5) are satisfied. For (4), use the Taylor expansion

    $$f(x+h) - f(x) = \int_0^1 Df(x+th)h\, dt.$$

16. Give the recursion formula for computing the values A_1, B_1 which appear in the Bairstow method.

17. (Tornheim 1964). Consider the scalar, multistep iteration function of $r+1$ variables

    $$\varphi(x_0, x_1, \ldots, x_r)$$

    and the iteration

    $$y_{i+1} := \varphi(y_i, y_{i-1}, \ldots, y_{i-r}), \qquad i = 0, 1, \ldots,$$

    where $y_0, y_{-1}, \ldots, y_{-r}$ are specified. Let φ have partial derivatives of at least order $r+1$. y^* is called a fixed point of φ if for all $k = 1, \ldots, r$ and arbitrary x_i, $i \neq k$, it follows that

    $$(*) \qquad y^* = \varphi(x_0, \ldots, x_{k-1}, y^*, x_{k+1}, \ldots, x_r).$$

    Show that

    (a) The partial derivatives

    $$D^s\varphi(x_0, \ldots, x_r) := \frac{\partial^{|s|}\varphi(x_0, \ldots, x_r)}{\partial x_0^{s_0}\, \partial x_s^{s_1} \ldots \partial x_r^{s_r}}, \qquad s = (s_0, \ldots, s_n), \quad |s| = \sum_{j=0}^r s_j,$$

    satisfy $D^s\varphi(y^*, \ldots, y^*) = 0$ if for some j, $0 \leqslant j \leqslant r$, $s_j = 0$. [Note that (*) holds identically in $x_0, \ldots, x_{k-1}, x_{k+1}, \ldots, x_r$ for all k.]

    (b) In a suitably small neighborhood of y^* the recursion

    $$(**) \qquad \varepsilon_{i+1} \leqslant c\varepsilon_i \varepsilon_{i-1} \ldots \varepsilon_{i-r}$$

    holds with $\varepsilon_i := |y_i - y^*|$ and with an appropriate constant c.

    Further, give:

    (c) the general solution of the recursion (**) and the local convergence order of the sequence y_i.

18. Prove (5.9.14).

References for Chapter 5

Baptist, P., Stoer, J.: On the relation between quadratic termination and convergence properties of minimization algorithms. Part II. Applications. *Numer. Math.* **28**, 367–391 (1977).

Bauer, F. L.: Beiträge zur Entwicklung numerischer Verfahren für programmgesteuerte Rechenanlagen. II. Direkte Faktorisierung eines Polynoms. *Bayer. Akad. Wiss. Math. Natur. Kl. S.B.* 163–203 (1956).

Brent, R. P.: *Algorithms for Minimization without Derivatives*. Englewood Cliffs, N.J.: Prentice-Hall 1973.

Broyden, C. G.: A class of methods for solving nonlinear simultaneous equations. *Math. Comput.* **19**, 577–593 (1965).

———: Quasi-Newton-methods and their application to function minimization. *Math. Comput.* **21**, 368–381 (1967).

———: The convergence of a class of double rank minimization algorithms. 2. The new algorithm. *J. Inst. Math. Appl.* **6**, 222–231 (1970).

———: Dennis, J. E., Moré, J. J.: On the local and superlinear convergence of quasi-Newton methods. *J. Inst. Math. Appl.* **12**, 223–245 (1973).

Collatz, L.: *Funktionalanalysis und numerische Mathematik*. Die Grundlehren der mathematischen Wissenschaften in Einzeldarstellungen. Bd. 120. Berlin, Heidelberg, New York: Springer 1968. English edition: *Functional Analysis and Numerical Analysis*. New York: Academic Press 1966.

Davidon, W. C.: Variable metric methods for minimization. Argonne National Laboratory Report ANL-5990, 1959.

———: Optimally conditioned optimization algorithms without line searches. *Math. Programming* **9**, 1–30 (1975).

Deuflhard, P.: A modified Newton method for the solution of ill-conditioned systems of nonlinear equations with applications to multiple shooting. *Numer. Math.* **22**, 289–315 (1974).

Dixon, L. C. W.: The choice of step length, a crucial factor in the performance of variable metric algorithms. In: *Numerical Methods for Nonlinear Optimization*. Edited by F. A. Lootsma. 149–170. New York: Academic Press 1971.

Fletcher, R., Powell, M. J. D.: A rapidly convergent descent method for minimization. *Comput. J.* **6**, 163–168 (1963).

Gill, P. E., Golub, G. H., Murray, W., Saunders, M. A.: Methods for modifying matrix factorizations. *Math. Comput.* **28**, 505–535 (1974).

Henrici, P.: *Applied and Computational Complex Analysis*. Vol. 1. New York: Wiley 1974.

Himmelblau, D. M.: *Applied Nonlinear Programming*. New York: McGraw-Hill 1972.

Householder, A. S.: *The Numerical Treatment of a Single Non-linear Equation*. New York: McGraw-Hill 1970.

Jenkins, M. A., Traub, J. F.: A three-stage variable-shift iteration for polynomial zeros and its relation to generalized Rayleigh iteration. *Numer. Math.* **14**, 252–263 (1970).

Luenberger, D. G.: *Introduction to Linear and Nonlinear Programming*. Reading, Mass.: Addison-Wesley 1973.

Maehly, H.: Zur iterativen Auflösung algebraischer Gleichungen. *Z. Angew. Math. Physik* **5**, 260–263 (1954).

Marden, M.: *Geometry of Polynomials*. Providence, R.I.: Amer. Math. Soc. 1966.

Nickel, K.: Die numerische Berechnung der Wurzeln eines Polynoms. *Numer. Math.* **9**, 80–98 (1966).

Oren, S. S., Luenberger, D. G.: Self-scaling variable metric (SSVM) algorithms. I. Criteria and sufficient conditions for scaling a class of algorithms. *Manage. Sci.* **20**, 845–862 (1974).

———, Spedicato, E.: Optimal conditioning of self-scaling variable metric algorithms. Stanford University Dept. of Engineering, Economic Systems Report ARG-MR 74-5, 1974.

Ortega, J. M., Rheinboldt, W. C.: *Iterative Solution of Non-linear Equations in Several Variables*. New York: Academic Press 1970.

Ostrowski, A. M.: *Solution of Equations in Euclidean and Banach Spaces*. New York: Academic Press 1973.

Peters, G., Wilkinson, J. H.: Eigenvalues of $Ax = \lambda Bx$ with band symmetric A and B. *Comput. J.* **12**, 398–404 (1969).

———, ———: Practical problems arising in the solution of polynomial equations. *J. Inst. Math. Appl.* **8**, 16–35 (1971).

Powell, M. J. D.: Some global convergence properties of a variable metric algorithm for minimization without exact line searches. In: *Proc. AMS Symposium on Nonlinear Programming 1975*. Amer. Math. Soc. 1976.

Stoer, J.: On the convergence rate of imperfect minimization algorithms in Broyden's β-class. *Math. Programming* **9**, 313–335 (1975).

———: On the relation between quadratic termination and convergence properties of minimization algorithms. Part I. Theory. *Numer. Math.* **28**, 343–366 (1977).

Tornheim, L.: Convergence of multipoint methods. *J. Assoc. Comput. Mach.* **11**, 210–220 (1964).

Traub, J. F.: *Iterative Methods for the Solution of Equations*. Englewood Cliffs, N.J.: Prentice-Hall 1964.

Wilkinson, J. H.: The evaluation of the zeros of ill-conditioned polynomials. Part I. *Numer. Math.* **1**, 150–180 (1959).

———: *Rounding Errors in Algebraic Processes*. Englewood Cliffs, N.J.: Prentice-Hall 1963.

———: *The Algebraic Eigenvalue Problem*. Oxford: Clarendon Press 1965.

6 Eigenvalue Problems

6.0 Introduction

Many practical problems in engineering and physics lead to eigenvalue problems. Typically, in all these problems, an overdetermined system of equations is given, say $n + 1$ equations for n unknowns $\xi_1, \ldots, \xi_n$ of the form

$$F(x; \lambda) :\equiv \begin{bmatrix} f_1(\xi_1, \ldots, \xi_n; \lambda) \\ \cdots\cdots\cdots\cdots \\ f_{n+1}(\xi_1, \ldots, \xi_n; \lambda) \end{bmatrix} = 0, \tag{6.0.1}$$

in which the functions f_i also depend on an additional parameter λ. Usually, (6.0.1) has a solution $x = [\xi_1, \ldots, \xi_n]^T$ only for specific values $\lambda = \lambda_i$, $i = 1, 2, \ldots$, of this parameter. These values λ_i are called *eigenvalues* of the eigenvalue problem (6.0.1), and a corresponding solution $x = x(\lambda_i)$ of (6.0.1) *eigensolution* belonging to the eigenvalue λ_i.

Eigenvalue problems of this general form occur, e.g., in the context of boundary value problems for differential equations (see Section 7.3.0). In this chapter we consider only the special class of *algebraic eigenvalue problems*, where all but one of the f_i in (6.0.1) depend linearly on x and λ, and which have the following form: Given real or complex $n \times n$ matrices A and B, find a number $\lambda \in \mathbb{C}$ such that the system of $n + 1$ equations

$$(6.0.2) \qquad \begin{aligned} (A - \lambda B)x &= 0, \\ x^H x &= 1 \end{aligned}$$

has a solution $x \in \mathbb{C}^n$. Clearly, this problem is equivalent to finding numbers $\lambda \in \mathbb{C}$ such that there is a *nontrivial* vector $x \in \mathbb{C}^n$, $x \neq 0$, with

(6.0.3) $$Ax = \lambda Bx.$$

For arbitrary A and B, this problem is still very general, and we treat it only briefly in Section 6.8. The main portion of Chapter 6 is devoted to the special case of (6.0.3) where $B := I$ is the identity matrix: For an $n \times n$ matrix A, find numbers $\lambda \in \mathbb{C}$ (the eigenvalues of A) and nontrivial vectors $x \in \mathbb{C}^n$ (the eigenvectors of A belonging to λ) such that

(6.0.4) $$Ax = \lambda x, \qquad x \neq 0.$$

Sections 6.1–6.4 provide the main theoretical results on the eigenvalue problem (6.0.4) for general matrices A. In particular, we describe various normal forms of a matrix A connected with its eigenvalues, additional results on the eigenvalue problem for important special classes of matrices A (such as Hermitian and normal matrices) and the basic facts on the singular values σ_i of a matrix A, i.e., the eigenvalues σ_i^2 of $A^H A$ and AA^H, respectively.

The methods for actually computing the eigenvalues and eigenvectors of a matrix A usually are preceded by a reduction step, in which the matrix A is transformed to a "similar" matrix B having the same eigenvalues as A. The matrix $B = (b_{ik})$ has a simpler structure than A (B is either a tridiagonal matrix, $b_{ik} = 0$ for $|i - k| > 1$, or a Hessenberg matrix, $b_{ik} = 0$ for $i \geqslant k + 2$), so that the standard methods for computing eigenvalues and eigenvectors are computationally less expensive when applied to B than when applied to A. Various reduction algorithms are described in Section 6.5 and its subsections.

The main algorithms for actually computing eigenvalues and eigenvectors are presented in Section 6.6, among others the LR algorithm of Rutishauser (Section 6.6.4) and the powerful QR algorithm of Francis (Section 6.6.6). Related to the QR algorithm is the method of Golub and Reinsch for computing the singular values of matrices, which is described in Section 6.7. After touching briefly on the more general eigenvalue problem (6.0.3) in Section 6.8, the chapter closes (Section 6.9) with a description of several useful estimates for eigenvalues. These may serve, e.g., to locate the eigenvalues of a matrix and to study their sensitivity with respect to small perturbations. A detailed treatment of all numerical aspects of the eigenvalue problem for matrices is given in the excellent monograph of Wilkinson (1965), and in Golub and van Loan (1983); the eigenvalue problem for symmetric matrices is treated in Parlett (1980). ALGOL programs for all algorithms described in this chapter are found in Wilkinson and Reinsch (1971), and FORTRAN programs in the "EISPACK Guide" of Smith et al. (1976) and its extension by Garbow et al. (1977).

6.1 Basic Facts on Eigenvalues

In the following we study the problem (6.0.4), i.e., given a real or complex $n \times n$ matrix A, find a number $\lambda \in \mathbb{C}$ such that the linear homogeneous system of equations

(6.1.1) $$(A - \lambda I)x = 0$$

has a nontrivial solution $x \neq 0$.

(6.1.2) Definition. A number $\lambda \in \mathbb{C}$ is called an *eigenvalue* of the matrix A if there is a vector $x \neq 0$ such that $Ax = \lambda x$. Every such vector is called a *(right) eigenvector of A associated with the eigenvalue* λ. The set of all eigenvalues is called the *spectrum* of A.

The set

$$L(\lambda) := \{x \mid (A - \lambda I)x = 0\}$$

forms a linear subspace of $\mathbb{C}^n$ of dimension

$$\rho(\lambda) = n - \operatorname{rank}(A - \lambda I),$$

and a number $\lambda \in \mathbb{C}$ is an eigenvalue of A precisely when $L(\lambda) \neq 0$, i.e., when $\rho(\lambda) > 0$ and thus $A - \lambda I$ is singular:

$$\det(A - \lambda I) = 0.$$

It is easily seen that $\varphi(\mu) := \det(A - \mu I)$ is a nth-degree polynomial of the form

$$\varphi(\mu) = (-1)^n(\mu^n + \alpha_{n-1}\mu^{n-1} + \cdots + \alpha_0).$$

It is called the

(6.1.3) *characteristic polynomial*

of the matrix A. Its zeros are the eigenvalues of A. If $\lambda_1, \ldots, \lambda_k$ are the distinct zeros of $\varphi(\mu)$, then φ can be represented in the form

$$\varphi(\mu) = (-1)^n(\mu - \lambda_1)^{\sigma_1}(\mu - \lambda_2)^{\sigma_2} \cdots (\mu - \lambda_k)^{\sigma_k}.$$

The integer σ_i, which we also denote by $\sigma(\lambda_i) = \sigma_i$, is called the *multiplicity* of the eigenvalue λ_i—more precisely, its *algebraic multiplicity*.

The eigenvectors associated with the eigenvalue λ are not uniquely determined: together with the zero vector, they fill precisely the linear subspace $L(\lambda)$ of $\mathbb{C}^n$. Thus,

(6.1.4). *If x and y are eigenvectors belonging to the eigenvalue λ of the matrix A, then so is every linear combination $\alpha x + \beta y \neq 0$.*

The integer $\rho(\lambda) = \dim L(\lambda)$ specifies the maximum number of linearly independent eigenvectors associated with the eigenvalue λ. It is therefore also called the

geometric multiplicity of the eigenvalue λ.

One should not confuse it with the algebraic multiplicity $\sigma(\lambda)$.

EXAMPLES. The diagonal matrix of order n,

$$D = \lambda I,$$

has the characteristic polynomial $\varphi(\mu) = \det(D - \mu I) = (\lambda - \mu)^n$. λ is the only eigenvalue, and every vector $x \in \mathbb{C}^n$, $x \neq 0$, is an eigenvector: $L(\lambda) = \mathbb{C}^n$; furthermore, $\sigma(\lambda) = n = \rho(\lambda)$. The nth-order matrix

(6.1.5)
$$C_n(\lambda) := \begin{bmatrix} \lambda & 1 & & & 0 \\ & \lambda & \cdot & & \\ & & \cdot & \cdot & \\ & & & \cdot & 1 \\ 0 & & & & \lambda \end{bmatrix}$$

also has the characteristic polynomial $\varphi(\mu) = (\lambda - \mu)^n$ and λ as its only eigenvalue, with $\sigma(\lambda) = n$. The rank of $C_n(\lambda) - \lambda I$, however, is now equal to $n - 1$; thus $\rho(\lambda) = n - (n - 1) = 1$, and

$$L(\lambda) = \{\alpha e_1 \mid \alpha \in \mathbb{C}\}, \qquad e_1 = \text{1st coordinate vector.}$$

Among further simple properties of eigenvalues we note:

(6.1.6). *Let $p(\mu) = \gamma_0 + \gamma_1 \mu + \cdots + \gamma_m \mu^m$ be an arbitrary polynomial, and A a matrix of order n. Defining the matrix $p(A)$ by*

$$p(A) := \gamma_0 I + \gamma_1 A + \cdots + \gamma_m A^m,$$

the matrix $p(A)$ has the eigenvector x corresponding to the eigenvalue $p(\lambda)$ if λ is an eigenvalue of A and x a corresponding eigenvector. In particular, αA has the eigenvalue $\alpha\lambda$, and $A + \tau I$ the eigenvalue $\lambda + \tau$.

PROOF. From $Ax = \lambda x$ one obtains immediately $A^2 x = A(Ax) = \lambda Ax = \lambda^2 x$, and in general $A^i x = \lambda^i x$. Thus,

$$\begin{aligned} p(A)x &= (\gamma_0 I + \gamma_1 A + \cdots + \gamma_m A^m)x \\ &= (\gamma_0 + \gamma_1 \lambda + \cdots + \gamma_m \lambda^m)x = p(\lambda)x. \end{aligned}$$ □

Furthermore, from

$$\det(A - \lambda I) = \det((A - \lambda I)^T) = \det(A^T - \lambda I),$$
$$\det(A^H - \bar{\lambda} I) = \det((A - \lambda I)^H) = \det\overline{((A - \lambda I)^T)} = \overline{\det(A - \lambda I)},$$

there follows:

(6.1.7). *If λ is an eigenvalue of A, then λ is also an eigenvalue of A^T, and $\bar{\lambda}$ an eigenvalue of A^H.*

Between the corresponding eigenvectors x, y, z,

$$A x = \lambda x,$$
$$A^T y = \lambda y,$$
$$A^H z = \bar{\lambda} z,$$

merely the trivial relationship $\bar{y} = z$ holds, in view of $A^H = \bar{A}^T$. In particular, there is no simple relationship, in general, between x and y, or x and z. Because of $y^T = z^H$ and $z^H A = \lambda z^H$, one calls z^H, or y^T, also a *left eigenvector* associated with the eigenvalue λ of A. Furthermore, if $x \neq 0$ is an eigenvector corresponding to the eigenvalue λ,

$$A x = \lambda x,$$

T an arbitrary nonsingular $n \times n$ matrix, and if one defines $y := T^{-1} x$, then

$$T^{-1} A T y = T^{-1} A x = \lambda T^{-1} x = \lambda y, \qquad y \neq 0,$$

i.e., y is an eigenvector of the transformed matrix

$$B := T^{-1} A T$$

associated with the same eigenvalue λ. Such transformations are called

similarity transformations,

and B is said to be similar to A, $A \sim B$. One easily shows that similarity of matrices is an equivalence relation, i.e.,

$$A \sim A,$$
$$A \sim B \quad \Rightarrow \quad B \sim A,$$
$$A \sim B, \;\; B \sim C \quad \Rightarrow \quad A \sim C.$$

Similar matrices have not only the same eigenvalues, but also the same characteristic polynomial. Indeed,

$$\begin{aligned} \det(T^{-1} A T - \mu I) &= \det(T^{-1}(A - \mu I) T) \\ &= \det(T^{-1}) \det(A - \mu I) \det(T) \\ &= \det(A - \mu I). \end{aligned}$$

Moreover, the integers $\rho(\lambda)$, $\sigma(\lambda)$ remain the same: For $\sigma(\lambda)$, this follows from the invariance of the characteristic polynomial; for $\rho(\lambda)$, from the fact that, T being nonsingular, the vectors $x_1, \ldots, x_\rho$ are linearly independent if and only if the corresponding vectors $y_i = T^{-1} x_i$, $i = 1, \ldots, \rho$, are linearly independent.

In the most important methods for calculating eigenvalues and eigenvectors of a matrix A, one first performs a sequence of similarity transformations

$$A^{(o)} := A,$$

$$A^{(i)} := T_i^{-1} A^{(i-1)} T_i, \qquad i = 1, 2, \ldots,$$

in order to gradually transform the matrix A into a matrix of simpler form, whose eigenvalues and eigenvectors can then be determined more easily.

6.2 The Jordan Normal Form of a Matrix

We remarked already in the previous section that for an eigenvalue λ of an $n \times n$ matrix A, the multiplicity $\sigma(\lambda)$ of λ as a zero of the characteristic polynomial need not coincide with $\rho(\lambda)$, the maximum number of linearly independent eigenvectors belonging to λ. It is possible, however, to prove the following inequality:

$$1 \leqslant \rho(\lambda) \leqslant \sigma(\lambda) \leqslant n. \tag{6.2.1}$$

PROOF. We prove only the nontrivial part $\rho(\lambda) \leqslant \sigma(\lambda)$. Let $\rho := \rho(\lambda)$, and let $x_1, \ldots, x_\rho$ be linearly independent eigenvectors associated with λ:

$$Ax_i = \lambda x_i, \qquad i = 1, \ldots, \rho.$$

We select $n - \rho$ additional linearly independent vectors $x_i \in \mathbb{C}^n$, $i = \rho + 1, \ldots, n$, such that the x_i, $i = 1, \ldots, n$, form a basis in $\mathbb{C}^n$. Then the square matrix $T := [x_1, \ldots, x_n]$ with columns x_i is nonsingular. For $i = 1, \ldots, \rho$, in view of $Te_i = x_i$, $e_i = T^{-1}x_i$, we now have

$$T^{-1}ATe_i = T^{-1}Ax_i = \lambda T^{-1}x_i = \lambda e_i.$$

$T^{-1}AT$, therefore, has the form

$$T^{-1}AT = \begin{array}{l} \left[\begin{array}{ccc|ccc} \lambda & & 0 & * & \cdots & * \\ & \ddots & & \vdots & & \vdots \\ 0 & & \lambda & * & \cdots & * \\ \hline & & & * & \cdots & * \\ & 0 & & \vdots & & \vdots \\ & & & * & \cdots & * \end{array}\right] \\ \;\underbrace{\hphantom{\lambda\quad\ddots\quad\lambda}}_{\rho} \end{array} = \left[\begin{array}{c|c} \lambda I & B \\ \hline 0 & C \end{array}\right],$$

and for the characteristic polynomial of A, or of $T^{-1}AT$, we obtain

$$\varphi(\mu) = \det(A - \mu I) = \det(T^{-1}AT - \mu I) = (\lambda - \mu)^\rho \det(C - \mu I).$$

φ is divisible by $(\lambda - \mu)^\rho$; hence λ is a zero of φ of multiplicity at least ρ.

□

In the example of the previous section we already introduced the $\nu \times \nu$ matrices [see (6.1.5)]

$$C_\nu(\lambda) = \begin{bmatrix} \lambda & 1 & & 0 \\ & \cdot & \cdot & \\ & & \cdot & \cdot \\ & & & \cdot \quad 1 \\ 0 & & & \lambda \end{bmatrix}$$

and showed that $1 = \rho(\lambda) < \sigma(\lambda) = \nu$ (if $\nu > 1$) for the (only) eigenvalue λ of these matrices. The unique eigenvector (up to scalar multiples) is e_1, and for the coordinate vectors e_i we have generally

(6.2.2)
$$\begin{aligned}(C_\nu(\lambda) - \lambda I)e_i &= e_{i-1}, \qquad i = \nu, \nu - 1, \ldots, 2,\\ (C_\nu(\lambda) - \lambda I)e_1 &= 0.\end{aligned}$$

Setting formally $e_k := 0$ for $k \le 0$, then for all $i, j \geqslant 1$,

$$(C_\nu(\lambda) - \lambda I)^i e_j = e_{j-i},$$

and thus

(6.2.3)
$$(C_\nu(\lambda) - \lambda I)^\nu = 0, \qquad (C_\nu(\lambda) - \lambda I)^{\nu - 1} \neq 0.$$

The significance of the matrices $C_\nu(\lambda)$ lies in the fact that they are used to build the so-called *Jordan normal form* J of a matrix. Indeed, the following fundamental theorem holds, which we state without proof:

(6.2.4) Theorem. *Let A be an arbitrary $n \times n$ matrix and $\lambda_1, \ldots, \lambda_k$ its distinct eigenvalues, with geometric and algebraic multiplicities $\rho(\lambda_i)$ and $\sigma(\lambda_i)$, respectively, $i = 1, \ldots, k$. Then for each of the eigenvalues λ_i, $i = 1, \ldots, k$, there exist $\rho(\lambda_i)$ natural numbers $\nu_j^{(i)}$, $j = 1, 2, \ldots, \rho(\lambda_i)$, with*

$$\sigma(\lambda_i) = \nu_1^{(i)} + \nu_2^{(i)} + \cdots + \nu_{\rho(\lambda_i)}^{(i)},$$

and there exists a nonsingular $n \times n$ matrix T, such that $J := T^{-1}AT$ has the following form:

(6.2.5)
$$J = \begin{bmatrix} C_{\nu_1^{(1)}}(\lambda_1) & & & & & 0 \\ & \ddots & & & & \\ & & C_{\nu_{\rho(\lambda[1])}^{(1)}}(\lambda_1) & & & \\ & & & \ddots & C_{\nu_1^{(k)}}(\lambda_k) & \\ & & & & & \ddots \\ 0 & & & & & C_{\nu_{\rho(\lambda[k])}^{(k)}}(\lambda_k) \end{bmatrix},$$

where, for typographical convenience, $\lambda[i] := \lambda_i$. The numbers $\nu_j^{(i)}$, $j = 1, \ldots, \rho(\lambda_i)$ (and with them, the matrix J) are uniquely determined up to order. J is called the Jordan normal form of A.

The matrix T, in general, is not uniquely determined.

If one partitions the matrix T columnwise, in accordance with the Jordan normal form J in (6.2.5),

$$T = [T_1^{(1)}, \ldots, T_{\rho(\lambda_1)}^{(1)}, \ldots, T_1^{(k)}, \ldots, T_{\rho(\lambda_k)}^{(k)}],$$

then from $T^{-1}AT = J$, and hence $AT = TJ$, there follow immediately the relations

$$(6.2.6) \quad AT_j^{(i)} = T_j^{(i)} C_{\nu_j^{(i)}}(\lambda_i), \qquad i = 1, 2, \ldots, k, \quad j = 1, 2, \ldots, \rho(\lambda_i).$$

Denoting the columns of the $n \times \nu_j^{(i)}$ matrix $T_j^{(i)}$ without further indices briefly by t_m, $m = 1, 2, \ldots, \nu_j^{(i)}$,

$$T_j^{(i)} = [t_1, t_2, \ldots, t_{\nu_j^{(i)}}],$$

it immediately follows from (6.2.6) and the definition of $C_{\nu_j^{(i)}}(\lambda_i)$ that

$$(A - \lambda_i I)[t_1, \ldots, t_{\nu_j^{(i)}}] = [t_1, \ldots, t_{\nu_j^{(i)}}] \begin{bmatrix} 0 & 1 & & 0 \\ & \cdot & \cdot & \\ & & \cdot & \cdot \\ & & & 1 \\ & & & \cdot \\ 0 & & & 0 \end{bmatrix},$$

or

$$(6.2.7) \quad \begin{aligned} (A - \lambda_i I)t_m &= t_{m-1}, \qquad m = \nu_j^{(i)}, \nu_j^{(i)} - 1, \ldots, 2, \\ (A - \lambda_i I)t_1 &= 0. \end{aligned}$$

In particular, t_1, the first column of $T_j^{(i)}$, is an eigenvector for the eigenvalue λ_i. The remaining t_m, $m = 2, 3, \ldots, \nu_j^{(i)}$, are called *principal vectors* corresponding to λ_i, and one sees that with each Jordan block $C_{\nu_j^{(i)}}(\lambda_i)$ there is associated an eigenvector and a set of principal vectors. Altogether, for an $n \times n$ matrix A, one can thus find a basis of $\mathbb{C}^n$ (namely, the columns of T) which consists entirely of eigenvectors and principal vectors of A.

The characteristic polynomials

$$(\lambda_i - \mu)^{\nu_j^{(i)}} = \det(C_{\nu_j^{(i)}}(\lambda_i) - \mu I)$$

of the individual Jordan blocks $C_{\nu_j^{(i)}}(\lambda_i)$ are called the

(6.2.8) *elementary divisors*

of A. Therefore, A has only linear elementary divisors precisely if $\nu_j^{(i)} = 1$ for all i and j, i.e., if the Jordan normal form is a diagonal matrix. One then calls A *diagonalizable* or also *normalizable*. This case is distinguished by the existence of a basis of $\mathbb{C}^n$ consisting solely of eigenvectors of A; principal vectors do not occur. Otherwise, one says that A has "higher," i.e., nonlinear elementary divisors.

From Theorem (6.2.4) there follows immediately:

(6.2.9) Theorem. *Every $n \times n$ matrix A with n distinct eigenvalues is diagonalizable.*

We will get to know further classes of diagonalizable matrices in Section 6.4.

Another extreme case occurs if with each of the distinct eigenvalues λ_i, $i = 1, \ldots, k$, of A there is associated only one Jordan block in the Jordan normal form J of (6.2.5). This is the case precisely if

$$\rho(\lambda_i) = 1 \quad \text{for } 1 = 1, 2, \ldots, k.$$

The matrix A is then called

(6.2.10) *nonderogatory*,

otherwise, *derogatory* (an $n \times n$ matrix with n distinct eigenvalues is thus both diagonalizable and nonderogatory). The class of nonderogatory matrices will be studied more fully in the next section.

A further important concept is that of the *minimal polynomial* of a matrix A. By this we mean the polynomial

$$\psi(\mu) = \gamma_0 + \gamma_1 \mu + \cdots + \gamma_{m-1} \mu^{m-1} + \mu^m$$

of smallest degree having the property

$$\psi(A) = 0.$$

The minimal polynomial can be read off at once from the Jordan normal form:

(6.2.11) Theorem. *Let A be an $n \times n$ matrix with the (distinct) eigenvalues $\lambda_1, \ldots, \lambda_k$ and with the Jordan normal form J of (6.2.5), and let $\tau_i := \max_{1 \leqslant j \leqslant \rho(\lambda_i)} \nu_j^{(i)}$. Then*

$$\psi(\mu) := (\mu - \lambda_1)^{\tau_1} (\mu - \lambda_2)^{\tau_2} \ldots (\mu - \lambda_k)^{\tau_k} \tag{6.2.12}$$

is the minimal polynomial of A. $\psi(\mu)$ divides every polynomial $\chi(\mu)$ with $\chi(A) = 0$.

Proof. We first show that all zeros of the minimal polynomial ψ of A, if it exists, are eigenvalues of A. Let, say, λ be a zero of ψ. Then

$$\psi(\mu) = (\mu - \lambda) g(\mu),$$

where the polynomial $g(\mu)$ has smaller degree than ψ, and hence by the definition of the minimal polynomial, $g(A) \neq 0$. There exists, therefore, a vector $z \neq 0$ with $x := g(A)z \neq 0$. Because of $\psi(A) = 0$ it then follows that

$$0 = \psi(A)z = (A - \lambda I) g(A)z = (A - \lambda I)x,$$

i.e., λ is an eigenvalue of A. If a minimal polynomial exists, it will thus have the form $\psi(\mu) = (\mu - \lambda_1)^{\tau_1}(\mu - \lambda_2)^{\tau_2} \ldots (\mu - \lambda_k)^{\tau_k}$ for certain τ_i. We wish to show now that $\tau_i := \max v_j^{(i)}$ will define a polynomial with $\psi(A) = 0$. With the notation of Theorem (6.2.4), indeed, $A = TJT^{-1}$ and thus $\psi(A) = T\psi(J)T^{-1}$. In view of the diagonal structure of J,

$$J = \operatorname{diag}(C_{v_1^{(1)}}(\lambda_1), \ldots, C_{v_{\rho(\lambda[k])}^{(k)}}(\lambda_k));$$

however, we now have

$$\psi(J) = \operatorname{diag}(\psi(C_{v_1^{(1)}}(\lambda_1)), \ldots, \psi(C_{v_{\rho(\lambda[k])}^{(k)}}(\lambda_k))).$$

Since $\psi(\mu) = (\mu - \lambda_i)^{\tau_i} g(\mu)$, there follows

$$\psi(C_{v_j^{(i)}}(\lambda_i)) = (C_{v_j^{(i)}}(\lambda_i) - \lambda_i I)^{\tau_i} g(C_{v_j^{(i)}}(\lambda_i)), \tag{6.2.13}$$

and thus, by virtue of $\tau_i \geqslant v_j^{(i)}$ and (6.2.3),

$$\psi(C_{v_j^{(i)}}(\lambda_i)) = 0.$$

Thus, $\psi(J) = 0$, and therefore also $\psi(A) = 0$.

At the same time one sees that none of the integers τ_i can be chosen smaller than $\max v_j^{(i)}$: If there were, say, $\tau_i < v_j^{(i)}$, then, by (6.2.3),

$$(C_{v_j^{(i)}}(\lambda_i) - \lambda_i I)^{\tau_i} \neq 0.$$

From $g(\lambda_i) \neq 0$ it would follow at once that

$$B := g(C_{v_j^{(i)}}(\lambda_i))$$

is nonsingular. Hence, by (6.2.13), also $\psi(C_{v_j^{(i)}}(\lambda_i)) \neq 0$, and neither $\psi(J)$ nor $\psi(A)$ would vanish. This shows that the specified polynomial is the minimal polynomial of A.

If, finally, $\chi(\mu)$ is a polynomial with $\chi(A) = 0$, then χ, with the aid of the minimal polynomial, can be written in the form

$$\chi(\mu) = g(\mu)\psi(\mu) + r(\mu),$$

where $\deg r < \deg \psi$. From $\chi(A) = \psi(A) = 0$ we thus get also $r(A) = 0$. Since ψ is the minimal polynomial of A, we must have identically $r(\mu) \equiv 0 : \psi$ is a divisor of χ. □

By (6.2.4), one has

$$\sigma(\lambda_i) = \sum_{j=1}^{\rho(\lambda_i)} v_j^{(i)} \geqslant \tau_i = \max_j v_j^{(i)},$$

i.e., the characteristic polynomial $\varphi(\mu) = \det(A - \mu I)$ of A is a multiple of the minimal polynomial. Equality $\sigma(\lambda_i) = \tau_i$, $i = 1, \ldots, k$, prevails precisely when A is nonderogatory. Thus,

(6.2.14) Corollary (Cayley–Hamilton). *The characteristic polynomial $\varphi(\mu)$ of a matrix A satisfies $\varphi(A) = 0$.*

(6.2.15) Corollary. *A matrix A is nonderogatory if and only if its minimal polynomial and characteristic polynomial coincide (up to a multiplicative constant).*

EXAMPLE. The Jordan matrix

$$J = \begin{bmatrix} 1 & 1 & & & & & & & \\ & 1 & 1 & & & & 0 & & \\ & & 1 & & & & & & \\ & & & 1 & 1 & & & & \\ & & & & 1 & & & & \\ & & & & & -1 & 1 & & \\ & & & & & & -1 & & \\ & 0 & & & & & & -1 & \\ & & & & & & & & -1 \end{bmatrix}$$

has the eigenvalues $\lambda_1 = 1$, $\lambda_2 = -1$ with multiplicities

$$\rho(\lambda_1) = 2, \qquad \rho(\lambda_2) = 3,$$

$$\sigma(\lambda_1) = 5, \qquad \sigma(\lambda_2) = 4.$$

Elementary divisors:

$$(1-\mu)^3,\ (1-\mu)^2,\ (-1-\mu)^2,\ (-1-\mu),\ (-1-\mu).$$

Characteristic polynomial:

$$\varphi(\mu) = (-1)^9(\mu - 1)^5(\mu + 1)^4.$$

Minimal polynomial:

$$\psi(\mu) = (\mu - 1)^3(\mu + 1)^2.$$

To $\lambda_1 = 1$ there correspond the linearly independent (right) eigenvectors e_1, e_4; to $\lambda_2 = -1$ the eigenvectors e_6, e_8, e_9.

6.3 The Frobenius Normal Form of a Matrix

In the previous section we studied the matrices $C_\nu(\lambda)$, which turned out to be the building blocks of the Jordan normal form of a matrix. Analogously, one builds up the *Frobenius normal form* (also called the *rational normal form*) of a matrix from *Frobenius matrices F* of the form

(6.3.1)
$$F = \begin{bmatrix} 0 & \cdots & \cdots & 0 & -\gamma_0 \\ 1 & \ddots & & 0 & -\gamma_1 \\ & \ddots & \ddots & \vdots & \vdots \\ & & \ddots & 0 & -\gamma_{m-2} \\ 0 & & & 1 & -\gamma_{m-1} \end{bmatrix},$$

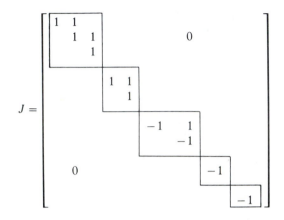

whose properties we now wish to discuss. One encounters matrices of this type in the study of *Krylov sequences* of vectors. By a Krylov sequence of vectors for the $n \times n$ matrix A and the initial vector $t_0 \in \mathbb{C}^n$ one means a sequence of vectors t_i, $i = 0, 1, \ldots, m-1$, with the following properties:

(6.3.2).

(a) $t_i = At_{i-1}$, $i \geqslant 1$,
(b) $t_0, t_1, \ldots, t_{m-1}$ are linearly independent,
(c) $t_m = At_{m-1}$ depends linearly on $t_0, t_1, \ldots, t_{m-1}$: there are constants γ_i with $t_m + \gamma_{m-1} t_{m-1} + \cdots + \gamma_0 t_0 = 0$.

The length m of the Krylov sequence of course depends on t_0. Clearly, $m \leqslant n$, since more than n vectors in $\mathbb{C}^n$ are always linearly dependent. If one forms the $n \times m$ matrix $T := [t_0, \ldots, t_{m-1}]$ and the matrix F of (6.3.1), then (6.3.2) is equivalent to

$$\operatorname{rank} T = m,$$

$$AT = A[t_0, \ldots, t_{m-1}] = [t_1, \ldots, t_m] \tag{6.3.3}$$
$$= [t_0, \ldots, t_{m-1}]F = TF.$$

Every eigenvalue of F is also eigenvalue of A: From $Fz = \lambda z$, $z \neq 0$, we indeed obtain for $x := Tz$, in view of (6.3.3),

$$x \neq 0 \quad \text{and} \quad Ax = ATz = TFz = \lambda Tz = \lambda x.$$

Moreover, we have:

(6.3.4) Theorem. *The matrix F of (6.3.1) is nonderogatory: The minimal polynomial of F is*

$$\psi(\mu) = \gamma_0 + \gamma_1 \mu + \cdots + \gamma_{m-1}\mu^{m-1} + \mu^m$$
$$= (-1)^m \det(F - \mu I).$$

Proof. Expanding $\varphi(\mu) := \det(F - \mu I)$ by the last column, the characteristic polynomial of F is found to be

$$\varphi(\mu) = \det \begin{bmatrix} -\mu & & & 0 & -\gamma_0 \\ 1 & -\mu & & & -\gamma_1 \\ & \ddots & \ddots & & \vdots \\ & & 1 & -\mu & -\gamma_{m-2} \\ 0 & & & 1 & -\gamma_{m-1} - \mu \end{bmatrix}$$

$$= (-1)^m(\gamma_0 + \gamma_1 \mu + \cdots + \gamma_{m-1}\mu^{m-1} + \mu^m).$$

By the results (6.2.12), (6.2.14) of the preceding section, the minimum polynomial $\psi(\mu)$ of T divides $\varphi(\mu)$. If we had $\deg \psi < m = \deg \varphi$, say

$$\psi(\mu) = \alpha_0 + \alpha_1 \mu + \cdots + \alpha_{r-1}\mu^{r-1} + \mu^r, \qquad r < m,$$

then from $\psi(F) = 0$ and $Fe_i = e_{i+1}$ for $1 \leqslant i \leqslant m - 1$, the following contradiction would result at once:

$$0 = \psi(F)e_1 = \alpha_0 e_1 + \alpha_1 e_2 + \cdots + \alpha_{r-1} e_r + e_{r+1}$$
$$= [\alpha_0, \alpha_1, \ldots, \alpha_{r-1}, 1, 0, \ldots, 0]^T \neq 0.$$

Thus, $\deg \psi = m$, and hence $\psi(\mu) = (-1)^m \varphi(\mu)$. Because of (6.2.15), the theorem is proved. □

Assuming the characteristic polynomial of F to have the zeros λ_i with multiplicities σ_i, $i = 1, \ldots, k$,

$$\psi(\mu) = \gamma_0 + \cdots + \gamma_{m-1}\mu^{m-1} + \mu^m = (\mu - \lambda_1)^{\sigma_1}(\mu - \lambda_2)^{\sigma_2} \cdots (\mu - \lambda_k)^{\sigma_k},$$

the Jordan normal form of F in (6.3.1), in view of (6.3.4), is given by

$$\begin{bmatrix} C_{\sigma_1}(\lambda_1) & & & 0 \\ & C_{\sigma_2}(\lambda_2) & & \\ & & \ddots & \\ 0 & & & C_{\sigma_k}(\lambda_k) \end{bmatrix}.$$

The significance of the Frobenius matrices lies in the fact that they furnish the building blocks of the so-called *Frobenius* or *rational normal form* of a matrix. Namely:

(6.3.5) Theorem. *For every $n \times n$ matrix A there is a nonsingular $n \times n$-matrix T with*

$$T^{-1}AT = \begin{bmatrix} F_1 & & & 0 \\ & F_2 & & \\ & & \ddots & \\ 0 & & & F_r \end{bmatrix}, \tag{6.3.6}$$

where the F_i are Frobenius matrices having the following properties:

(a) *If $\varphi_i(\mu) = \det(F_i - \mu I)$ is the characteristic polynomial of F_i, $i = 1, \ldots, r$, then $\varphi_i(\mu)$ is a divisor of $\varphi_{i-1}(\mu)$, $i = 2, 3, \ldots, r$.*
(b) *$\varphi_1(\mu)$, up to the multiplicative constant ± 1, is the minimal polynomial of A.*
(c) *The matrices F_i are uniquely determined by A.*

One calls (6.3.6) the *Frobenius normal form* of A.

Proof. This is easily done with the help of the Jordan normal form [see (6.2.4)]: We assume that J in (6.2.5) is the Jordan normal form of A. Without loss of generality, let the integers $v_i^{(j)}$ be ordered,

$$v_1^{(i)} \geqslant v_2^{(i)} \geqslant \cdots \geqslant v_{\rho(\lambda_i)}^{(i)}, \qquad i = 1, 2, \ldots, k. \tag{6.3.7}$$

Define the polynomials $\varphi_j(\mu)$, $j = 1, \ldots, r$, $r = \max_{1 \leqslant i \leqslant k} \rho(\lambda_i)$, by

$$\varphi_j(\mu) = (\lambda_1 - \mu)^{\nu_j^{(1)}}(\lambda_2 - \mu)^{\nu_j^{(2)}} \ldots (\lambda_k - \mu)^{\nu_j^{(k)}}$$

[using the convention $\nu_j^{(i)} := 0$ if $j > \rho(\lambda_i)$]. In view of (6.3.7), $\varphi_j(\mu)$ divides $\varphi_{j-1}(\mu)$ and $\pm\varphi_1(\mu)$ is the minimal polynomial of A. Now take as the Frobenius matrix F_j just the Frobenius matrix whose characteristic polynomial is $\varphi_j(\mu)$. Let S_i be the matrix that transforms F_i into its Jordan normal form J_i,

$$S_i^{-1} F_i S_i = J_i.$$

A Jordan normal form of A (the Jordan normal form is unique only up to permutations of the Jordan blocks) then is

$$J' = \begin{bmatrix} J_1 & & & \\ & J_2 & & \\ & & \ddots & \\ & & & J_r \end{bmatrix}$$

$$= \begin{bmatrix} S_1 & & & \\ & S_2 & & \\ & & \ddots & \\ & & & S_r \end{bmatrix}^{-1} \begin{bmatrix} F_1 & & & \\ & F_2 & & \\ & & \ddots & \\ & & & F_r \end{bmatrix} \begin{bmatrix} S_1 & & & \\ & S_2 & & \\ & & \ddots & \\ & & & S_r \end{bmatrix}.$$

According to Theorem (6.2.4) there is a matrix U with $U^{-1}AU = J'$. The matrix $T := US^{-1}$ with

$$S = \begin{bmatrix} S_1 & & & \\ & S_2 & & \\ & & \ddots & \\ & & & S_r \end{bmatrix}$$

transforms A into the desired form (6.3.6).

It is easy to convince oneself of the uniqueness of the F_i. □

EXAMPLE. For the matrix J in the example of Section 6.2 one has

$$\varphi_1(\mu) = (1 - \mu)^3(-1 - \mu)^2 = -(\mu^5 - \mu^4 - 2\mu^3 + 2\mu^2 + \mu - 1),$$
$$\varphi_2(\mu) = (1 - \mu)^2(-1 - \mu) = -(\mu^3 - \mu^2 - \mu + 1),$$
$$\varphi_3(\mu) = -(\mu + 1),$$

and there follows

$$F_1 = \begin{bmatrix} 0 & 0 & 0 & 0 & 1 \\ 1 & 0 & 0 & 0 & -1 \\ 0 & 1 & 0 & 0 & -2 \\ 0 & 0 & 1 & 0 & 2 \\ 0 & 0 & 0 & 1 & 1 \end{bmatrix}, \quad F_2 = \begin{bmatrix} 0 & 0 & -1 \\ .1 & 0 & 1 \\ 0 & 1 & 1 \end{bmatrix}, \quad F_3 = [-1].$$

The significance of the Frobenius normal form lies in its theoretical properties (Theorem (6.3.5)). Its practical importance for computing eigenvalues is very limited. For example, if the $n \times n$ matrix A is nonderogatory, then computing the Frobenius normal form F (6.3.1) is equivalent to the computation of the coefficients γ_i of the characteristic polynomial

$$\varphi(\mu) \equiv \det(A - \mu I) = (-1)^n(\mu^n + \gamma_{n-1}\mu^{n-1} + \cdots + \gamma_0),$$

which has the desired eigenvalues of A as zeros. But it is not advisable to first determine the γ_i in order to subsequently compute the eigenvalues as zeros of φ: In general, the zeros λ_j of φ react much more sensitively to small changes in the coefficients γ_i of φ than to small changes in the elements of the original matrix A (see Sections 5.8 and 6.9).

EXAMPLE. In Section 6.9, Theorem (6.9.7), it will be shown that the eigenvalue problem for Hermitian matrices $A = A^H$ is well conditioned in the following sense: For each eigenvalue $\lambda_i(A + \Delta A)$ of $A + \Delta A$ there is an eigenvalue $\lambda_j(A)$ of A such that

$$|\lambda_i(A + \Delta A) - \lambda_j(A)| \le \mathrm{lub}_2(\Delta A).$$

If the 20×20 matrix A has, say, the eigenvalues $\lambda_j = j$, $j = 1, 2, \dots, 20$, then $\mathrm{lub}_2(A) = 20$ because $A = A^H$ (see Exercise 8). Subjecting all elements of A to a relative error of at most eps, i.e., replacing A by $A + \Delta A$ with $|\Delta A| \le \mathrm{eps}|A|$, it follows that (see Exercise 11)

$$\begin{aligned} \mathrm{lub}_2(\Delta A) &\le \mathrm{lub}_2(|\Delta A|) \le \mathrm{lub}_2(\mathrm{eps}|A|) \\ &\le \mathrm{eps}\sqrt{20}\,\mathrm{lub}_2(A) < 90 \text{ eps}. \end{aligned}$$

On the other hand, a mere relative error $|\Delta\gamma_i| = \mathrm{eps}$ of the coefficients γ_i of the characteristic polynomial $\varphi(\mu) = (\mu - 1)(\mu - 2) \ldots (\mu - 20)$ produces tremendous changes in the zeros $\lambda_j = j$ of φ (see Example (1) of Section 5.8).

The situation is particularly bad for Hermitian matrices with clustered eigenvalues: the eigenvalue problem for these matrices is still well conditioned even for multiple eigenvalues, whereas multiple zeros of a polynomial φ are always ill-conditioned functions of the coefficients γ_i.

In addition, many methods for computing the coefficients γ_i of the characteristic polynomial are numerically unstable. As an example, we mention the method of Frazer, Duncan, and Collar, which is based on the observation (see (6.3.2) (c)) that the vector $c = [\gamma_0, \gamma_1, \dots, \gamma_{n-1}]^T$ is the solution of a linear system of equations $Tc = -t_n$ with the nonsingular matrix $T := [t_0, t_1, \dots, t_{n-1}]$, provided that $t_0, \dots, t_{n-1}$ is a Krylov sequence of length n. Unfortunately, the matrix T is in general ill-conditioned, $\mathrm{cond}(T) \gg 1$, so that the computed solution c can be highly incorrect (see Section 4.4). Indeed, as $k \to \infty$, the vectors $t_k = A^k t_0$ will, when scaled by suitable factors σ_k, converge toward a nonzero vector $t = \lim_k \sigma_k t_k$ that, in general, does not depend on the choice of the initial vector t_0. The columns of T tend, therefore, to become "more and more linear dependent" (see Section 6.6.3 and Exercise 12).

6.4 The Schur Normal Form of a Matrix; Hermitian and Normal Matrices; Singular Values of Matrices

If one does not admit arbitrary nonsingular matrices T in the similarity transformation $T^{-1}AT$, it is in general no longer possible to transform A to Jordan normal form. However, for unitary matrices T, i.e., matrices T with $T^H T = I$, one has the following result of Schur:

(6.4.1) Theorem. *For every $n \times n$ matrix A there is a unitary $n \times n$ matrix U with*

$$U^H A U = \begin{bmatrix} \lambda_1 & * & \cdots & * \\ & \lambda_2 & \ddots & \vdots \\ & & \ddots & * \\ 0 & & & \lambda_n \end{bmatrix}.$$

Here λ_i, $i = 1, \ldots, n$, are the (not necessarily distinct) eigenvalues of A.

PROOF. We use complete induction on n. For $n = 1$ the theorem is trivial. Suppose the theorem is true for matrices of order $n - 1$, and let A be an $n \times n$ matrix. Let λ_1 be any eigenvalue of A, and $x_1 \neq 0$ a corresponding eigenvector with $\|x_1\|_2^2 = x_1^H x_1 = 1$, $Ax_1 = \lambda_1 x_1$. Then one can find $n - 1$ additional vectors $x_2, \ldots, x_n$ such that $x_1, x_2, \ldots, x_n$ forms an orthonormal basis of $\mathbb{C}^n$, and the $n \times n$ matrix $X := [x_1, \ldots, x_n]$ with columns x_i is thus unitary, $X^H X = I$. Since

$$X^H A X e_1 = X^H A x_1 = \lambda_1 X^H x_1 = \lambda_1 e_1,$$

the matrix $X^H A X$ has the form

$$X^H A X = \left[\begin{array}{c|c} \lambda_1 & a \\ \hline 0 & A_1 \end{array}\right],$$

where A_1 is a matrix of order $n - 1$ and $a^H \in \mathbb{C}^{n-1}$. By the induction hypothesis, there exists a unitary $(n - 1) \times (n - 1)$ matrix U_1 such that

$$U_1^H A_1 U_1 = \begin{bmatrix} \lambda_2 & * & \cdot & \cdot & * \\ & \cdot & \cdot & & \cdot \\ & & \cdot & \cdot & \cdot \\ & & & \cdot & * \\ 0 & & & & \lambda_n \end{bmatrix}.$$

The matrix

$$U := X \cdot \left[\begin{array}{c|c} 1 & 0 \\ \hline 0 & U_1 \end{array}\right]$$

then is a unitary $n \times n$ matrix satisfying

$$
\begin{aligned}
U^H A U &= \begin{bmatrix} 1 & 0 \\ 0 & U_1^H \end{bmatrix} X^H A X \begin{bmatrix} 1 & 0 \\ 0 & U_1 \end{bmatrix} \\
&= \begin{bmatrix} 1 & 0 \\ 0 & U_1^H \end{bmatrix} \begin{bmatrix} \lambda_1 & a \\ 0 & A_1 \end{bmatrix} \begin{bmatrix} 1 & 0 \\ 0 & U_1 \end{bmatrix} \\
&= \begin{bmatrix} \lambda_1 & * & \cdot & \cdot & \cdot & * \\ & \cdot & \cdot & & & \cdot \\ & & \cdot & \cdot & & \cdot \\ & & & \cdot & \cdot & \cdot \\ & & & & \cdot & * \\ 0 & & & & & \lambda_n \end{bmatrix}.
\end{aligned}
$$

The fact that λ_i, $i = 1, \ldots, n$, are zeros of $\det(U^H A U - \mu I)$, and hence eigenvalues of A, is trivial. □

Now, if $A = A^H$ is a Hermitian matrix, then

$$(U^H A U)^H = U^H A^H U^{HH} = U^H A U$$

is again a Hermitian matrix. Thus, from (6.4.1), there follows immediately

(6.4.2) Theorem. *For every Hermitian $n \times n$ matrix $A = A^H$ there is a unitary matrix $U = [x_1 \ldots, x_n]$ with*

$$U^{-1} A U = U^H A U = \begin{bmatrix} \lambda_1 & & 0 \\ & \ddots & \\ 0 & & \lambda_n \end{bmatrix}.$$

The eigenvalues λ_i, $i = 1, \ldots, n$, of A are real. A is diagonalizable. The ith column x_i of U is an eigenvector belonging to the eigenvalue λ_i: $Ax_i = \lambda_i x_i$. A thus has n linearly independent pairwise orthogonal eigenvectors.

If the eigenvalues λ_i of a Hermitian $n \times n$ matrix $A = A^H$ are arranged in decreasing order,

$$\lambda_1 \geqslant \lambda_2 \geqslant \cdots \geqslant \lambda_n,$$

then λ_1 and λ_n can be characterized also in the following manner [see (6.9.14) for a generalization]:

$$\text{(6.4.3)} \qquad \lambda_1 = \max_{0 \neq x \in \mathbb{C}^n} \frac{x^H A x}{x^H x}, \qquad \lambda_n = \min_{0 \neq x \in \mathbb{C}^n} \frac{x^H A x}{x^H x}.$$

PROOF. If $U^H A U = \Lambda = \operatorname{diag}[\lambda_1, \ldots, \lambda_n]$, U unitary, then for all $x \neq 0$,

$$\frac{x^H A x}{x^H x} = \frac{(x^H U) U^H A U (U^H x)}{(x^H U)(U^H x)} = \frac{y^H \Lambda y}{y^H y} = \frac{\sum_i \lambda_i |\eta_i|^2}{\sum_i |\eta_i|^2} \leqslant \frac{\sum_i \lambda_1 |\eta_i|^2}{\sum_i |\eta_i|^2} = \lambda_1,$$

where $y := U^H x = [\eta_1, \ldots, \eta_n]^T \neq 0$. Taking for $x \neq 0$ in particular an eigenvector belonging to λ_1, $Ax = \lambda_1 x$, one gets $x^H Ax/x^H x = \lambda_1$, so that $\lambda_1 = \max_{0 \neq x \in \mathbb{C}^n} x^H Ax/x^H x$. The other assertion in (6.4.3) follows from what was just proved by replacing A with $-A$. □

From (6.4.3) and the definition (4.3.1) of a positive definite (positive semidefinite) matrix A, one obtains immediately

(6.4.4). *A Hermitian matrix A is positive definite (positive semidefinite) if and only if all eigenvalues of A are positive (nonnegative).*

A generalization of the notion of a Hermitian matrix is that of a normal matrix: An $n \times n$ matrix A is called *normal* if

$$A^H A = AA^H,$$

i.e., A commutes with A^H. For example, all Hermitian, diagonal, skew Hermitian, and unitary matrices are normal.

(6.4.5) Theorem. *An $n \times n$ matrix A is normal if and only if there exists a unitary matrix U such that*

$$U^{-1}AU = U^H AU = \begin{bmatrix} \lambda_1 & & 0 \\ & \ddots & \\ 0 & & \lambda_n \end{bmatrix}.$$

Normal matrices are diagonalizable and have n linearly independent pairwise orthogonal eigenvectors x_i $(i = 1, \ldots, n)$, $Ax_i = \lambda_i x_i$, namely the columns of the matrix $U = [x_1, \ldots, x_n]$.

Proof. By Schur's theorem (6.4.1), there exists a unitary matrix U with

$$U^H AU = \begin{bmatrix} \lambda_1 & * & \cdot & \cdot & \cdot & * \\ & \cdot & \cdot & & & \cdot \\ & & \cdot & \cdot & & \cdot \\ & & & \cdot & \cdot & \cdot \\ & & & & \cdot & * \\ 0 & & & & & \lambda_n \end{bmatrix} =: R = [r_{ik}].$$

From $A^H A = AA^H$ there now follows

$$\begin{aligned} R^H R &= U^H A^H U U^H AU = U^H A^H AU \\ &= U^H AA^H U = U^H AUU^H A^H U \\ &= RR^H. \end{aligned}$$

For the (1, 1) element of $R^H R = RR^H$ we thus obtain

$$\bar{\lambda}_1 \lambda_1 = |\lambda_1|^2 = |\lambda_1|^2 + \sum_{k=2}^{n} |r_{1k}|^2;$$

hence $r_{1k} = 0$ for $k = 2, \ldots, n$. In the same manner one shows that all nondiagonal elements of R vanish.

Conversely, if A is unitarily diagonalizable, $U^H AU = \text{diag}(\lambda_1, \ldots, \lambda_n) =: D$, $U^H U = I$, there follows at once

$$A^H A = UD^H U^H UDU^H = U|D|^2 U^H = UDU^H UD^H U^H = AA^H. \qquad \square$$

Given an arbitrary $m \times n$ matrix A, the $n \times n$ matrix $A^H A$ is positive semidefinite, since $x^H(A^H A)x = \|Ax\|_2^2 \geqslant 0$ for any $x \in \mathbb{C}^n$. Its eigenvalues $\lambda_1 \geqslant \lambda_2 \geqslant \cdots \geqslant \lambda_n \geqslant 0$ are nonnegative by (6.4.4) and can therefore be written in the form $\lambda_k = \sigma_k^2$ with $\sigma_k \geqslant 0$. The numbers $\sigma_1 \geqslant \cdots \geqslant \sigma_n \geqslant 0$ are called

(6.4.6) *singular values of A.*

Replacing the matrix A in (6.4.3) by $A^H A$, one obtains immediately

$$\text{(6.4.7)} \qquad \sigma_1 = \max_{0 \neq x \in \mathbb{C}^n} \frac{\|Ax\|_2}{\|x\|_2} = \text{lub}_2(A), \qquad \sigma_n = \min_{0 \neq x \in \mathbb{C}^n} \frac{\|Ax\|_2}{\|x\|_2}.$$

In particular, if $m = n$ and A is nonsingular, one has

$$\text{(6.4.8)} \qquad \begin{aligned} \frac{1}{\sigma_n} &= \max_{x \neq 0} \frac{\|x\|_2}{\|Ax\|_2} = \max_{y \neq 0} \frac{\|A^{-1}y\|_2}{\|y\|_2} = \text{lub}_2(A^{-1}), \\ \text{cond}_2(A) &= \text{lub}_2(A)\, \text{lub}_2(A^{-1}) = \sigma_1/\sigma_n. \end{aligned}$$

The smallest singular value σ_n of a square matrix A gives the distance of A to the "nearest" singular matrix:

(6.4.9) Theorem. *Let A and E be arbitrary $n \times n$ matrices and let A have the singular values $\sigma_1 \geqslant \sigma_2 \geqslant \cdots \geqslant \sigma_n \geqslant 0$. Then*

(a) $\text{lub}_2(E) \geqslant \sigma_n$ *if $A + E$ is singular.*
(b) *There is a matrix E with* $\text{lub}_2(E) = \sigma_n$ *such that $A + E$ is singular.*

PROOF. (a): Let $A + E$ be singular, thus $(A + E)x = 0$ for some $x \neq 0$. Then (6.4.7) gives

$$\sigma_n \|x\|_2 \leqslant \|Ax\|_2 = \|-Ex\|_2 \leqslant \text{lub}_2(E)\|x\|_2;$$

hence $\sigma_n \leqslant \text{lub}_2(E)$.

(b): If $\sigma_n = 0$, there is nothing to prove. For, by (6.4.7), one has $0 = \|Ax\|_2$ for some $x \neq 0$, so that A already is singular. Let, therefore, $\sigma_n > 0$. Because of (6.4.7), there exist vectors u, v such that

$$\|Au\|_2 = \sigma_n, \qquad \|u\|_2 = 1,$$

$$v := \frac{1}{\sigma_n} Au, \qquad \|v\|_2 = 1.$$

For the special $n \times n$ matrix $E := -\sigma_n v u^H$, one then has $(A + E)u = 0$, so that $A + E$ is singular, and moreover

$$\operatorname{lub}_2(E) = \sigma_n \max_{x \neq 0} \|v\|_2 \frac{|u^H x|}{\|x\|_2} = \sigma_n. \qquad \square$$

An arbitrary $m \times n$ matrix A can be transformed unitarily to a certain normal form in which the singular values of A appear:

(6.4.10) Theorem. *Let A be an arbitrary (complex) $m \times n$ matrix. Then:*

(a) *There exist a unitary $m \times m$ matrix U and a unitary $n \times n$ matrix V such that $U^H A V = \Sigma$ is an $m \times n$ "diagonal matrix" of the following form:*

$$\Sigma = \begin{bmatrix} D & 0 \\ 0 & 0 \end{bmatrix}, \qquad D := \operatorname{diag}(\sigma_1, \ldots, \sigma_r), \quad \sigma_1 \geqslant \sigma_2 \geqslant \cdots \geqslant \sigma_r > 0.$$

Here $\sigma_1, \ldots, \sigma_r$ are the nonvanishing singular values of A, and r is the rank of A.

(b) *The nonvanishing singular values of A^H are also precisely the numbers $\sigma_1, \ldots, \sigma_r$.*

The decomposition $A = U \Sigma V^H$ is called

(6.4.11) the *singular-value decomposition* of A.

PROOF. We show (a) by mathematical induction on m and n. For $m = 0$ or $n = 0$ there is nothing to prove. We assume that the theorem is true for $(m-1) \times (n-1)$ matrices and that A is an $m \times n$ matrix with $m \geqslant 1$, $n \geqslant 1$. Let σ_1 be the largest singular value of A. If $\sigma_1 = 0$, then by (6.4.7) also $A = 0$, and there is nothing to show. Let, therefore, $\sigma_1 > 0$, and let $x_1 \neq 0$ be an eigenvector of $A^H A$ for the eigenvalue σ_1^2, with $\|x_1\|_2 = 1$:

$$A^H A x_1 = \sigma_1^2 x_1. \tag{6.4.12}$$

Then one can find $n - 1$ additional vectors $x_2, \ldots, x_n \in \mathbb{C}^n$ such that the $n \times n$ matrix $X := [x_1, x_2, \ldots, x_n]$ with the columns x_i becomes unitary, $X^H X = I_n$. By virtue of $\|Ax_1\|_2^2 = x_1^H A^H A x_1 = \sigma_1^2 x_1^H x_1 = \sigma_1^2 > 0$, the vector $y_1 := (1/\sigma_1) A x_1 \in \mathbb{C}^m$ with $\|y_1\|_2 = 1$ is well defined, and one can find $m - 1$ additional vectors $y_2, \ldots, y_m \in \mathbb{C}^m$ such that the $m \times m$ matrix $Y = [y_1, y_2, \ldots, y_m]$ is likewise unitary, $Y^H Y = I_m$. Now from (6.4.12) and the definition of y_1, X, and Y, there follows at once, with $e_1 := [1, 0, \ldots, 0]^T \in \mathbb{C}^n$, $\bar{e}_1 := [1, 0, \ldots, 0]^T \in \mathbb{C}^m$, that

$$Y^H A X e_1 = Y^H A x_1 = \sigma_1 Y^H y_1 = \sigma_1 \bar{e}_1 \in \mathbb{C}^m$$

and

$$(Y^H A X)^H \bar{e}_1 = X^H A^H Y \bar{e}_1 = X^H A^H y_1 = \frac{1}{\sigma_1} X^H A^H A x_1$$
$$= \sigma_1 X^H x_1 = \sigma_1 e_1 \in \mathbb{C}^n,$$

so that the matrix $Y^H AX$ has the following form:

$$Y^H AX = \begin{bmatrix} \sigma_1 & 0 \\ 0 & \tilde{A} \end{bmatrix}.$$

Here $\tilde{A}$ is an $(m-1) \times (n-1)$ matrix.

By the induction hypothesis, there exist a unitary $(m-1) \times (m-1)$ matrix $\tilde{U}$ and a unitary $(n-1) \times (n-1)$ matrix $\tilde{V}$ such that

$$\tilde{U}^H \tilde{A} \tilde{V} = \tilde{\Sigma} = \begin{bmatrix} \tilde{D} & 0 \\ 0 & 0 \end{bmatrix}, \qquad \tilde{D} := \operatorname{diag}(\sigma_2, \ldots, \sigma_r), \quad \sigma_2 \geqslant \sigma_3 \geqslant \cdots \geqslant \sigma_r > 0,$$

with $\tilde{\Sigma}$ a $(m-1) \times (n-1)$ "diagonal matrix" of the form indicated. The $m \times m$ matrix

$$U := Y \cdot \begin{bmatrix} 1 & 0 \\ 0 & \tilde{U} \end{bmatrix}$$

is unitary, as is the $n \times n$ matrix

$$V := X \cdot \begin{bmatrix} 1 & 0 \\ 0 & \tilde{V} \end{bmatrix},$$

and one has

$$\begin{aligned} U^H AV &= \begin{bmatrix} 1 & 0 \\ 0 & \tilde{U}^H \end{bmatrix} Y^H AX \begin{bmatrix} 1 & 0 \\ 0 & \tilde{V} \end{bmatrix} = \begin{bmatrix} 1 & 0 \\ 0 & \tilde{U}^H \end{bmatrix} \begin{bmatrix} \sigma_1 & 0 \\ 0 & \tilde{A} \end{bmatrix} \begin{bmatrix} 1 & 0 \\ 0 & \tilde{V} \end{bmatrix} \\ &= \begin{bmatrix} \sigma_1 & 0 \\ 0 & \tilde{\Sigma} \end{bmatrix} = \begin{bmatrix} D & 0 \\ 0 & 0 \end{bmatrix} = \Sigma, \qquad D := \operatorname{diag}(\sigma_1, \ldots, \sigma_r), \end{aligned}$$

Σ being an $m \times n$ diagonal matrix with $\sigma_2 \geqslant \cdots \geqslant \sigma_r > 0$, $\sigma_1^2 = \lambda_{\max}(A^H A)$. Evidently, rank $A = r$, since rank $A =$ rank $U^H AV =$ rank Σ.

We must still prove that $\sigma_1 \geqslant \sigma_2$ and that the σ_i are the singular values of A. Now from $U^H AV = \Sigma$ there follows, for the $n \times n$ diagonal matrix $\Sigma^H \Sigma$,

$$\Sigma^H \Sigma = \operatorname{diag}(\sigma_1^2, \ldots, \sigma_r^2, 0, \ldots, 0) = V^H A^H U U^H AV = V^H (A^H A) V,$$

so that [see Theorem (6.4.2)] $\sigma_1^2, \ldots, \sigma_r^2$ are the nonvanishing eigenvalues of $A^H A$, and hence $\sigma_1, \ldots, \sigma_r$ are the nonvanishing singular values of A. Because of $\sigma_1^2 = \lambda_{\max}(A^H A)$, one also has $\sigma_1 \geqslant \sigma_2$. □

The unitary matrices U, V in the decomposition $U^H AV = \Sigma$ have the following meaning: The columns of U represent m orthonormal eigenvectors of the Hermitian $m \times m$ matrix AA^H, while those of V represent n orthonormal eigenvectors of the Hermitian $n \times n$ matrix $A^H A$. This follows at once from $U^H AA^H U = \Sigma\Sigma^H$, $V^H A^H AV = \Sigma^H \Sigma$, and Theorem (6.4.2). Finally we remark that the pseudoinverse A^+ (Section 4.8.5) of the $m \times n$ matrix A can be immediately obtained from the decomposition $U^H AV = \Sigma$: If

$$\Sigma = \begin{bmatrix} D & 0 \\ 0 & 0 \end{bmatrix}, \qquad D = \operatorname{diag}(\sigma_1, \ldots, \sigma_r), \quad \sigma_1 \geqslant \cdots \geqslant \sigma_r > 0,$$

then the $n \times m$ diagonal matrix

$$\Sigma^+ := \begin{bmatrix} D^{-1} & 0 \\ 0 & 0 \end{bmatrix}$$

is the pseudoinverse of Σ, and one verifies at once that the $n \times m$ matrix

$$A^+ := V\Sigma^+ U^H \tag{6.4.13}$$

satisfies the conditions of (4.8.5.1) for a pseudoinverse of A, so that A^+, in view of the uniqueness statement of Theorem (4.8.5.2), must be the pseudoinverse of A.

6.5 Reduction of Matrices to Simpler Form

The most common methods for determining the eigenvalues and eigenvectors of a dense matrix A proceed as follows. By means of a finite number of similarity transformations

$$A = A_0 \to A_1 \to \cdots \to A_m,$$
$$A_i = T_i^{-1} A_{i-1} T_i, \qquad i = 1, 2, \ldots, m,$$

one first transforms the matrix A into a matrix B of simpler form,

$$B = A_m = T^{-1}AT, \qquad T := T_1 T_2 \ldots T_m,$$

and then determines the eigenvalues λ and eigenvectors y of B, $By = \lambda y$. For $x := Ty = T_1 \ldots T_m y$, since $B = T^{-1}AT$, we then have

$$Ax = \lambda x,$$

i.e., to the eigenvalue λ of A there belongs the eigenvector x. The matrix B is chosen in such a way that

(1) the determination of the eigenvalues and eigenvectors of B is as simple as possible (i.e., requires as few operations as possible) and
(2) the eigenvalue problem for B is not (substantially) worse conditioned than that for A (i.e., small changes in the matrix B do not impair the eigenvalues of B, nor therefore those of A, substantially more than equally small changes in A).

In view of

$$B = T^{-1}AT,$$
$$B + \Delta B = T^{-1}(A + \Delta A)T, \qquad \Delta A := T\,\Delta B\,T^{-1},$$

given any vector norm $\|\cdot\|$ and corresponding matrix norm $\mathrm{lub}(\cdot)$, one gets the following estimates:

$$\mathrm{lub}(B) \leqslant \mathrm{cond}(T)\,\mathrm{lub}(A),$$
$$\mathrm{lub}(\Delta A) \leqslant \mathrm{cond}(T)\,\mathrm{lub}(\Delta B),$$

and hence

$$\frac{\operatorname{lub}(\Delta A)}{\operatorname{lub}(A)} \leqslant (\operatorname{cond}(T))^2 \frac{\operatorname{lub}(\Delta B)}{\operatorname{lub}(B)}.$$

For large $\operatorname{cond}(T) \gg 1$ the eigenvalue problem for B will be worse conditioned than that for A. Since

$$\operatorname{cond}(T) = \operatorname{cond}(T_1 \ldots T_m) \leqslant \operatorname{cond}(T_1) \ldots \operatorname{cond}(T_m),$$

well-conditioning will be insured if one chooses the matrices T_i such that $\operatorname{cond}(T_i)$ does not become too large. This is the case, in particular, for the maximum norm $\|x\|_\infty = \max_i |x_i|$ and elimination matrices of the form (see Section 4.1)

$$T_i = G_j = \begin{bmatrix} 1 & & & & & 0 \\ & \ddots & & & & \\ & & 1 & & & \\ & & l_{j+1,j} & \ddots & & \\ & & \vdots & & \ddots & \\ 0 & & l_{nj} & & 0 & 1 \end{bmatrix}, \qquad |l_{kj}| \leqslant 1,$$

(6.5.0.1)

$$G_j^{-1} = \begin{bmatrix} 1 & & & & & 0 \\ & \ddots & & & & \\ & & 1 & & & \\ & & -l_{j+1,j} & \ddots & & \\ & & \vdots & & \ddots & \\ 0 & & -l_{nj} & & 0 & 1 \end{bmatrix},$$

$$\operatorname{cond}(T_i) \leqslant 4,$$

and also for the Euclidean norm $\|x\|_2 = \sqrt{x^H x}$ and unitary matrices $T_i = U$ (e.g., Householder matrices) for which $\operatorname{cond}(T_i) = 1$. Reduction algorithms using either unitary matrices T_i or elimination matrices T_i as in (6.5.0.1) are described in the following sections. The "simple" terminal matrix $B = A_m$ that can be achieved in this manner, for arbitrary matrices, is an upper *Hessenberg* matrix, which has the following form:

$$B = \begin{bmatrix} * & \cdot & \cdot & \cdot & \cdot & \cdot & * \\ * & & & & & & \\ 0 & \cdot & & & & & \cdot \\ \cdot & \cdot & \cdot & & & & \cdot \\ \cdot & & \cdot & \cdot & & & \cdot \\ \cdot & & & \cdot & \cdot & & \\ 0 & \cdot & \cdot & \cdot & 0 & * & * \end{bmatrix}, \qquad b_{ik} = 0 \quad \text{for } k \leqslant i - 2.$$

For Hermitian matrices $A = A^H$ only unitary matrices T_i, $T_i^{-1} = T_i^H$, are used for the reduction. If A_{i-1} is Hermitian, then so is $A_i = T_i^{-1} A_{i-1} T_i$:

$$A_i^H = (T_i^H A_{i-1} T_i)^H = T_i^H A_{i-1}^H T_i = T_i^H A_{i-1} T_i = A_i.$$

For the terminal matrix B one thus obtains a Hermitian Hessenberg matrix, i.e., a (Hermitian) tridiagonal matrix, or Jacobi matrix:

$$B = \begin{bmatrix} \delta_1 & \bar{\gamma}_2 & & 0 \\ \gamma_2 & \ddots & \ddots & \\ & \ddots & \ddots & \bar{\gamma}_n \\ 0 & & \gamma_n & \delta_n \end{bmatrix}, \qquad \delta_i = \bar{\delta}_i.$$

6.5.1 Reduction of a Hermitian Matrix to Tridiagonal Form. The Method of Householder

In the method of Householder for the tridiagonalization of a Hermitian $n \times n$ matrix $A^H = A =: A_0$, one uses suitable Householder matrices (see Section 4.7)

$$T_i^H = T_i^{-1} = T_i = I - \beta_i u_i u_i^H$$

for the transformation

$$A_i = T_i^{-1} A_{i-1} T_i.$$

We assume that the matrix $A_{i-1} = [\alpha_{jk}]$ has already the following form:

$$(6.5.1.1) \qquad A_{i-1} = \left[\begin{array}{c|c|c} J_{i-1} & c & 0 \\ \hline c^H & \delta_i & a_i^H \\ \hline 0 & a_i & \tilde{A}_{i-1} \end{array}\right] = [\alpha_{jk}],$$

with

$$\left[\begin{array}{c|c} J_{i-1} & c \\ \hline c^H & \delta_i \end{array}\right] = \left[\begin{array}{ccccc|c} \delta_1 & \bar{\gamma}_2 & & & 0 & 0 \\ \gamma_2 & \ddots & \ddots & & & \vdots \\ \vdots & \ddots & \ddots & \ddots & & \vdots \\ \vdots & & \ddots & \ddots & \bar{\gamma}_{i-1} & 0 \\ 0 & \cdots & \cdots & \gamma_{i-1} & \delta_{i-1} & \bar{\gamma}_i \\ \hline 0 & \cdots & \cdots & 0 & \gamma_i & \delta_i \end{array}\right], \qquad a_i = \begin{bmatrix} \alpha_{i+1,i} \\ \vdots \\ \alpha_{ni} \end{bmatrix}.$$

According to Section 4.7, there is a Householder matrix $\tilde{T}_i$ of order $n - i$ such that

$$(6.5.1.2) \qquad \tilde{T}_i a_i = k \cdot e_1 \in \mathbb{C}^{n-i}.$$

$\tilde{T}_i$ has the form $\tilde{T}_i = I - \beta u u^H$, $u \in \mathbb{C}^{n-i}$, and is given by

$$\sigma := \|a_i\|_2 = \sqrt{\sum_{j=i+1}^{n} |\alpha_{ji}|^2}, \qquad \beta = \begin{cases} 1/(\sigma(\sigma + |\alpha_{i+1,i}|)) & \text{if } \sigma \neq 0, \\ 0 & \text{otherwise}, \end{cases}$$

(6.5.1.3)
$$k := -\sigma e^{i\varphi} \quad \text{if } \alpha_{i+1,i} = e^{i\varphi} |\alpha_{i+1,i}|,$$

$$u := \begin{bmatrix} e^{i\varphi}(\sigma + |\alpha_{i+1,i}|) \\ \alpha_{i+2,i} \\ \vdots \\ \alpha_{ni} \end{bmatrix}.$$

Then, for the unitary $n \times n$ matrix T_i, partitioned like (6.5.1.1),

$$T_i := \left[\begin{array}{c|c|c} I & 0 & 0 \\ \hline 0 & 1 & 0 \\ \hline 0 & 0 & \tilde{T}_i \end{array}\right] \}{i-1},$$

one clearly has $T_i^H = T_i^{-1} = T_i$ and, by (6.5.1.2),

$$T_i^{-1} A_{i-1} T_i = T_i A_{i-1} T_i = \left[\begin{array}{c|c|c} J_{i-1} & c & 0 \\ \hline c^H & \delta_i & a_i^H \tilde{T}_i \\ \hline 0 & \tilde{T}_i a_i & \tilde{T}_i \tilde{A}_{i-1} \tilde{T}_i \end{array}\right]$$

$$= \left[\begin{array}{cccc|c|c} \delta_1 & \bar{\gamma}_2 & & 0 & 0 & \\ \gamma_2 & \ddots & \ddots & & \vdots & 0 \\ & \ddots & \ddots & \bar{\gamma}_{i-1} & 0 & \\ 0 & & \gamma_{i-1} & \delta_{i-1} & \bar{\gamma}_i & \\ \hline 0 & \dots & 0 & \gamma_i & \delta_i & \bar{\gamma}_{i+1} 0 \dots 0 \\ \hline & & & & \gamma_{i+1} & \\ & & 0 & & 0 & \tilde{T}_i \tilde{A}_{i-1} \tilde{T}_i \\ & & & & \vdots & \\ & & & & 0 & \end{array}\right] =: A_i$$

with $\gamma_{i+1} := k$.

Since $\tilde{T}_i = I - \beta u u^H$, one can compute $\tilde{T}_i \tilde{A}_{i-1} \tilde{T}_i$ as follows:

$$\begin{aligned} \tilde{T}_i \tilde{A}_{i-1} \tilde{T}_i &= (I - \beta u u^H) \tilde{A}_{i-1} (I - \beta u u^H) \\ &= \tilde{A}_{i-1} - \beta \tilde{A}_{i-1} u u^H - \beta u u^H \tilde{A}_{i-1} + \beta^2 u u^H \tilde{A}_{i-1} u u^H. \end{aligned}$$

Introducing for brevity the vectors $p, q \in \mathbb{C}^{n-i}$,

$$p := \beta \tilde{A}_{i-1} u,$$

$$q := p - \frac{\beta}{2}(p^H u)u,$$

it follows immediately, in view of $\beta \geqslant 0$, $p^H u = \beta u^H \tilde{A}_{i-1} u = (p^H u)^H$, that

$$\begin{aligned} \tilde{T}_i \tilde{A}_{i-1} \tilde{T}_i &= \tilde{A}_{i-1} - pu^H - up^H + \beta u p^H u u^H \\ &= \tilde{A}_{i-1} - u\left[p - \frac{\beta}{2}(p^H u)u\right]^H - \left[p - \frac{\beta}{2}(p^H u)u\right]u^H \qquad (6.5.1.4) \\ &= \tilde{A}_{i-1} - uq^H - qu^H. \end{aligned}$$

The formulas (6.5.1.1)–(6.5.1.4) completely describe the ith transformation

$$A_i = T_i^{-1} A_{i-1} T_i.$$

Evidently,

$$B = A_{n-2} = \begin{bmatrix} \delta_1 & \bar{\gamma}_2 & & 0 \\ \gamma_2 & \ddots & \ddots & \\ & \ddots & \ddots & \bar{\gamma}_n \\ 0 & & \gamma_n & \delta_n \end{bmatrix}, \qquad \delta_i = \bar{\delta}_i,$$

is a Hermitian tridigonal matrix.

A formal ALGOL-like description of the Householder transformation for a real symmetric matrix $A = A^T = [a_{jk}]$ with $n \geqslant 2$ is

for $i := 1$ **step** 1 **until** $n - 2$ **do**
begin
  $\delta_i := a_{ii}$;
  $s := \oplus\sqrt{\sum_{j=i+1}^{n} |a_{ji}|^2}$; **if** $a_{i+1,i} < 0$ **then** $s := -s$;
  $\gamma_{i+1} := -s$; $e := s + a_{i+1,i}$;
  if $s = 0$ **then begin** $a_{ii} := 0$; **go to** MM **end**;
  $\beta := a_{ii} := 1/(s \times e)$;
  $u_{i+1} := a_{i+1,i} := e$;
  for $j := i + 2$ **step** 1 **until** n **do** $u_j := a_{ji}$;
  for $j := i + 1$ **step** 1 **until** n **do**
    $p_j = \left(\sum_{k=i+1}^{j} a_{jk} \times u_k + \sum_{k=j+1}^{n} a_{kj} \times u_k\right) \times \beta$;
  $sk := \left(\sum_{j=i+1}^{n} p_j \times u_j\right) \times \beta/2$;
  for $j := i + 1$ **step** 1 **until** n **do**
    $q_j := p_j - sk \times u_j$;
  for $j := i + 1$ **step** 1 **until** n **do**
    for $k := i + 1$ **step** 1 **until** j **do**
      $a_{jk} := a_{jk} - q_j \times u_k - u_j \times q_k$;

MM: **end**;

$$\delta_{n-1} := a_{n-1,n-1};\ \delta_n := a_{n,n};\ \gamma_n := a_{n,n-1};$$

This program takes advantage of the symmetry of A: Only the elements a_{jk} with $k \leqslant j$ need to be given. Moreover, the matrix A is overwritten with the essential elements β_i, u_i of the transformation matrix $\tilde{T}_i = I - \beta_i u_i u_i^H$, $i = 1, 2, \ldots, n-2$: Upon exiting from the program, the ith column of A will be occupied by the vector

$$\begin{bmatrix} a_{ii} \\ \hline a_{i+1,i} \\ \vdots \\ a_{ni} \end{bmatrix} := \begin{bmatrix} \beta_i \\ \hline u_i \end{bmatrix}, \qquad i = 1, 2, \ldots, n-2.$$

(The matrices T_i, $i = 1, 2, \ldots, n-2$, are needed for the "back transformation" of the eigenvectors: if y is an eigenvector of A_{n-2} for the eigenvalue λ,

$$A_{n-2}\, y = \lambda y,$$

then $x := T_1 T_2 \ldots T_{n-2}\, y$ is an eigenvector of A.)

Tested ALGOL programs for the Householder reduction and back transformation of eigenvectors can be found in Martin, Reinsch, and Wilkinson (1971); FORTRAN programs, in Smith et al. (1976).

Applying the transformations described above to an arbitrary non-Hermitian matrix A of order n, the formulas (6.5.1.2), (6.5.1.3) give rise to a chain of matrices A_i, $i = 0, 1, \ldots, n-2$, of the form

$$A_0 := A,$$

$$A_{i-1} = \left[\begin{array}{cccc|c|ccc}
* & \cdots & & & * & * & \cdots & * \\
* & \ddots & & & \vdots & \vdots & & \vdots \\
 & \ddots & \ddots & & \vdots & \vdots & & \vdots \\
0 & & * & * & * & * & \cdots & * \\ \hline
0 & & & * & * & * & \cdots & * \\ \hline
 & & & & * & * & \cdots & * \\
 & & & & \vdots & \vdots & & \vdots \\
\underbrace{0 \quad\quad\quad\quad}_{i-1} & & & & * & * & \cdots & *
\end{array}\right].$$

The first $i-1$ columns of A_{i-1} are already those of a Hessenberg matrix. A_{n-2} is a Hessenberg matrix. During the transition from A_{i-1} to A_i the elements α_{jk} of A_{i-1} with $j, k \leqslant i$ remain unchanged.

ALGOL programs for this algorithm can be found in Martin and Wilkinson (1971); FORTRAN programs, in Smith et al. (1976).

In Section 6.5.4 a further algorithm for the reduction of a general matrix A to Hessenberg form will be described which does not operate with unitary similarity transformations.

The numerical stability of the Householder reduction can be shown in the following way: Let $\bar{A}_i$ and $\bar{T}_i$ denote the matrices obtained if the algorithm is carried out in floating-point arithmetic with relative precision eps; let U_i denote the Householder matrix which, according to the rules of the algorithm, would have to be taken as transformation matrix for the transition $\bar{A}_{i-1} \to \bar{A}_i$ in exact arithmetic. Thus, U_i is an exact unitary matrix, while $\bar{T}_i$ is an approximate unitary matrix, namely the one obtained in place of U_i by computing U_i in floating-point arithmetic. The following relations thus hold:

$$\bar{T}_i = \mathrm{fl}(U_i),$$
$$\bar{A}_i = \mathrm{fl}(\bar{T}_i \bar{A}_{i-1} \bar{T}_i).$$

By means of the methods described in Section 1.3 one can now show [see, e.g., Wilkinson (1965)] that

$$\begin{aligned} &\mathrm{lub}_2(\bar{T}_i - U_i) \leqslant f(n)\,\mathrm{eps}, \\ (6.5.1.5)\qquad &\bar{A}_i = \mathrm{fl}(\bar{T}_i \bar{A}_{i-1} \bar{T}_i) = \bar{T}_i \bar{A}_{i-1} \bar{T}_i + G_i, \\ &\mathrm{lub}_2(G_i) \leqslant f(n)\,\mathrm{eps}\,\mathrm{lub}_2(\bar{A}_{i-1}), \end{aligned}$$

where $f(n)$ is a certain function [for which, generally, $f(n) = O(n^\alpha)$, $\alpha \approx 1$]. From (6.5.1.5), since $\mathrm{lub}_2(U_i) = 1$, $U_i^H = U_i^{-1} = U_i$ (U_i is a Householder matrix!), there follows at once

$$\mathrm{lub}_2(R_i) \leqslant f(n)\,\mathrm{eps}, \qquad R_i := \bar{T}_i - U_i,$$
$$\begin{aligned} \bar{A}_i &= U_i^{-1} \bar{A}_{i-1} U_i + R_i \bar{A}_{i-1} U_i + U_i \bar{A}_{i-1} R_i + R_i \bar{A}_{i-1} R_i + G_i \\ &= U_i^{-1} \bar{A}_{i-1} U_i + F_i, \end{aligned}$$

where

$$\mathrm{lub}_2(F_i) \leqslant \mathrm{eps} \cdot f(n)[3 + \mathrm{eps} \cdot f(n)]\, \mathrm{lub}_2(\bar{A}_{i-1}),$$

or, since $f(n)\,\mathrm{eps} \ll 3$, in first approximation:

$$\begin{aligned} &\mathrm{lub}_2(F_i) \mathrel{\dot{\leqslant}} 3\,\mathrm{eps} \cdot f(n)\, \mathrm{lub}_2(\bar{A}_{i-1}), \\ (6.5.1.6)\qquad &\mathrm{lub}_2(\bar{A}_i) \mathrel{\dot{\leqslant}} (1 + 3\,\mathrm{eps} \cdot f(n))\, \mathrm{lub}_2(\bar{A}_{i-1}) \\ &\qquad \mathrel{\dot{\leqslant}} (1 + 3\,\mathrm{eps} \cdot f(n))^i\, \mathrm{lub}_2(A). \end{aligned}$$

For the Hessenberg matrix $\bar{A}_{n-2}$, finally, since $A = \bar{A}_0$, one obtains

$$(6.5.1.7)\qquad \bar{A}_{n-2} = U_{n-2}^{-1} \cdots U_1^{-1}(A + F)U_1 \cdots U_{n-2},$$

where

$$F := \sum_{i=1}^{n-2} U_1 U_2 \ldots U_i F_i U_i^{-1} \ldots U_2^{-1} U_1^{-1}.$$

It thus follows from (6.5.1.6) that

$$\begin{aligned} \operatorname{lub}_2(F) &\leqslant \sum_{i=1}^{n-2} \operatorname{lub}_2(F_i) \\ &\dot{\leqslant} 3 \text{ eps} \cdot f(n) \operatorname{lub}_2(A) \sum_{i=1}^{n-2} (1 + 3 \text{ eps} \cdot f(n))^{i-1}, \end{aligned}$$

or, in first approximation,

(6.5.1.8) $$\operatorname{lub}_2(F) \dot{\leqslant} 3(n-2) f(n) \text{ eps} \operatorname{lub}_2(A).$$

Provided $nf(n)$ is not too large, the relations (6.5.1.7) and (6.5.1.8) show that the matrix $\bar{A}_{n-2}$ is exactly similar to the matrix $A + F$, which is only a slight perturbation of A, and that therefore the method is numerically stable.

6.5.2 Reduction of a Hermitian Matrix to Tridiagonal or Diagonal Form: The Methods of Givens and Jacobi

In Givens' method (1954), a precursor of the Householder method, the chain

$$A =: A_0 \to A_1 \to \cdots \to A_m, \qquad A_i = T_i^{-1} A_{i-1} T_i,$$

for the transformation of a Hermitian matrix A to tridiagonal form $B = A_m$ is constructed by means of unitary matrices $T_i = \Omega_{jk}$ of the form (φ, ψ real)

(6.5.2.1) $$\Omega_{jk} = \begin{bmatrix} 1 & & & & & & & & 0 \\ & \ddots & & & & & & & \\ & & 1 & & & & & & \\ & & & \cos\varphi & & & -e^{-i\psi}\sin\varphi & & \\ & & & & 1 & & & & \\ & & & & & \ddots & & & \\ & & & & & & 1 & & \\ & & & e^{i\psi}\sin\varphi & & & \cos\varphi & & \\ & & & & & & & 1 & \\ & & & & & & & & \ddots \\ 0 & & & & & & & & & 1 \end{bmatrix} \begin{matrix} \\ \\ \\ \leftarrow j \\ \\ \cdot \\ \\ \leftarrow k \\ \\ \\ \\ \end{matrix}$$

In order to describe the method of Givens we assume for simplicity that $A = A^H$ is real; we can choose in this case $\psi = 0$, and Ω_{jk} is orthogonal. Note that in the left multiplication $A \to \Omega_{jk}^{-1} A = \Omega_{jk}^H A$ only rows j and k of A undergo changes, while in the right multiplication $A \to A\Omega_{jk}$ only columns j and k change. We describe only the first transformation step $A = A_0 \to$

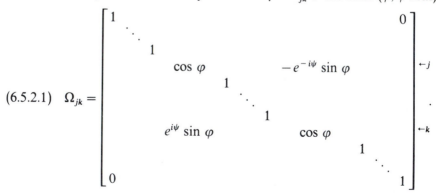

$T_1^{-1}A_0 =: A_0' \to A_0' T_1 = T_1^{-1} A_0 T_1 =: A_1$. In the half step $A_0 \to A_0'$ the matrix $T_1 = \Omega_{23}$, $T_1^{-1} = \Omega_{23}^H$ is chosen such (see Section 4.9) that the element of $A_0' = \Omega_{23}^H A_0$ in position (3, 1) is annihilated; in the subsequent right multiplication by Ω_{23}, $A_0 \to A_1 = A_0' \Omega_{23}$, the zero in position (3, 1) is preserved. Below is a sketch for a 4×4 matrix, where changing elements are denoted by $*$:

$$A_0 = \begin{bmatrix} \times & \times & \times & \times \\ \times & \times & \times & \times \\ \times & \times & \times & \times \\ \times & \times & \times & \times \end{bmatrix} \to \Omega_{23}^H A_0 = \begin{bmatrix} \times & \times & \times & \times \\ * & * & * & * \\ 0 & * & * & * \\ \times & \times & \times & \times \end{bmatrix} = A_0'$$

$$\to A_0' \Omega_{23} = \begin{bmatrix} \times & * & 0 & \times \\ \times & * & * & \times \\ 0 & * & * & \times \\ \times & * & * & \times \end{bmatrix} =: A_1.$$

Since with A_0, also A_1 is Hermitian, the transformation $A_0' \to A_1$ also annihilates the element in position (1, 3). After this, the element in position (4, 1) is transformed to zero by a Givens rotation $T_2 = \Omega_{24}$, etc. In general, one takes for T_i successively the matrices

$$\begin{aligned} &\Omega_{23}, \Omega_{24}, \ldots, \Omega_{2n}, \\ &\qquad \Omega_{34}, \ldots, \Omega_{3n}, \\ &\qquad\qquad \vdots \\ &\qquad\qquad \Omega_{n-1,n}, \end{aligned}$$

and chooses Ω_{jk}, $j = 2, 3, \ldots, n-1$, $k = j+1, j+2, \ldots, n$, so as to annihilate the element in position $(k, j-1)$. A comparison with Householder's method shows that this variant of Givens' method requires about twice as many operations. For this reason, Householder's method is usually preferred. There are, however, modern variants ("rational Givens transformations") which are comparable to the Householder method.

The method of Jacobi, too, employs similarity transformations with the special unitary matrices Ω_{jk} in (6.5.2.1); however, it no longer produces a finite sequence ending in a tridiagonal matrix, but an infinite sequence of matrices $A^{(i)}$, $i = 0, 1, 2, \ldots$, converging to a diagonal matrix

$$D = \begin{bmatrix} \lambda_1 & & 0 \\ & \ddots & \\ 0 & & \lambda_n \end{bmatrix}.$$

Here, the λ_i are just the eigenvalues of A. To explain the method, we again assume for simplicity that A is a *real* symmetric matrix. In the transformation step

$$A^{(i)} \to A^{(i+1)} = \Omega_{jk}^H A^{(i)} \Omega_{jk}$$

the quantities $c := \cos\varphi$, $s := \sin\varphi$ of the matrix Ω_{jk}, $j < k$, in (6.5.2.1) are now determined so that $a'_{jk} = 0$ (we denote the elements of $A^{(i)}$ by a_{rs}, those of $A^{(i+1)}$ by a'_{rs}):

$$A^{(i+1)} = \begin{bmatrix} a'_{11} & \cdots & a'_{1j} & \cdots & a'_{1k} & \cdots & a'_{1n} \\ \vdots & & \vdots & & \vdots & & \vdots \\ a'_{j1} & \cdots & a'_{jj} & \cdots & 0 & \cdots & a'_{jn} \\ \vdots & & \vdots & & \vdots & & \vdots \\ a'_{k1} & \cdots & 0 & \cdots & a'_{kk} & \cdots & a'_{kn} \\ \vdots & & \vdots & & \vdots & & \vdots \\ a'_{n1} & \cdots & a'_{nj} & \cdots & a'_{nk} & \cdots & a'_{nn} \end{bmatrix}.$$

Only the entries in frames are changed, according to the formulas

$$\begin{aligned} &\left.\begin{aligned} a'_{rj} &= a'_{jr} = ca_{rj} + sa_{rk} \\ a'_{rk} &= a'_{kr} = -sa_{rj} + ca_{rk} \end{aligned}\right\} \quad \text{for } r \neq j, k, \\ &a'_{jj} = c^2 a_{jj} + s^2 a_{kk} + 2csa_{jk}, \\ &a'_{jk} = a'_{kj} = -cs(a_{jj} - a_{kk}) + (c^2 - s^2)a_{jk} = 0, \\ &a'_{kk} = s^2 a_{jj} + c^2 a_{kk} - 2csa_{jk}. \end{aligned} \tag{6.5.2.2}$$

From this, one obtains for the angle φ the defining equation

$$\tan 2\varphi = \frac{2a_{jk}}{a_{jj} - a_{kk}}, \qquad |\varphi| \leqslant \frac{\pi}{4}.$$

By means of trigonometric identities, one can compute from this the quantities c and s and, by (6.5.2.2), the a'_{rs}.

It is recommended, however, that the following numerically more stable formulas be used [see Rutishauser (1971), where an ALGOL program can also be found]. First compute the quantity $\vartheta := \cot 2\varphi$ from

$$\vartheta := \frac{a_{jj} - a_{kk}}{2a_{jk}},$$

and then $t := \tan\varphi$ as the root of smallest modulus of the quadratic equation

$$t^2 + 2t\vartheta - 1 = 0,$$

that is,

$$t = \frac{s(\vartheta)}{|\vartheta| + \sqrt{1 + \vartheta^2}}, \qquad s(\vartheta) := \begin{cases} 1 & \text{if } \vartheta \geqslant 0, \\ -1 & \text{otherwise,} \end{cases}$$

or $t := 1/2\vartheta$ if $|\vartheta|$ is so large that ϑ^2 would overflow. Obtain the quantities c and s from

$$c := \frac{1}{\sqrt{1+t^2}}, \qquad s := tc.$$

Finally compute the number $\tau := \tan(\varphi/2)$ from

$$\tau := \frac{s}{1+c},$$

and with the aid of s, t, and τ rewrite the formulas (6.5.2.2) in a numerically more stable way as

$$\left.\begin{aligned} a'_{rj} = a'_{jr} &:= a_{rj} + s\cdot(a_{rk} - \tau a_{rj}) \\ a'_{rk} = a'_{kr} &:= a_{rk} - s\cdot(a_{rj} + \tau a_{rk}) \end{aligned}\right\} \quad \text{for } r \neq j, k,$$

$$a'_{jj} := a_{jj} + t a_{jk},$$

$$a'_{jk} := a'_{kj} := 0,$$

$$a'_{kk} := a_{kk} - t a_{jk}.$$

For the proof of convergence, one considers

$$S(A^{(i)}) := \sum_{j \neq k} |a_{jk}|^2, \qquad S(A^{(i+1)}) = \sum_{j \neq k} |a'_{jk}|^2,$$

the sums of the squares of the off-diagonal elements of $A^{(i)}$ and $A^{(i+1)}$, respectively. For these one finds, by (6.5.2.2),

$$0 \leqslant S(A^{(i+1)}) = S(A^{(i)}) - 2|a_{jk}|^2 < S(A^{(i)}) \quad \text{if } a_{jk} \neq 0.$$

The sequence of nonnegative numbers $S(A^{(i)})$ therefore decreases monotonically, and thus converges. One can show that $\lim_{i\to\infty} S(A^{(i)}) = 0$ (i.e., the $A^{(i)}$ converge to a diagonal matrix), provided the transformations Ω_{jk} are executed in a suitable order, namely row-wise,

$$\begin{matrix} \Omega_{12}, \Omega_{13}, \ldots, \Omega_{1n} \\ \Omega_{23}, \ldots, \Omega_{2n} \\ \vdots \\ \Omega_{n-1,n} \end{matrix}$$

and, in this order, cyclically repeated. Under these conditions one can even prove quadratic convergence of the Jacobi method, if A has only simple eigenvalues:

$$S(A^{(i+N)}) \leqslant \frac{S(A^{(i)})^2}{\delta} \quad \text{with } N := \frac{n(n-1)}{2},$$

$$\delta := \min_{i \neq j} |\lambda_i(A) - \lambda_j(A)| > 0$$

[for the proof, see Wilkinson (1962); further literature: Rutishauser (1971); Schwarz, Rutishauser, and Stiefel (1972), Parlett (1980)].

In spite of this rapid convergence and the additional advantage that an orthogonal system of eigenvectors of A can easily be obtained from the Ω_{jk} employed, it is more advantageous in practical situations, particularly for large n, to reduce the matrix A to a tridiagonal matrix J by means of the Householder method (see Section 6.5.1) and to compute the eigenvalues and eigenvectors of J by the QR method, since this method converges cubically. This all the more so if A has already the form of a band matrix: in the QR method this form is preserved; the Jacobi method destroys it.

We remark that Eberlein developed a method for non-Hermitian matrices similar to Jacobi's. An ALGOL program for this method, and further details, can be found in Eberlein (1971).

6.5.3 Reduction of a Hermitian Matrix to Tridiagonal Form: The Method of Lanczos

Krylov sequences of vectors q, Aq, A^2q, ... belonging to an $n \times n$ matrix A and a starting vector $q \in \mathbb{C}^n$ were already used for the derivation of the Frobenius normal form of a general matrix in Section 6.3. They also play an important role in the method of Lanczos (1950) for reducing a Hermitian matrix to tridiagonal form. Closely related to such a sequence of vectors is a sequence of subspaces of $\mathbb{C}^n$

$$K_i(q, A) := \text{span}[q, Aq, \ldots, A^{i-1}q], \quad i \geq 1, \quad K_0(q, A) := \{0\},$$

called *Krylov spaces*: $K_i(q, A)$ is the subspace spanned by the first i vectors of the sequence $\{A^i q\}_{i \geq 0}$. As in Section 6.3, we denote by m the largest index i for which $q, Aq, \ldots, A^{i-1}q$ are still linearly independent, that is, $\dim K_i(q, A) = i$. Then $m \leq n$, $A^m q \in K_m(q, A)$, the vectors $q, Aq, \ldots, A^{m-1}q$ form a basis of $K_m(q, A)$, and therefore $AK_m(q, A) \subset K_m(q, A)$: the Krylov space $K_m(q, A)$ is A-invariant and the map $x \mapsto \phi(x) := Ax$ describes a linear map of $K_m(q, A)$ into itself.

In Section 6.3 we arrived at the Frobenius matrix (6.3.1) when the map ϕ was described with respect to the basis $q, Aq, \ldots, A^{m-1}q$ of $K_m(q, A)$. The idea of the Lanczos method is closely related: Here, the map ϕ is described with respect to a special *orthonormal* basis $q_1, q_2, \ldots, q_m$ of $K_m(q, A)$, where the q_j are chosen such that for all $i = 1, 2, \ldots, m$, the vectors $q_1, q_2, \ldots, q_i$ form an orthonormal basis of $K_i(q, A)$. If $A = A^H$ is a Hermitian $n \times n$ matrix, then such a basis is easily constructed for a given starting vector q. We assume $q \neq 0$ in order to exclude the trivial case and suppose in addition that $\|q\| = 1$, where $\|\cdot\|$ is the Euclidean norm. Then there is a three-term recursion formula for the vectors q_i (similar recursions are known for orthogonal polynomials, cf. Theorem (3.6.3))

(6.5.3.1a)
$$q_1 := q, \qquad \gamma_1 q_0 := 0$$
$$Aq_i = \gamma_i q_{i-1} + \delta_i q_i + \gamma_{i+1} q_{i+1} \quad \text{for } i \geq 1,$$

where

(6.5.3.1b)
$$\delta_i := q_i^H A q_i$$
$$\gamma_{i+1} := \|r_i\| \qquad \text{with } r_i := Aq_i - \delta_i q_i - \gamma_i q_{i-1}$$
$$q_{i+1} := r_i/\gamma_{i+1}, \qquad \text{if } \gamma_{i+1} \neq 0.$$

Here, all coefficients γ_i, δ_i are real. The recursion breaks off with the first index $i =: i_0$ with $\gamma_{i+1} = 0$, and then the following holds

$$i_0 = m = \max_i \dim K_i(q, A).$$

Proof. We show (6.5.3.1) by induction over i. Clearly, since $\|q\| = 1$, the vector $q_1 := q$ provides an orthonormal basis for $K_1(q, A)$. Assume now that for some $j \geq 1$ vectors $q_1, \ldots, q_j$ are given, so that (6.5.3.1) and

$$\operatorname{span}[q_1, \ldots, q_i] = K_i(q, A)$$

holds for all $i \leq j$, and that $r_i \neq 0$ in (6.5.3.1b) for all $i < j$. We show first that these statements are also true for $j + 1$, if $r_j \neq 0$. In fact, then $\gamma_{j+1} \neq 0$, δ_j and q_{j+1} are well defined by (6.5.3.1b), and $\|q_{j+1}\| = 1$. The vector q_{j+1} is orthogonal to all q_i with $i \leq j$: This holds for $i = j$, because $\gamma_{j+1} \neq 0$, because

$$Aq_j = \gamma_j q_{j-1} + \delta_j q_j + \gamma_{j+1} q_{j+1}$$

from the definition of δ_j, and using the induction hypothesis

$$\gamma_{j+1} q_j^H q_{j+1} = q_j^H A q_j - \delta_j q_j^H q_j = 0.$$

For $i = j - 1$, the same reasoning and $A = A^H$ first give

$$\gamma_{j+1} q_{j-1}^H q_{j+1} = q_{j-1}^H A q_j - \gamma_j q_{j-1}^H q_{j-1} = (Aq_{j-1})^H q_j - \gamma_j.$$

The orthogonality of the q_i for $i \leq j$ and $Aq_{j-1} = \gamma_{j-1} q_{j-2} + \delta_{j-1} q_{j-1} + \gamma_j q_j$ then imply $(Aq_{j-1})^H q_j = \bar{\gamma}_j = \gamma_j$, and therefore $q_{j-1}^T q_{j+1} = 0$. For $i < j - 1$ we get the same result with the aid of $Aq_i = \gamma_i q_{i-1} + \delta_i q_i + \gamma_{i+1} q_{i+1}$:

$$\gamma_{j+1} q_i^H q_{j+1} = q_i^H A q_j = (Aq_i)^H q_j = 0.$$

Finally, since $\operatorname{span}[q_1, \ldots, q_i] = K_i(q, A) \subset K_j(q, A)$ for $i \leq j$, we also have

$$Aq_j \in K_{j+1}(q, A),$$

which implies by (6.5.3.1b)

$$q_{j+1} \in \operatorname{span}[q_{j-1}, q_j, Aq_j] \subset K_{j+1}(q, A),$$

and therefore $\operatorname{span}[q_1, \ldots, q_{j+1}] \subset K_{j+1}(q, A)$. Since the orthonormal vectors

$q_1, \ldots, q_{j+1}$ are linearly independent and $\dim K_{j+1}(q, A) \le j + 1$ we obtain

$$K_{j+1}(q, A) = \operatorname{span}[q_1, \ldots, q_{j+1}].$$

This also shows $j + 1 \le m = \max_i \dim K_i(q, A)$, and $i_0 \le m$ for the break-off index i_0 of (6.5.3.1). On the other hand, by the definition of i_0

$$Aq_{i_0} \in \operatorname{span}[q_{i_0-1}, q_{i_0}] \subset \operatorname{span}[q_1, \ldots, q_{i_0}] = K_{i_0}(q, A),$$

so that, because

$$Aq_i \in \operatorname{span}[q_1, \ldots, q_{i+1}] = K_{i+1}(q, A) \subset K_{i_0}(q, A) \quad \text{for } i < i_0,$$

we get the A-invariance of $K_{i_0}(q, A)$, $AK_{i_0}(q, A) \subset K_{i_0}(q, A)$. Therefore $i_0 \ge m$, since $K_m(q, A)$ is the first A-invariant subspace among the $K_i(q, A)$. This finally shows $i_0 = m$, and the proof is complete. □

The recursion (6.5.3.1) can be written in terms of the matrices

$$Q_i := [q_1, \ldots, q_i], \qquad J_i := \begin{bmatrix} \delta_1 & \gamma_2 & & 0 \\ \gamma_2 & \delta_2 & \ddots & \\ & \ddots & \ddots & \gamma_i \\ 0 & & \gamma_i & \delta_i \end{bmatrix}, \qquad 1 \le i \le m,$$

as a matrix equation

$$\begin{aligned} AQ_i &= Q_i J_i + [0, \ldots, 0, \gamma_{i+1} q_{i+1}] \\ &= Q_i J_i + \gamma_{i+1} q_{i+1} e_i^T, \qquad i = 1, 2, \ldots, m, \end{aligned}$$

where $e_i := [0, \ldots, 0, 1]^T \in \mathbb{R}^i$ is the ith axis vector of $\mathbb{R}^i$. This equation is easily verified by comparing the jth columns, $j = 1, \ldots, i$, on both sides. Note, that the $n \times i$ matrices Q_i have orthonormal columns, $Q_i^H Q_i = I_i$ ($:= i \times i$ identity matrix) and the J_i are real symmetric tridiagonal matrices. Since $i = m$ is the first index with $\gamma_{m+1} = 0$, the matrix J_m is irreducible, and the preceding matrix equation reduces to (cf. (6.3.3))

$$AQ_m = Q_m J_m$$

where $Q_m^H Q_m = I_m$. Any eigenvalue of J_m is also an eigenvalue of A, since $J_m z = \lambda z$, $z \ne 0$ implies $x := Q_m z \ne 0$ and

$$Ax = AQ_m z = Q_m J_m z = \lambda Q_m z = \lambda x.$$

If $m = n$, i.e., if the method does not terminate prematurely with an $m < n$, then Q_n is a unitary matrix, and the tridiagonal matrix $J_n = Q_n^H A Q_n$ is unitarily similar to A.

Given any vector $q =: q_1$ with $\|q\| = 1$, the method of Lanczos consists of computing the numbers γ_i, δ_i, $i = 1, 2, \ldots, m$, ($\gamma_1 := 0$), and the tridiagonal matrix J_m by means of (6.5.3.1). Subsequently, one may apply the methods of Section 6.6 to compute the eigenvalues and eigenvectors of J_m (and thereby those of A). Concerning the implementation of the method, the following remarks are in order:

1. The number of operations can be reduced by introducing an auxiliary vector defined by

$$u_i := Aq_i - \gamma_i q_{i-1}.$$

Then $r_i = u_i - \delta_i q_i$, and the number

$$\delta_i = q_i^H A q_i = q_i^H u_i$$

can also be computed from u_i, since $q_i^H q_{i-1} = 0$.

2. It is not necessary to store the vectors q_i if one is not interested in the eigenvectors of A: In order to carry out (6.5.3.1) only two auxiliary vectors v, $w \in \mathbb{C}^n$ are needed, where initially $v := q$ is the given starting vector with $\|q\| = 1$. Within the following program, which implements the Lanczos algorithm for a given Hermitian $n \times n$ matrix $A = A^H$, v_k and w_k, $k = 1, \ldots, n$, denote the components of v and w, respectively:

```
    w := 0; γ_1 := 1; i := 1;
1:  if γ_i ≠ 0 then
    begin if i ≠ 1 then
              for k := 1 step 1 until n do
              begin t := v_k; v_k := w_k/γ_i; w_k := −γ_i t end;
          w := Av + w; δ_i := v^H w; w := w − δ_i v;
          m := i; i := i + 1; γ_i := √(w^H w);
          goto 1;
    end;
```

Each step $i \to i + 1$ requires about $5n$ scalar multiplications and one multiplication of the matrix A with a vector. Therefore, the method is inexpensive if A is sparse, so that it is particularly valuable for solving the eigenvalue problem for large sparse matrices $A = A^H$.

3. In theory, the method is finite: it stops with the first index $i = m \le n$ with $\gamma_{i+1} = 0$. However, because of the influence of roundoff, one will rarely find a computed $\gamma_{i+1} = 0$ in practice. Yet, it is usually not necessary to perform many steps of the method until one finds a zero or a very small γ_{i+1}: The reason is that, under weak assumptions, the largest and smallest eigenvalues of J_i converge very rapidly with increasing i toward the largest and smallest eigenvalues of A (Kaniel-Paige theory: see Kaniel (1966), Paige (1971), and Saad (1980)). Therefore, if one is only interested in the extreme eigenvalues of A (which is quite frequently the case in applications), only relatively few steps of Lanczos' method are necessary to find a J_i, $i \ll n$, with extreme eigenvalues that already approximate the extreme eigenvalues of A to machine precision.

4. The method of Lanczos will generate orthogonal vectors q_i only in theory: In practice, due to roundoff, the vectors $\tilde{q}_i$ actually computed become less and less orthogonal as i increases. This defect could be corrected by *reorthogonalizing* a newly computed vector $\hat{q}_{i+1}$ with respect to *all* previous

vectors $\tilde{q}_j$, $j \le i$, that is, by replacing $\hat{q}_{i+1}$ by

$$\tilde{q}_{i+1} := \hat{q}_{i+1} - \sum_{j=1}^{i} (\tilde{q}_j^H \hat{q}_{i+1})\tilde{q}_j.$$

However, reorthogonalization is quite expensive: The vectors $\tilde{q}_i$ have to be stored, and step i of the Lanczos method now requires $O(i \cdot n)$ operations instead of $O(n)$ operations as before. But it is possible to avoid a full reorthogonalization to some extent and still obtain very good approximations for the eigenvalues of A in spite of the difficulties mentioned. Details can be found in the following literature, which also contains a systematic investigation of the interesting numerical properties of the Lanczos method: Paige (1971), Parlett and Scott (1979), and Cullum and Willoughby (1985), where one can also find programs.

6.5.4 Reduction to Hessenberg Form

It was already observed in Section 6.5.1 that one can transform a given $n \times n$ matrix A by means of $n - 2$ Householder matrices T_i similarly to Hessenberg form B,

$$A := A_0 \to A_1 \to \cdots \to A_{n-2} = B, \qquad A_i = T_i^{-1} A_{i-1} T_i.$$

We now wish to describe a second algorithm of this kind, in which one uses as transformation matrices T_i permutation matrices

$$P_{rs} = \begin{bmatrix} 1 &&&&&&&&& \\ & \ddots &&&&&&&& \\ && 1 &&&&&&& \\ &&& 0 &&&& 1 && \\ &&&& 1 &&&&& \\ &&&&& \ddots &&&& \\ &&&&&& 1 &&& \\ &&& 1 &&&& 0 && \\ &&&&&&&& 1 & \\ &&&&&&&&& \ddots & \\ &&&&&&&&&& 1 \end{bmatrix} \begin{matrix} \\ \\ \\ \leftarrow r \\ \\ \\ \\ \leftarrow s \\ \\ \\ \\ \end{matrix}$$

and elimination matrices of the form

$$G_j = \begin{bmatrix} 1 &&&&& \\ & \ddots &&&& \\ && 1 &&& \\ && l_{j+1,j} & 1 && \\ && \vdots && \ddots & \\ && l_{nj} &&& 1 \end{bmatrix}, \qquad |l_{ij}| \leqslant 1.$$

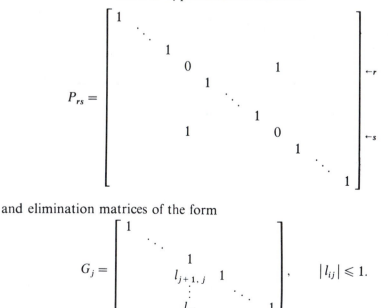

These matrices have the property

$$P_{rs}^{-1} = P_{rs},$$

$$(6.5.4.1) \qquad G_j^{-1} = \begin{bmatrix} 1 & & & & & \\ & \ddots & & & & \\ & & 1 & & & \\ & & -l_{j+1,j} & 1 & & \\ & & \vdots & & \ddots & \\ & & -l_{nj} & & & 1 \end{bmatrix}.$$

A left multiplication $P_{rs}^{-1}A$ of A by $P_{rs}^{-1} = P_{rs}$ has the effect of interchanging rows r and s of A, whereas a right multiplication AP_{rs} interchanges columns r and s of A. A left multiplication $G_j^{-1}A$ of A by G_j^{-1} has the effect of subtracting l_{rj} times row j from row r of the matrix A for $r = j+1, j+2, \ldots, n$, while a right multiplication AG_j means that l_{rj} times column r is added to column j of A for $r = j+1, j+2, \ldots, n$.

In order to successively transform A to Hessenberg form by means of similarity transformations of the type considered, we proceed as follows: We set $A = A_0$, and we assume inductively that A_{i-1} is already a matrix whose first $i-1$ columns have Hessenberg form:

$$(6.5.4.2) \qquad A_{i-1} = \left[\begin{array}{cccccc|c|ccc} * & \cdot & \cdot & \cdot & \cdot & * & * & \cdot & \cdot & * \\ * & \cdot & & & & \cdot & \cdot & & & \cdot \\ & \cdot & \cdot & & & \cdot & \cdot & & & \cdot \\ & & \cdot & \cdot & & \cdot & \cdot & & & \cdot \\ & & & \cdot & \cdot & \cdot & \cdot & & & \cdot \\ 0 & & & & * & * & * & \cdot & \cdot & * \\ \hline 0 & \cdot & \cdot & \cdot & 0 & * & * & \cdot & \cdot & * \\ \hline & & & & & & \cdot & & & \cdot \\ & & & & & & \cdot & & & \cdot \\ & & & & & & \cdot & & & \cdot \\ 0 & & & & & & * & \cdot & \cdot & * \end{array}\right]$$

$$\underbrace{\qquad\qquad\qquad}_{i-1}$$

$$= \left[\begin{array}{c|c|c} B_{i-1} & d & \hat{A}_{i-1} \\ \hline c & \delta_i & b \\ \hline 0 & a & \tilde{A}_{i-1} \end{array}\right], \qquad a = \begin{bmatrix} \alpha_{i+1,i} \\ \vdots \\ \alpha_{ni} \end{bmatrix}.$$

In order to compute a matrix $A_i = T_i^{-1} A_{i-1} T_i$ of analogous form, one chooses as matrix T_i a matrix of the type

$$T_i = P_{r,i+1} G_{i+1}, \quad T_i^{-1} = G_{i+1}^{-1} P_{r,i+1}^{-1}, \qquad r \geqslant i+1,$$

in fact such that (cf. Section 4.1 on Gauss elimination)

$$T_i^{-1}\begin{bmatrix} d \\ \\ \delta_i \\ \\ a \end{bmatrix} = \begin{bmatrix} d \\ \\ \delta_i \\ \\ \bar{a} \end{bmatrix}, \qquad \bar{a} = \begin{bmatrix} * \\ 0 \\ \vdots \\ \vdots \\ 0 \end{bmatrix}.$$

To this end, one must first determine the absolutely largest component of a,

$$|\alpha_{ri}| = \max_{i+1 \leqslant j \leqslant n} |\alpha_{ji}|, \qquad r \geqslant i+1.$$

Thereupon, one interchanges rows r and $i+1$, as well as columns r and $i+1$ of A_{i-1}, i.e., one computes

$$A' := P_{r,\,i+1}^{-1} A_{i-1} P_{r,\,i+1},$$

so that now the dominant element of a is in the first position. Denoting the resulting matrix by $A' = [\alpha'_{jk}]$, one proceeds to form the matrix G_{i+1} by means of the quantities

$$l_{j,\,i+1} := \begin{cases} \dfrac{\alpha'_{ji}}{\alpha'_{i+1,\,i}} & \text{if } \alpha'_{i+1,\,i} \neq 0 \\ 0 & \text{otherwise} \end{cases} \qquad (j = i+2, \ldots, n).$$

For $j = i+2, i+3, \ldots, n$, one now subtracts $l_{j,\,i+1}$ times row $i+1$ from row j of the matrix A' and then adds $l_{j,\,i+1}$ times column j to column $i+1$. As a result one obtains the matrix

$$\begin{aligned} A_i &= G_{i+1}^{-1} A' G_{i+1} \\ &= T_i^{-1} A_{i-1} T_i, \qquad T_i = P_{r,\,i+1} G_{i+1}, \end{aligned}$$

which has the desired form. Note that the transformations just described leave the elements of the submatrix

$$\begin{bmatrix} B_{i-1} & d \\ c & \delta_i \end{bmatrix}$$

[see (6.5.4.2)] of A_{i-1} unchanged.

After $n-2$ steps of this type one ends up with a Hessenberg matrix $B = A_{n-2}$.

ALGOL programs for this algorithm and for the back transformation of the eigenvectors can be found in Martin and Wilkinson (1971); FORTRAN programs, in Smith et al. (1976).

The numerical stability of this method can be examined as follows: Let $\bar{A}_i$ and $\bar{T}_i$ be the matrices actually obtained in place of the A_i, T_i during the course of the algorithm in a floating-point computation. In view of (6.5.4.1), no further rounding errors are committed in the computation of $\bar{T}_i^{-1}$ from

$\bar{T}_i$,

$$\bar{T}_i^{-1} = \text{fl}(\bar{T}_i^{-1}),$$

and by definition of $\bar{A}_i$ one has

(6.5.4.3) $$\bar{A}_i = \text{fl}(\bar{T}_i^{-1}\bar{A}_{i-1}\bar{T}_i) = \bar{T}_i^{-1}\bar{A}_{i-1}\bar{T}_i + R_i.$$

With the methods of Section 1.3 one can derive for the error matrix R_i estimates of the following form:

(6.5.4.4) $$\text{lub}_\infty(R_i) \leqslant f(n) \text{ eps} \cdot \text{lub}_\infty(\bar{A}_{i-1}),$$

where $f(n)$ is a certain function of n with $f(n) = O(n^\alpha)$, $\alpha \approx 1$ (cf. Exercise 13). It finally follows from (6.5.4.3), since $A = \bar{A}_0$, that

(6.5.4.5) $$\bar{A}_{n-2} = \bar{T}_{n-2}^{-1} \ldots \bar{T}_1^{-1}(A + F)\bar{T}_1 \ldots \bar{T}_{n-2},$$

where

(6.5.4.6) $$F = \sum_{i=1}^{n-2} \bar{T}_1\bar{T}_2 \ldots \bar{T}_i R_i \bar{T}_i^{-1} \ldots \bar{T}_2^{-1}\bar{T}_1^{-1}.$$

This shows that $\bar{A}_{n-2}$ is exactly similar to a perturbed initial matrix $A + F$. The smaller F is compared to A, the more stable is the method.

For the matrices $\bar{T}_i$ we now have, in view of (6.5.4.1) and $|l_{ij}| \leqslant 1$,

$$\text{lub}_\infty(\bar{T}_i) \leqslant 2, \qquad \text{lub}_\infty(\bar{T}_i^{-1}) \leqslant 2,$$

so that, rigorously,

(6.5.4.7) $$\text{lub}_\infty(\bar{T}_1 \ldots \bar{T}_i) \leqslant 2^i.$$

As a consequence, taking note of $\bar{A}_i \approx \bar{T}_i^{-1} \ldots \bar{T}_1^{-1}A\bar{T}_1 \ldots \bar{T}_i$, (6.5.4.4), and (6.5.4.6), one can get the estimates

(6.5.4.8)
$$\text{lub}_\infty(\bar{A}_i) \mathrel{\dot{\leqslant}} C_i \,\text{lub}_\infty(A), \qquad C_i := 2^{2i},$$
$$\text{lub}_\infty(F) \mathrel{\dot{\leqslant}} Kf(n) \text{ eps} \cdot \text{lub}_\infty(A), \qquad K := \sum_{i=1}^{n-2} 2^{4i-2},$$

with a factor $K = K(n)$ which grows rapidly with n. Fortunately, the bound (6.5.4.7) in most practical cases is much too pessimistic. It is already rather unlikely that

$$\text{lub}_\infty(\bar{T}_1 \ldots \bar{T}_i) \geqslant 2i \quad \text{for } i \geqslant 2.$$

In all these cases the constants C_i and K in (6.5.4.8) can be replaced by substantially smaller constants, which means that in most cases F is small compared to A and the method is numerically stable.

For non-Hermitian matrices Householder's reduction method (see Section 6.5.1) requires about twice as many operations as the method described in this section. Since in most practical cases this method is not essentially less stable than the Householder method, it is preferred in practice (tests even show that the Householder method, due to the larger number of arithmetic operations, often produces somewhat less accurate results).

6.6 Methods for Determining the Eigenvalues and Eigenvectors

In this section we first describe how classical methods for the determination of zeros of polynomials (see Sections 5.5, 5.6) can be used to determine the eigenvalues of Hermitian tridiagonal and Hessenberg matrices.

We then note some iterative methods for the solution of the eigenvalue problem. A prototype of these methods is the simple vector iteration, in which one iteratively computes a specified eigenvalue and corresponding eigenvector of a matrix. A refinement of this method is Wielandt's inverse iteration method, by means of which all eigenvalues and eigenvectors can be determined, provided one knows sufficiently accurate approximations for the eigenvalues. To these iterative methods, finally, belong also the *LR* and the *QR* method for the calculation of all eigenvalues. The last two methods, especially the *QR* method, are the best methods currently known for the solution of the eigenvalue problem.

6.6.1 Computation of the Eigenvalues of a Hermitian Tridiagonal Matrix

For determining the eigenvalues of a Hermitian tridiagonal matrix

(6.6.1.1) $$J = \begin{bmatrix} \delta_1 & \bar{\gamma}_2 & & 0 \\ \gamma_2 & \ddots & \ddots & \\ & \ddots & \ddots & \bar{\gamma}_n \\ 0 & & \gamma_n & \delta_n \end{bmatrix}, \qquad \delta_i = \bar{\delta}_i,$$

there are (in addition to the most important method, the *QR* method, which will be described in Section 6.6.6) two obvious methods which we now wish to discuss. Without restricting generality, let J be an irreducible tridiagonal matrix, i.e., $\gamma_i \neq 0$ for all i. Otherwise, J would decompose into irreducible tridiagonal matrices $J^{(i)}$, $i = 1, \ldots, k$,

$$J = \begin{bmatrix} J^{(1)} & & & 0 \\ & J^{(2)} & & \\ & & \ddots & \\ 0 & & & J^{(k)} \end{bmatrix};$$

but the eigenvalues of J are just the eigenvalues of the $J^{(i)}$, $i = 1, \ldots, k$, so that it suffices to consider irreducible matrices J. The characteristic polynomial $\varphi(\mu)$ of J can easily be computed by recursion: indeed, letting

$$p_i(\mu) := \det(J_i - \mu I), \qquad J_i := \begin{bmatrix} \delta_1 & \bar{\gamma}_2 & & 0 \\ \gamma_2 & \ddots & \ddots & \\ & \ddots & \ddots & \bar{\gamma}_i \\ 0 & & \gamma_i & \delta_i \end{bmatrix},$$

and expanding $\det(J_i - \mu I)$ by the last column, one finds the recurrence relations

$$(6.6.1.2)\qquad \begin{aligned} p_0(\mu) &:= 1, \\ p_1(\mu) &= \delta_1 - \mu, \\ p_i(\mu) &= (\delta_i - \mu)p_{i-1}(\mu) - |\gamma_i|^2 p_{i-2}(\mu), \qquad i = 2, 3, \ldots, n, \\ \varphi(\mu) &\equiv p_n(\mu). \end{aligned}$$

Since δ_i and $|\gamma_i|^2$ are real, the polynomials $p_i(\mu)$ form a Sturm sequence [see Theorem (5.6.5)], provided that $\gamma_i \neq 0$ for $i = 2, \ldots, n$.

It is possible, therefore, to determine the eigenvalues of J with the bisection method described in Section 5.6. This method is recommended especially if one wants to compute not all, but only certain prespecified eigenvalues of J. It can further be recommended on account of its numerical stability when some eigenvalues of J are lying very close to each other [see Barth, Martin, and Wilkinson (1971) for an ALGOL program].

Since the eigenvalues of J are all real and simple, they can also be determined with Newton's method, say, in Maehly's version [see (5.5.13)], at least if the eigenvalues are not excessively clustered. The values $p_n(\lambda^{(j)})$, $p_n'(\lambda^{(j)})$ of the characteristic polynomial and its derivative required for Newton's method can be computed recursively: $p_n(\lambda^{(j)})$ by means of (6.6.1.2), and $p_n'(\lambda^{(j)})$ by means of the following formulas which are obtained by differentiating (6.6.1.2):

$$\begin{aligned} p_0'(\mu) &= 0, \\ p_1'(\mu) &= -1, \\ p_i'(\mu) &= -p_{i-1}(\mu) + (\delta_i - \mu)p_{i-1}'(\mu) - |\gamma_i|^2 p_{i-2}'(\mu), \qquad i = 2, \ldots, n. \end{aligned}$$

A starting value $\lambda^{(0)} \geq \max_i \lambda_i$ for Newton's method can be obtained from

(6.6.1.3) Theorem. *The eigenvalues λ_j of the matrix J in* (6.6.1.1) *satisfy the inequality*

$$|\lambda_j| \leq \max_{1 \leq i \leq n} \{|\gamma_i| + |\delta_i| + |\gamma_{i+1}|\}, \qquad \gamma_1 := \gamma_{n+1} := 0.$$

PROOF. For the maximum norm $\|x\|_\infty = \max|x_i|$ one has

$$\text{lub}_\infty(J) = \max_{1 \leqslant i \leqslant n} \{|\gamma_i| + |\delta_i| + |\gamma_{i+1}|\},$$

and from $Jx = \lambda_j x$, $x \neq 0$, there follows at once

$$|\lambda_j| \, \|x\|_\infty = \|Jx\|_\infty \leqslant \text{lub}_\infty(J) \cdot \|x\|_\infty, \qquad \|x\|_\infty \neq 0,$$

and hence $|\lambda_j| \leqslant \text{lub}_\infty(J)$. □

6.6.2 Computation of the Eigenvalues of a Hessenberg Matrix. The Method of Hyman

Besides the QR method (see Section 6.6.6), which is used most frequently in practice, one can in principle use all the methods of Chapter 5 to determine the zeros of the characteristic polynomial $p(\mu) = \det(B - \mu I)$ of a Hessenberg matrix $B = [b_{ik}]$. For this, as for example in the application of Newton's method, one must evaluate the values $p(\mu)$ and $p'(\mu)$ for given μ. The following method for computing these quantities is due to Hyman:

We assume that B is irreducible, i.e., that $b_{i,i-1} \neq 0$ for $i = 2, \dots, n$. For fixed μ one can then determine numbers $\alpha, x_1, \dots, x_{n-1}$ such that $x = [x_1, \dots, x_{n-1}, x_n]^T$, $x_n := 1$, is a solution of the system of equations

$$(B - \mu I)x = \alpha e_1,$$

or, written out in full,

$$\begin{aligned} (b_{11} - \mu)x_1 + b_{12}x_2 + \cdots + b_{1n}x_n &= \alpha, \\ b_{21}x_1 + (b_{22} - \mu)x_2 + \cdots + b_{2n}x_n &= 0, \\ &\vdots \\ b_{n,n-1}x_{n-1} + (b_{nn} - \mu)x_n &= 0. \end{aligned} \tag{6.6.2.1}$$

Starting with $x_n = 1$, one can indeed determine x_{n-1} from the last equation, x_{n-2} from the second to last equation, ..., x_1 from the second equation, and finally α from the first equation. The numbers x_i and α of course depend on μ. By interpreting (6.6.2.1) as an equation for x, given α, it follows by Cramer's rule that

$$1 = x_n = \frac{\alpha(-1)^{n-1} b_{21} b_{32} \dots b_{n,n-1}}{\det(B - \mu I)},$$

or

$$\alpha = \alpha(\mu) = \frac{(-1)^{n-1}}{b_{21} b_{32} \dots b_{n,n-1}} \det(B - \mu I). \tag{6.6.2.2}$$

Apart from a constant factor, $\alpha = \alpha(\mu)$ is thus identical with the characteristic polynomial of B. By differentiation with respect to μ, letting $x_i' := x_i'(\mu)$ and observing $x_n \equiv 1$, $x_n' \equiv 0$, one further obtains from (6.6.2.1) the formulas

$$\begin{aligned}
(b_{11} - \mu)x_1' - x_1 + b_{12}x_2' + \cdots + b_{1,n-1}x_{n-1}' &= \alpha', \\
b_{21}x_1' + (b_{22} - \mu)x_2' - x_2 + \cdots + b_{2,n-1}x_{n-1}' &= 0, \\
\vdots \qquad & \quad \vdots \\
b_{n,n-1}x_{n-1}' - x_n &= 0,
\end{aligned}$$

which, together with (6.6.2.1) and starting with the last equation, can be solved recursively for the x_i, x_i', α, and finally α'. In this manner one can compute $\alpha = \alpha(\mu)$ and $\alpha'(\mu)$ for every μ, and thus apply Newton's method to compute the zeros of $\alpha(\mu)$, i.e., by (6.6.2.2), the eigenvalues of B.

6.6.3 Simple Vector Iteration and Inverse Iteration of Wielandt

A precursor of all iterative methods for determining eigenvalues and eigenvectors of a matrix A is the simple vector iteration: Starting with an arbitrary initial vector $t_0 \in \mathbb{C}^n$, one forms the sequence of vectors $\{t_i\}$ with

$$t_i = At_{i-1}, \qquad i = 1, 2, \ldots.$$

Then

$$t_i = A^i t_0.$$

In order to examine the convergence of this sequence, we first assume A to be a diagonalizable $n \times n$ matrix having eigenvalues λ_i,

$$|\lambda_1| \geqslant |\lambda_2| \geqslant \cdots \geqslant |\lambda_n|.$$

We further assume that there is no eigenvalue λ_j different from λ_1 with $|\lambda_j| = |\lambda_1|$, i.e., there is an integer $r > 0$ such that

(6.6.3.1)
$$\begin{gathered}
\lambda_1 = \lambda_2 = \cdots = \lambda_r, \\
|\lambda_1| = \cdots = |\lambda_r| > |\lambda_{r+1}| \geqslant \cdots \geqslant |\lambda_n|.
\end{gathered}$$

The matrix A, being diagonalizable, has n linearly independent eigenvectors x_i, $Ax_i = \lambda_i x_i$, which form a basis of $\mathbb{C}^n$. We can therefore write t_0 in the form

(6.6.3.2)
$$t_0 = \rho_1 x_1 + \cdots + \rho_n x_n.$$

For t_i there follows the representation

(6.6.3.3)
$$t_i = A^i t_0 = \rho_1 \lambda_1^i x_1 + \cdots + \rho_n \lambda_n^i x_n.$$

Assuming now further that t_0 satisfies

$$\rho_1 x_1 + \cdots + \rho_r x_r \neq 0$$

—this makes precise the requirement that t_0 be "sufficiently general"—it follows from (6.6.3.3) and (6.6.3.1) that

$$\text{(6.6.3.4)} \quad \frac{1}{\lambda_1^i} t_i = \rho_1 x_1 + \cdots + \rho_r x_r + \rho_{r+1}\left(\frac{\lambda_{r+1}}{\lambda_1}\right)^i x_{r+1} + \cdots + \rho_n\left(\frac{\lambda_n}{\lambda_1}\right)^i x_n,$$

and thus, because $|\lambda_j/\lambda_1| < 1$ for $j \geqslant r+1$,

$$\text{(6.6.3.5)} \quad \lim_{i\to\infty} \frac{1}{\lambda_1^i} t_i = \rho_1 x_1 + \cdots + \rho_r x_r.$$

Normalizing the $t_i = [\tau_1^{(i)}, \ldots, \tau_n^{(i)}]^T$ in any way, for example, by letting

$$z_i := \frac{1}{\tau_{j_i}^{(i)}} t_i, \qquad |\tau_{j_i}^{(i)}| = \max_s |\tau_s^{(i)}|,$$

it follows from (6.6.3.5) that

$$\text{(6.6.3.6)} \quad \lim_{i\to\infty} \frac{\tau_{j_i}^{(i+1)}}{\tau_{j_i}^{(i)}} = \lambda_1, \qquad \lim_{i\to\infty} z_i = \alpha(\rho_1 x_1 + \cdots + \rho_r x_r),$$

where $\alpha \neq 0$ is a normalization constant. Under the stated assumptions the method thus furnishes both the dominant eigenvalue λ_1 of A and an eigenvector belonging to λ_1, namely the vector $z = \alpha(\rho_1 x_1 + \cdots + \rho_r x_r)$, and we say that "the vector iteration converges toward λ_1 and a corresponding eigenvector."

Note that for $r = 1$ (λ_1 a simple eigenvalue) the limit vector z is independent of the choice of t_0 (provided only that $\rho_1 \neq 0$). If λ_1 is a multiple dominant eigenvalue ($r > 1$), then the eigenvector z obtained depends on the ratios $\rho_1 : \rho_2 : \ldots : \rho_r$, and thus on the initial vector t_0. In addition, it can be seen from (6.6.3.4) that we have linear convergence with convergence factor $|\lambda_{r+1}/\lambda_1|$: the method converges better for smaller $|\lambda_{r+1}/\lambda_1|$. The proof of convergence, at the same time, shows that the method does not always converge to λ_1 and an associated eigenvector, but may converge toward an eigenvalue λ_k and an eigenvector belonging to λ_k, if in the decomposition (6.6.3.2) of t_0 one has

$$\rho_1 = \cdots = \rho_{k-1} = 0, \qquad \rho_k \neq 0$$

(and there is no eigenvalue which differs from λ_k and has the same modulus as λ_k). This statement, however, has only theoretical significance, for even if initially one has exactly $\rho_1 = 0$ for t_0, due to the effect of rounding errors we will get for the computed $\bar{t}_1 = \mathrm{fl}(At_0)$, in general,

$$\bar{t}_1 = \varepsilon\lambda_1 x_1 + \bar{\rho}_2\lambda_2 x_2 + \cdots + \bar{\rho}_n\lambda_n x_n$$

with a small $\varepsilon \neq 0$ and $\bar{\rho}_i \approx \rho_i$, $i = 2, \ldots, n$, so that the method eventually still converges to λ_1 and a corresponding eigenvector.

Suppose now that A is a nondiagonalizable matrix with a uniquely determined dominant eigenvalue λ_1 (i.e., $|\lambda_1| = |\lambda_i|$ implies $\lambda_1 = \lambda_i$). Replacing (6.6.3.2) by a representation of t_0 as a linear combination of eigenvectors and principal vectors of A, one can show in the same way that for "sufficiently general" t_0 the vector iteration converges to λ_1 and an associated eigenvector.

For practical computation the simple vector iteration is only conditionally useful, since it converges slowly if the moduli of the eigenvalues are not sufficiently separated, and moreover furnishes only one eigenvalue and associated eigenvector. These disadvantages are avoided in the *inverse iteration* (also called *fractional iteration*) of Wielandt. Here one assumes that a good approximation λ is already known for one of the eigenvalues $\lambda_1, \ldots, \lambda_n$ of A, say λ_j:

(6.6.3.7) $$|\lambda_j - \lambda| \ll |\lambda_k - \lambda| \quad \text{for all } \lambda_k \neq \lambda_j.$$

Starting with a "sufficiently general" initial vector $t_0 \in \mathbb{C}^n$, one then forms the vectors t_i, $i = 1, 2, \ldots$, according to

(6.6.3.8) $$(A - \lambda I)t_i = t_{i-1}.$$

If $\lambda \neq \lambda_i$, $i = 1, \ldots, n$, then $(A - \lambda I)^{-1}$ exists and (6.6.3.8) is equivalent to

$$t_i = (A - \lambda I)^{-1} t_{i-1},$$

i.e., to the simple vector iteration with the matrix $(A - \lambda I)^{-1}$ and the eigenvalues $1/(\lambda_k - \lambda)$, $k = 1, 2, \ldots, n$. Because of (6.6.3.7) one has

$$\left|\frac{1}{\lambda_j - \lambda}\right| \gg \left|\frac{1}{\lambda_k - \lambda}\right| \quad \text{for } \lambda_k \neq \lambda_j.$$

Assuming A again diagonalizable, with eigenvectors x_i, it follows from $t_0 = \rho_1 x_1 + \cdots + \rho_n x_n$ (if λ_j is a simple eigenvalue) that

$$t_i = (A - \lambda I)^{-i} t_0 = \sum_{k=1}^{n} \frac{\rho_k}{(\lambda_k - \lambda)^i} x_k,$$

(6.6.3.9) $$(\lambda_j - \lambda)^i t_i = \rho_j x_j + \sum_{k \neq j} \left(\frac{\lambda_j - \lambda}{\lambda_k - \lambda}\right)^i \rho_k x_k,$$

$$\lim_{i \to \infty} (\lambda_j - \lambda)^i t_i = \rho_j x_j.$$

The smaller $|\lambda_j - \lambda|/|\lambda_k - \lambda|$ for $\lambda_k \neq \lambda_j$ (i.e., the better the approximation λ), the better will be the convergence.

The relations (6.6.3.9) may give the impression that an initial vector t_0 is the more suitable for inverse iteration the more closely it agrees with the eigenvector x_j, to whose eigenvalue λ_j one knows a good approximation λ. They further seem to suggest that with the choice $t_0 \approx x_j$ the "accuracy" of the t_i increases uniformly with i. This impression is misleading, and true in general only for the well-conditioned eigenvalues λ_j of a matrix A (see Section 6.9), i.e., for the eigenvalues λ_j which under a small perturbation of A change only a little:

$$\lambda_j(A + \Delta A) - \lambda_j(A) = O(\text{eps}) \quad \text{if} \quad \frac{\text{lub}(\Delta A)}{\text{lub}(A)} = O(\text{eps}).$$

According to Section 6.9 all eigenvalues of symmetric matrices A are well conditioned. A poor condition for λ_j must be expected in those cases in which λ_j is a multiple zero of the characteristic polynomial of A and A has nonlinear elementary divisors, or, what often occurs in practice, when λ_j belongs to a cluster of eigenvalues whose eigenvectors are almost linearly dependent.

Before we study in an example the possible complications for the inverse iteration caused by an ill-conditioned λ_j, let us clarify what we mean by a *numerically acceptable* eigenvalue λ (in the context of the machine precision eps used) and a corresponding eigenvector $x_\lambda \neq 0$ of a matrix A.

The number λ is called a *numerically acceptable eigenvalue* of A if there exists a small matrix ΔA with

$$\frac{\operatorname{lub}(\Delta A)}{\operatorname{lub}(A)} = O(\text{eps})$$

such that λ is an exact eigenvalue of $A + \Delta A$. Such numerically acceptable eigenvalues λ are produced by every numerically stable method for the determination of eigenvalues.

The vector $x_\lambda \neq 0$ is called a *numerically acceptable eigenvector* corresponding to a given approximation λ of an eigenvalue of A if there exists a small matrix $\Delta_1 A$ with

$$\text{(6.6.3.10)} \qquad (A + \Delta_1 A - \lambda I)x_\lambda = 0, \qquad \frac{\operatorname{lub}(\Delta_1 A)}{\operatorname{lub}(A)} = O(\text{eps}).$$

EXAMPLE 1. For the matrix

$$A = \begin{bmatrix} 1 & 1 \\ 0 & 1 \end{bmatrix}$$

the number $\lambda := 1 + \sqrt{\text{eps}}$ is a numerically acceptable eigenvalue, for λ is an exact eigenvalue of

$$A + \Delta A \quad \text{with } \Delta A = \begin{bmatrix} 0 & 0 \\ \text{eps} & 0 \end{bmatrix}.$$

Although

$$x_\lambda := \begin{bmatrix} 1 \\ 0 \end{bmatrix}$$

is an exact eigenvector of A, it is not, in the sense of the above definition, a numerically acceptable eigenvector corresponding to the approximation $\lambda = 1 + \sqrt{\text{eps}}$, because every matrix

$$\Delta_1 A = \begin{bmatrix} \alpha & \beta \\ \gamma & \delta \end{bmatrix}$$

with

$$(A + \Delta_1 A - \lambda I)x_\lambda = 0$$

has the form

$$\Delta_1 A = \begin{bmatrix} \sqrt{\text{eps}} & \beta \\ 0 & \delta \end{bmatrix}, \qquad \beta, \delta \text{ arbitrary.}$$

For all these matrices, $\text{lub}_\infty(\Delta_1 A) \geqslant \sqrt{\text{eps}} \gg O(\text{eps} \cdot \text{lub}(A))$.

It is thus possible to have the following paradoxical situation: Let λ_0 be an exact eigenvalue of a matrix, x_0 a corresponding exact eigenvector, and λ a numerically acceptable approximation of λ_0. Then x_0 need not be a numerically acceptable eigenvector to the given approximation λ of λ_0.

If one has such a numerically acceptable eigenvalue λ of A, then with inverse iteration one merely tries to find a corresponding numerically acceptable eigenvector x_λ. A vector t_j, $j \geqslant 1$, found by means of inverse iteration from an initial vector t_0 can be accepted as such an x_λ, provided

$$\text{(6.6.3.11)} \qquad \frac{\|t_{j-1}\|}{\|t_j\|} = O(\text{eps} \cdot \text{lub}(A)).$$

For then, in view of $(A - \lambda I)t_j = t_{j-1}$, one has

$$(A + \Delta A - \lambda I)t_j = 0$$

with

$$\Delta A := -\frac{t_{j-1} t_j^H}{t_j^H t_j},$$

$$\text{lub}_2(\Delta A) = \frac{\|t_{j-1}\|_2}{\|t_j\|_2} = O(\text{eps} \cdot \text{lub}_2(A)).$$

It is now surprising that for ill-conditioned eigenvalues we can have an iterate t_j which is numerically acceptable, while its successor t_{j+1} is not:

EXAMPLE 2. The matrix

$$A = \begin{bmatrix} \eta & 1 \\ \eta & \eta \end{bmatrix}, \quad \eta = O(\text{eps}), \qquad \text{lub}_\infty(A) = 1 + |\eta|,$$

has the eigenvalues $\lambda_1 = \eta + \sqrt{\eta}$, $\lambda_2 = \eta - \sqrt{\eta}$ with corresponding eigenvectors

$$x_1 = \begin{bmatrix} 1 \\ \sqrt{\eta} \end{bmatrix}, \qquad x_2 = \begin{bmatrix} 1 \\ -\sqrt{\eta} \end{bmatrix}.$$

Since η is small, A is very ill-conditioned and λ_1, λ_2 form a cluster of eigenvalues whose eigenvectors x_1, x_2 are almost linearly dependent [e.g., the slightly perturbed matrix

$$\tilde{A} := \begin{bmatrix} 0 & 1 \\ 0 & 0 \end{bmatrix} = A + \Delta A, \qquad \frac{\text{lub}_\infty(\Delta A)}{\text{lub}_\infty(A)} = \frac{2|\eta|}{1 + |\eta|} = O(\text{eps}),$$

has the eigenvalue $\lambda(\tilde{A}) = 0$ with $\min|\lambda(\tilde{A}) - \lambda_i| \geqslant \sqrt{|\eta|} - |\eta| \gg O(\text{eps})$]. The number $\lambda = 0$ is a numerically acceptable eigenvalue of A to the eigenvector

$$x_\lambda = \begin{bmatrix} 1 \\ 0 \end{bmatrix}.$$

Indeed,

$$(A + \Delta A - 0 \cdot I)x_\lambda = 0 \quad \text{for } \Delta A := \begin{bmatrix} -\eta & 0 \\ -\eta & 0 \end{bmatrix},$$

$$\frac{\text{lub}_\infty(\Delta A)}{\text{lub}_\infty(A)} = \frac{|\eta|}{1 + |\eta|} = O(\text{eps}).$$

Taking this numerically acceptable eigenvector

$$x_\lambda = \begin{bmatrix} 1 \\ 0 \end{bmatrix}$$

as the initial vector t_0 for the inverse iteration, and $\lambda = 0$ as an approximate eigenvalue of A, one obtains after the first step

$$t_1 := \frac{-1}{1 - \eta} \begin{bmatrix} 1 \\ -1 \end{bmatrix}.$$

The vector t_1, however, unlike t_0, is no longer numerically acceptable as an eigenvector of A, every matrix $\Delta_1 A$ with

$$(A + \Delta_1 A - 0 \cdot I)t_1 = 0$$

having the form

$$\Delta_1 A = \begin{bmatrix} \alpha & \beta \\ \gamma & \delta \end{bmatrix} \quad \text{with } \alpha = 1 + \beta - \eta,\ \gamma = \delta.$$

These matrices $\Delta_1 A$ all satisfy $\text{lub}_\infty(\Delta_1 A) \geqslant 1 - |\eta|$; there is no small matrix among them with $\text{lub}_\infty(\Delta_1 A) = O(\text{eps})$.

On the other hand, every initial vector t_0 of the form

$$t_0 = \begin{bmatrix} \tau \\ 1 \end{bmatrix}$$

with $|\tau|$ not too large produces for t_1 a numerically acceptable eigenvector. For example, if

$$t_0 = \begin{bmatrix} 1 \\ 1 \end{bmatrix},$$

the vector

$$t_1 = \frac{1}{\eta} \begin{bmatrix} 1 \\ 0 \end{bmatrix}$$

is numerically acceptable, as we just showed.

Similar circumstances to those in this example prevail in general whenever the eigenvalue considered has nonlinear elementary divisors. Suppose λ_j is an eigenvalue of A for which there are elementary divisors of degrees at most k [$k = \tau_j$; see (6.2.11)]. Through a numerically stable method one can in general [see (6.9.11)] obtain only an approximate value λ within the error $\lambda - \lambda_j = O(\text{eps}^{1/k})$ [we assume for simplicity $\text{lub}(A) = 1$].

Let now $x_0, x_1, \ldots, x_{k-1}$ be a chain of principal vectors (see Sections 6.2, 6.3) for the eigenvalue λ_j,

$$(A - \lambda_j I)x_i = x_{i-1} \text{ for } i = k - 1, \ldots, 0 \qquad (x_{-1} := 0).$$

It then follows immediately, for $\lambda \neq \lambda_i$, $i = 1, \ldots, n$, that

$$(A - \lambda I)^{-1}[A - \lambda I + (\lambda - \lambda_j)I]x_i = (A - \lambda I)^{-1}x_{i-1},$$

$$(A - \lambda I)^{-1}x_i = \frac{1}{\lambda_j - \lambda}x_i + \frac{1}{\lambda - \lambda_j}(A - \lambda I)^{-1}x_{i-1},$$

$$i = k - 1, \ldots, 0,$$

and from this, by induction,

$$(A - \lambda I)^{-1}x_{k-1} = \frac{1}{\lambda_j - \lambda}x_{k-1} - \frac{1}{(\lambda_j - \lambda)^2}x_{k-2} + \cdots \pm \frac{1}{(\lambda_j - \lambda)^k}x_0.$$

For the initial vector $t_0 := x_{k-1}$, we thus obtain for the corresponding t_1

$$\|t_1\| = O((\lambda_j - \lambda)^{-k}) = O\left(\frac{1}{\text{eps}}\right),$$

and therefore $\|t_0\|/\|t_1\| = O(\text{eps})$, so that t_1 is acceptable as an eigenvector of A [see (6.6.3.11)]. If, however, we had taken as initial vector $t_0 := x_0$, the exact eigenvector of A, we would get

$$t_1 = \frac{1}{\lambda_j - \lambda}t_0$$

and thus

$$\frac{\|t_0\|}{\|t_1\|} = O(\text{eps}^{1/k}).$$

Since $\text{eps}^{1/k} \gg \text{eps}$, the vector t_1 is not a numerically acceptable eigenvector to the approximate value λ at hand (cf. Example 1).

For this reason, one applies inverse iteration in practice only in a rather rudimentary form: Having computed (numerically acceptable) approximations λ for the exact eigenvalues of A by means of a numerically stable method, one tentatively determines, for a few different initial vectors t_0, the corresponding t_1 and accepts as eigenvector that vector t_1 for which $\|t_0\|/\|t_1\|$ best represents a quantity of $O(\text{eps} \cdot \text{lub}(A))$.

Since the step $t_0 \to t_1$ requires the solution of a system of linear equations (6.6.3.8), one applies inverse iteration in practice only for tridiagonal and Hessenberg matrices A. To solve the system of equations (6.6.3.8) ($A - \lambda I$ is almost singular), one decomposes the matrix $A - \lambda I$ into the product of a lower and an upper triangular matrix, L and R, respectively. To insure numerical stability one must use *partial pivoting*, i.e., one determines a permutation matrix P and matrices L, R with (see Section 4.1)

$P(A - \lambda I) = LR$, $L =$ 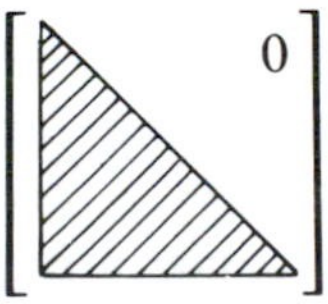, $R =$ 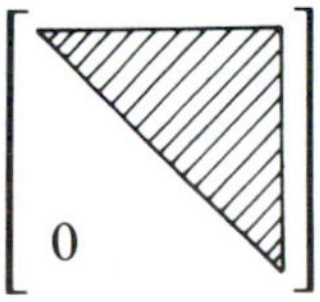

$|l_{ik}| \leq 1.$

The solution t_1 of (6.6.3.8) is then obtained from the two triangular systems of equations

$$\begin{aligned} Lz &= Pt_0, \\ Rt_1 &= z. \end{aligned} \tag{6.6.3.12}$$

For tridiagonal and Hessenberg matrices A, the matrix L is very sparse: In each column of L at most two elements are different from zero. R for Hessenberg matrices A is an upper triangular matrix; for tridiagonal matrices A, a matrix of the form

$$R = \begin{bmatrix} * & * & * & & & 0 \\ & \cdot & \cdot & \cdot & & \\ & & \cdot & \cdot & \cdot & \\ & & & \cdot & \cdot & \cdot \\ & & & & \cdot & \cdot & * \\ & & & & & \cdot & * \\ 0 & & & & & & * \end{bmatrix},$$

so that in each case the solution of (6.6.3.12) requires only a few operations. Besides, $\|z\| \approx \|t_0\|$, by the first relation in (6.6.3.12). The work can therefore be simplified further by determining t_1 only from $Rt_1 = z$, where one tries to choose the vector z so that $\|z\|/\|t_1\|$ becomes as small as possible.

An ALGOL program for the computation of the eigenvectors of a symmetric tridiagonal matrix using inverse iteration can be found in Peters and Wilkinson (1971a); FORTRAN programs, in Smith et al. (1976).

6.6.4 The *LR* and *QR* Methods

The *LR* method of Rutishauser (1958) and the *QR* method of Francis (1961/62) and Kublanovskaja (1961) are also iterative methods for the computation of the eigenvalues of an $n \times n$ matrix A. We first describe the historically earlier *LR* method. Here, one generates a sequence of matrices A_i, beginning with $A_1 := A$, according to the following rules: Use the Gauss elimination method (see Section 4.1) to represent A_i as the product of a lower triangular matrix $L_i = [l_{jk}]$ with $l_{jj} = 1$ and an upper triangular matrix R_i

$$A_i =: L_i R_i, \qquad L_i = \begin{bmatrix} 1 & & 0 \\ \vdots & \ddots & \\ * & \cdots & 1 \end{bmatrix}, \qquad R_i = \begin{bmatrix} * & \cdots & * \\ & \ddots & \vdots \\ 0 & & * \end{bmatrix}. \tag{6.6.4.1}$$

Thereafter, let

$$A_{i+1} := R_i L_i =: L_{i+1} R_{i+1}, \qquad i = 1, 2, \ldots.$$

From Section 4.1 we know that an arbitrary matrix A_i does not always have a factorization $A_i = L_i R_i$. We assume, however, for the following discussion that A_i can be so factored. We begin by showing the following.

(6.6.4.2) Theorem. *If all decompositions $A_i = L_i R_i$ exist, then*

(a) *A_{i+1} is similar to A_i:*

$$A_{i+1} = L_i^{-1} A_i L_i, \qquad i = 1, 2, \ldots$$

(b) $A_{i+1} = (L_1 L_2 \ldots L_i)^{-1} A_1 (L_1 L_2 \ldots L_i), i = 1, 2, \ldots.$

(c) *for the lower triangular matrix $T_i := L_1 \ldots L_i$ and the upper triangular matrix $U_i := R_i \ldots R_1$ one has*

$$A^i = A_1^i = T_i U_i, \qquad i = 1, 2, \ldots.$$

PROOF. (a) From $A_i = L_i R_i$ we get

$$L_i^{-1} A_i L_i = R_i L_i =: A_{i+1}.$$

(b) follows immediately from (a).

To prove (c), note that (b) implies

$$L_1 \ldots L_i A_{i+1} = A_1 L_1 \ldots L_i, \qquad i = 1, 2, \ldots.$$

It follows for $i = 1, 2, \ldots$ that

$$\begin{aligned} T_i U_i &= L_1 \ldots L_{i-1}(L_i R_i) R_{i-1} \ldots R_1 \\ &= L_1 \ldots L_{i-1} A_i R_{i-1} \ldots R_1 \\ &= A_1 L_1 \ldots L_{i-1} R_{i-1} \ldots R_1 \\ &= A_1 T_{i-1} U_{i-1}. \end{aligned}$$

With this, the theorem is proved. □

It is possible to show, under certain conditions, that the matrices A_i converge to an upper triangular matrix A_∞ with the eigenvalues λ_j of A as diagonal elements, $\lambda_j = (A_\infty)_{jj}$. However, we wish to analyze the convergence only for the QR method, which is closely related to the LR method, because the LR method has a serious drawback: it breaks down if one of the matrices A_i does not have a triangular decomposition, and even if the decomposition $A_i = L_i R_i$ exists, the problem of computing L_i and R_i may be ill conditioned.

In order to avoid these difficulties (and to make the method numerically stable), one could think of forming the LR decomposition with partial pivoting:

$$\begin{aligned} &P_i A_i =: L_i R_i, \qquad P_i \text{ permutation matrix}, P_i^{-1} = P_i^T, \\ &A_{i+1} := R_i P_i^T L_i = L_i^{-1}(P_i A_i P_i^T) L_i. \end{aligned}$$

There are examples, however, in which the modified process fails to converge

$$A_1 = \begin{bmatrix} 1 & 3 \\ 2 & 0 \end{bmatrix}, \qquad \lambda_1 = 3, \qquad \lambda_2 = -2$$

$$P_1 = \begin{bmatrix} 0 & 1 \\ 1 & 0 \end{bmatrix}, \qquad P_1 A_1 = \begin{bmatrix} 2 & 0 \\ 1 & 3 \end{bmatrix} = \begin{bmatrix} 1 & 0 \\ 1/2 & 1 \end{bmatrix} \begin{bmatrix} 2 & 0 \\ 0 & 3 \end{bmatrix} = L_1 R_1$$

$$A_2 = R_1 P_1^T L_1 = \begin{bmatrix} 1 & 2 \\ 3 & 0 \end{bmatrix}$$

$$P_2 = \begin{bmatrix} 0 & 1 \\ 1 & 0 \end{bmatrix}, \qquad P_2 A_2 = \begin{bmatrix} 3 & 0 \\ 1 & 2 \end{bmatrix} = \begin{bmatrix} 1 & 0 \\ 1/3 & 1 \end{bmatrix} \begin{bmatrix} 3 & 0 \\ 0 & 2 \end{bmatrix} = L_2 R_2$$

$$A_3 = R_2 P_2^T L_2 = \begin{bmatrix} 1 & 3 \\ 2 & 0 \end{bmatrix} \equiv A_1.$$

The *LR* method without pivoting, on the other hand, would converge for this example.

These difficulties of the *LR* method are avoided in the *QR* method of Francis (1961/62), which can be regarded as a natural modification of the *LR* algorithm. Formally the *QR* method is obtained if one replaces the *LR* decompositions in (6.6.4.1) by *QR* decompositions (see Section 4.7). Thus, again beginning with $A_1 := A$, the *QR* method generates matrices Q_i, R_i, and A_i according to the following rules:

$$A_i =: Q_i R_i, \qquad Q_i^H Q_i = I, \qquad R_i = \begin{bmatrix} * & \cdots & * \\ & \ddots & \vdots \\ 0 & & * \end{bmatrix}$$

(6.6.4.3)

$$A_{i+1} := R_i Q_i.$$

Here, the matrices are factored into a product of a unitary matrix Q_i and an upper triangular matrix R_i. Note that a *QR* decomposition of A_i always exists and that it can be computed with the methods of Section 4.7 in a numerically stable way: There are $n-1$ Householder matrices $H_j^{(i)}$, $j = 1, \ldots, n-1$, satisfying

$$H_{n-1}^{(i)} \ldots H_1^{(i)} A_i = R_i,$$

where R_i is upper triangular. Then, since $(H_j^{(i)})^H = (H_j^{(i)})^{-1} = H_j^{(i)}$ the matrix

$$Q_i := H_1^{(i)} \ldots H_{n-1}^{(i)}$$

is unitary and satisfies $A_i = Q_i R_i$, so that A_{i+1} can be computed as

$$A_{i+1} = R_i Q_i = R_i H_1^{(i)} \ldots H_{n-1}^{(i)}.$$

Note further that the *QR* decomposition of a matrix is not unique. If S is an arbitrary diagonal unitary matrix, i.e., a *phase matrix* of the form

$$S = \operatorname{diag}(e^{i\phi_1}, e^{i\phi_2}, \ldots, e^{i\phi_n}),$$

then $Q_i S_i$ and $S_i^H R_i$ are unitary and upper triangular, respectively, and $(Q_i S)(S^H R_i) = Q_i R_i$. In analogy to Theorem (6.6.4.2), the following holds.

(6.6.4.4) Theorem. *The matrices* A_i, Q_i, R_i *in* (6.6.4.3) *and*

$$P_i := Q_1 Q_2 \ldots Q_i, \qquad U_i := R_i R_{i-1} \ldots R_1,$$

satisfy:

(a) A_{i+1} *is unitarily similar to* A_i, $A_{i+1} = Q_i^H A_i Q_i$.
(b) $A_{i+1} = (Q_1 \ldots Q_i)^H A_1 (Q_1 \ldots Q_i) = P_i^H A_1 P_i$.
(c) $A^i = P_i U_i$.

The proof is carried out in much the same way as for Theorem (6.6.4.2) and is left to the reader.

In order to make the convergence properties (see (6.6.4.12)) of the QR method plausible, we show that this method can be regarded as a natural generalization of both the simple and inverse vector iteration (see Section 6.6.3). As in the proof of (6.6.4.2), starting with $P_0 := I$, we obtain from (6.6.4.4) the relation

$$\text{(6.6.4.5)} \qquad P_i R_i = A P_{i-1}, \qquad i \geq 1.$$

If we partition the matrices P_i and R_i as follows

$$P_i = [P_i^{(r)}, \hat{P}_i^{(r)}], \qquad R_i = \begin{bmatrix} R_i^{(r)} & * \\ 0 & \hat{R}_i^{(r)} \end{bmatrix},$$

so that $P_i^{(r)}$ is a $n \times r$ matrix and $R_i^{(r)}$ an $r \times r$ matrix, then (6.6.4.5) implies

$$\text{(6.6.4.6)} \qquad A P_{i-1}^{(r)} = P_i^{(r)} R_i^{(r)} \qquad \text{for } i \geq 1,\ 1 \leq r \leq n.$$

Denoting the linear subspace, which is spanned by the orthonormal columns of $P_i^{(r)}$ by $\mathscr{P}_i^{(r)} := R(P_i^{(r)}) = \{P_i^{(r)} z \mid z \in \mathbb{C}^r\}$, we obtain

$$\mathscr{P}_i^{(r)} \supset A \mathscr{P}_{i-1}^{(r)} = R(P_i^{(r)} R_i^{(r)}), \qquad \text{for } i \geq 1,\ 1 \leq r \leq n,$$

with equality holding, when A is nonsingular, implying that all R_i, $R_i^{(r)}$ will also be nonsingular, that is, the QR method can be regarded as a *subspace iteration*. The special case $r = 1$ in (6.6.4.6) is equivalent to the usual vector iteration (see 6.6.3) with the starting vector $e_1 = P_0^{(1)}$: If we write $p_i := P_i^{(1)}$ then

$$\text{(6.6.4.7)} \qquad r_{11}^{(i)} p_i = A p_{i-1}, \qquad \|p_i\| = 1,\ i \geq 1,$$

as follows from (6.6.4.6); also, $R_i^{(1)} = r_{11}^{(i)}$, and $\|P_i^{(1)}\| = 1$. The convergence of the vectors p_i can be investigated as in 6.6.3: We assume for simplicity that A is diagonalizable and has the eigenvalues λ_i with

$$|\lambda_1| > |\lambda_2| \geq \cdots \geq |\lambda_n|.$$

Further let $X = (x_1, \ldots, x_n) = (x_{ik})$, $Y := X^{-1} = (y_1, \ldots, y_n)^T = (y_{ik})$ be matrices with

$$\text{(6.6.4.8}a\text{)} \qquad A = XDY, \qquad D = \begin{bmatrix} \lambda_1 & & 0 \\ & \ddots & \\ 0 & & \lambda_n \end{bmatrix}$$

so that x_i and y_i^T are a right and left eigenvector belonging to λ_i:

$$\text{(6.6.4.8}b\text{)} \quad Ax_i = \lambda_i x_i, \qquad y_i^T A = \lambda_i y_i^T, \qquad y_i^T x_k = \begin{cases} 1 & \text{for } i = k \\ 0 & \text{otherwise.} \end{cases}$$

If $\rho_1 = y_{11} \neq 0$ holds in the decomposition of e_1

$$e_1 = \rho_1 x_1 + \cdots + \rho_n x_n, \qquad \rho_i = y_i^T e_1 = y_{i1},$$

then by the results of 6.6.3 the ordinary vector iteration, $t_k := A^k e_1$ satisfies:

$$\text{(6.6.4.9)} \qquad \lim_{k\to\infty} \frac{1}{\lambda_1^k} t_k = \rho_1 x_1.$$

On the other hand (6.6.4.7) implies

$$A^i e_1 = r_{11}^{(1)} r_{11}^{(2)} \ldots r_{11}^{(i)} p_i.$$

Since $\|p_i\| = 1$, there are phase factors $\sigma_k = e^{i\phi_k}$, $|\sigma_k| = 1$, so that

$$\lim_i \sigma_i p_i = \hat{x}_1, \qquad \lim r_{11}^{(i)} \frac{\sigma_{i-1}}{\sigma_i} = \lambda_1, \qquad \text{where } \hat{x}_1 := x_1/\|x_1\|.$$

Thus, the $r_{11}^{(i)}$ and p_i "essentially converge" (i.e., up to a phase factor) toward λ_1 and $\hat{x}_1$ as i tends to ∞. By the results of 6.6.3 the speed of convergence is determined by the factor $|\lambda_2/\lambda_1| < 1$,

$$\text{(6.6.4.10)} \qquad \|\sigma_i p_i - \hat{x}_1\| = O\left(\left|\frac{\lambda_2}{\lambda_1}\right|^i\right).$$

Using (6.6.4.10) and $A_{i+1} = P_i^H A P_i$ (Theorem (6.6.4.4) (b)) it is easy to see that the first column $A_i e_1$ of A_i converges to the vector $\lambda_1 e_1 = [\lambda_1, 0, \ldots, 0]^T$ as $i \to \infty$, and the smaller $|\lambda_2/\lambda_1| < 1$ is, the faster the error $\|A_i e_1 - \lambda_1 e_1\| =$

$O((|\lambda_2|/|\lambda_1|)^i)$ tends to zero. We note that the condition $\rho_1 = y_{11} \neq 0$, which ensures the convergence, is satisfied if the matrix Y has a decomposition $Y = L_Y R_Y$ into a lower triangular matrix L_Y with $(L_Y)_{jj} = 1$ and an upper triangular matrix R_Y.

It will turn out to be particularly important that the QR method is also related to the inverse vector iteration (see 6.6.3) if A is nonsingular. Since $P_i^H P_i = I$, we obtain from (6.6.4.5) $P_{i-1}^H A^{-1} = R_i^{-1} P_i^H$ or

$$A^{-H} P_{i-1} = P_i R_i^{-H}.$$

Denote the $n \times (n - r + 1)$ matrix consisting of the last $n - r + 1$ columns of P_i by $\hat{P}_i^{(r)}$, denote the subspace generated by the columns of $\hat{P}_i^{(r)}$ by $\hat{\mathscr{P}}_i^{(r)} = R(\hat{P}_i^{(r)})$, and denote the upper triangular matrix consisting of the last $n - r + 1$ rows and columns of R_i by $\hat{R}_i^{(r)}$. Then the last equation can be written as a vector space iteration with the inverse $A^{-H} = (A^H)^{-1}$ of the matrix A^H:

$$\left.\begin{aligned} A^{-H}\hat{P}_{i-1}^{(r)} &= \hat{P}_i^{(r)}(\hat{R}_i^{(r)})^{-H} \\ A^{-H}\hat{\mathscr{P}}_{i-1}^{(r)} &= \hat{\mathscr{P}}_i^{(r)} \end{aligned}\right\} \quad \text{for } i \geq 1,\ 1 \leq r \leq n.$$

In the special case $r = n$ this iteration reduces to an ordinary inverse vector iteration for the last column $\hat{p}_i := \hat{P}_i^{(n)}$ of the matrices P_i:

$$A^{-H}\hat{p}_{i-1} = \hat{p}_i \cdot \bar{\rho}_{nn}^{(i)}, \qquad \rho_{nn}^{(i)} := (R_i^{-1})_{nn}, \qquad \|\hat{p}_i\| = 1,$$

arising from the starting vector $\hat{p}_0 = e_n$. Convergence of the iteration can be investigated as in Section 6.6.3 and as previously. Again, the discussion is easy if A is diagonalizable and A^{-H} has a unique eigenvalue of maximal modulus, i.e., if the eigenvalues λ_i of A now satisfy $|\lambda_1| \geq \cdots \geq |\lambda_{n-1}| > |\lambda_n| > 0$ (note that A^{-H} has the eigenvalues $\bar{\lambda}_j^{-1}$, $j = 1, \ldots, n$). Let x_i and y_i^T be the right and left eigenvector of A belonging to the eigenvalue λ_i, and consider (6.6.4.8). If the starting vector $\hat{p}_0 = e_n$ is sufficiently general, the vectors $\hat{p}_i$ will "essentially" converge toward a normalized eigenvector u, $\|u\| = 1$, of A^{-H} for its largest eigenvalue $\bar{\lambda}_n^{-1}$, $A^{-H}u = \bar{\lambda}_n^{-1}u$, or equivalently $u^H A = \lambda_n u^H$. Therefore the last rows $e_n^T A_i$ of the matrices A_i will converge to $\lambda_n e_n^T = [0, \ldots, 0, \lambda_n]$ as $i \to \infty$.

The convergence speed is now determined by the quotient $|\lambda_n/\lambda_{n-1}| < 1$,

$$\text{(6.6.4.11)} \qquad \|e_n^T A_i - [0, \ldots, 0, \lambda_n]\| = O\left(\left|\frac{\lambda_n}{\lambda_{n-1}}\right|^i\right),$$

since A^{-H} has the eigenvalues $\bar{\lambda}_j^{-1}$ satisfying $|\lambda_n^{-1}| > |\lambda_{n-1}^{-1}| \geq \cdots \geq |\lambda_1^{-1}|$ by hypothesis. For the formal convergence analysis we have to write the starting vector $\hat{p}_0 = e_n$ as a linear combination

$$e_n = \rho_1 \bar{y}_1 + \cdots + \rho_n \bar{y}_n.$$

of the right eigenvectors $\bar{y}_j$ of the matrix A^{-H}, $A^{-H}\bar{y}_j = \bar{\lambda}_j^{-1}\bar{y}_j$. Convergence is ensured if the coefficient ρ_n associated with the eigenvalue $\bar{\lambda}_n^{-1}$ of A^{-H} largest in modulus is $\neq 0$. Now, $I = X^H Y^H = [x_1, \ldots, x_n]^H[\bar{y}_1, \ldots, \bar{y}_n]$ implies $\rho_n = x_n^H e_n$, so that $\rho_n \neq 0$ if the element $x_{nn} = e_n^T x_n = \bar{\rho}_n$ of the matrix X is nonzero. We note that this is the case if the matrix $Y = L_Y R_Y$ has an LR decomposition: the reason is that $X = R_Y^{-1} L_Y^{-1}$, since $X = Y^{-1}$ and R_Y^{-1} and L_Y^{-1} are upper and lower triangular matrices, which implies $x_{nn} = e_n^T R_Y^{-1} L_Y^{-1} e_n \neq 0$. This analysis motivates part of the following theorem, which assures the convergence of the QR method if *all* eigenvalues of A have different moduli.

(6.6.4.12) Theorem. *Let $A =: A_1$ be an $n \times n$ matrix satisfying the following hypotheses*

(1) *The eigenvalues λ_i of A have distinct moduli:*

$$|\lambda_1| > |\lambda_2| > \cdots > |\lambda_n|.$$

(2) *The matrix Y with $A = XDY$, $X = Y^{-1}$, $D = \operatorname{diag}(\lambda_1, \ldots, \lambda_n) =$ the Jordan normal form of A, has a triangular decomposition*

$$Y = L_Y R_Y, \qquad L_Y = \begin{bmatrix} 1 & & 0 \\ \vdots & \ddots & \\ x & \cdots & 1 \end{bmatrix}, \qquad R_Y = \begin{bmatrix} x & \cdots & x \\ & \ddots & \vdots \\ 0 & & x \end{bmatrix}.$$

Then the matrices A_i, Q_i, R_i of the QR method (6.6.4.3) *have the following convergence properties: There are phase matrices*

$$S_i = \operatorname{diag}(\sigma_1, \ldots, \sigma_n), \qquad |\sigma_k| = 1,$$

such that $\lim_i S_{i-1}^H Q_i S_i = I$ *and*

$$\lim_{i\to\infty} S_i^H R_i S_{i-1} = \lim_{i\to\infty} S_{i-1}^H A_i S_{i-1} = \begin{bmatrix} \lambda_1 & x & \cdots & x \\ & \lambda_2 & \ddots & \vdots \\ & & \ddots & x \\ 0 & & & \lambda_n \end{bmatrix}.$$

In particular, $\lim_{i\to\infty} a_{jj}^{(i)} = \lambda_j$, $j = 1, \ldots, n$, *where* $A_i = (a_{jk}^{(i)})$.

Remark. Hypothesis (2) is used only to ensure that the diagonal elements of A_i converge to the eigenvalues λ_j in their natural order, $|\lambda_1| > |\lambda_2| > \cdots > |\lambda_n|$.

PROOF [following Wilkinson (1965)]. We carry out the proof under the additional hypothesis $\lambda_n \neq 0$ so that D^{-1} exists. Then from $X^{-1} = Y$

$$\begin{aligned} A^i &= XD^iY \\ &= Q_X R_X D^i L_Y R_Y \\ &= Q_X R_X (D^i L_Y D^{-i}) D^i R_Y, \end{aligned} \tag{6.6.4.13}$$

where $Q_X R_X = X$ is a QR decomposition of the nonsingular matrix X into a unitary matrix Q_X and a (nonsingular) upper triangular matrix R_X. Now $D^i L_Y D^{-i} = [l_{jk}^{(i)}]$ is lower triangular, with

$$l_{jk}^{(i)} = \left(\frac{\lambda_j}{\lambda_k}\right)^i l_{jk}, \qquad L_Y =: [l_{jk}], \qquad l_{jk} = \begin{cases} 1 & \text{for } j = k \\ 0 & \text{for } j < k \end{cases}.$$

Since $|\lambda_j| < |\lambda_k|$ for $j > k$ it follows that $\lim_i l_{jk}^{(i)} = 0$ for $j > k$, and thus

$$D^i L_Y D^{-i} = I + E_i, \qquad \lim_{i \to \infty} E_i = 0.$$

Here the speed of convergence depends on the separation of the moduli of the eigenvalues. Next, we obtain from (6.6.4.13)

$$\begin{aligned} A^i &= Q_X R_X (I + E_i) D^i R_Y \\ &= Q_X (I + R_X E_i R_X^{-1}) R_X D^i R_Y \qquad (6.6.4.14) \\ &= Q_X (I + F_i) R_X D^i R_Y \end{aligned}$$

with $F_i := R_X E_i R_X^{-1}$, $\lim_i F_i = 0$. Now the positive-definite matrix

$$(I + F_i)^H (I + F_i) \equiv I + H_i, \qquad H_i := F_i^H + F_i + F_i^H F_i$$

with $\lim_i H_i = 0$, has a uniquely determined Cholesky decomposition (Theorem (4.3.3))

$$I + H_i = \tilde{R}_i^H \tilde{R}_i,$$

where $\tilde{R}_i$ is an upper triangular matrix with positive diagonal elements. Clearly, the Cholesky factor $\tilde{R}_i$ depends continuously on the matrix $I + H_i$, as is shown by the formulas of the Cholesky method. Therefore $\lim_i H_i = 0$ implies $\lim_i \tilde{R}_i = I$. Also the matrix

$$\tilde{Q}_i := (I + F_i) \tilde{R}_i^{-1}$$

is unitary:

$$\begin{aligned} \tilde{Q}_i^H \tilde{Q}_i &= \tilde{R}_i^{-H} (I + F_i)^H (I + F_i) \tilde{R}_i^{-1} = \tilde{R}_i^{-H} (I + H_i) \tilde{R}_i^{-1} \\ &= \tilde{R}_i^{-H} (\tilde{R}_i^H \tilde{R}_i) \tilde{R}_i^{-1} = I. \end{aligned}$$

Therefore, the matrix $I + F_i$ has the QR decomposition $I + F_i = \tilde{Q}_i \tilde{R}_i$, with $\lim_i \tilde{Q}_i = \lim_i (I + F_i) \tilde{R}_i^{-1} = I$, $\lim_i \tilde{R}_i = I$. Thus, by (6.6.4.14)

$$A^i = (Q_X \tilde{Q}_i)(\tilde{R}_i R_X D^i R_Y),$$

where $Q_X \tilde{Q}_i$ is unitary, and $\tilde{R}_i R_X D^i R_Y$ is an upper triangular matrix.

On the other hand, by Theorem (6.6.4.4) (c), the matrix A^i has the QR decomposition

$$A^i = P_i U_i, \qquad P_i := Q_1 \dots Q_i, \qquad U_i := R_i \dots R_1.$$

Since the QR decomposition for nonsingular A is unique up to a rescaling of the columns (rows) of Q (resp. R) by phase factors $\sigma = e^{i\phi}$, there are phase matrices

$$S_i = \operatorname{diag}(\sigma_1^{(i)}, \dots, \sigma_n^{(i)}), \qquad |\sigma_k^{(i)}| = 1$$

with

$$P_i = Q_X \tilde{Q}_i S_i^H, \qquad U_i = S_i \tilde{R}_i R_X D^i R_Y, \qquad i \geq 1,$$

and it follows that

$$\lim_i P_i S_i = \lim_i Q_X \tilde{Q}_i = Q_X$$

$$Q_i = P_{i-1}^{-1} P_i = S_{i-1} \tilde{Q}_{i-1}^H \tilde{Q}_i S_i^H$$

$$\lim_i S_{i-1}^H Q_i S_i = I$$

$$\begin{aligned} R_i = U_i U_{i-1}^{-1} &= S_i \tilde{R}_i R_X D^i R_Y \cdot R_Y^{-1} D^{-i+1} R_X^{-1} \tilde{R}_{i-1}^{-1} S_{i-1}^H \\ &= S_i \tilde{R}_i R_X D R_X^{-1} \tilde{R}_{i-1}^{-1} S_{i-1}^H \end{aligned}$$

$$S_i^H R_i S_{i-1} = \tilde{R}_i R_X D R_X^{-1} \tilde{R}_{i-1}^{-1}, \qquad \lim_i S_i^H R_i S_{i-1} = R_X D R_X^{-1},$$

and finally, by $A_i = Q_i R_i$,

$$\lim_i S_{i-1}^H A_i S_{i-1} = \lim_i S_{i-1}^H Q_i S_i S_i^H R_i S_{i-1} = R_X D R_X^{-1}.$$

The proof is now complete, since the matrix $R_X D R_X^{-1}$ is upper triangular and has diagonal D

$$R_X D R_X^{-1} = \begin{bmatrix} \lambda_1 & * & \cdots & * \\ & \lambda_2 & \ddots & \vdots \\ & & \ddots & * \\ 0 & & & \lambda_n \end{bmatrix}.$$

It can be seen from the proof that the convergence of the Q_i, R_i, and A_i is linear, and improves with decreasing "converging factors" $|\lambda_j/\lambda_k|$, $j > k$, i.e., with improved separation of the eigenvalues in absolute value. The hypotheses of the theorem can be weakened, e.g., our analysis that led to estimates (6.6.4.10) and (6.6.4.11) already showed that the first column and the last row of A_i converge under weaker conditions on the separation of the eigenvalues. In particular, the strong hypotheses of the theorem are violated in the important case of a real matrix A with a pair λ_r, $\lambda_{r+1} = \bar{\lambda}_r$ of conjugate complex eigenvalues. Assuming, for example, that

$$|\lambda_1| > \cdots > |\lambda_r| = |\lambda_{r+1}| > \cdots > |\lambda_n|$$

and that the remaining hypotheses of Theorem (6.6.4.12) are satisfied, the following can still be shown for the matrices $A_i = [a_{jk}^{(i)}]$.

(6.6.4.15). (a) $\lim_i a_{jk}^{(i)} = 0$ *for all* $(j, k) \neq (r+1, r)$ *with* $j > k$.
(b) $\lim_i a_{jj}^{(i)} = \lambda_j$ *for* $j \neq r, r+1$.
(c) *Although the* 2×2 *matrices*

$$\begin{bmatrix} a_{rr}^{(i)} & a_{r,r+1}^{(i)} \\ a_{r+1,r}^{(i)} & a_{r+1,r+1}^{(i)} \end{bmatrix}$$

diverge in general as $i \to \infty$*, their eigenvalues converge toward* λ_r *and* λ_{r+1}.

In other words, the convergence of the matrices A_i takes place in the positions denoted by λ_j and 0 of the following figure, and the eigenvalues of the 2×2 matrix denoted by $*$ converge

$$A_i \xrightarrow[i\to\infty]{} \begin{bmatrix} \lambda_1 & x & \cdots & x & x & x & x & \cdots & x \\ 0 & \lambda_2 & \ddots & & & & & & \vdots \\ & \ddots & \ddots & x & & & & & \vdots \\ & & 0 & \lambda_{r-1} & x & x & & & \vdots \\ & & & 0 & * & * & & & \vdots \\ & & & & * & * & x & & \vdots \\ & & & & & 0 & \lambda_{r+2} & \ddots & \vdots \\ & & & & & & \ddots & \ddots & x \\ 0 & & & & & & & 0 & \lambda_n \end{bmatrix}.$$

For a detailed investigation of the convergence of the QR method, the reader is referred to the following literature: Parlett (1967), Parlett and Poole (1973), and Golub and Van Loan (1983).

6.6.5 The Practical Implementation of the QR Method

In its original form (6.6.4.3), the QR method has some series drawbacks that make it hardly competitive with the methods considered so far:

(a) The method is expensive: a complete step $A_i \to A_{i+1}$ for a dense $n \times n$ matrix A requires $O(n^3)$ operations.

(b) Convergence is very slow if some quotients $|\lambda_j/\lambda_k|$ of A are close to 1. However, there are remedies for these disadvantages.

(a) Because of the amount of work, one applies the QR method only to *reduced* matrices A, namely, matrices of Hessemberg form, or in the case of Hermitian matrices, to Hermitian tridiagonal matrices (i.e., Hermitian-Hessenberg matrices). A general matrix A, therefore, must first be reduced to one of these forms by means of the methods described in Section 6.5. For this procedure to make sense, one must show that these special matrices are invariant under the QR transformation: if A_i is a (possibly Hermitian) Hessenberg matrix then so is A_{i+1}. This invariance is easily established. From Theorem (6.6.4.4) (a), the matrix $A_{i+1} = Q_i^H A_i Q_i$ is unitarily similar to A_i, so that A_{i+1} is Hermitian if A_i is. If A_i is an $n \times n$ Hessenberg matrix, then A_{i+1} can be computed as follows: First, reduce the subdiagonal elements of A_i to 0 by means of suitable Givens matrices of type $\Omega_{12}, \ldots, \Omega_{n-1,n}$ (see Section 4.9)

$$\Omega_{n-1,n} \cdots \Omega_{23}\Omega_{12}A_i = R_i = \begin{bmatrix} * & \cdots & * \\ & \ddots & \vdots \\ 0 & & * \end{bmatrix}$$

$$A_i = Q_i R_i, \qquad Q_i := \Omega_{12}^H \Omega_{23}^H \ldots \Omega_{n-1,n}^H,$$

and then compute A_{i+1} by means of

$$A_{i+1} = R_i Q_i = R_i \Omega_{12}^H \Omega_{23}^H \ldots \Omega_{n-1,n}^H.$$

Because of the special structure of the $\Omega_{j,j+1}$, the upper triangular matrix R_i is transformed by the postmultiplications with the $\Omega_{j,j+1}^H$ into a matrix A_{i+1} of Hessenberg form. Note that A_{i+1} can be computed from A_i in one sweep if the matrix multiplications are carried out in the following order

$$A_{i+1} = (\Omega_{n-1,n} \cdots (\Omega_{23}((\Omega_{12}A_i)\Omega_{12}^H))\Omega_{23}^H \ldots)\Omega_{n-1,n}^H.$$

One verifies easily that it takes only $O(n^2)$ operations to transform an $n \times n$ Hessenberg matrix A_i into A_{i+1} in this way, and only $O(n)$ operations in the Hermitian case, where A_i is a Hermitian tridiagonal matrix.

We therefore assume for the following discussion that A and thus all A_i are (possibly Hermitian) Hessenberg matrices. We may also assume that the Hessenberg matrices A_i are irreducible, i.e., their subdiagonal elements $a_{j,j-1}^{(i)}$, $j = 2, \ldots, n$, are nonzero. Otherwise A_i has the form

$$A_i = \begin{bmatrix} A_i' & * \\ 0 & A_i'' \end{bmatrix},$$

where A_i', A_i'' are Hessenberg matrices of order lower than n. Since the eigenvalues of A_i are just the eigenvalues of A_i' and A_i'', it is sufficient to determine the eigenvalues of the matrices A_i', A_i'' separately. So, the eigenvalue problem for A_i can be reduced to the eigenvalue problem for smaller irreducible matrices.

Basically, the QR method for Hessenberg matrices runs as follows: The A_i are computed according to (6.6.4.3) until one of the last two subdiagonal elements $a_{n,n-1}^{(i)}$ and $a_{n-1,n-2}^{(i)}$ of the Hessenberg matrix A_i (which converge to 0 in general, see (6.6.4.12), (6.6.4.15)) become negligible, that is,

$$\min\{|a_{n,n-1}^{(i)}|, |a_{n-1,n-2}^{(i)}|\} \le \text{eps}(|a_{nn}^{(i)}| + |a_{n-1,n-1}^{(i)}|),$$

eps being, say, the relative machine precision. In case $a_{n,n-1}^{(i)}$ is negligible, $a_{nn}^{(i)}$ is a numerically acceptable eigenvalue (see Section 6.6.3) of A, because it is the exact eigenvalue of a Hessenberg matrix $\tilde{A}_i$ close to A_i, $\|\tilde{A}_i - A_i\| \le \text{eps}\|A_i\|$: $\tilde{A}_i$ is obtained from A_i by replacing $a_{n,n-1}^{(i)}$ by 0. In case $a_{n-1,n-2}^{(i)}$ is negligible, the eigenvalues of the 2×2 matrix

$$\begin{bmatrix} a_{n-1,n-1}^{(i)} & a_{n-1,n}^{(i)} \\ a_{n,n-1}^{(i)} & a_{nn}^{(i)} \end{bmatrix}$$

are two numericably acceptable eigenvalues of A_i, since they are the exact eigenvalues of a Hessenberg matrix $\tilde{A}_i$ close to A_i, which is obtained by replacing the small element $a_{n-1,n-2}^{(i)}$ by 0. If $a_{nn}^{(i)}$ is taken as an eigenvalue, one crosses out the last row and column, and if the eigenvalues of the 2×2 matrix are taken, one removes the last two rows and columns of A_i, and continues the algorithm with the remaining matrix.

By (6.6.4.4) the QR method generates matrices that are unitarily similar to each other. We note an interesting property of "almost" irreducible Hessenberg matrices that are unitarily similar to each other, a property that will be important for the implicit shift techniques to be discussed later.

(6.6.5.1) Theorem. *Let $Q = [q_1, \ldots, q_n]$ and $U = [u_1, \ldots, u_n]$ be unitary matrices with columns q_i and u_i, and suppose that $H = [h_{jk}] := Q^H AQ$ and $K := U^H AU$ are both Hessenberg matrices that are similar to the same matrix A using Q and U, respectively, for unitary similarity transformations. Assume further that $h_{i,i-1} \ne 0$ for $i \le n-1$ and that $u_1 = \sigma_1 q_1$, $|\sigma_1| = 1$. Then there is a phase matrix $S = \operatorname{diag}(\sigma_1, \ldots, \sigma_n)$, $|\sigma_k| = 1$, such that $U = QS$ and $K = S^H HS$.*

That is, if H is "almost" irreducible and Q and U have "essentially" the same first column, then all columns of these matrices are "essentially" the same, and the Hessenberg matrices K and H agree up to phase factors, $K = S^H HS$.

PROOF. In terms of the unitary matrix $V = [v_1, \ldots, v_n] := U^H Q$, we have $H = V^H KV$ and therefore

$$KV = VH.$$

A comparison of the $(i-1)$-th column on both sides shows

$$h_{i,i-1} v_i = K v_{i-1} - \sum_{j=1}^{i-1} h_{j,i-1} v_j, \qquad 2 \le i \le n.$$

With $h_{i,i-1} \neq 0$ for $2 \le i \le n-1$, this recursion permits one to compute v_i, $i \le n-1$, from v_1. Since $v_1 = U^H q_1 = \sigma_1 U^H u_1 = \sigma_1 e_1$ and K is a Hessenberg matrix we find that the matrix $V = [v_1, \ldots, v_n]$ is upper triangular. As V is also unitary, $VV^H = I$, V is a phase matrix, which we denote by $S^H := V$. The theorem now follows from $K = VHV^H$. □

(b) The slow convergence of the QR method can be improved substantially by means of so-called *shift techniques*. We know from the relationship between the QR method and inverse vector iteration (see (6.6.4.11)) that, for nonsingular A, the last row $e_n^T A_i$ of the matrices A_i converges in general to $\lambda_n e_n^T$ as $i \to \infty$, if the eigenvalues λ_i of A satisfy

$$|\lambda_1| \ge |\lambda_2| \ge \cdots \ge |\lambda_{n-1}| > |\lambda_n| > 0.$$

In particular, the convergence speed of the error $\|e_n^T A_i - \lambda_n e_n^T\| = O(|\lambda_n/\lambda_{n-1}|^i)$ is determined by the quotient $|\lambda_n/\lambda_{n-1}|$. This suggests that we may accelerate the convergence by the same techniques used with inverse vector iteration (see Section 6.6.3), namely, by applying the QR method not to A but to a suitable *shifted matrix* $\tilde{A} = A - kI$. Here, the *shift parameter* k is chosen as a close approximation to some eigenvalue of A, so that, perhaps after a reordering of the eigenvalues,

$$|\lambda_1 - k| \ge |\lambda_2 - k| \ge \cdots \ge |\lambda_{n-1} - k| \gg |\lambda_n - k| > 0.$$

Then the elements $\tilde{a}_{n,n-1}^{(i)}$ of the matrices $\tilde{A}_i$ associated with $\tilde{A}$ will converge much faster to 0, namely, as

$$\left| \frac{\lambda_n - k}{\lambda_{n-1} - k} \right|^i \ll 1.$$

Note that if A is a Hessenberg matrix, so also are $\tilde{A} = A - kI$ and all $\tilde{A}_i$, hence the last row of $\tilde{A}_i$ has the form

$$e_n^T \tilde{A}_i = [0, \ldots, 0, \tilde{a}_{n,n-1}^{(i)}, \tilde{a}_{nn}^{(i)}].$$

More generally, if we choose a new shift parameter in each step, the QR *method with shifts* is obtained:

$$\begin{aligned} &A_1 := A \\ (6.6.5.2) \qquad &A_i - k_i I =: Q_i R_i \qquad (QR \text{ decomposition}) \\ &A_{i+1} := R_i Q_i + k_i I. \end{aligned}$$

The matrices A_i are still unitarily similar to each other, since

$$A_{i+1} = Q_i^H (A_i - k_i I) Q_i + k_i I = Q_i^H A_i Q_i.$$

In addition, it can be shown as in the proof of Theorem (6.6.4.4) (see Exercise 20)

(6.6.5.3)
$$A_{i+1} = P_i^H A P_i$$
$$(A - k_1 I) \dots (A - k_i I) = P_i U_i,$$

where again $P_i := Q_1 Q_2 \dots Q_i$ and $U_i = R_i R_{i-1} \dots R_1$. Moreover,

(6.6.5.4)
$$\begin{aligned} A_{i+1} &= R_i A_i R_i^{-1} \\ &= U_i A U_i^{-1} \end{aligned}$$

holds if all R_j are nonsingular. Also, A_{i+1} will be a Hessenberg matrix, but this assertion can be sharpened.

(6.6.5.5) Theorem. *Let A_i be an irreducible Hessenberg matrix of order n. If k_i is not an eigenvalue of A_i then A_{i+1} will also be an irreducible Hessenberg matrix. Otherwise, A_{i+1} will have the form*

$$A_{i+1} = \begin{bmatrix} \tilde{A}_{i+1} & * \\ 0 & k_i \end{bmatrix},$$

where $\tilde{A}_{i+1}$ is an irreducible Hessenberg matrix of order $(n-1)$.

PROOF. It follows from (6.6.5.2) that

(6.6.5.6)
$$R_i = Q_i^H (A_i - k_i I).$$

If k_i is not an eigenvalue of A_i, then R_i is nonsingular by (6.6.5.2), and therefore (6.6.5.4) yields

$$a_{j+1,j}^{(i+1)} = e_{j+1}^T A_{i+1} e_j = e_{j+1}^T R_i A_i R_i^{-1} e_j = r_{j+1,j+1}^{(i)} a_{j+1,j}^{(i)} (r_{jj}^{(i)})^{-1} \neq 0,$$

so that A_{i+1} is irreducible, too.

On the other hand, if k_i is an eigenvalue of A_i, then R_i is singular. But since A_i is irreducible, the first $n-1$ columns of $A_i - k_i I$ are linearly independent, and therefore, by (6.6.5.6) so also are the first $n-1$ columns of R_i. Hence, by the singularity of R_i, the last row of R_i must vanish

$$R_i = \begin{bmatrix} \tilde{R}_i & * \\ 0 & 0 \end{bmatrix},$$

with $\tilde{R}_i$ being a nonsingular upper triangular matrix of order $(n-1)$. Because of the irreducibility of $A_i - k_i I$ and the fact that

$$A_i - k_i I = Q_i R_i = Q_i \begin{bmatrix} \tilde{R}_i & * \\ 0 & 0 \end{bmatrix},$$

Q_i is also an irreducible Hessenberg matrix. Therefore, since

$$A_{i+1} = \begin{bmatrix} \tilde{R}_i & * \\ 0 & 0 \end{bmatrix} Q_i + k_i I = \begin{bmatrix} \tilde{A}_{i+1} & * \\ 0 & k_i \end{bmatrix},$$

the submatrix $\tilde{A}_{i+1}$ is an irreducible Hessenberg matrix of order $n-1$, and A_{i+1} has the structure asserted in the theorem. □

We now turn to the problem of choosing the shift parameters k_i appropriately. To simplify the discussion, we consider only the case of real matrices A, which is the most important case in practice, and we assume that A is a Hessenberg matrix. The earlier analysis suggests choosing close approximations to the eigenvalues of A as shifts k_i, and the problem is finding such approximations. But Theorems (6.6.4.12) and (6.6.4.15) are helpful here: According to them we have $\lim_i a_{nn}^{(i)} = \lambda_n$, if A has a unique eigenvalue λ_n of smallest absolute value. Therefore, for large i, the last diagonal element $a_{nn}^{(i)}$ of A_i will, in general, be a good approximation for λ_n, which suggests the choice

(6.6.5.7a) $$k_i := a_{nn}^{(i)},$$

if the convergence of the $a_{nn}^{(i)}$ has stabilized, say, if

$$\left| 1 - \frac{a_{nn}^{(i-1)}}{a_{nn}^{(i)}} \right| \le \eta < 1$$

for a small η (even the choice $\eta = 1/3$ leads to acceptable results). It is possible to show for Hessenberg matrices under weak conditions that the choice (6.6.5.7a) ensures

$$|a_{n,n-1}^{(i+1)}| \le C\varepsilon^2$$

if $|a_{n,n-1}^{(i)}| \le \varepsilon$ and ε is sufficiently small, so that the QR method then converges locally at a quadratic rate. By direct calculation, this is easily verified for real 2×2 matrices

$$A = \begin{bmatrix} a & b \\ \varepsilon & 0 \end{bmatrix}$$

with $a \ne 0$ and small $|\varepsilon|$: Using the shift $k_1 = a_{nn} = 0$ one finds the matrix

$$\tilde{A} = \begin{bmatrix} \tilde{a} & \tilde{b} \\ \tilde{c} & \tilde{d} \end{bmatrix} \qquad \text{with } |\tilde{c}| = \frac{|b|\varepsilon^2}{a^2 + \varepsilon^2},$$

as the QR successor of A, so that indeed $|\tilde{c}| = O(\varepsilon^2)$ for small $|\varepsilon|$. For symmetric matrices, $b = \varepsilon$, the same example shows

$$|\tilde{c}| = \frac{|\varepsilon^3|}{a^2 + \varepsilon^2},$$

so that one can expect local convergence at a cubic rate in this case. In fact, for general Hermitian tridiagonal matrices and shift strategy (6.6.5.7a), local cubic convergence was shown by Wilkinson [for details see Wilkinson (1965), p. 548, and Exercise 23].

A more general shift strategy, which takes account of both (6.6.4.12) and (6.6.4.15), is the following.

(6.6.5.7.b) *Choose k_i as the eigenvalue λ of the 2×2 matrix*

$$\begin{bmatrix} a_{n-1,n-1}^{(i)} & a_{n-1,n}^{(i)} \\ a_{n,n-1}^{(i)} & a_{nn}^{(i)} \end{bmatrix},$$

for which $|a_{nn}^{(i)} - \lambda|$ is smallest.

This strategy also allows for the possibility that A_i has several eigenvalues of equal absolute value. But it may happen that k_i is a complex number even for real A_i, if A_i is nonsymmetric, a case that will considered later. For real symmetric matrices Wilkinson (1968) even showed the following global convergence theorem.

(6.6.5.8) Theorem. *If the QR method with shift strategy* (6.6.5.7b) *is applied to a real, irreducible, symmetric tridiagonal $n \times n$ matrix in all iterations, then the elements $a_{n,n-1}^{(i)}$ of the ith iteration matrix A_i converge to zero at least quadratically as $i \to \infty$, while $a_{nn}^{(i)}$ converges at least quadratically toward an eigenvalue of A. Disregarding rare exceptions, the convergence rate is even cubic.* □

Numerical Example. The spectrum of the matrix

(6.6.5.9)
$$A = \begin{bmatrix} 12 & 1 & & & \\ 1 & 9 & 1 & & \\ & 1 & 6 & 1 & \\ & & 1 & 3 & 1 \\ & & & 1 & 0 \end{bmatrix}$$

lies symmetrically around 6; in particular, 6 is an eigenvalue of A. For the QR method with shift strategy (b) we display in the following the elements $a_{n,n-1}^{(i)}$, $a_{nn}^{(i)}$, as well as the shift parameters k_i:

i	$a_{54}^{(i)}$	$a_{55}^{(i)}$	k_i
1	1	0	$-.302\ 775\ 637\ 732_{10}0$
2	$-.454\ 544\ 295\ 102_{10}-2$	$-.316\ 869\ 782\ 391_{10}0$	$-.316\ 875\ 874\ 226_{10}0$
3	$+.106\ 774\ 452\ 090_{10}-9$	$-.316\ 875\ 952\ 616_{10}0$	$-.316\ 875\ 952\ 619_{10}0$
4	$+.918\ 983\ 519\ 419_{10}-22$	$\boxed{-.316\ 875\ 952\ 617_{10}0 = \lambda_5}$	

Processing the 4 × 4 matrix further, we find

i	$a_{43}^{(i)}$	$a_{44}^{(i)}$	k_i
4	$+.143\ 723\ 850\ 633_{10}0$	$+.299\ 069\ 135\ 875_{10}1$	$+.298\ 389\ 967\ 722_{10}1$
5	$-.171\ 156\ 231\ 712_{10}-5$	$+.298\ 386\ 369\ 683_{10}1$	$+.298\ 386\ 369\ 682_{10}1$
6	$-.111\ 277\ 687\ 663_{10}-17$	$\boxed{+.298\ 386\ 369\ 682_{10}1 = \lambda_4}$	

Processing the 3 × 3-matrix further, we get

i	$a_{32}^{(i)}$	$a_{33}^{(i)}$	k_i
6	$+.780\ 088\ 052\ 879_{10}-1$	$+.600\ 201\ 597\ 254_{10}1$	$+.600\ 000\ 324\ 468_{10}1$
7	$-.838\ 854\ 980\ 961_{10}-7$	$+.599\ 999\ 999\ 996_{10}1$	$+.599\ 999\ 999\ 995_{10}1$
8	$+.127\ 181\ 135\ 623_{10}-19$	$\boxed{+.599\ 999\ 999\ 995_{10}1 = \lambda_3}$	

The remaining 2 × 2-matrix has the eigenvalues

$$\boxed{\begin{array}{l} +.901\ 613\ 630\ 314_{10}1 = \lambda_2 \\ +.123\ 168\ 759\ 526_{10}2 = \lambda_1 \end{array}}$$

This ought to be compared with the result of eleven QR steps *without* shifts:

k	$a_{k,k-1}^{(12)}$	$a_{kk}^{(12)}$
1		$+.123\ 165\ 309\ 125_{10}2$
2	$+.337\ 457\ 586\ 637_{10}-1$	$+.901\ 643\ 819\ 611_{10}1$
3	$+.114\ 079\ 951\ 421_{10}-1$	$+.600\ 004\ 307\ 566_{10}1$
4	$+.463\ 086\ 759\ 853_{10}-3$	$+.298\ 386\ 376\ 789_{10}1$
5	$-.202\ 188\ 244\ 733_{10}-10$	$-.316\ 875\ 952\ 617_{10}0$

Such convergence behavior is to be expected, since

$$a_{54}^{(i)} = O((\lambda_5/\lambda_4)^i), \qquad |\lambda_5/\lambda_4| \approx 0.1.$$

Further processing the 4 × 4 matrix requires as many as 23 iterations in order to achieve

$$a_{43}^{(35)} = +0.487\ 637\ 464\ 425_{10}-10.$$

The fact that the eigenvalues in this example come out ordered in size is accidental: With the use of shifts this is not to be expected in general.

When the QR step (6.6.5.2) with shifts for symmetric tridiagonal matrices is implemented in practice, there is a danger of losing accuracy when subtracting and adding $k_i I$ in (6.6.5.2), in particular, if $|k_i|$ is large. Using Theorem (6.6.5.1), however, there is a way to compute A_i without subtracting and adding $k_i I$ explicitly. Here, we use the fact that the first column q_1 of the matrix $Q_i := [q_1, \ldots, q_n]$ can be computed from the first column of $A_i - k_i I$: at the beginning of this section we have seen that

$$Q_i = \Omega_{12}^H \Omega_{23}^H \ldots \Omega_{n-1,n}^H,$$

where the $\Omega_{j,j+1}$ are Givens matrices. Note that, for $j > 1$, the first column of $\Omega_{j,j+1}$ equals the first unit column e_1, so that $q_1 = Q_i e_1 = \Omega_{12}^H e_1$ is also the first column of Ω_{12}^H. Matrix Ω_{12} is by definition the Givens matrix that annihilates the first subdiagonal element of $A_i - k_i I$, and it is therefore fully determined by the first column of that matrix. The matrix $B = \Omega_{12}^H A_i \Omega_{12}$, then has the form

$$B = \begin{bmatrix} x & x & x & & & 0 \\ x & x & x & & & \\ x & x & x & x & & \\ & & x & x & \ddots & \\ & & & \ddots & \ddots & x \\ 0 & & & & x & x \end{bmatrix},$$

that is, it is symmetric and, except for the potentially nonzero elements $b_{13} = b_{31}$, tridiagonal. In analogy to the procedure described in Section 6.5.1, one can find a sequence of Givens transformations

$$B \to \tilde{\Omega}_{23}^H B \tilde{\Omega}_{23} \to \cdots \to \tilde{\Omega}_{n-1,n}^H \ldots \tilde{\Omega}_{23}^H B \tilde{\Omega}_{23} \ldots \tilde{\Omega}_{n-1,n} = C,$$

which transform B into a unitarily similar matrix C that is symmetric and tridiagonal. First, $\tilde{\Omega}_{23}$ is chosen so as to annihilate the elements $b_{31} = b_{13}$ of B. The corresponding Givens transformation will produce nonzero elements in the positions (4, 2) and (2, 4) of $\tilde{\Omega}_{23}^H B \tilde{\Omega}_{23}$. These, in turn, are annihilated by a proper choice of $\tilde{\Omega}_{34}$, and so on.

So, for $n = 5$ the matrices $B \to \cdots \to C$ have the following structure, where again 0 and $*$ denote new zero and potentially nonzero elements, respectively.

$$B = \begin{bmatrix} x & x & x & & \\ x & x & x & & \\ x & x & x & x & \\ & & x & x & x \\ & & & x & x \end{bmatrix} \xrightarrow{\tilde{\Omega}_{23}} \begin{bmatrix} x & x & 0 & & \\ x & x & x & * & \\ 0 & x & x & x & \\ & * & x & x & x \\ & & & x & x \end{bmatrix}$$

$$\xrightarrow{\tilde{\Omega}_{34}} \begin{bmatrix} x & x & & & \\ x & x & x & 0 & \\ & x & x & x & * \\ & 0 & x & x & x \\ & & * & x & x \end{bmatrix} \xrightarrow{\tilde{\Omega}_{45}} \begin{bmatrix} x & x & & & \\ x & x & x & & \\ & x & x & x & 0 \\ & & x & x & x \\ & & 0 & x & x \end{bmatrix} = C.$$

In terms of the unitary matrix $U := \Omega_{12}\tilde{\Omega}_{23} \ldots \tilde{\Omega}_{n-1,n}$, we have $C = U^H A_i U$ since $B = \Omega_{12}^H A_i \Omega_{12}$. The matrix U has the same first column as the unitary matrix Q_i, which transforms A_i into $A_{i+1} = Q_i^H A_i Q_i$ according to (6.6.5.2), namely, $q_1 = \Omega_{12} e_1$. If A_i is irreducible then by Theorem (6.6.5.5) the elements $a_{j,j-1}^{(i+1)}$ of $A_{i+1} = [a_{jk}^{(i+1)}]$ are nonzero for $j = 2, 3, \ldots, n-1$. Thus, by Theorem (6.6.5.1), the matrices C and $A_{i+1} = SCS^H$ are essentially the same, $S = \operatorname{diag}(\pm 1, \ldots, \pm 1)$, $U = QS$. Since in the explicit QR step (6.6.5.2), the matrix Q_i is unique only up to a rescaling $Q_i \to Q_i S$ by a phase matrix S, we have found an alternate algorithm to determine A_{i+1}, namely, by computing C. This technique, which avoids the explicit subtraction and addition of $k_i I$, is called the *implicit shift technique* for the computation of A_{i+1}.

Now, let A be a real *nonsymmetric* Hessenberg matrix. The shift strategy (6.6.5.7b) then leads to a complex shift k_i, even if the matrix A_i is real, whenever the 2×2 matrix in (6.6.5.7b) has two conjugate complex eigenvalues k_i and $\bar{k}_i$. In this case, it is possible to avoid complex calculations, if one combines the two QR steps $A_i \to A_{i+1} \to A_{i+2}$ by choosing the shift k_i in the first, and the shift $k_{i+1} := \bar{k}_i$ in the second. It is easy to verify that for real A_i the matrix A_{i+2} is real again, even though A_{i+1} may be complex. Following an elegant proposal by Francis (1961/62) one can compute A_{i+2} directly from A_i using only real arithmetic, provided the real Hessenberg matrix A_i is irreducible: For such matrices A_i this method permits the direct calculation of the subsequence

$$A = A_1 \to A_3 \to A_5 \to \cdots$$

of the sequence of matrices generated by the QR method with shifts. It is in this context that implicit shift techniques were used first.

In this text, we describe only the basic ideas of Francis' technique. Explicit formulas can be found, for instance, in Wilkinson (1965). For simplicity, we omit the iteration index and write $A = [a_{jk}]$ for the real irreducible Hessenberg matrix A_i, k for k_i, and $\tilde{A}$ for A_{i+2}. By (6.6.5.2), (6.6.5.3) there are a unitary matrix Q and an upper triangular matrix R such that $\tilde{A} = Q^H A Q$ is a real Hessenberg matrix and $(A - kI)(A - \bar{k}I) = QR$. We try to find a unitary matrix U having essentially the same first column as Q, $Qe_1 = \pm Ue_1$, so that $U^H A U =: K$ is also a Hessenberg matrix. If k is not an eigenvalue of A, then Theorems (6.6.5.1) and (6.6.5.5) show again that the matrices $\tilde{A}$ and K differ only by a phase transformation: $K = S^H \tilde{A} S$, $S = \operatorname{diag}(\pm 1, \ldots, \pm 1)$.

This can be achieved as follows: The matrix

$$B := (A - kI)(A - \bar{k}I) = A^2 - (k + \bar{k})A + k\bar{k}I$$

is real, and its first column $b = Be_1$, which is readily computed from A and k, has the form $b = [x, x, x, 0, \ldots, 0]^T$, since A is a Hessenberg matrix. Choose P (see Section 4.7) as the Householder matrix that transforms b into a multiple of e_1, $Pb = \mu_1 e_1$. Then the first column of P has the same structure as b, that is, $Pe_1 = [x, x, x, 0, \ldots, 0]^T$, because $P^2 b = b = \mu_1 P e_1$. Since $B = QR$, $B = P^2 B = P[\mu_1 e_1, *, \ldots, *]$, the matrix P has essentially the same first column $Pe_1 = \pm Qe_1$ as the matrix Q. Next, one computes the matrix $P^H A P =$

PAP, which, by the structure of P (note that $Pe_k = e_k$ for $k \geq 4$) is a matrix of the form

$$PAP = \begin{bmatrix} x & \cdots & & & & & x \\ x & x & & & & & \vdots \\ * & x & x & & & & \vdots \\ * & * & x & x & & & \vdots \\ & & & x & x & & \vdots \\ & & & & \ddots & \ddots & \vdots \\ & & & & & x & x \end{bmatrix},$$

in other words, except for few additional subdiagonal elements in positions (3, 1), (4, 1), and (4, 2), it is again a Hessenberg matrix. Applying Householder's method (see the end of Section 6.5.4), PAP is then transformed into a genuine Hessenberg matrix K using unitary similarity transformations

$$PAP \to T_1 PAPT_1 \to \cdots \to T_{n-2} \ldots T_1 PAPT_1 \ldots T_{n-2} =: K.$$

The Householder matrices T_k, $k = 1, \ldots, n-2$, according to Section 6.5.1, differ from the unit matrix only in three columns, $T_k e_i = e_i$ for $i \notin \{k+1, k+2, k+3\}$.

For $n = 6$ one obtains a sequence of matrices of the following form (0 and $*$ again denote new zero and nonzero elements, respectively):

$$PAP = \begin{bmatrix} x & \cdots & & & & x \\ x & x & & & & \vdots \\ x & x & x & & & \vdots \\ x & x & x & x & & \vdots \\ & & & x & x & \vdots \\ & & & & x & x \end{bmatrix} \xrightarrow{T_1} \begin{bmatrix} x & \cdots & & & & x \\ x & x & & & & \vdots \\ 0 & x & x & & & \vdots \\ 0 & x & x & x & & \vdots \\ & * & * & x & x & \vdots \\ & & & & x & x \end{bmatrix} \xrightarrow{T_2}$$

$$\xrightarrow{T_2} \begin{bmatrix} x & \cdots & & & & x \\ x & x & & & & \vdots \\ & x & x & & & \vdots \\ & 0 & x & x & & \vdots \\ & 0 & x & x & x & \vdots \\ & & * & * & x & x \end{bmatrix} \xrightarrow{T_3} \begin{bmatrix} x & \cdots & & & & x \\ x & x & & & & \vdots \\ & x & x & & & \vdots \\ & & x & x & & \vdots \\ & & 0 & x & x & \vdots \\ & & 0 & x & x & x \end{bmatrix} \xrightarrow{T_4}$$

$$\xrightarrow{T_4} \begin{bmatrix} x & \cdots & & & & x \\ x & x & & & & \vdots \\ & x & x & & & \vdots \\ & & x & x & & \vdots \\ & & & x & x & \vdots \\ & & & 0 & x & x \end{bmatrix} = K.$$

As $T_k e_1 = e_1$ for all k, the unitary matrix $U := PT_1 \dots T_{n-2}$ with $U^H AU = K$ has the same first column as P, $Ue_1 = PT_1 \dots T_{n-2}e_1 = Pe_1$. Thus, by Theorem (6.6.5.1), the matrix K is (essentially) identical to $\tilde{A}$.

Note that it is not necessary for the practical implementation of QR method to compute and store the product matrix $P_i := Q_1 \dots Q_i$, if one wishes to determine only the eigenvalues of A. This is no longer true in the Hermitian case if one also wants to find a set of eigenvectors that are orthogonal. The reason is that for Hermitian A

$$\lim_i P_i^H AP_i = D = \operatorname{diag}(\lambda_1, \dots, \lambda_n),$$

which follows from (6.6.5.3), (6.6.4.12), so that for large i the jth column $p_j^{(i)} = P_i e_j$ of P_i will be a very good approximation to an eigenvector of A for the eigenvalue λ_j, and these approximations $p_j^{(i)}$, $j = 1, \dots, n$, are orthogonal to each other, since P_i is unitary. If A is not Hermitian, it is better to determine the eigenvectors by the inverse iteration of Wielandt (see Section 6.6.3), where the usually excellent approximations to the eigenvalues that have been found by the QR method can be used.

The QR method converges very quickly, as is also shown by the example given earlier. For symmetric matrices, experience suggests that the QR method is about four times as fast as the widely used Jacobi method if eigenvalues and eigenvectors are to be computed, and about ten times as fast if only the eigenvalues are desired. There are many programs for the QR method. ALGOL programs can be found in the handbook of Wilkinson and Reinsch (1971) (see the contributions by Bowdler, Martin, Peters, Reinsch, and Wilkinson), FORTRAN programs are in Smith et al. (1976).

6.7 Computation of the Singular Values of a Matrix

The singular values (6.4.6) and the singular-value decomposition (6.4.11) of an $m \times n$ matrix A can be computed rapidly, and in a numerically stable manner, by a method due to Golub and Reinsch (1971), which is closely related to the QR method. We assume, without loss of generality, that $m \geqslant n$ (otherwise, replace A by A^H). The decomposition (6.4.11) can then be written in the form

$$\text{(6.7.1)} \quad A = U\begin{bmatrix} D \\ 0 \end{bmatrix}V^H, \qquad D := \operatorname{diag}(\sigma_1, \dots, \sigma_n), \quad \sigma_1 \geqslant \sigma_2 \geqslant \cdots \geqslant \sigma_n \geqslant 0,$$

where U is a unitary $m \times m$ matrix, V is a unitary $n \times n$ matrix, and $\sigma_1, \dots, \sigma_n$ are the singular values of A, i.e., σ_i^2 are the eigenvalues of $A^H A$. In principle, therefore, one could determine the singular values by solving the eigenvalue problem for the Hermitian matrix $A^H A$, but this approach can be subject to loss of accuracy: for the matrix

$$A := \begin{bmatrix} 1 & 1 \\ \varepsilon & 0 \\ 0 & \varepsilon \end{bmatrix}, \qquad |\varepsilon| < \sqrt{\text{eps}}, \quad \text{eps} = \text{machine precision},$$

for example, the matrix $A^H A$ is given by

$$A^H A = \begin{bmatrix} 1 + \varepsilon^2 & 1 \\ 1 & 1 + \varepsilon^2 \end{bmatrix},$$

and A has the singular values $\sigma_1(A) = \sqrt{2 + \varepsilon^2}$, $\sigma_2(A) = |\varepsilon|$. In floating-point arithmetic, with precision eps, instead of $A^H A$ one obtains the matrix

$$\text{fl}(A^H A) =: B = \begin{bmatrix} 1 & 1 \\ 1 & 1 \end{bmatrix},$$

with eigenvalues $\tilde{\lambda}_1 = 2$ and $\tilde{\lambda}_2 = 0$; $\sigma_2(A) = |\varepsilon|$ does *not* agree to machine precision with $\sqrt{\tilde{\lambda}_2} = 0$.

In the method of Golub and Reinsch one first reduces A, in a preliminary step, unitarily to bidiagonal form. By the Householder method (Section 4.7), one begins with determining an $m \times m$ Householder matrix P_1 which annihilates the subdiagonal elements in the first column of A. One thus obtains a matrix $A' = P_1 A$ of the form shown in the following sketch for $m = 5$, $n = 4$, where changing elements are denoted by *:

$$A = \begin{bmatrix} x & x & x & x \\ x & x & x & x \\ x & x & x & x \\ x & x & x & x \\ x & x & x & x \end{bmatrix} \to P_1 A = A' = \begin{bmatrix} * & * & * & * \\ 0 & * & * & * \\ 0 & * & * & * \\ 0 & * & * & * \\ 0 & * & * & * \end{bmatrix}.$$

Then, as in Section 4.7, one determines an $n \times n$ Householder matrix Q_1 of the form

$$Q_1 = \begin{bmatrix} 1 & 0 \\ 0 & \tilde{Q} \end{bmatrix},$$

such that the elements in the positions $(1, 3), \ldots, (1, n)$ in the first row of $A'' := A'Q_1$ vanish:

$$A' = \begin{bmatrix} x & x & x & x \\ 0 & x & x & x \\ 0 & x & x & x \\ 0 & x & x & x \\ 0 & x & x & x \end{bmatrix} \to A'Q_1 = A'' = \begin{bmatrix} x & * & 0 & 0 \\ 0 & * & * & * \\ 0 & * & * & * \\ 0 & * & * & * \\ 0 & * & * & * \end{bmatrix}.$$

In general, one obtains in the first step a matrix A'' of the form

$$A'' = \left[\begin{array}{c|c} q_1 & a_1 \\ \hline 0 & \bar{A} \end{array}\right], \qquad a_1 := [e_2, 0, \ldots, 0] \in \mathbb{C}^{n-1},$$

with an $(m-1) \times (n-1)$ matrix $\bar{A}$. One now treats the matrix $\bar{A}$ in the same manner as A, etc., and in this way, after n reduction steps, obtains an $m \times n$ bidiagonal matrix $J^{(0)}$ of the form

$$J^{(0)} = \begin{bmatrix} J_0 \\ 0 \end{bmatrix}, \qquad J_0 = \begin{bmatrix} q_1^{(0)} & e_2^{(0)} & & & 0 \\ & q_2^{(0)} & e_3^{(0)} & & \\ & & \ddots & \ddots & \\ & & & \ddots & e_n^{(0)} \\ 0 & & & & q_n^{(0)} \end{bmatrix},$$

where $J^{(0)} = P_n P_{n-1} \ldots P_1 A Q_1 Q_2 \ldots Q_{n-2}$ and P_i, Q_i are certain Householder matrices.

Since $Q := Q_1 Q_2 \ldots Q_{n-2}$ is unitary and $J^{(0)H} J^{(0)} = J_0^H J_0 = Q^H A^H A Q$, the matrices J_0 and A have the same singular values; likewise, with $P := P_1 P_2 \ldots P_n$, the decomposition

$$\left(P \begin{bmatrix} G & 0 \\ 0 & I_{m-n} \end{bmatrix}\right) \begin{bmatrix} D \\ 0 \end{bmatrix} (H^H Q^H)$$

is the singular-value decomposition (6.7.1) of A if $J_0 = GDH^H$ is the singular-value decomposition of J_0. It suffices therefore to consider the problem for $n \times n$ bidiagonal matrices J.

In the first step, J is multiplied on the right by a certain Givens reflection of the type T_{12}, the choice of which we want to leave open for the moment. Thereby, J changes into a matrix $J^{(1)} = JT_{12}$ of the following form (sketch for $n = 4$):

$$J = \begin{bmatrix} x & x & & \\ & x & x & \\ & & x & x \\ & & & x \end{bmatrix} \to JT_{12} = \begin{bmatrix} * & * & & \\ \circledast & * & x & \\ & & x & x \\ & & & x \end{bmatrix} = J^{(1)}.$$

The subdiagonal element in position (2, 1) of $J^{(1)}$, generated in this way, is again annihilated through left multiplication by a suitable Givens reflection of the type S_{12}. One thus obtains the matrix $J^{(2)} = S_{12} J^{(1)}$:

$$J^{(1)} = \begin{bmatrix} x & x & & \\ x & x & x & \\ & & x & x \\ & & & x \end{bmatrix} \to S_{12} J^{(1)} = \begin{bmatrix} * & * & \circledast & \\ 0 & * & * & \\ & & x & x \\ & & & x \end{bmatrix} = J^{(2)}.$$

The element in position (1, 3) of $J^{(2)}$ is annihilated by multiplication from the right with a Givens reflection of the type T_{23},

$$J^{(2)} = \begin{bmatrix} x & x & x & \\ & x & x & \\ & & x & x \\ & & & x \end{bmatrix} \to J^{(2)}T_{23} = \begin{bmatrix} x & * & 0 & \\ & * & * & \\ & \circledast & * & x \\ & & & x \end{bmatrix} = J^{(3)}.$$

The subdiagonal element in position (3, 2) is now annihilated by left multiplication with a Givens reflection S_{23}, which generates an element in position (2, 4), etc. In the end, for each Givens reflection T_{12} one can thereby determine a sequence of further Givens reflections $S_{i,i+1}$, $T_{i,i+1}$, $i = 1, 2, \ldots, n-1$, in such a manner that the matrix

$$\bar{J} := S_{n-1,n} S_{n-2,n-1} \cdots S_{12} J T_{12} T_{23} \cdots T_{n-1,n}$$

is again a bidiagonal matrix. The matrices $S := S_{12} S_{23} \cdots S_{n-1,n}$, $T := T_{12} T_{23} \cdots T_{n-1,n}$ are unitary, so that, by virtue of $\bar{J} = S^H J T$, the matrix $\bar{M} := \bar{J}^H \bar{J} = T^H J^H S S^H J T = T^H M T$ is a tridiagonal matrix which is unitarily similar to the tridiagonal matrix $M := J^H J$.

Up to now, the choice of T_{12} was open. We show that T_{12} can be so chosen that $\bar{M}$ becomes precisely the matrix which is produced from the tridiagonal matrix M by the QR method with shift k. To show this, we may assume without loss of generality that $M = J^H J$ is irreducible, for otherwise the eigenvalue problem for M could be reduced to the eigenvalue problem for irreducible tridiagonal matrices of smaller order than that of M. We already know (see (6.6.5.2)) that the QR method, with shift k, applied to the tridiagonal matrix M yields another tridiagonal matrix $\tilde{M}$ according to the following rules:

$$\text{(6.7.2)} \qquad \begin{aligned} M - kI &=: \tilde{T}R, \qquad R \text{ upper triangular}, \qquad \tilde{T} \text{ unitary} \\ \tilde{M} &:= R\tilde{T} + kI. \end{aligned}$$

Both tridiagonal matrices $\tilde{M} = \tilde{T}^H M \tilde{T}$ and $\bar{M} = T^H M T$ are unitarily similar to M, where the unitary matrix $\tilde{T} = \tilde{T}(k)$ depends, of course, on the shift k. Now, Theorem (6.6.5.5) shows that for all shifts k the matrix $\tilde{M} = [\tilde{m}_{ij}]$ is almost irreducible, that is, $\tilde{m}_{j,j-1} \neq 0$ for $j = 2, 3, \ldots, n-1$, since M is irreducible. Hence we have the situation of Theorem (6.6.5.1): If the matrices T and $\tilde{T} = \tilde{T}(k)$ have the same first column, then according to this theorem there is a phase matrix $\Delta = \operatorname{diag}(e^{i\phi_1}, e^{i\phi_2}, \ldots, e^{i\phi_n})$, $\phi_1 = 0$, so that $\tilde{M} = \Delta^H \bar{M} \Delta$, that is, $\tilde{M}$ is essentially equal to M. We show next that it is in fact possible to choose T_{12} in such a way that T and $\tilde{T}(k)$ have the same first column.

The first column t_1 of $T = T_{12} T_{23} \cdots T_{n-1,n}$ is precisely the first column of T_{12}, since (with $e_1 := [1, 0, \ldots, 0]^T \in \mathbb{C}^n$)

$$t_1 = Te_1 = T_{12} \cdots T_{n-1,n} e_1 = T_{12} \cdots T_{n-2,n-1} e_1 = \cdots = T_{12} e_1,$$

and has the form $t_1 = [c, s, 0, \ldots, 0]^T \in \mathbb{R}^n$, $c^2 + s^2 = 1$. On the other hand, the tridiagonal matrix $M - kI$, by virtue of $M = J^H J$, has the form

$$\text{(6.7.3)}\quad M - kI = \begin{bmatrix} \delta_1 & \bar{\gamma}_2 & & & 0 \\ \gamma_2 & \delta_2 & \cdot & & \\ & \cdot & \cdot & \cdot & \\ & & \cdot & \cdot & \cdot \\ & & & \cdot & \bar{\gamma}_n \\ 0 & & & \gamma_n & \delta_n \end{bmatrix}, \qquad \delta_1 = |q_1|^2 - k, \quad \gamma_2 = \bar{e}_2 q_1,$$

if

$$J = \begin{bmatrix} q_1 & e_2 & & & 0 \\ & q_2 & e_3 & & \\ & & \cdot & \cdot & \\ & & & \cdot & \cdot \\ & & & & e_n \\ 0 & & & & q_n \end{bmatrix}.$$

A unitary matrix $\tilde{T}$ with (6.7.2), $\tilde{T}^H(M - kI) = R$ upper triangular, can be determined as a product of $n - 1$ Givens reflections of the type $\tilde{T}_{i,i+1}$ (see Section 4.9), $\tilde{T} = \tilde{T}_{12}\tilde{T}_{23} \cdots \tilde{T}_{n-1,n}$ which successively annihilate the subdiagonal elements of $M - kI$. The matrix $\tilde{T}_{12}$, in particular, has the form

$$\text{(6.7.4)}\quad \tilde{T}_{12} = \begin{bmatrix} \tilde{c} & \tilde{s} & & 0 \\ \tilde{s} & -\tilde{c} & & \\ & & 1. & \\ & & & \ddots \\ 0 & & & 1 \end{bmatrix}, \qquad \begin{bmatrix} \tilde{c} \\ \tilde{s} \end{bmatrix} = \alpha \begin{bmatrix} \delta_1 \\ \gamma_2 \end{bmatrix} = \alpha \begin{bmatrix} |q_1|^2 - k \\ \bar{e}_2 q_1 \end{bmatrix}, \qquad \tilde{c}^2 + \tilde{s}^2 = 1,$$

where $\tilde{s} \neq 0$ if M is irreducible.

The first column $\tilde{t}_1$ of $\tilde{T}$ agrees with the first column of $\tilde{T}_{12}$, which is given by $\tilde{c}$, $\tilde{s}$ in (6.7.4). Since $\tilde{M} = \tilde{T}^H M \tilde{T}$ is a tridiagonal matrix, it follows from Francis's observation that $\bar{M}$ essentially agrees with $\tilde{M}$ (up to scaling factors), provided one chooses as first column t_1 of T_{12} (through which T_{12} is determined) precisely the first column of $\tilde{T}_{12}$ (if the tridiagonal matrix M is not reducible).

Accordingly, a typical step $J \to \bar{J}$ of the method of Golub and Reinsch consists in first determining a real shift parameter k, for example by means of one of the strategies in Section 6.6.6; then choosing $T_{12} := \tilde{T}_{12}$with the aid of the formulas (6.7.4); and finally determining, as described above, further Givens matrices $T_{i,i+1}$, $S_{i,i+1}$ such that

$$\bar{J} = S_{n-1,n} \cdots S_{23} S_{12} J T_{12} T_{23} \cdots T_{n-1,n}$$

again becomes a bidiagonal matrix. Through this *implicit* handling of the shifts, one avoids in particular the loss of accuracy which in (6.7.2) would occur, say for large k, if the subtraction $M \to M - kI$ and subsequent addition $R\tilde{T} \to R\tilde{T} + kI$ were executed explicitly. The convergence properties of the method are of course the same as those of the QR method: in particular, using appropriate shift strategies [see Theorem (6.6.6.5)], one has cubic convergence as a rule.

An ALGOL program for this method can be found in Golub and Reinsch (1971).

6.8 Generalized Eigenvalue Problems

In applications one frequently encounters eigenvalue problems of the following form: For given $n \times n$ matrices A, B, numbers λ are to be found such that there exists a vector $x \neq 0$ with

(6.8.1) $$Ax = \lambda Bx.$$

For nonsingular matrices B, this is equivalent to the classical eigenvalue problem

(6.8.2) $$B^{-1}Ax = \lambda x$$

for the matrix $B^{-1}A$ (a similar statement holds if A^{-1} exists). Now usually, in applications, the matrices A and B are real symmetric, and in addition B is positive definite. Although in general $B^{-1}A$ is not symmetric, it is still possible to reduce (6.8.1) to a classical eigenvalue problem for symmetric matrices: If

$$B = LL^T$$

is the Cholesky decomposition of the positive definite matrix B, then L is nonsingular and $B^{-1}A$ similar to the matrix $G := L^{-1}A(L^{-1})^T$,

$$L^T(B^{-1}A)(L^T)^{-1} = L^T(L^T)^{-1}L^{-1}A(L^{-1})^T = L^{-1}A(L^{-1})^T = G.$$

But now, the matrix G, like A, is symmetric. The eigenvalues λ of (6.8.1) are thus precisely the eigenvalues of the symmetric matrix G.

The computation of G can be simplified as follows: One first computes

$$F := A(L^{-1})^T$$

by solving $FL^T = A$, and then obtains

$$G = L^{-1}F$$

from the equation $LG = F$. In view of the symmetry of G it suffices to determine the elements below the diagonal of G. For this, knowledge of the lower triangle of F (f_{ik} with $k \leqslant i$) is sufficient. It suffices, therefore, to compute only these elements of F from the equation $FL^T = A$.

Together with the Cholesky decomposition $B = LL^T$, which requires about $\frac{1}{6}n^3$ multiplications, the computation of $G = L^{-1}A(L^{-1})^T$ from A, B thus costs about $\frac{2}{3}n^3$ multiplications.

For additional methods, see Martin and Wilkinson (1971) and Peters and Wilkinson (1970). A more recent method for the solution of the generalized eigenvalue problem (6.8.1) is the QZ method of Moler and Stewart (1973).

6.9 Estimation of Eigenvalues

Using the concepts of vector and matrix norm developed in Section 4.4, we now wish to give some simple estimates for the eigenvalues of a matrix. We assume that for $x \in \mathbb{C}^n$

$$\|x\|$$

is a given vector norm and

$$\operatorname{lub}(A) = \max_{x \neq 0} \frac{\|Ax\|}{\|x\|}$$

the associated matrix norm. In particular, we employ the maximum norm

$$\|x\|_\infty = \max_i |x_i|, \qquad \operatorname{lub}_\infty(A) = \max_i \sum_k |a_{ik}|.$$

We distinguish between two types of eigenvalue estimations:

(1) exclusion theorems,
(2) inclusion theorems.

Exclusion theorems give domains in the complex plane which contain *no* eigenvalue (or whose complement contains *all* eigenvalues); inclusion theorems give domains in which there lies *at least* one eigenvalue.

An exclusion theorem of the simplest type is

(6.9.1) Theorem (Hirsch). *For all eigenvalues λ of A one has*

$$|\lambda| \leqslant \operatorname{lub}(A).$$

PROOF. If x is an eigenvector to the eigenvalue λ, then from

$$Ax = \lambda x, \qquad x \neq 0,$$

there follows

$$\|\lambda x\| = |\lambda| \cdot \|x\| \leqslant \operatorname{lub}(A) \cdot \|x\|,$$
$$|\lambda| \leqslant \operatorname{lub}(A). \qquad \square$$

If λ_i are the eigenvalues of A, then

$$\rho(A) := \max_{1 \leqslant i \leqslant n} |\lambda_i|$$

is called the *spectral radius* of A. By (6.9.1) we have $\rho(A) \leqslant \operatorname{lub}(A)$ for every vector norm.

(6.9.2) Theorem.

(a) *For every matrix A and every $\varepsilon > 0$ there exists a vector norm such that*

$$\operatorname{lub}(A) \leqslant \rho(A) + \varepsilon.$$

(b) *If every eigenvalue λ of A with $|\lambda| = \rho(A)$ has only linear elementary divisors, then there exists a vector norm such that*

$$\operatorname{lub}(A) = \rho(A).$$

PROOF. (a): Given A, there exists a nonsingular matrix T such that

$$TAT^{-1} = J$$

is the Jordan normal form of A [see (6.2.5)], i.e., J is made up of diagonal blocks of the form

$$C_\nu(\lambda_i) = \begin{bmatrix} \lambda_i & 1 & & & 0 \\ & \cdot & \cdot & & \\ & & \cdot & \cdot & \\ & & & \cdot & 1 \\ 0 & & & & \lambda_i \end{bmatrix}.$$

By means of the transformation $J \to D_\varepsilon^{-1} J D_\varepsilon$, with diagonal matrix

$$D_\varepsilon := \operatorname{diag}(1, \varepsilon, \varepsilon^2, \ldots, \varepsilon^{n-1}), \qquad \varepsilon > 0,$$

one reduces the $C_\nu(\lambda_i)$ to the form

$$\begin{bmatrix} \lambda_i & \varepsilon & & & 0 \\ & \cdot & \cdot & & \\ & & \cdot & \cdot & \\ & & & \cdot & \varepsilon \\ 0 & & & & \lambda_i \end{bmatrix}.$$

From this it follows immediately that

$$\operatorname{lub}_\infty(D_\varepsilon^{-1} J D_\varepsilon) = \operatorname{lub}_\infty(D_\varepsilon^{-1} TAT^{-1} D_\varepsilon) \leqslant \rho(A) + \varepsilon.$$

Now the following is true in general: If S is a nonsingular matrix and $\|\cdot\|$ a vector norm, then $p(x) := \|Sx\|$ is also a vector norm, and $\operatorname{lub}_p(A) = \operatorname{lub}(SAS^{-1})$. For the norm $p(x) := \|D_\varepsilon^{-1} Tx\|_\infty$ it then follows that

$$\operatorname{lub}_p(A) = \operatorname{lub}_\infty(D_\varepsilon^{-1} TAT^{-1} D_\varepsilon) \leqslant \rho(A) + \varepsilon.$$

(b): Let the eigenvalues λ_i of A be ordered as follows:

$$\rho(A) = |\lambda_1| = \cdots = |\lambda_s| > |\lambda_{s+1}| \geqslant \cdots \geqslant |\lambda_n|.$$

Then, by assumption, for $1 \leqslant i \leqslant s$ each Jordan box $C_\nu(\lambda_i)$ in J has dimension 1, i.e., $C_\nu(\lambda_i) = [\lambda_i]$. Choosing

$$\varepsilon = \rho(A) - |\lambda_{s+1}|,$$

we therefore have

$$\mathrm{lub}_\infty(D_\varepsilon^{-1} TAT^{-1} D_\varepsilon) = \rho(A).$$

For the norm $p(x) := \|D_\varepsilon^{-1} Tx\|$ it follows, as in (a), that

$$\mathrm{lub}_p(A) = \rho(A). \qquad \square$$

A better estimate than (6.9.1) is given by the following theorem [cf. Bauer and Fike (1960)]:

(6.9.3) Theorem. *If B is an arbitrary $n \times n$ matrix, then for all eigenvalues λ of A one has*

$$1 \leqslant \mathrm{lub}((\lambda I - B)^{-1}(A - B)) \leqslant \mathrm{lub}((\lambda I - B)^{-1})\, \mathrm{lub}(A - B)$$

unless λ is also an eigenvalue of B.

PROOF. If x is an eigenvector of A for the eigenvalue λ, then from the identity

$$(A - B)x = (\lambda I - B)x$$

it follows immediately, if λ is not an eigenvalue of B, that

$$(\lambda I - B)^{-1}(A - B)x = x,$$

and hence

$$\mathrm{lub}[(\lambda I - B)^{-1}(A - B)] \geqslant 1. \qquad \square$$

Choosing in particular

$$B = A_D := \begin{bmatrix} a_{11} & & 0 \\ & \ddots & \\ 0 & & a_{nn} \end{bmatrix},$$

the diagonal of A, and taking the maximum norm, it follows that

$$\mathrm{lub}_\infty[(\lambda I - A_D)^{-1}(A - A_D)] = \max_{1 \leqslant i \leqslant n} \frac{1}{|\lambda - a_{ii}|} \sum_{\substack{k=1 \\ k \neq i}}^{n} |a_{ik}|.$$

From Theorem (6.9.3) we now get

(6.9.4) Theorem (Gershgorin). *The union of all discs*

$$K_i := \left\{ \mu \in \mathbb{C} \;\middle|\; |\mu - a_{ii}| \leqslant \sum_{\substack{k=1 \\ k \neq i}}^{n} |a_{ik}| \right\}$$

contains all eigenvalues of the $n \times n$ matrix $A = [a_{ik}]$.

Since the disc K_i has center at a_{ii} and radius $\sum_{k=1, k \neq i}^{n} |a_{ik}|$, this estimate will be sharper as A deviates less from a diagonal matrix.

EXAMPLE 1.

$$A = \begin{bmatrix} 1 & 0.1 & -0.1 \\ 0 & 2 & 0.4 \\ -0.2 & 0 & 3 \end{bmatrix},$$

$$K_1 = \{\mu \mid |\mu - 1| \leqslant 0.2\}$$

$$K_2 = \{\mu \mid |\mu - 2| \leqslant 0.4\}$$

$$K_3 = \{\mu \mid |\mu - 3| \leqslant 0.2\}$$

(See Figure 10.)

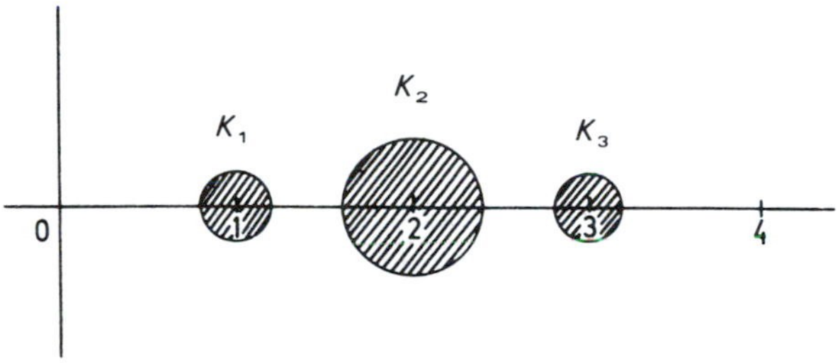

Figure 10 Gershgorin circles.

The preceding theorem can be sharpened as follows:

(6.9.5) Corollary. *If the union $M_1 = \bigcup_{j=1}^{k} K_{i_j}$ of k discs K_{i_j}, $j = 1, \ldots, k$, and the union M_2 of the remaining discs are disjoint, then M_1 contains exactly k eigenvalues of A and M_2 exactly $n - k$ eigenvalues.*

PROOF. If $A = A_D + R$, for $t \in [0, 1]$ let

$$A_t := A_D + tR.$$

Then

$$A_0 = A_D, \qquad A_1 = A.$$

The eigenvalues of A_t are continuous functions of t. Applying the theorem of Gershgorin to A_t, one finds that for $t = 0$ there are exactly k eigenvalues of A_0 in M_1 and $n - k$ in M_2 (counting multiple eigenvalues according to their multiplicities as zeros of the characteristic polynomial). Since for $0 \leqslant t \leqslant 1$ all eigenvalues of A_t likewise must lie in these discs, it follows for reasons of continuity that also k eigenvalues of A lie in M_1 and the remaining $n - k$ in M_2. □

Since A and A^T have the same eigenvalues, one can apply (6.9.4), (6.9.5) to A as well as to A^T and thus, in general, obtain more information about the location of the eigenvalues.

It is often possible to improve the estimates of Gershgorin's theorem by first applying a similarity transformation $A \to D^{-1}AD$ to A with a diagonal matrix $D = \operatorname{diag}(d_1, \ldots, d_n)$. For the eigenvalues of $D^{-1}AD$, and thus of A, one thus obtains the discs

$$K_i = \left\{\mu \,\middle|\, |\mu - a_{ii}| \leqslant \sum_{\substack{k=1 \\ k \neq i}}^{n} \left| \frac{a_{ik} d_k}{d_i} \right| =: \rho_i \right\}.$$

By a suitable choice of D one can often substantially reduce the radius ρ_i of a disc (the remaining discs, as a rule, become larger) in such a way, indeed, that the disc K_i in question remains disjoint with the other discs $K_j, j \neq i$. K_i then contains exactly one eigenvalue of A.

EXAMPLE 2.

$$A = \begin{bmatrix} 1 & \varepsilon & \varepsilon \\ \varepsilon & 2 & \varepsilon \\ \varepsilon & \varepsilon & 2 \end{bmatrix},$$

$$K_1 = \{\mu \mid |\mu - 1| \leqslant 2\varepsilon\},$$
$$K_2 = K_3 = \{\mu \mid |\mu - 2| \leqslant 2\varepsilon\},$$
$$0 < \varepsilon \ll 1.$$

Transformation with $D = \operatorname{diag}(1, k\varepsilon, k\varepsilon)$, $k > 0$, yields

$$A' = D^{-1}AD = \begin{bmatrix} 1 & k\varepsilon^2 & k\varepsilon^2 \\ 1/k & 2 & \varepsilon \\ 1/k & \varepsilon & 2 \end{bmatrix}.$$

For A' we have

$$\rho_1 = 2k\varepsilon^2, \qquad \rho_2 = \rho_3 = \frac{1}{k} + \varepsilon.$$

The discs K_1 and $K_2 = K_3$ for A' are disjoint if

$$\rho_1 + \rho_2 = 2k\varepsilon^2 + \frac{1}{k} + \varepsilon < 1.$$

For this to be true we must clearly have $k > 1$. The optimal value $\tilde{k}$, for which K_1 and K_2 touch one another, is obtained from $\rho_1 + \rho_2 = 1$. One finds

$$\tilde{k} = \frac{2}{1 - \varepsilon + \sqrt{(1 - \varepsilon)^2 - 8\varepsilon^2}} = 1 + \varepsilon + O(\varepsilon^2),$$

and thus

$$\rho_1 = 2\tilde{k}\varepsilon^2 = 2\varepsilon^2 + O(\varepsilon^3).$$

Through the transformation $A \to A'$ the radius ρ_1 of K_1 can thus be reduced from the initial 2ε to about $2\varepsilon^2$.

The estimate of Theorem (6.9.3) can also be interpreted as a perturbation theorem, indicating how much the eigenvalues of A can deviate from the eigenvalues of B. In order to show this, let us assume that B is normalizable:

$$B = P\Lambda_B P^{-1}, \qquad \Lambda_B = \operatorname{diag}(\lambda_1(B), \ldots, \lambda_n(B)).$$

It then follows, if λ is no eigenvalue of B, that

$$\begin{aligned}
\operatorname{lub}((\lambda I - B)^{-1}) &= \operatorname{lub}(P(\lambda I - \Lambda_B)^{-1}P^{-1}) \\
&\leqslant \operatorname{lub}((\lambda I - \Lambda_B)^{-1}) \operatorname{lub}(P) \operatorname{lub}(P^{-1}) \\
&= \max_{1 \leqslant i \leqslant n} \frac{1}{|\lambda - \lambda_i(B)|} \operatorname{cond}(P) \\
&= \frac{1}{\min\limits_{1 \leqslant i \leqslant n} |\lambda - \lambda_i(B)|} \operatorname{cond}(P).
\end{aligned}$$

This estimate is valid for all norms $\|\cdot\|$ satisfying, like the maximum norm and the Euclidean norm,

$$\operatorname{lub}(D) = \max_{1 \leqslant i \leqslant n} |d_i|$$

for all diagonal matrices $D = \operatorname{diag}(d_1, \ldots, d_n)$. Such norms are called *absolute*; they are also characterized by the condition $\| |x| \| = \|x\|$ for all $x \in \mathbb{C}^n$ [see Bauer, Stoer, and Witzgall (1961)].

From (6.9.3) we thus obtain [cf. Bauer and Fike (1960), Householder (1964)]:

(6.9.6) Theorem. *If B is a diagonalizable $n \times n$ matrix, $B = P\Lambda_B P^{-1}$, $\Lambda_B = \operatorname{diag}(\lambda_1(B), \ldots, \lambda_n(B))$, and A an arbitrary $n \times n$ matrix, then for each eigenvalue $\lambda(A)$ there is an eigenvalue $\lambda_i(B)$ such that*

$$|\lambda(A) - \lambda_i(B)| \leqslant \operatorname{cond}(P) \operatorname{lub}(A - B).$$

Here cond *and* lub *are to be formed with reference to an absolute norm $\|\cdot\|$.*

This estimate shows that for the sensitivity of the eigenvalues of a matrix B under perturbations, the controlling factor is the condition

$$\operatorname{cond}(P)$$

of the matrix P, not the condition of B. But the columns of P are just the (right) eigenvectors of B. For Hermitian, and more generally, normal matrices B one can choose for P a unitary matrix [Theorem (6.4.5)]. With respect to the Euclidean norm $\|x\|_2$ one therefore has $\operatorname{cond}_2(P) = 1$, and thus:

(6.9.7) Theorem. *If B is a normal $n \times n$ matrix and A an arbitrary $n \times n$ matrix, then for each eigenvalue $\lambda(A)$ of A there is an eigenvalue $\lambda(B)$ of B such that*

$$|\lambda(A) - \lambda(B)| \leqslant \operatorname{lub}_2(A - B).$$

The eigenvalue problem for Hermitian matrices is thus always well conditioned.

The estimates in (6.9.6), (6.9.7) are of a global nature. We now wish to examine in a first approximation the sensitivity of a fixed eigenvalue λ of A under small perturbations $A \to A + \varepsilon C$, $\varepsilon \to 0$. We limit ourselves to the case in which the eigenvalue λ under study is a simple zero of the characteristic polynomial of A. To λ then belong uniquely determined (up to a constant factor) right and left eigenvectors x and y, respectively:

$$Ax = \lambda x, \quad y^H A = \lambda y^H, \qquad x \neq 0, \quad y \neq 0.$$

For these, one has $y^H x \neq 0$, as one easily shows with the aid of the Jordan normal form of A, which contains only one Jordan block, of order one, for the eigenvalue λ.

(6.9.8) Theorem. *Let λ be a simple zero of the characteristic polynomial of the $n \times n$ matrix A, and x and y^H corresponding right and left eigenvectors of A, respectively,*

$$Ax = \lambda x, \quad y^H A = \lambda y^H, \qquad x \neq 0, \quad y \neq 0,$$

and let C be an arbitrary $n \times n$ matrix. Then there exists a function $\lambda(\varepsilon)$ which is analytic for ε sufficiently small, $|\varepsilon| < \varepsilon_0$, $\varepsilon_0 > 0$, such that

$$\lambda(0) = \lambda, \qquad \lambda'(0) = \frac{y^H C x}{y^H x},$$

and $\lambda(\varepsilon)$ is a simple zero of the characteristic polynomial of $A + \varepsilon C$. One thus has, in first approximation,

$$\lambda(\varepsilon) \doteq \lambda + \varepsilon \frac{y^H C x}{y^H x}.$$

Proof. The characteristic polynomial of the matrix $A + \varepsilon C$,

$$\varphi_\varepsilon(\mu) = \det(A + \varepsilon C - \mu I),$$

is an analytic function of ε and μ. Let K be a circle about λ,

$$K = \{\mu \mid |\mu - \lambda| = r\},$$

which does not contain any eigenvalues of A other than λ. Then

$$\inf_{\mu \in K} |\varphi_0(\mu)| =: m > 0.$$

Since $\varphi_\varepsilon(\mu)$ depends continuously on ε, there is an $\varepsilon_0 > 0$ such that also

(6.9.9) $$\inf_{\mu \in K} |\varphi_\varepsilon(\mu)| > 0 \quad \text{for all } |\varepsilon| < \varepsilon_0.$$

According to a well-known theorem in complex variables, the number of zeros of $\varphi_\varepsilon(\mu)$ within K is given by

$$\nu(\varepsilon) = \frac{1}{2\pi i} \oint_K \frac{\varphi_\varepsilon'(\mu)}{\varphi_\varepsilon(\mu)} \, d\mu.$$

In view of (6.9.9), $\nu(\varepsilon)$ is continuous for $|\varepsilon| \leqslant \varepsilon_0$; hence $1 = \nu(0) = \nu(\varepsilon)$ for $|\varepsilon| \leqslant \varepsilon_0$, ν being integer valued. The simple zero $\lambda(\varepsilon)$ of $\varphi_\varepsilon(\mu)$ inside of K, according to another theorem in complex variables, admits the representation

$$\lambda(\varepsilon) = \frac{1}{2\pi i} \oint_K \frac{\mu \varphi_\varepsilon'(\mu)}{\varphi_\varepsilon(\mu)} \, d\mu. \tag{6.9.10}$$

For $|\varepsilon| \leqslant \varepsilon_0$ the integrand of (6.9.10) is an analytic function of ε and therefore also $\lambda(\varepsilon)$, according to a well-known theorem on the interchangeability of differentiation and integration. For the simple eigenvalue $\lambda(\varepsilon)$ of $A + \varepsilon C$ one can choose right and left eigenvectors $x(\varepsilon)$ and $y(\varepsilon)$,

$$(A + \varepsilon C)x(\varepsilon) = \lambda(\varepsilon)x(\varepsilon), \qquad y(\varepsilon)^H(A + \varepsilon C) = \lambda(\varepsilon)y(\varepsilon)^H,$$

in such a way that $x(\varepsilon)$ and $y(\varepsilon)$ are analytic functions of ε for $|\varepsilon| \leqslant \varepsilon_0$. We may put, e.g., $x(\varepsilon) = [\xi_1(\varepsilon), \ldots, \xi_n(\varepsilon)]^T$ with

$$\xi_i(\varepsilon) = (-1)^i \det B_{1i},$$

where B_{1i} is the $(n-1) \times (n-1)$ matrix obtained by deleting row 1 and column i in the matrix $A + \varepsilon C - \lambda(\varepsilon)I$. From

$$(A + \varepsilon C - \lambda(\varepsilon)I)x(\varepsilon) = 0,$$

differentiating with respect to ε at $\varepsilon = 0$, one obtains

$$(C - \lambda'(0)I)x(0) + (A - \lambda(0)I)x'(0) = 0,$$

from which, in view of $y(0)^H(A - \lambda(0)I) = 0$,

$$y(0)^H(C - \lambda'(0)I)x(0) = 0,$$

and hence, since $y(0)^H x(0) \neq 0$,

$$\lambda'(0) = \frac{y^H C x}{y^H x}, \qquad y = y(0), \quad x = x(0),$$

as was to be shown. □

Denoting, for the Euclidean norm $\|\cdot\|_2$, by

$$\cos(x, y) := \frac{y^H x}{\|x\|_2 \, \|y\|_2}$$

the cosine of the angle between x and y, the preceding result implies the estimate

$$\begin{aligned} |\lambda'(0)| &= \frac{|y^H C x|}{\|y\|_2 \, \|x\|_2 \, |\cos(x, y)|} \leqslant \frac{\|Cx\|_2}{\|x\|_2 \, |\cos(x, y)|} \\ &\leqslant \frac{\operatorname{lub}_2(C)}{|\cos(x, y)|}. \end{aligned}$$

The sensitivity of λ will thus increase with decreasing $|\cos(x, y)|$. For Hermitian matrices we always have $x = y$ (up to constant multiples), and hence $|\cos(x, y)| = 1$. This is in harmony with Theorem (6.9.7), according to which the eigenvalues of Hermitian matrices are relatively insensitive to perturbations.

Theorem (6.9.8) asserts that an eigenvalue λ of A which is only a simple zero of the characteristic polynomial is relatively insensitive to perturbations $A \to A + \varepsilon C$, in the sense that for the corresponding eigenvalue $\lambda(\varepsilon)$ of $A + \varepsilon C$ there exist a constant K and an $\varepsilon_0 > 0$ such that

$$|\lambda(\varepsilon) - \lambda| \leqslant K \cdot |\varepsilon| \quad \text{for } |\varepsilon| \leqslant \varepsilon_0.$$

However, for ill-conditioned eigenvalues λ, i.e., if the corresponding left and right eigenvectors are almost orthogonal, the constant K is very large.

This statement remains valid if the eigenvalue λ is a multiple zero of the characteristic polynomial and has only linear elementary divisors. If to λ there belong nonlinear elementary divisors, however, the statement becomes false. The following can be shown in this case (cf. Exercise 29): Let $(\mu - \lambda)^{\nu_1}$, $(\mu - \lambda)^{\nu_2}$, ..., $(\mu - \lambda)^{\nu_\rho}$, $\nu_1 \geqslant \nu_2 \geqslant \cdots \geqslant \nu_\rho$, be the elementary divisors belonging to the eigenvalue λ of A. Then the matrix $A + \varepsilon C$, for sufficiently small ε, has eigenvalues $\lambda_i(\varepsilon)$, $i = 1, \ldots, \sigma$, $\sigma := \nu_1 + \cdots + \nu_\rho$, satisfying, with some constant K,

$$(6.9.11) \qquad |\lambda_i(\varepsilon) - \lambda| \leqslant K |\varepsilon|^{1/\nu_1} \quad \text{for } i = 1, \ldots, \sigma, \ |\varepsilon| \leqslant \varepsilon_0.$$

This has the following numerical consequence: If in the practical computation of the eigenvalues of a matrix A [with $\text{lub}(A) \approx 1$] one applies a stable method, then the rounding errors committed during the course of the method can be interpreted as having the effect of producing *exact* results not for A, but for a perturbed initial matrix $A + \Delta A$, $\text{lub}(\Delta A) = O(\text{eps})$. If to the eigenvalue λ_i of A there belong only linear elementary divisors, then the computed approximation $\tilde{\lambda}_i$ of λ_i agrees with λ_i up to an error of the order eps. If however λ_i has elementary divisors of order at most ν, one must expect an error of order of magnitude $\text{eps}^{1/\nu}$ for $\tilde{\lambda}_i$.

For the derivation of a typical inclusion theorem, we limit ourselves to the case of the Euclidean norm

$$\|x\|_2 = \sqrt{x^H x} = \sqrt{\sum_i |x_i|^2},$$

and begin by proving the formula

$$\min_{x \neq 0} \frac{\|Dx\|_2}{\|x\|_2} = \min_i |d_i|$$

for a diagonal matrix $D = \text{diag}(d_1, \ldots, d_n)$. Indeed, for all $x \neq 0$,

$$\frac{\|Dx\|_2^2}{\|x\|_2^2} = \frac{\sum_i |x_i|^2 \, |d_i|^2}{\sum_i |x_i|^2} \geqslant \min |d_i|^2.$$

If $|d_j| = \min_i |d_i|$, then the lower bound is attained for $x = e_j$ (the jth unit vector). Furthermore, if A is a normal matrix, i.e., a matrix which can be transformed to diagonal form by a unitary matrix U,

$$A = U^H D U, \qquad D = \operatorname{diag}(d_1, \ldots, d_n), \quad d_i = \lambda_i(A),$$

and if $f(\lambda)$ is an arbitrary polynomial, then

$$f(A) = U^H f(D) U,$$

and from the unitary invariance of $\|x\|_2 = \|Ux\|_2$ it follows at once, for all $x \neq 0$, that

$$\begin{aligned} \frac{\|f(A)x\|_2}{\|x\|_2} &= \frac{\|U^H f(D)Ux\|_2}{\|x\|_2} = \frac{\|f(D)Ux\|_2}{\|Ux\|_2} \\ &\geqslant \min_{1 \leqslant i \leqslant n} |f(d_i)| = \min_{1 \leqslant i \leqslant n} |f(\lambda_i(A))|. \end{aligned}$$

We therefore have the following [see, e.g., Householder (1964)]:

(6.9.12) Theorem. *If A is a normal matrix, $f(\lambda)$ an arbitrary polynomial, and $x \neq 0$ an arbitrary vector, then there is an eigenvalue $\lambda(A)$ of A such that*

$$|f(\lambda(A))| \leqslant \frac{\|f(A)x\|_2}{\|x\|_2}.$$

In particular, choosing for f the linear polynomial

$$f(\lambda) \equiv \lambda - \frac{x^H A x}{x^H x} \equiv \lambda - \frac{\mu_{01}}{\mu_{00}},$$

where

$$\mu_{ik} := x^H (A^H)^i A^k x, \qquad i, k = 0, 1, 2, \ldots,$$

we immediately get, in view of $\mu_{ik} = \bar{\mu}_{ki}$,

$$\begin{aligned} \|f(A)x\|_2^2 &= x^H \left(A^H - \frac{\bar{\mu}_{01}}{\mu_{00}} I \right) \left(A - \frac{\mu_{01}}{\mu_{00}} I \right) x \\ &= \mu_{11} - \frac{\mu_{10}\mu_{01}}{\mu_{00}} - \frac{\mu_{10}\mu_{01}}{\mu_{00}} + \frac{\mu_{01}\mu_{10}}{\mu_{00}^2} \mu_{00} \\ &= \mu_{11} - \frac{\mu_{01}\mu_{10}}{\mu_{00}}. \end{aligned}$$

We thus have:

(6.9.13) Theorem (Weinstein). *If A is normal and $x \neq 0$ an arbitrary vector, then the disc*

$$\left\{ \lambda \,\middle|\, \left| \lambda - \frac{\mu_{01}}{\mu_{00}} \right| \leqslant \sqrt{\frac{\mu_{11} - (\mu_{10}\mu_{01}/\mu_{00})}{\mu_{00}}} \right\}$$

contains at least one eigenvalue of A.

The quotient $\mu_{01}/\mu_{00} = x^H Ax/x^H x$, incidentally, is called the

Rayleigh quotient

of A belonging to x. The last theorem is used especially in connection with the vector iteration: If x is approximately equal to x_1, the eigenvector corresponding to the eigenvalue λ_1,

$$x \approx x_1, \qquad Ax_1 = \lambda_1 x_1,$$

then the Rayleigh quotient $x^H Ax/x^H x$ belonging to x is in general a very good approximation to λ_1. Theorem (6.9.13) then shows how much at most this approximate value $x^H Ax/x^H x$ deviates from an eigenvalue of A.

The set

$$G[A] = \left\{ \frac{x^H Ax}{x^H x} \,\middle|\, x \neq 0 \right\}$$

of all Rayleigh quotients is called the *field of values* of the matrix A. Choosing for x an eigenvector of A, it follows at once that $G[A]$ contains the eigenvalues of A. Hausdorff, furthermore, has shown that $G[A]$ is always convex. For normal matrices

$$A = U^H \Lambda U, \qquad \Lambda = \begin{bmatrix} \lambda_1 & & 0 \\ & \ddots & \\ 0 & & \lambda_n \end{bmatrix}, \quad U^H U = I,$$

one even has

$$\begin{aligned} G[A] &= \left\{ \frac{x^H U^H \Lambda U x}{x^H U^H U x} \,\middle|\, x \neq 0 \right\} \\ &= \left\{ \frac{y^H \Lambda y}{y^H y} \,\middle|\, y \neq 0 \right\} \\ &= \left\{ \mu \,\middle|\, \mu = \sum_{i=1}^{n} \tau_i \lambda_i,\ \tau_i \geqslant 0,\ \sum_{i=1}^{n} \tau_i = 1 \right\}. \end{aligned}$$

That is, for normal matrices, $G[A]$ is the convex hull of the eigenvalues of A.

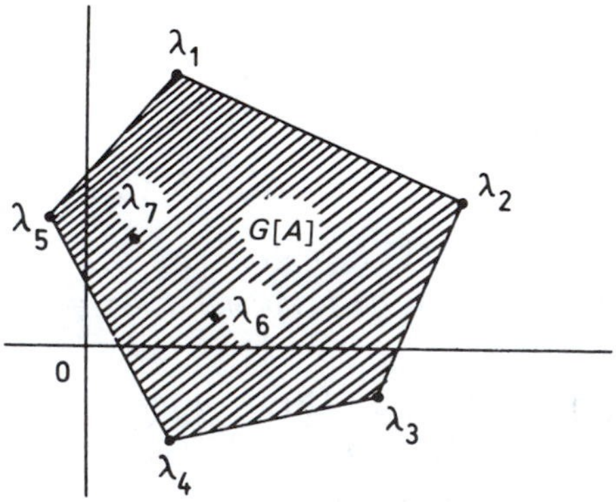

Figure 11 Field of values.

For a Hermitian matrix $H = H^H = U^H \Lambda U$ with eigenvalues $\lambda_1 \geqslant \lambda_2 \geqslant \cdots \geqslant \lambda_n$, we thus recover the result (6.4.3), which characterizes λ_1 and λ_n by extremal properties. The remaining eigenvalues of H, too, can be characterized similarly, as is shown by the following theorem:

(6.9.14) Theorem (Courant, Weyl). *For the eigenvalues $\lambda_1 \geqslant \lambda_2 \geqslant \cdots \geqslant \lambda_n$ of an $n \times n$ Hermitian matrix H, one has for $i = 0, 1, \ldots, n-1$*

$$\lambda_{i+1} = \min_{y_1, \ldots, y_i \in \mathbb{C}^n} \max_{\substack{x \in \mathbb{C}^n:\, x^H y_1 = \cdots = x^H y_i = 0 \\ x \neq 0}} \frac{x^H H x}{x^H x}.$$

Proof. For arbitrary $y_1, \ldots, y_i \in \mathbb{C}^n$ define $\mu(y_1, \ldots, y_i)$ by

$$\mu(y_1, \ldots, y_i) := \max_{\substack{x \in \mathbb{C}^n:\, x^H y_1 = \cdots = x^H y_i = 0 \\ x^H x = 1}} x^H H x.$$

Further let $x_1, \ldots, x_n$ be a set of n orthonormal eigenvectors of H for the eigenvalues λ_j [see Theorem (6.4.2)]: $Hx_j = \lambda_j x_j$, $x_j^H x_k = \delta_{jk}$ for $j, k = 1, 2, \ldots, n$. For $y_j := x_j$, $j = 1, \ldots, i$, all $x \in \mathbb{C}^n$ with $x^H y_j = 0$, $j = 1, \ldots, i$, $x^H x = 1$, can then be represented in the form

$$x = \rho_{i+1} x_{i+1} + \cdots + \rho_n x_n, \qquad \sum_{k > i} |\rho_k|^2 = 1,$$

so that, since $\lambda_k \leqslant \lambda_{i+1}$ for $k \geqslant i + 1$, one has for such x

$$\begin{aligned} x^H H x &= (\rho_{i+1} x_{i+1} + \cdots + \rho_n x_n)^H H (\rho_{i+1} x_{i+1} + \cdots + \rho_n x_n) \\ &= |\rho_{i+1}|^2 \lambda_{i+1} + \cdots + |\rho_n|^2 \lambda_n \\ &\leqslant \lambda_{i+1} (|\rho_{i+1}|^2 + \cdots + |\rho_n|^2) = \lambda_{i+1}, \end{aligned}$$

where equality holds for $x := x_{i+1}$. Hence,

$$\mu(x_1, \ldots, x_i) = \lambda_{i+1}.$$

On the other hand, for arbitrary $y_1, \ldots, y_i \in \mathbb{C}^n$, the subspaces $E := \{x \in \mathbb{C}^n \mid x^H y_j = 0 \text{ for } j \leqslant i\}$, $F := \{\sum_{j \leqslant i+1} \rho_j x_j \mid \rho_j \in \mathbb{C}\}$ have dimensions $\dim E \geqslant n - i$, $\dim F = i + 1$, so that $\dim E \cap F \geqslant 1$, and there exists a vector $x_0 \in E \cap F$ with $x_0^H x_0 = 1$. Therefore, because $x_0 = \rho_1 x_1 + \cdots + \rho_{i+1} x_{i+1} \in F$,

$$\begin{aligned} \mu(y_1, \ldots, y_i) \geqslant x_0^H H x_0 &= |\rho_1|^2 \lambda_1 + \cdots + |\rho_{i+1}|^2 \lambda_{i+1} \\ &\geqslant (|\rho_1|^2 + \cdots + |\rho_{i+1}|^2) \lambda_{i+1} = \lambda_{i+1}. \end{aligned}$$ □

Defining, for an arbitrary matrix A,

$$H_1 := \tfrac{1}{2}(A + A^H), \qquad H_2 := \frac{1}{2i}(A - A^H),$$

then H_1, H_2 are Hermitian and

$$A = H_1 + iH_2.$$

(H_1, H_2 are also denoted by Re A and Im A, respectively; it should be noted that the elements of Re A are not real in general.)

For every eigenvalue λ of A, one has, on account of $\lambda \in G[A]$ and (6.4.3),

$$\operatorname{Re} \lambda \leqslant \max_{x \neq 0} \operatorname{Re} \frac{x^H A x}{x^H x} = \max_{x \neq 0} \frac{1}{x^H x} \frac{1}{2} (x^H A x + x^H A^H x)$$

$$= \max_{x \neq 0} \frac{x^H H_1 x}{x^H x} = \lambda_{\max}(H_1),$$

$$\operatorname{Im} \lambda \leqslant \max_{x \neq 0} \operatorname{Im} \frac{x^H A x}{x^H x} = \lambda_{\max}(H_2).$$

By estimating Re λ, Im λ analogously from below, one obtains

(6.9.15) Theorem (Bendixson). *Decomposing an arbitrary matrix A into $A = H_1 + iH_2$, where H_1 and H_2 are Hermitian, then for every eigenvalue λ of A one has*

$$\lambda_{\min}(H_1) \leqslant \operatorname{Re} \lambda \leqslant \lambda_{\max}(H_1),$$

$$\lambda_{\min}(H_2) \leqslant \operatorname{Im} \lambda \leqslant \lambda_{\max}(H_2).$$

The estimates of this section lead immediately to estimates for the zeros of a polynomial

$$p(\lambda) = a_n \lambda^n + \cdots + a_0, \qquad a_n \neq 0.$$

We need only observe that to p there corresponds the Frobenius matrix

$$F = \begin{bmatrix} 0 & & & & -\gamma_0 \\ 1 & \cdot & & & -\gamma_1 \\ & & \cdot & & \cdot \\ & & & \cdot & \cdot \\ 0 & & & 1 & -\gamma_{n-1} \end{bmatrix}, \quad \text{with } \gamma_i = \frac{a_i}{a_n},$$

which has the characteristic polynomial $(1/a_n)(-1)^n p(\lambda)$. In particular, from the estimate (6.9.1) of Hirsch, with $\operatorname{lub}_\infty(A) = \max_i \sum_k |a_{ik}|$, applied to F and F^T, one obtains the following estimates for all zeros λ_i of $p(\lambda)$, respectively:

(a) $$|\lambda_i| \leqslant \max\left\{\left|\frac{a_0}{a_n}\right|, \max_{1 \leqslant k \leqslant n-1}\left(1 + \left|\frac{a_k}{a_n}\right|\right)\right\},$$

(b) $$|\lambda_i| \leqslant \max\left\{1, \sum_{k=0}^{n-1} \left|\frac{a_k}{a_n}\right|\right\}.$$

EXAMPLE 3. For $p(\lambda) = \lambda^3 - 2\lambda^2 + \lambda - 1$ one obtains

(a) $|\lambda_i| \leqslant \max\{1, 2, 3\} = 3$,
(b) $|\lambda_i| \leqslant \max\{1, 1 + 1 + 2\} = 4$.

In this case (a) gives a better estimate.

Exercises for Chapter 6

1. For the matrix

$$A = \left[\begin{array}{ccc|cc|c|c} 2 & 1 & & & & & 0 \\ & 2 & 1 & & & & \\ & & 2 & & & & \\ \hline & & & 2 & 1 & & \\ & & & & 2 & & \\ \hline & & & & & 1 & \\ \hline 0 & & & & & & 0 \end{array}\right]$$

find the characteristic polynomial, the minimal polynomial, a system of eigenvectors and principal vectors, and the Frobenius normal form.

2. How many distinct (i.e., not mutually similar) 6×6 matrices are there whose characteristic polynomial is

$$p(\lambda) = (3 - \lambda)^4(1 - \lambda)^2?$$

3. What are the properties of the eigenvalues of

   positive definite/semidefinite,
   orthogonal/unitary,
   real skew-symmetric $(A^T = -A)$

   matrices? Determine the minimal polynomial of a projection matrix $A = A^2$.

4. For the matrices

   (a) $A = uv^T$, $u, v \in \mathbb{R}^n$,
   (b) $H = I - 2ww^H$, $w^H w = 1$, $w \in \mathbb{C}^n$,
   (c) $P = \begin{bmatrix} 0 & 0 & 0 & 1 \\ 1 & 0 & 0 & 0 \\ 0 & 1 & 0 & 0 \\ 0 & 0 & 1 & 0 \end{bmatrix}$,

   determine the eigenvalues λ_i, the multiplicities σ_i and ρ_i, the characteristic polynomial $\varphi(\mu)$, the minimal polynomial $\psi(\mu)$, a system of eigenvectors and principal vectors, and a Jordan normal form J.

5. For $u, v, w, z \in \mathbb{R}^n$, $n > 2$, let

$$A := uv^T + wz^T.$$

   (a) With λ_1, λ_2 the eigenvalues of

$$\tilde{A} := \begin{bmatrix} v^T u & v^T w \\ z^T u & z^T w \end{bmatrix},$$

   show that A has the eigenvalues $\lambda_1, \lambda_2, 0$.
   (b) How many eigenvectors and principal vectors can A have? What types of Jordan normal form J of A are possible? Determine J in particular for $A = 0$.

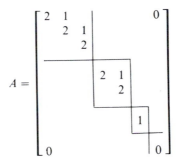

6. (a) Show: λ is an eigenvalue of A if and only if $-\lambda$ is an eigenvalue of B, where

$$A := \begin{bmatrix} \delta_1 & \gamma_2 & & & 0 \\ \beta_2 & \delta_2 & \gamma_3 & & \\ & \ddots & \ddots & \ddots & \\ & & \ddots & \ddots & \gamma_n \\ 0 & & & \beta_n & \delta_n \end{bmatrix}, \qquad B := \begin{bmatrix} -\delta_1 & \gamma_2 & & & 0 \\ \beta_2 & -\delta_2 & \gamma_3 & & \\ & \ddots & \ddots & \ddots & \\ & & \ddots & \ddots & \gamma_n \\ 0 & & & \beta_n & -\delta_n \end{bmatrix}.$$

(b) Suppose the real symmetric tridiagonal matrix

$$A = \begin{bmatrix} \delta_1 & \gamma_2 & & & 0 \\ \gamma_2 & \delta_2 & \ddots & & \\ & \ddots & \ddots & \ddots & \\ & & \ddots & \ddots & \gamma_n \\ 0 & & & \gamma_n & \delta_n \end{bmatrix}$$

satisfies

$$\delta_i = -\delta_{n+1-i}, \qquad i = 1, \ldots, n,$$

$$\gamma_i = \gamma_{n+2-i}, \qquad i = 2, \ldots, n.$$

Show: If λ is an eigenvalue of A, then so is $-\lambda$. What does this imply for the eigenvalues of the matrix (6.6.6.6)?

(c) Show that the eigenvalues of the matrix

$$A = \begin{bmatrix} 0 & \bar{\gamma}_2 & & & 0 \\ \gamma_2 & 0 & \bar{\gamma}_3 & & \\ & \gamma_3 & \ddots & \ddots & \\ & & \ddots & \ddots & \bar{\gamma}_n \\ 0 & & & \gamma_n & 0 \end{bmatrix}$$

are symmetric with respect to the origin, and that

$$\det A = \begin{cases} (-1)^k |\gamma_2|^2 \, |\gamma_4|^2 \cdots |\gamma_n|^2 & \text{if } n \text{ is even, } n = 2k, \\ 0 & \text{otherwise.} \end{cases}$$

7. Let C be a real $n \times n$ matrix and

$$f(x) := \frac{x^T C x}{x^T x} \qquad \text{for } x \in \mathbb{R}^n,\ x \neq 0.$$

Show: f is stationary at $\tilde{x} \neq 0$ precisely if $\tilde{x}$ is an eigenvector of $\frac{1}{2}(C + C^T)$ with $f(\tilde{x})$ as corresponding eigenvalue.

8. Show:

(a) If A is normal, and has the eigenvalues λ_i, $|\lambda_1| \geqslant \cdots \geqslant |\lambda_n|$, and the singular values σ_i, then

$$\sigma_i = |\lambda_i|, \qquad i = 1, \ldots, n,$$

$$\operatorname{lub}_2(A) = |\lambda_1| = \rho(A),$$

$$\operatorname{cond}_2(A) = \frac{|\lambda_1|}{|\lambda_n|} = \rho(A)\rho(A^{-1}) \quad (\text{if } A^{-1} \text{ exists}).$$

(b) For every $n \times n$ matrix,

$$(\operatorname{lub}_2(A))^2 = \operatorname{lub}_2(A^H A).$$

9. Show: If A is a normal $n \times n$ matrix with the eigenvalues λ_i,

$$|\lambda_1| \geqslant \cdots \geqslant |\lambda_n|,$$

and if U, V are unitary, then for the eigenvalues μ_i of UAV one has

$$|\lambda_1| \geqslant |\mu_i| \geqslant |\lambda_n|.$$

10. Let B be a $n \times m$ matrix. Prove:

$$M = \left[\begin{array}{c|c} I_n & B \\ \hline B^H & I_m \end{array}\right] \text{ positive definite} \quad \Leftrightarrow \quad \rho(B^H B) < 1$$

[I_n, I_m unit matrices, $\rho(B^H B) =$ spectral radius of $B^H B$].

11. For $n \times n$ matrices A, B show:

(a) $|A| \leqslant |B| \Rightarrow \operatorname{lub}_2(|A|) \leqslant \operatorname{lub}_2(|B|)$,
(b) $\operatorname{lub}_2(A) \leqslant \operatorname{lub}_2(|A|) \leqslant \sqrt{n}\, \operatorname{lub}_2(A)$.

12. The content of this exercise is the assertion in Section 6.3 that a matrix is ill conditioned if two column vectors are "almost linearly dependent." For the matrix $A = [a_1, \ldots, a_n]$, $a_i \in \mathbb{R}^n$, $i = 1, \ldots, n$, let

$$|a_1^T a_2| \geqslant \|a_1\|_2 \, \|a_2\|_2 (1 - \varepsilon), \qquad 0 < \varepsilon < 1$$

(i.e., a_1 and a_2 form an angle α with $1 - \varepsilon \leqslant |\cos \alpha| \leqslant 1$). Show: Either A is singular or $\operatorname{cond}_2(A) \geqslant 1/\sqrt{\varepsilon}$.

13. Estimation of the function $f(n)$ in formula (6.5.4.4) for floating-point arithmetic with relative precision eps: Let A be a $n \times n$ matrix in floating-point representation and G_j an elimination matrix of the type (6.5.4.1) with the essential elements l_{ij}. The l_{ij} are formed by division of elements of A and are thus subject to a relative error of at most eps. With the methods of Section 1.3, derive the following estimates (higher powers eps^i, $i \geqslant 2$, are to be neglected):

(a) $\operatorname{lub}_\infty[\operatorname{fl}(G_j^{-1} A) - G_j^{-1} A] \mathrel{\dot{\leqslant}} 4 \text{ eps } \operatorname{lub}_\infty(A)$
(b) $\operatorname{lub}_\infty[\operatorname{fl}(A G_j) - A G_j] \mathrel{\dot{\leqslant}} (n - j + 2) \text{ eps } \operatorname{lub}_\infty(A)$
(c) $\operatorname{lub}_\infty[\operatorname{fl}(G_j^{-1} A G_j) - G_j^{-1} A G_j] \mathrel{\dot{\leqslant}} 2(n - j + 6) \text{ eps } \operatorname{lub}_\infty(A)$.

14. Convergence behavior of the vector iteration: Let A be a real symmetric $n \times n$ matrix having the eigenvalues λ_i with

$$|\lambda_1| > |\lambda_2| \geqslant \cdots \geqslant |\lambda_n|$$

and the corresponding eigenvectors $x_1, \ldots, x_n$ with $x_i^T x_k = \delta_{ik}$. Starting with an initial vector y_0 for which $x_1^T y_0 \neq 0$, suppose one computes

$$y_{k+1} := \frac{1}{\|A y_k\|} A y_k \quad \text{for } k = 0, 1, 2, \ldots$$

with an arbitrary vector norm $\|\cdot\|$, and concurrently the quantities

$$q_{ki} := \frac{(A y_k)_i}{(y_k)_i}, \quad 1 \leqslant i \leqslant n, \quad \text{in case } (y_k)_i \neq 0,$$

and the Rayleigh quotient

$$r_k := \frac{y_k^T A y_k}{y_k^T y_k}.$$

Prove:

(a) $q_{ki} = \lambda_1[1 + O((\lambda_2/\lambda_1)^k)]$ for all i with $(x_1)_i \neq 0$,
(b) $r_k = \lambda_1[1 + O((\lambda_2/\lambda_1)^{2k})]$.

15. In the real matrix

$$A = A^T = \begin{bmatrix} -9 & * & * & * & * \\ * & 0 & * & * & * \\ * & * & 1 & * & * \\ * & * & * & 4 & * \\ * & * & * & * & 21 \end{bmatrix},$$

stars represent elements of modulus $\leq \frac{1}{4}$. Suppose the vector iteration is carried out with A and the initial vector $y_0 = e_5$.

(a) Show that e_5 is an "appropriate" initial vector, i.e., the sequence y_k in Exercise 14 does indeed converge toward the eigenvector belonging to the dominant eigenvalue of A.
(b) Estimate how many correct decimal digits r_{k+5} gains compared to r_k.

16. Prove: $\text{lub}_\infty(F) < 1 \Rightarrow A := I + F$ admits a triangular factorization $A = L \cdot R$.

17. Let the matrix A be nonsingular and admit a triangular factorization $A = L \cdot R$ $(l_{ii} = 1)$. Show:
(a) L and R are uniquely determined.
(b) If A is an upper Hessenberg matrix, then

$$L = \begin{bmatrix} 1 & & & 0 \\ * & 1 & & \\ & * & \ddots & \\ & & \ddots & \ddots \\ 0 & & * & 1 \end{bmatrix}, \qquad RL \text{ upper Hessenberg.}$$

(c) If A is tridiagonal, then L is as in (b) and

$$R = \begin{bmatrix} * & * & & 0 \\ & \ddots & \ddots & \\ & & & * \\ 0 & & & * \end{bmatrix}, \qquad RL \text{ tridiagonal.}$$

18. (a) Which upper triangular matrices are at the same time unitary; which ones real orthogonal?
(b) In what do different QR factorizations of a nonsingular matrix differ from one another? Is the answer valid also for singular matrices?

19. Consider an upper triangular matrix

$$R = \begin{bmatrix} \lambda_1 & * \\ 0 & \lambda_2 \end{bmatrix} \qquad \text{with } \lambda_1 \neq \lambda_2$$

and determine a Givens rotation Ω so that

$$\Omega^T R \Omega = \begin{bmatrix} \lambda_2 & * \\ 0 & \lambda_1 \end{bmatrix}.$$

Hint: Ωe_1 is an eigenvector of R corresponding to the eigenvalue λ_2.

20. QR method with shifts: Prove formula (6.6.5.2).

21. Let A be a normal $n \times n$ matrix with the eigenvalues $\lambda_1, \ldots, \lambda_n$, $A = QR$, $Q^H Q = I$, $R = [r_{ik}]$ upper triangular. Prove:

$$\min_i |\lambda_i| \leqslant |r_{jj}| \leqslant \max_i |\lambda_i|, \qquad j = 1, \ldots, n.$$

22. Compute a QR step with the matrix $A = \begin{bmatrix} 2 & \varepsilon \\ \varepsilon & 1 \end{bmatrix}$

    (a) without shift
    (b) with shift $k = 1$, i.e., following strategy (a) of Section 6.6.6.

23. Effect of a QR step with shift for tridiagonal matrices:

$$A = \begin{bmatrix} \delta_1 & \gamma_2 & & 0 \\ \gamma_2 & \delta_2 & \ddots & \\ & \ddots & \ddots & \gamma_n \\ 0 & & \gamma_n & \delta_n \end{bmatrix} = \left[\begin{array}{ccc|c} & & & 0 \\ & & & \vdots \\ & B & & 0 \\ & & & \gamma_n \\ \hline 0 & \cdots & 0 \quad \gamma_n & \delta_n \end{array}\right], \qquad \gamma_i \neq 0, \quad i = 2, \ldots, n.$$

Suppose we carry out a QR step with A, using the shift parameter $k = \delta_n$,

$$A - \delta_n I = QR \rightarrow RQ + \delta_n I =: A' = \begin{bmatrix} \delta'_1 & \gamma'_2 & & 0 \\ \gamma'_2 & \delta'_2 & \ddots & \\ & \ddots & \ddots & \gamma'_n \\ 0 & & \gamma'_n & \delta'_n \end{bmatrix}.$$

Prove: If $d := \min_i |\lambda_i(B) - \delta_n| > 0$, then

$$|\gamma'_n| \leqslant \frac{|\gamma_n|^3}{d^2}, \qquad |\delta'_n - \delta_n| \leqslant \frac{|\gamma_n|^2}{d}.$$

Hint: Q is a product of suitable Givens rotations; apply Exercise 21.
Example: What does one obtain for

$$A = \begin{bmatrix} 5 & 1 & & & \\ 1 & 5 & 1 & & \\ & 1 & 5 & 1 & \\ & & 1 & 5 & 0.1 \\ & & & 0.1 & 1 \end{bmatrix}?$$

24. Is shift strategy (a) in the QR method meaningful for real tridiagonal matrices of the type

$$\begin{bmatrix} \delta & \gamma_2 & & & 0 \\ \gamma_2 & \delta & \gamma_3 & & \\ & \gamma_3 & \delta & \ddots & \\ & & \ddots & \ddots & \gamma_n \\ 0 & & & \gamma_n & \delta \end{bmatrix}?$$

Answer this question with the aid of Exercise 6 (c) and the following fact: If A is real symmetric and tridiagonal, and if all diagonal elements are zero, then the same is true after one QR step.

25. For the matrix

$$A = \begin{bmatrix} 5.2 & 0.6 & 2.2 \\ 0.6 & 6.4 & 0.5 \\ 2.2 & 0.5 & 4.7 \end{bmatrix}$$

compute an upper bound for $\operatorname{cond}_2(A)$, using estimates of the eigenvalues by the method of Gershgorin.

26. Estimate the eigenvalues of the following matrices as accurately as possible:

(a) The matrix A in Exercise 15.

(b) $\begin{bmatrix} 1 & 10^{-3} & 10^{-4} \\ 10^{-3} & 2 & 10^{-3} \\ 10^{-4} & 10^{-3} & 3 \end{bmatrix}.$

Hint: Use the method of Gershgorin in conjunction with a transformation $A \to D^{-1}AD$, D a suitable diagonal matrix.

27. (a) Let A, B be Hermitian square matrices and

$$H = \begin{bmatrix} A & C \\ C^H & B \end{bmatrix}.$$

Show: For every eigenvalue $\lambda(B)$ of B there is an eigenvalue $\lambda(H)$ of H such that

$$|\lambda(H) - \lambda(B)| \leqslant \sqrt{\operatorname{lub}_2(C^H C)}.$$

(b) Apply (a) to the practically important case in which H is a Hermitian, "almost reducible" tridiagonal matrix of the form

$$H = \left[\begin{array}{cccc|cccc} * & * & & & & & & 0 \\ * & \ddots & \ddots & & & & & \\ & \ddots & \ddots & * & & & & \\ & & * & * & \varepsilon & & & \\ \hline & & & \varepsilon & * & * & & \\ & & & & * & \ddots & \ddots & \\ & & & & & \ddots & \ddots & * \\ 0 & & & & & & * & * \end{array}\right], \qquad \varepsilon \text{ small.}$$

How can the eigenvalues of H be estimated in terms of those of A and B?

28. Show: If $A = [a_{ik}]$ is Hermitian, then for every diagonal element a_{ii} there exists an eigenvalue $\lambda(A)$ of A such that

$$|\lambda(A) - a_{ii}| \leqslant \sqrt{\sum_{j \neq i} |a_{ij}|^2}.$$

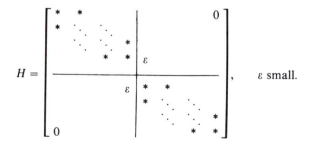

29.

(a) For the $\nu \times \nu$ matrix

$$C_\nu(\lambda) = \begin{bmatrix} \lambda & 1 & & & 0 \\ & \lambda & 1 & & \\ & & \ddots & \ddots & \\ & & & \ddots & 1 \\ 0 & & & & \lambda \end{bmatrix},$$

prove with the aid of Gershgorin's theorem and appropriate scaling with diagonal matrices: The eigenvalues $\lambda_i(\varepsilon)$ of the perturbed matrix $C_\nu(\lambda) + \varepsilon F$ satisfy, for ε sufficiently small,

$$|\lambda_i(\varepsilon) - \lambda| \leqslant K|\varepsilon^{1/\nu}|,$$

with some constant K. Through a special choice of the matrix F, show that the case $\lambda_i(\varepsilon) - \lambda = O(\varepsilon^{1/\nu})$ does indeed occur.

(b) Prove the result (6.9.11).

Hint: Transform A to Jordan normal form.

30. Obtain the field of values $G[A] = \{x^H A x \mid x^H x = 1\}$ for

$$A = \begin{bmatrix} 1 & 1 & 0 & 0 \\ 1 & 1 & 0 & 0 \\ 0 & 0 & -1 & 1 \\ 0 & 0 & -1 & -1 \end{bmatrix}.$$

31.

$$A := \begin{bmatrix} 2 & 0 \\ 2 & 2 \end{bmatrix}, \quad B := \begin{bmatrix} i & i \\ -i & i \end{bmatrix}, \quad C := \begin{bmatrix} 2 & 5+i & 1-2i \\ 5+i & 1+4i & 3 \\ 1+2i & 1 & -i \end{bmatrix} (i^2 = -1).$$

Decompose these matrices into $H_1 + iH_2$ with H_1, H_2 Hermitian.

32. For the eigenvalues of

$$A = \begin{bmatrix} 3. & 1.1 & -0.1 \\ 0.9 & 15.0 & -2.1 \\ 0.1 & -1.9 & 19.5 \end{bmatrix}$$

use the theorems of Gershgorin and Bendixson to determine inclusion regions which are as small as possible.

References for Chapter 6

Barth, W., Martin, R. S., Wilkinson, J. H.: Calculation of the eigenvalues of a symmetric tridiagonal matrix by the method of bisection. Contribution II/5 in Wilkinson and Reinsch (1971).

Bauer, F. L., Fike, C. T.: Norms and exclusion theorems. *Numer. Math.* **2**, 137–141 (1960).

———, Stoer, J., Witzgall, C.: Absolute and monotonic norms. *Numer. Math.* **3**, 257–264 (1961).

Bowdler, H., Martin, R. S., Wilkinson, J. H.: The *QR* and *QL* algorithms for symmetric matrices. Contribution II/3 in Wilkinson and Reinsch (1971).

Bunse, W., Bunse-Gerstner, A.: *Numerische Lineare Algebra.* Stuttgart: Teubner (1985).

Cullum, J., Willoughby, R. A.: *Lanczos Algorithms for Large Symmetric Eigenvalue Computations. Vol. I: Theory, Vol. II: Programs.* Progress in Scientific Computing, Vol. 3, 4. Basel: Birkhäuser (1985).

Eberlein, P. J.: Solution to the complex eigenproblem by a norm reducing Jacobi type method. Contribution II/17 In Wilkinson and Reinsch (1971).

Francis, J. F. G.: The *QR* transformation. A unitary analogue to the *LR* transformation. I. *Computer J.* **4**, 265–271 (1961/62). The *QR* transformation. II. *ibid.*, 332–345 (1961/62).

Garbow, B. S., et al.: *Matrix Eigensystem Routines*—EISPACK *Guide Extension.* Lecture Notes in Computer Science 51. Berlin, Heidelberg, New York: Springer-Verlag (1977).

Givens, J. W.: Numerical computation of the characteristic values of a real symmetric matrix. Oak Ridge National Laboratory Report ORNL-1574 (1954).

Golub, G. H., Reinsch, C.: Singular value decomposition and least squares solution. Contribution I/10 in Wilkinson and Reinsch (1971).

———, Van Loan, C. F.: *Matrix Computations.* Baltimore: The Johns-Hopkins University Press (1983).

———, Wilkinson, J. H.: Ill-conditioned eigensystems and the computation of the Jordan canonical form. *SIAM Review* **18**, 578–619 (1976).

Householder, A. S.: *The Theory of Matrices in Numerical Analysis.* New York: Blaisdell (1964).

Kaniel, S.: Estimates for some computational techniques in linear algebra. *Math. Comp.* **20**, 369–378 (1966).

Kielbasinski, A., Schwetlick, H.: *Numerische Lineare Algebra.* Thun, Frankfurt/M.: Deutsch (1988).

Kublanovskaya, V. N.: On some algorithms for the solution of the complete eigenvalue problem. *Ž. Vyčisl. Mat. i Mat. Fiz.* **1**, 555–570 (1961).

Lanczos, C.: An iteration method for the solution of the eigenvalue problem of linear differential and integral operators. *J. Res. Nat. Bur. Stand.* **45**, 255–282 (1950).

Martin, R. S., Peters, G., Wilkinson, J. H.: The *QR* algorithm for real Hessenberg matrices. Contribution II/14 in Wilkinson and Reinsch (1971).

———, Reinsch, C., Wilkinson. J. H.: Householder's tridiagonalization of a symmetric matrix. Contribution I/2 in Wilkinson and Reinsch (1971).

———, Wilkinson, J. H.: Reduction of the symmetric eigenproblem $Ax = \lambda Bx$ and related problems to standard form. Contribution II/10 in Wilkinson and Reinsch (1971).

———, Wilkinson, J. H.: Similarity reduction of a general matrix to Hessenberg form. Contribution II/13 in Wilkinson and Reinsch (1971).

Moler, C. B., Stewart, G. W.: An algorithm for generalized matrix eigenvalue problems. *SIAM J. Numer. Anal.* **10**, 241–256 (1973).

Paige, C. C.: The computation of eigenvalues and eigenvectors of very large sparse matrices. Ph.D. thesis, London University (1971).

Parlett, B. N.: Convergence of the QR algorithm. *Numer. Math.* **7**, 187–193 (1965) (corr. in **10**, 163–164 (1967)).

———. *The Symmetric Eigenvalue Problem.* Englewood Cliffs, N.J.: Prentice-Hall (1980).

———, Poole, W. G.: A geometric theory for the QR, LU and power iterations. *SIAM J. Numer. Anal.* **10**. 389–412 (1973).

———, Scott, D. S.: The Lanczos algorithm with selective orthogonalization. *Math Comp.* **33**, 217–238 (1979).

Peters, G., Wilkinson J. H.: $Ax = \lambda Bx$ and the generalized eigenproblem. *SIAM J. Numer. Anal.* **7**, 479–492 (1970).
———, ———. Eigenvectors of real and complex matrices by *LR* and *QR* triangularizations. Contribution II/15 in Wilkinson and Reinsch (1971).
———, ———. The calculation of specified eigenvectors by inverse iteration. Contribution II/18 in Wilkinson and Reinsch (1971).
Rutishauser, H.: Solution of eigenvalue problems with the *LR*-transformation. *Nat. Bur. Standards Appl. Math. Ser.* **49**, 47–81 (1958).
———. The Jacobi method for real symmetric matrices. Contribution II/1 in Wilkinson and Reinsch (1971).
Saad, Y.: On the rates of convergence of the Lanczos and the block Lanczos methods. *SIAM J. Num. Anal.* **17**, 687–706 (1980).
Schwarz, H. R., Rutishauser, H., Stiefel, E.: *Numerik symmetrischer Matrizen*. 2d ed. Stuttgart: Teubner (1972). (English translation: Englewood Cliffs, N.J.: Prentice-Hall (1973.))
Smith, B. T., et al.: *Matrix Eigensystems Routines*—EISPACK *Guide*. Lecture Notes in Computer Science 6, 2d ed. Berlin, Heidelberg, New York: Springer-Verlag (1976).
Stewart, G. W.: *Introduction to matrix computations*. New York, London: Academic Press (1973).
Wilkinson, J. H.: Note on the quadratic convergence of the cyclic Jacobi process. *Numer. Math.* **4**, 296–300 (1962).
———: *The Algebraic Eigenvalue Problem*. Oxford: Clarendon Press (1965).
———: Global convergence of tridiagonal *QR* algorithm with origin shifts. *Linear Algebra and Appl.* **1**, 409–420 (1968).
———, Reinsch, C.: *Linear Algebra, Handbook for Automatic Computation*, Vol. II. Berlin, Heidelberg, New York: Springer-Verlag (1971).

7 Ordinary Differential Equations

7.0 Introduction

Many problems in applied mathematics lead to ordinary differential equations. In the simplest case one seeks a differentiable function $y = y(x)$ of one real variable x, whose derivative $y'(x)$ is to satisfy an equation of the form $y'(x) = f(x, y(x))$, or more briefly,

(7.0.1)
$$y' = f(x, y);$$

one then speaks of an *ordinary differential equation.* In general there are infinitely many different functions y which are solutions of (7.0.1). Through additional requirements one can single out certain solutions from the set of all solutions. Thus, in an *initial-value problem,* one seeks a solution y of (7.0.1) which for given x_0, y_0 satisfies an *initial condition* of the form

(7.0.2)
$$y(x_0) = y_0.$$

More generally, one also considers *systems of n ordinary differential equations*

$$\begin{aligned} y_1'(x) &= f_1(x, y_1(x), \ldots, y_n(x)),\\ y_2'(x) &= f_2(x, y_1(x), \ldots, y_n(x)),\\ &\vdots\\ y_n'(x) &= f_n(x, y_1(x), \ldots, y_n(x)) \end{aligned}$$

for n unknown real functions $y_i(x)$, $i = 1, \ldots, n$, of a real variable. Such systems can be written analogously to (7.0.1) in vector form:

$$\text{(7.0.3)} \quad y' = f(x, y), \quad y' := \begin{bmatrix} y_1' \\ \vdots \\ y_n' \end{bmatrix}, \quad f(x, y) := \begin{bmatrix} f_1(x, y_1, \ldots, y_n) \\ \vdots \\ f_n(x, y_1, \ldots, y_n) \end{bmatrix}.$$

To the initial condition (7.0.2) there now corresponds a condition of the form

$$\text{(7.0.4)} \qquad y(x_0) = y_0 = \begin{bmatrix} y_{10} \\ \vdots \\ y_{n0} \end{bmatrix}.$$

In addition to *ordinary differential equations of first order* (7.0.1), (7.0.3), in which there occur only first derivatives of the unknown function $y(x)$, there are *ordinary differential equations of mth order* of the form

$$\text{(7.0.5)} \qquad y^{(m)}(x) = f(x, y(x), y'(x), \ldots, y^{(m-1)}(x)).$$

By introducing auxiliary functions

$$\begin{aligned} z_1(x) &:= y(x), \\ z_2(x) &:= y'(x), \\ &\vdots \\ z_m(x) &:= y^{(m-1)}(x), \end{aligned}$$

(7.0.5) can always be transformed into an equivalent system of first-order differential equations,

$$\text{(7.0.6)} \qquad z' = \begin{bmatrix} z_1' \\ \vdots \\ z_{m-1}' \\ z_m' \end{bmatrix} = \begin{bmatrix} z_2 \\ \vdots \\ z_m \\ f(x, z_1, z_2, \ldots, z_m) \end{bmatrix}.$$

By an initial-value problem for the ordinary differential equation of mth order (7.0.5) one means the problem of finding an m times differentiable function $y(x)$ which satisfies (7.0.5) and initial conditions of the form

$$y^{(i)}(x_0) = y_{i0}, \qquad i = 0, 1, \ldots, m-1.$$

Initial-value problems will be treated in Section 7.2.

Besides initial-value problems for systems of ordinary differential equations, *boundary-value problems* also frequently occur in practice. Here, the desired solution $y(x)$ of the differential equation (7.0.3) has to satisfy a *boundary condition* of the form

$$\text{(7.0.7)} \qquad r(y(a), y(b)) = 0,$$

where $a \neq b$ are two *different* numbers and

$$r(u, v) := \begin{bmatrix} r_1(u_1, \ldots, u_n, v_1, \ldots, v_n) \\ \vdots \\ r_n(u_1, \ldots, u_n, v_1, \ldots, v_n) \end{bmatrix}$$

is a vector of n given functions r_i of $2n$ variables $u_1, \ldots, u_n, v_1, \ldots, v_n$. Problems of this kind will be considered in Sections 7.3, 7.4, and 7.5.

In the methods which are to be discussed in the following, one does not construct a closed-form expression for the desired solution $y(x)$—this is not even possible, in general—but in correspondence to certain discrete abscissae x_i, $i = 0, 1, \ldots$, one determines approximate values $\eta_i := \eta(x_i)$ for the exact values $y_i := y(x_i)$. The discrete abscissae are often equidistant, $x_i = x_0 + ih$. For the approximate values η_i we then also write more precisely $\eta(x_i; h)$, since the η_i, like the x_i, depend on the *stepsize* h used. An important problem, for a given method, will be to examine whether, and how fast, $\eta(x; (x - x_0)/n)$ converges to $y(x)$ as $n \to \infty$, i.e., $h \to 0$.

For a detailed treatment of numerical methods for solving initial- and boundary-value problems we refer to the literature: In addition to the classical book by Henrici (1963), and the more recent exposition by Hairer et al. (1987, 1991), we mention the books by Gear (1971), Grigorieff (1972, 1977), Keller (1968), Shampine and Gordon (1975), and Stetter (1973).

7.1 Some Theorems from the Theory of Ordinary Differential Equations

For later use we list a few results—some without proof—from the theory of ordinary differential equations. We assume throughout that [see (7.0.3)]

$$y' = f(x, y)$$

is a system of n ordinary differential equations, $\|\cdot\|$ a norm on $\mathbb{R}^n$, and $\|A\|$ an associated consistent multiplicative matrix norm with $\|I\| = 1$ [see (4.4.8)]. It can then be shown [see, e.g., Henrici (1962), Theorem 3.1] that the initial-value problem (7.0.3), (7.0.4)—and thus in particular (7.0.1), (7.0.2)—has exactly one solution, provided f satisfies a few simple regularity conditions:

(7.1.1) Theorem. *Let f be defined and continuous on the strip $S := \{(x, y) \mid a \leqslant x \leqslant b,\ y \in \mathbb{R}^n\}$, a, b finite. Further, let there be a constant L such that*

(7.1.2) $$\|f(x, y_1) - f(x, y_2)\| \leqslant L\|y_1 - y_2\|$$

for all $x \in [a, b]$ and all $y_1, y_2 \in \mathbb{R}^n$ ("Lipschitz condition"). Then for every $x_0 \in [a, b]$ and every $y_0 \in \mathbb{R}^n$ there exists exactly one function $y(x)$ such that

(a) $y(x)$ *is continuous and continuously differentiable for* $x \in [a, b]$;
(b) $y'(x) = f(x, y(x))$ *for* $x \in [a, b]$;
(c) $y(x_0) = y_0$.

From the mean-value theorem it easily follows that the Lipschitz condition is satisfied if the partial derivatives $\partial f_i / \partial y_j$, $i, j = 1, \ldots, n$, exist on the strip S and are continuous and bounded there. For later, we denote by

$$F_N(a, b) \tag{7.1.3}$$

the set of functions f for which all partial derivatives up to and including order N exist on the strip $S = \{(x, y) \mid a \leqslant x \leqslant b, \ y \in \mathbb{R}^n\}$, a, b finite, and are continuous and bounded there. The functions $f \in F_1(a, b)$ thus fulfill the assumptions of (7.1.1).

In applications, f is usually continuous on S and also continuously differentiable there, but the derivatives $\partial f_i / \partial y_j$ are often unbounded on S. Then, while the initial-value problem (7.0.3), (7.0.4) is still solvable, the solution may be defined only in a certain neighborhood $U(x_0)$ of the initial point and not on all of $[a, b]$ [see, e.g., Henrici (1962)].

Example. The initial-value problem

$$y' = y^2, \qquad y(0) = 1$$

has the solution $y(x) = 1/(1 - x)$, which is defined only for $x < 1$.

(7.1.1) is the fundamental existence and uniqueness theorem for the initial-value problem given in (7.0.3), (7.0.4).

We now show that the solution of an initial-value problem depends continuously on the initial value:

(7.1.4) Theorem. *Let the function* $f: S \to \mathbb{R}^n$ *be continuous on the strip* $S = \{(x, y) \mid a \leqslant x \leqslant b, \ y \in \mathbb{R}^n\}$ *and satisfy the Lipschitz condition*

$$\| f(x, y_1) - f(x, y_2) \| \leqslant L \| y_1 - y_2 \|$$

for all $(x, y_i) \in S$, $i = 1, 2$. *Let* $a \leqslant x_0 \leqslant b$. *Then for the solution* $y(x; s)$ *of the initial-value problem*

$$y' = f(x, y), \qquad y(x_0; s) = s$$

there holds the estimate

$$\| y(x; s_1) - y(x; s_2) \| \leqslant e^{L|x - x_0|} \| s_1 - s_2 \|$$

for $a \leqslant x \leqslant b$.

Proof. By definition of $y(x; s)$ one has

$$y(x; s) = s + \int_{x_0}^{x} f(t, y(t; s)) \, dt$$

for $a \leqslant x \leqslant b$. It follows that

$$y(x; s_1) - y(x; s_2) = s_1 - s_2 + \int_{x_0}^{x} [f(t, y(t; s_1)) - f(t, y(t; s_2))]\, dt,$$

and thus

$$\|y(x; s_1) - y(x; s_2)\| \leqslant \|s_1 - s_2\| + L\left|\int_{x_0}^{x} \|y(t; s_1) - y(t, s_2)\|\, dt\right|. \tag{7.1.5}$$

For the function

$$\Phi(x) := \int_{x_0}^{x} \|y(t; s_1) - y(t; s_2)\|\, dt$$

one has $\Phi'(x) = \|y(x; s_1) - y(x; s_2)\|$ and thus, by (7.1.5), for $x \geqslant x_0$

$$\alpha(x) \leqslant \|s_1 - s_2\| \quad \text{with } \alpha(x) := \Phi'(x) - L\Phi(x).$$

The initial-value problem

$$\Phi'(x) = \alpha(x) + L\Phi(x), \qquad \Phi(x_0) = 0 \tag{7.1.6}$$

has for $x \geqslant x_0$ the solution

$$\Phi(x) = e^{L(x - x_0)} \int_{x_0}^{x} \alpha(t) e^{-L(t - x_0)}\, dt. \tag{7.1.7}$$

On account of $\alpha(x) \leqslant \|s_1 - s_2\|$, one thus obtains the estimate

$$\begin{aligned} 0 \leqslant \Phi(x) &\leqslant e^{L(x - x_0)} \|s_1 - s_2\| \int_{x_0}^{x} e^{-L(t - x_0)}\, dt \\ &= \frac{1}{L} \|s_1 - s_2\| [e^{L(x - x_0)} - 1] \quad \text{for } x \geqslant x_0. \end{aligned}$$

The desired result finally follows for $x \geqslant x_0$:

$$\begin{aligned} \|y(x; s_1) - y(x; s_2)\| &= \Phi'(x) = \alpha(x) + L\Phi(x) \\ &\leqslant \|s_1 - s_2\| e^{L|x - x_0|}. \end{aligned}$$

If $x < x_0$ one proceeds similarly. □

The preceding theorem can be sharpened: Under additional assumptions the solution of the initial-value problem actually depends on the initial value in a continuously differentiable manner. We have the following:

(7.1.8) Theorem. *If in addition to the assumptions of Theorem* (7.1.4) *the Jacobian matrix* $D_y f(x, y) = [\partial f_i / \partial y_j]$ *exists on* S *and is continuous and bounded there,*

$$\|D_y f(x, y)\| \leqslant L \quad \textit{for } (x, y) \in S,$$

then the solution $y(x; s)$ of $y' = f(x, y)$, $y(x_0; s) = s$ is continuously differentiable for all $x \in [a, b]$ and all $s \in \mathbb{R}^n$. The derivative

$$Z(x; s) := D_s y(x; s) = \left[\frac{\partial y(x; s)}{\partial \sigma_1}, \dots, \frac{\partial y(x; s)}{\partial \sigma_n}\right], \qquad s = [\sigma_1, \dots, \sigma_n]^T,$$

is the solution of the initial-value problem ($Z' = D_x Z$)

(7.1.9) $$Z' = D_y f(x, y(x; s))Z, \qquad Z(x_0; s) = I.$$

Note that Z', Z, and $D_y f(x, y(x; s))$ are $n \times n$ matrices. (7.1.9) thus describes an initial-value problem for a system of n^2 differential equations. Formally, (7.1.9) can be obtained by differentiating with respect to s the identities

$$y'(x; s) = f(x, y(x; s)), \qquad y(x_0; s) = s.$$

A proof of Theorem (7.1.8) can be found, e.g., in Coddington and Levinson (1955).

For many purposes it is important to estimate the growth of the solution Z of (7.1.9). Suppose, then, that $T(x)$ is an $n \times n$ matrix, and the $n \times n$ matrix $Y(x)$ solution of the linear initial-value problem

(7.1.10) $$Y' = T(x)Y, \qquad Y(a) = I.$$

One can then show:

(7.1.11) Theorem. *If $T(x)$ is continuous on $[a, b]$, and $k(x) := \|T(x)\|$, then the solution $Y(x)$ of (7.1.10) satisfies*

$$\|Y(x) - I\| \leqslant \exp\left(\int_a^x k(t)\,dt\right) - 1, \qquad x \geqslant a.$$

PROOF. By definition of $Y(x)$ one has

$$Y(x) = I + \int_a^x T(t)Y(t)\,dt.$$

Letting

$$\varphi(x) := \|Y(x) - I\|,$$

by virtue of $\|Y(x)\| \leqslant \varphi(x) + \|I\| = \varphi(x) + 1$ there follows for $x \geqslant a$ the estimate

(7.1.12) $$\varphi(x) \leqslant \int_a^x k(t)(\varphi(t) + 1)\,dt.$$

Now let $c(x)$ be defined by

(7.1.13) $$\int_a^x k(t)(\varphi(t) + 1)\,dt = c(x)\exp\left(\int_a^x k(t)\,dt\right) - 1, \qquad c(a) = 1.$$

Through differentiation of (7.1.13) [$c(x)$ is clearly differentiable] one obtains

$$k(x)(\varphi(x)+1) = c'(x)\exp\left(\int_a^x k(t)\,dt\right) + k(x)c(x)\exp\left(\int_a^x k(t)\,dt\right)$$
$$= c'(x)\exp\left(\int_a^x k(t)\,dt\right) + k(x)\cdot\left[1+\int_a^x k(t)(\varphi(t)+1)\,dt\right],$$

from which, because of $k(x) \geqslant 0$ and (7.1.12), it follows that

$$c'(x)\exp\left(\int_a^x k(t)\,dt\right) + k(x)\int_a^x k(t)(\varphi(t)+1)\,dt = k(x)\varphi(x)$$
$$\leqslant k(x)\int_a^x k(t)(\varphi(t)+1)\,dt.$$

One thus obtains finally

$$c'(x) \leqslant 0$$

and therefore

$$(7.1.14) \qquad c(x) \leqslant c(a) = 1 \cdot \text{ for } x \geqslant a.$$

The assertion of the theorem now follows immediately from (7.1.12)–(7.1.14). □

7.2 Initial-Value Problems

7.2.1 One-Step Methods: Basic Concepts

As one can already surmise from Section 7.1, the methods and results for initial-value problems for systems of ordinary differential equations of first order are essentially independent of the number n of unknown functions. In the following we therefore limit ourselves to the case of only *one* ordinary differential equation of first order for only *one* unknown function (i.e., $n = 1$). The results, however, are valid, as a rule, also for systems (i.e., $n > 1$), provided quantities such as y and $f(x, y)$ are interpreted as vectors, and $|\cdot|$ as norm $\|\cdot\|$. For the following, we assume that the initial-value problem under consideration is always uniquely solvable.

A first numerical method for the solution of the initial-value problem

$$(7.2.1.1) \qquad \text{(I)} \qquad y' = f(x, y), \qquad y(x_0) = y_0$$

is suggested by the following simple observation: Since $f(x, y(x))$ is just the slope $y'(x)$ of the desired exact solution $y(x)$ of (7.2.1.1), one has for $h \neq 0$ approximately

$$\frac{y(x+h) - y(x)}{h} \approx f(x, y(x)),$$

or

$$y(x+h) \approx y(x) + hf(x, y(x)). \tag{7.2.1.2}$$

Once a steplength $h \neq 0$ is chosen, starting with the given initial values x_0, $y_0 = y(x_0)$, one thus obtains at equidistant points $x_i = x_0 + ih$, $i = 1, 2, \ldots,$ approximations η_i to the values $y_i = y(x_i)$ of the exact solution $y(x)$ as follows:

(7.2.1.3)

$$\begin{aligned} &\eta_0 := y_0; \\ &\text{for } i = 0, 1, 2, \ldots: \\ &\eta_{i+1} := \eta_i + hf(x_i, \eta_i), \qquad x_{i+1} := x_i + h. \end{aligned}$$

One arrives at the *polygon method* of Euler shown in Figure 12.

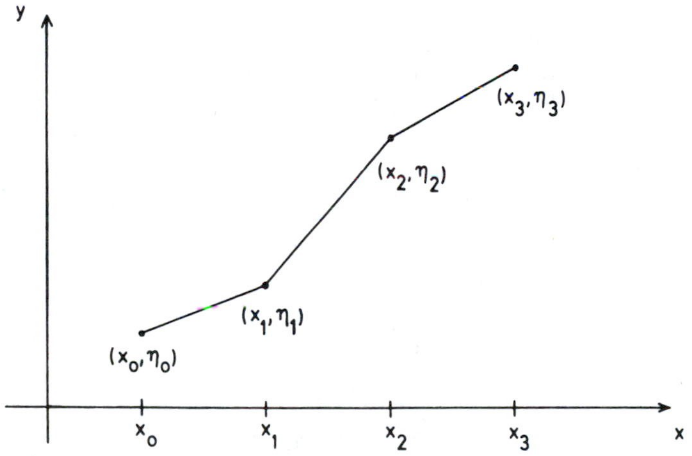

Figure 12 Euler's method.

Evidently, the approximate values η_i depend on the stepsize h. To indicate this, we also write more precisely $\eta(x_i; h)$ instead of η_i. The "approximate solution" $\eta(x; h)$ is thus defined only for

$$x \in R_h := \{x_0 + ih \mid i = 0, 1, 2, \ldots\},$$

or, alternatively, for

$$h \in H_x := \left\{ \frac{x - x_0}{n} \,\middle|\, n = 1, 2, \ldots \right\};$$

in fact, it is defined recursively by [cf. (7.2.1.3)]

$$\eta(x_0; h) := y_0,$$
$$\eta(x + h; h) := \eta(x; h) + h \cdot f(x, \eta(x; h)).$$

Euler's method is a typical *one-step method*. In general, such methods are given by a function

$$\Phi(x, y; h; f).$$

Starting with the initial values x_0, y_0 of the initial-value problem (I) of (7.2.1.1), one now obtains approximate values η_i for the quantities $y_i := y(x_i)$ of the exact solution $y(x)$ by means of

$$\begin{aligned} &\eta_0 := y_0; \\ &\text{for } i = 0, 1, 2, \ldots: \\ &\eta_{i+1} := \eta_i + h\Phi(x_i, \eta_i; h; f), \\ &x_{i+1} := x_i + h. \end{aligned} \tag{7.2.1.4}$$

In the method of Euler, for example, one has $\Phi(x, y; h; f) := f(x, y)$; here, Φ is independent of h.

For simplicity, the argument f in the function Φ will from now on be omitted. As in Euler's method (see above) we also write more precisely $\eta(x_i; h)$ instead of η_i, in order to indicate the dependence of the approximate values on the stepsize h used.

Let now x and y be arbitrary, but fixed, and let $z(t)$ be the exact solution of the initial-value problem

$$z'(t) = f(t, z(t)), \qquad z(x) = y, \tag{7.2.1.5}$$

with initial values x, y. Then the function

$$\Delta(x, y; h; f) := \begin{cases} \dfrac{z(x + h) - y}{h} & \text{if } h \neq 0, \\ f(x, y) & \text{if } h = 0 \end{cases} \tag{7.2.1.6}$$

represents the difference quotient of the exact solution $z(t)$ of (7.2.1.5) for stepsize h, while $\Phi(x, y; h)$ is the difference quotient for stepsize h of the approximate solution of (7.2.1.5) produced by Φ. As in Φ, we shall also drop the argument f in Δ.

The magnitude of the difference

$$\tau(x, y; h) := \Delta(x, y; h) - \Phi(x, y; h)$$

indicates how well the exact solution of the differential equation obeys the equation of the one-step method: it is a measure of the quality of the approximation method. One calls $\tau(x, y; h)$ the *local discretization error* at the point (x, y) of the method in question. For a reasonable one-step method one will require that

$$\lim_{h \to 0} \tau(x, y; h) = 0.$$

In as much as $\lim_{h \to 0} \Delta(x, y; h) = f(x, y)$, this is equivalent to

$$\lim_{h \to 0} \Phi(x, y; h) = f(x, y). \tag{7.2.1.7}$$

One calls Φ, and the associated one-step method, *consistent* if (7.2.1.7) holds for all $x \in [a, b]$, $y \in \mathbb{R}$, and $f \in F_1(a, b)$ [see (7.1.3)].

EXAMPLE. Euler's method, $\Phi(x, y; h) := f(x, y)$, is obviously consistent. The result can be sharpened: If f has sufficiently many continuous partial derivatives, it is possible to say with what order $\tau(x, y; h)$ goes to zero as $h \to 0$. For this, expand the solution $z(t)$ of (7.2.1.5) into a Taylor series about the point $t = x$:

$$z(x + h) = z(x) + hz'(x) + \frac{h^2}{2} z''(x) + \cdots + \frac{h^p}{p!} z^{(p)}(x + \theta h), \qquad 0 < \theta < 1.$$

We now have, because $z(x) = y$, $z'(t) = f(t, z(t))$,

$$\begin{aligned} z''(x) &= \frac{d}{dt} f(t, z(t))\Big|_{t=x} = f_x(t, z(t))\Big|_{t=x} + f_y(t, z(t))z'(t)\Big|_{t=x} \\ &= f_x(x, y) + f_y(x, y)f(x, y), \\ z'''(x) &= f_{xx}(x, y) + 2f_{xy}(x, y)f(x, y) + f_{yy}(x, y)f(x, y)^2 + f_y(x, y)z''(x), \end{aligned}$$

etc., and thus

$$\begin{aligned} \Delta(x, y; h) &= z'(x) + \frac{h}{2!} z''(x) + \cdots + \frac{h^{p-1}}{p!} z^{(p)}(x + \theta h) \\ &= f(x, y) + \frac{h}{2}[f_x(x, y) + f_y(x, y)f(x, y)] + \cdots. \end{aligned} \tag{7.2.1.8}$$

For Euler's method, $\Phi(x, y; h) := f(x, y)$, it follows that

$$\begin{aligned} \tau(x, y; h) &= \Delta(x, y; h) - \Phi(x, y; h) \\ &= \frac{h}{2}[f_x(x, y) + f_y(x, y)f(x, y)] + \cdots \\ &= O(h). \end{aligned}$$

Generally, one speaks of a *method of order p* if

$$\tau(x, y; h) = O(h^p) \tag{7.2.1.9}$$

for all $x \in [a, b]$, $y \in \mathbb{R}$, and $f \in F_p(a, b)$.

Euler's method thus is a method of order 1.

The preceding example shows how to obtain methods of order greater than 1. Simply take for $\Phi(x, y; h)$ sections of the Taylor series (7.2.1.8) of $\Delta(x, y; h)$. For example,

$$\Phi(x, y; h) := f(x, y) + \frac{h}{2}[f_x(x, y) + f_y(x, y)f(x, y)] \tag{7.2.1.10}$$

produces a method of order 2. The higher-order methods so obtained, however, are hardly useful, since in every step $(x_i, \eta_i) \to (x_{i+1}, \eta_{i+1})$ one must compute not only f, but also the partial derivatives f_x, f_y, etc.

Simpler methods of higher order can be obtained, e.g., by means of the construction

$$\Phi(x, y; h) := a_1 f(x, y) + a_2 f(x + p_1 h, y + p_2 hf(x, y)), \tag{7.2.1.11}$$

in which the constants a_1, a_2, p_1, p_2 are so chosen that the Taylor expansion of $\Delta(x, y; h) - \Phi(x, y; h)$ in powers of h starts with the largest possible power. For $\Phi(x, y; h)$ in (7.2.1.11), the Taylor expansion is

$$\Phi(x, y; h) = (a_1 + a_2)f(x, y) + a_2 h[p_1 f_x(x, y) + p_2 f_y(x, y)f(x, y)] + O(h^2).$$

Comparison with (7.2.1.8) yields for a second-order method the conditions

$$a_1 + a_2 = 1, \qquad a_2 p_1 = \tfrac{1}{2}, \qquad a_2 p_2 = \tfrac{1}{2}.$$

One solution of these equations is

$$a_1 = \tfrac{1}{2}, \qquad a_2 = \tfrac{1}{2}, \qquad p_1 = 1, \qquad p_2 = 1,$$

and one obtains the method of Heun (1900):

$$\Phi(x, y; h) := \tfrac{1}{2}[f(x, y) + f(x + h, y + hf(x, y))], \tag{7.2.1.12}$$

which requires only two evaluations of f per step. Another solution is

$$a_1 = 0, \qquad a_2 = 1, \qquad p_1 = \tfrac{1}{2}, \qquad p_2 = \tfrac{1}{2},$$

leading to the *modified Euler method* [Collatz (1960)]

$$\Phi(x, y; h) := f\left(x + \frac{h}{2}, y + \frac{h}{2} f(x, y)\right), \tag{7.2.1.13}$$

which again is of second order and requires two evaluations of f per step.

The *Runge–Kutta method* [Runge (1895), Kutta (1901)] is obtained from a construction which is somewhat more general than (7.2.1.11). It has the form

$$\Phi(x, y; h) := \tfrac{1}{6}[k_1 + 2k_2 + 2k_3 + k_4], \tag{7.2.1.14}$$

where

$$\begin{aligned} k_1 &:= f(x, y), \\ k_2 &:= f(x + \tfrac{1}{2}h, y + \tfrac{1}{2}hk_1), \\ k_3 &:= f(x + \tfrac{1}{2}h, y + \tfrac{1}{2}hk_2), \\ k_4 &:= f(x + h, y + hk_3). \end{aligned}$$

Through Taylor expansion in h one finds, for $f \in F_4(a, b)$,

$$\Phi(x, y; h) - \Delta(x, y; h) = O(h^4).$$

The Runge–Kutta method, therefore, is a method of fourth order. It requires four evaluations of f per step.

If $f(x, y)$ does not depend on y, then the solution of the initial-value problem

$$y' = f(x), \qquad y(x_0) = y_0$$

is just the integral $y(x) = y_0 + \int_{x_0}^{x} f(t)\,dt$. The method of Heun then corresponds to the approximation of $y(x)$ by means of trapezoidal sums, the modified Euler method to the midpoint rule, and the Runge–Kutta method to Simpson's rule (see Section 3.1).

Further methods of this type, also of order greater than 4, are given in Butcher (1964), Fehlberg (1964, 1966, 1969), and Shanks (1966); a general exposition of one-step methods is found in the recent book by Hairer, Nørsett, and Wanner (1987), and in Grigorieff (1972) and Stetter (1973).

7.2.2 Convergence of One-Step Methods

In this section we wish to examine the convergence behavior as $h \to 0$ of an approximate solution $\eta(x; h)$ furnished by a one-step method. We assume that $f \in F_1(a, b)$ and denote by $y(x)$ the exact solution of the initial-value problem

$$\text{(I)} \qquad y' = f(x, y), \qquad y(x_0) = y_0.$$

Let $\Phi(x, y; h)$ define a one-step method,

$$\begin{aligned} &\eta_0 := y_0; \\ &\text{for } i = 0, 1, 2, \ldots: \\ &\eta_{i+1} := \eta_i + h\Phi(x_i, \eta_i; h), \\ &x_{i+1} := x_i + h, \end{aligned}$$

which for $x \in R_h := \{x_0 + ih \mid i = 0, 1, 2, \ldots\}$ produces the approximate solution $\eta(x; h)$:

$$\eta(x; h) := \eta_i \quad \text{if} \quad x = x_0 + ih.$$

We are interested in the behavior of the *global discretization error*

$$e(x; h) := \eta(x; h) - y(x)$$

for fixed x and $h \to 0$, $h \in H_x := \{(x - x_0)/n \mid n = 1, 2, \ldots\}$. Since $e(x; h)$, like $\eta(x; h)$, is only defined for $h \in H_x$, this means a study of the convergence of

$$e(x; h_n), \quad h_n := \frac{x - x_0}{n}, \qquad \text{as } n \to \infty.$$

We say that the one-step method is *convergent* if

(7.2.2.1) $$\lim_{n\to\infty} e(x; h_n) = 0$$

for all $x \in [a, b]$ and all $f \in F_1(a, b)$.

We will see that methods of order $p > 0$ [cf. (7.2.1.9)] are convergent, and even that

$$e(x; h_n) = O(h_n^p).$$

The order of the global discretization error is thus equal to the order of the local discretization error.

We begin by showing the following:

(7.2.2.2) Lemma. *If the numbers* ξ_i *satisfy estimates of the form*

$$|\xi_{i+1}| \leqslant (1+\delta)|\xi_i| + B, \qquad \delta > 0, \quad B \geqslant 0, \quad i = 0, 1, 2, \ldots,$$

then

$$|\xi_n| \leqslant e^{n\delta}|\xi_0| + \frac{e^{n\delta}-1}{\delta} B.$$

PROOF. From the assumptions we get immediately

$$\begin{aligned}
|\xi_1| &\leqslant (1+\delta)|\xi_0| + B,\\
|\xi_2| &\leqslant (1+\delta)^2|\xi_0| + B(1+\delta) + B,\\
&\;\;\vdots\\
|\xi_n| &\leqslant (1+\delta)^n|\xi_0| + B[1 + (1+\delta) + (1+\delta)^2 + \cdots + (1+\delta)^{n-1}]\\
&= (1+\delta)^n|\xi_0| + B\frac{(1+\delta)^n - 1}{\delta}\\
&\leqslant e^{n\delta}|\xi_0| + B\frac{e^{n\delta}-1}{\delta},
\end{aligned}$$

since $0 < 1 + \delta \leqslant e^\delta$ for $\delta > -1$. □

With this, we can prove the following main theorem:

(7.2.2.3) Theorem. *Consider, for* $x_0 \in [a, b]$, $y_0 \in \mathbb{R}$, *the initial-value problem*

(I) $$y' = f(x, y), \qquad y(x_0) = y_0,$$

having the exact solution $y(x)$. *Let the function* Φ *be continuous on* $G := \{(x, y, h) \mid a \leqslant x \leqslant b,\ |y - y(x)| \leqslant \gamma,\ 0 \leqslant |h| \leqslant h_0\}$, $h_0 > 0$, $\gamma > 0$, *and let there exist positive constants M and N such that*

(7.2.2.4) $$|\Phi(x, y_1; h) - \Phi(x, y_2; h)| \leqslant M|y_1 - y_2|$$

for all $(x, y_i, h) \in G$, $i = 1, 2$, *and*

(7.2.2.5) $$|\tau(x, y(x); h)| = |\Delta(x, y(x); h) - \Phi(x, y(x); h)| \leqslant N|h|^p, \quad p > 0,$$

for all $x \in [a, b]$, $|h| \leqslant h_0$. *Then there exists an* $\bar{h}$, $0 < \bar{h} \leqslant h_0$, *such that for the global discretization error* $e(x; h) = \eta(x; h) - y(x)$,

$$|e(x; h_n)| \leqslant |h_n|^p N \frac{e^{M|x - x_0|} - 1}{M}$$

for all $x \in [a, b]$ *and all* $h_n = (x - x_0)/n$, $n = 1, 2, \ldots$, *with* $|h_n| \leqslant \bar{h}$. *If* $\gamma = \infty$, *then* $\bar{h} = h_0$.

PROOF. The function

$$\tilde{\Phi}(x, y; h) := \begin{cases} \Phi(x, y; h) & \text{if } (x, y, h) \in G, \\ \Phi(x, y(x) + \gamma; h) & \text{if } x \in [a, b],\ |h| \leqslant h_0,\ y \geqslant y(x) + \gamma, \\ \Phi(x, y(x) - \gamma; h) & \text{if } x \in [a, b],\ |h| \leqslant h_0,\ y \leqslant y(x) - \gamma \end{cases}$$

is evidently continuous on $\tilde{G} := \{(x, y, h) \mid x \in [a, b], y \in \mathbb{R}, |h| \leqslant h_0\}$ and satisfies the condition

(7.2.2.6) $$|\tilde{\Phi}(x, y_1; h) - \tilde{\Phi}(x, y_2; h)| \leqslant M|y_1 - y_2|$$

for all $(x, y_i, h) \in \tilde{G}$, $i = 1, 2$, and, because $\tilde{\Phi}(x, y(x); h) = \Phi(x, y(x); h)$, also the condition

(7.2.2.7) $$|\Delta(x, y(x); h) - \tilde{\Phi}(x, y(x); h)| \leqslant N|h|^p \quad \text{for } x \in [a, b],\ |h| \leqslant h_0.$$

Let the one-step method generated by $\tilde{\Phi}$ furnish the approximate values $\tilde{\eta}_i := \tilde{\eta}(x_i; h)$ for $y_i := y(x_i)$, $x_i = x_0 + ih$:

$$\tilde{\eta}_{i+1} = \tilde{\eta}_i + h\tilde{\Phi}(x_i, \tilde{\eta}_i; h).$$

In view of

$$y_{i+1} = y_i + h\Delta(x_i, y_i; h),$$

one obtains for the error $\tilde{e}_i := \tilde{\eta}_i - y_i$, by subtraction, the recurrence formula

(7.2.2.8)

$$\tilde{e}_{i+1} = \tilde{e}_i + h[\tilde{\Phi}(x_i, \tilde{\eta}_i; h) - \tilde{\Phi}(x_i, y_i; h)] + h[\tilde{\Phi}(x_i, y_i; h) - \Delta(x_i, y_i; h)].$$

Now from (7.2.2.6), (7.2.2.7) it follows that

$$|\tilde{\Phi}(x_i, \tilde{\eta}_i; h) - \tilde{\Phi}(x_i, y_i; h)| \leqslant M|\tilde{\eta}_i - y_i| = M|\tilde{e}_i|,$$
$$|\Delta(x_i, y_i; h) - \tilde{\Phi}(x_i, y_i; h)| \leqslant N|h|^p,$$

and hence from (7.2.2.8) we have the recursive estimate

$$|\tilde{e}_{i+1}| \leqslant (1 + |h|M)|\tilde{e}_i| + N|h|^{p+1}.$$

Lemma (7.2.2.2), since $\tilde{e}_0 = \tilde{\eta}_0 - y_0 = 0$, gives

$$(7.2.2.9) \qquad |\tilde{e}_k| \leqslant N|h|^p \frac{e^{k|h|M} - 1}{M}.$$

Now let $x \in [a, b]$, $x \neq x_0$, be fixed and $h := h_n = (x - x_0)/n$, $n > 0$ an integer. Then $x_n = x_0 + nh = x$, and from (7.2.2.9) with $k = n$, since $\tilde{e}(x; h_n) = \tilde{e}_n$, it follows at once that

$$(7.2.2.10) \qquad |\tilde{e}(x; h_n)| \leqslant N|h_n|^p \frac{e^{M|x-x_0|} - 1}{M}$$

for all $x \in [a, b]$ and h_n with $|h_n| \leqslant h_0$. Since $|x - x_0| \leqslant |b - a|$ and $\gamma > 0$, there exists an $\bar{h}$, $0 < \bar{h} \leqslant h_0$, such that $|\tilde{e}(x; h_n)| \leqslant \gamma$ for all $x \in [a, b]$, $|h_n| \leqslant \bar{h}$, i.e., for the one-step method generated by Φ,

$$\eta_0 = y_0,$$
$$\eta_{i+1} = \eta_i + \Phi(x_i, \eta_i; h),$$

we have for $|h| \leqslant \bar{h}$, according to the definition of $\tilde{\Phi}$,

$$\tilde{\eta}_i = \eta_i, \qquad \tilde{e}_i = e_i, \quad \text{and} \quad \tilde{\Phi}(x_i, \tilde{\eta}_i; h) = \Phi(x_i, \eta_i; h).$$

The assertion of the theorem,

$$|e(x; h_n)| \leqslant |h_n|^p N \frac{e^{M|x-x_0|} - 1}{M},$$

thus follows for all $x \in [a, b]$ and all $h_n = (x - x_0)/n$, $n = 1, 2, \ldots$, with $|h_n| \leqslant \bar{h}$. □

From the preceding theorem it follows in particular that methods of order $p > 0$ which in the neighborhood of the exact solution satisfy a Lipschitz condition of the form (7.2.2.4) are convergent in the sense (7.2.2.1). Observe that the condition (7.2.2.4) is fulfilled, e.g., if $(\partial/\partial y)\Phi(x, y; h)$ exists and is continuous in a domain G of the form stated in the theorem.

Theorem (7.2.2.3) also provides an upper bound for the discretization error, which in principle can be evaluated if one knows M and N. One could use it, e.g., to determine the steplength h which is required to compute $y(x)$ within an error ε, given x and $\varepsilon > 0$. Unfortunately, in practice this is doomed by the fact that the constants M and N are not easily accessible, since an estimation of M and N is only possible via estimates of higher derivatives of f. Already in the simple Euler's method, $\Phi(x, y; h) := f(x, y)$, e.g., one has [see (7.2.1.8) f.]

$$N \approx \tfrac{1}{2}|f_x(x, y(x)) + f_y(x, y(x))f(x, y(x))|,$$
$$M \approx \left|\frac{\partial\Phi}{\partial y}\right| = |f_y(x, y)|.$$

For the Runge–Kutta method, one would already have to estimate derivatives of f of the fourth order.

7.2.3 Asymptotic Expansions for the Global Discretization Error of One-Step Methods

It may be conjectured from Theorem (7.2.2.3) that the approximate solution $\eta(x; h)$, furnished by a method of order p, possesses an asymptotic expansion in powers of h of the form

$$(7.2.3.1) \qquad \eta(x; h) = y(x) + e_p(x)h^p + e_{p+1}(x)h^{p+1} + \cdots$$

for all $h = h_n = (x - x_0)/n$, $n = 1, 2, \ldots$, with certain coefficient functions $e_i(x)$, $i = p, p + 1, \ldots$, that are independent of h. This is indeed true for general one-step methods of order p, provided only that $\Phi(x, y; h)$ and f satisfy certain additional regularity conditions. One has [see Gragg (1963)]

(7.2.3.2) Theorem. *Let $f(x, y) \in F_{N+2}(a, b)$ [cf. (7.1.3)] and let $\eta(x; h)$ be the approximate solution obtained by a one-step method of order p, $p \leqslant N$, to the solution $y(x)$ of the initial value problem*

$$(\text{I}) \qquad y' = f(x, y), \qquad y(x_0) = y_0, \quad x_0 \in [a, b].$$

Then $\eta(x; h)$ has an asymptotic expansion of the form

$$(7.2.3.3) \qquad \begin{aligned} \eta(x; h) = y(x) + h^p e_p(x) + h^{p+1} e_{p+1}(x) + \cdots + h^N e_N(x) \\ + h^{N+1} E_{N+1}(x; h) \quad \text{with } e_k(x_0) = 0, \quad k = p, p + 1, \ldots \end{aligned}$$

which is valid for all $x \in [a, b]$ and all $h = h_n = (x - x_0)/n$, $n = 1, 2, \ldots$. The functions $e_i(x)$ therein are independent of h, and the remainder term $E_{N+1}(x; h)$ is bounded for fixed x and all $h = h_n = (x - x_0)/n$, $n = 1, 2, \ldots$.

The following elegant proof is due to Hairer and Lubich (1984). Suppose that the one-step method is given by $\Phi(x, y, h)$. Since the method has order p, and $f \in F_{N+2}$, there follows (see Section 7.2.1)

$$\begin{aligned} y(x + h) &- y(x) - h\Phi(x, y(x); h) \\ &= d_{p+1}(x)h^{p+1} + \cdots + d_{N+1}(x)h^{N+1} + O(h^{N+2}). \end{aligned}$$

First, we use only

$$y(x + h) - y(x) - h\Phi(x, y(x); h) = d_{p+1}(x)h^{p+1} + O(h^{p+2})$$

and show that there is a differentiable function $e_p(x)$ such that

$$\eta(x; h) - y(x) = e_p(x)h^p + O(h^{p+1}).$$

To this end, we consider the function

$$\hat{\eta}(x; h) := \eta(x; h) - e_p(x)h^p,$$

where the choice of e_p is still left open. It is easy to see that $\hat{\eta}$ can be considered as the result of another one-step method,

$$\hat{\eta}(x + h; h) = \hat{\eta}(x; h) + h\hat{\Phi}(x, \hat{\eta}(x); h),$$

if $\hat{\Phi}$ is defined by

$$\hat{\Phi}(x, y; h) := \Phi(x, y + e_p(x)h^p; h) - (e_p(x + h) - e_p(x))h^{p-1}.$$

By Taylor expansion with respect to h, we find

$$\begin{aligned} &y(x + h) - y(x) - h\hat{\Phi}(x, y(x); h) \\ &\quad = [d_{p+1}(x) - f_y(x, y(x))e_p(x) - e'_p(x)]h^{p+1} + O(h^{p+2}). \end{aligned}$$

Hence, the one-step method belonging to $\hat{\Phi}$ has order $p + 1$ if e_p is taken as the solution of the initial value problem

$$e'_p(x) = d_{p+1}(x) - f_y(x, y(x))e_p(x), \qquad e_p(x_0) = 0.$$

With this choice of e_p, Theorem (7.2.2.3) applied to $\hat{\Phi}$ then shows that

$$\hat{\eta}(x; h) - y(x) = \eta(x; h) - y(x) - e_p(x)h^p = O(h^{p+1}).$$

A repetition of these arguments with $\hat{\Phi}$ in place of Φ completes the proof of the theorem. □

Asymptotic laws of the type (7.2.3.1) or (7.2.3.3) are significant in practice for two reasons. In the first place, one can use them to estimate the global discretization error $e(x; h)$. Suppose the method of order p has an asymptotic expansion of the form (7.2.3.1), so that

$$e(x; h) = \eta(x; h) - y(x) = h^p e_p(x) + O(h^{p+1}).$$

Having found the approximate value $\eta(x; h)$ with stepsize h, one computes for the same x, but with another stepsize (say $h/2$), the approximation $\eta(x; h/2)$. For sufficiently small h [and $e_p(x) \neq 0$] one then has in first approximation

(7.2.3.4) $$\eta(x; h) - y(x) \doteq e_p(x) \cdot h^p,$$

(7.2.3.5) $$\eta\left(x; \frac{h}{2}\right) - y(x) \doteq e_p(x) \cdot \left(\frac{h}{2}\right)^p.$$

Subtracting the second equation from the first gives

$$\eta(x; h) - \eta\left(x; \frac{h}{2}\right) \doteq e_p(x) \cdot \left(\frac{h}{2}\right)^p (2^p - 1),$$

$$e_p(x)\left(\frac{h}{2}\right)^p \doteq \frac{\eta(x; h) - \eta(x; h/2)}{2^p - 1},$$

and one obtains, by substitution in (7.2.3.5),

$$\eta\left(x; \frac{h}{2}\right) - y(x) \doteq \frac{\eta(x; h) - \eta(x; h/2)}{2^p - 1}. \tag{7.2.3.6}$$

For the Runge–Kutta method one has $p = 4$ and obtains the frequently used formula

$$\eta\left(x; \frac{h}{2}\right) - y(x) \doteq \frac{\eta(x; h) - \eta(x; h/2)}{15}.$$

The other, more important, significance of asymptotic expansions lies in the fact that they justify the application of extrapolation methods (see Section 3.4). Since a little later [see (7.2.12.7) f.] we will get to know a discretization method for which the asymptotic expansion of $\eta(x; h)$ contains only even powers of h and which, therefore, is more suitable for extrapolation algorithms (see Section 3.5) than Euler's method, we defer the description of extrapolation algorithms to Section 7.2.14.

7.2.4 The Influence of Rounding Errors in One-Step Methods

If a one-step method

(7.2.4.1)
$$\begin{aligned}
&\eta_0 := y_0;\\
&\text{for } i = 0, 1, 2, \ldots:\\
&\eta_{i+1} := \eta_i + h\Phi(x_i, \eta_i; h),\\
&x_{i+1} := x_i + h
\end{aligned}$$

is executed in floating-point arithmetic (t decimal digits) with relative precision eps $= 5 \times 10^{-t}$, then instead of the η_i one obtains other numbers $\tilde{\eta}_i$, which satisfy a recurrence formula of the form

(7.2.4.2)
$$\begin{aligned}
&\tilde{\eta}_0 := y_0;\\
&\text{for } i = 0, 1, 2, \ldots:\\
&c_i := \mathrm{fl}(\Phi(x_i, \tilde{\eta}_i; h)),\\
&d_i := \mathrm{fl}(hc_i),\\
&\tilde{\eta}_{i+1} := \mathrm{fl}(\tilde{\eta}_i + d_i) = \tilde{\eta}_i + h\Phi(x_i, \tilde{\eta}_i; h) + \varepsilon_{i+1},
\end{aligned}$$

where the total rounding error ε_{i+1}, in first approximation, is made up of three components:

$$\varepsilon_{i+1} \doteq h\Phi(x_i, \tilde{\eta}_i; h)(\alpha_{i+1} + \mu_{i+1}) + \tilde{\eta}_{i+1}\sigma_{i+1}.$$

Here

$$\alpha_{i+1} = \frac{\mathrm{fl}(\Phi(x_i, \tilde{\eta}_i; h)) - \Phi(x_i, \tilde{\eta}_i; h)}{\Phi(x_i, \tilde{\eta}_i; h)}$$

is the relative rounding error committed in the floating-point computation of Φ, μ_{i+1} the relative rounding error committed in the computation of the product hc_i, and σ_{i+1} the relative rounding error which occurs in the addition $\tilde{\eta}_i + d_i$. Normally, in practice, the stepsize h is so small that $|h\Phi(x_i, \tilde{\eta}_i; h)| \ll |\tilde{\eta}_i|$, and if $|\alpha_{i+1}| \leqslant \text{eps}$ and $|\mu_{i+1}| \leqslant \text{eps}$, one thus has $\varepsilon_{i+1} \doteq \tilde{\eta}_{i+1}\sigma_{i+1}$, i.e., the influence of rounding errors is determined primarily by the addition error σ_{i+1}.

Remark. It is natural, therefore, to reduce the influence of rounding errors by carrying out the addition in double precision ($2t$ decimal places). Denoting by $\mathrm{fl}_2(a + b)$ a double-precision addition, by $\tilde{\eta}_i$ a double-precision number ($2t$ decimal places), and by $\bar{\eta}_i := \mathrm{rd}_1(\tilde{\eta}_i)$ the number $\tilde{\eta}_i$ rounded to single precision, then the algorithm, instead of (7.2.4.2), now runs as follows,

(7.2.4.3)
$$\begin{aligned}
&\tilde{\eta}_0 := y_0;\\
&\text{for } i = 0, 1, 2, \ldots:\\
&\bar{\eta}_i := \mathrm{rd}_1(\tilde{\eta}_i),\\
&c_i := \mathrm{fl}(\Phi(x_i, \bar{\eta}_i; h)),\\
&d_i := \mathrm{fl}(hc_i),\\
&\tilde{\eta}_{i+1} := \mathrm{fl}_2(\tilde{\eta}_i + d_i).
\end{aligned}$$

Let us now briefly estimate the total influence of all rounding errors ε_i. For this, let $y_i = y(x_i)$ be the values of the exact solution of the initial-value problem, $\eta_i = \eta(x_i; h)$ the discrete solutions produced by the one-step method (7.2.4.1) in exact arithmetic, and finally $\tilde{\eta}_i$ the approximate values of η_i actually obtained in t-digit floating-point arithmetic. The latter satisfy relations of the form

(7.2.4.4)
$$\begin{aligned}
&\tilde{\eta}_0 = y_0;\\
&\text{for } i = 0, 1, 2, \ldots:\\
&\tilde{\eta}_{i+1} = \tilde{\eta}_i + h\Phi(x_i, \tilde{\eta}_i; h) + \varepsilon_{i+1}.
\end{aligned}$$

For simplicity, we also assume

$$|\varepsilon_{i+1}| \leqslant \varepsilon \quad \text{for all } i \geqslant 0.$$

We assume further that Φ satisfies a Lipschitz condition of the form (7.2.2.4),

$$|\Phi(x, y_1; h) - \Phi(x, y_2; h)| \leqslant M|y_1 - y_2|.$$

Then, for the error $r(x_i; h) := r_i := \tilde{\eta}_i - \eta_i$, there follows by subtraction of (7.2.4.1) from (7.2.4.4)

$$r_{i+1} = r_i + h(\Phi(x_i, \tilde{\eta}_i; h) - \Phi(x_i, \eta_i; h)) + \varepsilon_{i+1},$$

and thus

(7.2.4.5) $$|r_{i+1}| \leqslant (1 + |h|M)|r_i| + \varepsilon.$$

Since $r_0 = 0$, Lemma (7.2.2.2) gives

$$|r(x; h)| \leqslant \frac{\varepsilon}{|h|} \frac{e^{M|x-x_0|} - 1}{M}$$

for all $x \in [a, b]$ and $h = h_n = (x - x_0)/n$, $n = 1, 2, \ldots$. It follows, therefore, that for a method of order p the total error

$$v(x_i; h) := v_i := \tilde{\eta}_i - y_i = (\tilde{\eta}_i - \eta_i) + (\eta_i - y_i) = r(x_i; h) + e(x_i; h),$$

under the assumptions of Theorem (7.2.2.3), obeys the estimate

(7.2.4.6) $$|v(x; h)| \leqslant \left[N|h|^p + \frac{\varepsilon}{|h|}\right] \frac{e^{M|x-x_0|} - 1}{M}$$

for all $x \in [a, b]$ and for all sufficiently small $h := h_n = (x - x_0)/n$.

This formula reveals that, on account of the influence of rounding errors, the total error $v(x; h)$ begins to increase again, once h is reduced beyond a certain critical value. The following table shows this behavior. For the initial-value problem

$$y' = -200xy^2, \qquad y(-1) = \tfrac{1}{101},$$

with exact solution $y(x) = 1/(1 + 100x^2)$, an approximate value $\eta(0; h)$ for $y(0) = 1$ has been computed by the Runge–Kutta method in 12-digit arithmetic:

h	10^{-2}	0.5×10^{-2}	10^{-3}	0.5×10^{-3}
$v(0; h)$	-0.276×10^{-4}	-0.178×10^{-5}	-0.229×10^{-7}	-0.192×10^{-7}

h	10^{-4}	0.5×10^{-4}	10^{-5}
$v(0; h)$	-0.478×10^{-6}	-0.711×10^{-6}	-0.227×10^{-5}

The appearance of the term $\varepsilon/|h|$ in (7.2.4.6) becomes plausible if one considers that the number of steps to get from x_0 to x, using steplength h, is just $(x - x_0)/h$ and that all essentially independent rounding errors were assumed equal to ε. Nevertheless, the estimate is much too coarse to be practically significant.

7.2.5 Practical Implementation of One-Step Methods

In practice, initial-value problems present themselves mostly in the following form: What is desired is the value which the exact solution $y(x)$ assumes for a certain $x \neq x_0$. It is tempting to compute this solution approximately by means of a one-step method in a single step, i.e., choosing stepsize $\bar{h} = x - x_0$. For large $x - x_0$ this of course leads to a large discretization error $e(x; \bar{h})$; the choice made for $\bar{h}$ would be entirely inadequate. Normally, therefore, one will introduce suitable intermediate points x_i, $i = 1, \ldots, k-1$, $x_0 < x_1 < \cdots < x_k = x$, and, beginning with x_0, $y_0 = y(x_0)$, compute successive approximate values of $y(x_i)$: Having determined an approximation $\bar{y}(x_i)$ of $y(x_i)$, one computes $\bar{y}(x_{i+1})$ by applying a step of the method with stepsize $h_i := x_{i+1} - x_i$,

$$\begin{aligned}\bar{y}(x_{i+1}) &= \bar{y}(x_i) + h_i \Phi(x_i, \bar{y}(x_i); h_i),\\ x_{i+1} &= x_i + h_i.\end{aligned}$$

There again arises, however, the problem of how the stepsizes h_i are to be chosen. Since the amount of work involved in the method is proportional to the number of individual steps, one will attempt to choose the stepsizes h_i as large as possible. On the other hand, they must not be chosen too large if one wants to keep the discretization error small. In principle, one has the following problem: For given x_0, y_0, determine a stepsize h as large as possible, but such that the discretization error $e(x_0 + h; h)$ after one step with this stepsize still remains below a certain tolerance ε. This tolerance ε should not be selected smaller than $K \cdot \text{eps}$, i.e., $\varepsilon \geqslant K \cdot \text{eps}$, where eps is the relative machine precision and K a bound for the solution $y(x)$ in the region under consideration,

$$K \approx \max\{|y(x)| : x \in [x_0, x_0 + h]\}.$$

A choice of h corresponding to $\varepsilon = K \cdot \text{eps}$ then guarantees that the approximate solution $\eta(x_0 + h; h)$ obtained agrees with the exact solution $y(x_0 + h)$ to machine precision. Such a stepsize $h = h(\varepsilon)$, with

$$|e(x_0 + h; h)| \approx \varepsilon, \qquad \varepsilon \geqslant K \cdot \text{eps},$$

can be found approximately with the methods of Section 7.2.3: For a method of order p one has in first approximation

$$e(x; h) \doteq e_p(x) h^p. \tag{7.2.5.1}$$

Now $e_p(x_0) = 0$; thus in first approximation,

$$e_p(x) \doteq (x - x_0)e_p'(x_0). \tag{7.2.5.2}$$

Therefore $|e(x_0 + h; h)| \doteq \varepsilon$ will hold if

$$\varepsilon \doteq |e_p(x_0 + h)h^p| \doteq |h^{p+1}e_p'(x_0)|. \tag{7.2.5.3}$$

If we know $e_p'(x_0)$, we can compute from this the appropriate value of h. An approximate value for $e_p'(x_0)$, however, can be obtained from (7.2.3.6). Using first the stepsize H to compute $\eta(x_0 + H; H)$ and $\eta(x_0 + H; H/2)$, one then has by (7.2.3.6)

$$e\left(x_0 + H; \frac{H}{2}\right) \doteq \frac{\eta(x_0 + H; H) - \eta(x_0 + H; H/2)}{2^p - 1}. \tag{7.2.5.4}$$

By (7.2.5.1), (7.2.5.2), on the other hand,

$$\begin{aligned} e\left(x_0 + H; \frac{H}{2}\right) &\doteq e_p(x_0 + H)\left(\frac{H}{2}\right)^p \\ &\doteq e_p'(x_0)H\left(\frac{H}{2}\right)^p. \end{aligned}$$

From (7.2.5.4) thus follows the estimate

$$e_p'(x_0) \doteq \frac{1}{H^{p+1}}\frac{2^p}{2^p - 1}\left[\eta(x_0 + H; H) - \eta\left(x_0 + H; \frac{H}{2}\right)\right].$$

Equation (7.2.5.3) therefore yields for h the formula

$$\frac{H}{h} \doteq \sqrt[p+1]{\frac{2^p}{2^p - 1}\frac{|\eta(x_0 + H; H) - \eta(x_0 + H; H/2)|}{\varepsilon}}, \tag{7.2.5.5}$$

which can be used in the following way: Choose a stepsize H; compute $\eta(x_0 + H; H)$, $\eta(x_0 + H; H/2)$, and h from (7.2.5.5). If $H/h \gg 2$, then by (7.2.5.4) the error $e(x_0 + H; H/2)$ is much larger than the prescribed ε. It is expedient, therefore, to reduce H. Replace H by $2h$; with the new H compute again $\eta(x_0 + H; H)$, $\eta(x_0 + H; H/2)$, and from (7.2.5.5) the corresponding h, until finally $|H/h| \leqslant 2$. Once this is the case, one accepts $\eta(x_0 + H; H/2)$ as an approximation for $y(x_0 + H)$ and proceeds to the next integration step, replacing x_0, y_0, and H by the new starting values $x_0 + H$, $\eta(x_0 + H; H/2)$, and $2h$:

(7.2.5.6)

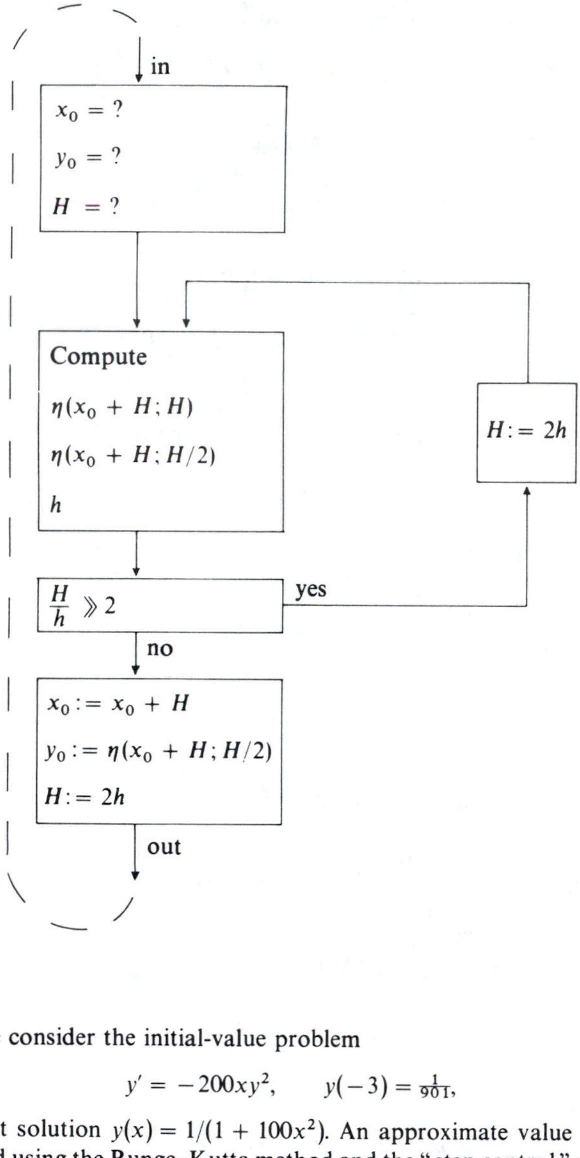

EXAMPLE. We consider the initial-value problem

$$y' = -200xy^2, \qquad y(-3) = \tfrac{1}{901},$$

with the exact solution $y(x) = 1/(1 + 100x^2)$. An approximate value η for $y(0) = 1$ was computed using the Runge–Kutta method and the "step control" procedure just discussed. The criterion $H/h \gg 2$ in (7.2.5.6) was replaced by the test $H/h \geqslant 3$. Computation in 12 digits yields:

$\eta - y(0)$	Number of required Runge–Kutta steps	Smallest stepsize H observed
-0.13585×10^{-6}	1476	$0.1226 \cdots \times 10^{-2}$

For fixed stepsize h the Runge–Kutta method yields:

h	$\eta(0; h) - y(0)$	Number of Runge–Kutta steps
$\frac{3}{1476} = 0.2032 \cdots \times 10^{-2}$	-0.5594×10^{-6}	1476
$0.1226 \cdots \times 10^{-2}$	-0.4052×10^{-6}	$\frac{3}{0.1226 \cdots \times 10^{-2}} = 2446$

A fixed choice of stepsize thus yields worse results with the same, or even greater, computational effort. In the first case, the stepsize $h = 0.2032 \cdots \times 10^{-2}$ is probably too large in the "critical" region near 0; the discretization error becomes too large. The stepsize $h = 0.1226 \cdots \times 10^{-2}$, on the other hand, will be too small in the "harmless" region from -3 to close to 0; one takes unnecessarily many steps and thereby commits more rounding errors.

Our discussion shows that in order to be able to make statements about the size of an approximately optimal steplength h, there must be available two approximate values $\eta(x_0 + H; H)$ and $\eta(x_0 + H; H/2)$—obtained from the same discretization method—for the exact solution $y(x_0 + H)$. Efficient methods for controlling the stepsize, however, can be obtained in still another manner. Instead of comparing two approximations (with different H) from the same discretization method, one takes, following an idea of Fehlberg's, two approximations (with the same H) which originate from (different) discretization methods of orders p and $p + 1$. This gives rise to so-called *Runge–Kutta–Fehlberg methods*. The explicit realization of the idea will be explained in the example of Runge–Kutta–Fehlberg methods of orders 2 and 3.

Let

$$\text{(7.2.5.7)} \qquad \begin{aligned} \bar{y}_{i+1} &= \bar{y}_i + h\Phi_{\mathrm{I}}(x_i, \bar{y}_i; h), \\ \hat{y}_{i+1} &= \bar{y}_i + h\Phi_{\mathrm{II}}(x_i, \bar{y}_i; h), \end{aligned}$$

where we assume, according to (7.2.2.5),

$$\text{(7.2.5.8)} \qquad \begin{aligned} |\Delta(x, y(x); h) - \Phi_{\mathrm{I}}(x, y(x); h)| &\leqslant N_{\mathrm{I}} h^2, \\ |\Delta(x, y(x); h) - \Phi_{\mathrm{II}}(x, y(x); h)| &\leqslant N_{\mathrm{II}} h^3. \end{aligned}$$

The Φ_{I}, Φ_{II} are constructed as follows:

$$\text{(7.2.5.9)} \qquad \begin{aligned} \Phi_{\mathrm{I}}(x, y; h) &= \sum_{k=0}^{2} c_k f_k(x, y; h), \\ \Phi_{\mathrm{II}}(x, y; h) &= \sum_{k=0}^{3} \hat{c}_k f_k(x, y; h), \end{aligned}$$

with

(7.2.5.10) $$f_k = f_k(x, y; h) = f\left(x + \alpha_k h,\ y + h \sum_{l=0}^{k-1} \beta_{kl} f_l\right), \qquad k = 0, 1, 2, 3.$$

Note that in the computation of Φ_{II} the function values f_0, f_1, f_2 used in method I are again utilized, so that only one additional evaluation, f_3, is required.

The constants α_k, β_{kl}, c_k, and $\hat{c}_k$ are now determined, as in (7.2.1.11), so that both relations in (7.2.5.8) are satisfied. Using the additional condition $\alpha_k = \sum_{j=0}^{k-1} \beta_{kj}$, $k = 1, 2, 3$, this yields the equations

(7.2.5.11) $$\begin{aligned} &\sum_{k=0}^{2} c_k - 1 = 0, && \sum_{k=1}^{2} \alpha_k c_k - \frac{1}{2} = 0, \\ &\sum_{k=0}^{3} \hat{c}_k - 1 = 0, && \sum_{k=1}^{3} \alpha_k \hat{c}_k - \frac{1}{2} = 0, \\ &\sum_{k=1}^{3} \alpha_k^2 \hat{c}_k - \frac{1}{3} = 0, && \sum_{k=2}^{3} P_{k1} \hat{c}_k - \frac{1}{6} = 0, \end{aligned}$$

$$P_{21} := \alpha_1 \beta_{21}, \qquad P_{31} := \alpha_1 \beta_{31} + \alpha_2 \beta_{32}.$$

The system of equations admits infinitely many solutions. One can therefore impose additional requirements to enhance economy: for example, make the value f_3 from the ith step reusable as f_0 in the $(i + 1)$st step. Since

$$f_3 = f(x + \alpha_3 h,\ y + h(\beta_{30} f_0 + \beta_{31} f_1 + \beta_{32} f_2))$$

and

$$\text{“new”}\ f_0 = f(x + h,\ y + h\Phi_{\text{I}}),$$

this implies

$$\alpha_3 = 1, \qquad \beta_{30} = c_0, \qquad \beta_{31} = c_1, \qquad \beta_{32} = c_2.$$

Further requirements can be made concerning the magnitudes of the coefficients in the error terms $\Delta - \Phi_{\text{I}}$, $\Delta - \Phi_{\text{II}}$. However, we will not pursue this any further and refer, instead, to the papers of Fehlberg (1964, 1966, 1969).

For the set of coefficients, one then finds

k	α_k	β_{kl}			c_k	$\hat{c}_k$
		$l = 0$	$l = 1$	$l = 2$		
0	0	0	—	—	$\frac{214}{891}$	$\frac{533}{2106}$
1	$\frac{1}{4}$	$\frac{1}{4}$	—	—	$\frac{1}{33}$	0
2	$\frac{27}{40}$	$-\frac{189}{800}$	$\frac{729}{800}$	—	$\frac{650}{891}$	$\frac{800}{1053}$
3	1	$\frac{214}{891}$	$\frac{1}{33}$	$\frac{650}{891}$	—	$-\frac{1}{78}$

Step control is now accomplished as follows: Consider the difference $\bar{y}_{i+1} - \hat{y}_{i+1}$. From (7.2.5.7) it follows that

$$\bar{y}_{i+1} - \hat{y}_{i+1} = h[\Phi_{\mathrm{I}}(x_i, \bar{y}_i; h) - \Phi_{\mathrm{II}}(x_i, \bar{y}_i; h)], \tag{7.2.5.12}$$

and from (7.2.5.8),

$$\begin{aligned} \Phi_{\mathrm{I}} - \Delta &= h^2 C_{\mathrm{I}}(x) + \text{terms of higher order}, \\ \Phi_{\mathrm{II}} - \Delta &= h^3 C_{\mathrm{II}}(x) + \text{terms of higher order}. \end{aligned} \tag{7.2.5.13}$$

Hence,

$$\bar{y}_{i+1} - \hat{y}_{i+1} = h^3 C_{\mathrm{I}}(x_i) + \text{terms of higher order}. \tag{7.2.5.14}$$

Suppose the integration from x_i to x_{i+1} was successful, i.e., for given error tolerance $\varepsilon > 0$, it achieved

$$|\bar{y}_{i+1} - \hat{y}_{i+1}| \leqslant \varepsilon.$$

Neglecting the terms of higher order, one then has also

$$|C_{\mathrm{I}}(x_i)h^3| \leqslant \varepsilon.$$

If we wish the "new" stepsize $h_{\text{new}} = x_{i+2} - x_{i+1}$ to be successful, we must have

$$|C_{\mathrm{I}}(x_{i+1})h_{\text{new}}^3| \leqslant \varepsilon.$$

Now, up to errors of the first order, we have

$$C_{\mathrm{I}}(x_i) \doteq C_{\mathrm{I}}(x_{i+1}).$$

For $C_{\mathrm{I}}(x_i)$, however, one has from (7.2.5.14) the approximation

$$|C_{\mathrm{I}}(x_i)| \doteq \frac{|\bar{y}_{i+1} - \hat{y}_{i+1}|}{|h^3|}.$$

This yields the approximate relation

$$\frac{|\bar{y}_{i+1} - \hat{y}_{i+1}|}{|h^3|} |h_{\text{new}}^3| \leqslant \varepsilon,$$

which can be used to estimate the new stepsize,

$$h_{\text{new}} \doteq h \sqrt[3]{\frac{\varepsilon}{|\bar{y}_{i+1} - \hat{y}_{i+1}|}}. \tag{7.2.5.15}$$

Pairs of methods of the type considered are also called *embedded methods.* Clearly, the idea expressed in (7.2.5.7), (7.2.5.9), and (7.2.5.10) can be generalized to construct embedded methods of orders higher than 2 and 3, say, to pairs of methods of order p and $p + 1$: Then $\bar{y}_{i+1}$ and $\hat{y}_{i+1}$ (see (7.2.5.7)) will be defined by functions of the form

$$\Phi_I(x, y; h) := \sum_{k=0}^{p} c_k f_k(x, y; h),$$

$$\Phi_{II}(x, y; h) := \sum_{k=0}^{p+1} \hat{c}_k f_k(x, y; h),$$

where

$$f_k = f_k(x, y; h) := f\left(x + \alpha_k h,\ y + h \sum_{l=0}^{k-1} \beta_{kl} f_l\right), \qquad k = 0, 1, \ldots, p+1$$

and the coefficients α_k, β_{kl}, c_k, and $\hat{c}_k$ are determined such that Φ_I and Φ_{II} generate one-step methods of order p and $p+1$, respectively.

The following coefficients for a pair of embedded methods of order 4 and 5 were given by Dormand and Prince (1980):

k	α_k	β_{k0}	β_{k1}	β_{k2}	β_{k3}	β_{k4}	β_{k5}	c_k	$\hat{c}_k$
0	0							$\frac{35}{384}$	$\frac{5179}{57600}$
1	$\frac{1}{5}$	$\frac{1}{5}$						0	0
2	$\frac{3}{10}$	$\frac{3}{40}$	$\frac{9}{40}$					$\frac{500}{1113}$	$\frac{7571}{16695}$
3	$\frac{4}{5}$	$\frac{44}{45}$	$-\frac{56}{15}$	$\frac{32}{9}$				$\frac{125}{192}$	$\frac{393}{640}$
4	$\frac{8}{9}$	$\frac{19372}{6561}$	$-\frac{25360}{2187}$	$\frac{64448}{6561}$	$-\frac{212}{729}$			$-\frac{2187}{6784}$	$-\frac{92097}{339200}$
5	1	$\frac{9017}{3168}$	$-\frac{355}{33}$	$\frac{46732}{5247}$	$\frac{49}{176}$	$-\frac{5103}{18656}$		$\frac{11}{84}$	$\frac{187}{2100}$
6	1	$\frac{35}{384}$	0	$\frac{500}{1113}$	$\frac{125}{192}$	$-\frac{2187}{6784}$	$\frac{11}{84}$	0	$\frac{1}{40}$

Of course, if $\bar{y}_{i+1}$ and $\hat{y}_{i+1}$ are obtained from a pair of methods of order p and $p+1$, then we have to replace (7.2.5.15) by

(7.2.5.16)
$$h_{\text{new}} \doteq h \sqrt[p+1]{\frac{\varepsilon}{|\bar{y}_{i+1} - \hat{y}_{i+1}|}}.$$

Step-size control according to this formula is relatively cheap: The computation of $\bar{y}_{i+1}$ and $\hat{y}_{i+1}$ in (7.2.5.16) requires only $p+1$ evaluations of the right-hand side $f(x, y)$ of the differential equation. Compare this with the analogous relation (7.2.5.5): Since the computation of $\eta(x_0 + H; H/2)$ necessitates further evaluation of f (three times as many, altogether), step-size control according to (7.2.5.16) is more efficient than according to (7.2.5.5).

We finally remark that many authors, based on extensive numerical experimentation, recommend a modification of (7.2.5.16), viz.,

(7.2.5.17)
$$h_{\text{new}} \doteq \alpha h \sqrt[p]{\frac{\varepsilon |h|}{|\bar{y}_{i+1} - \hat{y}_{i+1}|}},$$

where α is a suitable adjustment factor: $\alpha \approx 0.9$.

7.2.6 Multistep Methods: Examples

In a multistep method for the solution of the initial-value problem

$$\text{(I)} \qquad y' = f(x, y), \qquad y(x_0) = y_0,$$

one computes an approximate value η_{j+r} of $y(x_{j+r})$ from $r \geqslant 2$ given approximate values η_k of $y(x_k)$, $k = j, j+1, \ldots, j+r-1$, at equidistant points $x_k = x_0 + kh$:

$$\text{(7.2.6.1)} \qquad \begin{aligned} &\text{for } j = 0, 1, 2, \ldots: \\ &\eta_j, \eta_{j+1}, \ldots, \eta_{j+r-1} \Rightarrow \eta_{j+r}. \end{aligned}$$

To initiate such methods, it is of course necessary that r starting values η_0, $\eta_1, \ldots, \eta_{r-1}$ be at our disposal; these must be obtained by other means, e.g., with the aid of a one-step method.

We begin by introducing some examples of multistep methods. A class of such methods can be derived from the formula

$$\text{(7.2.6.2)} \qquad y(x_{p+k}) - y(x_{p-j}) = \int_{x_{p-j}}^{x_{p+k}} f(t, y(t))\, dt,$$

which is obtained by integrating the identity $y'(x) = f(x, y(x))$. As in the derivation of the Newton–Cotes formulas (see Section 3.1), one now replaces the integrand in (7.2.6.2) by an interpolating polynomial $P_q(x)$ with

(1) $\deg P_q(x) \leqslant q$,
(2) $P_q(x_i) = f(x_i, y(x_i))$, $i = p, p-1, \ldots, p-q$.

We assume here that the x_i are equidistant and $h := x_{i+1} - x_i$. With the abbreviation $y_i := y(x_i)$, and using the Lagrange interpolation formula (2.1.1.4),

$$P_q(x) = \sum_{i=0}^{q} f(x_{p-i}, y_{p-i}) L_i(x), \qquad L_i(x) := \prod_{\substack{l=0 \\ l \neq i}}^{q} \frac{x - x_{p-l}}{x_{p-i} - x_{p-l}},$$

one obtains the approximate formula

$$\text{(7.2.6.3)} \qquad \begin{aligned} y_{p+k} - y_{p-j} &\approx \sum_{i=0}^{q} f(x_{p-i}, y_{p-i}) \int_{x_{p-j}}^{x_{p+k}} L_i(x)\, dx \\ &= h \sum_{i=0}^{q} \beta_{qi} f(x_{p-i}, y_{p-i}) \end{aligned}$$

with

$$\text{(7.2.6.4)} \qquad \begin{aligned} \beta_{qi} &:= \frac{1}{h} \int_{x_{p-j}}^{x_{p+k}} L_i(x)\, dx \\ &= \int_{-j}^{k} \prod_{\substack{l=0 \\ l \neq i}}^{q} \frac{s+l}{-i+l}\, ds, \qquad i = 0, 1, \ldots, q. \end{aligned}$$

Replacing in (7.2.6.3) the y_i by approximate values η_i and $\approx$ by the equality sign, one obtains the formula

$$\eta_{p+k} = \eta_{p-j} + h \sum_{i=0}^{q} \beta_{qi} f_{p-i}, \qquad f_i := f(x_i, \eta_i).$$

For different choices of k, j, and q, one obtains different multistep methods.

For $k = 1$, $j = 0$, and $q = 0, 1, 2, \ldots$, one obtains the *Adams-Bashforth methods*:

$$\eta_{p+1} = \eta_p + h[\beta_{q0} f_p + \beta_{q1} f_{p-1} + \cdots + \beta_{qq} f_{p-q}]$$

(7.2.6.5)
$$\text{with } \beta_{qi} := \int_0^1 \prod_{\substack{l=0 \\ l \neq i}}^{q} \frac{s+l}{-i+l}\, ds, \qquad i = 0, 1, \ldots, q.$$

Comparison with (7.2.6.1) shows that here $r = q + 1$. A few numerical values:

β_{qi} \ i	0	1	2	3	4
β_{0i}	1				
$2\beta_{1i}$	3	-1			
$12\beta_{2i}$	23	-16	5		
$24\beta_{3i}$	55	-59	37	-9	
$720\beta_{4i}$	1901	-2774	2616	-1274	251

For $k = 0$, $j = 1$ and $q = 0, 1, 2, \ldots$, one obtains the *Adams-Moulton formulas*:

$$\eta_p = \eta_{p-1} + h[\beta_{q0} f_p + \beta_{q1} f_{p-1} + \cdots + \beta_{qq} f_{p-q}].$$

Replacing p by $p + 1$ gives

(7.2.6.6)
$$\eta_{p+1} = \eta_p + h[\beta_{q0} f(x_{p+1}, \eta_{p+1}) + \beta_{q1} f_p + \cdots + \beta_{qq} f_{p+1-q}]$$
$$\text{with } \beta_{qi} := \int_{-1}^{0} \prod_{\substack{l=0 \\ l \neq i}}^{q} \frac{s+l}{-i+l}\, ds.$$

At first sight it appears that (7.2.6.6) no longer has the form (7.2.6.1), since η_{p+1} occurs on both the left-hand and the right-hand side of the equation in (7.2.6.6). Thus for given $\eta_p, \eta_{p-1}, \ldots, \eta_{p+1-q}$ Equation (7.2.6.6) represents, in general, a nonlinear equation for η_{p+1}, and the Adams–Moulton method is an *implicit method*. The following iterative method for determining η_{p+1}

suggests itself naturally:

(7.2.6.7)

$$\eta_{p+1}^{(i+1)} = \eta_p + h[\beta_{q0} f(x_{p+1}, \eta_{p+1}^{(i)}) + \beta_{q1} f_p + \cdots + \beta_{qq} f_{p+1-q}],$$
$$i = 0, 1, 2, \ldots .$$

This iteration has the form $\eta_{p+1}^{(i+1)} = \Psi(\eta_{p+1}^{(i)})$, and with the methods of Section 5.2 it can easily be shown that for sufficiently small $|h|$ the mapping $z \to \Psi(z)$ is contractive (see Exercise 10) and thus has a fixed point η_{p+1}, $\eta_{p+1} = \Psi(\eta_{p+1})$, which solves (7.2.6.6). This solution of course depends on $x_p, \eta_p, \eta_{p-1}, \ldots, \eta_{p+1-q}$, and h, and the Adams–Moulton method therefore is indeed a multistep method of the type (7.2.6.1). Here $r = q$.

For given $\eta_p, \eta_{p-1}, \ldots, \eta_{p+1-q}$ a good initial value $\eta_{p+1}^{(0)}$ for the iteration (7.2.6.7) can be found, e.g., with the aid of the Adams–Bashforth method (7.2.6.5). For this reason, one also calls explicit methods like the Adams–Bashforth method *predictor methods*, and implicit methods like the Adams–Moulton method *corrector methods* [through the iteration (7.2.6.7) one "corrects" $\eta_{p+1}^{(i)}$].

A few numerical values for the coefficients occurring in (7.2.6.6):

β_{qi} \ i	0	1	2	3	4
β_{0i}	1				
$2\beta_{1i}$	1	1			
$12\beta_{2i}$	5	8	−1		
$24\beta_{3i}$	9	19	−5	1	
$720\beta_{4i}$	251	646	−264	106	−19

In the *method of Nyström* one chooses $k = 1$, $j = 1$ in (7.2.6.2) and thus obtains

(7.2.6.8) $$\eta_{p+1} = \eta_{p-1} + h[\beta_{q0} f_p + \beta_{q1} f_{p-1} + \cdots + \beta_{qq} f_{p-q}]$$

with

$$\beta_{qi} := \int_{-1}^{1} \prod_{\substack{l=0 \\ l \neq i}}^{q} \frac{s+l}{-i+l}\, ds, \qquad i = 0, 1, \ldots, q.$$

This is a predictor method, which has obviously the form (7.2.6.1) with $r = q + 1$.

Worthy of note is the special case $q = 0$. Here $\beta_{00} = \int_{-1}^{1} 1\, ds = 2$, and (7.2.6.8) becomes

(7.2.6.9) $$\eta_{p+1} = \eta_{p-1} + 2hf_p.$$

This is the so-called *midpoint rule*, which corresponds in the approximation of an integral to "rectangular sums".

The *methods of Milne* are corrector methods. They are obtained from (7.2.6.2) by taking $k = 0$, $j = 2$, and by replacing p with $p + 1$:

$$\eta_{p+1} = \eta_{p-1} + h[\beta_{q0} f(x_{p+1}, \eta_{p+1}) + \beta_{q1} f_p + \cdots + \beta_{qq} f_{p+1-q}],$$

(7.2.6.10)
$$\beta_{qi} := \int_{-2}^{0} \prod_{\substack{l=0 \\ l \neq i}}^{q} \frac{s+l}{-i+l}\, ds, \qquad i = 0, 1, \ldots, q.$$

Analogously to (7.2.6.7), one solves (7.2.6.10) also by iteration.

7.2.7 General Multistep Methods

All multistep methods discussed in Section 7.2.6, as well as the one-step methods of Section 7.2.1, can be written in the following form:

(7.2.7.1)
$$\eta_{j+r} + a_{r-1}\eta_{j+r-1} + \cdots + a_0 \eta_j = hF(x_j; \eta_{j+r}, \eta_{j+r-1}, \ldots, \eta_j; h; f).$$

Generally, a multistep method given by (7.2.7.1) is called an *r-step method*. In the examples considered in Section 7.2.6 the function F, in addition, depends linearly on f as follows:

$$F(x_j; \eta_{j+r}, \eta_{j+r-1}, \ldots, \eta_j; h; f) = b_r f(x_{j+r}, \eta_{j+r}) + \cdots + b_0 f(x_j, \eta_j).$$

The b_i, $i = 0, 1, \ldots, r$, are certain constants. One then speaks of a *linear r-step method*; such methods will be further discussed in Section 7.2.11.

For the Adams–Bashforth method (7.2.6.5), for example, $r = q + 1$,

$$a_q = -1, \qquad a_{q-1} = \cdots = a_0 = 0, \qquad b_{q+1} = 0,$$

and

$$b_{q-i} = \beta_{qi} = \int_{0}^{1} \prod_{\substack{l=0 \\ l \neq i}}^{q} \frac{s+l}{-i+l}\, ds, \qquad i = 0, 1, \ldots, q.$$

Any r initial values $\eta_0, \ldots, \eta_{r-1}$ define, through (7.2.7.1), a unique sequence η_j, $j \geqslant 0$. For the initial values η_i one chooses, as well as possible, certain approximations for the exact solution $y_i = y(x_i)$ of (7.2.1.1) at $x_i = x_0 + ih$, $i = 0, 1, \ldots, r - 1$. These can be obtained, e.g., by means of appropriate one-step methods. We denote the errors in the starting values by

$$\varepsilon_i := \eta_i - y(x_i), \qquad i = 0, 1, \ldots, r - 1.$$

Further errors, e.g. rounding errors in the evaluation of F, occur during the execution of (7.2.7.1). We want to include the influence of these errors in our study also, and therefore consider, more generally than (7.2.7.1), the following recurrence formulas:

(7.2.7.2)

$$\begin{aligned} \eta_0 &:= y_0 + \varepsilon_0, \\ &\vdots \\ \eta_{r-1} &:= y_{r-1} + \varepsilon_{r-1}; \end{aligned}$$

for $j = 0, 1, 2, \ldots$:

$$\eta_{j+r} + a_{r-1}\eta_{j+r-1} + \cdots + a_0\eta_j := hF(x_j; \eta_{j+r}, \eta_{j+r-1}, \ldots, \eta_j; h; f) + h\varepsilon_{j+r}.$$

The solution η_i of (7.2.7.2) depends on h and the ε_j, and represents a function

$$\eta(x; \varepsilon; h)$$

which, like the error function $\varepsilon = \varepsilon(x; h)$, is defined only for $x \in R_h = \{x_0 + ih \mid i = 0, 1, \ldots\}$, or $h \in H_x = \{(x - x_0)/n, n = 1, 2, \ldots\}$, through

$$\eta(x_i; \varepsilon; h) := \eta_i, \qquad x_i := x_0 + ih,$$
$$\varepsilon(x_i; h) := \varepsilon_i.$$

As in one-step methods, one can define the local discretization error $\tau(x, y; h)$ of a multistep method (7.2.7.1) at the point x, y. Indeed, let $f \in F_1(a, b)$, $x \in [a, b]$, $y \in \mathbb{R}$, and let $z(t)$ be the solution of the initial-value problem

$$z'(t) = f(t, z(t)), \qquad z(x) = y.$$

Then the *local discretization error* $\tau(x, y; h)$ is defined to be the quantity

(7.2.7.3)

$$\begin{aligned} \tau(x, y; h) := \frac{1}{h}\Bigg[& z(x + rh) + \sum_{i=0}^{r-1} a_i z(x + ih) \\ & - hF(x; z(x + rh), z(x + (r-1)h), \ldots, z(x); h; f)\Bigg]. \end{aligned}$$

The local discretization error $\tau(x, y; h)$ thus indicates how well the exact solution of a differential equation satisfies the recurrence formula (7.2.7.1). From a reasonable method one expects that this error will become small for small $|h|$. Generalizing (7.2.1.7), one thus defines the consistency of multistep methods by:

(7.2.7.4) Definition. The multistep method is called *consistent* if for each $f \in F_1(a, b)$ there exists a function $\sigma(h)$ with $\lim_{h\to 0} \sigma(h) = 0$ such that

$$|\tau(x, y; h)| \leqslant \sigma(h) \quad \text{for all } x \in [a, b],\ y \in \mathbb{R}. \tag{7.2.7.5}$$

One speaks of a *method of order p* if for $f \in F_p(a, b)$,

$$\sigma(h) = O(h^p).$$

EXAMPLE. For the midpoint rule (7.2.6.9) one has, in view of $z'(t) = f(t, z(t))$, $z(x) = y$,

$$\begin{aligned}\tau(x, y; h) &= \frac{1}{h}[z(x + 2h) - z(x) - 2hf(x + h, z(x + h))]\\ &= \frac{1}{h}[z(x + 2h) - z(x) - 2hz'(x + h)].\end{aligned}$$

Through Taylor expansion in h one finds

$$\begin{aligned}\tau(x, y; h) &= \frac{1}{h}\left[z(x) + 2hz'(x) + 2h^2z''(x) + \frac{8h^3}{6}z'''(x)\right.\\ &\qquad \left.- z(x) - 2h\left(z'(x) + hz''(x) + \frac{h^2}{2}z'''(x)\right)\right] + O(h^3)\\ &= \frac{h^2}{3}z'''(x) + O(h^3).\end{aligned}$$

The method is thus consistent and of second order.

The order of the methods in Section 7.2.6 can be determined more conveniently by means of the error estimates for interpolating polynomials [e.g., (2.1.4.1)] and Newton–Cotes formulas (3.1.1).

For $f(x, y) :\equiv 0$ and $z(x) :\equiv y$, a consistent method gives

$$\tau(x, y; h) = \frac{1}{h}[y(1 + a_{r-1} + \cdots + a_0) - hF(x; y, y, \ldots, y; h; 0)],$$

$$|\tau(x, y; h)| \leqslant \sigma(h), \qquad \lim_{h\to 0} \sigma(h) = 0.$$

For continuous $F(x; y, y, \ldots, y; \cdot; 0)$, since y is arbitrary, it follows that

$$1 + a_{r-1} + \cdots + a_0 = 0. \tag{7.2.7.6}$$

We will often, in the following, impose on F the condition

$$F(x; u_r, u_{r-1}, \ldots, u_0; h; 0) \equiv 0 \tag{7.2.7.7}$$

for all $x \in [a, b]$, all h, and all u_i. For linear multistep methods, (7.2.7.7) is certainly always fulfilled. (7.2.7.7), together with (7.2.7.6), guarantees that the exact solution $y(x) \equiv y_0$ of the trivial differential equation $y' = 0$, $y(x_0) = y_0$, is also exact solution of (7.2.7.2) if $\varepsilon_i = 0$ for all i.

Since the approximate solution $\eta(x; \varepsilon; h)$ furnished by a multistep method (7.2.7.2) also depends on the errors ε_i, the definition of convergence is more complicated than for one-step methods. One cannot expect, of course, that the *global discretization error*

$$e(x; \varepsilon; h) := \eta(x; \varepsilon; h) - y(x)$$

for fixed x and for $h = h_n = (x - x_0)/n$, $n = 1, 2, \ldots$ will converge toward 0, unless the errors $\varepsilon(x; h)$ also become arbitrarily small as $h \to 0$. One therefore defines:

(7.2.7.8) Definition. The multistep method given by (7.2.7.2) is called *convergent* if

$$\lim_{n\to\infty} \eta(x; \varepsilon; h_n) = y(x), \qquad h_n := \frac{x - x_0}{n}, \quad n = 1, 2, \ldots,$$

for all $x \in [a, b]$, all $f \in F_1(a, b)$, and all functions $\varepsilon(z; h)$ for which there is a $\rho(h)$ such that

$$\text{(7.2.7.9)} \qquad \begin{aligned} |\varepsilon(z; h)| &\leq \rho(h) \quad \text{for all } z \in R_h, \\ \lim_{h\to 0} \rho(h) &= 0. \end{aligned}$$

7.2.8 An Example of Divergence

The results of Section 7.2.2, in particular Theorem (7.2.2.3), may suggest that multistep methods, too, converge faster with increasing order p of the local discretization error [see (7.2.7.4)]. The point of the following example is to show that this conjecture is false. At the same time the example furnishes a method for constructing multistep methods of maximum order.

Suppose we construct a multistep method of the type (7.2.7.1) with $r = 2$, having the form

$$\eta_{j+2} + a_1\eta_{j+1} + a_0\eta_j = h[b_1 f(x_{j+1}, \eta_{j+1}) + b_0 f(x_j, \eta_j)].$$

The constants a_0, a_1, b_0, b_1 are to be determined so as to yield a method of maximum order. If $z'(t) = f(t, z(t))$, then for the local discretization error $\tau(x, y; h)$ of (7.2.7.3) we have

$$h\tau(x, y; h) = z(x + 2h) + a_1 z(x + h) + a_0 z(x) - h[b_1 z'(x + h) + b_0 z'(x)].$$

Now expand the right-hand side in a Taylor series in h,

$$h\tau(x, y; h) = z(x)[1 + a_1 + a_0] + hz'(x)[2 + a_1 - b_1 - b_0] + h^2 z''(x)[2 + \tfrac{1}{2}a_1 - b_1] + h^3 z'''(x)[\tfrac{4}{3} + \tfrac{1}{6}a_1 - \tfrac{1}{2}b_1] + O(h^4).$$

The coefficients a_0, a_1, b_0, b_1 are to be determined in such a way as to annihilate as many h-powers as possible, thus producing a method which has the largest possible order.

This leads to the equations

$$\begin{aligned} 1 + a_1 + a_0 &= 0, \\ 2 + a_1 - b_1 - b_0 &= 0, \\ 2 + \tfrac{1}{2}a_1 - b_1 &= 0, \\ \tfrac{4}{3} + \tfrac{1}{6}a_1 - \tfrac{1}{2}b_1 &= 0, \end{aligned}$$

with the solution $a_1 = 4$, $a_0 = -5$, $b_1 = 4$, $b_0 = 2$, and hence to the method

$$\eta_{j+2} + 4\eta_{j+1} - 5\eta_j = h[4f(x_{j+1}, \eta_{j+1}) + 2f(x_j, \eta_j)]$$

of order 3 [since $h\tau(x, y; h) = O(h^4)$, i.e., $\tau(x, y; h) = O(h^3)$].

If one tries to use this method to solve the initial-value problem

$$y' = -y, \qquad y(0) = 1,$$

with the exact solution $y(x) = e^{-x}$, computation in 10-digit arithmetic for $h = 10^{-2}$, even taking as starting values the exact (within machine precision) values $\eta_0 := 1$, $\eta_1 := e^{-h}$, produces the results given in the following table.

j	$\eta_j - y_j$	$-\dfrac{x_j^4}{216}\dfrac{(-5)^j}{j^4}e^{3x_j/5}$ †
2	$-.164 \times 10^{-8}$	$-.753 \times 10^{-9}$
3	$+.501 \times 10^{-8}$	$+.378 \times 10^{-8}$
4	$-.300 \times 10^{-7}$	$-.190 \times 10^{-7}$
5	$+.144 \times 10^{-6}$	$+.958 \times 10^{-7}$
⋮	⋮	⋮
96	$-.101 \times 10^{58}$	$-.668 \times 10^{57}$
97	$+.512 \times 10^{58}$	$+.336 \times 10^{58}$
98	$-.257 \times 10^{59}$	$-.169 \times 10^{59}$
99	$+.129 \times 10^{60}$	$+.850 \times 10^{59}$
100	$-.652 \times 10^{60}$	$-.427 \times 10^{60}$

† (Cf. (7.2.8.3))

How does one account for this wildly oscillating behavior of the η_j? If we assume that the exact values $\eta_0 := 1$, $\eta_1 := e^{-h}$ are used as starting values and no rounding errors are committed during the execution of the method ($\varepsilon_j = 0$ for all j), we obtain a sequence of numbers η_j with

$$\eta_0 = 1,$$
$$\eta_1 = e^{-h},$$
$$\eta_{j+2} + 4\eta_{j+1} - 5\eta_j = h[-4\eta_{j+1} - 2\eta_j] \quad \text{for } j = 0, 1, \ldots,$$

or

(7.2.8.1) $$\eta_{j+2} + 4(1+h)\eta_{j+1} + (-5+2h)\eta_j = 0 \quad \text{for } j = 0, 1, \ldots .$$

Such *difference equations* have special solutions of the form $\eta_j = \lambda^j$. Upon substituting this expression in (7.2.8.1), one finds for λ the equation

$$\lambda^j[\lambda^2 + 4(1+h)\lambda + (-5+2h)] = 0,$$

which, aside from the trivial solution $\lambda = 0$, has the solutions

$$\lambda_1 = -2 - 2h + 3\sqrt{1 + \tfrac{2}{3}h + \tfrac{4}{9}h^2},$$
$$\lambda_2 = -2 - 2h - 3\sqrt{1 + \tfrac{2}{3}h + \tfrac{4}{9}h^2}.$$

For small h one has

$$\sqrt{1 + \tfrac{2}{3}h + \tfrac{4}{9}h^2} = 1 + \tfrac{1}{3}h + \tfrac{1}{6}h^2 - \tfrac{1}{18}h^3 + \tfrac{1}{216}h^4 + O(h^5);$$

hence

(7.2.8.2) $$\begin{aligned} \lambda_1 &= 1 - h + \tfrac{1}{2}h^2 - \tfrac{1}{6}h^3 + \tfrac{1}{72}h^4 + O(h^5), \\ \lambda_2 &= -5 - 3h + O(h^2). \end{aligned}$$

One can now show that every solution η_j of (7.2.8.1) can be written as a linear combination

$$\eta_j = \alpha\lambda_1^j + \beta\lambda_2^j$$

of the two particular solutions λ_1^j, λ_2^j just found [see (7.2.9.9)]. The constants α and β, in our case, are determined by the initial conditions $\eta_0 = 1$, $\eta_1 = e^{-h}$, which lead to the following system of equations for α, β:

$$\eta_0 = \alpha + \beta = 1,$$
$$\eta_1 = \alpha\lambda_1 + \beta\lambda_2 = e^{-h}.$$

The solution is given by

$$\alpha = \frac{\lambda_2 - e^{-h}}{\lambda_2 - \lambda_1}, \qquad \beta = \frac{e^{-h} - \lambda_1}{\lambda_2 - \lambda_1}.$$

From (7.2.8.2) one easily verifies

$$\alpha = 1 + O(h^2), \qquad \beta = -\tfrac{1}{216}h^4 + O(h^5).$$

For fixed $x \neq 0$, $h = h_n = x/n$, $n = 0, 1, 2, \ldots$, one therefore obtains for the approximate solution $\eta_n = \eta(x; h_n)$

$$\begin{aligned}\eta(x; h_n) &= \alpha\lambda_1^n + \beta\lambda_2^n \\ &= \left[1 + O\left(\left(\frac{x}{n}\right)^2\right)\right]\left[1 - \frac{x}{n} + O\left(\left(\frac{x}{n}\right)^2\right)\right]^n \\ &\quad - \frac{1}{216}\frac{x^4}{n^4}\left[1 + O\left(\frac{x}{n}\right)\right]\left[-5 - 3\frac{x}{n} + O\left(\left(\frac{x}{n}\right)^2\right)\right]^n.\end{aligned}$$

The first term tends to e^{-x} as $n \to \infty$; the second term for $n \to \infty$ behaves like

(7.2.8.3) $$-\frac{x^4}{216}\frac{(-5)^n}{n^4}e^{3x/5}.$$

Since $\lim_{n\to\infty} 5^n/n^4 = \infty$, this term oscillates more and more violently as $n \to \infty$. This explains the oscillatory behavior and divergence of the method. As is easily seen, the reason for this behavior lies in the fact that -5 is a root of the quadratic equation $\mu^2 + 4\mu - 5 = 0$. It is to be expected that also in the general case (7.2.7.2) the zeros of the polynomial $\psi(\mu) = \mu^r + a_{r-1}\mu^{r-1} + \cdots + a_0$ play a significant role for the convergence of the method.

7.2.9 Linear Difference Equations

In the following section we need a few simple results on linear difference equations. By a *linear homogeneous difference equation of order r* one means an equation of the form

(7.2.9.1)

$$u_{j+r} + a_{r-1}u_{j+r-1} + a_{r-2}u_{j+r-2} + \cdots + a_0 u_j = 0, \qquad j = 0, 1, 2, \ldots.$$

For every set of starting values $u_0, u_1, \ldots, u_{r-1}$ one can evidently determine exactly one sequence of numbers u_j, $j = 0, 1, 2, \ldots$, which solves (7.2.9.1).

In applications to multistep methods a point of interest is the growth behavior of the u_n as $n \to \infty$, in dependence of the starting values $u_0, u_1, \ldots, u_{r-1}$. In particular, one would like to know conditions guaranteeing that

(7.2.9.2) $$\lim_{n\to\infty} \frac{u_n}{n} = 0 \quad \text{for all real starting values } u_0, u_1, \ldots u_{r-1}.$$

Since the solutions $u_n = u_n(U_0)$, $U_0 := [u_0, u_1, \ldots, u_{r-1}]^T$, obviously depend linearly on the starting vector,

$$u_n(\alpha U_0 + \beta V_0) = \alpha u_n(U_0) + \beta u_n(V_0),$$

the restriction to real starting vectors $U_0 \in \mathbb{R}^r$ is unnecessary, and (7.2.9.2) is equivalent to

$$\text{(7.2.9.3)} \quad \lim_{n\to\infty} \frac{u_n}{n} = 0 \quad \text{for all (complex) starting values } u_0, u_1, \ldots u_{r-1}.$$

With the difference equation (7.2.9.1) one associates the polynomial

$$\text{(7.2.9.4)} \qquad \psi(\mu) := \mu^r + a_{r-1}\mu^{r-1} + \cdots + a_0 .$$

One now says that (7.2.9.1) satisfies the

(7.2.9.5) *stability condition*

if for every zero λ of $\psi(\mu)$ one has $|\lambda| \leqslant 1$, and further $\psi(\lambda) = 0$ and $|\lambda| = 1$ together imply that λ is a simple zero of ψ.

(7.2.9.6) Theorem. *The stability condition* (7.2.9.5) *is necessary and sufficient for* (7.2.9.3).

PROOF. (1) Assume (7.2.9.3), and let λ be a zero of ψ in (7.2.9.4). Then the sequence $u_j := \lambda^j$, $j = 0, 1, \ldots,$ is a solution of (7.2.9.1). For $|\lambda| > 1$ the sequence $u_n/n = \lambda^n/n$ diverges, so that from (7.2.9.3) it follows at once that $|\lambda| \leqslant 1$. Let now λ be a multiple zero of ψ with $|\lambda| = 1$. Then

$$\psi'(\lambda) = r\lambda^{r-1} + (r-1)a_{r-1}\lambda^{r-2} + \cdots + 1 \cdot a_1 = 0.$$

The sequence $u_j := j\lambda^j$, $j \geqslant 0$, is thus a solution of (7.2.9.1):

$$\begin{aligned} &u_{j+r} + a_{r-1}u_{j+r-1} + \cdots + a_0 u_j \\ &= j\lambda^j(\lambda^r + a_{r-1}\lambda^{r-1} + \cdots + a_0) + \lambda^{j+1}(r\lambda^{r-1} + (r-1)a_{r-1}\lambda^{r-2} + \cdots + a_1) \\ &= 0. \end{aligned}$$

Since $u_n/n = \lambda^n$ does not converge to zero as $n \to \infty$, λ must be a simple zero.

(2) Conversely, let the stability condition (7.2.9.5) be satisfied. We first use the fact that, with the abbreviations

$$U_j := \begin{bmatrix} u_j \\ u_{j+1} \\ \cdot \\ \cdot \\ \cdot \\ \cdot \\ u_{j+r-1} \end{bmatrix} \in \mathbb{C}^r, \qquad A := \begin{bmatrix} 0 & 1 & 0 & \cdot & \cdot & \cdot & 0 \\ \cdot & & \ddots & \ddots & & & \cdot \\ \cdot & & & \ddots & \ddots & & \cdot \\ \cdot & & & & \ddots & \ddots & \cdot \\ \cdot & & & & & \ddots & 0 \\ 0 & & \cdot & \cdot & \cdot & 0 & 1 \\ -a_0 & & \cdot & \cdot & \cdot & & -a_{r-1} \end{bmatrix},$$

the difference equation (7.2.9.1) is equivalent to the recurrence formula

$$U_{j+1} = AU_j.$$

A is a Frobenius matrix with the characteristic polynomial $\psi(\mu)$ in (7.2.9.4) [Theorem (6.3.4)]. Therefore, if the stability condition (7.2.9.5) is satisfied, one can choose, by Theorem (6.9.2), a norm $\|\cdot\|$ on $\mathbb{C}^r$ such that for the corresponding matrix norm $\operatorname{lub}(A) \leqslant 1$. It thus follows, for all $U_0 \in \mathbb{C}^r$, that

$$(7.2.9.7) \qquad \|U_n\| = \|A^n U_0\| \leqslant \|U_0\| \quad \text{for } n = 0, 1, \ldots.$$

Since on $\mathbb{C}^r$ all norms are equivalent [Theorem (4.4.6)], there exists a $k > 0$ with $(1/k)\|U\| \leqslant \|U\|_\infty \leqslant k\|U\|$, and one obtains from (7.2.9.7) in particular

$$\|U_n\|_\infty \leqslant k^2 \|U_0\|_\infty, \qquad n = 0, 1, \ldots,$$

i.e., one has $\lim_{n\to\infty} (1/n)\|U_n\|_\infty = 0$, and hence (7.2.9.3). □

In the proof of the preceding theorem we exploited the fact that the zeros λ_i of ψ furnish particular solutions of (7.2.9.1) of the form $u_j := \lambda_i^j$, $j = 0, 1, \ldots$. The following theorem shows that one can similarly represent all solutions of (7.2.9.1) in terms of the zeros of ψ:

(7.2.9.8) Theorem. *Let the polynomial*

$$\psi(\mu) := \mu^r + a_{r-1}\mu^{r-1} + \cdots + a_0$$

have the k distinct zeros λ_i, $i = 1, 2, \ldots, k$, *with multiplicities* σ_i, $i = 1, 2, \ldots, k$, *and let* $a_0 \neq 0$. *Then for arbitrary polynomials* $p_i(t)$ *with* $\deg p_i < \sigma_i$, $i = 1, 2, \ldots, k$, *the sequence*

$$(7.2.9.9) \qquad u_j := p_1(j)\lambda_1^j + p_2(j)\lambda_2^j + \cdots + p_k(j)\lambda_k^j, \qquad j = 0, 1, \ldots,$$

is a solution of the difference equation (7.2.9.1). *Conversely, every solution of* (7.2.9.1) *can be uniquely represented in the form* (7.2.9.9).

PROOF. We only show the first part of the theorem. Since with $\{u_j\}$, $\{v_j\}$, also $\{u_j + v_j\}$ is a solution of (7.2.9.1), it suffices to show that for a σ-fold zero λ of ψ the sequence

$$u_j := p(j)\lambda^j, \qquad j = 0, 1, \ldots,$$

is a solution of (7.2.9.1) if $p(t)$ is an arbitrary polynomial with $\deg p < \sigma$. For fixed $j \geqslant 0$, let us represent $p(j + t)$ with the aid of Newton's interpolation formula (2.1.3.1) in the following form:

$$p(j + t) = \alpha_0 + \alpha_1 t + \alpha_2 t(t - 1) + \cdots + \alpha_r t(t - 1) \ldots (t - r + 1),$$

where $\alpha_\sigma = \alpha_{\sigma+1} = \cdots = \alpha_r = 0$, because deg $p < \sigma$. With the notation $a_r := 1$, we thus have

$$
\begin{aligned}
u_{j+r} &+ a_{r-1} u_{j+r-1} + \cdots + a_0 u_j \\
&= \lambda^j \sum_{\rho=0}^{r} a_\rho \lambda^\rho p(j+\rho) \\
&= \lambda^j \sum_{\rho=0}^{r} a_\rho \lambda^\rho \left[\alpha_0 + \sum_{\tau=1}^{r} \alpha_\tau \rho(\rho-1)\ldots(\rho-\tau+1)\right] \\
&= \lambda^j[\alpha_0 \psi(\lambda) + \alpha_1 \lambda \psi'(\lambda) + \cdots + \alpha_{\sigma-1} \lambda^{\sigma-1} \psi^{(\sigma-1)}(\lambda)] \\
&= 0,
\end{aligned}
$$

since λ is a σ-fold zero of ψ and thus $\psi^{(\tau)}(\lambda) = 0$ for $0 \leqslant \tau \leqslant \sigma - 1$. This proves the first part of the theorem.

Now a polynomial $p(t) = c_0 + c_1 t + \cdots + c_{\sigma-1} t^{\sigma-1}$ of degree $<\sigma$ is precisely determined by its σ coefficients c_m, $m = 0, 1, \ldots, \sigma - 1$, so that by $\sigma_1 + \sigma_2 + \cdots + \sigma_k = r$ the representation (7.2.9.9) contains a total of r free parameters, namely the coefficients of the $p_i(t)$. The second part of the theorem therefore asserts that by suitably choosing these r parameters one can obtain every solution of (7.2.9.1), i.e., that for every choice of initial values $u_0, u_1, \ldots, u_{r-1}$ there is a unique solution to the following system of r linear equations for the r unknown coefficients of the $p_i(t)$, $i = 1, \ldots, k$:

$$p_1(j)\lambda_1^j + p_2(j)\lambda_2^j + \cdots + p_k(j)\lambda_k^j = u_j, \qquad j = 0, 1, \ldots, r-1.$$

The proof of this, although elementary, is tedious. We therefore omit it.

□

7.2.10 Convergence of Multistep Methods

We now want to use the results of the previous section in order to study the convergence behavior of the multistep method (7.2.7.2). It turns out that for a consistent method [see (7.2.7.4)] the stability condition (7.2.9.5) is necessary and sufficient for the convergence of the method, provided F satisfies certain additional regularity conditions [see (7.2.10.3)].

In connection with the stability condition (7.2.9.5), note from (7.2.7.6) that for a consistent method $\psi(\mu) = \mu^r + a_{r-1}\mu^{r-1} + \cdots + a_0$ has $\lambda = 1$ as a zero.

We begin by showing that the stability condition is necessary for convergence:

(7.2.10.1) Theorem. *If the multistep method* (7.2.7.2) *is convergent* [*Definition* (7.2.7.8)] *and F obeys the condition* (7.2.7.7), $F(x; u_r, u_{r-1}, \ldots, u_0; h; 0) \equiv 0$, *then the stability condition* (7.2.9.5) *holds.*

PROOF. Since the method (7.2.7.2) is supposed to be convergent, in the sense of (7.2.7.8), for the integration of the differential equation $y' \equiv 0$, $y(x_0) = 0$, with the exact solution $y(x) \equiv 0$, it must produce an approximate solution $\eta(x; \varepsilon; h)$ satisfying

$$\lim_{h \to 0} \eta(x; \varepsilon; h) = 0$$

for all $x \in [a, b]$ and all ε with $|\varepsilon(z; h)| \leqslant \rho(h)$, $\rho(h) \to 0$ for $h \to 0$.

Let $x \neq x_0$, $x \in [a, b]$, be given. For $h = h_n = (x - x_0)/n$, $n = 1, 2, \ldots$, it follows that $x_n = x$ and $\eta(x; \varepsilon; h_n) = \eta_n$, where η_n, in view of $F(x; u_r, \ldots, u_0; h; 0) \equiv 0$, is determined by the recurrence formula

$$\eta_i = \varepsilon_i, \qquad i = 0, 1, \ldots, r-1,$$

$$\eta_{j+r} + a_{r-1}\eta_{j+r-1} + \cdots + a_0\eta_j = h_n\varepsilon_{j+r}, \qquad j = 0, 1, \ldots, n-r,$$

with $\varepsilon_i := \varepsilon(x_0 + ih_n; h_n)$. We choose $\varepsilon_{j+r} := 0$, $j = 0, 1, \ldots, n-r$, $\varepsilon_i := h_n u_i$, $i = 0, 1, \ldots, r-1$, with arbitrary constants $u_0, u_1, \ldots, u_{r-1}$. With

$$\rho(h) := |h| \max_{0 \leqslant i \leqslant r-1} |u_i|,$$

one then has

$$|\varepsilon_i| \leqslant \rho(h_n), \qquad i = 0, 1, \ldots, n,$$

and

$$\lim_{h \to 0} \rho(h) = 0.$$

Now, $\eta_n = h_n u_n$, where u_n is obtained recursively from $u_0, u_1, \ldots, u_{r-1}$ via the difference equation

$$u_{j+r} + a_{r-1}u_{j+r-1} + \cdots + a_0 u_j = 0, \qquad j = 0, 1, \ldots, n-r.$$

Since, by assumption, the method is convergent, we have

$$\lim_{n \to \infty} \eta_n = (x - x_0) \lim_{n \to \infty} \frac{u_n}{n} = 0,$$

i.e., (7.2.9.2) follows, the $u_0, u_1, \ldots, u_{r-1}$ having been chosen arbitrarily. The assertion now follows from Theorem (7.2.9.6). □

We now impose on F the additional condition of being "Lipschitz continuous" in the following sense: For every function $f \in F_1(a, b)$ there exist constants $h_0 > 0$ and M (which may depend on f) such that

(7.2.10.2)

$$|F(x; u_r, u_{r-1}, \ldots, u_0; h; f) - F(x; v_r, v_{r-1}, \ldots, v_0; h; f)| \leqslant M \sum_{i=0}^{r} |u_i - v_i|$$

for all $x \in [a, b]$, $|h| \leqslant h_0$, $u_i, v_i \in \mathbb{R}$ [cf., the analogous condition (7.2.2.4)].

We then show that for consistent methods the stability condition is also sufficient for convergence:

(7.2.10.3) Theorem. *Let the multistep method* (7.2.7.2) *be consistent in the sense of* (7.2.7.4), *and F satisfy the conditions* (7.2.7.7) *and* (7.2.10.2). *Then the method is convergent in the sense of* (7.2.7.8) *for all* $f \in F_1(a, b)$ *if and only if the stability condition* (7.2.9.5) *holds.*

PROOF. The fact that the stability condition is necessary for convergence follows from (7.2.10.1). In order to show that it is also sufficient under the stated assumptions, one proceeds, analogously to the proof of Theorem (7.2.2.3), in the following way: Let $y(x)$ be the exact solution of $y' = f(x, y)$, $y(x_0) = y_0$, $y_i := y(x_i)$, $x_i = x_0 + ih$, and η_i the solution of (7.2.7.2):

$$\eta_i = y_i + \varepsilon_i, \qquad i = 0, \ldots, r-1,$$
$$\eta_{j+r} + a_{r-1}\eta_{j+r-1} + \cdots + a_0\eta_j = hF(x_j; \eta_{j+r}, \ldots, \eta_j; h; f) + h\varepsilon_{j+r},$$

for $j = 0, 1, \ldots$, with $|\varepsilon_i| \leqslant \rho(h)$, $\lim_{h \to 0} \rho(h) = 0$.

For the error $e_i := \eta_i - y_i$ one has

$$\text{(7.2.10.4)} \qquad \begin{aligned} &e_i = \varepsilon_i, \qquad i = 0, \ldots, r-1, \\ &e_{j+r} + a_{r-1}e_{j+r-1} + \cdots + a_0 e_j = c_{j+r}, \qquad j = 0, 1, \ldots, \end{aligned}$$

where

$$\begin{aligned} c_{j+r} :=\; & h[F(x_j; \eta_{j+r}, \ldots, \eta_j; h; f) - F(x_j; y_{j+r}, \ldots, y_j; h; f)] \\ & + h(\varepsilon_{j+r} - \tau_{j+r}), \\ \tau_{j+r} :=\; & \tau(x_j, y_j; h). \end{aligned}$$

By virtue of the consistency (7.2.7.4), one has for a suitable function $\sigma(h)$,

$$|\tau_{j+r}| \leqslant \sigma(h), \qquad \lim_{h \to 0} \sigma(h) = 0,$$

and by (7.2.10.2),

$$\text{(7.2.10.5)} \qquad |c_{j+r}| \leqslant |h| M \sum_{i=0}^{r} |e_{j+i}| + |h| \cdot [\rho(h) + \sigma(h)].$$

With the help of the vectors

$$E_j := \begin{bmatrix} e_j \\ e_{j+1} \\ \cdot \\ \cdot \\ \cdot \\ e_{j+r-1} \end{bmatrix}, \qquad B := \begin{bmatrix} 0 \\ 0 \\ \vdots \\ 0 \\ 1 \end{bmatrix} \in \mathbb{R}^r$$

and the matrix

$$A := \begin{bmatrix} 0 & 1 & \cdot & \cdot & \cdot & 0 \\ \cdot & \cdot & \cdot & & & \cdot \\ \cdot & & & \cdot & \cdot & \cdot \\ \cdot & & & & \cdot & \cdot \\ 0 & \cdot & \cdot & \cdot & 0 & 1 \\ -a_0 & \cdot & \cdot & \cdot & \cdot & -a_{r-1} \end{bmatrix},$$

(7.2.10.4) can be written equivalently in the form

(7.2.10.6) $$E_{j+1} = AE_j + c_{j+r}B, \qquad E_0 := \begin{bmatrix} \varepsilon_0 \\ \vdots \\ \varepsilon_{r-1} \end{bmatrix}.$$

Since by assumption the stability condition (7.2.9.5) is valid, one can choose, by Theorem (6.9.2), a norm $\|\cdot\|$ on $\mathbb{C}^r$ with $\operatorname{lub}(A) \leqslant 1$. Now all norms on $\mathbb{C}^r$ are equivalent [see (4.4.6)], i.e., there is a constant $k > 0$ such that

$$\frac{1}{k}\|E_j\| \leqslant \sum_{i=0}^{r-1} |e_{j+i}| \leqslant k\|E_j\|.$$

Since

$$\sum_{i=0}^{r-1} |e_{j+i}| \leqslant k\|E_j\|, \qquad \sum_{i=1}^{r} |e_{j+i}| \leqslant k\|E_{j+1}\|,$$

one thus obtains from (7.2.10.5)

$$|c_{j+r}| \leqslant |h| Mk(\|E_j\| + \|E_{j+1}\|) + |h|[\rho(h) + \sigma(h)].$$

Using (7.2.10.6) and $\|B\| \leqslant k$, it follows that

(7.2.10.7)

$$(1 - |h|Mk^2)\|E_{j+1}\| \leqslant (1 + |h|Mk^2)\|E_j\| + k|h|[\rho(h) + \sigma(h)],$$
$$j = 0, 1, \ldots,$$
$$\|E_0\| \leqslant kr\rho(h).$$

For $|h| \leqslant 1/2Mk^2$, we now have $1 - |h|Mk^2 \geqslant \frac{1}{2}$ and

$$\frac{1 + |h|Mk^2}{1 - |h|Mk^2} \leqslant 1 + 4|h|Mk^2.$$

(7.2.10.7), therefore, yields for $|h| \leqslant 1/2Mk^2$

$$\|E_{j+1}\| \leqslant (1 + 4|h|Mk^2)\|E_j\| + 2k|h|[\rho(h) + \sigma(h)], \qquad j = 0, 1, \ldots,$$
$$\|E_0\| \leqslant kr\rho(h).$$

Lemma (7.2.2.2) shows that

$$\|E_n\| \leqslant e^{4n|h|Mk^2} kr\rho(h) + [\rho(h) + \sigma(h)]\frac{e^{4n|h|Mk^2} - 1}{2Mk},$$

i.e., for $x \neq x_0$, $h = h_n = (x - x_0)/n$, $|h_n| \leqslant 1/2Mk^2$, one has

$$\|E_n\| \leqslant e^{4Mk^2|x-x_0|}kr\rho(h_n) + [\rho(h_n) + \sigma(h_n)]\frac{e^{4Mk^2|x-x_0|} - 1}{2Mk}.$$

There exist, therefore, constants C_1 and C_2 independent of h such that

(7.2.10.8) $$|e_n| = |\eta(x; \varepsilon; h_n) - y(x)| \leqslant C_1\rho(h_n) + C_2\sigma(h_n)$$

for all sufficiently large n. Convergence of the method now follows in view of $\lim_{h\to 0}\rho(h) = \lim_{h\to 0}\sigma(h) = 0$. □

From (7.2.10.8) one immediately obtains the following:

(7.2.10.9) Corollary. *If in addition to the assumptions of Theorem* (7.2.10.3) *the multistep method is a method of order* p [see (7.2.7.4)], $\sigma(h) = O(h^p)$, *and* $f \in F_p(a, b)$, *then the global discretization error also satisfies*

$$\eta(x; \varepsilon; h_n) - y(x) = O(h_n^p)$$

for all $h_n = (x - x_0)/n$, *n sufficiently large, provided the errors*

$$\varepsilon_i = \varepsilon(x_0 + ih_n; h_n), \qquad i = 0, 1, \ldots, n,$$

obey an estimate

$$|\varepsilon_i| \leqslant \rho(h_n), \qquad i = 0, \ldots, n,$$

with

$$\rho(h_n) = O(h_n^p) \quad \text{as } n \to \infty.$$

7.2.11 Linear Multistep Methods

In the following sections we assume that in (7.2.7.2), aside from the starting errors ε_i, $0 \leqslant i \leqslant r - 1$, there are no further rounding errors: $\varepsilon_j = 0$ for $j \geqslant r$. Since it will always be clear from the context what starting values are being used and hence what the starting errors are, we will simplify further by writing $\eta(x; h)$ in place of $\eta(x; \varepsilon; h)$ for the approximate solution produced by the multistep method (7.2.7.2).

The most commonly used multistep methods are linear ones. For these, the function $F(x_j; \eta_{j+r}, \eta_{j+r-1}, \ldots, \eta_j; h; f)$ in (7.2.7.1) has the following form:

(7.2.11.1)

$$F(x_j; \eta_{j+r}, \ldots, \eta_j; h; f)$$
$$\equiv b_r f(x_{j+r}, \eta_{j+r}) + b_{r-1} f(x_{j+r-1}, \eta_{j+r-1}) + \cdots + b_0 f(x_j, \eta_j),$$

where $x_{j+i} := x_j + ih$, $i = 0, 1, \ldots, r$. A linear multistep method therefore is determined by listing the coefficients $a_0, \ldots, a_{r-1}, b_0, \ldots, b_r$. By means of the recurrence formula

$$\eta_{j+r} + a_{r-1}\eta_{j+r-1} + \cdots + a_0\eta_j = h[b_r f(x_{j+r}, \eta_{j+r}) + \cdots + b_0 f(x_j, \eta_j)],$$
$$x_i := x_0 + ih,$$

for each set of starting values $\eta_0, \eta_1, \ldots, \eta_{r-1}$ and for each (sufficiently small) stepsize $h \neq 0$, the method produces approximate values η_j for the values $y(x_j)$ of the exact solution $y(x)$ of an initial-value problem $y' = f(x, y)$, $y(x_0) = y_0$.

If $b_r \neq 0$, one deals with a corrector method; if $b_r = 0$, with a predictor method.

For $f \in F_1(a, b)$ every linear multistep method evidently satisfies the conditions (7.2.10.2) and (7.2.7.7). According to Theorem (7.2.10.1), therefore, the stability condition (7.2.9.5) for the polynomial

$$\psi(\mu) := \mu^r + a_{r-1}\mu^{r-1} + \cdots + a_0$$

is necessary for convergence [see (7.2.7.8)] of these methods. By Theorem (7.2.10.3), the stability condition for ψ together with consistency, (7.2.7.4), is also sufficient for convergence.

To check consistency, by Definition (7.2.7.4) one has to examine the behavior of the expression

(7.2.11.2)

$$\begin{aligned} L[z(x); h] &:= z(x + rh) + \sum_{i=0}^{r-1} a_i z(x + ih) - h \sum_{i=0}^{r} b_i f(x + ih, z(x + ih)) \\ &\equiv z(x + rh) + \sum_{i=0}^{r-1} a_i z(x + ih) - h \sum_{i=0}^{r} b_i z'(x + ih) \\ &\equiv h \cdot \tau(x, y; h) \end{aligned}$$

for the solution $z(t)$ with $z'(t) = f(t, z(t))$, $z(x) = y$, $x \in [a, b]$, $y \in \mathbb{R}$. Assuming that $z(t)$ is sufficiently often differentiable (this is the case if f has sufficiently many continuous partial derivatives), one finds by Taylor expansion of $L[z(x); h]$ in powers of h,

$$\begin{aligned} L[z(x); h] &= C_0 z(x) + C_1 h z'(x) + \cdots + C_q h^q z^{(q)}(x)(1 + O(h)) \\ &= h\tau(x, y; h). \end{aligned}$$

Here, the C_i are independent of $z(x)$ and h, and one has in particular

$$\begin{aligned} C_0 &= a_0 + a_1 + \cdots + a_{r-1} + 1, \\ C_1 &= a_1 + 2a_2 + \cdots + (r-1)a_{r-1} + r \cdot 1 - (b_0 + b_1 + \cdots + b_r). \end{aligned}$$

In terms of the polynomial $\psi(\mu)$ and the additional polynomial

(7.2.11.3) $$\chi(\mu) := b_0 + b_1\mu + \cdots + b_r\mu^r, \qquad \mu \in \mathbb{C},$$

one can write C_0 and C_1 in the form

$$C_0 = \psi(1),$$
$$C_1 = \psi'(1) - \chi(1).$$

Now

$$\tau(x, y; h) = \frac{1}{h} L[z(x); h] = \frac{C_0}{h} z(x) + C_1 z'(x) + O(h),$$

and, according to Definition (7.2.7.4), a consistent multistep method requires

$$C_0 = C_1 = 0,$$

i.e., a consistent linear multistep method has at least order 1.

In general, it has order p [see Definition (7.2.7.4)], for $f \in F_p(a, b)$, if

$$C_0 = C_1 = \cdots = C_p = 0, \qquad C_{p+1} \neq 0.$$

In addition to Theorems (7.2.10.1) and (7.2.10.3), we now have for linear multistep methods the following:

(7.2.11.4) Theorem. *A linear multistep method which is convergent is also consistent.*

PROOF. Consider the initial-value problem

$$y' = 0, \qquad y(0) = 1$$

with the exact solution $y(x) \equiv 1$. For starting values $\eta_i := 1$, $i = 0, 1, \ldots, r-1$, the method produces values η_{j+r}, $j = 0, 1, \ldots$, with

(7.2.11.5) $$\eta_{j+r} + a_{r-1}\eta_{j+r-1} + \cdots + a_0\eta_j = 0.$$

Letting $h_n := x/n$, one gets $\eta(x; h_n) = \eta_n$, and in view of the convergence of the method,

$$\lim_{n\to\infty} \eta(x; h_n) = \lim_{n\to\infty} \eta_n = y(x) = 1.$$

For $j \to \infty$ it thus follows at once from (7.2.11.5) that

$$C_0 = 1 + a_{r-1} + \cdots + a_0 = 0.$$

In order to show $C_1 = 0$, we utilize the fact that the method must converge also for the initial-value problem

$$y' = 1, \qquad y(0) = 0$$

with the exact solution $y(x) \equiv x$. We already know that $C_0 = \psi(1) = 0$. By Theorem (7.2.10.1) the stability condition (7.2.9.5) holds, hence $\lambda = 1$ is only a simple zero of ψ, i.e., $\psi'(1) \neq 0$; the constant

$$K := \frac{\chi(1)}{\psi'(1)}$$

is thus well defined. With the starting values

$$\eta_j := jhK, \qquad j = 0, 1, \ldots, r-1,$$

we have for the initial-value problem $y' = 1$, $y(0) = 0$, in view of $y(x_j) = x_j = jh$,

$$\eta_j = y(x_j) + \varepsilon_j \qquad \text{with} \qquad \varepsilon_j := jh(K-1), \qquad j = 0, 1, \ldots, r-1,$$

whereby

$$\lim_{h \to 0} \varepsilon_j = 0 \quad \text{for } j = 0, 1, \ldots, r-1.$$

The method, with these starting values, yields a sequence η_j for which

$$\text{(7.2.11.6)} \quad \eta_{j+r} + a_{r-1}\eta_{j+r-1} + \cdots + a_0\eta_j = h(b_0 + b_1 + \cdots + b_r) = h\chi(1).$$

Through substitution in (7.2.11.6), and observing $C_0 = 0$, one easily sees that

$$\eta_j = jhK \quad \text{for } all\ j.$$

Now, $\eta_n = \eta(x; h_n)$, $h_n := x/n$. By virtue of the convergence of the method, therefore,

$$x = y(x) = \lim_{n \to \infty} \eta(x; h_n) = \lim_{n \to \infty} \eta_n = \lim_{n \to \infty} nh_n K = xK.$$

Consequently $K = 1$, and thus,

$$C_1 = \psi'(1) - \chi(1) = 0.$$

□

Together with Theorems (7.2.10.1), (7.2.10.3) this yields the following:

(7.2.11.7) Theorem. *A linear multistep method is convergent for all $f \in F_1(a, b)$ if and only if it satisfies the stability condition* (7.2.9.5) *for ψ and is consistent* [i.e., $\psi(1) = 0$, $\psi'(1) - \chi(1) = 0$].

The following theorem gives a convenient means of determining the order of a linear multistep method:

(7.2.11.8) Theorem. *A linear multistep method is a method of order p if and only if the function $\varphi(\mu) := [\psi(\mu)/\ln \mu] - \chi(\mu)$ has $\mu = 1$ as a p-fold zero.*

PROOF. Put $z(x) := e^x$ in $L[z(x); h]$ of (7.2.11.2). Then, for a method of order p,

$$L[e^x; h] = C_{p+1} h^{p+1} e^x (1 + O(h)).$$

On the other hand,

$$L[e^x; h] = e^x[\psi(e^h) - h\chi(e^h)].$$

We thus have a method of order p precisely if

$$\varphi(e^h) = \frac{1}{h}[\psi(e^h) - h\chi(e^h)] = C_{p+1}h^p(1 + O(h)),$$

i.e., if $h = 0$ is a p-fold zero of $\varphi(e^h)$, or, in other words, if $\varphi(\mu)$ has the p-fold zero $\mu = 1$. □

This theorem suggests the following procedure: For given constants a_0, $a_1, \ldots, a_{r-1}$, suppose additional constants $b_0, b_1, \ldots, b_r$ are to be determined so that the resulting multistep method has the largest possible order. To this end, the function $\psi(\mu)/\ln \mu$, which for consistent methods is holomorphic in a neighborhood of $\mu = 1$, is expanded in a Taylor series about $\mu = 1$:

(7.2.11.9)
$$\frac{\psi(\mu)}{\ln \mu} = c_0 + c_1(\mu - 1) + c_2(\mu - 1)^2 + \cdots + c_{r-1}(\mu - 1)^{r-1} + c_r(\mu - 1)^r + \cdots.$$

Choosing

(7.2.11.10)
$$\begin{aligned}\chi(\mu) &:= c_0 + c_1(\mu - 1) + \cdots + c_r(\mu - 1)^r \\ &= b_0 + b_1\mu + \cdots + b_r\mu^r\end{aligned}$$

then gives rise to a corrector method of order at least $r + 1$. Taking

$$\begin{aligned}\chi(\mu) &:= c_0 + c_1(\mu - 1) + \cdots + c_{r-1}(\mu - 1)^{r-1} \\ &= b_0 + b_1\mu + \cdots + b_{r-1}\mu^{r-1} + 0 \cdot \mu^r\end{aligned}$$

one obtains a predictor method of order at least r.

In order to achieve methods of still higher order, one could think of further determining the constants $a_0, \ldots, a_{r-1}$ in such a way that in (7.2.11.9)

(7.2.11.11)
$$\begin{aligned}\psi(1) &= 1 + a_{r-1} + \cdots + a_0 = 0, \\ c_{r+1} &= c_{r+2} = \cdots = c_{2r-1} = 0.\end{aligned}$$

The choice (7.2.11.10) for $\chi(\mu)$ would then lead to a corrector method of order $2r$. Unfortunately, the methods so obtained are no longer convergent, since the polynomials ψ for which (7.2.11.11) holds no longer satisfy the stability condition (7.2.9.5). Dahlquist (1956, 1959) was able to show that an r-step method which satisfies the stability condition (7.2.9.5) has order

$$p \leqslant \begin{cases} r + 1 & \text{if } r \text{ is odd,} \\ r + 2 & \text{if } r \text{ is even} \end{cases}$$

(cf. Section 7.2.8).

EXAMPLE. The consistent method of maximum order, for $r = 2$, is obtained by setting

$$\psi(\mu) = \mu^2 - (1 + a)\mu + a = (\mu - 1)(\mu - a).$$

Taylor expansion of $\psi(\mu)/\ln \mu$ about $\mu = 1$ yields

$$\frac{\psi(\mu)}{\ln \mu} = 1 - a + \frac{3 - a}{2}(\mu - 1) + \frac{a + 5}{12}(\mu - 1)^2 - \frac{1 + a}{24}(\mu - 1)^3 + \cdots.$$

Putting

$$\chi(\mu) := 1 - a + \frac{3 - a}{2}(\mu - 1) + \frac{a + 5}{12}(\mu - 1)^2,$$

the resulting linear multistep method has order 3 for $a \neq -1$, and order 4 for $a = -1$. Since $\psi(\mu) = (\mu - 1)(\mu - a)$, the stability condition (7.2.9.5) is satisfied only for $-1 \leqslant a < 1$. In particular, for $a = 0$, one obtains

$$\psi(\mu) = \mu^2 - \mu, \qquad \chi(\mu) = \tfrac{1}{12}(5\mu^2 + 8\mu - 1).$$

This is just the Adams–Moulton method (7.2.6.6) for $q = 2$, which therefore has order 3. For $a = -1$ one obtains

$$\psi(\mu) = \mu^2 - 1, \qquad \chi(\mu) = \tfrac{1}{3}\mu^2 + \tfrac{4}{3}\mu + \tfrac{1}{3},$$

which corresponds to Milne's method (7.2.6.10) for $q = 2$ and has order 4 (see also Exercise 11).

One should not overlook that for multistep methods of order p the integration error is of order $O(h^p)$ only if the solution $y(x)$ of the differential equation is at least $p + 1$ times differentiable $[f \in F_p(a, b)]$.

7.2.12 Asymptotic Expansions of the Global Discretization Error for Linear Multistep Methods

In analogy to Section 7.2.3, one can also try to find asymptotic expansions in the stepsize h for approximate solutions generated by multistep methods. There arise, however, a number of difficulties.

To begin with, the approximate solution $\eta(x; h)$, and thus certainly also its asymptotic expansion (if it exists), will depend on the starting values used. Furthermore, it is not necessarily true that an asymptotic expansion of the form [cf. (7.2.3.3)]

(7.2.12.1)

$$\eta(x; h) = y(x) + h^p e_p(x) + h^{p+1} e_{p+1}(x) + \cdots + h^N e_N(x) + h^{N+1} E_{N+1}(x; h)$$

exists for all $h = h_n = (x - x_0)/n$, with functions $e_i(x)$ independent of h and a remainder term $E_{N+1}(x; h)$ which is bounded in h for each x.

We show this for a simple linear multistep method, the midpoint rule [see (7.2.6.9)], i.e.,

(7.2.12.2) $\quad \eta_{j+1} = \eta_{j-1} + 2hf(x_j, \eta_j), \quad x_j = x_0 + jh, \qquad j = 1, 2, \ldots.$

We will use this method to treat the initial-value problem

$$y' = -y, \qquad x_0 = 0, \qquad y_0 = y(0) = 1,$$

whose exact solution is $y(x) = e^{-x}$. We take the starting values

$$\eta_0 = 1,$$
$$\eta_1 = 1 - h$$

[η_1 is the approximate value for $y(x_1) = e^{-h}$ obtained by Euler's polygon method (7.2.1.3)]. Beginning with these starting values, the sequence $\{\eta_j\}$—and hence the function $\eta(x; h)$ for all $x \in R_h = \{x_j = jh \mid j = 0, 1, 2, \ldots\}$—is then defined by (7.2.12.2) through

$$\eta(x; h) := \eta_j = \eta_{x/h} \quad \text{if } x = x_j = jh.$$

According to (7.2.12.2), since $f(x_j, \eta_j) = -\eta_j$, the η_j satisfy the following difference equation:

$$\eta_{j+1} + 2h\eta_j - \eta_{j-1} = 0, \qquad j = 1, 2, \ldots.$$

With the help of Theorem (7.2.9.8), the η_j can be obtained explicitly: The polynomial

$$\mu^2 + 2h\mu - 1$$

has the zeros

$$\lambda_1 = \lambda_1(h) = -h + \sqrt{1 + h^2} = \sqrt{1 + h^2}\left(1 - \frac{h}{\sqrt{1 + h^2}}\right),$$

$$\lambda_2 = \lambda_2(h) = -h - \sqrt{1 + h^2} = -\sqrt{1 + h^2}\left(1 + \frac{h}{\sqrt{1 + h^2}}\right).$$

Therefore, by (7.2.9.8),

$$\text{(7.2.12.3)} \qquad \eta_j = c_1\lambda_1^j + c_2\lambda_2^j, \qquad j = 0, 1, 2, \ldots,$$

where the constants c_1, c_2 can be determined by means of the starting values $\eta_0 = 1$, $\eta_1 = 1 - h$. One finds

$$\eta_0 = 1 = c_1 + c_2,$$
$$\eta_1 = 1 - h = c_1\lambda_1 + c_2\lambda_2,$$

and thus

$$c_1 = c_1(h) = \frac{\lambda_2 - (1 - h)}{\lambda_2 - \lambda_1} = \frac{1 + \sqrt{1 + h^2}}{2\sqrt{1 + h^2}},$$

$$c_2 = c_2(h) = \frac{1 - h - \lambda_1}{\lambda_2 - \lambda_1} = \frac{h^2}{2}\,\frac{1}{\sqrt{1 + h^2} + 1 + h^2}.$$

Consequently, for $x \in R_h$, $h \neq 0$,

(7.2.12.4) $$\eta(x; h) := \eta_{x/h} = c_1(h)[\lambda_1(h)]^{x/h} + c_2(h)[\lambda_2(h)]^{x/h}.$$

One easily verifies that

$$\varphi_1(h) := c_1(h) \cdot [\lambda_1(h)]^{x/h}$$

is an analytic function of h in $|h| < 1$. Furthermore,

$$\varphi_1(h) = \varphi_1(-h),$$

since, evidently, $c_1(-h) = c_1(h)$ and $\lambda_1(-h) = 1/\lambda_1(h)$. The second term in (7.2.12.4) exhibits more complicated behavior. We have

$$c_2(h)[\lambda_2(h)]^{x/h} = (-1)^{x/h}\varphi_2(h),$$

with

$$\varphi_2(h) = c_2(h)[\lambda_1(-h)]^{x/h} = c_2(h)[\lambda_1(h)]^{-x/h}$$

an analytic function in $|h| < 1$. As above, one sees that $\varphi_2(-h) = \varphi_2(h)$. It follows that φ_1 and φ_2, for $|h| < 1$, have convergent power-series expansions of the form

$$\varphi_1(h) = u_0(x) + u_1(x)h^2 + u_2(x)h^4 + \cdots,$$
$$\varphi_2(h) = v_0(x) + v_1(x)h^2 + v_2(x)h^4 + \cdots,$$

with certain analytic functions $u_j(x)$, $v_j(x)$. The first terms in these series can easily be found from the explicit formulas for $c_i(h)$, $\lambda_i(h)$:

$$u_0(x) = e^{-x}, \qquad u_1(x) = \frac{e^{-x}}{4}[-1 + 2x],$$

$$v_0(x) = 0, \qquad v_1(x) = \frac{e^x}{4}.$$

Therefore, $\eta(x; h)$ has an expansion of the form

(7.2.12.5)

$$\eta(x; h) = y(x) + \sum_{k=1}^{\infty} h^{2k}[u_k(x) + (-1)^{x/h}v_k(x)] \quad \text{for all } h = \frac{x}{n},\ n = 1, 2, \ldots.$$

Because of the oscillating term $(-1)^{x/h}$, which depends on h, this is *not* an asymptotic expansion of the form (7.2.12.1).

Restricting the choice of h in such a way that x/h is always even, or always odd, one obtains true asymptotic expansions

(7.2.12.6)

$$\eta(x; h) = y(x) + \sum_{k=1}^{\infty} h^{2k}[u_k(x) + v_k(x)] \quad \text{for all } h = \frac{x}{2n},\ n = 1, 2, \ldots,$$

$$\eta(x; h) = y(x) + \sum_{k=1}^{\infty} h^{2k}[u_k(x) - v_k(x)] \quad \text{for all } h = \frac{x}{2n-1},\ n = 1, 2, \ldots.$$

Computing the starting value η_1 with the Runge–Kutta method (7.2.1.14) instead of Euler's method, one obtains the starting values

$$\eta_0 := 1,$$

$$\eta_1 := 1 - h + \frac{h^2}{2} - \frac{h^3}{6} + \frac{h^4}{24}.$$

For c_1 and c_2 one finds in the same way as before

$$c_1 = c_1(h) = \frac{1}{2\sqrt{1+h^2}}\left[1 + \sqrt{1+h^2} + \frac{h^2}{2} - \frac{h^3}{6} + \frac{h^4}{24}\right],$$

$$c_2 = c_2(h) = \frac{\sqrt{1+h^2} - 1 - \frac{h^2}{2} + \frac{h^3}{6} - \frac{h^4}{24}}{2\sqrt{1+h^2}}.$$

Since $c_1(h)$ and $c_2(h)$, and thus $\eta(x; h)$, are no longer even functions of h, there will be no expansion of the form (7.2.12.5) for $\eta(x; h)$, but merely an expansion of the type

$$\eta(x; h) = y(x) + \sum_{k=2}^{\infty} h^k[\tilde{u}_k(x) + (-1)^{x/h}\tilde{v}_k(x)] \quad \text{for } h = \frac{x}{n},\ n = 1, 2, \ldots.$$

The form of the asymptotic expansion thus depends critically on the starting values used.

In general, we have the following theorem due to Gragg (1965) [see Hairer and Lubich (1984) for a short proof]:

(7.2.12.7) Theorem. *Let $f \in F_{2N+2}(a, b)$, and $y(x)$ be the exact solution of the initial-value problem*

(I) $$y' = f(x, y), \quad y(x_0) = y_0, \qquad x_0 \in [a, b].$$

For $x \in R_h = \{x_0 + ih \mid i = 0, 1, 2, \ldots\}$ let $\eta(x; h)$ be defined by

$$\eta(x_0; h) := y_0,$$

(7.2.12.8) $$\eta(x_0 + h; h) := y_0 + hf(x_0, y_0),$$

$$\eta(x + h; h) := \eta(x - h; h) + 2hf(x, \eta(x; h)).$$

Then $\eta(x; h)$ has an expansion of the form

(7.2.12.9)

$$\eta(x; h) = y(x) + \sum_{k=1}^{N} h^{2k}[u_k(x) + (-1)^{(x-x_0)/h}v_k(x)] + h^{2N+2}E_{2N+2}(x; h),$$

valid for $x \in [a, b]$ and all $h = (x - x_0)/n$, $n = 1, 2, \ldots$. The functions $u_k(x)$, $v_k(x)$ are independent of h. The remainder term $E_{2N+2}(x; h)$ for fixed x remains bounded for all $h = (x - x_0)/n$, $n = 1, 2, \ldots$.

We note here explicitly that Theorem (7.2.12.7) holds also for systems of differential equations (7.0.3): f, y_0, η, y, u_k, v_k, etc. are then to be understood as vectors.

Under the assumptions of Theorem (7.2.12.7) the error $e(x; h) := \eta(x; h) - y(x)$ in first approximation is equal to

$$h^2[u_1(x) + (-1)^{(x-x_0)/h} v_1(x)].$$

In view of the term $(-1)^{(x-x_0)/h}$, it shows an oscillating behavior:

$$e(x \pm h; h) \doteq h^2[u_1(x) - (-1)^{(x-x_0)/h} v_1(x)].$$

One says, for this reason, that the midpoint rule is "weakly unstable". The principal oscillating term $(-1)^{(x-x_0)/h} v_1(x)$ can be removed by a trick. Set [Gragg (1965)]

(7.2.12.10) $\quad S(x; h) := \frac{1}{2}[\eta(x; h) + \eta(x - h; h) + hf(x, \eta(x; h))],$

where $\eta(x; h)$ is defined by (7.2.12.9). On account of (7.2.12.8), one now has

$$\eta(x + h; h) = \eta(x - h; h) + 2hf(x, \eta(x; h)),$$

and hence also

(7.2.12.11) $\quad S(x; h) = \frac{1}{2}[\eta(x; h) + \eta(x + h; h) - hf(x, \eta(x; h))].$

Addition of (7.2.12.10) and (7.2.12.11) yields

(7.2.12.12) $\quad S(x; h) = \frac{1}{2}[\eta(x; h) + \frac{1}{2}\eta(x - h; h) + \frac{1}{2}\eta(x + h; h)].$

By (7.2.12.12), one thus obtains for S an expansion of the form

$$\begin{aligned} S(x; h) = \tfrac{1}{2}\Big\{ & y(x) + \tfrac{1}{2}[y(x + h) + y(x - h)] \\ & + \sum_{k=1}^{N} h^{2k}[u_k(x) + \tfrac{1}{2}(u_k(x + h) + u_k(x - h)) \\ & + (-1)^{(x-x_0)/h}(v_k(x) - \tfrac{1}{2}(v_k(x + h) + v_k(x - h)))]\Big\} + O(h^{2N+2}). \end{aligned}$$

Expanding $y(x \pm h)$ and the coefficient functions $u_k(x \pm h)$, $v_k(x \pm h)$ in Taylor series in h, one finally obtains for $S(x; h)$ an expansion of the form

(7.2.12.13)

$$\begin{aligned} S(x; h) = y(x) & + h^2[u_1(x) + \tfrac{1}{4}y''(x)] \\ & + \sum_{k=2}^{N} h^{2k}[\tilde{u}_k(x) + (-1)^{(x-x_0)/h}\tilde{v}_k(x)] + O(h^{2N+2}), \end{aligned}$$

in which the leading error term no longer contains an oscillating factor.

7.2.13 Practical Implementation of Multistep Methods

The unpredictable behavior of solutions of differential equations forces the numerical integration to proceed with stepsizes which, in general, must vary from point to point if a prescribed error bound is to be maintained. Multistep methods which use equidistant nodes x_i and a fixed order, therefore, are not very suitable in practice. Every change of the steplength requires a recomputation of starting data (cf. Section 7.2.6) or entails a complicated interpolation process, which severely reduces the efficiency of these integration methods.

In order to construct efficient multistep methods, the equal spacing of the nodes x_i must be given up. We briefly sketch the construction of such methods. Our point of departure is again Equation (7.2.6.2), which was obtained by formal integration of $y' = f(x, y)$:

$$y(x_{p+k}) - y(x_{p-j}) = \int_{x_{p-j}}^{x_{p+k}} f(t, y(t))\, dt.$$

As in (7.2.6.3) we replace the integrand by an interpolating polynomial Q_q of degree q, but use here Newton's interpolation formula (2.1.3.4), which, for the purpose of step control, offers some advantages. With the interpolating polynomial one first obtains an approximation formula

$$\eta_{p+k} - \eta_{p-j} = \int_{x_{p-j}}^{x_{p+k}} Q_q(x)\, dx,$$

and from this the recurrence formula

$$\eta_{p+k} - \eta_{p-j} = \sum_{i=0}^{q} f[x_p, \ldots, x_{p-i}] \int_{x_{p-j}}^{x_{p+k}} \bar{Q}_i(x)\, dx,$$

with

$$\bar{Q}_i(x) = (x - x_p) \ldots (x - x_{p-i+1}), \qquad \bar{Q}_0(x) \equiv 1.$$

In the case $k = 1$, $j = 0$, $q = 0, 1, 2, \ldots$, one obtains an "explicit" formula (predictor)

$$\eta_{p+1} = \eta_p + \sum_{i=0}^{q} g_i f[x_p, \ldots, x_{p-i}], \tag{7.2.13.1}$$

where

$$g_i = \int_{x_p}^{x_{p+1}} \bar{Q}_i(x)\, dx.$$

For $k = 0$, $j = 1$, $q = 0, 1, 2, \ldots$, and replacing p by $p + 1$, there results an "implicit" formula (corrector)

$$\eta_{p+1} = \eta_p + \sum_{i=0}^{q} g_i^* f[x_{p+1}, \ldots, x_{p-i+1}], \tag{7.2.13.2}$$

with

$$g_i^* = \int_{x_p}^{x_{p+1}} (x - x_{p+1}) \dots (x - x_{p+2-i})\, dx, \qquad i > 0,$$

$$g_0^* = \int_{x_p}^{x_{p+1}} dx.$$

The approximation $\eta_{p+1} = \eta_{p+1}^{(0)}$ from the predictor formula can be used, as in Section 7.2.6, as "starting value" for the iteration on the corrector formula. However, we carry out only one iteration step, and denote the resulting approximation by $\eta_{p+1}^{(1)}$. From the difference of the two approximate values we can again derive statements about the error and stepsize. Subtracting the predictor formula from the corrector formula gives

$$\eta_{p+1}^{(1)} - \eta_{p+1}^{(0)} = \int_{x_p}^{x_{p+1}} (Q_q^{(1)}(x) - Q_q^{(0)}(x))\, dx.$$

$Q_q^{(1)}(x)$ is the interpolation polynomial of the corrector formula (7.2.13.2), involving the points

$$(x_{p-q+1}, f_{p-q+1}), \dots, (x_p, f_p), (x_{p+1}, f_{p+1}^{(1)})$$

with

$$f_{p+1}^{(1)} = f(x_{p+1}, \eta_{p+1}^{(0)}).$$

$Q_q^{(0)}(x)$ is the interpolation polynomial of the predictor formula, with points $(x_{p-q}, f_{p-q}), \dots, (x_p, f_p)$. Defining $f_{p+1}^{(0)} := Q_q^{(0)}(x_{p+1})$, then $Q_q^{(0)}(x)$ is also determined uniquely by the points

$$(x_{p-q+1}, f_{p-q+1}), \dots, (x_p, f_p), (x_{p+1}, f_{p+1}^{(0)}).$$

The difference $Q_q^{(1)}(x) - Q_q^{(0)}(x)$ therefore vanishes at the nodes $x_{p-q+1}, \dots, x_p$, and at x_{p+1} has the value

$$f_{p+1}^{(1)} - f_{p+1}^{(0)}.$$

Therefore,

$$\eta_{p+1}^{(1)} - \eta_{p+1}^{(0)} = C_q \cdot (f_{p+1}^{(1)} - f_{p+1}^{(0)}). \tag{7.2.13.3}$$

Now let $y(x)$ denote the exact solution of the differential equation $y' = f(x, y)$ with initial value $y(x_p) = \eta_p$. We assume that the "past" approximate values $\eta_p, \eta_{p-1}, \dots, \eta_{p-q}$ are "exact," and we make the additional assumption that $f(x, y(x))$ is exactly represented by the polynomial $Q_{q+1}^{(1)}(x)$ [with interpolation points $(x_{p-q}, f_{p-q}), \dots, (x_p, f_p), (x_{p+1}\, f_{p+1}^{(1)})$]. Then

$$\begin{aligned} y(x_{p+1}) &= \eta_p + \int_{x_p}^{x_{p+1}} f(x, y(x))\, dx \\ &= \eta_p + \int_{x_p}^{x_{p+1}} Q_{q+1}^{(1)}(x)\, dx. \end{aligned}$$

For the error

$$E_q := y(x_{p+1}) - \eta_{p+1}^{(1)},$$

one obtains

$$\begin{aligned} E_q &= \eta_p + \int_{x_p}^{x_{p+1}} Q_{q+1}^{(1)}(x)\,dx - \eta_p - \int_{x_p}^{x_{p+1}} Q_q^{(1)}(x)\,dx \\ &= \int_{x_p}^{x_{p+1}} (Q_{q+1}^{(1)}(x) - Q_q^{(0)}(x))\,dx - \int_{x_p}^{x_{p+1}} (Q_q^{(1)}(x) - Q_q^{(0)}(x))\,dx. \end{aligned}$$

In the first term, $Q_q^{(0)}(x)$ may be replaced by $Q_{q+1}^{(0)}(x)$ if one takes as interpolation points

$$(x_{p-q}, f_{p-q}), \ldots, (x_p, f_p), (x_{p+1}, f_{p+1}^{(0)}).$$

In analogy to (7.2.13.3) one then obtains

$$\begin{aligned} E_q &= C_{q+1}(f_{p+1}^{(1)} - f_{p+1}^{(0)}) - C_q(f_{p+1}^{(1)} - f_{p+1}^{(0)}) \\ &= (C_{q+1} - C_q)(f_{p+1}^{(1)} - f_{p+1}^{(0)}). \end{aligned}$$

If now the nodes $x_{p+1}, x_p, \ldots, x_{p-q}$ are equidistant and $h := x_{j+1} - x_j$, one has for the error

$$E_q = y(x_{p+1}) - \eta_{p+1}^{(1)} = O(h^{q+2}).$$

The further development now follows the previous pattern (cf. Section 7.2.5). Let ε be a prescribed error bound. The "old" step h_{old} is accepted as "successful" if

$$|E_q| = |C_{q+1} - C_q| \cdot |f_{p+1}^{(1)} - f_{p+1}^{(0)}| \doteq |C h_{\text{old}}^{q+2}| \leqslant \varepsilon.$$

The new step h_{new} is considered "successful" if we can make

$$|C \cdot h_{\text{new}}^{q+2}| \leqslant \varepsilon.$$

Elimination of C again yields

$$h_{\text{new}} \doteq h_{\text{old}} \left(\frac{\varepsilon}{|E_q|} \right)^{1/(q+2)}.$$

This strategy can still be combined with a change of q (variable-order method). One computes the three quantities

$$\left(\frac{\varepsilon}{|E_{q-1}|} \right)^{1/(q+1)}, \quad \left(\frac{\varepsilon}{|E_q|} \right)^{1/(q+2)}, \quad \left(\frac{\varepsilon}{|E_{q+1}|} \right)^{1/(q+3)}$$

and determines their maximum. If the first quantity is the maximum, one reduces q by 1. If the second is the maximum, one holds on to q. If the third is the maximum, one increases q by 1. The important point to note is that the quantities E_{q-1}, E_q, E_{q+1} can be computed recursively from the table of divided differences. One has

$$E_{q-1} = g_{q-1,2} f^{(1)}[x_{p+1}, x_p, \ldots, x_{p-q+1}],$$
$$E_q = g_{q,2} f^{(1)}[x_{p+1}, x_p, \ldots, x_{p-q}],$$
$$E_{q+1} = g_{q+1,2} f^{(1)}[x_{p+1}, x_p, \ldots, x_{p-q-1}],$$

where $f^{(1)}[x_{p+1}, x_p, \ldots, x_{p-i}]$ are the divided differences relative to the interpolation points $(x_{p+1}, f^{(1)}_{p+1})$, $(x_p, f_p), \ldots, (x_{p-i}, f_{p-i})$, and g_{ij} is defined by

$$g_{ij} = \int_{x_p}^{x_{p+1}} \bar{Q}_i(x)(x - x_{p+1})^{j-1}\, dx, \qquad i, j \geqslant 1.$$

The quantities g_{ij} satisfy the recursion

$$g_{ij} = (x_{p+1} - x_{p+1-i})g_{i-1,j} + g_{i-1,j+1},$$
$$j = 1, 2, \ldots, q + 2 - i, \quad i = 2, 3, \ldots, q,$$

with

$$g_{1j} = \frac{(-(x_{p+1} - x_p))^{j+1}}{j(j+1)}, \qquad j = 1, \ldots, q + 1.$$

The method discussed is "self-starting"; one begins with $q = 0$, then raises q in the next step to $q = 1$, etc. The "initialization" (cf. Section 7.2.6) required for multistep methods with equidistant nodes and fixed q becomes unnecessary.

For a more detailed study we refer to the relevant literature, for example, Hairer, Nørsett, and Wanner (1987); Shampine and Gordon (1975); Gear (1971); and Krogh (1974).

7.2.14 Extrapolation Methods for the Solution of the Initial-Value Problem

As discussed in Section 3.4, asymptotic expansions suggest the application of extrapolation methods. Particularly effective extrapolation methods result from discretization methods which have asymptotic expansions containing only even powers of h. Observe that this is the case for the midpoint rule or the modified midpoint rule [see (7.2.12.8), (7.2.12.9) or (7.2.12.10), (7.2.12.13)].

In practice, one uses especially Gragg's function $S(x; h)$ of (7.2.12.10), whose definition we repeat here because of its importance: Given the triple

(f, x_0, y_0), a real number H, and a natural number $n > 0$, define $\bar{x} := x_0 + H$, $h := H/n$. For the initial-value problem

$$y' = f(x, y), \qquad y(x_0) = y_0,$$

with exact solution $y(x)$, one then defines the function value $S(\bar{x}; h)$ in the following way:

$$\begin{aligned} &\eta_0 := y_0, \\ &\eta_1 := \eta_0 + hf(x_0, \eta_0), \qquad x_1 := x_0 + h, \\ &\text{for } j = 1, 2, \ldots, n-1: \\ &\quad \eta_{j+1} := \eta_{j-1} + 2hf(x_j, \eta_j), \qquad x_{j+1} := x_j + h, \\ &S(\bar{x}; h) := \tfrac{1}{2}[\eta_n + \eta_{n-1} + hf(x_n, \eta_n)]. \end{aligned} \tag{7.2.14.1}$$

In an extrapolation method for the computation of $y(\bar{x})$ (see Sections 3.4, 3.5), one then has to select a sequence of natural numbers

$$F = \{n_0, n_1, n_2, \ldots\}, \qquad 0 < n_0 < n_1 < n_2 < \cdots,$$

and for $h_i := H/n_i$ to compute the values $S(\bar{x}; h_i)$. Because of the oscillation term $(-1)^{(x-x_0)/h}$ in (7.2.12.13), however, F must contain only even or only odd numbers. Usually, one takes the sequence

$$F = \{2, 4, 6, 8, 12, 16, \ldots\}, \qquad n_i := 2n_{i-2} \quad \text{for } i \geqslant 3. \tag{7.2.14.2}$$

As described in Section 3.4, beginning with the 0th column, one then computes by means of interpolation formulas a tableau of numbers T_{ik}, one upward diagonal after another:

$$\begin{array}{llll} S(\bar{x}; h_0) =: T_{00} & & & \\ & T_{11} & & \\ S(\bar{x}; h_1) =: T_{10} & & T_{22} & \\ & T_{21} & & T_{33}. \\ S(\bar{x}; h_2) =: T_{20} & & T_{32} & \\ & T_{31} & & \\ S(\bar{x}; h_3) =: T_{30} & & & \end{array} \tag{7.2.14.3}$$

Here

$$T_{ik} := \tilde{T}_{ik}(0)$$

is just the value of the interpolating polynomial (rational functions would be better)

$$\tilde{T}_{ik}(h) = a_0 + a_1 h^2 + \cdots + a_k h^{2k}$$

of degree k in h^2 with $\tilde{T}_{ik}(h_j) = S(\bar{x}; h_j)$ for $j = i, i-1, \ldots, i-k$. As shown in Section 3.5, each column of (7.2.14.3) converges to $y(\bar{x})$:

$$\lim_{i \to \infty} T_{ik} = y(\bar{x}) \quad \text{for } k = 0, 1, \ldots.$$

In particular, for k fixed, the T_{ik} for $i \to \infty$ converge to $y(\bar{x})$ like a method of order $2k+2$. We have in first approximation, by virtue of (7.2.12.13) [see (3.5.6)],

$$T_{ik} - y(\bar{x}) \doteq (-1)^k h_i^2 h_{i-1}^2 \ldots h_{i-k}^2[\tilde{u}_{k+1}(\bar{x}) + \tilde{v}_{k+1}(\bar{x})].$$

One can further, as described in Section 3.5, exploit the monotonic behavior of the T_{ik} in order to compute explicit estimates of the error $T_{ik} - y(\bar{x})$.

Once a sufficiently accurate $T_{ik} =: \bar{y}$ is found, $\bar{y}$ will be accepted as an approximate value for $y(\bar{x})$. Thereafter, one can compute in the same way an approximation to $y(\bar{\bar{x}})$ at an additional point $\bar{\bar{x}} = \bar{x} + \bar{H}$ by replacing x_0, y_0, H by $\bar{x}$, $\bar{y}$, $\bar{H}$ and by solving the new initial-value problem again, just as described.

We wish to emphasize that the extrapolation method can be applied also for the solution of an initial-value problem (7.0.3), (7.0.4) for systems of n ordinary differential equations. In this case, $f(x, y)$ and $y(x)$ are vectors of functions, and y_0, η_i, and $S(\bar{x}; h)$ in (7.2.14.1) are vectors in $\mathbb{R}^n$. The asymptotic expansions (7.2.12.9) and (7.2.12.13) remain valid, and they mean that each component of $S(\bar{x}; h) \in \mathbb{R}^n$ has an asymptotic expansion of the form indicated. The elements T_{ik} of (7.2.14.3), likewise, are vectors in $\mathbb{R}^n$ and are computed as before, component by component, from the corresponding components of $S(\bar{x}; h_i)$.

In the practical realization of the method, the problem arises how large the basic steps H are to be chosen. If one chooses H too large, one must compute a very large tableau (7.2.14.3) before a sufficiently accurate T_{ik} is found; k is a large integer, and to find T_{ik} one has to compute $S(\bar{x}; h_{i-k+j})$ for $j = 0, 1, \ldots, k$. The computation of $S(\bar{x}; h_j)$ requires $n_j + 1$ evaluations of the right-hand side $f(x, y)$ of the differential equation.

For the sequence F in (7.2.14.2) chosen above, the numbers $s_i := \sum_{j=0}^{i}(n_j + 1)$, and thus the expenditure of work for a tableau with $i+1$ upward diagonals grows rapidly with i: one has $s_{i+1} \approx 1.4 s_i$.

If, on the other hand, the step H is too small, one takes unnecessarily small, and hence too many, integration steps $(x_0, y(x_0)) \to (x_0 + H, y(x_0 + H))$. It is therefore very important for the efficiency of the method that one incorporate into the method, as in Section 7.2.5, a mechanism for estimating a reasonable stepsize H. This mechanism must accomplish two things:

(1) It must guarantee that any stepsize H which is too large is reduced, so that no unnecessarily large tableau is constructed.
(2) It should propose to the user of the method (of the program) a reasonable stepsize $\bar{H}$ for the next integration step.

However, we don't want to enter into further discussion of such mechanisms and only remark that one can proceed, in principle, very much as in Section 7.2.5.

An ALGOL program for the solution of initial-value problems by means of extrapolation methods can be found in Bulirsch and Stoer (1966).

7.2.15 Comparison of Methods for Solving Initial-Value Problems

The methods we have described fall into three classes:

(1) one-step methods,
(2) multistep methods,
(3) extrapolation methods.

All methods allow a change of the steplength in each integration step; an adjustment of the respective stepsizes in each of them does not run into any inherent difficulties. The modern multistep methods are not employed with fixed orders; nor are the extrapolation methods. In extrapolation methods, for example, one can easily increase the order by appending another column to the tableau of extrapolated values. One-step methods of the Runge–Kutta–Fehlberg type, by construction, are tied to a fixed order, although methods of variable orders can also be constructed with correspondingly more complicated procedures. Research on this is currently in progress.

In an attempt to find out the advantages and disadvantages of the various integration schemes, computer programs for the methods mentioned above have been prepared with the utmost care, and extensive numerical experiments have been conducted with a large number of differential equations. The result may be described roughly as follows:

The least amount of computation, measured in evaluations of the right-hand side of the differential equation, is required by the multistep methods. In a predictor method the right-hand side of the differential equation must be evaluated only *once* per step, while in a corrector method this number is equal to the (generally small) number of iterations. The expense caused by the step control in multistep methods, however, destroys this advantage to a large extent. Multistep methods have the largest amount of overhead time. Advantages accrue particularly in cases where the right-hand side of the differential equation is built in a very complicated manner (large amount of computational work to evaluate the right-hand side). In contrast, extrapolation methods have the least amount of overhead time, but on the other hand sometimes do not react as "sensitively" as one-step methods or multistep methods to changes in the prescribed accuracy tolerance ε: often results are furnished with more correct digits than necessary. The reliability of extrapolation methods is quite high, but for modest accuracy requirements they no longer work economically (are too expensive).

For modest accuracy requirements, Runge–Kutta–Fehlberg methods with low orders of approximation p are to be preferred. Runge–Kutta–Fehlberg methods of certain orders sometimes react less sensitively to discontinuities in the right-hand side of the differential equation than multistep or extrapolation methods. It is true that in the absence of special precautions, the accuracy at a discontinuity in Runge–Kutta–Fehlberg methods is drastically reduced at first, but afterward these methods continue to function without disturbances. In certain practical problems, this can be an advantage.

None of the methods is endowed with such advantages that it can be preferred over all others (assuming that for all methods computer programs are used that were developed with the greatest care). The question of which method ought to be employed for a particular problem depends on so many factors that it cannot be discussed here in full detail; we must refer for this to the original papers, e.g., Clark (1968), Crane and Fox (1969), Hull et al. (1972), Shampine et al. (1976), Diekhoff et al. (1977).

7.2.16 Stiff Differential Equations

In the upper atmosphere, ozone decays under the influence of the radiation of the sun. This process is described by the formulas

$$O_3 + O_2 \underset{k_2}{\overset{k_1}{\rightleftharpoons}} O + 2O_2; \qquad O_3 + O \overset{k_3}{\rightarrow} 2O_2$$

of chemical reaction kinetics. The kinetic parameters $k_j, j = 1, 2, 3$ are either known from direct measurements or are determined indirectly from the observed concentrations as functions of time by solving an "inverse problem." If we denote by $y_1 = [O_3]$, $y_2 = [O]$, and $y_3 = [O_2]$ the concentration of the reacting gases, then according to a simple model of physical chemistry, the reaction can be described by the following set of ordinary differential equation [cf. Willoughby (1974)]:

$$\begin{aligned}
\dot{y}_1 &= -k_1 y_1 y_3 + k_2 y_2 y_3^2 - k_3 y_1 y_2, \\
\dot{y}_2 &= k_1 y_1 y_3 - k_2 y_2 y_3^2 - k_3 y_1 y_2, \\
\dot{y}_3 &= -k_1 y_1 y_3 + k_2 y_2 y_3^2 + k_3 y_1 y_2.
\end{aligned}$$

We simplify this system by assuming that the concentration of molecular oxygen $[O_2]$ is constant, i.e., $\dot{y}_3 = 0$, and consider the situation where initially the concentration $[O]$ of the radical O is zero. Then, after inserting the appropriate constants and rescaling of the k_j, one obtains the initial value problem

$$\begin{aligned}
\dot{y}_1 &= -y_1 - y_1 y_2^2 + 294 y_2, \qquad & y_1(0) &= 1, \\
\dot{y}_2 &= (y_1 - y_1 y_2)/98 - 3y_2, \qquad & y_2(0) &= 0,
\end{aligned} \qquad t \geq 0.$$

Typical for chemical kinetics are the widely different velocities (time scales) of the various elementary reactions. This results in reaction constants k_j and coefficients of the differential equations of very different orders of magnitudes. A linearization therefore gives linear differential equations with a matrix having a large spread of eigenvalues [see the Gershgorin theorem (6.9.4)]. One therefore has to expect a solution structure characterized by "boundary layers" and "asymptotic phases." The integration of systems of this kind gives rise to peculiar difficulties. The following example will serve as an illustration [cf. Grigorieff (1972, 1977)].

Suppose we are given the system

(7.2.16.1)
$$y_1' = \frac{\lambda_1 + \lambda_2}{2} y_1 + \frac{\lambda_1 - \lambda_2}{2} y_2,$$
$$y_2' = \frac{\lambda_1 - \lambda_2}{2} y_1 + \frac{\lambda_1 + \lambda_2}{2} y_2,$$

with constants λ_1, λ_2; $\lambda_i < 0$. The general solution is

(7.2.16.2)
$$\left.\begin{aligned} y_1(x) &= C_1 e^{\lambda_1 x} + C_2 e^{\lambda_2 x} \\ y_2(x) &= C_1 e^{\lambda_1 x} - C_2 e^{\lambda_2 x} \end{aligned}\right\} \quad x \geqslant 0,$$

where C_1, C_2 are constants of integration. If (7.2.16.1) is integrated by Euler's method [cf. (7.2.1.3)], the numerical solution can be represented in "closed form" as follows,

(7.2.16.3)
$$\eta_{1i} = C_1(1 + h\lambda_1)^i + C_2(1 + h\lambda_2)^i,$$
$$\eta_{2i} = C_1(1 + h\lambda_1)^i - C_2(1 + h\lambda_2)^i.$$

Evidently, the approximations converge to zero as $i \to \infty$ only if the steplength h is chosen small enough to have

(7.2.16.4)
$$|1 + h\lambda_1| < 1 \quad \text{and} \quad |1 + h\lambda_2| < 1.$$

Now let $|\lambda_2|$ be large compared to $|\lambda_1|$. Since $\lambda_2 < 0$, the influence of the component $e^{\lambda_2 x}$ in (7.2.16.2) is negligibly small in comparison with $e^{\lambda_1 x}$. Unfortunately, this is not true for the numerical integration. In view of (7.2.16.4), indeed, the steplength $h > 0$ must be chosen so small that

$$h < \frac{2}{|\lambda_2|}.$$

For example, if $\lambda_1 = -1$, $\lambda_2 = -1000$, we must have $h < 0.002$. Thus, even though e^{-1000x} contributes practically nothing to the solution, the factor 1000 in the exponent severely limits the step size. This behavior in the numerical solution is referred to as "stiffness." Euler's method (7.2.1.3) is hardly suitable for the numerical integration of such systems; the same can be said for the Runge–Kutta–Fehlberg methods, multistep methods, and extrapolation methods, discussed earlier.

Appropriate methods for integrating stiff differential equations can be derived from so-called *implicit* methods. As an example, we may take the *implicit Euler method*,

(7.2.16.5)
$$\eta_{i+1} = \eta_i + hf(x_{i+1}, \eta_{i+1}), \qquad i = 0, 1, 2, \ldots.$$

The "new" approximation η_{i+1} will have to be determined by iteration. The computational effort thus increases considerably.

One finds that many methods, applied with constant steplength $h > 0$ to the linear system of differential equations

(7.2.16.6)
$$y' = Ay, \qquad y(0) = y_0,$$
$$A \text{ a constant } n \times n \text{ matrix},$$

will produce a sequence η_i of approximation vectors for the solution $y(x_i)$ which satisfy a recurrence formula

(7.2.16.7)
$$\eta_{i+1} = g(hA)\eta_i.$$

The function $g(z)$ depends only on the method employed and is usually a rational function in which it is permissible to substitute a matrix for the argument.

EXAMPLE 1. For the explicit Euler method (7.2.1.3) one finds

$$\eta_{i+1} = \eta_i + hA\eta_i = (1 + hA)\eta_i, \quad \text{whence } g(z) = 1 + z;$$

for the implicit Euler method (7.2.16.5),

$$\eta_{i+1} = \eta_i + hA\eta_{i+1}, \quad \eta_{i+1} = (1 - hA)^{-1}\eta_i, \quad \text{whence } g(z) = \frac{1}{1-z}.$$

If one assumes that the matrix A in (7.2.16.6) has only eigenvalues λ_j with $\operatorname{Re} \lambda_j < 0$ [cf. (7.2.16.1)], then the solution $y(x)$ of (7.2.16.6) converges to zero as $x \to \infty$, while the discrete solution $\{\eta_i\}$, by virtue of (7.2.16.7), converges to zero as $i \to \infty$ only for those stepsizes $h > 0$ for which $|g(h\lambda_j)| < 1$ for all eigenvalues λ_j of A.

Inasmuch as the presence of eigenvalues λ_j with $\operatorname{Re} \lambda_j \ll 0$ will *not* force the employment of small stepsizes $h > 0$, a method will be suitable for the integration of stiff differential equations if it is absolutely stable in the following sense:

(7.2.16.8) Definition. A method (7.2.16.7) is called *absolutely stable* if $|g(z)| < 1$ for all z with $\operatorname{Re} z < 0$.

A more accurate description of the behavior of a method (7.2.16.7) is provided by its *region of absolute stability*, by which we mean the set

(7.2.16.9)
$$\mathfrak{M} = \{z \in \mathbb{C} \mid |g(z)| < 1\}.$$

The larger the intersection $\mathfrak{M} \cap \mathbb{C}_-$ of $\mathfrak{M}$ with the left half-plane $\mathbb{C}_- = \{z \mid \operatorname{Re} z < 0\}$, the more suitable the method is for the integration of stiff differential equations; it is absolutely stable if $\mathfrak{M}$ contains $\mathbb{C}_-$.

EXAMPLE 2. The region of absolute stability for the explicit Euler method is

$$\{z \mid |1 + z| < 1\};$$

for the implicit Euler method,

$$\{z \mid |1 - z| > 1\}.$$

The implicit Euler method is thus absolutely stable; the explicit Euler method is not.

Taking A-stability into account, one can then develop one-step, multistep, and extrapolation methods as in the previous Sections (7.2.1), (7.2.9), and (7.2.14). All these methods are implicit or semi-implicit, since only these methods have a proper rational stability function. All explicit methods considered earlier lead to polynomial stability functions, and hence cannot be A-stable. The implicit character of all stable methods for solving stiff differential equations implies that one has to solve a linear system of equations in each step at least once (semi-implicit methods), sometimes even repeatedly, resulting in Newton-type iterative methods. In general, the matrix E of these linear equations contains the Jacobian matrix $f_y = f_y(x, y)$ and usually has the form $E = I - h\gamma f_y$ for some number γ.

Extrapolation methods

In order to derive an extrapolation method for solving a stiff system of the form* $y' = f(y)$, we note that the stiff part of the solution $y(t)$ can be factored off near $t = x$ by putting $c(t) := e^{-A(t-x)}y(t)$, where $A := f_y(y(x))$. Then

$$c'(x) = \bar{f}(y(x)), \qquad \bar{f}(y) := f(y) - Ay,$$

and Euler's method (7.2.1.2), resp., the midpoint rule (7.2.6.9), lead to the approximations

$$\begin{aligned} c(x+h) &\approx y(x) + h\bar{f}(y(x)), \\ c(x+h) &\approx c(x-h) + 2h\bar{f}(y(x)). \end{aligned}$$

Using that $c(x \pm h) = e^{\mp Ah}y(x \pm h) \approx (I \mp Ah)y(x \pm h)$, we obtain a *semi-implicit midpoint rule* [cf. (7.2.12.8)]

$$\begin{aligned} \eta(x_0; h) &:= y_0, \qquad A := f_y(y_0), \\ \eta(x_0 + h; h) &:= (I - hA)^{-1}[y_0 + h\bar{f}(y_0)], \\ \eta(x + h; h) &:= (I - hA)^{-1}[(I + hA)\eta(x - h; h) + 2h\bar{f}(\eta(x; h))] \end{aligned}$$

for the computation of an approximate solution $\eta(x; h) \approx y(x)$ of the initial value problem $y' = f(y)$, $y(x_0) = y_0$. This rule was used by Bader and Deuflhard (1983) as a basis for extrapolation methods, and they applied it with great success to problems of chemical reaction kinetics.

One-step methods

In analogy to the Runge-Kutta-Fehlberg methods [see Equation (7.2.5.7) ff.], Kaps and Rentrop (1979) construct for stiff autonomous systems $y' = f(y)$ methods that are distinguished by a simple structure, efficiency, and

* Every differential equation can be reduced to this *autonomous* form: $\tilde{y}' = \tilde{f}(\tilde{x}, \tilde{y})$ is equivalent to $\begin{bmatrix} \tilde{y} \\ \tilde{x} \end{bmatrix}' = \begin{bmatrix} \tilde{f}(\tilde{x}, \tilde{y}) \\ 1 \end{bmatrix}$.

robust step control. They have been tested numerically for stiffness ratios as large as

$$\left|\frac{\lambda_{\max}}{\lambda_{\min}}\right| = 10^7 \quad \text{(with 12-digit computation).}$$

As in (7.2.5.7) one takes

$$\begin{aligned} \bar{y}_{i+1} &= \bar{y}_i + h\Phi_{\rm I}(\bar{y}_i; h), \\ \hat{y}_{i+1} &= \bar{y}_i + h\Phi_{\rm II}(\bar{y}_i; h), \end{aligned} \tag{7.2.16.10}$$

with

$$\begin{aligned} |\Delta(x, y(x); h) - \Phi_{\rm I}(y(x); h)| &\leqslant N_{\rm I} h^3, \\ |\Delta(x, y(x); h) - \Phi_{\rm II}(y(x); h) &\leqslant N_{\rm II} h^4 \end{aligned} \tag{7.2.16.11}$$

and

$$\Phi_{\rm I}(y; h) = \sum_{k=1}^{3} c_k f_k^*(y; h),$$

$$\Phi_{\rm II}(y; h) = \sum_{k=1}^{4} \hat{c}_k f_k^*(y; h),$$

where

$$\begin{aligned} f_k^* &= f_k^*(y; h) \\ &= f\left(y + h \sum_{l=1}^{4} \beta_{kl} f_l^*\right) + hf'(y) \sum_{l=1}^{4} \gamma_{kl} f_l^*, \qquad k = 1, 2, 3, 4. \end{aligned} \tag{7.2.16.12}$$

For specified constants, the f_k^* must be determined iteratively from these systems. The constants obey equations similar to those in (7.2.5.11). Kaps and Rentrop provide the following values:

$$\begin{aligned}
\gamma &= 0.220\,428\,410, \\
\gamma_{21} &= 0.822\,867\,461, \\
\gamma_{31} &= 0.695\,700\,194, & \gamma_{32} &= 0, \\
\gamma_{41} &= 3.904\,813\,42, & \gamma_{42} &= 0, & \gamma_{43} &= 1, \\
\beta_{21} &= -0.554\,591\,416, \\
\beta_{31} &= 0.252\,787\,696, & \beta_{32} &= 1, \\
\beta_{41} &= \beta_{31}, & \beta_{42} &= \beta_{32}, & \beta_{43} &= 0, \\
\hat{c}_1 &= 0.545\,211\,088, & \hat{c}_2 &= 0.301\,486\,480, \\
\hat{c}_3 &= 0.177\,064\,668, & \hat{c}_4 &= -0.237\,622\,363 \times 10^{-1},
\end{aligned}$$

$$c_1 = -0.162\,871\,035, \qquad c_2 = 1.182\,153\,60,$$
$$c_3 = -0.192\,825\,995 \times 10^{-1}.$$

Here,

$$\gamma = \gamma_{ii}, \qquad i = 1, 2, 3, 4,$$
$$\gamma_{ij} = 0 \quad \text{for} \quad i < j,$$
$$\beta_{ij} = 0 \quad \text{for} \quad i \leqslant j.$$

Step control is accomplished as in (7.2.5.17):

$$h_{\text{new}} = 0.9h \sqrt[3]{\frac{\varepsilon |h|}{|\hat{y}_{i+1} - \bar{y}_{i+1}|}}.$$

Multistep methods

It was shown by Dahlquist (1963) that there are no A-stable multistep methods of order $r > 2$ and that the implicit trapezoidal rule

$$\eta_{n+1} = \eta_n + \frac{h}{2}(f(x_n, \eta_n) + f(x_{n+1}, \eta_{n+1})), \qquad n > 0,$$

where η_1 is given by the implicit Euler method (7.2.16.5), is the A-stable method of order 2 with an error coefficient $c = -\frac{1}{12}$ of smallest modulus.

Gear (1971) also showed that the *BDF-methods* up to order $r = 6$ have good stability properties: Their stability regions (7.2.16.9) contain subsets of $\mathbb{C}_- = \{z | \operatorname{Re} z < 0\}$ having the shape sketched in Figure 13.

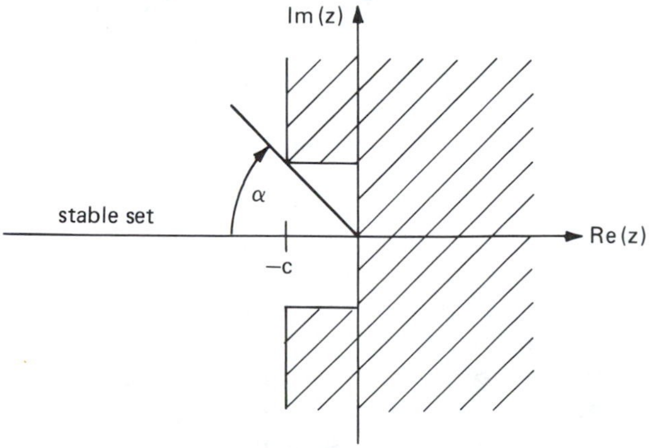

Figure 13

These multistep methods all belong to the choice [see (7.2.11.3)]

$$\chi(\mu) = b_r \mu^r$$

and can be represented by backward difference formulas, which explains their name. Coefficients of the standard representation

$$\eta_{j+r} + a_{r-1}\eta_{j+r-1} + \cdots + a_0\eta_j = hb_r f(x_{j+r}, \eta_{j+r})$$

are listed in the following table taken from Gear (1971) and Lambert (1973). In this table, $\alpha = \alpha_r$ and $c = c_r$ are the parameters indicated in Figure 13 that correspond to the particular *BDF*-method. Byrne and Hindmarsh (1987) report very favorable numerical results for these methods.

More details on stiff differential equations can be found in, e.g., Grigorieff (1972, 1977); Enright, Hull, and Lindberg (1975); and Willoughby (1974). For a recent comprehensive treatment, see Hairer and Wanner (1991).

r	α_r	c_r	b_r	a_0	a_1	a_2	a_3	a_4	a_5
1	90°	0	1	-1					
2	90°	0	$\frac{2}{3}$	$\frac{1}{3}$	$-\frac{4}{3}$				
3	88°	0.1	$\frac{6}{11}$	$-\frac{2}{11}$	$\frac{9}{11}$	$-\frac{18}{11}$			
4	73°	0.7	$\frac{12}{25}$	$\frac{3}{25}$	$-\frac{16}{25}$	$\frac{36}{25}$	$-\frac{48}{25}$		
5	51°	2.4	$\frac{60}{137}$	$-\frac{12}{137}$	$\frac{75}{137}$	$-\frac{200}{137}$	$\frac{300}{137}$	$-\frac{300}{137}$	
6	18°	6.1	$\frac{60}{147}$	$\frac{10}{147}$	$-\frac{72}{147}$	$\frac{225}{147}$	$-\frac{400}{147}$	$\frac{450}{147}$	$-\frac{360}{147}$

7.2.17 Implicit Differential Equations. Differential-Algebraic Equations

So far we have only considered explicit ordinary differential equations $y' = f(x, y)$ (7.0.1). For many modern applications this is too restrictive, and it is important, for reasons of efficiency, to deal with *implicit systems*

$$F(x, y, y') = 0$$

directly without transforming them to explicit form, which may not even be possible.

Large-scale examples of this type abound in many applications, e.g., in the design of efficient microchips for modern computers. Thousands of transistors are densely packed on these chips, so that their economical design is not possible without using efficient numerical techniques. Such a chip can be modeled as a large complicated electrical network. Kirchhoff's laws for the electrical potentials and currents in the nodes of this network lead to a huge system of ordinary differential equations for these potentials and currents as functions of time t. By solving this system numerically, the behavior of these chips, and their electrical properties, can be checked before their actual production, that is, such a chip can be simulated by a computer [see, e.g., Bank, Bulirsch, and Merten (1990); Horneber (1985)].

The resulting differential equations are implicit, and the following types, depending on the complexity of the model for a single transistor, can be distinguished:

(7.2.17.1) Linear implicit systems
$$C\dot{U}(t) = BU(t) + f(t),$$

(7.2.17.2) Linear-implicit nonlinear systems
$$C\dot{U}(t) = f(t, U(t)),$$

(7.2.17.3) Quasilinear-implicit systems
$$C(U)\dot{U}(t) = f(t, U(t)),$$

(7.2.17.4) General implicit systems
$$F(t, U(t), \dot{Q}(t)) = 0$$
$$Q(t) = C(U)U(t).$$

The vector U, which may have a very large number of components, describes the electrical potentials of the nodes of the network. The matrix C contains the capacities of the network, which may depend on the voltages, $C = C(U)$. Usually, this matrix is singular and very sparse.

There are already algorithms for solving systems of type (7.2.17.2) [cf. Deuflhard, Hairer, and Zugck (1987); Petzold (1982); Rentrop (1985); Rentrop, Roche, and Steinebach (1989)]. But the efficient and reliable solution of systems of the types (7.2.17.3) and (7.2.17.4) is still the subject of research [see, e.g., Hairer, Lubich, and Roche (1989); Petzold (1982)].

We now describe some basic differences between initial value problems for implicit and explicit systems (we again denote the independent variable by x)

(7.2.17.5) $$y' = f(x, y(x)), \qquad y(x_0) = y_0$$

treated so far.

Concerning systems of type (7.2.17.1),

(7.2.17.6) $$Ay'(x) = By(x) + f(x),$$
$$y(x_0) = y_0,$$

where $y(x) \in R^n$ and A, B are $n \times n$ matrices, the following cases can be distinguished:

1st Case. A regular, B arbitrary.

The formal solution of the associated homogeneous system is given by

$$y(x) = e^{(x-x_0)A^{-1}B}y_0.$$

Even though the system can be reduced to explicit form $y' = A^{-1}By + A^{-1}f$, this is not feasible for large sparse A in practice, since then the inverse A^{-1} usually is a large full matrix and $A^{-1}B$ would be hard to compute and store.

2nd Case. A singular, B regular.
Then (7.2.17.6) is equivalent to

$$B^{-1}Ay'(x) = y(x) + B^{-1}f(x).$$

The Jordan normal form J of the singular matrix $B^{-1}A = TJT^{-1}$ has the structure

$$J = \begin{bmatrix} W & O \\ O & N \end{bmatrix},$$

where W contains all Jordan blocks belonging to the nonzero eigenvalues of $B^{-1}A$, and N the Jordan blocks corresponding to the eigenvalue zero.

We say that N has *index* k if N is nilpotent of order k, i.e., $N^k = 0$, $N^j \neq 0$ for $j < k$. The transformation to Jordan normal form decouples the system: In terms of

$$\begin{pmatrix} u \\ v \end{pmatrix} := T^{-1}y, \qquad \begin{pmatrix} p(x) \\ q(x) \end{pmatrix} := T^{-1}B^{-1}f(x)$$

we obtain the equivalent system

$$Wu'(x) = u(x) + p(x),$$
$$Nv'(x) = v(x) + q(x).$$

The partial system for u' has the same structure as in Case 1, so it has a unique solution for any initial value y_0. This is not true for the second system for v': First, this system can only be solved under additional *smoothness assumption* for f:

$$f \in C^{k-1}[x_0, x_{\text{end}}],$$

where k is the index of N. Then

$$\begin{aligned} v(x) &= -q(x) + Nv'(x) \\ &= -(q(x) + Nq'(x)) + N^2q''(x) \\ &\;\vdots \\ &= -(q(x) + Nq'(x) + \cdots + N^{k-1}q^{(k-1)}(x)). \end{aligned}$$

Since $N^k = 0$, this resolution chain for v finally terminates, proving that $v(x)$ is entirely determined by $q(x)$ and its derivatives. Therefore, the following principal differences to (7.2.17.5) can be noted:

(1) The index of the Jordan normal form of $B^{-1}A$ determines the smoothness requirements for $f(x)$.

(2) Not all components of the initial value y_0 can be chosen arbitrarily: $v(x_0)$ is already determined by $q(x)$, and therefore by $f(x)$, and its derivatives at $x = x_0$. For the solvability of the system, the initial values $y_0 = y(x_0)$ have to satisfy a *consistency condition*, and one speaks of *consistent initial values* in

this context. However, the computation of consistent initial values may be difficult in practice.

3rd Case. A and B singular.

Here it is necessary to restrict the investigation to "meaningful" systems. Since there are matrices (the zero matrix $A := B := 0$ provides a trivial example) for which (7.2.17.6) has no unique solution, one considers only pairs of matrices (A, B) that lead to uniquely solvable systems. It is possible to characterize such pairs by means of the concept of a *regular matrix pencil*: These are pairs (A, B) of matrices such that $\lambda A + B$ is regular for some $\lambda \in \mathbb{C}$. Since then $\det(\lambda A + B)$ is a nonvanishing polynomial in λ of degree $\leq n$, there are at most n numbers λ, namely, the eigenvalues of the generalized eigenvalue problem (see 6.8) $Bx = \lambda Ax$, for which $\lambda A + B$ is singular.

It is possible to show for regular matrix pencils that (7.2.17.6) has at most one solution. Also, there is a closed formula for this solution in terms of the *Drazin-inverse* of matrices [see, e.g., Wilkinson (1982) or Gantmacher (1969)]. We will not pursue this analysis further, as it does not lead to essentially new phenomena compared with Case 2.

If the capacity matrices C of the systems (7.2.17.1) and (7.2.17.2) are singular, then, after a suitable transformation, these systems decompose into a differential equation and a nonlinear system of equations (we denote the independent variable again by x):

$$\begin{aligned} y'(x) &= f(x, y(x), z(x)), \\ 0 &= g(x, y(x), z(x)) \in \mathbb{R}^{n_2}, \qquad (7.2.17.7) \\ y(x_0) \in \mathbb{R}^{n_1},\ & z(x_0) \in \mathbb{R}^{n_2}\text{: consistent initial values.} \end{aligned}$$

Such a decomposed system is called a *differential-algebraic system* [see, e.g., Griepentrog and März (1986), Hairer and Wanner (1991)]. By the implicit function theorem, it has a unique local solution if the Jacobian

$$\left(\frac{\partial g}{\partial z}\right) \quad \text{is regular } (\textit{Index-1 assumption}).$$

Differential-algebraic systems typically occur in models of the dynamics of multibody systems; however, the Index-1 assumption may not always be satisfied [see Gear (1988)].

Numerical methods

The differential-algebraic system (7.2.17.7) can be viewed as a differential equation on a manifold. Therefore, such systems can be successfully solved by methods combining solvers for differential equations (see Section 7.2), for nonlinear equations (see Section 5.4), and continuation methods.

Extrapolation and Runge-Kutta type methods can be derived by imbedding the nonlinear equation of (7.2.17.7) formally into a differential equation

$$\varepsilon z'(x) = g(x, y(x), z(x))$$

and then considering the limiting case $\varepsilon = 0$. Efficient and reliable one-step and extrapolation methods for solving differential-algebraic systems are described in Deuflhard, Hairer, and Zugck (1987). The method (7.2.16.10)–(7.2.16.12) for solving stiff systems can be modified into a method for solving implicit systems of the form (7.2.17.2). The following method of order 4 with steps size control is described in Rentrop, Roche, and Steinebach (1989): In the notation of Section 7.2.16, and for an autonomous implicit system $Cy' = f(y)$, the method has the form [cf. (7.2.16.12)]

$$Cf_k^* = hf\left(y + \sum_{l=1}^{k-1} \beta_{kl} f_l^*\right) + hf'(y) \sum_{l=1}^{k} \gamma_{kl} f_l^*, \qquad k = 1, 2, \ldots, 5,$$

where

$$\gamma = 0.5, \qquad \gamma_{ll} = \gamma, \qquad l = 1, \ldots, 5;$$

$$\begin{array}{llll}
\gamma_{21} = 4.0 & & & \\
\gamma_{31} = 0.46\overline{296} & \gamma_{32} = -1.5 & & \\
\gamma_{41} = 2.208\overline{3} & \gamma_{42} = -1.8958\overline{3} & \gamma_{43} = -2.25 & \\
\gamma_{51} = -12.0069\overline{4} & \gamma_{52} = 7.30324\overline{074} & \gamma_{53} = 19.0 & \gamma_{54} = -15.\overline{851} \\
\beta_{21} = 0 & & & \\
\beta_{31} = 0.25 & \beta_{32} = 0.25 & & \\
\beta_{41} = -0.005208\overline{3} & \beta_{42} = 0.192708\overline{3}\ldots & \beta_{43} = 0.5625 & \\
\beta_{51} = 1.200810\overline{185} & \beta_{52} = -1.950810\overline{185} & \beta_{53} = 0.25 & \beta_{54} = 1.0 \\
c_1 = 0.53\overline{8} & c_2 = -0.13\overline{148} & c_3 = -0.2 & c_4 = 0.5\overline{925} \\
c_5 = 0.2 & & & \\
\hat{c}_1 = 0.4523\overline{148} & \hat{c}_2 = -0.1560\overline{185} & \hat{c}_3 = 0 & \hat{c}_4 = 0.5\overline{037} \\
\hat{c}_5 = 0.2. & & &
\end{array}$$

Multistep methods can also be used for the numerical solution of differential-algebraic systems. Here, we only refer to the program package DASSL [see Petzold (1982)] for solving implicit systems of the form

$$F(x, y(x), y'(x)) = 0.$$

There, the derivative $y'(x)$ is replaced by a BDF-formula, which leads to a nonlinear system of equations. Concerning details of solution strategies and their implementation, we refer to the literature, e.g., Byrne and Hindmarsh (1987), Gear and Petzold (1984), and Petzold (1982).

7.3 Boundary-Value Problems

7.3.0 Introduction

More general than initial-value problems are boundary-value problems. In these one seeks a solution $y(x)$ of a system of n ordinary differential equations,

$$(7.3.0.1a)\quad y' = f(x, y), \qquad y = \begin{bmatrix} y_1 \\ \vdots \\ y_n \end{bmatrix}, \quad f(x, y) = \begin{bmatrix} f_1(x, y_1, \ldots, y_n) \\ \vdots \\ f_n(x, y_1, \ldots, y_n) \end{bmatrix},$$

satisfying a boundary condition of the form

$$(7.3.0.1b)\qquad Ay(a) + By(b) = c.$$

Here, $a \neq b$ are given numbers, A, B square matrices of order n, and c a vector in $\mathbb{R}^n$. In practice, the boundary conditions are usually *separated*:

$$(7.3.0.1b')\qquad A_1 y(a) = c_1, \qquad B_2 y(b) = c_2,$$

i.e., in (7.3.0.1b) the rows of the matrix $[A, B, c]$ can be permuted such that for the rearranged matrix $[\bar{A}, \bar{B}, \bar{c}]$,

$$[\bar{A}, \bar{B}, \bar{c}] = \left[\begin{array}{c|c|c} A_1 & 0 & c_1 \\ \hline 0 & B_2 & c_2 \end{array}\right].$$

The boundary conditions (7.3.0.1b) are linear (more precisely, affine) in $y(a)$, $y(b)$.

Occasionally, in practice, one encounters also nonlinear boundary conditions of the type

$$(7.3.0.1b'')\qquad r(y(a), y(b)) = 0,$$

which are formed by means of a vector r of n functions r_i, $i = 1, \ldots, n$, of $2n$ variables:

$$r(u, v) \equiv \begin{bmatrix} r_1(u_1, \ldots, u_n, v_1, \ldots, v_n) \\ \vdots \\ r_n(u_1, \ldots, u_n, v_1, \ldots, v_n) \end{bmatrix}.$$

Even separated linear boundary-value problems are still very general. Thus, initial-value problems, e.g., can be thought of as special boundary-value problems of this type (with $A = I$, $c = y_0$, $B = 0$).

Whereas initial-value problems are normally uniquely solvable [see Theorem (7.1.1)], boundary-value problems can also have no solution or several solutions.

EXAMPLE. The differential equation

$$w'' + w = 0 \tag{7.3.0.2a}$$

for the real function $w: \mathbb{R} \to \mathbb{R}$, with the notation $y_1(x) := w(x)$, $y_2(x) := w'(x)$, can be written in the form (7.3.0.1a),

$$\begin{bmatrix} y_1 \\ y_2 \end{bmatrix}' = \begin{bmatrix} y_2 \\ -y_1 \end{bmatrix}.$$

It has the general solution

$$w(x) = c_1 \sin x + c_2 \cos x, \qquad c_1, c_2 \text{ arbitrary}.$$

The special solution $w(x) := \sin x$ is the only solution satisfying the boundary conditions

$$w(0) = 0, \qquad w(\pi/2) = 1. \tag{7.3.0.2b}$$

All functions $w(x) := c_1 \sin x$, with c_1 arbitrary, satisfy the boundary conditions

$$w(0) = 0, \qquad w(\pi) = 0, \tag{7.3.0.2c}$$

while there is no solution $w(x)$ of (7.3.0.2a) obeying the boundary conditions

$$w(0) = 0, \qquad w(\pi) = 1. \tag{7.3.0.2d}$$

(Observe that all boundary conditions (7.3.0.2b–d) have the form (7.3.0.1b′) with $A_1 = B_2 = [1, 0]$.)

The preceding example shows that there will be no theorem, such as (7.1.1), of similar generality, for the existence and uniqueness of solutions to boundary-value problems (see, in this connection, Section 7.3.3).

Many practically important problems can be reduced to boundary-value problems (7.3.0.1). Such is the case, e.g., for

eigenvalue problems for differential equations,

in which the right-hand side f of a system of n differential equations depends on a parameter λ,

$$y' = f(x, y, \lambda), \tag{7.3.0.3a}$$

and one has to satisfy $n + 1$ boundary conditions of the form

(7.3.0.3b)

$$r(y(a), y(b), \lambda) = 0, \qquad r(u, v, \lambda) = \begin{bmatrix} r_1(u_1, \ldots, u_n, v_1, \ldots, v_n, \lambda) \\ \vdots \\ r_{n+1}(u_1, \ldots, u_n, v_1, \ldots, v_n, \lambda) \end{bmatrix}.$$

The problem (7.3.0.3) is overdetermined and therefore, in general, has no solution for an arbitrary choice of λ. The eigenvalue problem in (7.3.0.3) consists in determining those numbers λ_i, the "eigenvalues" of (7.3.0.3), for which (7.3.0.3) does have a solution. Through the introduction of an additional function

$$y_{n+1}(x) := \lambda$$

and an additional differential equation

$$y'_{n+1}(x) = 0,$$

(7.3.0.3) is seen to be equivalent to the problem

$$\bar{y}' = \bar{f}(x, \bar{y}), \qquad \bar{r}(\bar{y}(a), \bar{y}(b)) = 0,$$

which now has the form (7.3.0.1), with

$$\bar{y} := \begin{bmatrix} y \\ y_{n+1} \end{bmatrix}, \qquad \bar{f}(x, \bar{y}) := \begin{bmatrix} f(x, y, y_{n+1}) \\ 0 \end{bmatrix},$$

$$\bar{r}(u_1, \ldots, u_n, u_{n+1}, v_1, \ldots, v_n, v_{n+1}) := r(u_1, \ldots, u_n, v_1, \ldots, v_n, v_{n+1}).$$

Furthermore, so-called

boundary-value problems with free boundary

are also reducible to (7.3.0.1). In these problems, only one boundary abscissa, a, is prescribed, while b is to be determined so that the system of n ordinary differential equations

(7.3.0.4a) $$y' = f(x, y)$$

has a solution y satisfying $n + 1$ boundary conditions

(7.3.0.4b) $$r(y(a), y(b)) = 0, \qquad r(u, v) = \begin{bmatrix} r_1(u, v) \\ \vdots \\ r_{n+1}(u, v) \end{bmatrix}.$$

Here, in place of x, one introduces a new independent variable t and a constant $z_{n+1} := b - a$, yet to be determined, by means of

$$x - a = tz_{n+1}, \qquad 0 \leqslant t \leqslant 1,$$

$$\dot{z}_{n+1} = \frac{dz_{n+1}}{dt} = 0.$$

[Instead of this choice of parameters, any substitution of the form $x - a = \Phi(t, z_{n+1})$ with $\Phi(1, z_{n+1}) = z_{n+1}$ is also suitable.] For $z(t) := y(a + tz_{n+1})$, where $y(x)$ is a solution of (7.3.0.4), one then obtains

$$\dot{z}(t) = D_t z(t) = D_x y(a + tz_{n+1})z_{n+1} = f(a + tz_{n+1}, z(t))z_{n+1},$$

and (7.3.0.4) is thus equivalent to a boundary-value problem of the type (7.3.0.1) for the functions $z_i(t)$, $i = 1, 2, \ldots, n+1$,

$$\text{(7.3.0.5)} \qquad \begin{bmatrix} \dot{z}_1 \\ \vdots \\ \dot{z}_n \\ \dot{z}_{n+1} \end{bmatrix} = \begin{bmatrix} z_{n+1} f_1(a + t z_{n+1}, z_1, \ldots, z_n) \\ \vdots \\ z_{n+1} f_n(a + t z_{n+1}, z_1, \ldots, z_n) \\ 0 \end{bmatrix},$$

$$r_i(z_1(0), \ldots, z_n(0), z_1(1), \ldots, z_n(1)) = 0, \qquad i = 1, 2, \ldots, n+1.$$

7.3.1 The Simple Shooting Method

We want to explain the *simple shooting method* first by means of an example. Suppose we are given the boundary-value problem

$$\text{(7.3.1.1)} \qquad \begin{gathered} w'' = f(x, w, w'), \\ w(a) = \alpha, \qquad w(b) = \beta, \end{gathered}$$

with separated boundary conditions. The initial-value problem

$$\text{(7.3.1.2)} \qquad w'' = f(x, w, w'), \qquad w(a) = \alpha, \quad w'(a) = s$$

in general has a uniquely determined solution $w(x) \equiv w(x; s)$ which of course depends on the choice of the initial value s for $w'(a)$. To solve the boundary-value problem (7.3.1.1), we must determine $s = \bar{s}$ so as to satisfy the second boundary condition, $w(b) = w(b; \bar{s}) = \beta$. In other words: one has to find a zero $\bar{s}$ of the function $F(s) := w(b; s) - \beta$. For every argument s the function $F(s)$ can be computed. For this, one has to determine (e.g., with the methods of Section 7.2) the value $w(b) = w(b; s)$ of the solution $w(x; s)$ of the initial-value problem (7.3.1.2) at the point $x = b$. The computation of $F(s)$ thus amounts to the solution of an initial-value problem.

For determining a zero $\bar{s}$ of $F(s)$ one can use, in principle, any method of Chapter 5. If one knows, e.g., values $s^{(0)}$, $s^{(1)}$ with

$$F(s^{(0)}) < 0, \qquad F(s^{(1)}) > 0$$

(see Figure 14), one can compute $\bar{s}$ by means of a simple *bisection method* (see Section 5.6).

Since $w(b; s)$, and hence $F(s)$, are in general [see Theorem (7.1.8)] continuously differentiable functions of s, one can also use Newton's method to determine $\bar{s}$. Starting with an initial approximation $s^{(0)}$, one then has to iteratively compute values $s^{(i)}$ according to the prescription

$$\text{(7.3.1.3)} \qquad s^{(i+1)} = s^{(i)} - \frac{F(s^{(i)})}{F'(s^{(i)})}.$$

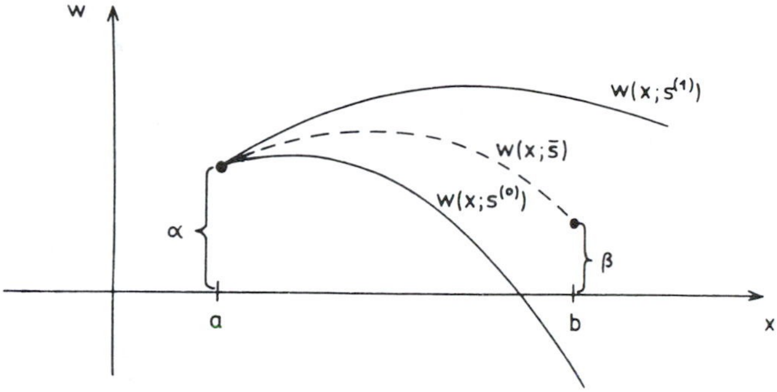

Figure 14 Simple shooting.

$w(b; s^{(i)})$, and thus $F(s^{(i)})$, can be determined by solving the initial-value problem

$$(7.3.1.4) \qquad w'' = f(x, w, w'), \qquad w(a) = \alpha, \quad w'(a) = s^{(i)}.$$

The value of the derivative of F,

$$F'(s) = \frac{\partial}{\partial s} w(b; s),$$

for $s = s^{(i)}$ can be obtained, e.g., by adjoining an additional initial-value problem: With the aid of (7.1.9) one easily verifies that the function $v(x) :\equiv v(x; s) = (\partial/\partial s) w(x; s)$ satisfies

$$(7.3.1.5) \quad v'' = f_w(x, w, w')v + f_{w'}(x, w, w')v', \qquad v(a) = 0, \quad v'(a) = 1.$$

Because of the partial derivatives $f_w, f_{w'}$, the initial-value problem (7.3.1.5) is in general substantially more complex than (7.3.1.4). For this reason, one often replaces the derivative $F'(s^{(i)})$ in Newton's formula (7.3.1.3) by a difference quotient $\Delta F(s^{(i)})$,

$$\Delta F(s^{(i)}) := \frac{F(s^{(i)} + \Delta s^{(i)}) - F(s^{(i)})}{\Delta s^{(i)}},$$

where $\Delta s^{(i)}$ is chosen "sufficiently" small. $F(s^{(i)} + \Delta s^{(i)})$ is computed, like $F(s^{(i)})$, by solving an initial-value problem. The following difficulties then arise:

If $\Delta s^{(i)}$ is chosen too large, $\Delta F(s^{(i)})$ is a poor approximation to $F'(s^{(i)})$ and the iteration

$$(7.3.1.3a) \qquad s^{(i+1)} = s^{(i)} - \frac{F(s^{(i)})}{\Delta F(s^{(i)})}$$

converges toward $\bar{s}$ considerably more slowly (if at all) than (7.3.1.3). If $\Delta s^{(i)}$ is chosen too small, then $F(s^{(i)} + \Delta s^{(i)}) \approx F(s^{(i)})$, and the subtraction $F(s^{(i)} + \Delta s^{(i)}) - F(s^{(i)})$ is subject to cancellation, so that even small errors in the calculation of $F(s^{(i)})$ and $F(s^{(i)} + \Delta s^{(i)})$ strongly impair the result $\Delta F(s^{(i)})$.

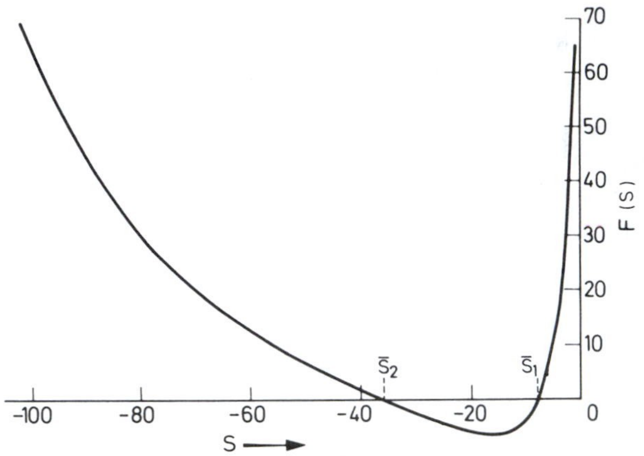

Figure 15 Graph of $F(s) = w(1; s) - 1$.

The solution of the initial-value problems (7.3.1.4), i.e., the calculation of F, therefore has to be carried out as accurately as possible. The relative error of $F(s^{(i)})$ and $F(s^{(i)} + \Delta s^{(i)})$ is only allowed to have the order of magnitude of the machine precision eps. Such accuracy can be attained with extrapolation methods (see Section 7.2.14). If $\Delta s^{(i)}$ is then further chosen so that in t-digit computation $F(s^{(i)})$ and $F(s^{(i)} + \Delta s^{(i)})$ have approximately the first $t/2$ digits in common, $\Delta s^{(i)} \, \Delta F(s^{(i)}) \approx \sqrt{\text{eps}}\, F(s^{(i)})$, then the effects of cancellation are still tolerable. As a rule, this is the case with the choice $\Delta s^{(i)} = \sqrt{\text{eps}}\, s^{(i)}$.

Example. Consider the boundary-value problem

$$w'' = \tfrac{3}{2}w^2,$$

$$w(0) = 4, \qquad w(1) = 1.$$

Following (7.3.1.2), one finds the solution of the initial-value problem

$$w'' = \tfrac{3}{2}w^2,$$

$$w(0; s) = 4, \qquad w'(0; s) = s.$$

The graph of $F(s) := w(1; s) - 1$ is shown in Figure 15. It is seen that $F(s)$ has two zeros $\bar{s}_1$, $\bar{s}_2$. The iteration according to (7.3.1.3a) yields

$$\bar{s}_1 = -8.000\,000\,0000,$$

$$\bar{s}_2 = -35.858\,548\,7278.$$

Figure 16 shows the graphs of the two solutions $w_i(x) = w(x; \bar{s}_i)$, $i = 1, 2$, of the boundary-value problem. The solutions were computed to about ten decimal digits. Both solutions, incidentally, can be expressed in closed form by

$$w(x, \bar{s}_1) = \frac{4}{(1+x)^2},$$

$$w(x, \bar{s}_2) = C_1^2\left(\frac{1 - \operatorname{cn}(C_1 x - C_2 | k^2)}{1 + \operatorname{cn}(C_1 x - C_2 | k^2)} - \frac{1}{\sqrt{3}}\right),$$

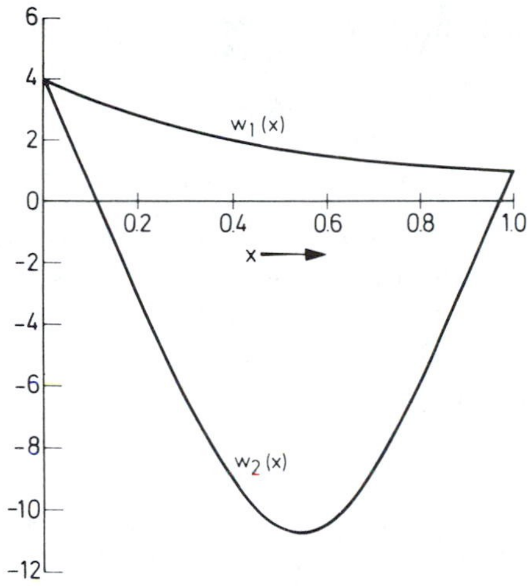

Figure 16 The solutions w_1 and w_2.

where $\mathrm{cn}(\xi \,|\, k^2)$ denotes the Jacobian elliptic function with modulus

$$k = \frac{\sqrt{2 + \sqrt{3}}}{2}.$$

From the theory of elliptic functions, and by using an iterative method of Section 5.2, one obtains for the constants of integration C_1, C_2 the values

$$C_1 = 4.303\ 10990\ \ldots,$$

$$C_2 = 2.334\ 64196\ \ldots.$$

For the solution of a general boundary-value problem (7.3.0.1a), (7.3.0.1b″) involving n unknown functions $y_i(x)$, $i = 1, \ldots, n$,

(7.3.1.6) $$y' = f(x, y), \quad r(y(a), y(b)) = 0, \qquad y = [y_1, \ldots, y_n]^T,$$

where $f(x, y)$ and $r(u, v)$ are vectors of n functions, one proceeds as in the example above. One tries again to determine a starting vector $s \in \mathbb{R}^n$ for the initial-value problem

(7.3.1.7) $$y' = f(x, y), \qquad y(a) = s$$

in such a way that the solution $y(x) = y(x; s)$ obeys the boundary conditions of (7.3.1.6),

$$r(y(a; s), y(b; s)) \equiv r(s, y(b; s)) = 0.$$

One thus has to find a solution $s = [\sigma_1, \sigma_2, \ldots, \sigma_n]^T$ of the equation

(7.3.1.8) $$F(s) = 0, \qquad F(s) :\equiv r(s, y(b; s)).$$

This can be done, e.g., by means of the general Newton's method (5.1.6)

(7.3.1.9) $$s^{(i+1)} = s^{(i)} - DF(s^{(i)})^{-1} \cdot F(s^{(i)}).$$

In each iteration step, therefore, one has to compute $F(s^{(i)})$, the Jacobian matrix

$$DF(s^{(i)}) = \left[\frac{\partial F_j}{\partial \sigma_k}\right]_{s=s^{(i)}},$$

and the solution $d^{(i)} := s^{(i)} - s^{(i+1)}$ of the linear system of equations $DF(s^{(i)})d^{(i)} = F(s^{(i)})$. For the computation of $F(s^{(i)}) = r(s^{(i)}, y(b; s^{(i)}))$ one must determine $y(b; s^{(i)})$, i.e., solve the initial value problem (7.3.1.7) for $s = s^{(i)}$. For the computation of $DF(s^{(i)})$ one observes

$$DF(s) = D_u r(s, y(b; s)) + D_v r(s, y(b; s)) \cdot Z(b; s), \tag{7.3.1.10}$$

with the matrices

$$\begin{aligned} D_u r(u, v) &= \left[\frac{\partial r_i(u, v)}{\partial u_j}\right], \qquad D_v r(u, v) = \left[\frac{\partial r_i(u, v)}{\partial v_j}\right], \\ Z(b; s) &= D_s y(b; s) = \left[\frac{\partial y_i(b; s)}{\partial \sigma_j}\right]. \end{aligned} \tag{7.3.1.11}$$

In the case of nonlinear functions $r(u, v)$, however, one will not compute $DF(s)$ by means of these complicated formulas, but instead will approximate by means of difference quotients. Thus, $DF(s)$ will be approximated by the matrix

$$\Delta F(s) = [\Delta_1 F(s), \ldots, \Delta_n F(s)],$$

where

(7.3.1.12)
$$\Delta_j F(s) = \frac{1}{\Delta\sigma_j}(F(\sigma_1, \ldots, \sigma_j + \Delta\sigma_j, \ldots, \sigma_n) - F(\sigma_1, \ldots, \sigma_j, \ldots, \sigma_n)).$$

In view of $F(s) = r(s, y(b; s))$, the calculation of $\Delta_j F(s)$ of course will require that $y(b; s) = y(b; \sigma_1, \ldots, \sigma_n)$ and $y(b; \sigma_1, \ldots, \sigma_j + \Delta\sigma_j, \ldots, \sigma_n)$ be determined through the solution of the corresponding initial-value problems.

For linear boundary conditions (7.3.0.1b),

$$r(u, v) \equiv Au + Bv - c, \qquad D_u r = A, \quad D_v r = B,$$

the formulas simplify somewhat. One has

$$\begin{aligned} F(s) &\equiv As + By(b; s) - c, \\ DF(s) &\equiv A + BZ(b; s). \end{aligned}$$

In this case, in order to form $DF(s)$, one needs to determine the matrix

$$Z(b; s) = \left[\frac{\partial y(b; s)}{\partial \sigma_1}, \ldots, \frac{\partial y(b; s)}{\partial \sigma_n}\right].$$

As just described, the jth column $\partial y(b; s)/\partial \sigma_j$ of $Z(b; s)$ is replaced by a difference quotient

$$\Delta_j y(b; s) := \frac{1}{\Delta \sigma_j}(y(b; \sigma_1, \ldots, \sigma_j + \Delta\sigma_j, \ldots, \sigma_n) - y(b; \sigma_1, \ldots, \sigma_j, \ldots, \sigma_n)).$$

One obtains the approximation

(7.3.1.13)

$$\Delta F(s) = A + B\,\Delta y(b; s), \qquad \Delta y(b; s) = [\Delta_1 y(b; s), \ldots, \Delta_n y(b; s)].$$

Therefore, to carry out the approximate Newton's method

(7.3.1.14) $$s^{(i+1)} = s^{(i)} - \Delta F(s^{(i)})^{-1} F(s^{(i)}),$$

the following has to be done:

(0) Choose a starting vector $s^{(0)}$.

For $i = 0, 1, 2, \ldots$:

(1) Determine $y(b; s^{(i)})$ by solving the initial-value problem (7.3.1.7) for $s = s^{(i)}$, and compute $F(s^{(i)}) = r(s^{(i)}, y(b; s^{(i)}))$.
(2) Choose (sufficiently small) numbers $\Delta\sigma_j \neq 0, j = 1, \ldots, n$, and determine $y(b; s^{(i)} + \Delta\sigma_j e_j)$ by solving the n initial-value problems (7.3.1.7) for $s = s^{(i)} + \Delta\sigma_j e_j = [\sigma_1^{(i)}, \ldots, \sigma_j^{(i)} + \Delta\sigma_j, \ldots, \sigma_n^{(i)}]^T, j = 1, \ldots, n$.
(3) Compute $\Delta F(s^{(i)})$ by means of (7.3.1.12) [or (7.3.1.13)] and also the solution $d^{(i)}$ of the system of linear equations $\Delta F(s^{(i)})\, d^{(i)} = F(s^{(i)})$, and put $s^{(i+1)} := s^{(i)} - d^{(i)}$.

In each step of the method one thus has to solve $n + 1$ initial-value problems and an nth-order system of linear equations.

In view of the mere local convergence of the (approximate) Newton's method (7.3.1.14), the method will in general diverge unless the starting vector $s^{(0)}$ is already sufficiently close to a solution $\bar{s}$ of $F(s) = 0$ [see Theorem (5.3.2)]. Since, as a rule, such initial values are not known, the method in the form (7.3.1.9), or (7.3.1.14), is not very useful in practice. For this reason, one replaces (7.3.1.9), or (7.3.1.14), by the modified Newton method [see (5.4.2.4)], which usually converges even for starting vectors $s^{(0)}$ that are not particularly good (if the boundary-value problem is at all solvable).

7.3.2 The Simple Shooting Method for Linear Boundary-Value Problems

By substituting $\Delta F(s^{(i)})$ for $DF(s^{(i)})$ in (7.3.1.9), one generally loses the (local) quadratic convergence of Newton's method. The substitute method (7.3.1.14) as a rule converges only linearly (locally), the rate of convergence being larger for better approximations $\Delta F(s^{(i)})$ to $DF(s^{(i)})$.

In the special case of *linear boundary-value problems*, one now has $DF(s) = \Delta F(s)$ for all s (and for arbitrary choice of the $\Delta\sigma_j$), so that (7.3.1.9) and (7.3.1.14) become identical. By a linear boundary-value problem one means a problem in which $f(x, y)$ is an affine function in y and the boundary conditions (7.3.0.1b) are linear, i.e.,

(7.3.2.1)
$$y' = T(x)y + g(x),$$
$$Ay(a) + By(b) = c,$$

with an $n \times n$ matrix $T(x)$, a function $g : \mathbb{R} \to \mathbb{R}^n$, $c \in \mathbb{R}^n$, and constant $n \times n$ matrices A and B. We assume in the following that $T(x)$ and $g(x)$ are continuous functions on $[a, b]$. By $y(x; s)$ we again denote the solution of the initial-value problem

(7.3.2.2)
$$y' = T(x)y + g(x), \qquad y(a; s) = s.$$

For $y(x; s)$ one can give an explicit formula,

(7.3.2.3)
$$y(x; s) = Y(x)s + y(x; 0),$$

where the $n \times n$ matrix $Y(x)$ is the solution of the initial-value problem

$$Y' = T(x)Y, \qquad Y(a) = I.$$

Denoting the right-hand side of (7.3.2.3) by $u(x; s)$, one indeed has

$$\begin{aligned} u(a; s) &= Y(a)s + y(a; 0) = Is + 0 = s, \\ D_x u(x; s) &= u'(x; s) = Y'(x)s + y'(x; 0) \\ &= T(x)Y(x)s + T(x)y(x; 0) + g(x) \\ &= T(x)u(x; s) + g(x), \end{aligned}$$

i.e., $u(x; s)$ is a solution of (7.3.2.2). Since, under the assumptions on $T(x)$ and $g(x)$ made above, the initial-value problem has a unique solution, it follows that $u(x; s) = y(x; s)$. Using (7.3.2.3), one obtains for the function $F(s)$ in (7.3.1.8)

(7.3.2.4)
$$F(s) = As + By(b; s) - c = [A + BY(b)]s + By(b; 0) - c.$$

Thus, $F(s)$ is also an affine function of s. Consequently [cf. (7.3.1.13)],

$$DF(s) = \Delta F(s) = A + BY(b) = \Delta F(0).$$

The solution $\bar{s}$ of $F(s) = 0$ [assuming the existence of $\Delta F(0)^{-1}$] is given by

$$\begin{aligned} \bar{s} &= -[A + BY(b)]^{-1}[By(b; 0) - c] \\ &= 0 - \Delta F(0)^{-1}F(0), \end{aligned}$$

or, slightly more generally, by

(7.3.2.5)
$$\bar{s} = s^{(0)} - \Delta F(s^{(0)})^{-1}F(s^{(0)}),$$

where $s^{(0)} \in \mathbb{R}^n$ is arbitrary. In other words, the solution $\bar{s}$ of $F(s) = 0$, and hence the solution of the linear boundary-value problem (7.3.2.1), will be produced by the method (7.3.1.14) in *one* iteration step, initiated with an arbitrary starting vector $s^{(0)}$.

7.3.3 An Existence and Uniqueness Theorem for the Solution of Boundary-Value Problems

Under very restrictive conditions one can show the unique solvability of certain boundary-value problems. To this end, we consider in the following boundary-value problems with nonlinear boundary conditions:

$$\text{(7.3.3.1)} \qquad \begin{aligned} y' &= f(x, y), \\ r(y(a), y(b)) &= 0. \end{aligned}$$

The problem given in (7.3.3.1) is solvable precisely if the function $F(s)$ in (7.3.1.8) has a zero $\bar{s}$:

$$\text{(7.3.3.2)} \qquad F(\bar{s}) = r(\bar{s}, y(b; \bar{s})) = 0.$$

The latter is certainly true if one can find a nonsingular $n \times n$ matrix Q such that

$$\text{(7.3.3.3)} \qquad \Phi(s) := s - QF(s)$$

is a contractive mapping in $\mathbb{R}^n$; the zero $\bar{s}$ of $F(s)$ is then a fixed point of Φ, $\Phi(\bar{s}) = \bar{s}$.

With the help of Theorem (7.1.8), we can now prove the following result, which for linear boundary conditions is due to Keller (1968).

(7.3.3.4) Theorem. *For the boundary-value problem* (7.3.3.1), *let the following assumptions be satisfied:*

(1) *f and $D_y f$ are continuous on $S := \{(x, y) \mid a \leqslant x \leqslant b, y \in \mathbb{R}^n\}$.*
(2) *There is a $k \in C[a, b]$ with $\|D_y f(x, y)\| \leqslant k(x)$ for all $(x, y) \in S$.*
(3) *The matrix*

$$P(u, v) := D_u r(u, v) + D_v r(u, v)$$

admits for all $u, v \in \mathbb{R}^n$ a representation of the form

$$P(u, v) = P_0(I + M(u, v))$$

with a constant nonsingular matrix P_0 and a matrix $M = M(u, v)$, and there are constants μ and m with

$$\|M(u, v)\| \leqslant \mu < 1, \qquad \|P_0^{-1} D_v r(u, v)\| \leqslant m$$

for all $u, v \in \mathbb{R}^n$.

(4) *There is a number* $\lambda > 0$ *with* $\lambda + \mu < 1$ *such that*

$$\int_a^b k(t)\,dt \leqslant \ln\left(1 + \frac{\lambda}{m}\right).$$

Then the boundary-value problem (7.3.3.1) *has exactly one solution* $y(x)$.

PROOF. We show that with a suitable choice of Q, namely $Q := P_0^{-1}$, the function $\Phi(s)$ in (7.3.3.3) satisfies

(7.3.3.5) $$\|D_s\Phi(s)\| \leqslant K < 1 \quad \text{for all } s \in \mathbb{R}^n,\ K := \lambda + \mu < 1.$$

From this it then follows at once that

$$\|\Phi(s_1) - \Phi(s_2)\| \leqslant K\|s_1 - s_2\| \quad \text{for all } s_1, s_2 \in \mathbb{R}^n,$$

i.e., Φ is a contractive mapping which, according to Theorem (5.2.3), has exactly one fixed point $\bar{s} = \Phi(\bar{s})$, which is a zero of $F(s)$, on account of the nonsingularity of Q.

For $\Phi(s) := s - P_0^{-1}r(s, y(b; s))$ one has

(7.3.3.6)

$$\begin{aligned} D_s\Phi(s) &= I - P_0^{-1}[D_u r(s, y(b; s)) + D_v r(s, y(b; s))Z(b; s)] \\ &= I - P_0^{-1}[P(s, y(b; s)) + D_v r(s, y(b; s))(Z(b; s) - I)] \\ &= I - P_0^{-1}[P_0(I + M) + D_v r(Z - I)] \\ &= -M(s, y(b; s)) - P_0^{-1}D_v r(s, y(b; s))(Z(b; s) - I), \end{aligned}$$

where the matrix

$$Z(x; s) := D_s y(x; s)$$

is the solution of the initial-value problem

$$Z' = T(x)Z, \qquad Z(a; s) = I, \quad T(x) := D_y f(x, y(x; s))$$

[see (7.1.8), (7.1.9)]. From Theorem (7.1.11), by virtue of assumption (2), there thus follows for Z the estimate

$$\|Z(b; s) - I\| \leqslant \exp\left(\int_a^b k(t)\,dt\right) - 1,$$

and further, from (7.3.3.6) and assumptions (3) and (4),

$$\begin{aligned} \|D_s\Phi(s)\| &\leqslant \mu + m\left[\exp\left(\int_a^b k(t)\,dt\right) - 1\right] \\ &\leqslant \mu + m\left[1 + \frac{\lambda}{m} - 1\right] \\ &= \mu + \lambda < 1. \end{aligned}$$

The theorem is now proved. □

Remark. The conditions of the theorem are only sufficient and are also very restrictive. Assumption (3), for example, already in the case $n = 2$, is not satisfied for such simple boundary conditions as

$$y_1(a) = c_1, \qquad y_1(b) = c_2$$

(see Exercise 20). Even though some of the assumptions, for example (3), can be weakened, one nevertheless obtains only theorems whose conditions are very rarely satisfied in practice.

7.3.4 Difficulties in the Execution of the Simple Shooting Method

From a method for solving the boundary-value problem

$$y' = f(x, y), \qquad r(y(a), y(b)) = 0$$

one will require that for each x_0 in the region of definition of a solution y it will produce an approximate value for $y(x_0)$. In the shooting method discussed up to now, only the initial value $y(a) = \bar{s}$ will be determined. One could believe that the problem is thus solved, since the value $y(x_0)$ of the solution at every other point x_0 can be (approximately) determined by treating the initial-value problem

$$\text{(7.3.4.1)} \qquad y' = f(x, y), \qquad y(a) = \bar{s},$$

say, with the methods of Section 7.2. This, however, is only true in principle. In practice, there often accrue considerable inaccuracies if the solution $y(x) = y(x; \bar{s})$ of (7.3.4.1) depends very sensitively on $\bar{s}$. An example will demonstrate this.

EXAMPLE 1. The linear system of differential equations

$$\text{(7.3.4.2)} \qquad \begin{bmatrix} y_1 \\ y_2 \end{bmatrix}' = \begin{bmatrix} 0 & 1 \\ 110 & 1 \end{bmatrix} \begin{bmatrix} y_1 \\ y_2 \end{bmatrix}$$

has the general solution

$$\text{(7.3.4.3)} \qquad y(x) = \begin{bmatrix} y_1(x) \\ y_2(x) \end{bmatrix} = c_1 e^{-10x} \begin{bmatrix} 1 \\ -10 \end{bmatrix} + c_2 e^{11x} \begin{bmatrix} 1 \\ 11 \end{bmatrix},$$

$$c_1, c_2 \text{ arbitrary.}$$

Let $y(x; s)$ be the solution of (7.3.4.2) satisfying the initial condition

$$y(0) = s = \begin{bmatrix} s_1 \\ s_2 \end{bmatrix}.$$

One verifies at once that

$$\text{(7.3.4.4)} \qquad y(x; s) = \frac{11s_1 - s_2}{21} e^{-10x} \begin{bmatrix} 1 \\ -10 \end{bmatrix} + \frac{10s_1 + s_2}{21} e^{11x} \begin{bmatrix} 1 \\ 11 \end{bmatrix}.$$

We now wish to determine the solution $y(x)$ of (7.3.4.2) which satisfies the linear separated boundary conditions

$$\text{(7.3.4.5)} \qquad \begin{bmatrix} 1 & 0 \\ 0 & 0 \end{bmatrix} y(0) + \begin{bmatrix} 0 & 0 \\ 1 & 0 \end{bmatrix} y(10) - \begin{bmatrix} 1 \\ 1 \end{bmatrix} = \begin{bmatrix} 0 \\ 0 \end{bmatrix},$$

or simpler,

$$y_1(0) = 1, \qquad y_1(10) = 1.$$

The exact solution is found from (7.3.4.3) to be

$$\text{(7.3.4.6)} \qquad y(x) = \frac{e^{110} - 1}{e^{110} - e^{-100}} e^{-10x} \begin{bmatrix} 1 \\ -10 \end{bmatrix} + \frac{1 - e^{-100}}{e^{110} - e^{-100}} e^{11x} \begin{bmatrix} 1 \\ 11 \end{bmatrix}.$$

The initial value $\bar{s} = y(0)$ of the exact solution is

$$\bar{s} = \begin{bmatrix} 1 \\ -10 + \dfrac{21(1 - e^{-100})}{e^{110} - e^{-100}} \end{bmatrix}.$$

In the computation of $\bar{s}$, e.g., in 10-digit floating-point arithmetic, instead of $\bar{s}$ one obtains at best an approximate value $\tilde{s}$ of the form

$$\tilde{s} = \mathrm{fl}(\bar{s}) = \begin{bmatrix} 1(1 + \varepsilon_1) \\ -10(1 + \varepsilon_2) \end{bmatrix},$$

with $|\varepsilon_i| \leqslant \mathrm{eps} = 10^{-10}$. Let, e.g., $\varepsilon_1 = 0$, $\varepsilon_2 = -10^{-10}$. To the initial value

$$\tilde{s} = \begin{bmatrix} 1 \\ -10 + 10^{-9} \end{bmatrix},$$

however, there belongs by (7.3.4.4) an exact solution $y(x; \tilde{s})$ with

$$y_1(10; \tilde{s}) \approx \frac{10^{-9}}{21} e^{110} \approx 2.8 \times 10^{37}.$$

The above example shows that even the computation of the initial value $\bar{s} = y(a)$ to full machine accuracy does not guarantee that additional values $y(x)$ can be determined accurately. On account of (7.3.4.4) one has in this example, for $x > 0$ sufficiently large,

$$\|y(x; s_1) - y(x; s_2)\| = O(e^{11x}) \cdot \|s_1 - s_2\|,$$

i.e., the influence of inaccurate initial data grows exponentially with x.

Theorem (7.1.4) shows that this is true in general. For the solution $y(x; s)$ of the initial-value problem $y' = f(x, y)$, $y(a; s) = s$ one has

$$\|y(x; s_1) - y(x; s_2)\| \leqslant \|s_1 - s_2\| e^{L|x-a|},$$

provided the hypotheses of Theorem (7.1.4) are satisfied.

This estimate, however, also shows that the influence of inaccurate initial data $s = y(a)$ can be made arbitrarily small by a reduction of the interval

$|x - a|$. The dilemma would thus be eliminated if one knew the values $\bar{s}_k = y(x_k)$, $k = 1, 2, \ldots, m$, of the solution $y(x)$ at *several* sufficiently close points x_k with

$$a = x_1 < x_2 < \cdots < x_m = b.$$

Choosing x_k such that $e^{L|x_{k+1} - x_k|}$ is not too large, one can determine $y(\xi)$ for each $\xi \in [a, b]$, say $\xi \in [x_k, x_{k+1}]$, to sufficient accuracy by solving the initial-value problem

$$y' = f(x, y), \qquad y(x_k) = \bar{s}_k$$

over the small interval $[x_k, \xi]$ with the methods of Section 7.2. The problem thus arises of how the value $y(x_k) = \bar{s}_k$ can be computed for a particular x_k. For this, however, as for the initial value $\bar{s} = y(a)$, one can again apply the simple shooting method. Indeed, denoting by $y(x; x_k, s)$ the solution of the initial-value problem

$$\text{(7.3.4.7)} \qquad y' = f(x, y), \qquad y(x_k) = s,$$

$\bar{s}_k$ is now a zero of the function

$$F^{(k)}(s) :\equiv r(y(a; x_k, s), y(b; x_k, s)).$$

This zero can again be determined iteratively ($s_k^{(i)} \to s_k^{(i+1)}$, $\lim_{i \to \infty} s_k^{(i)} = \bar{s}_k$), for example, by means of the (modified) Newton's method, as described in Section 7.3.1. The only difference from the simple shooting method consists in the fact that in each Newton step $s_k^{(i)} \to s_k^{(i+1)}$ for the determination of $\bar{s}_k$ one now has to compute *two* solution values, namely $y(a; x_k, s_k^{(i)})$ and $y(b; x_k, s_k^{(i)})$, through the solution of initial-value problems of the type (7.3.4.7). For $k \geqslant 2$ the value $s_k^{(0)} := y(x_k; x_{k-1}, \bar{s}_{k-1})$ suggests itself as the starting value $s_k^{(0)}$ for the Newton iteration, since theoretically $s_k^{(0)} = \bar{s}_k$ [provided that $\bar{s}_{k-1}$ is exactly equal to $y(x_{k-1})$ and $y(x_k; x_{k-1}, \bar{s}_{k-1})$ is computed without rounding errors]. Provided only that $|x_k - x_{k-1}|$ is sufficiently small, the effect on $s_k^{(0)}$ of errors in $\bar{s}_{k-1}$, according to Theorem (7.1.4), is kept within bounds, $s_k^{(0)} \approx \bar{s}_k$.

This, however, does not remove all difficulties. In the computation of the $\bar{s}_k$, $k = 1, 2, \ldots, m$, the following problem, in fact, frequently arises, which severely restricts the practical significance of the simple shooting method: The function f in the differential equation $y' = f(x, y)$, to be sure, often has continuous partial derivatives with respect to y for $\bar{x} \in [a, b]$, but $\|D_y f(x, y)\|$ is unbounded on $S = \{(x, y) \mid a \leqslant x \leqslant b,\ y \in \mathbb{R}^n\}$. In this case, the solution $y(x) = y(x; \bar{x}, s)$ of the initial-value problem $y' = f(x, y)$, $y(\bar{x}) = s$, need only be defined in a certain neighborhood $U_s(\bar{x})$ of $\bar{x}$, whose size depends on s. Thus, $y(a; \bar{x}, s)$ and $y(b; \bar{x}, s)$ possibly exist only for the values s in a small set M. Besides, M as a rule is not known. The simple shooting method, therefore, will always break down if one chooses for the starting value in Newton's method a vector $s^{(0)} \notin M$.

EXAMPLE 2. Consider the boundary-value problem [cf. Troesch (1960), (1976)]

(7.3.4.8) $$y'' = \lambda \sinh \lambda y,$$

(7.3.4.9) $$y(0) = 0, \qquad y(1) = 1$$

(λ a fixed parameter).

In order that the problem can be treated with the simple shooting method, one must first "estimate" the initial slope $y'(0) = s$. In the numerical integration of the initial-value problem (7.3.4.8) with $\lambda = 5$, $y(0) = 0$, $y'(0) = s$, it turns out that the solution $y(x; s)$ depends extremely sensitively on s: For $s = 0.1, 0.2, \ldots$ the computation breaks down even before the right-hand boundary ($x = 1$) is reached, due to exponent overflow, i.e., $y(x; s)$ has a singular point (which depends on s) at some $x_s \leqslant 1$. The effect of the initial slope $y'(0) = s$ upon the location of the singularity can be estimated in this example:

$$y'' = \lambda \sinh \lambda y$$

possesses the first integral

(7.3.4.10) $$\frac{(y')^2}{2} = \cosh \lambda y + C.$$

The conditions $y(0) = 0$, $y'(0) = s$ define the constant of integration,

$$C = \frac{s^2}{2} - 1.$$

Integration of (7.3.4.10) leads to

$$x = \frac{1}{\lambda}\int_0^{\lambda y} \frac{d\eta}{\sqrt{s^2 + 2\cosh \eta - 2}}.$$

The *singular* point is then given by

$$x_s = \frac{1}{\lambda}\int_0^{\infty} \frac{d\eta}{\sqrt{s^2 + 2\cosh \eta - 2}}.$$

For the approximate evaluation of the integral we decompose the interval of integration,

$$\int_0^\infty = \int_0^\varepsilon + \int_\varepsilon^\infty, \qquad \varepsilon > 0 \text{ arbitrary},$$

and estimate the partial integrals separately. We find

$$\begin{aligned}\int_0^\varepsilon \frac{d\eta}{\sqrt{s^2 + 2\cosh\eta - 2}} &= \int_0^\varepsilon \frac{d\eta}{\sqrt{s^2 + \eta^2 + \eta^4/12 + \cdots}} \leqslant \int_0^\varepsilon \frac{d\eta}{\sqrt{s^2 + \eta^2}} \\ &= \ln\left(\frac{\varepsilon}{|s|} + \sqrt{1 + \frac{\varepsilon^2}{s^2}}\right),\end{aligned}$$

and

$$\begin{aligned}\int_\varepsilon^\infty \frac{d\eta}{\sqrt{s^2 + 2\cosh\eta - 2}} &= \int_\varepsilon^\infty \frac{d\eta}{\sqrt{s^2 + 4\sinh^2(\eta/2)}} \leqslant \int_\varepsilon^\infty \frac{d\eta}{2\sinh(\eta/2)} \\ &= -\ln(\tanh(\varepsilon/4)).\end{aligned}$$

One thus gets the estimate

$$x_s \leqslant \frac{1}{\lambda} \ln \left(\frac{\dfrac{\varepsilon}{|s|} + \sqrt{1 + \dfrac{\varepsilon^2}{s^2}}}{\tanh \dfrac{\varepsilon}{4}} \right) =: H(\varepsilon, s).$$

For each $\varepsilon > 0$ the quantity $H(\varepsilon, s)$ is an upper bound for x_s; therefore, in particular,

$$x_s \leqslant H(\sqrt{|s|}, s) \quad \text{for all } s \neq 0.$$

The asymptotic behavior of $H(\sqrt{|s|}, s)$ for $s \to 0$ can easily be found: For small $|s|$, in fact, we have in first approximation

$$\tanh\left(\frac{\sqrt{|s|}}{4}\right) \doteq \frac{\sqrt{|s|}}{4}, \qquad \frac{1}{\sqrt{|s|}} + \sqrt{1 + \frac{1}{|s|}} \doteq \frac{2}{\sqrt{|s|}},$$

so that, asymptotically for $s \to 0$,

(7.3.4.11) $$x_s \leqslant H(\sqrt{|s|}, s) \doteq \frac{1}{\lambda} \ln\left(\frac{2/\sqrt{|s|}}{\sqrt{|s|}/4}\right) = \frac{1}{\lambda} \ln \frac{8}{|s|}.$$

[One can even show (see below) that, asymptotically for $s \to 0$,

(7.3.4.12) $$x_s \doteq \frac{1}{\lambda} \ln \frac{8}{|s|}$$

holds.]

If, in the shooting process, one is to arrive at the right-hand boundary $x = 1$, the value of $|s|$ can thus be chosen at most so large that

$$1 \mathrel{\dot{\leqslant}} \frac{1}{\lambda} \ln \frac{8}{|s|}, \quad \text{i.e., } |s| \mathrel{\dot{\leqslant}} 8e^{-\lambda};$$

for $\lambda = 5$, one obtains the small domain

(7.3.4.13) $$|s| \mathrel{\dot{\leqslant}} 0.05.$$

For the "connoisseur" we add the following: The initial-value problem

(7.3.4.14) $$y'' = \lambda \sinh \lambda y, \qquad y(0) = 0, \quad y'(0) = s$$

has the exact solution

$$y(x; s) = \frac{2}{\lambda} \sinh^{-1}\left(\frac{s}{2} \frac{\operatorname{sn}(\lambda x \mid k^2)}{\operatorname{cn}(\lambda x \mid k^2)}\right), \qquad k^2 = 1 - \frac{s^2}{4};$$

here, sn and cn are the Jacobian elliptic functions with modulus k, which here depends on the initial slope s. If $K(k^2)$ denotes the quarter period of cn, then cn has a zero at

(7.3.4.15) $$x_s = \frac{K(k^2)}{\lambda}$$

and therefore $y(x; s)$ has a logarithmic singularity there. But $K(k^2)$ has the expansion

$$K(k^2) = \ln \frac{4}{\sqrt{1 - k^2}} + \frac{1}{4}\left(\ln \frac{4}{\sqrt{1 - k^2}} - 1\right)(1 - k^2) + \cdots,$$

or, rewritten in terms of s,

$$K(k^2) = \ln \frac{8}{|s|} + \frac{s^2}{16}\left(\ln \frac{8}{|s|} - 1\right) + \cdots,$$

from which (7.3.4.12) follows.

For the solution of the actual boundary-value problem, i.e.,

$$y(1; s) = 1,$$

one finds for $\lambda = 5$ the value

$$s = 4.575\,04614 \times 10^{-2}$$

[cf. (7.3.4.13)]. This value is obtained from a relation between Jacobi functions combined with an iteration method of Section 5.1. For completeness, we add still another relation, which can be derived from the theory of Jacobi functions. The solution of the boundary-value problem (7.3.4.8), (7.3.4.9) has a logarithmic singularity at

(7.3.4.16) $$x_s \doteq 1 + \frac{1}{\lambda \cosh(\lambda/2)}.$$

For $\lambda = 5$ it follows that

$$x_s \doteq 1.0326 \ldots;$$

the singularity of the exact solution lies in the immediate neighborhood of the right-hand boundary point. The example sufficiently illuminates the difficulties that can arise in the numerical solution of boundary value problems.

For a direct numerical solution of the problem without knowledge of the theory of elliptic functions, see Section 7.3.6.

A further example for the occurrence of singularities can be found in Exercise 19.

7.3.5 The Multiple Shooting Method

The multiple shooting method has been described repeatedly in the literature, for example in Keller (1968), Osborne (1969), and Bulirsch (1971). A FORTRAN program can be found in Oberle and Grimm (1989).

In a multiple shooting method, the values

$$\bar{s}_k = y(x_k), \qquad k = 1, 2, \ldots, m,$$

of the exact solution $y(x)$ of a boundary-value problem

(7.3.5.1) $$y' = f(x, y), \qquad r(y(a), y(b)) = 0,$$

at several points

$$a = x_1 < x_2 < \cdots < x_m = b$$

are computed simultaneously by iteration. To this end, let $y(x; x_k, s_k)$ be the solution of the initial-value problem

$$y' = f(x, y), \qquad y(x_k) = s_k.$$

The problem now consists in determining the vectors s_k, $k = 1, 2, \ldots, m$, in such a way that the function

$$y(x) := y(x; x_k, s_k) \quad \text{for } x \in [x_k, x_{k+1}), \qquad k = 1, 2, \ldots, m-1,$$
$$y(b) := s_m,$$

pieced together by the $y(x; x_k, s_k)$, is continuous, and thus a solution of the differential equation $y' = f(x, y)$, and in addition satisfies the boundary conditions $r(y(a), y(b)) = 0$ (see Figure 17).

This yields the following nm conditions:

$$\text{(7.3.5.2)} \qquad \begin{aligned} y(x_{k+1}; x_k, s_k) &= s_{k+1}, \qquad k = 1, 2, \ldots, m-1, \\ r(s_1, s_m) &= 0 \end{aligned}$$

for the nm unknown components σ_{kj}, $j = 1, 2, \ldots, n$, $k = 1, 2, \ldots, m$, of the s_k:

$$s_k = [\sigma_{k1}, \sigma_{k2}, \ldots, \sigma_{kn}]^T.$$

Altogether, (7.3.5.2) represents a system of equations of the form

$$\text{(7.3.5.3)} \quad F(s) := \begin{bmatrix} F_1(s_1, s_2) \\ F_2(s_2, s_3) \\ \vdots \\ F_{m-1}(s_{m-1}, s_m) \\ F_m(s_1, s_m) \end{bmatrix} := \begin{bmatrix} y(x_2; x_1, s_1) - s_2 \\ y(x_3; x_2, s_2) - s_3 \\ \vdots \\ y(x_m; x_{m-1}, s_{m-1}) - s_m \\ r(s_1, s_m) \end{bmatrix} = 0$$

in the unknowns

$$s = \begin{bmatrix} s_1 \\ \vdots \\ s_m \end{bmatrix}.$$

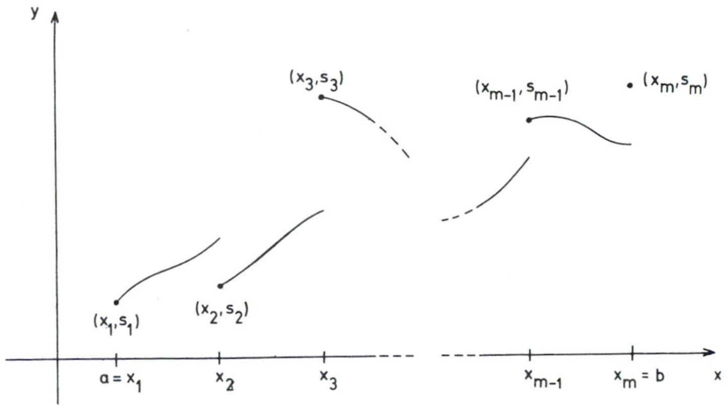

Figure 17 Multiple shooting.

It can be solved iteratively with the help of Newton's method,

$$s^{(i+1)} = s^{(i)} - [DF(s^{(i)})]^{-1} F(s^{(i)}), \qquad i = 0, 1, \ldots. \tag{7.3.5.4}$$

In order to still induce convergence if at all possible, even for a poor choice of the starting vector $s^{(0)}$, instead of (7.3.5.4) one takes in practice the modified Newton method (5.4.2.4) (see Section 7.3.6 for further indications concerning the implementation of the method). In each step of the method one must compute $F(s)$ and $DF(s)$ for $s = s^{(i)}$. For the computation of $F(s)$ one has to determine the values $y(x_{k+1}; x_k, s_k)$ for $k = 1, 2, \ldots, m-1$ by solving the initial-value problems

$$y' = f(x, y), \qquad y(x_k) = s_k,$$

and to compute $F(s)$ according to (7.3.5.3). The Jacobian matrix

$$DF(s) = [D_{s_k} F_i(s)]_{i,\,k=1,\ldots,m},$$

in view of the special structure of the F_i in (7.3.5.3), has the form

$$DF(s) = \begin{bmatrix} G_1 & -I & 0 & & 0 \\ 0 & G_2 & -I & \ddots & \\ & \ddots & \ddots & \ddots & 0 \\ 0 & & \ddots & G_{m-1} & -I \\ A & 0 & & 0 & B \end{bmatrix}, \tag{7.3.5.5}$$

where the $n \times n$ matrices A, B, G_k, $k = 1, \ldots, m-1$, in turn are Jacobian matrices,

(7.3.5.6)

$$\begin{aligned} G_k &:\equiv D_{s_k} F_k(s) \equiv D_{s_k} y(x_{k+1}; x_k, s_k), \qquad k = 1, 2, \ldots, m-1, \\ B &:\equiv D_{s_m} F_m(s) \equiv D_{s_m} r(s_1, s_m), \\ A &:\equiv D_{s_1} F_m(s) \equiv D_{s_1} r(s_1, s_m). \end{aligned}$$

As already described in the simple shooting method, it is expedient in practice to again replace the differential quotients in the matrices A, B, G_k by difference quotients, which can be computed by solving additional $(m-1)n$ initial-value problems (n initial-value problems for each matrix $G_1, \ldots, G_{m-1}$). The computation of $s^{(i+1)}$ from $s^{(i)}$ according to (7.3.5.4) can be carried out as follows: With the abbreviations

$$\begin{bmatrix} \Delta s_1 \\ \vdots \\ \Delta s_m \end{bmatrix} := s^{(i+1)} - s^{(i)}, \qquad F_k := F_k(s_k^{(i)}, s_{k+1}^{(i)}), \tag{7.3.5.7}$$

(7.3.5.4) (or the slightly modified substitute problem) is equivalent to the following system of linear equations:

$$
\begin{aligned}
G_1\,\Delta s_1 - \Delta s_2 &= -F_1,\\
G_2\,\Delta s_2 - \Delta s_3 &= -F_2,\\
&\ \vdots\\
G_{m-1}\,\Delta s_{m-1} - \Delta s_m &= -F_{m-1},\\
A\,\Delta s_1 + B\,\Delta s_m &= -F_m.
\end{aligned}
\tag{7.3.5.8}
$$

Beginning with the first equation, one can express all Δs_k successively in terms of Δs_1. One thus finds

$$
\begin{aligned}
\Delta s_2 &= G_1\,\Delta s_1 + F_1,\\
&\ \vdots\\
\Delta s_m &= G_{m-1}G_{m-2}\ldots G_1\,\Delta s_1 + \sum_{j=1}^{m-1}\left(\prod_{l=j+1}^{m-1} G_l\right)F_j,
\end{aligned}
\tag{7.3.5.9}
$$

and from this finally, by means of the last equation,

$$(A + BG_{m-1}G_{m-2}\ldots G_1)\,\Delta s_1 = w, \tag{7.3.5.10}$$

where $w = -(F_m + BF_{m-1} + BG_{m-1}F_{m-2} + \cdots + BG_{m-1}G_{m-2}\ldots G_2F_1)$.

This is a system of linear equations for the unknown vector Δs_1, which can be solved by means of Gauss elimination. Once Δs_1 is determined, one obtains $\Delta s_2, \Delta s_3, \ldots, \Delta s_m$ successively from (7.3.5.8) and $s^{(i+1)}$ from (7.3.5.7).

We further remark that under the assumptions of Theorem (7.3.3.4) one can show that the matrix $A + BG_{m-1}\ldots G_1$ in (7.3.5.10) is nonsingular. Furthermore, there again will be a nonsingular $nm \times nm$ matrix Q such that the function

$$\Phi(s) := s - QF(s)$$

is contracting, and thus has exactly one fixed point $\bar{s}$ with $F(\bar{s}) = 0$.

It is seen, in addition, that $F(s)$, and essentially also $DF(s)$, is defined for all vectors

$$s = \begin{bmatrix} s_1 \\ \vdots \\ s_m \end{bmatrix} \in M := M^{(1)} \times M^{(2)} \times \cdots \times M^{(m-1)} \times \mathbb{R}^n,$$

so that the iteration (7.3.5.4) of the multiple shooting method can be executed for $s \in M$. Here $M^{(k)}$, $k = 1, 2, \ldots, m-1$, is the set of all vectors s_k for which the solution $y(x; x_k, s_k)$ exists (at least) on the small interval $[x_k, x_{k+1}]$. This set $M^{(k)}$ includes the set M_k of all s_k for which $y(x; x_k, s_k)$ exists on all of $[a, b]$. Now Newton's method for computing $\bar{s}_k$ by means of the simple shooting method can only be executed for $s_k \in M_k \subset M^{(k)}$. This shows that the demands of the multiple shooting method upon the quality of the starting vectors in Newton's method are considerably more modest than those of the simple shooting method.

7.3.6 Hints for the Practical Implementation of the Multiple Shooting Method

The multiple shooting method in the form described in the previous section is still rather expensive. The iteration in the modified Newton's method (5.4.2.4), for example, has the form

$$s^{(i+1)} = s^{(i)} - \lambda_i d^{(i)}, \qquad d^{(i)} := [\Delta F(s^{(i)})]^{-1} F(s^{(i)}), \tag{7.3.6.1}$$

and in each step one has to compute at least the approximation $\Delta F(s^{(i)})$ of the Jacobian matrix $DF(s^{(i)})$ by forming appropriate difference quotients. The computation of $\Delta F(s^{(i)})$ alone amounts to the solution of n initial-value problems. This enormous amount of work can be substantially reduced by means of the techniques described in Section 5.4.3, recomputing the matrix $\Delta F(s^{(i)})$ only occasionally, and employing the rank-one procedure of Broyden in all remaining iteration steps in order to obtain approximations for $\Delta F(s^{(i)})$.

We next wish to consider the problem of how to choose the intermediate points x_k, $k = 1, \ldots, m$, in $[a, b]$, provided an approximate solution (*starting trajectory*) $\eta(x)$ is known for the boundary-value problem. Put $x_1 := a$. Having already chosen $x_i (< b)$, integrate the initial-value problem

$$\eta_i' = f(x, \eta_i), \qquad \eta_i(x_i) = \eta(x_i)$$

by means of the methods of Section 7.2, and terminate the integration at the first point $x = \xi$ for which the solution $\|\eta_i(\xi)\|$ becomes "too large" compared to $\|\eta(\xi)\|$, say $\|\eta_i(\xi)\| \geqslant 2\|\eta(\xi)\|$; then put $x_{i+1} := \xi$.

EXAMPLE 1. Consider the boundary-value problem

$$y'' = 5 \sinh 5y, \qquad y(0) = 0, \quad y(1) = 1 \tag{7.3.6.2}$$

(cf. Example 2 of Section 7.3.4).

For the starting trajectory $\eta(x)$ we take the straight line connecting the two boundary points $[\eta(x) \equiv x]$.

The subdivision $0 = x_1 < x_2 < \cdots < x_m = 1$ found by the computer is shown in Figure 18. Starting with this subdivision and the values $s_k := \eta(x_k)$, $s_k' := \eta'(x_k)$, the multiple shooting method yields the solution to about 9 digits in 7 iterations.

The problem suggests a "more advantageous" starting trajectory: the *linearized* problem

$$y'' = 5 \cdot 5y, \qquad y(0) = 0, \quad y(1) = 1$$

has the solution $y(x) = (\sinh 5x)/(\sinh 5)$. This function would have been a somewhat "better" starting trajectory.

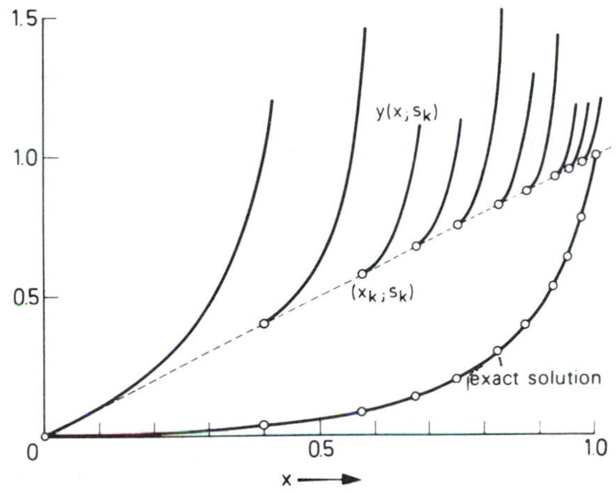

Figure 18 Subdivision in multiple shooting.

In some practical problems the right-hand side $f(x, y)$ of the differential equation is only piecewise continuous on $[a, b]$ as a function of x, or else only piecewise continuously differentiable. In these cases one ought to be careful to include the points of discontinuity among the subdivision points x_i; otherwise there are convergence difficulties (see Example 2).

There remains the problem of how to find a first approximation $\eta(x)$ for a boundary-value problem. In many cases one knows the qualitative behavior of the solution, e.g. on physical grounds, so that one can easily find at least a rough approximation. Usually this also suffices for convergence, as the modified Newton's method does not impose particularly high demands on the quality of the starting vector.

In complicated cases the so-called *continuation method* (homotopy method) is usually effective. Here one gradually feels one's way, via the solution of "neighboring" problems, toward the solution of the actually posed problem: Almost all problems contain certain parameters α,

$$\text{(7.3.6.3)} \qquad P_\alpha: \qquad y' = f(x, y; \alpha), \qquad r(y(a), y(b); \alpha) = 0,$$

and the problem consists in determining the solution $y(x)$, which of course depends also on α, for a certain value $\alpha = \bar{\alpha}$. It is usually true that for a certain other value $\alpha = \alpha_0$ the boundary-value problem P_α is simpler, so that at least a good approximation $\eta_0(x)$ is known for the solution $y(x; \alpha_0)$ of P_{α_0}. The continuation method now consists in the following: Starting with $\eta_0(x)$, one first determines the solution $y(x; \alpha_0)$ of P_{α_0}. One then chooses a finite sequence of sufficiently close numbers ε_i with $0 < \varepsilon_1 < \varepsilon_2 < \cdots < \varepsilon_l = 1$, puts $\alpha_i := \alpha_0 + \varepsilon_i(\bar{\alpha} - \alpha_0)$, and for $i = 0, 1, 2, \ldots, l - 1$ takes the solution $y(x; \alpha_i)$ of P_{α_i} as the starting approximation for the determination of the solution $y(x; \alpha_{i+1})$ of $P_{\alpha_{i+1}}$. In $y(x; \alpha_l) = y(x; \bar{\alpha})$ one then finally has the desired solution of $P_{\bar{\alpha}}$. For the method to work, it is important that one not take too large steps $\varepsilon_i \to \varepsilon_{i+1}$, and that "natural" parameters α, which are intrinsic to the problem, be chosen: The introduction of artificial parameters, say constructs of the type

$$\bar{f}(x, y; \alpha) = \alpha f(x, y) + (1 - \alpha)g(x, y), \qquad \bar{\alpha} = 1, \quad \alpha_0 = 0,$$

where $f(x, y)$ is the given right-hand side of the differential equation and $g(x, y)$ an arbitrarily chosen "simple" function which has nothing to do with the problem, does not succeed in critical cases.

A more refined variant of the continuation method, which has proved to be useful in the context of multiple shooting methods, is described in Deuflhard (1979).

EXAMPLE 2 (A singular boundary-value problem). The steady-state temperature distribution in the interior of a cylinder of radius 1 is described by the solution $y(x; \alpha)$ of the nonlinear boundary-value problem [cf., e.g., Na and Tang (1969)]

$$y'' = -\frac{y'}{x} - \alpha e^y, \tag{7.3.6.4}$$
$$y'(0) = y(1) = 0.$$

Here, α is a "natural" parameter with

$$\alpha \doteq \frac{\text{heat generation}}{\text{conductivity}}, \qquad 0 < \alpha \leqslant 0.8.$$

For the numerical computation of the solution for $\alpha = 0.8$ one could use the subdivision

$$\alpha = 0, 0.1, \ldots, 0.8$$

and, beginning with the explicitly known solution $y(x; 0) \equiv 0$, construct the homotopy chain for the computation of $y(x; 0.8)$. The problem, however, exhibits one further difficulty: The right-hand side of the differential equation has a singularity at $x = 0$. While it is true that $y'(0) = 0$ and the solution $y(x; \alpha)$ is defined, indeed analytic, on the whole interval $0 \leqslant x \leqslant 1$, there nevertheless arise considerable convergence difficulties. The reason is to be found in the following: A backward analysis (cf. Section 1.3) shows that the result $\tilde{y}$ of the numerical computation can be interpreted as the exact solution of the boundary-value problem (7.3.6.4) with slightly perturbed boundary data

$$\tilde{y}'(0) = \varepsilon_1, \qquad \tilde{y}(1) = \varepsilon_2$$

[see, e.g., Babuška et al. (1966)]. This solution $\tilde{y}(x; \alpha)$ is "near" the solution $y(x; \alpha)$, but in contrast to $y(x; \alpha)$ has only a continuous derivative at $x = 0$, because

$$\lim_{x \to 0} \tilde{y}'' = -\lim_{x \to 0} \frac{\varepsilon_1}{x} = \pm\infty.$$

Since the order of convergence of every numerical method depends not only on the method itself, but also (as is often overlooked) on the existence and boundedness of the higher derivatives of the solution, the order of convergence in the present example will be considerably reduced (to essentially 1). These difficulties, and others as well, are encountered in many practical boundary-value problems. The cause of failure is often not recognized, and the "blame" put on the method or the computer.

Through a simple artifice these difficulties can be avoided. The "neighboring" solutions with $y'(0) \neq 0$ are filtered out by means of a power series expansion about $x = 0$:

$$(7.3.6.5) \qquad y(x) = y(0) + \frac{x^2}{2!} y^{(2)}(0) + \frac{x^3}{3!} y^{(3)}(0) + \frac{x^4}{4!} y^{(4)}(0) + \cdots.$$

The coefficients $y^{(i)}(0)$, $i = 2, 3, 4, \ldots$, can all be expressed in terms of $\lambda := y(0)$, an unknown constant; through substitution in the differential equation (7.3.6.4) one finds

$$(7.3.6.6) \qquad y^{(2)}(x) = -\left(y^{(2)}(0) + \frac{x}{2!} y^{(3)}(0) + \frac{x^2}{3!} y^{(4)}(0) + \cdots\right) - \alpha e^{y(x)},$$

from which, for $x \to 0$,

$$y^{(2)}(0) = -y^{(2)}(0) - \alpha e^{y(0)}, \quad \text{i.e.,} \; y^{(2)}(0) = -\tfrac{1}{2}\alpha e^{\lambda}.$$

Differentiation gives

$$y^{(3)}(x) = -\left(\tfrac{1}{2}y^{(3)}(0) + \tfrac{1}{3}xy^{(4)}(0) + \cdots\right) - \alpha y'(x)e^{y(x)}$$

with

$$y^{(3)}(0) = 0.$$

Further,

$$y^{(4)}(x) = -\left(\tfrac{1}{3}y^{(4)}(0) + x(\cdots)\right) - \alpha\{(y'(x))^2 + y^{(2)}(x)\}e^{y(x)}$$

and

$$y^{(4)}(0) = \tfrac{3}{8}\alpha^2 e^{2\lambda}.$$

One can show that $y^{(5)}(0) = 0$, and in general, $y^{(2i+1)}(0) = 0$.

The singular boundary-value problem can now be treated as follows. For $y(x; \alpha)$ one utilizes in the neighborhood of $x = 0$ the power-series representation (7.3.6.5), and at a sufficient distance from the singularity $x = 0$ the differential equation (7.3.6.4) itself—for example,

$$(7.3.6.7) \qquad y''(x) = \begin{cases} -\tfrac{1}{2}\alpha e^{\lambda}[1 - \tfrac{3}{8}x^2\alpha e^{\lambda}] & \text{if } 0 \leqslant x \leqslant 10^{-2}, \\ -\dfrac{y'(x)}{x} - \alpha e^{y(x)} & \text{if } 10^{-2} \leqslant x \leqslant 1. \end{cases}$$

The error is of the order of magnitude 10^{-8}. Now the right-hand side still contains the unknown parameter $\lambda = y(0)$. As shown in Section 7.3.0, however, this can be interpreted as an extended boundary-value problem. One puts

$$\begin{aligned} y_1(x) &:= y(x), \\ y_2(x) &:= y'(x), \\ y_3(x) &:= y(0) = \lambda, \end{aligned}$$

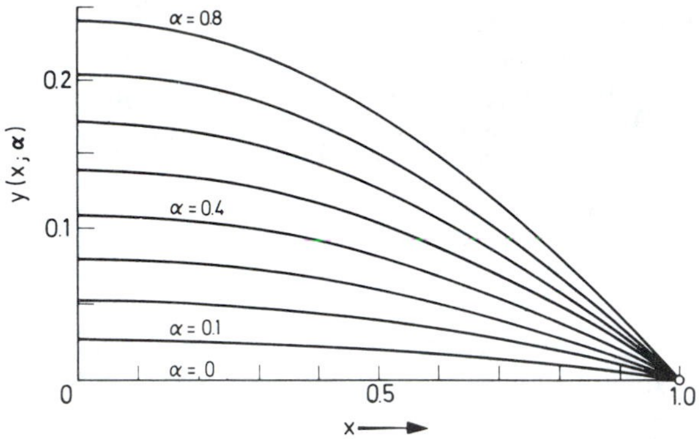

Figure 19 The solution $y(x; \alpha)$ for $\alpha = 0(0.1)0.8$.

and obtains the system of differential equations on $0 \leqslant x \leqslant 1$,

$$y_1' = y_2,$$

(7.3.6.8)
$$y_2' = \begin{cases} -\frac{1}{2}\alpha e^{y_3}[1 - \frac{3}{8}x^2\alpha e^{y_3}] & \text{if } 0 \leqslant x \leqslant 10^{-2}, \\ -\dfrac{y_2}{x} - \alpha e^{y_1} & \text{if } 10^{-2} \leqslant x \leqslant 1, \end{cases}$$

$$y_3' = 0$$

with the boundary conditions

(7.3.6.9)
$$r = \begin{bmatrix} y_2(0) \\ y_1(1) \\ y_3(0) - y_1(0) \end{bmatrix} = 0.$$

In the multiple shooting method one chooses the "patch point" 10^{-2} as one of the subdivision points, say,

$$x_1 = 0, \quad x_2 = 10^{-2}, \quad x_3 = 0.1, \ldots, \quad x_{11} = 0.9, \quad x_{12} = 1.$$

There is no need to worry about the smoothness of the solution at the patch point; it is ensured by construction. With the use of the homotopy [see (7.3.6.3)], beginning with $y(x; \alpha_0) \equiv 0$, the solution $y(x; \alpha_k)$ is easily obtained from $y(x; \alpha_{k-1})$ in only 3 iterations to an accuracy of about 6 digits (total computing time on the CDC 3600 computer for all 8 trajectories: 40 seconds). The results are shown in Figure 19.

7.3.7 An Example: Optimal Control Program for a Lifting Reentry Space Vehicle

The following extensive problem originates in space travel. This concluding example is to illustrate how nonlinear boundary-value problems are handled and solved in practice, because experience shows that the actual solution of such real-life problems causes the greatest difficulties to the beginner. The numerical solution was carried out with the multiple shooting method; a program can be found in Oberle and Grimm (1989).

The coordinates of the trajectory of an Apollo type vehicle satisfy, during the flight of the vehicle through the earth's atmosphere, the following differential equations:

$$\dot{v} = V(v, \gamma, \xi, u) = -\frac{S\rho v^2}{2m} C_D(u) - \frac{g \sin \gamma}{(1+\xi)^2},$$

$$\dot{\gamma} = \Gamma(v, \gamma, \xi, u) = \frac{S\rho v}{2m} C_L(u) + \frac{v \cos \gamma}{R(1+\xi)} - \frac{g \cos \gamma}{v(1+\xi)^2},$$

(7.3.7.1)

$$\dot{\xi} = \Xi(v, \gamma, \xi, u) = \frac{v \sin \gamma}{R},$$

$$\dot{\zeta} = Z(v, \gamma, \xi, u) = \frac{v}{1+\xi} \cos \gamma.$$

The meanings are: v: velocity; γ: flight-path angle; h: altitude above the earth's surface; R: earth radius; $\xi = h/R$: normalized altitude; ζ: distance on the earth's surface; $\rho = \rho_0 e^{-\beta R \xi}$: atmospheric density; $C_D(u) = 1.174 - 0.9 \cdot \cos u$: aerodynamical drag coefficient; $C_L(u) = 0.6 \sin u$: aerodynamical lift coefficient; u: control parameter, which can be chosen arbitrarily as a function of time; g: gravitational acceleration; S/m: (frontal area)/(mass of vehicle). Numerical values are: $R = 209$ ($= 209_{10}5$ ft); $\beta = 4.26$; $\rho_0 = 2.704_{10}{-3}$; $g = 3.2172_{10}{-4}$; $S/m = 53{,}200$ (1 ft = 0.3048 m). (See Figure 20.)

The differential equations have been simplified somewhat by assuming (1) a spherical earth at rest, (2) a flight trajectory in a great circle plane, (3) that vehicle and astronauts can be subjected to unlimited deceleration. The right-hand sides of the differential equations nevertheless contain all terms which are essential physically. The largest effect is produced by the terms mul-

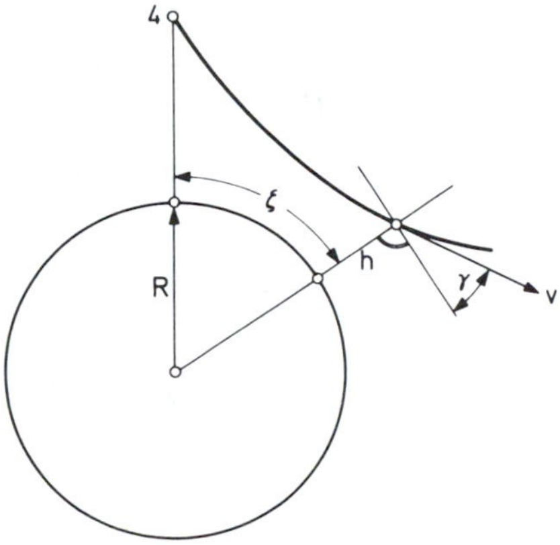

Figure 20 The coordinates of the trajectory.

tiplied by $C_D(u)$ and $C_L(u)$, respectively; these are the atmospheric forces, which, in spite of the low air density ρ, are particularly significant by virtue of the high speed of the vehicle; they can be influenced via the parameter u (angle of attack). The terms multiplied by g are the gravitational forces of the earth acting on the vehicle; the remaining terms result from the choice of the coordinate system.

During the flight through the earth's atmosphere the vehicle is heated up considerably. The total stagnation point convective heating per unit area is given by the integral

$$J = \int_0^T \dot{q}\, dt, \qquad \dot{q} = 10v^3\sqrt{\rho}. \tag{7.3.7.2}$$

The range of integration is from $t = 0$, the time when the vehicle hits the 400,000 ft atmospheric level, until a time instant T. The vehicle is to be maneuvered into an initial position favorable for the final splashdown in the Pacific. Through the freely disposable parameter u the maneuver is to be executed in such a way that the heating J becomes minimal and that the following boundary conditions are satisfied: Data at the moment of entry:

$$\begin{aligned} v(0) &= 0.36 \ (= 36{,}000 \text{ ft/sec}), \\ \gamma(0) &= -8.1^\circ \frac{\pi}{180^\circ}, \\ \xi(0) &= \frac{4}{R} \qquad [h(0) = 400{,}000 \text{ ft}]. \end{aligned} \tag{7.3.7.3}$$

Data at the end of maneuver:

$$\begin{aligned} v(T) &= 0.27 \ (=27{,}000 \text{ ft/sec}), \\ \gamma(T) &= 0, \\ \xi(T) &= \frac{2.5}{R} \qquad [h(T) = 250{,}000 \text{ ft}]. \end{aligned} \tag{7.3.7.4}$$

The terminal time T is free. ζ is ignored in the optimization process.

The calculus of variations [cf., e.g., Hestenes (1966)] now teaches the following: Form, with parameters (Lagrange multipliers), the expression

$$H := 10v^3\sqrt{\rho} + \lambda_v V + \lambda_\gamma \Gamma + \lambda_\xi \Xi, \tag{7.3.7.5}$$

where λ_v, λ_γ, λ_ξ satisfy the three differential equations

$$\begin{aligned} \dot{\lambda}_v &= -\frac{\partial H}{\partial v}, \\ \dot{\lambda}_\gamma &= -\frac{\partial H}{\partial \gamma}, \\ \dot{\lambda}_\xi &= -\frac{\partial H}{\partial \xi}. \end{aligned} \tag{7.3.7.6}$$

The optimal control u is then given by

(7.3.7.7)

$$\sin u = \frac{-0.6\lambda_\gamma}{\alpha}, \quad \cos u = \frac{-0.9v\lambda_v}{\alpha}, \qquad \alpha = \sqrt{(0.6\lambda_\gamma)^2 + (0.9v\lambda_v)^2}.$$

Note that (7.3.7.6), by virtue of (7.3.7.7), is nonlinear in λ_v, λ_γ.

Since the terminal time T is not subject to any condition, the further boundary condition must be satisfied:

(7.3.7.8)
$$H\bigg|_{t=T} = 0.$$

The problem is thus reduced to a boundary-value problem for the six differential equations (7.3.7.1), (7.3.7.6) with the seven boundary conditions (7.3.7.3), (7.3.7.4), (7.3.7.8). We are thus dealing with a *free* boundary-value problem [cf. (7.3.0.4a,b)]. A closed-form solution is impossible; one must use numerical methods.

It would now be wrong to construct a starting trajectory $\eta(x)$ without reference to reality. The unexperienced should not be misled by the innocent-looking form of the right-hand side of the differential equation (7.3.7.1): During the numerical integration one quickly observes that v, γ, ξ, λ_v, λ_γ, λ_ξ depend in an extremely sensitive way on the initial data. The solution has moving singularities which lie in the immediate neighborhood of the initial point of integration [see for this the comparatively trivial example (7.3.6.2)]. This sensitivity is a consequence of the effect of atmospheric forces, and the physical interpretation of the singularity is a "crash" of the vehicle or a "hurling back" into space. As can be shown by an *a posteriori* calculation, there exist differentiable solutions of the boundary-value problem only for an extremely narrow domain of boundary data. This is the mathematical formulation of the danger involved in the reentry maneuver.

Construction of a Starting Trajectory. For aerodynamic reasons the graph of the control parameter u will have the shape depicted in Figure 21 (information from space engineers). This function can be approximated, e.g., by

(7.3.7.9)
$$u = p_1 \operatorname{erf}(p_2(p_3 - \tau)),$$

where

$$\tau = \frac{t}{T}, \qquad 0 < \tau < 1,$$

$$\operatorname{erf} x = \frac{2}{\sqrt{\pi}} \int_0^x e^{-\sigma^2}\, d\sigma,$$

and p_1, p_2, p_3 are, for the moment, unknown constants.

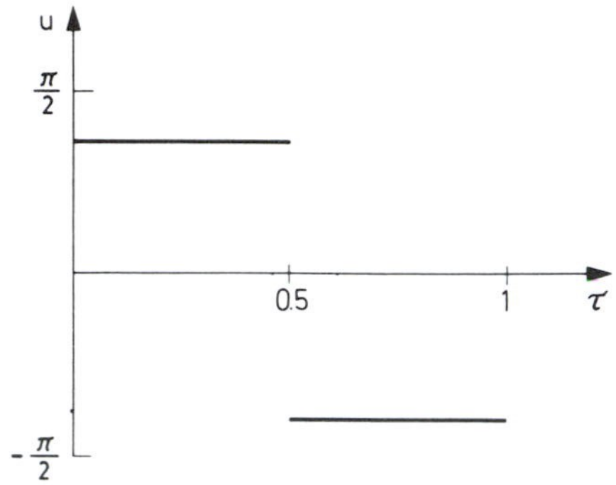

Figure 21 Control parameter (empirical).

For the determination of the p_i one solves the following auxiliary boundary-value problem: the differential equations (7.3.7.1) with u from (7.3.7.9) and in addition

$$\dot{p}_1 = 0,\quad \dot{p}_2 = 0,\quad \dot{p}_3 = 0, \tag{7.3.7.10}$$

with boundary conditions (7.3.7.3), (7.3.7.4) and $T = 230$.

The boundary value $T = 230$ sec is an estimate for the duration of the reentry maneuver. With the relatively poor approximations

$$p_1 = 1.6, \qquad p_2 = 4., \qquad p_3 = 0.5,$$

the auxiliary boundary-value problem can be solved even with the simple shooting method, provided one integrates in the backward direction (initial point $\tau = 1$, terminal point $\tau = 0$). The result after 11 iterations is

$$p_1 = 1.09835, \qquad p_2 = 6.48578, \qquad p_3 = 0.347717. \tag{7.3.7.11}$$

Solution of the Actual Boundary-Value Problem. With the "nonoptimal" control function u from (7.3.7.9), (7.3.7.11) one obtains approximations to $v(t)$, $\gamma(t)$, $\xi(t)$ by integrating (7.3.7.1). This "incomplete" starting trajectory can be made to a "complete" one as follows: since $\cos u > 0$, it follows from (7.3.7.7) that $\lambda_v < 0$; we choose

$$\lambda_v \equiv -1.$$

Further, from

$$\tan u = \frac{6\lambda_\gamma}{9v\lambda_v}$$

there follows an approximation for λ_γ. An approximation for λ_ξ can be obtained from the relation $H \equiv 0$, where H is given by (7.3.7.5), because $H = \text{const}$ is a first integral of the equations of motion.

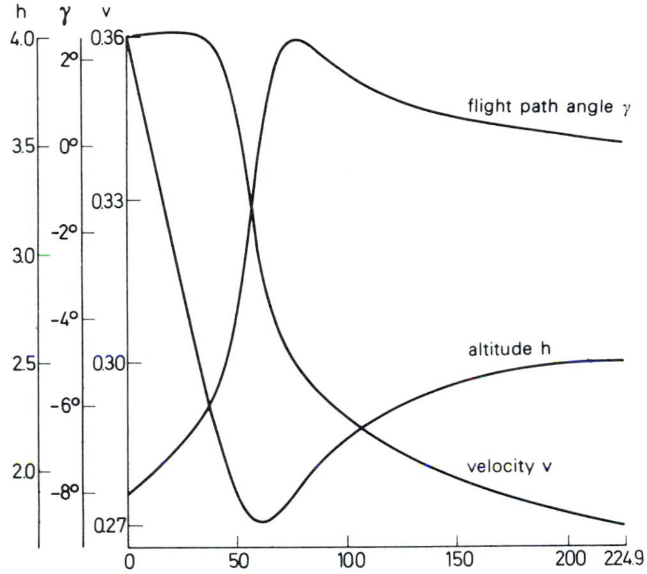

Figure 22 The trajectories h, γ, v.

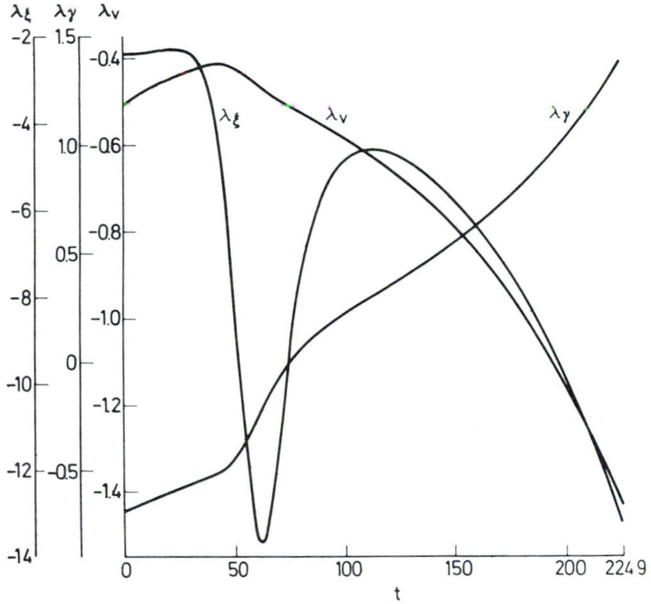

Figure 23 The trajectories of the adjoint variables λ_ξ, λ_γ, λ_v.

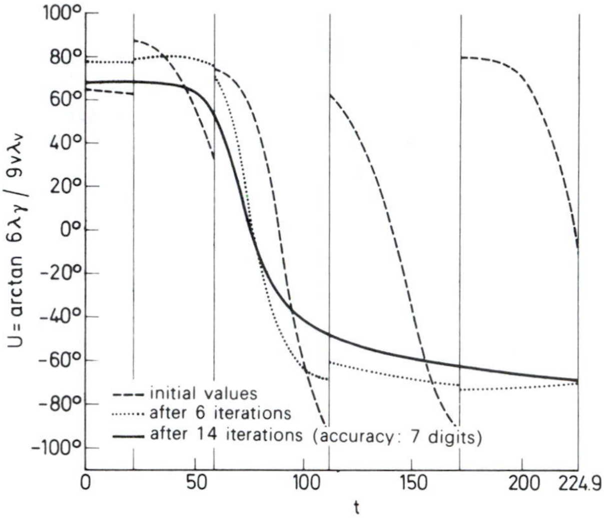

Figure 24 Successive approximations to the control u.

With this approximate trajectory for $v, \gamma, \ldots, \lambda_\xi, T$ ($T = 230$) and the techniques of Section 7.3.6, one can now determine the subdivision points for the multiple shooting method ($m = 6$ suffices).

Figures 22 and 23 show the result of the computation (after 14 iterations); the total time of the optimal reentry maneuver comes out to be $T = 224.9$ sec; the accuracy of the results is about 7 digits.

Figure 24 shows the behavior of the control $u = \tan^{-1}(6\lambda_\gamma/9v\lambda_v)$ during the course of the iteration. It can be seen how the initially large jumps at the subdivision points of the multiple shooting method are "flattened out."

The following table shows the convergence behavior of the modified Newton's method [see Sections 5.4.2 and 5.4.3].

Error $\|F(s^{(i)})\|^2$ [cf. (7.3.5.3)]	Steplength in (5.4.3.5) λ	Error $\|F(s^{(i)})\|^2$ [cf. (7.3.5.3)]	Steplength in (5.4.3.5) λ
5×10^2			
3×10^2	0.250	1×10^{-1}	1.000
6×10^4	0.500 (trial)	6×10^{-2}	1.000
7×10^2	0.250 (trial)	3×10^{-2}	1.000
2×10^2	0.125	1×10^{-2}	1.000
1×10^2	0.125	1×10^{-3}	1.000
8×10^1	0.250	4×10^{-5}	1.000
1×10^1	0.500	3×10^{-7}	1.000
1×10^0	1.000	1×10^{-9}	1.000

7.3.8 The Limiting Case $m \to \infty$ of the Multiple Shooting Method (General Newton's Method, Quasilinearization)

If the subdivision of the interval $[a, b]$ is made finer and finer ($m \to \infty$), the multiple shooting method converges toward the general Newton's method for boundary-value problems [cf., e.g., Collatz (1966)], also known as *quasilinearization*.

In this method an approximation $\eta(x)$ for the exact solution $y(x)$ of the *nonlinear* boundary-value problem (7.3.5.1) is improved by solving *linear* boundary-value problems. Using Taylor expansion of $f(x, y(x))$ and $r(y(a), y(b))$ about $\eta(x)$ one finds in first approximation (cf. the derivation of Newton's method in Section 5.1)

$$y'(x) = f(x, y(x)) \doteq f(x, \eta(x)) + D_y f(x, \eta(x))(y(x) - \eta(x)),$$
$$0 = r(y(a), y(b)) \doteq r(\eta(a), \eta(b)) + A(y(a) - \eta(a)) + B(y(b) - \eta(b)),$$

where $A := D_u r(\eta(a), \eta(b))$, $B := D_v r(\eta(a), \eta(b))$. One may expect, therefore, that the solution $\bar{\eta}(x)$ of the linear boundary-value problem

$$\text{(7.3.8.1)} \qquad \begin{aligned} \bar{\eta}' &= f(x, \eta(x)) + D_y f(x, \eta(x))(\bar{\eta} - \eta(x)), \\ A(\bar{\eta}(a) - \eta(a)) &+ B(\bar{\eta}(b) - \eta(b)) = -r(\eta(a), \eta(b)), \end{aligned}$$

will be a better solution of (7.3.5.1) than $\eta(x)$. For later use we introduce the correction function $\Delta\eta(x) := \bar{\eta}(x) - \eta(x)$, which by definition is solution of the boundary-value problem

$$\text{(7.3.8.2)} \qquad \begin{aligned} (\Delta\eta)' &= f(x, \eta(x)) - \eta'(x) + D_y f(x, \eta(x))\, \Delta\eta, \\ A\, \Delta\eta(a) &+ B\, \Delta\eta(b) = -r(\eta(a), \eta(b)), \end{aligned}$$

so that

$$\text{(7.3.8.3)} \quad \Delta\eta(x) = \Delta\eta(a) + \int_a^x [D_y f(t, \eta(t))\, \Delta\eta(t) + f(t, \eta(t)) - \eta'(t)]\, dt.$$

Replacing η in (7.3.8.1) by $\bar{\eta}$, one obtains a further approximation $\bar{\bar{\eta}}$ for y, etc. In spite of the simple derivation, this iteration method has serious disadvantages which cast some doubt on its practical value:

(1) The vector functions $\eta(x)$, $\bar{\eta}(x)$, ... must be stored over their entire range $[a, b]$.
(2) The matrix function $D_y f(x, y)$ must be computed in explicit analytic form and must likewise be stored over its entire range $[a, b]$. The realization of both these requirements is next to impossible in the problems that currently arise in practice (e.g., f with 25 components, 500–1000 arithmetic operations per evaluation of f).

We now wish to show that the multiple shooting method as $m \to \infty$ converges to the method (7.3.8.1) in the following sense: Let $\eta(x)$ be a sufficiently smooth function on $[a, b]$ and let further (7.3.8.1) be uniquely solvable with the solution $\bar{\eta}(x)$, $\Delta\eta(x) := \bar{\eta}(x) - \eta(x)$. Then the following is true: If in the multiple shooting method (7.3.5.3)–(7.3.5.10) one chooses any subdivision $a = x_1 < x_2 < \cdots < x_m = b$ of fineness $h := \max_k |x_{k+1} - x_k|$ and the starting vector

$$s = \begin{bmatrix} s_1 \\ s_2 \\ \vdots \\ s_m \end{bmatrix} \quad \text{with } s_k := \eta(x_k),$$

then the solution of (7.3.5.8) will yield a vector of corrections

$$\text{(7.3.8.4)} \qquad \Delta s = \begin{bmatrix} \Delta s_1 \\ \Delta s_2 \\ \vdots \\ \Delta s_m \end{bmatrix} \quad \text{with } \max_k \|\Delta s_k - \Delta\eta(x_k)\| = O(h).$$

For the proof we assume for simplicity $|x_{k+1} - x_k| = h$, $k = 1, 2, \ldots, m - 1$. We want to show that to each h there exists a differentiable function $\Delta\bar{s}: [a, b] \to \mathbb{R}^n$ such that $\max_k \|\Delta s_k - \Delta\bar{s}(x_k)\| = O(h)$, $\Delta\bar{s}(a) = \Delta s_1$, and $\max_{x \in [a, b]} \|\Delta\bar{s}(x) - \Delta\eta(x)\| = O(h)$.

To this end, we first show that the products

$$G_{k-1} G_{k-2} \ldots G_{j+1} = G_{k-1} G_{k-2} \ldots G_1 (G_j G_{j-1} \ldots G_1)^{-1}$$

appearing in (7.3.5.9), (7.3.5.10) have simple limits as $h \to 0$ ($hk = \text{const}$, $hj = \text{const}$). The representation (7.3.5.9) for Δs_k can be written in the form

(7.3.8.5)

$$\Delta s_k = F_{k-1} + G_{k-1} \ldots G_1 \left[\Delta s_1 + \sum_{j=1}^{k-2} (G_j \ldots G_1)^{-1} F_j \right], \qquad 2 \leqslant k \leqslant m.$$

According to (7.3.5.3), F_k is given by

$$F_k = y(x_{k+1}; x_k, s_k) - s_{k+1}, \qquad 1 \leqslant k \leqslant m - 1;$$

thus

$$F_k = s_k + \int_{x_k}^{x_{k+1}} f(t, y(t; x_k, s_k))\, dt - s_{k+1},$$

and with the help of the mean-value theorem,

$$\text{(7.3.8.6)} \quad F_k = [f(\tau_k, y(\tau_k; x_k, s_k)) - \eta'(\tilde{\tau}_k)]h, \qquad x_k < \tau_k, \tilde{\tau}_k < x_{k+1}.$$

Let now $Z(x)$ be the solution of the initial-value problem

$$\text{(7.3.8.7)} \qquad Z' = D_y f(x, \eta(x)) Z, \qquad Z(a) = I,$$

and the matrix $\bar{Z}_k$ solution of

$$\bar{Z}_k' = D_y f(x, \eta(x)) \bar{Z}_k, \qquad \bar{Z}_k(x_k) = I, \qquad 1 \leqslant k \leqslant m - 1,$$

so that

$$\bar{Z}_k(x) = I + \int_{x_k}^{x} D_y f(t, \eta(t))\bar{Z}_k(t)\,dt. \tag{7.3.8.8}$$

For the matrices $Z_k := \bar{Z}_k(x_{k+1})$ one now shows easily by induction on k that

$$Z(x_{k+1}) = Z_k Z_{k-1} \ldots Z_1. \tag{7.3.8.9}$$

Indeed, since $\bar{Z}_1 = Z$, this is true for $k = 1$. If the assertion is true for $k - 1$, then the function

$$\bar{Z}(x) := \bar{Z}_k(x) Z_{k-1} \ldots Z_1$$

satisfies the differential equation $\bar{Z}' = D_y f(x, \eta(x))\bar{Z}$ and the initial conditions $\bar{Z}(x_k) = \bar{Z}_k(x_k)Z_{k-1} \ldots Z_1 = Z_{k-1} \ldots Z_1 = Z(x_k)$. By uniqueness of the solution of initial-value problems there follows $\bar{Z}(x) = Z(x)$, hence (7.3.8.9).

By Theorem (7.1.8) we further have for the matrices $\bar{G}_k(x) := D_{s_k} y(x; x_k, s_k)$,

$$\begin{aligned} \bar{G}_k(x) &= I + \int_{x_k}^{x} D_y f(t, y(t; x_k, s_k))\bar{G}_k(t)\,dt, \\ G_k &= \bar{G}_k(x_{k+1}). \end{aligned} \tag{7.3.8.10}$$

With the abbreviation

$$\varphi(x) := \|\bar{Z}_k(x) - \bar{G}_k(x)\|,$$

upon subtraction of (7.3.8.8) and (7.3.8.10), one obtains the estimate

$$\begin{aligned} \varphi(x) \leqslant & \int_{x_k}^{x} \|D_y f(t, \eta(t)) - D_y f(t, y(t; x_k, s_k))\| \cdot \|\bar{Z}_k(t)\|\,dt \\ & + \int_{x_k}^{x} \|D_y f(t, y(t; x_k, s_k))\|\,\varphi(t)\,dt. \end{aligned} \tag{7.3.8.11}$$

If $D_y f(t, y)$, for all $t \in [a, b]$, is uniformly Lipschitz continuous in y, one easily shows for $x_k \leqslant t \leqslant x_{k+1}$ that

$$\|D_y f(t, \eta(t)) - D_y f(t, y(t; x_k, s_k))\| \leqslant L\|\eta(t) - y(t; x_k, s_k)\| = O(h),$$

as well as the uniform boundedness of $\|\bar{Z}_k(t)\|$, $\|D_y f(t, y(t; x_k, s_k))\|$ for $t \in [a, b]$, $k = 1, 2, \ldots, m - 1$. From (7.3.8.11) and the proof techniques of Theorems (7.1.4), (7.1.11), one then obtains an estimate of the form

$$\varphi(x) = O(h^2) \quad \text{for } x \in [x_k, x_{k+1}]$$

—in particular, for $x = x_{k+1}$,

$$\|Z_k - G_k\| \leqslant c_1 h^2, \qquad 1 \leqslant k \leqslant m - 1, \tag{7.3.8.12}$$

with a constant c_1 independent of k and h.

From the identity

$$Z_k Z_{k-1} \ldots Z_1 - G_k G_{k-1} \ldots G_1 = (Z_k - G_k) Z_{k-1} \ldots Z_1 + G_k(Z_{k-1} - G_{k-1}) Z_{k-2} \ldots Z_1 + \cdots + G_k G_{k-1} \ldots G_2 (Z_1 - G_1)$$

and the further estimates

$$\|G_k\| \leqslant 1 + c_2 h, \quad \|Z_k\| \leqslant 1 + c_2 h, \qquad 1 \leqslant k \leqslant m-1,$$

c_2 independent of k and h, which follow from Theorem (7.1.11), we thus obtain, in view of $kh \leqslant b - a$ and (7.3.8.9),

$$\|Z(x_{k+1}) - G_k G_{k-1} \ldots G_1\| \leqslant c_1 h^2 \cdot k(1 + c_2 h)^{k-1} \leqslant Dh, \tag{7.3.8.13}$$
$$1 \leqslant k \leqslant m-1,$$

with a constant D independent of k and h.

From (7.3.8.5), (7.3.8.6), (7.3.8.9), (7.3.8.12), (7.3.8.13) one obtains

(7.3.8.14)

$$\begin{aligned}\Delta s_k = {} & (Z(x_k) + O(h^2)) \\ & \times \left[\Delta s_1 + \sum_{j=1}^{k-2} (Z(x_{j+1}) + O(h^2))^{-1} (f(\tau_j, y(\tau_j; x_j, s_j)) - \eta'(\tilde{\tau}_j))h\right] + O(h) \\ = {} & Z(x_k)\, \Delta s_1 + Z(x_k) \int_a^{x_k} Z(t)^{-1} [f(t, \eta(t)) - \eta'(t)]\, dt + O(h) \\ = {} & \Delta \bar{s}(x_k) + O(h).\end{aligned}$$

Here, $\Delta \bar{s}(x)$ is the function

$$\begin{aligned}\Delta \bar{s}(x) &:= Z(x)\, \Delta s_1 + Z(x) \int_a^x Z(t)^{-1} [f(t, \eta(t)) - \eta'(t)]\, dt, \\ \Delta \bar{s}(a) &= \Delta s_1.\end{aligned} \tag{7.3.8.15}$$

Note that $\Delta \bar{s}$ depends also on h, since Δs_1 depends on h. Evidently, $\Delta \bar{s}$ is differentiable with respect to x, and by (7.3.8.7) we have

$$\begin{aligned}\Delta \bar{s}'(x) &= Z'(x) \left[\Delta s_1 + \int_a^x Z(t)^{-1} [f(t, \eta(t)) - \eta'(t)]\, dt\right] + f(x, \eta(x)) - \eta'(x) \\ &= D_y f(x, \eta(x))\, \Delta \bar{s}(x) + f(x, \eta(x)) - \eta'(x),\end{aligned}$$

so that

$$\Delta \bar{s}(x) = \Delta \bar{s}(a) + \int_a^x [D_y f(t, \eta(t))\, \Delta \bar{s}(t) + f(t, \eta(t)) - \eta'(t)]\, dt.$$

By subtracting this equation from (7.3.8.3), we obtain for the difference

$$\theta(x) := \Delta \eta(x) - \Delta \bar{s}(x)$$

the equation

$$\theta(x) = \theta(a) + \int_a^x D_y f(t, \eta(t))\theta(t)\,dt, \tag{7.3.8.16}$$

and further, with the aid of (7.3.8.7),

$$\theta(x) = Z(x)\theta(a), \quad \|\theta(x)\| \leqslant K\|\theta(a)\| \quad \text{for } a \leqslant x \leqslant b, \tag{7.3.8.17}$$

with a suitable constant K.

In view of (7.3.8.14) and (7.3.8.17) it suffices to show $\|\theta(a)\| = O(h)$ in order to complete the proof of (7.3.8.4). With the help of (7.3.5.10) and (7.3.8.15) one now shows in the same way as (7.3.8.14) that $\Delta s_1 = \Delta\bar{s}(a)$ satisfies an equation of the form

$$\begin{aligned}
&[A + B(Z(b) + O(h^2))]\,\Delta s_1 \\
&\qquad = -F_m - BZ(b)\int_a^b Z(t)^{-1}(f(t, \eta(t)) - \eta'(t))\,dt + O(h) \\
&\qquad = -F_m - B[\Delta\bar{s}(b) - Z(b)\,\Delta s_1] + O(h).
\end{aligned}$$

From $F_m = r(s_1, s_m) = r(\eta(a), \eta(b))$ it follows that

$$A\,\Delta\bar{s}(a) + B\,\Delta\bar{s}(b) = -r(\eta(a), \eta(b)) + O(h).$$

Subtraction of (7.3.8.2), in view of (7.3.8.17), yields

$$A\theta(a) + B\theta(b) = [A + BZ(b)]\theta(a) = O(h),$$

and thus $\theta(a) = O(h)$, since (7.3.8.1) by assumption is uniquely solvable and hence $A + BZ(b)$ is nonsingular. Because of (7.3.8.14), (7.3.8.17), this proves (7.3.8.4). □

7.4 Difference Methods

The basic idea underlying all difference methods is to replace the differential quotients in a differential equation by suitable difference quotients and to solve the discrete equations that are so obtained.

We illustrate this with the following simple boundary-value problem of second order for a function $y : [a, b] \to \mathbb{R}$:

$$\begin{aligned}
-y'' + q(x)y &= g(x), \\
y(a) = \alpha, \qquad y(b) &= \beta.
\end{aligned} \tag{7.4.1}$$

Under the assumptions that $q, g \in C[a, b]$ (i.e., q and g are continuous functions on $[a, b]$) and $q(x) \geqslant 0$ for $x \in [a, b]$, it can be shown that (7.4.1) has a unique solution $y(x)$.

In order to discretize (7.4.1), we subdivide $[a, b]$ into $n+1$ equal subintervals,

$$a = x_0 < x_1 < \cdots < x_n < x_{n+1} = b, \qquad x_j = a + jh, \quad h := \frac{b-a}{n+1},$$

and, with the abbreviation $y_j := y(x_j)$, replace the differential quotient $y_i'' = y''(x_i)$ for $i = 1, 2, \ldots, n$ by the second difference quotient

$$\Delta^2 y_i := \frac{y_{i+1} - 2y_i + y_{i-1}}{h^2}.$$

We now estimate the error $\tau_i(y) := y''(x_i) - \Delta^2 y_i$. We assume that y is four times continuously differentiable on $[a, b]$, $y \in C^4[a, b]$. Then, by Taylor expansion of $y(x_i \pm h)$ about x_i, one finds

$$y_{i\pm 1} = y_i \pm h y_i' + \frac{h^2}{2!} y_i'' \pm \frac{h^3}{3!} y_i''' + \frac{h^4}{4!} y^{(4)}(x_i \pm \theta_i^{\pm} h), \qquad 0 < \theta_i^{\pm} < 1,$$

and thus

$$\Delta^2 y_i = y_i'' + \frac{h^2}{24}[y^{(4)}(x_i + \theta_i^+ h) + y^{(4)}(x_i - \theta_i^- h)].$$

Since $y^{(4)}(x)$ is still continuous, it follows that

$$\text{(7.4.2)} \quad \tau_i(y) = y''(x_i) - \Delta^2 y_i = -\frac{h^2}{12} y^{(4)}(x_i + \theta_i h) \quad \text{for some } |\theta_i| < 1.$$

Because of (7.4.1), the values $y_i = y(x_i)$ satisfy the equations

$$y_0 = \alpha,$$

$$\text{(7.4.3)} \quad \frac{2y_i - y_{i-1} - y_{i+1}}{h^2} + q(x_i)y_i = g(x_i) + \tau_i(y), \qquad i = 1, 2, \ldots, n,$$

$$y_{n+1} = \beta.$$

With the abbreviations $q_i := q(x_i)$, $g_i := g(x_i)$, the vectors

$$\bar{y} := \begin{bmatrix} y_1 \\ y_2 \\ \vdots \\ y_n \end{bmatrix}, \qquad \tau(y) := \begin{bmatrix} \tau_1(y) \\ \tau_2(y) \\ \vdots \\ \tau_n(y) \end{bmatrix}, \qquad k := \begin{bmatrix} g_1 + \dfrac{\alpha}{h^2} \\ g_2 \\ \vdots \\ g_{n-1} \\ g_n + \dfrac{\beta}{h^2} \end{bmatrix},$$

and the symmetric $n \times n$ tridiagonal matrix

$$(7.4.4) \quad A := \frac{1}{h^2}\begin{bmatrix} 2+q_1h^2 & -1 & & & 0 \\ -1 & 2+q_2h^2 & -1 & & \\ & -1 & \ddots & \ddots & \\ & & \ddots & \ddots & -1 \\ 0 & & & -1 & 2+q_nh^2 \end{bmatrix},$$

the equations (7.4.3) are equivalent to

$$(7.4.5) \qquad A\bar{y} = k + \tau(y).$$

The difference method now consists in dropping the error term $\tau(y)$ in (7.4.5) and taking the solution $u = [u_1, \ldots, u_n]^T$ of the resulting system of linear equations,

$$(7.4.6) \qquad Au = k,$$

as an approximation to $\bar{y}$.

We first want to show some properties of the matrix A in (7.4.4). We will write $A \leqslant B$ for two $n \times n$ matrices if $a_{ij} \leqslant b_{ij}$ for $i, j = 1, 2, \ldots, n$. We now have the following:

(7.4.7) Theorem. *If $q_i \geqslant 0$ for $i = 1, \ldots, n$, then A in* (7.4.4) *is positive definite, and* $0 \leqslant A^{-1} \leqslant \tilde{A}^{-1}$ *with the positive definite $n \times n$ matrix*

$$(7.4.8) \qquad \tilde{A} := \frac{1}{h^2}\begin{bmatrix} 2 & -1 & & \\ -1 & \ddots & \ddots & \\ & \ddots & \ddots & -1 \\ & & -1 & 2 \end{bmatrix}.$$

PROOF. We begin by showing that A is positive definite. For this, we consider the $n \times n$ matrix $A_n := h^2\tilde{A}$. According to Gershgorin's theorem (6.9.4), we have for the eigenvalues the estimate $|\lambda_i - 2| \leqslant 2$, and hence $0 \leqslant \lambda_i \leqslant 4$. If $\lambda_i = 0$ were an eigenvalue of A_n, it would follow that $\det A_n = 0$; however, $\det A_n = n + 1$, as one easily verifies by means of the recurrence formula $\det A_{n+1} = 2 \det A_n - \det A_{n-1}$ (which is obtained at once by expanding $\det A_{n+1}$ along the first row). Neither is $\lambda_i = 4$ an eigenvalue of A_n, since otherwise $A_n - 4I$ would be singular. But

$$A_n - 4I = \begin{bmatrix} -2 & -1 & & \\ -1 & \ddots & \ddots & \\ & \ddots & \ddots & -1 \\ & & -1 & -2 \end{bmatrix} = -DA_nD^{-1},$$

$$D := \operatorname{diag}(1, -1, 1, \ldots, \pm 1),$$

so that $|\det(A_n - 4I)| = |\det A_n|$, and with A_n, also $A_n - 4I$ is nonsingular. We thus obtain the estimate $0 < \lambda_i < 4$ for the eigenvalues of A_n. This shows in particular that A_n is positive definite. By virtue of

$$z^H Az = z^H \tilde{A} z + \sum_{i=1}^{n} q_i |z_i|^2, \quad z^H \tilde{A} z > 0 \quad \text{for } z \neq 0, \text{ and } q_i \geqslant 0,$$

it follows immediately that $z^H Az > 0$ for $z \neq 0$; hence the positive definiteness of A. Theorem (4.3.2) shows the existence of A^{-1} and $\tilde{A}^{-1}$, and it only remains to prove the inequality $0 \leqslant A^{-1} \leqslant \tilde{A}^{-1}$. To this end, we consider the matrices D, $\tilde{D}$, J, $\tilde{J}$ with

$$\begin{aligned} h^2 A &= D(I - J), & D &= \operatorname{diag}(2 + q_1 h^2, \ldots, 2 + q_n h^2), \\ h^2 \tilde{A} &= \tilde{D}(I - \tilde{J}), & \tilde{D} &= 2I. \end{aligned}$$

Since $q_i \geqslant 0$, we obviously have the inequalities

$$0 \leqslant \tilde{D} \leqslant D,$$

$$(7.4.9) \quad 0 \leqslant J = \begin{bmatrix} 0 & \frac{1}{2 + q_1 h^2} & & 0 \\ \frac{1}{2 + q_2 h^2} & \ddots & \ddots & \\ & \ddots & \ddots & \frac{1}{2 + q_{n-1} h^2} \\ 0 & & \frac{1}{2 + q_n h^2} & 0 \end{bmatrix}$$

$$\leqslant \tilde{J} = \begin{bmatrix} 0 & \frac{1}{2} & & 0 \\ \frac{1}{2} & \ddots & \ddots & \\ & \ddots & \ddots & \frac{1}{2} \\ 0 & & \frac{1}{2} & 0 \end{bmatrix}.$$

In view of $\tilde{J} = \frac{1}{2}(-A_n + 2I)$ and the estimate $0 < \lambda_i < 4$ for the eigenvalues of A_n, we have $-1 < \mu_i < 1$ for the eigenvalues μ_i of $\tilde{J}$, i.e., $\rho(\tilde{J}) < 1$ for the spectral radius of $\tilde{J}$. From Theorem (6.9.2) there easily follows the convergence of the series

$$(0 \leqslant) I + \tilde{J} + \tilde{J}^2 + \tilde{J}^3 + \cdots = (I - \tilde{J})^{-1}.$$

Since $0 \leqslant J \leqslant \tilde{J}$, we then also have convergence in

$$(0 \leqslant) I + J + J^2 + J^3 + \cdots = (I - J)^{-1} \leqslant (I - \tilde{J})^{-1},$$

and since, by (7.4.9), $0 \leqslant D^{-1} \leqslant \tilde{D}^{-1}$, we get

$$0 \leqslant (h^2 A)^{-1} = (I - J)^{-1} D^{-1} \leqslant (I - \tilde{J})^{-1} \tilde{D}^{-1} = (h^2 \tilde{A})^{-1},$$

as was to be shown. □

From the above theorem it follows in particular that the system of equations (7.4.6) has a solution (if $q(x) \geqslant 0$ for $x \in [a, b]$), which can easily be found, e.g., by means of the Cholesky method (see Section 4.3). Since A is a tridiagonal matrix, the number of operations for the solution of (7.4.6) is proportional to n.

We now wish to derive an estimate for the errors

$$y_i - u_i$$

of the approximations u_i obtained from (7.4.6) for the exact solutions $y_i = y(x_i)$, $i = 1, \ldots, n$.

(7.4.10) Theorem. *Let the boundary-value problem* (7.4.1) *have a solution* $y(x) \in C^4[a, b]$, *and let* $|y^{(4)}(x)| \leqslant M$ *for* $x \in [a, b]$. *Also, let* $q(x) \geqslant 0$ *for* $x \in [a, b]$ *and* $u = [u_1, \ldots, u_n]^T$ *be the solution of* (7.4.6). *Then, for* $i = 1, 2, \ldots, n$,

$$|y(x_i) - u_i| \leqslant \frac{Mh^2}{24}(x_i - a)(b - x_i).$$

Proof. Because of (7.4.5) and (7.4.6) we have for $\bar{y} - u$ the equation

$$A(\bar{y} - u) = \tau(y).$$

Using the notation

$$|y| := \begin{bmatrix} |y_1| \\ \vdots \\ |y_n| \end{bmatrix} \quad \text{for } y \in \mathbb{R}^n,$$

we obtain from Theorem (7.4.7) and the representation (7.4.2) of $\tau(y)$,

$$\text{(7.4.11)} \qquad |\bar{y} - u| = |A^{-1}\tau(y)| \leqslant \tilde{A}^{-1}|\tau(y)| \leqslant \frac{Mh^2}{12}\tilde{A}^{-1}e,$$

where $e = [1, 1, \ldots, 1]^T$. The vector $\tilde{A}^{-1}e$ can be obtained at once by the following observation: The special boundary-value problem

$$-y'' = 1, \qquad y(a) = y(b) = 0,$$

of the type (7.4.1) has the exact solution $y(x) = \frac{1}{2}(x - a)(b - x)$. For this boundary-value problem, however, we have $\tau(y) = 0$ by (7.4.2), and the discrete solution u of (7.4.6) coincides with the exact solution $\bar{y}$ of (7.4.5). In addition, for this special boundary-value problem the matrix A in (7.4.4) is just the matrix $\tilde{A}$ in (7.4.8), and moreover $k = e$. We thus have $\tilde{A}^{-1}e = u$, $u_i = \frac{1}{2}(x_i - a)(b - x_i)$. Together with (7.4.11) this yields the assertion of the theorem. □

Under the assumptions of Theorem (7.4.10), the errors go to zero like h^2: the difference method has order 2. The method of Störmer and Numerov, which descretizes the differential equation $y'' = f(x, y)$ by

$$y_{i+1} - 2y_i + y_{i-1} = \frac{h^2}{12}(f_{i+1} + 10f_i + f_{i-1})$$

and leads to tridiagonal matrices, has order 4. All these methods can also be applied to nonlinear boundary value problems

$$y'' = f(x, y), \qquad y(a) = \alpha, \qquad y(b) = \beta.$$

We then obtain a system of nonlinear equations for the approximations $u_i \approx y(x_i)$, which in general can be solved only iteratively. At any rate, one obtains only methods of low order. To achieve high accuracy, one has to use a very fine subdivision of the interval $[a, b]$; in contrast, e.g., with the multiple shooting method (see Section 7.6 for comparative examples).

For an example of the application of difference methods to partial differential equations, see Section 8.4.

7.5 Variational Methods

Variational methods (Rayleigh–Ritz–Galerkin methods) are based on the fact that the solutions of some important types of boundary-value problems possess certain minimality properties. We want to explain these methods for the following simple boundary-value problem for a function $u : [a, b] \to \mathbb{R}$,

$$\text{(7.5.1)} \qquad \begin{gathered} -(p(x)u'(x))' + q(x)u(x) = g(x, u(x)), \\ u(a) = \alpha, \qquad u(b) = \beta. \end{gathered}$$

Note that the problem (7.5.1) is somewhat more general than (7.4.1).

Under the assumptions

$$\text{(7.5.2)} \qquad \begin{aligned} &p \in C^1[a, b], && p(x) \geqslant p_0 > 0, \\ &q \in C[a, b], && q(x) \geqslant 0, \\ &g \in C^1([a, b] \times \mathbb{R}), && g_u(x, u) \leqslant \lambda_0, \end{aligned}$$

with λ_0 the smallest eigenvalue of the eigenvalue problem

$$-(pz')' - (\lambda - q)z = 0, \qquad z(a) = z(b) = 0,$$

it is known that (7.5.1) always has exactly one solution. For the following we therefore assume (7.5.2) and make the simplifying assumption $g(x, u(x)) = g(x)$ (no u-dependence of the right-hand side).

If $u(x)$ is the solution of (7.5.1), then $y(x) := u(x) - l(x)$ with

$$l(x) := \alpha\frac{b - x}{b - a} + \beta\frac{a - x}{a - b}, \qquad l(a) = \alpha, \quad l(b) = \beta,$$

is the solution of a boundary-value problem of the form

$$-(py')' + qy = f, \qquad (7.5.3)$$
$$y(a) = 0, \qquad y(b) = 0,$$

with vanishing boundary values. Without loss of generality, we can thus consider, instead of (7.5.1), problems of the form (7.5.3). With the help of the *differential operator*

$$L(v) :\equiv -(pv')' + qv \qquad (7.5.4)$$

associated with (7.5.3), we want to formulate the problem (7.5.3) somewhat differently. The operator L maps the set

$$D_L := \{v \in C^2[a, b] \mid v(a) = 0,\ v(b) = 0\}$$

of all real functions that are twice continuously differentiable on $[a, b]$ and satisfy the boundary conditions $v(a) = v(b) = 0$ into the set $C[a, b]$ of continuous functions on $[a, b]$. The boundary-value problem (7.5.3) is thus equivalent to finding a solution of

$$L(y) = f, \qquad y \in D_L. \qquad (7.5.5)$$

Evidently, D_L is a real vector space and L a linear operator on D_L: for $u, v \in D_L$, also $\alpha u + \beta v$ belongs to D_L, and one has $L(\alpha u + \beta v) = \alpha L(u) + \beta L(v)$ for all real numbers α, β. On the set $L_2(a, b)$ of all square-integrable functions on $[a, b]$ we now introduce a bilinear form and a norm by means of the definition

$$(u, v) := \int_a^b u(x)v(x)\,dx, \qquad \|u\|_2 := (u, u)^{1/2}. \qquad (7.5.6)$$

The differential operator L in (7.5.4) has a few properties which are important for the understanding of the variational methods. One has the following:

(7.5.7) Theorem. *L is a symmetric operator on D_L, i.e., we have*

$$(u, L(v)) = (L(u), v) \quad \textit{for all } u, v \in D_L.$$

PROOF. Through integration by parts one finds

$$\begin{aligned}
(u, L(v)) &= \int_a^b u(x)[-(p(x)v'(x))' + q(x)v(x)]\,dx \\
&= -u(x)p(x)v'(x)\Big|_a^b + \int_a^b [p(x)u'(x)v'(x) + q(x)u(x)v(x)]\,dx \\
&= \int_a^b [p(x)u'(x)v'(x) + q(x)u(x)v(x)]\,dx,
\end{aligned}$$

since $u(a) = u(b) = 0$ for $u \in D_L$. For reasons of symmetry it follows likewise that

$$(7.5.8) \qquad (L(u), v) = \int_a^b [p(x)u'(x)v'(x) + q(x)u(x)v(x)]\, dx;$$

hence the assertion. □

The right-hand side of (7.5.8) is not only defined for $u, v \in D_L$. Indeed, let $D := \{u \in \mathscr{K}^1(a, b) \mid u(a) = u(b) = 0\}$ be the set of all functions u that are absolutely continuous on $[a, b]$ with $u(a) = u(b) = 0$, for which u' on $[a, b]$ (exists almost everywhere and) is square integrable [see Definition (2.4.1.3)]. In particular, all piecewise continuously differentiable functions satisfying the boundary conditions belong to D. D is again a real vector space with $D \supseteq D_L$. The right-hand side of (7.5.8) defines on D the symmetric bilinear form

$$(7.5.9) \qquad [u, v] := \int_a^b [p(x)u'(x)v'(x) + q(x)u(x)v(x)]\, dx,$$

which for $u, v \in D_L$ coincides with $(u, L(v))$. As above, one shows for $y \in D_L$, $u \in D$, through integration by parts, that

$$(7.5.10) \qquad (u, L(y)) = [u, y].$$

Relative to the scalar product introduced on D_L by (7.5.6), L is a *positive definite operator* in the following sense:

(7.5.11) Theorem. *Under the assumptions* (7.5.2) *one has*

$$[u, u] = (u, L(u)) > 0 \quad \text{for all } u \neq 0,\ u \in D_L.$$

One even has the estimate

$$(7.5.12) \qquad \gamma \|u\|_\infty^2 \leqslant [u, u] \leqslant \Gamma \|u'\|_\infty^2 \quad \text{for all } u \in D$$

with the norm $\|u\|_\infty := \sup_{a \leqslant x \leqslant b} |u(x)|$ *and the constants*

$$\gamma := \frac{p_0}{b - a}, \qquad \Gamma := \|p\|_\infty (b - a) + \|q\|_\infty (b - a)^3.$$

PROOF. In view of $\gamma > 0$ it suffices to show (7.5.12). For $u \in D$ we have, because $u(a) = 0$,

$$u(x) = \int_a^x u'(\xi)\, d\xi \quad \text{for } x \in [a, b].$$

The Schwarz inequality yields the estimate

$$[u(x)]^2 \leqslant \int_a^x 1^2\, d\xi \cdot \int_a^x [u'(\xi)]^2\, d\xi = (x-a)\int_a^x [u'(\xi)]^2\, d\xi$$
$$\leqslant (b-a)\int_a^b [u'(\xi)]^2\, d\xi,$$

and thus

$$\|u\|_\infty^2 \leqslant (b-a)\int_a^b [u'(x)]^2\, dx \leqslant (b-a)^2\|u'\|_\infty^2. \tag{7.5.13}$$

Now, by virtue of the assumption (7.5.2), we have $p(x) \geqslant p_0 > 0$, $q(x) \geqslant 0$ for $x \in [a, b]$; from (7.5.9) and (7.5.13) it thus follows that

$$[u, u] = \int_a^b (p(x)[u'(x)]^2 + q(x)[u(x)]^2)\, dx$$
$$\geqslant p_0 \int_a^b [u'(x)]^2\, dx$$
$$\geqslant \frac{p_0}{b-a}\|u\|_\infty^2.$$

By (7.5.13), finally, we also have

$$[u, u] = \int_a^b (p(x)[u'(x)]^2 + q(x)[u(x)]^2)\, dx$$
$$\leqslant \|p\|_\infty (b-a)\|u'\|_\infty^2 + \|q\|_\infty (b-a)\|u\|_\infty^2$$
$$\leqslant \Gamma\|u'\|_\infty^2,$$

as was to be shown. □

In particular, from (7.5.11) we may immediately deduce the uniqueness of the solution y of (7.5.3) or (7.5.5). If $L(y_1) = L(y_2) = f$, $y_1, y_2 \in D_L$, then $L(y_1 - y_2) = 0$, and hence $0 = (y_1 - y_2, L(y_1 - y_2)) \geqslant \gamma\|y_1 - y_2\|_\infty^2 \geqslant 0$, which yields at once $y_1 = y_2$.

We now define for $u \in D$ a quadratic functional by

$$F(u) := [u, u] - 2(u, f); \tag{7.5.14}$$

here f is the right-hand side of (7.5.3) or (7.5.5). F associates with each function $u \in D$ a real number $F(u)$. Fundamental for variational methods is the observation that the function F attains its smallest value exactly for the solution y of (7.5.5):

(7.5.15) Theorem. *Let $y \in D_L$ be the solution of* (7.5.5). *Then*

$$F(u) > F(y)$$

for all $u \in D$, $u \neq y$.

PROOF. We have $L(y) = f$ and therefore, by (7.5.10) and the definition of F, for $u \neq y$, $u \in D$,

$$\begin{aligned} F(u) &= [u, u] - 2(u, f) = [u, u] - 2(u, L(y)) \\ &= [u, u] - 2[u, y] + [y, y] - [y, y] \\ &= [u - y, u - y] - [y, y] \\ &> -[y, y] = F(y), \end{aligned}$$

since, by Theorem (7.5.11), $[u - y, u - y] > 0$ for $u \neq y$. □

As a side result, we note the identity

(7.5.16) $$[u - y, u - y] = F(u) + [y, y] \quad \text{for all } u \in D.$$

Theorem (7.5.15) suggests approximating the desired solution y by minimizing $F(u)$ approximately. Such an approximate minimum of F may be obtained systematically as follows: One chooses a finite-dimensional subspace S of D, $S \subset D$. If $\dim S = m$, then relative to a basis $u_1, \ldots, u_m$ of S, every $u \in S$ admits a representation of the form

(7.5.17) $$u = \delta_1 u_1 + \cdots + \delta_m u_m, \qquad \delta_i \in \mathbb{R}.$$

One then determines the minimum $u_S \in S$ of F on S,

(7.5.18) $$F(u_S) = \min_{u \in S} F(u),$$

and takes u_S to be an approximation for the exact solution y of (7.5.5), which according to (7.5.15) minimizes F on the whole space D. For the computation of the approximation u_S consider the following representation of $F(u)$, $u \in S$, obtained via (7.5.17),

$$\begin{aligned} \Phi(\delta_1, \delta_2, \ldots, \delta_m) &:\equiv F(\delta_1 u_1 + \cdots + \delta_m u_m) \\ &= \left[\sum_{i=1}^m \delta_i u_i, \sum_{k=1}^m \delta_k u_k\right] - 2\left(\sum_{k=1}^m \delta_k u_k, f\right) \\ &= \sum_{i,k=1}^m [u_i, u_k]\delta_i \delta_k - 2\sum_{k=1}^m (u_k, f)\delta_k. \end{aligned}$$

With the help of the vectors δ, φ and the $m \times m$ matrix A,

(7.5.19) $$\delta := \begin{bmatrix} \delta_1 \\ \vdots \\ \delta_m \end{bmatrix}, \qquad \varphi := \begin{bmatrix} (u_1, f) \\ \vdots \\ (u_m, f) \end{bmatrix}, \qquad A := \begin{bmatrix} [u_1, u_1] & \ldots & [u_1, u_m] \\ \vdots & & \vdots \\ [u_m, u_1] & \ldots & [u_m, u_m] \end{bmatrix},$$

one obtains for the quadratic function $\Phi : \mathbb{R}^m \to \mathbb{R}$

(7.5.20) $$\Phi(\delta) = \delta^T A \delta - 2\varphi^T \delta.$$

The matrix A is positive definite, since A is symmetric by (7.5.9), and for all vectors $\delta \neq 0$ one also has $u := \delta_1 u_1 + \cdots + \delta_m u_m \neq 0$ and thus, by Theorem (7.5.11),

$$\delta^T A \delta = \sum_{i,k} \delta_i \delta_k [u_i, u_k] = [u, u] > 0.$$

The system of linear equations

$$A\delta = \varphi, \tag{7.5.21}$$

therefore, has a unique solution $\delta = \bar{\delta}$, which can be computed by means of the Cholesky method (see Section 4.3). In view of the identity

$$\begin{aligned} \Phi(\delta) = \delta^T A \delta - 2\varphi^T \delta &= \delta^T A \delta - 2\bar{\delta}^T A \delta + \bar{\delta}^T A \bar{\delta} - \bar{\delta}^T A \bar{\delta} \\ &= (\delta - \bar{\delta})^T A (\delta - \bar{\delta}) - \bar{\delta}^T A \bar{\delta} \\ &= (\delta - \bar{\delta})^T A (\delta - \bar{\delta}) + \Phi(\bar{\delta}) \end{aligned}$$

and $(\delta - \bar{\delta})^T A (\delta - \bar{\delta}) > 0$ for $\delta \neq \bar{\delta}$, it follows at once that $\Phi(\delta) > \Phi(\bar{\delta})$ for $\delta \neq \bar{\delta}$, and consequently that the function

$$u_S := \bar{\delta}_1 u_1 + \cdots + \bar{\delta}_m u_m$$

belonging to $\bar{\delta}$ furnishes the minimum (7.5.18) of $F(u)$ on S. With y the solution of (7.5.5), it follows immediately from (7.5.16), by virtue of $F(u_S) = \min_{u \in S} F(u)$, that

$$[u_S - y, u_S - y] = \min_{u \in S} [u - y, u - y]. \tag{7.5.22}$$

We want to use this relation to estimate the error $\|u_S - y\|_\infty$. One has the following:

(7.5.23) Theorem. *Let y be the exact solution of* (7.5.3), (7.5.5). *Let $S \subset D$ be a finite-dimensional subspace of D, and let $F(u_S) = \min_{u \in S} F(u)$. Then the estimate*

$$\|u_S - y\|_\infty \leqslant C \|u' - y'\|_\infty$$

holds for all $u \in S$. Here $C = \sqrt{\Gamma/\gamma}$, where Γ, γ are the constants of Theorem (7.5.11).

PROOF. (7.5.12) and (7.5.22) yield immediately, for arbitrary $u \in S \subseteq D$,

$$\gamma \|u_S - y\|_\infty^2 \leqslant [u_S - y, u_S - y] \leqslant [u - y, u - y] \leqslant \Gamma \|u' - y'\|_\infty^2.$$

From this, the assertion follows. □

Every upper bound for $\inf_{u \in S} [u - y, u - y]$, or more weakly for

$$\inf_{u \in S} \|u' - y'\|_\infty,$$

immediately gives rise to an estimate for $\|u_S - y\|_\infty$. We wish to indicate such a bound for an important special case. We choose for the subspace S of D the set

$$S = \mathrm{Sp}_\Delta := \{S_\Delta \mid S_\Delta(a) = S_\Delta(b) = 0\}$$

of all cubic spline functions S_Δ [see Definition (2.4.1.1)] which belong to a fixed subdivision of the interval $[a, b]$,

$$\Delta : a = x_0 < x_1 < x_2 < \cdots < x_n = b,$$

and vanish at the end points a, b. Evidently, $\mathrm{Sp}_\Delta \subseteq D_L \subseteq D$. We denote by $\|\Delta\|$ the width of the largest subinterval of the subdivision Δ,

$$\|\Delta\| := \max_{1 \leqslant i \leqslant n} (x_i - x_{i-1}),$$

and further set

$$K := \max_{1 \leqslant i \leqslant n} \frac{\|\Delta\|}{x_i - x_{i-1}}$$

(for equidistant subdivisions, $K = 1$).

The spline function $u := S_\Delta$ with

$$u(x_i) = y(x_i), \qquad i = 0, 1, \ldots, n,$$
$$u'(\xi) = y'(\xi) \quad \text{for } \xi = a, b,$$

where y is the exact solution of (7.5.3), (7.5.5), clearly belongs to $S = \mathrm{Sp}_\Delta$.

From Theorem (2.4.3.3) one gets the estimate

$$\|u' - y'\|_\infty \leqslant \tfrac{7}{4} K \|y^{(4)}\|_\infty \cdot \|\Delta\|^3,$$

provided $y \in C^4[a, b]$. Together with Theorem (7.5.23) this gives the following result:

(7.5.24) Theorem. *Let the exact solution y of (7.5.3) belong to $C^4[a, b]$, and let the assumptions (7.5.2) be satisfied. Let $S := \mathrm{Sp}_\Delta$, and u_S be the spline function for which*

$$F(u_S) = \min_{u \in S} F(u).$$

Then there exist constants K and C which can be determined independently of y such that

$$\|u_S - y\|_\infty \leqslant \tfrac{7}{4} KC \|y^{(4)}\|_\infty \cdot \|\Delta\|^3.$$

By the footnote following Theorem (2.4.3.3), the estimate can be improved:

$$\|u_S - y\|_\infty \leqslant \tfrac{1}{24} C \|y^{(4)}\|_\infty \cdot \|\Delta\|^3.$$

The error bound thus goes to zero like the third power of the fineness of Δ; in this respect, therefore, the method is superior to the difference method of the previous section [see Theorem (7.4.10)].

For the practical implementation of the variational method, say in the case $S = Sp_\Delta$, one first has to select a basis for Sp_Δ. One easily sees that $m := \dim Sp_\Delta = n + 1$ [according to (2.4.1.2) the spline function $S_\Delta \in Sp_\Delta$ is uniquely determined by $n + 1$ conditions

$$S_\Delta(x_i) = y_i, \qquad i = 1, 2, \ldots, n-1,$$

$$S'_\Delta(a) = y'_0, \qquad S'_\Delta(b) = y'_n,$$

for arbitrary y_i, y'_0, y'_n]. As in Exercise 31, Chapter 2, one can find a basis of spline functions $S_0, S_1, \ldots, S_n$ in Sp_Δ such that

(7.5.25) $\quad S_j(x) = 0 \quad$ for $x \leqslant \max(x_0, x_{j-2})$ and $x \geqslant \min(x_n, x_{j+2})$.

This basis has the advantage that the corresponding matrix A in (7.5.19) is a *band matrix* of the form

$$\text{(7.5.26)} \qquad A = ([S_i, S_k]) = \begin{bmatrix} x & x & x & x & & & 0 \\ x & \cdot & & & \cdot & & \\ x & & \cdot & & & \cdot & \\ x & & & \cdot & & & \cdot \\ & \cdot & & & \cdot & & x \\ & & \cdot & & & \cdot & x \\ & & & \cdot & & \cdot & x \\ 0 & & & x & x & x & x \end{bmatrix},$$

since, by virtue of (7.5.9), (7.5.25),

$$[S_i, S_k] = \int_a^b [p(x)S'_i(x)S'_k(x) + q(x)S_i(x)S_k(x)]\, dx = 0$$

whenever $|i - k| \geqslant 4$. Once the components of the matrix A and of the vector φ in (7.5.19) for this basis $S_0, \ldots, S_n$ have been found by integration, one solves the system of linear equations (7.5.21) for δ and so obtains the approximation u_S for the exact solution y.

Of course, instead of Sp_Δ one can also choose other spaces $S \subseteq D$. For example, one could take for S the set

$$S = \left\{ P \,\middle|\, P(x) = (x - a)(x - b) \sum_{i=0}^{n-2} a_i x^i,\ a_i \text{ arbitrary} \right\}$$

of all polynomials of degree at most n which vanish at a and b. The matrix A in this case would in general no longer be a band matrix (and besides would be very ill-conditioned if the special polynomials

$$P_i(x) := (x - a)(x - b)x^i, \qquad i = 0, 1, \ldots, n-2,$$

were chosen as basis in S).

Similarly to the spline functions, one can also choose for S the set [see (2.1.5.11)]

$$S = H_\Delta^{(m)},$$

where Δ is again a subdivision $a = x_0 < x_1 < \cdots < x_n = b$ and $H_\Delta^{(m)}$ consists of all functions $u \in C^{m-1}[a, b]$ which in each subinterval $[x_i, x_{i+1}]$, $i = 0, 1, \ldots, n-1$, coincide with a polynomial of degree $\leqslant 2m - 1$ ("Hermite function space"). Here again, through a suitable choice of the basis $\{u_i\}$ in $H_\Delta^{(m)}$, one can assure that the matrix $A = ([u_i, u_k])$ is a band matrix. By appropriate choices of the parameter m one can even obtain methods of order higher than 3 in this way [cf. Theorem (7.5.24)]. For $S = H_\Delta^{(m)}$, analogously to Theorem (7.5.24), using the estimate (2.1.5.15) instead of (2.4.3.3), one obtains

$$\|u_S - y\|_\infty \leqslant \frac{C}{2^{2m-2}(2m-2)!} \|y^{(2m)}\|_\infty \|\Delta\|^{2m-1},$$

provided $y \in C^{2m}[a, b]$; we have a method of order at least $2m - 1$. [Proofs of these and similar estimates, as well as the generalization of the variational methods to certain nonlinear boundary-value problems, can be found in Ciarlet, Schultz, and Varga (1967).] Observe, however, that with increasing m the matrix A becomes more and more difficult to compute.

We further point out that the variational method can be applied to considerably more general boundary-value problems than (7.5.3), e.g., to partial differential equations. The important prerequisites for the crucial theorems (7.5.15), (7.5.23) and the solvability of (7.5.21) are essentially only the symmetry of L and the possibility of estimates of the form (7.5.12) (see Section 7.7 for applications to the Dirichlet problem).

The bulk of the work with methods of this type consists in computing the coefficients of the system of linear equations (7.5.21) by means of integration, having made a decision concerning an appropriate basis $u_1, \ldots, u_m$ in S. For boundary-value problems in ordinary differential equations these methods as a rule are too expensive and too inaccurate, and are not competitive with the multiple shooting method (see the following section for comparative results). Its value becomes evident only in boundary-value problems for partial differential equations.

Finite-dimensional spaces $S \subset D$ of functions satisfying the boundary conditions of the boundary value problem (7.5.5) also play a role in *collocation methods*. The idea of these methods is quite natural: One tries to approximate the solution $y(x)$ of (7.5.5) by a function $u(x)$ in S represented by $u = \delta_1 u_1 + \cdots + \delta_m u_m$ in terms of a basis $\{u_j\}$ of S. To this end, one selects m different *collocation points* $x_j \in (a, b)$, $j = 1, 2, \ldots, m$ and determines $u \in S$ such that

(7.5.27a) $$(Lu)(x_j) = f(x_j), \qquad j = 1, 2, \ldots, m,$$

that is, the differential equation $L(u) = f$ is to be satisfied exactly at the collocation points. Of course, this is equivalent to solving the following linear equations for the coefficients δ_k, $k = 1, 2, \ldots, m$:

(7.5.27b)
$$\sum_{k=1}^{m} L(u_k)(x_j)\delta_k = f(x_j), \qquad j = 1, 2, \ldots, m.$$

There are many possibilities of implementing such methods, namely, by different choices of S, of bases of S, and of collocation points x_j. With a suitable choice, one may obtain very efficient methods.

For instance, it is advantageous for the basis functions u_j, $j = 1, 2, \ldots, m$ to have compact support, like the B-splines [see 2.4.4 and (7.5.25)]. Then the matrix $A = [L(u_k)(x_j)]$ of the linear system (7.5.27) is a band matrix. In the so-called *spectral methods*, one chooses spaces S of trigonometric polynomials (2.3.1.1) and (2.3.1.2), which are spanned by finitely many simple trigonometric functions, say, e^{ikx}, $k = 0, \pm 1, \pm 2, \ldots$. Then, with a suitable choice of the collocation points x_j, one can often solve (7.5.27) with the aid of fast Fourier transforms (see 2.3.2).

For illustration, consider the simple boundary value problem

$$-y''(x) = f(x), \qquad y(0) = y(\pi) = 0,$$

of the form (7.5.3). Here, all functions $u(x) = \sum_{k=1}^{m} \delta_k \sin kx$ satisfy the boundary conditions. If we choose the collocation points $x_j := j\pi/(m+1)$, $j = 1, 2, \ldots, m$, then by (7.5.27) the numbers $\gamma_k := \delta_k k^2$ will solve the trigonometric interpolation problem (see 2.3.1)

$$\sum_{k=1}^{m} \gamma_k \sin \frac{kj\pi}{m+1} = f_j := f(x_j), \qquad j = 1, 2, \ldots, m.$$

Therefore, the γ_k can be computed by using fast Fourier transforms.

These methods play an important role for the solution of initial-boundary value problems for partial differential equations. For a detailed exposition, the reader is referred to the literature, e.g., Gottlieb and Orszag (1977), Canuto et al. (1987), and Boyd (1989).

7.6 Comparison of the Methods for Solving Boundary-Value Problems for Ordinary Differential Equations

The methods described in the previous sections, namely

(1) the simple shooting method,
(2) the multiple shooting method,

(3) the difference method,
(4) the variational method,

will now be compared by means of the following example. We consider the linear boundary-value problem with linear separated boundary conditions

$$-y'' + 400y = -400\cos^2 \pi x - 2\pi^2 \cos 2\pi x, \qquad y(0) = y(1) = 0, \tag{7.6.1}$$

with the exact solution (see Figure 23)

$$y(x) = \frac{e^{-20}}{1 + e^{-20}} e^{20x} + \frac{1}{1 + e^{-20}} e^{-20x} - \cos^2 \pi x. \tag{7.6.2}$$

Although this problem must be considered very simple compared with most problems in practice, the following peculiarities are present which lead us to expect difficulties in its solution:

(1) The homogeneous differential equation $-y'' + 400y = 0$ associated with (7.6.1) has solutions of the form $y(x) = ce^{\pm 20x}$ which grow or decay at a rapid exponential rate. This leads to difficulties in the simple shooting methods.
(2) The derivatives $y^{(i)}(x)$, $i = 1, 2, \ldots$, of the exact solution (7.6.2) are very large for $x \approx 0$ and $x \approx 1$. The error estimates (7.4.10) for the difference method, and (7.5.24) for the variational method, thus suggest large errors.

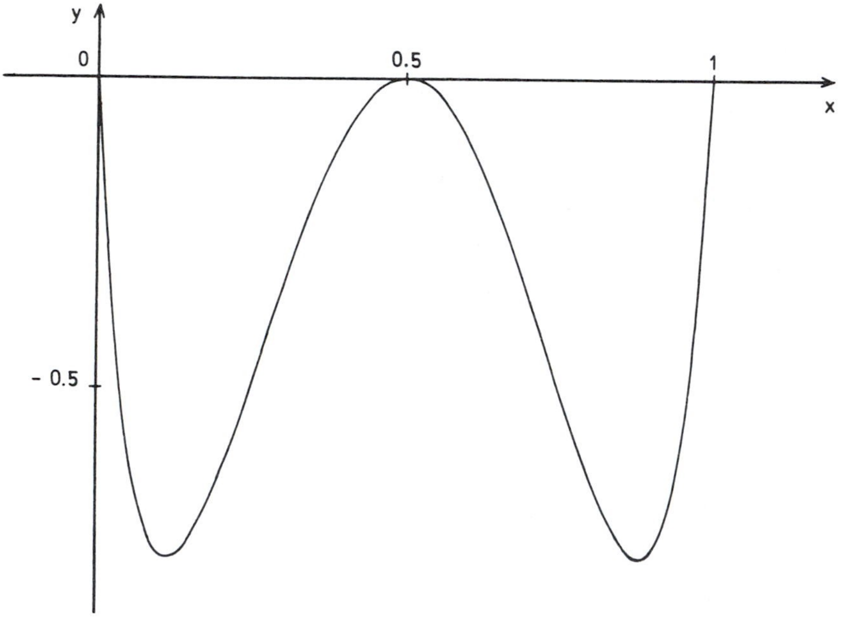

Figure 25 Exact solution of (7.6.1).

On a 12-digit computer the following results were found [in evaluating the error, especially the relative error, one ought to observe the behavior of the solution (see Figure 25): $\max_{0 \leqslant x \leqslant 1} |y(x)| = 0.77\ldots$, $y(0) = y(1) = 0$, $y(0.5) = 0.907\,998\,593\,379 \times 10^{-4}$]:

(a) *Simple shooting method:* The initial values

$$y(0) = 0, \qquad y'(0) = -20\frac{1 - e^{-20}}{1 + e^{-20}} = -19.999\,999\,9176\ldots$$

of the exact solution were found, after one iteration, with a relative error $\leqslant 3.2 \times 10^{-11}$. If the initial-value problem corresponding to (7.6.1) is then solved by means of an extrapolation method [see Section 7.2.14; ALGOL procedure *Diffsys* in Bulirsch and Stoer (1966)], taking as initial values the exact initial values $y(0)$, $y'(0)$ rounded to machine precision, one obtains instead of the exact solution $y(x)$ in (7.6.2) an approximate solution $\tilde{y}(x)$ having the absolute error $\Delta y = \tilde{y}(x) - y(x)$ and relative error $\varepsilon_y(x) = [\tilde{y}(x) - y(x)]/y(x)$ as shown in the table below. Here the effect of the exponentially growing solution $y(x) = ce^{20x}$ of the homogeneous problem is clearly noticeable (cf. Section 7.3.4).

(b) *Multiple shooting method*: To reduce the effect of the exponentially growing solution $y(x) = ce^{20x}$ of the homogeneous problem, we chose m large, $m = 21$:

$$0 = x_1 < x_2 < \cdots < x_{21} = 1, \qquad x_k = \frac{k-1}{20}.$$

x	$\lvert\Delta y(x)\rvert$[a]	$\lvert\varepsilon_y(x)\rvert$
0.1	1.9×10^{-11}	2.5×10^{-11}
0.2	1.5×10^{-10}	2.4×10^{-10}
0.3	1.1×10^{-9}	3.2×10^{-9}
0.4	8.1×10^{-9}	8.6×10^{-8}
0.5	6.0×10^{-8}	6.6×10^{-4}
0.6	4.4×10^{-7}	4.7×10^{-5}
0.7	3.3×10^{-6}	9.6×10^{-6}
0.8	2.4×10^{-5}	3.8×10^{-5}
0.9	1.8×10^{-4}	2.3×10^{-4}
1.0	1.3×10^{-3}	∞

[a] $\max_{x \in [0,1]} |\Delta y(x)| \approx 1.3 \times 10^{-3}$.

x	$\lvert\Delta y(x)\rvert$[a]	$\lvert\varepsilon_y(x)\rvert$
0.1	9.2×10^{-13}	1.2×10^{-12}
0.2	2.7×10^{-12}	4.3×10^{-12}
0.3	4.4×10^{-13}	1.3×10^{-12}
0.4	3.4×10^{-13}	3.6×10^{-12}
0.5	3.5×10^{-13}	3.9×10^{-9}
0.6	1.3×10^{-12}	1.4×10^{-11}
0.7	1.8×10^{-12}	5.3×10^{-12}
0.8	8.9×10^{-13}	1.4×10^{-12}
0.9	9.2×10^{-13}	1.2×10^{-12}
1.0	5.0×10^{-12}	∞

[a] $\max_{x \in [0,1]} \lvert\Delta y(x)\rvert \approx 5 \times 10^{-12}$.

Three iterations of the multiple shooting method [program in Oberle and Grimm (1989)] gave the absolute and relative errors in the above table.

(c) *Difference method*: The method of Section 7.4 for stepsizes $h = 1/(n+1)$ produces the following absolute errors $\Delta y = \max_{0 \leqslant i \leqslant n+1} \lvert\Delta y(x_i)\rvert$, $x_i = ih$ [powers of 2 were chosen for h in order that the matrix A in (7.4.4) could be calculated exactly on the computer]:

h	Δy
2^{-4}	2.0×10^{-2}
2^{-6}	1.4×10^{-3}
2^{-8}	9.0×10^{-5}
2^{-10}	5.6×10^{-6}

These errors are in harmony with the estimate of Theorem (7.4.10). Halving the step h reduces the error to $\frac{1}{4}$ of the old value. In order to achieve errors comparable to those of the multiple shooting method, $\Delta y \approx 5 \times 10^{-12}$, one would have to choose $h \approx 10^{-6}$!

(d) *Variational method*: In the method of Section 7.5 we choose for the spaces S spaces Sp_Δ of cubic spline functions corresponding to equidistant subdivisions

$$\Delta: \qquad 0 = x_0 < x_1 < \cdots < x_n = 1, \qquad x_i = ih, \qquad h = \frac{1}{n}.$$

The following maximum absolute errors $\Delta y = \|u_S - y\|_\infty$ [cf. Theorem (7.5.24)] were found:

h	Δy
1/10	6.0×10^{-3}
1/20	5.4×10^{-4}
1/30	1.3×10^{-4}
1/40	4.7×10^{-5}
1/50	2.2×10^{-5}
1/100	1.8×10^{-6}

In order to achieve errors of the order of magnitude $\Delta y \approx 5 \times 10^{-12}$ as in the multiple shooting method, one would have to choose $h \approx 10^{-4}$. Merely to compute the band matrix A in (7.5.26), this would require about 4×10^4 integrations for this stepsize.

These results clearly demonstrate the superiority of the multiple shooting method, even for simple linear separated boundary-value problems. Difference methods and variational methods are feasible, even here, only if the solution need not be computed very accurately. For the same accuracy, difference methods require the solution of larger systems of equations than variational methods. This advantage of the variational methods, however, is of importance only for smaller stepsizes h, and its significance is considerably reduced by the fact that the computation of the coefficients in the systems of equations is much more involved than in the simple difference methods. For the treatment of nonlinear boundary-value problems for ordinary differential equations the only feasible methods, effectively, are the multiple shooting method and its modifications.

7.7 Variational Methods for Partial Differential Equations. The Finite-Element Method

The methods described in Section 7.5 can also be used to solve boundary-value problems for partial differential equations. We want to explain this for a *Dirichlet boundary-value problem* in $\mathbb{R}^2$. We are given a (open bounded) region $\Omega \subseteq \mathbb{R}^2$ with boundary $\partial\Omega$. We seek a function $y : \bar{\Omega} \to \mathbb{R}$, $\bar{\Omega} := \Omega \cup \partial\Omega$, such that

$$(7.7.1) \qquad \begin{aligned} -\left[\frac{\partial^2 y(x)}{\partial x_1^2} + \frac{\partial^2 y(x)}{\partial x_2^2}\right] + c(x)y(x) = f(x) \quad &\text{for } x = (x_1, x_2) \in \Omega, \\ y(x) = 0 \qquad &\text{for } x \in \partial\Omega. \end{aligned}$$

Here $c, f: \bar{\Omega} \to \mathbb{R}$ are given continuous functions with $c(x) \geqslant 0$ for $x \in \bar{\Omega}$. To simplify the discussion, we assume that (7.7.1) has a solution $y \in C^2(\bar{\Omega})$, i.e., y has continuous derivatives $D_i^\alpha y := \partial^\alpha y/\partial x_i^\alpha$ on Ω for $\alpha \leqslant 2$ which are continuously extendable to $\bar{\Omega}$. As domain D_L of the differential operator $L(v) := -\Delta v + c \cdot v$ associated to (7.7.1) we accordingly take $D_L := \{v \in C^2(\bar{\Omega}) \mid v(x) = 0 \text{ for } x \in \partial\Omega\}$. Thus, the problem is to find a solution of

$$\text{(7.7.2)} \qquad L(v) = f, \qquad v \in D_L.$$

We further assume that the region Ω is such that the integral theorems of Gauss and Green apply, and, moreover, that each line $x_i = \text{const.}$, $i = 1, 2$, intersects Ω in at most finitely many segments. With the abbreviations

$$(u, v) := \int_\Omega u(x)v(x)\,dx, \quad \|u\|_2 := (u, u)^{1/2} \qquad (dx = dx_1 dx_2)$$

we then have [cf. Theorem (7.5.7)]:

(7.7.3) Theorem. *L is a symmetric operator on D_L:*

$$\text{(7.7.4)} \qquad \begin{aligned}(u, L(v)) = (L(u), v) = \int_\Omega [D_1 u(x) D_1 v(x) + D_2 u(x) D_2 v(x) \\ + c(x)u(x)v(x)]\,dx\end{aligned}$$

for all $u, v \in D_L$.

PROOF. One of Green's formulas is

$$-\int_\Omega u\,\Delta v\,dx = \int_\Omega \left(\sum_{i=1}^{2} D_i u\, D_i v\right) dx - \int_{\partial\Omega} u \frac{\partial v}{\partial \nu}\,d\omega.$$

Here $\partial v/\partial \nu$ is the differential quotient in the direction of the outward normal, and $d\omega$ the line element of $\partial\Omega$. Since $u \in D_L$, we have $u(x) = 0$ for $x \in \partial\Omega$; thus $\int_{\partial\Omega} u(\partial v/\partial \nu)\,d\omega = 0$. From this, the assertion of the theorem follows at once. □

The right-hand side of (7.7.4) again defines a bilinear form [cf. (7.5.9)]

$$\text{(7.7.5)} \qquad [u, v] := \int_\Omega \left(\sum_{i=1}^{2} D_i u D_i v + cuv\right) dx,$$

and there holds an analog of Theorem (7.5.11):

(7.7.6) Theorem. *There are constants $\gamma > 0$, $\Gamma > 0$ such that*

$$\gamma\|u\|^2_{W^{(1)}} \leqslant [u, u] \leqslant \Gamma\left(\sum_{i=1}^{2}\|D_i u\|_2^2\right) \quad \textit{for all } u \in D_L. \tag{7.7.7}$$

Here

$$\|u\|^2_{W^{(1)}} := (u, u) + (D_1 u, D_1 u) + (D_2 u, D_2 u)$$

is the so-called "Sobolev norm."

PROOF. The region Ω is bounded. There exists, therefore, a square Ω_1 with side length a which contains Ω in its interior. Without loss of generality, let the origin be a corner of Ω_1 (Figure 26).

Now let $u \in D_L$. Since $u(x) = 0$ for $x \in \partial\Omega$, we can extend u continuously onto Ω_1 by letting $u(x) := 0$ for $x \in \Omega_1 \backslash \Omega$. By our assumption on Ω, $D_1 u(t_1, \,.)$ is piecewise continuous, so that by $u(0, x_2) = 0$

$$u(x_1, x_2) = \int_0^{x_1} D_1 u(t_1, x_2)\, dt_1 \quad \text{for all } x \in \Omega_1.$$

The Schwarz inequality therefore gives

$$\begin{aligned}[u(x_1, x_2)]^2 &\leqslant x_1 \int_0^{x_1} [D_1 u(t_1, x_2)]^2\, dt_1 \\ &\leqslant a \int_0^{a} [D_1 u(t_1, x_2)]^2\, dt_1 \quad \text{for } x \in \Omega_1.\end{aligned}$$

Integrating this inequality over Ω_1, one obtains

$$\begin{aligned}\int_{\Omega_1} [u(x_1, x_2)]^2\, dx_1\, dx_2 &\leqslant a^2 \int_{\Omega_1} [D_1 u(t_1, t_2)]^2\, dt_1\, dt_2 \\ &\leqslant a^2 \int_{\Omega_1} [(D_1 u)^2 + (D_2 u)^2]\, dx.\end{aligned} \tag{7.7.8}$$

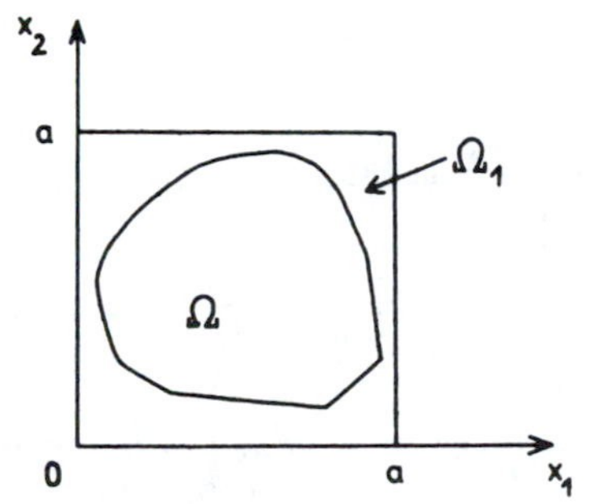

Figure 26 The regions Ω and Ω_1.

Since $u(x) = 0$ for $x \in \Omega_1 \setminus \Omega$, one can restrict the integration to Ω, and because $c(x) \geqslant 0$ for $x \in \bar{\Omega}$, it follows immediately that

$$a^2[u, u] \geqslant a^2 \sum_{i=1}^{2} \|D_i u\|_2^2 \geqslant \|u\|_2^2,$$

and from this finally

$$\gamma \|u\|_{W^{(1)}}^2 \leqslant [u, u] \quad \text{with } \gamma := \frac{1}{a^2 + 1}.$$

Again from (7.7.8) it follows that

$$\int_\Omega cu^2 \, dx \leqslant C \int_\Omega u^2 \, dx \leqslant Ca^2 \sum_{i=1}^{2} \|D_i u\|_2^2, \qquad C := \max_{x \in \Omega} |c(x)|,$$

and thus

$$[u, u] \leqslant \Gamma \sum_{i=1}^{2} \|D_i u\|_2^2 \quad \text{with } \Gamma := 1 + Ca^2.$$

This proves the theorem. □

As in Section 7.5, one can find a larger vector space $D \supseteq D_L$ of functions such that Definition (7.7.5) of $[u, v]$ remains meaningful for $u, v \in D$, the statement (7.7.7) of the previous theorem remains valid for $u \in D$, and

(7.7.9) $$[u, y] = (u, L(y)) \quad \text{for } y \in D_L \text{ and } u \in D.$$

In what follows, we are not interested in what the largest possible set D with this property looks like;* we are only interested in special finite-dimensional vector spaces S of functions for which the validity of (7.7.7) and (7.7.9) for $u \in D^S$, D^S the span of S and D_L, can easily be established individually.

Example. Let the set Ω be supplied with a triangulation $\mathcal{T} = \{T_1, \ldots, T_k\}$, i.e., $\bar{\Omega}$ is the union of finitely many triangles $T_i \in \mathcal{T}$, $\bar{\Omega} = \bigcup_{i=1}^{k} T_i$, such that any two triangles in $\mathcal{T}$ either are disjoint or have exactly one vertex or one side in common (Figure 27).

By S_i $(i \geqslant 1)$ we denote the set of all functions $u : \bar{\Omega} \to \mathbb{R}$ with the following properties:

(1) u is continuous on $\bar{\Omega}$.
(2) $u(x) = 0$ for $x \in \partial\Omega$.
(3) On each triangle T of the triangulation $\mathcal{T}$ of Ω the function u coincides with a polynomial of degree i,

$$u(x_1, x_2) = \sum_{j+k \leqslant i} a_{jk} x_1^j x_2^k.$$

* One obtains such a D by "completion" of the vector space $C_0^1(\Omega)$, equipped with the norm $\|\cdot\|_{W^{(1)}}$, of all once continuously differentiable functions φ on Ω, whose "support" $\operatorname{supp}(\varphi) := \overline{\{x \in \Omega \mid \varphi(x) \neq 0\}}$ is compact and contained in Ω.

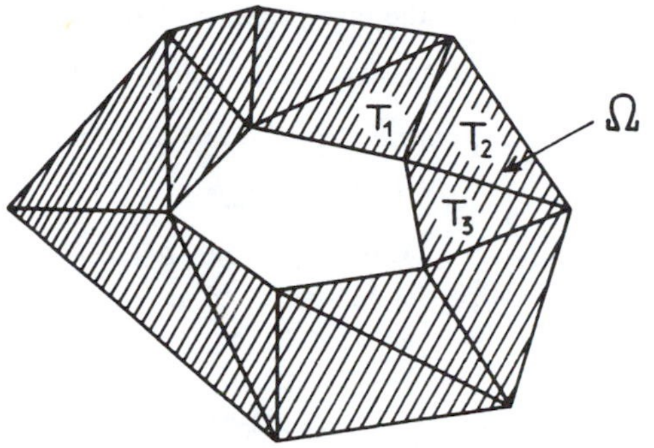

Figure 27 Triangulation of Ω.

Evidently, each S_i is a real vector space. Even though $u \in S_i$ need not even be continuously differentiable in Ω, it is nevertheless continuously differentiable in the interior of each triangle $T \in \mathscr{T}$. Definition (7.7.5) is meaningful for all $u, v \in D^{S_i}$; likewise, one sees immediately that (7.7.7) is valid for all $u \in D^{S_i}$. The function spaces $S = S_i$, $i = 1, 2, 3$—in particular S_2 and S_3—are used in the finite-element method in conjunction with the Rayleigh–Ritz method.

Exactly as in Section 7.5 [cf. (7.5.14), (7.5.15), (7.5.22)] one shows the following:

(7.7.10) Theorem. *Let y be the solution of* (7.7.2), *and S a finite-dimensional space of functions for which* (7.7.7), (7.7.9) *hold for all $u \in D^S$. Let $F(u)$ be defined by*

(7.7.11) $$F(u) := [u, u] - 2(u, f) \quad \text{for } u \in D^S.$$

Then:

(a) *$F(u) > F(y)$ for all $u \neq y$.*
(b) *There exists a $u_S \in S$ with $F(u_S) = \min_{u \in S} F(u)$.*
(c) *For this u_S one has*

$$[u_S - y, u_S - y] = \min_{u \in S} [u - y, u - y].$$

The approximate solution u_S can be represented as in (7.5.17)–(7.5.21) and determined by solving a linear system of equations of the form (7.5.21). For this, one must choose a basis in S and calculate the coefficients (7.5.19) of the system of linear equations (7.5.21).

A practically useful basis for S_2, e.g., is obtained in the following way: Let $\mathscr{P}$ be the set of all vertices and midpoints of sides of the triangles T_i, $i = 1, 2, \ldots, k$, of the triangulation $\mathscr{T}$ which are *not* located on the boundary $\partial\Omega$. One can show, then, that for each $P \in \mathscr{P}$ there exists exactly one function $u_P \in S_2$ satisfying

$$u_P(P) = 1, \qquad u_P(Q) = 0 \qquad \text{for } Q \neq P,\ Q \in \mathscr{P}.$$

In addition, these functions have the nice property that

$$u_P(x) = 0 \quad \text{for all } x \in T \quad \text{with } P \notin T;$$

it follows from this that in the matrix A of (7.5.21), all those elements vanish,

$$[u_P, u_Q] = 0,$$

for which there is no triangle $T \in \mathscr{T}$ to which both P and Q belong: the matrix A is sparse. Bases with similar properties can be found also for the spaces S_1 and S_3 [see Zlámal (1968)].

The error estimate of Theorem (7.5.23), too, carries over.

(7.7.12) Theorem. *Let y be the exact solution of (7.7.1), (7.7.2), and S a finite-dimensional space of functions such that (7.7.7), (7.7.9) holds for all $u \in D^S$. For the approximate solution u_S with $F(u_S) = \min_{u \in S} F(u)$ one then has the estimate*

$$\|u_S - y\|_{W^{(1)}} \leqslant \inf_{u \in S} \sqrt{\frac{\Gamma}{\gamma} \sum_{j=1}^{2} \|D_j u - D_j y\|_2^2}.$$

We now wish to indicate upper bounds for

$$\inf_{u \in S} \sqrt{\sum_{j=1}^{2} \|D_j u - D_j y\|_2^2}$$

in the case of the spaces $S = S_i$, $i = 1, 2, 3$, defined in the above example of triangulated regions Ω. The following theorem holds:

(7.7.13) Theorem. *Let Ω be a triangulated region with triangulation $\mathscr{T}$. Let h be the maximum side length, and θ the smallest angle, of all triangles occurring in $\mathscr{T}$. Then the following is true:*

(1) *If $y \in C^2[\bar{\Omega}]$ and*

$$\left|\frac{\partial^2 y(x_1, x_2)}{\partial x_i\, \partial x_j}\right| \leqslant M_2 \quad \textit{for all } 1 \leqslant i, j \leqslant 2, \qquad x \in \Omega,$$

then there exists a function $\tilde{u} \in S_1$ with

(a) $|\tilde{u}(x) - y(x)| \leqslant M_2 h^2$ *for all* $x \in \bar{\Omega}$,

(b) $|D_j \tilde{u}(x) - D_j y(x)| \leqslant 6M_2 h/\sin\theta$ *for* $j = 1, 2$ *and for all* $x \in T^0$, $T \in \mathscr{T}$.

(2) *If $y \in C^3[\bar{\Omega}]$ and*

$$\left|\frac{\partial^3 y(x_1, x_2)}{\partial x_i\, \partial x_j\, \partial x_k}\right| \leqslant M_3 \quad \textit{for all } 1 \leqslant i, j, k \leqslant 2, \qquad x \in \Omega,$$

then there exists a $\tilde{u} \in S_2$ with

(a) $|\tilde{u}(x) - y(x)| \leqslant M_3 h^3$ *for* $x \in \bar{\Omega}$,

(b) $|D_j \tilde{u}(x) - D_j y(x)| \leqslant 2M_3 h^2/\sin\theta$ *for* $j = 1, 2$, $x \in T^0$, $T \in \mathscr{T}$.

(3) *If* $y \in C^4[\bar{\Omega}]$ *and*

$$\left| \frac{\partial^4 y(x_1, x_2)}{\partial x_i \, \partial x_j \, \partial x_k \, \partial x_l} \right| \leqslant M_4 \quad \text{for } 1 \leqslant i, j, k, l \leqslant 4, \qquad x \in \Omega,$$

then there exists a $\tilde{u} \in S_3$ *with*
(a) $|\tilde{u}(x) - y(x)| \leqslant 3M_4 h^4/\sin\theta$ *for* $x \in \bar{\Omega}$,
(b) $|D_j\tilde{u}(x) - D_j y(x)| \leqslant 5M_4 h^3/\sin\theta$ *for* $j = 1, 2$, $x \in T^0$, $T \in \mathscr{T}$.

[Here, T^0 means the interior of the triangle T. The limitation $x \in T^0$, $T \in \mathscr{T}$ in the estimates (b) is necessary because the first derivatives $D_j \tilde{u}$ may have jumps along the sides of the triangles T.]

Since for $u \in S_i$, $i = 1, 2, 3$, the function $D_j u$ is only piecewise continuous, and since

$$\begin{aligned} \|D_j u - D_j y\|_2^2 &= \sum_{T \in \mathscr{T}} \int_T [D_j u(x) - D_j y(x)]^2 \, dx \\ &\leqslant C \max_{\substack{x \in T^0 \\ T \in \mathscr{T}}} |D_j u(x) - D_j y(x)|^2, \qquad C := \int_\Omega 1 \, dx, \end{aligned}$$

from each of the estimates given under (b) there immediately follows, via Theorem (7.7.12), an upper bound for the Sobolev norm $\|u_S - y\|_{W^{(1)}}$ of the error $u_S - y$ for the special function spaces $S = S_i$, $i = 1, 2, 3$, provided only that the exact solution y satisfies the differentiability assumptions of (7.7.13):

$$\|u_{S_1} - y\|_{W^{(1)}} \leqslant C_1 M_2 \frac{h}{\sin\theta},$$

$$\|u_{S_2} - y\|_{W^{(1)}} \leqslant C_2 M_3 \frac{h^2}{\sin\theta},$$

$$\|u_{S_3} - y\|_{W^{(1)}} \leqslant C_3 M_4 \frac{h^3}{\sin\theta}.$$

The constants C_i are independent of the triangulation $\mathscr{T}$ of Ω and of y, and can be specified explicitly; the constants M_i, h, θ have the same meaning as in (7.7.13). These estimates show, e.g. for $S = S_3$, that the error (i.e., its Sobolev norm) goes to zero like the third power of the fineness h of the triangulation.

We will not prove Theorem (7.7.13) here. Proofs can be found, e.g., in Zlàmal (1968) and in Ciarlet and Wagschal (1971). We only indicate that the special functions $\tilde{u} \in S_i$, $i = 1, 2, 3$, of the theorem are obtained by interpolation of y: For $S = S_2$, e.g., one obtains the function $\tilde{u} \in S_2$ by interpolation of y at the points $P \in \mathscr{P}$ of the triangulation (see above):

$$\tilde{u}(x) = \sum_{P \in \mathscr{P}} y(P) u_P(x).$$

An important advantage of the finite-element method lies in the fact that boundary-value problems can be treated also for relatively complicated regions Ω, provided Ω can still be triangulated. We remark, in addition, that not only can problems in $\mathbb{R}^2$ of the special form (7.7.1) be attacked by these methods, but also boundary-value problems in higher-dimensional spaces and for more general differential operators than the Laplace operator.

A detailed exposition of the finite element method and its practical implementation can be found in Strang and Fix (1973), Oden and Reddy (1976), Schwarz (1988), and Ciarlet and Lions (1991).

Exercises for Chapter 7

1. Let A be a real diagonalizable $n \times n$ matrix with the real eigenvalues $\lambda_1, \ldots, \lambda_n$ and the eigenvectors $c_1, \ldots, c_n$. How can the solution set of the system

$$y' = Ay$$

be described in terms of the λ_k and c_k? How is the special solution $y(x)$ obtained with $y(0) = y_0$, $y_0 \in \mathbb{R}^n$?

2. Determine the solution of the initial-value problem

$$y' = Jy, \qquad y(0) = y_0$$

with $y_0 \in \mathbb{R}^n$ and the $n \times n$ matrix

$$J = \begin{bmatrix} \lambda & 1 & & 0 \\ & \ddots & \ddots & \\ & & \ddots & 1 \\ 0 & & & \lambda \end{bmatrix}, \qquad \lambda \in \mathbb{R}.$$

Hint: Seek the kth component of the solution vector $y(x)$ in the form $y_k(x) = p_k(x)e^{\lambda x}$, p_k a polynomial of degree $\leqslant n - k$.

3. Consider the initial-value problem

$$y' = x - x^3, \qquad y(0) = 0.$$

Suppose we use Euler's method with stepsize h to compute approximate values $\eta(x_j; h)$ for $y(x_j)$, $x_j = jh$. Find an explicit formula for $\eta(x_j; h)$ and $e(x_j; h) = \eta(x_j; h) - y(x_j)$, and show that $e(x; h)$, for x fixed, goes to zero as $h = x/n \to 0$.

4. The initial-value problem

$$y' = \sqrt{y}, \qquad y(0) = 0$$

has the nontrivial solution $y(x) = (x/2)^2$. Application of Euler's method however yields $\eta(x; h) = 0$ for all x and $h = x/n$, $n = 1, 2, \ldots$. Explain this paradox.

5. Let $\eta(x; h)$ be the approximate solution furnished by Euler's method for the initial-value problem

$$y' = y, \qquad y(0) = 1.$$

(a) One has $\eta(x; h) = (1 + h)^{x/h}$.
(b) Show that $\eta(x; h)$ has the expansion

$$\eta(x; h) = \sum_{i=0}^{\infty} \tau_i(x) h^i \quad \text{with } \tau_0(x) = e^x,$$

which converges for $|h| < 1$; the $\tau_i(x)$ here are analytic functions independent of h.
(c) Determine $\tau_i(x)$ for $i = 1, 2, 3$.
(d) The $\tau_i(x)$, $i \geqslant 1$, are the solutions of the initial-value problems

$$\tau_i'(x) = \tau_i(x) - \sum_{k=1}^{i} \frac{\tau_{i-k}^{(k+1)}(x)}{(k+1)!}, \qquad \tau_i(0) = 0.$$

6. Show that the modified Euler method furnishes the exact solution of the differential equation $y' = -2ax$.

7. Show that the one-step method given by

$$\Phi(x, y; h) := \tfrac{1}{6}[k_1 + 4k_2 + k_3],$$
$$k_1 := f(x, y),$$
$$k_2 := f\left(x + \frac{h}{2}, y + \frac{h}{2} k_1\right),$$
$$k_3 := f(x + h, y + h(-k_1 + 2k_2))$$

("simple Kutta formula") is of third order.

8. Consider the one-step method given by

$$\Phi(x, y; h) := f(x, y) + \frac{h}{2} g(x + \tfrac{1}{3}h, y + \tfrac{1}{3}hf(x, y)),$$

where

$$g(x, y) := \frac{\partial}{\partial x} f(x, y) + \left(\frac{\partial}{\partial y} f(x, y)\right) \cdot f(x, y).$$

Show that it is a method of order 3.

9. What does the general solution of the difference equation

$$u_{j+2} = u_{j+1} + u_j, \qquad j \geqslant 0,$$

look like? (For $u_0 = 0$, $u_1 = 1$ one obtains the "Fibonacci sequence".)

10. In the methods of Adams–Moulton type, given approximate values $\eta_{p-j}, \ldots, \eta_p$ for $y(x_{p-j}), \ldots, y(x_p)$, one computes an approximate value η_{p+1} for $y(x_{p+1})$, $x_{p+1} \in [a, b]$, through the following iterative process [see (7.2.6.7)]:

$\eta_{p+1}^{(0)}$ arbitrary;

for $i = 0, 1, \ldots$:

$$\eta_{p+1}^{(i+1)} := \Psi(\eta_{p+1}^{(i)}) := \eta_p + h[\beta_{q0} f(x_{p+1}, \eta_{p+1}^{(i)}) + \beta_{q1} f_p + \cdots + \beta_{qq} f_{p+1-q}].$$

Show: For $f \in F_1(a, b)$ there exists an $h_0 > 0$ such that for all $|h| \leqslant h_0$ the sequence $\{\eta_{p+1}^{(i)}\}$ converges toward an η_{p+1} with $\eta_{p+1} = \Psi(\eta_{p+1})$.

11. Use the error estimates for interpolation by polynomials, or for Newton–Cotes formulas, to show that the Adams–Moulton method for $q = 2$ is of third order and the Milne method for $q = 2$ of fourth order.

12. For $q = 1$ and $q = 2$ determine the coefficients β_{qi} in Nyströms's formulas

$$\eta_{p+1} = \eta_{p-1} + h[\beta_{10} f_p + \beta_{11} f_{p-1}],$$

$$\eta_{p+1} = \eta_{p-1} + h[\beta_{20} f_p + \beta_{21} f_{p-1} + \beta_{22} f_{p-2}].$$

13. Check whether or not the linear multistep method

$$\eta_p - \eta_{p-4} = \frac{h}{3}[8f_{p-1} - 4f_{p-2} + 8f_{p-3}]$$

is convergent.

14. Let $\psi(1) = 0$, and assume that for the coefficients in

$$\frac{\psi(\mu)}{\mu - 1} = \gamma_{r-1}\mu^{r-1} + \gamma_{r-2}\mu^{r-2} + \cdots + \gamma_1 \mu + \gamma_0$$

one has $|\gamma_{r-1}| > |\gamma_{r-2}| \geqslant \cdots \geqslant |\gamma_0|$. Does $\psi(\mu)$ satisfy the stability condition?

15. Determine α, β, and γ such that the linear multistep method

$$\eta_{j+4} - \eta_j + \alpha(\eta_{j+3} - \eta_{j+1}) = h[\beta(f_{j+3} - f_{j+1}) + \gamma f_{j+2}]$$

has order 3. Is the method thus found stable?

16. Consider the predictor method given by

$$\eta_{j+2} + a_1 \eta_{j+1} + a_0 \eta_j = h[b_0 f(x_j, \eta_j) + b_1 f(x_{j+1}, \eta_{j+1})].$$

(a) Determine a_0, b_0, and b_1 as a function of a_1 such that the method has order at least 2.
(b) For which a_1-values is the method thus found stable?
(c) What special methods are obtained for $a_1 = 0$ and $a_1 = -1$?
(d) Can a_1 be so chosen that there results a stable method of order 3?

17. Suppose the method

$$\eta_{j+2} + 9\eta_{j+1} - 10\eta_j = \frac{h}{2}[13f_{j+1} + 9f_j]$$

is applied to the initial-value problem

$$y' = 0, \qquad y(0) = c.$$

Let the starting values be $\eta_0 = c$ and $\eta_1 = c + \text{eps}$ (eps = machine precision). What values η_j are to be expected for arbitrary stepsize h?

18. For the solution of the boundary-value problem

$$y'' = 100y, \qquad y(0) = 1, \qquad y(3) = e^{-30},$$

consider the initial-value problem

$$y'' = 100y, \qquad y(0) = 1, \qquad y'(0) = s,$$

with solution $y(x; s)$, and determine $s = \bar{s}$ iteratively such that $y(3; \bar{s}) = e^{-30}$. Assume further that $\bar{s}$ is computed only within a relative error ε, i.e., instead of $\bar{s}$ one obtains $\bar{s}(1 + \varepsilon)$. How large is $y(3; \bar{s}(1 + \varepsilon))$? In this case, is the simple shooting method (as described above) a suitable method for the solution of the boundary-value problem?

19. Consider the initial-value problem

$$y' = \kappa(y + y^3), \qquad y(0) = s.$$

(a) Determine the solution $y(x; s)$ of this problem.
(b) In which x-neighborhood $U_s(0)$ of 0 is $y(x; s)$ defined?
(c) For a given $b \neq 0$ find a $k > 0$ such that $y(b; s)$ exists for all $|s| < k$.

20. Show that assumption (3) of Theorem (7.3.3.4) in the case $n = 2$ is not satisfied for the boundary conditions

$$y_1(a) = c_1, \qquad y_1(b) = c_2.$$

21. Under the assumptions of Theorem (7.3.3.4) prove that $E := A + BG_{m-1} \dots G_1$ in (7.3.5.10) is nonsingular.
Hint: $E = P_0(I + H)$, $H = M + P_0^{-1} D_v r \cdot (Z - I)$.

22. Show by an error analysis (backward analysis) that the solution vector of (7.3.5.10), (7.3.5.9), subject to rounding errors, can be interpreted as the *exact* result of (7.3.5.8) with *slightly* perturbed right-hand sides $\tilde{F}_j$ and perturbed last equation

$$(A + E_1)\,\Delta s_1 + E_2\,\Delta s_2 + \cdots + E_{m-1}\,\Delta s_{m-1} + (B + E_m)\,\Delta s_m = -\tilde{F}_m,$$

$\operatorname{lub}(E_j)$ small.

23. Let $DF(s)$ be the matrix in (7.3.5.5). Prove:

$$\det(DF(s)) = \det(A + BG_{m-1} \dots G_1).$$

For the inverse $(DF(s))^{-1}$ find explicitly a decomposition in block matrices R, S, T:

$$(DF(s))^{-1} = RST$$

with S block-diagonal, R unit lower block triangular, and S, T chosen so that $ST \cdot DF(s)$ is unit lower block triangular.

24. Let $\Delta \in \mathbb{R}^n$ be arbitrary. Prove: If $\Delta s_1 := \Delta$ and Δs_j, $j = 2, \dots, m$ is obtained from (7.3.5.9), then

$$\Delta s = \begin{bmatrix} \Delta s_1 \\ \vdots \\ \Delta s_m \end{bmatrix}$$

satisfies

$$(DF(s) \cdot F)^T\,\Delta s = -F^T F + F_m^T((A + BG_{m-1} \dots G_1)\,\Delta - w).$$

Thus, if Δ is a solution of (7.3.5.10), then always

$$(DF(s) \cdot F)^T \, \Delta s < 0.$$

25. Consider the boundary-value problem

$$y' = f(x, y)$$

with the *separated* boundary conditions

$$\begin{bmatrix} y_1(a) - \alpha_1 \\ \vdots \\ y_k(a) - \alpha_k \\ y_{k+1}(b) - \beta_{k+1} \\ \vdots \\ y_n(b) - \beta_n \end{bmatrix} = 0.$$

There are thus k initial values and $n - k$ terminal values which are known. The information can be used to reduce the dimension of the system of equations (7.3.5.8) (k components of Δs_1 and $n - k$ components of Δs_m are zero). For $m = 3$ construct the system of equations for the corrections Δs_i. Integrate (using the given information) from $x_1 = a$ to x_2 as well as from $x_3 = b$ to x_2 ("counter shooting"; see Figure 28).

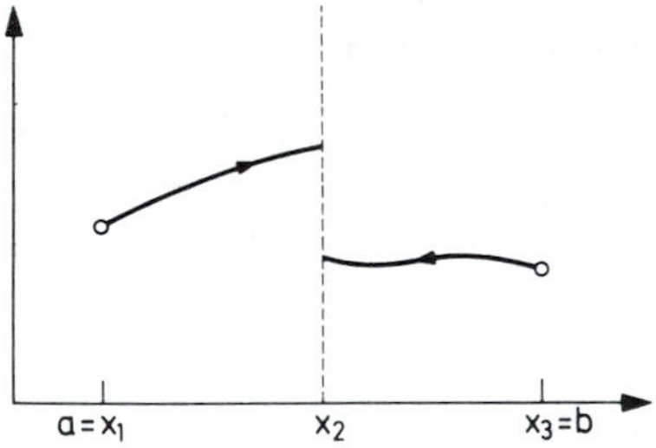

Figure 28 Counter shooting.

26. The steady concentration of a substrate in a spherical cell of radius 1 in an enzyme catalyzed reaction is described by the solution $y(x; \alpha)$ of the following singular boundary-value problem [cf. Keller (1968)]:

$$y'' = -\frac{2y'}{x} + \frac{y}{\alpha(y + k)},$$

$$y'(0) = 0, \qquad y(1) = 1,$$

α, k parameters, $k = 0.1$, $10^{-3} \leqslant \alpha \leqslant 10^{-1}$. Although f is singular at $x = 0$, there exists a solution $y(x)$ which is analytic for small $|x|$. In spite of this, every numerical integration method fails near $x = 0$. Help yourself as follows: By using the symmetries of $y(x)$, expand $y(x)$ as in Example 2, Section 7.3.6, into a Taylor series about 0 up to and including the term x^6; thereby express all coefficients in terms of $\lambda := y(0)$. By means of $p(x; \lambda) = y(0) + y'(0)x + \cdots + y^{(6)}(0)x^6/6!$ one obtains a modified problem

$$(*) \qquad y'' = F(x, y, y') = \begin{cases} \dfrac{d^2p(x; \lambda)}{dx^2} & \text{if } 0 \leqslant x \leqslant 10^{-2}, \\ f(x, y, y') & \text{if } 10^{-2} < x \leqslant 1, \end{cases}$$

$$y(0) = \lambda, \qquad y'(0) = 0, \qquad y(1) = 1,$$

which is better suited for numerical solution. Consider $(*)$ an eigenvalue problem for the eigenvalue λ, and formulate $(*)$ as in Section 7.3.0 as a boundary-value problem (without eigenvalue). For checking purposes we give the derivatives:

$$y(0) = \lambda, \qquad y^{(2)}(0) = \frac{\lambda}{3\alpha(\lambda + k)}, \qquad y^{(4)}(0) = \frac{k\lambda}{5\alpha^2(\lambda + k)^3},$$

$$y^{(6)}(0) = \frac{(3k - 10\lambda)k\lambda}{21\alpha^3(\lambda + k)^5}, \qquad y^{(i)}(0) = 0 \quad \text{for } i = 1, 3, 5.$$

27. Consider the boundary-value problem

$$y'' - p(x)y' - q(x)y = r(x), \qquad y(0) = \alpha, \quad y(b) = \beta$$

with $q(x) \geqslant q_0 > 0$ for $a \leqslant x \leqslant b$. We are seeking approximate values u_i for the exact values $y(x_i)$, $i = 1, 2, \ldots, n$, at $x_i = a + ih$, $h = (b - a)/(n + 1)$. Replacing $y'(x_i)$ by $(u_{i+1} - u_{i-1})/2h$ and $y''(x_i)$ by $(u_{i-1} - 2u_i + u_{i+1})/h^2$ for $i = 1, 2, \ldots, n$, and putting $u_0 = \alpha$, $u_{n+1} = \beta$, one obtains from the differential equation a system of equations for the vector $u = [u_1, \ldots, u_n]^T$

$$Au = k, \qquad A \text{ an } n \times n \text{ matrix}, \quad k \in \mathbb{R}^n.$$

(a) Determine A and k.
(b) Show that the system of equations is uniquely solvable for h sufficiently small.

28. Consider the boundary-value problem

$$y'' = g(x), \qquad y(0) = y(1) = 0,$$

with $g \in C[0, 1]$.
(a) Show:

$$y(x) = \int_0^1 G(x, \xi)g(\xi)\,d\xi$$

with

$$G(x, \xi) := \begin{cases} \xi(x - 1) & \text{for } 0 \leqslant \xi \leqslant x \leqslant 1, \\ x(\xi - 1) & \text{for } 0 \leqslant x \leqslant \xi \leqslant 1. \end{cases}$$

(b) Replacing $g(x)$ by $g(x) + \Delta g(x)$ with $|\Delta g(x)| \leqslant \varepsilon$ for all x, the solution $y(x)$ changes to $y(x) + \Delta y(x)$. Prove:

$$|\Delta y(x)| \leqslant \frac{\varepsilon}{2} x(1 - x) \quad \text{for } 0 \leqslant x \leqslant 1.$$

(c) The difference method described in Section 7.4 yields the system of equations [see (7.4.6)]

$$Au = k,$$

the solution vector $u = [u_1, \ldots, u_n]^T$ of which provides approximate values u_i for $y(x_i)$, $x_i = i/(n+1)$ and $i = 1, 2, \ldots, n$. Replacing k by $k + \Delta k$ with $|\Delta k_i| \leqslant \varepsilon$, $i = 1, 2, \ldots, n$, the vector u changes to $u + \Delta u$. Show:

$$|\Delta u_i| \leqslant \frac{\varepsilon}{2} x_i(1 - x_i), \qquad i = 1, 2, \ldots, n.$$

29. Let

$$D := \{u \mid u(0) = 0,\ u \in C^2[0, 1]\}$$

and

$$F(u) := \int_0^1 \{\tfrac{1}{2}(u'(x))^2 + f(x, u(x))\}\, dx + p(u(1))$$

with $u \in D$, $f_{uu}(x, u) \geqslant 0$, $p''(u) \geqslant 0$. Prove: If $y(x)$ is solution of

$$y'' - f_u(x, y) = 0,$$

$$y(0) = 0, \qquad y'(1) + p'(y(1)) = 0,$$

then

$$F(y) < F(u), \qquad u \in D, \quad u \neq y,$$

and vice versa.

30.

(a) Let T be a triangle in $\mathbb{R}^2$ with the vertices P_1, P_3, and P_5; let further $P_2 \in \overline{P_1 P_3}$, $P_4 \in \overline{P_3 P_5}$, and $P_6 \in \overline{P_5 P_1}$ be different from P_1, P_3, P_5. Then for arbitrary real numbers $y_1, \ldots, y_6$ there exists exactly one polynomial of degree at most 2,

$$u(x_1, x_2) = \sum_{0 \leqslant j+k \leqslant 2} a_{jk} x_1^j x_2^k,$$

which assumes the values y_i at the points P_i, $i = 1, \ldots, 6$.
Hint: It evidently suffices to show this for a single triangle with suitably chosen coordinates.

(b) Is the statement analogous to (a) valid for arbitrary position of the P_i?

(c) Let T_1, T_2 be two triangles of a triangulation with a common side g and u_1, u_2 polynomials of degree at most 2 on T_1 and T_2, respectively. Show: If u_1 and u_2 agree in three distinct points of g, then u_2 is a continuous extension of u_1 onto T_2 (i.e., u_1 and u_2 agree on all of g).

References for Chapter 7

Babuška, I., Prager, M., Vitásek, E.: *Numerical Processes in Differential Equations.* New York: Interscience (1966).

Bader, G., Deuflhard, P.: A semi-implicit midpoint rule for stiff systems of ordinary systems of differential equations. *Numer. Math.* **41**, 373–398 (1983).

Bank, R. E., Bulirsch, R., Merten, K.: Mathematical modelling and simulation of electrical circuits and semiconductor devices. *ISNM* 93, Basel: Birkhäuser (1990).

Boyd, J. P.: *Chebyshev and Fourier Spectral Methods*. Berlin, Heidelberg, New York: Springer (1989).

Bulirsch, R.: Die Mehrzielmethode zur numerischen Lösung von nichtlinearen Randwertproblemen und Aufgaben der optimalen Steuerung. Report of the Carl-Cranz-Gesellschaft (1971).

———, Stoer, J.: Numerical treatment of ordinary differential equations by extrapolation methods. *Numer. Math.* **8**, 1–13 (1966).

Butcher, J. C.: On Runge-Kutta processes of high order. *J. Austral. Math. Soc.* **4**, 179–194 (1964).

Byrne, D. G., Hindmarsh, A. C.: Stiff o.d.e.-solvers: A review of current and coming attractions. *J. Comp. Phys.* **70**, 1–62 (1987).

Ciarlet, P. G., Lions, J. L. (Eds.): *Handbook of Numerical Analysis. Vol. II. Finite Element Methods (Part 1)*. Amsterdam: North Holland (1991).

———, Schultz, M. H., Varga, R. S.: Numerical methods of high order accuracy for nonlinear boundary value problems. *Numer. Math.* **9**, 394–430 (1967).

———, Wagschal, C.: Multipoint Taylor formulas and applications to the finite element method. *Numer. Math.* **17**, 84–100 (1971).

Canuto, C., Hussaini, M. Y., Quarteroni, A., Zang, T. A.: *Spectral Methods in Fluid Dynamics*. Berlin, Heidelberg, New York: Springer (1987).

Coddington, E. A., Levinson, N.: *Theory of Ordinary Differential Equations*. New York: McGraw-Hill (1955).

Collatz, L.: *The Numerical Treatment of differential Equations*. Berlin: Springer (1960).

———: *Functional Analysis and Numerical Mathematics*. New York: Academic Press (1966).

Dahlquist, G.: Convergence and stability in the numerical integration of ordinary differential equations. *Math. Scand.* **4**, 33–53 (1956).

———: Stability and error bounds in the numerical integration of ordinary differential equations. *Trans. Roy. Inst. Tech.* (Stockholm), No. 130 (1959).

———: A special stability problem for linear multistep methods. *BIT* **3**, 27–43 (1963).

Deuflhard, P.: A stepsize control for continuation methods and its special application to multiple shooting techniques. *Numer. Math.* **33**, 115–146 (1979).

———, Hairer, E., Zugck, J.: One-step and extrapolation methods for differential-algebraic systems. *Numer. Math.* **51**, 501–516 (1987).

Diekhoff, H.-J., Lory, P., Oberle, H. J., Pesch, H.-J., Rentrop, P., Seydel, R.: Comparing routines for the numerical solution of initial value problems of ordinary differential equations in multiple shooting. *Numer. Math.* **27**, 449–469 (1977).

Dormand, J. R., Prince, P. J.: A family of embedded Runge-Kutta formulae. *J. Comp. Appl. Math.* **6**, 19–26 (1980).

Enright, W. H., Hull, T. E., Lindberg, B.: Comparing numerical methods for stiff systems of ordinary differential equations. *BIT* **55**, 10–48 (1975).

Fehlberg, E.: New high-order Runge-Kutta formulas with stepsize control for systems of first- and second-order differential equations. *Z. Angew. Math. Mech.* **44**, T17–T29 (1964).

———: New high–order Runge–Kutta formulas with an arbitrary small truncation error. *Z. Angew. Math. Mech.* **46**, 1–16 (1966).

———: Klassische Runge–Kutta Formeln fünfter und siebenter Ordnung mit Schrittweiten–Kontrolle. *Computing* **4**, 93–106 (1969).

Gantmacher, R. R.: *Matrizenrechnung II*. Berlin: VEB Deutscher Verlag der Wissenschaften (1969).

Gear, C. W.: *Numerical Initial Value Problems in Ordinary Differential Equations*. Englewood Cliffs, N.J.: Prentice-Hall (1971).

———: Differential algebraic equation index transformations. *SIAM J. Sci. Statist. Comput.* **9**, 39–47 (1988).

———, Petzold, L. R.: ODE methods for the solution of differential/algebraic systems. *SIAM J. Numer. Anal.* **21**, 716–728 (1984).

Gottlieb, D., Orszag, S. A.: *Numerical Analysis of Spectral Methods: Theory and Applications.* Philadelphia: SIAM (1977).

Gragg, W.: Repeated extrapolation to the limit in the numerical solution of ordinary differential equations. Thesis, UCLA (1963).

———: On extrapolation algorithms for ordinary initial value problems. *J. SIAM Numer. Anal. Ser. B* **2**, 384–403 (1965).

Griepentrog, E., März, R.: *Differential-Algebraic Equations and Their Numerical Treatment.* Leipzig: Teubner (1986).

Grigorieff, R. D.: *Numerik gewöhnlicher Differentialgleichungen 1, 2.* Stuttgart: Teubner (1972, 1977).

Hairer, E., Lubich, Ch.: Asymptotic expansions of the global error of fixed step-size methods. *Numer. Math.* **45**, 345–360 (1984).

———, ———, Roche, M.: The numerical solution of differential-algebraic systems by Runge–Kutta methods. In: *Lecture Notes in Mathematics* 1409. Berlin, Heidelberg, New York: Springer (1989).

———, Nørsett, S. P., Wanner, G.: *Solving Ordinary Differential Equations I. Nonstiff Problems.* Berlin, Heidelberg, New York: Springer (1987).

———, Wanner, G.: *Solving Ordinary Differential Equations II. Stiff and Differential-Algebraic Problems.* Berlin, Heidelberg, New York: Springer (1991).

Henrici, P.: *Discrete Variable Methods in Ordinary Differential Equations.* New York: John Wiley (1962).

Hestenes, M. R.: *Calculus of Variations and Optimal Control Theory.* New York: John Wiley (1966).

Heun, K.: Neue Methode zur Approximativen Integration der Differentialgleichungen einer unabhängigen Variablen. *Z. Math. Phys.* **45**, 23–38 (1900).

Horneber, E. H.: Simulation elektrischer Schaltungen auf dem Rechner. Fachberichte Simulation, Bd. 5. Berlin, Heidelberg, New York: Springer (1985).

Hull, E. E., Enright, W. H., Fellen, B. M., Sedgwick, A. E.: Comparing numerical methods for ordinary differential equations. *SIAM J. Numer. Anal.* **9**, 603–637 (1972). [Errata, *ibid.* **11**, 681 (1974).]

Isaacson, E., Keller, H. B.: *Analysis of Numerical Methods.* New York: John Wiley (1966).

Kaps, P., Rentrop, P.: Generalized Runge–Kutta methods of order four with stepsize control for stiff ordinary differential equations. *Numer. Math.* **33**, 55–68 (1979).

Keller, H. B.: *Numerical Methods for Two-Point Boundary-Value Problems.* London: Blaisdell (1968).

Krogh, F. T.: Changing step size in the integration of differential equations using modified divided differences. In: *Proceedings of the Conference on the Numerical Solution of Ordinary Differential Equations*, 22–71. Lecture Notes in Mathematics 362. Berlin, Heidelberg, New York: Springer (1974).

Kutta, W.: Beitrag zur näherungsweisen Integration totaler Differentialgleichungen. *Z. Math. Phys.* **46**, 435–453 (1901).

Lambert, J. D.: *Computational Methods in Ordinary Differential Equations.* London, New York, Sydney, Toronto: John Wiley (1973).

Na, T. Y., Tang, S. C.: A method for the solution of conduction heat transfer with non-linear heat generation. *Z. Angew. Math. Mech.* **49**, 45–52 (1969).

Oberle, H. J., Grimm, W.: BNDSCO—A program for the numerical solution of optimal control problems. Internal Report No. 515-89/22, Institute for Flight Systems Dynamics, DLR, Oberpfaffenhofen, Germany (1989).

Oden, J. T., Reddy, J. N.: *An Introduction to the Mathematical Theory of Finite Elements*. New York: John Wiley (1976).
Osborne, M. R.: On shooting methods for boundary value problems. *J. Math. Anal. Appl.* **27**, 417–433 (1969).
Petzold, L. R.: A description of DASSL—A differential algebraic solver. *IMACS Trans. Sci. Comp.* **1**, 65ff., Ed. H. Stepleman. Amsterdam: North Holland (1982).
Rentrop, P.: Partitioned Runge–Kutta methods with stiffness detection and step-size control. *Numer. Math.* **47**, 545–564 (1985).
———, Roche, M., Steinebach, G.: The application of Rosenbrock-Wanner type methods with stepsize control in differential-algebraic equations. *Numer. Math.* **55**, 545–563 (1989).
Runge, C.: Über die numerische Auflösung von Differentialgleichungen. *Math. Ann.* **46**, 167–178 (1895).
Schwarz, H. R.: *Finite Element Methods*. London, New York: Academic Press (1988).
Shampine, L. F., Gordon, M. K.: *Computer Solution of Ordinary Differential Equations. The Initial Value Problem*. San Francisco: Freeman and Company (1975).
———, Watts, H. A., Davenport, S. M.: Solving nonstiff ordinary differential equations—The state of the art. *SIAM Review* **18**, 376–411 (1976).
Shanks, E. B.: Solution of differential equations by evaluation of functions. *Math. Comp.* **20**, 21–38 (1966).
Stetter, H. J.: *Analysis of Discretization Methods for Ordinary Differential Equations*. Berlin, Heidelberg, New York: Springer (1973).
Strang, G., Fix, G. J.: *An Analysis of the Finite Element Method*. Englewood Cliffs, N.J.: Prentice-Hall (1973).
Troesch, B. A.: Intrinsic difficulties in the numerical solution of a boundary value problem. Report NN-142, TRW, Inc. Redondo Beach, CA (1960).
———: A simple approach to a sensitive two-point boundary value problem. *J. Computational Phys.* **21**, 279–290 (1976).
Wilkinson, J. H.: Note on the practical significance of the Drazin inverse. In: L. S. Campbell, Ed. *Recent Applications of Generalized Inverses*. Pitman Publ. **66**, 82–89 (1982).
Willoughby, R. A.: *Stiff Differential Systems*. New York: Plenum Press (1974).
Zlámal, M.: On the finite element method. *Numer. Math.* **12**, 394–409 (1968).

8 Iterative Methods for the Solution of Large Systems of Linear Equations. Some Further Methods

8.0 Introduction

Many problems in practice require the solution of very large systems of linear equations $Ax = b$ in which the matrix A, fortunately, is sparse, i.e., has relatively few nonvanishing elements. Systems of this type arise, e.g., in the application of difference methods or finite-element methods to the approximate solution of boundary-value problems in partial differential equations. The usual elimination methods (see Chapter 4) cannot normally be applied here, since without special precautions they tend to lead to the formation of more or less dense intermediate matrices, making the number of arithmetic operations necessary for the solution much too large, even for present-day computers, not to speak of the fact that the intermediate matrices no longer fit into the usually available computer memory.

For these reasons, researchers have long since moved to *iterative methods* for solving such systems of equations. In these methods, beginning with an initial vector $x^{(0)}$, one generates a sequence of vectors

$$x^{(0)} \to x^{(1)} \to x^{(2)} \to \cdots$$

which converge toward the desired solution x. The common feature of all these methods is the fact that an individual iteration step $x^{(i)} \to x^{(i+1)}$ requires an amount of work which is comparable to the multiplication of A with a vector—a very modest amount if A is sparse. For this reason, one can, with a reasonable amount of work, still carry out a relatively large number of iterations. This is necessary, if for no other reasons than the fact that these methods converge only linearly, and very slowly at that. Iterative methods are therefore usually inferior to the elimination methods if A is a small matrix (a 100×100 matrix is small in this sense) or not a sparse matrix.

Somewhat outside of this framework are the so-called *method of iterative refinement* [see (8.1.9)–(8.1.11)] and the *conjugate-gradient method* (see Section 8.7). Iterative refinement is used relatively frequently and serves to iteratively improve the accuracy of an approximate solution $\tilde{x}$ of a system of equations which has been obtained by an elimination method in machine arithmetic. The general characteristics of iterative methods, stated above, also apply to the conjugate-gradient method, with the one exception that the method in exact arithmetic terminates with the exact solution x after a finite number of steps. With regard to the applicability of the conjugate-gradient method, however, the same remarks apply as for the true iterative methods. For this reason we treat this method in the present chapter.

For the solution of certain important special large systems of equations, in addition, there are some direct methods which furnish the solution in a finite number of steps and are superior to iterative methods: one of the earliest of these methods, the *algorithm of Buneman*, is described in Section 8.8.

It should be pointed out that, most recently, a number of additional general techniques have been developed for treating large systems of equations $Ax = b$ with A a sparse matrix. These techniques concern themselves with appropriate ways of storing such matrices, and with the determination of suitable permutation matrices P_1, P_2 such that in the application of elimination methods to the equivalent system $P_1 A P_2 y = P_1 b$, $y = P_2^{-1} x$, the intermediate matrices generated remain as sparse as possible. A method of this type was described in Section 4.A. For a relevant exposition, we refer the reader to the literature cited there.

A detailed treatment of iterative methods can be found in the fundamental book of Varga (1962), and also in Young (1971).

8.1 General Procedures for the Construction of Iterative Methods

Let a nonsingular $n \times n$ matrix A be given, and a system of linear equations

$$Ax = b \tag{8.1.1}$$

with the exact solution $x := A^{-1}b$. We consider iterative methods of the form (cf. Chapter 5)

$$x^{(i+1)} = \Phi(x^{(i)}), \qquad i = 0, 1, \ldots. \tag{8.1.2}$$

With the help of an arbitrary nonsingular $n \times n$ matrix B such iteration algorithms can be obtained from the equation

$$Bx + (A - B)x = b,$$

by putting

(8.1.3) $$Bx^{(i+1)} + (A - B)x^{(i)} = b,$$

or solved for $x^{(i+1)}$,

(8.1.4) $$x^{(i+1)} = x^{(i)} - B^{-1}(Ax^{(i)} - b) = (I - B^{-1}A)x^{(i)} + B^{-1}b.$$

Such iteration methods, in this generality, were first considered by Wittmeyer (1936).

Note that (8.1.4) is identical with the following special vector iteration (see Section 6.6.3):

$$\begin{bmatrix} 1 \\ x^{(i+1)} \end{bmatrix} = W \begin{bmatrix} 1 \\ x^{(i)} \end{bmatrix}, \qquad W := \left[\begin{array}{c|c} 1 & 0 \\ \hline B^{-1}b & I - B^{-1}A \end{array}\right],$$

where the matrix W of order $n+1$, in correspondence to the eigenvalue $\lambda_0 := 1$, has the left eigenvector $[1, 0]$ and the right eigenvector $\left[\begin{smallmatrix} 1 \\ x \end{smallmatrix}\right]$, $x := A^{-1}b$. According to the results of Section 6.6.3, the sequence $\left[\begin{smallmatrix} 1 \\ x^{(i)} \end{smallmatrix}\right]$ will converge to $\left[\begin{smallmatrix} 1 \\ x \end{smallmatrix}\right]$ only if $\lambda_0 = 1$ is a simple dominant eigenvalue of W, i.e., if

$$\lambda_0 = 1 > |\lambda_1| \geqslant \cdots \geqslant |\lambda_n|,$$

the remaining eigenvalues $\lambda_1, \ldots, \lambda_n$ of W (these are the eigenvalues of $I - B^{-1}A$) being smaller in absolute value than 1.

Each choice of a nonsingular matrix B leads to a potential iterative method (8.1.4). The better B satisfies the following conditions, the more useful the method will be:

(1) the system of equations (8.1.3) is easily solved for $x^{(i+1)}$,
(2) the eigenvalues of $I - B^{-1}A$ have moduli which are as small as possible.

The better B agrees with A, the more likely the latter will be true. These questions of optimality and convergence will be examined in the next sections. Here, we only wish to indicate a few important special iterative methods (8.1.3) obtained by different choices of B. We introduce, for this purpose, the following standard decomposition of A:

(8.1.5) $$A = D - E - F,$$

with

$$D = \begin{bmatrix} a_{11} & & 0 \\ & \ddots & \\ 0 & & a_{nn} \end{bmatrix},$$

$$E = -\begin{bmatrix} 0 & \cdots & \cdots & 0 \\ a_{21} & \ddots & & \vdots \\ \vdots & \ddots & \ddots & \vdots \\ a_{n1} & \cdots & a_{n,n-1} & 0 \end{bmatrix}, \qquad F = -\begin{bmatrix} 0 & a_{12} & \cdots & a_{1n} \\ \vdots & \ddots & \ddots & \vdots \\ \vdots & & \ddots & a_{n-1,n} \\ 0 & \cdots & \cdots & 0 \end{bmatrix},$$

as well as the abbreviations

(8.1.6) $L := D^{-1}E, \qquad U := D^{-1}F, \qquad J := L + U, \qquad H := (I - L)^{-1}U,$

assuming $a_{ii} \neq 0$ for $i = 1, 2, \ldots, n$.

(1) In the *Jacobi method* or *total-step method* one chooses

(8.1.7) $$B := D, \qquad I - B^{-1}A = J.$$

One thus obtains from (8.1.3) the iteration prescription

$$a_{jj}x_j^{(i+1)} + \sum_{k \neq j} a_{jk}x_k^{(i)} = b_j, \qquad j = 1, 2, \ldots, n, \quad i = 0, 1, \ldots,$$

where $x^{(i)} := [x_1^{(i)}, \ldots, x_n^{(i)}]^T$.

(2) In the *Gauss–Seidel method* or *single-step method* one chooses

(8.1.8) $$B := D - E, \qquad I - B^{-1}A = (I - L)^{-1}U = H.$$

One thus obtains for (8.1.3)

$$\sum_{k<j} a_{jk}x_k^{(i+1)} + a_{jj}x_j^{(i+1)} + \sum_{k>j} a_{jk}x_k^{(i)} = b_j,$$
$$j = 1, 2, \ldots, n, \qquad i = 0, 1, \ldots.$$

(3) The *method of iterative refinement* is a special case in itself. Here, the following situation is assumed. As the result of an elimination method for the solution of $Ax = b$ one obtains, owing to rounding errors, a (generally) good approximate solution $x^{(0)}$ for the exact solution x and a lower and upper triangular matrix $\bar{L}$ and $\bar{R}$, respectively, such that $\bar{L}\bar{R} \approx A$ (see Section 4.5). The approximate solution $x^{(0)}$ can then subsequently be improved by means of an iterative method of the form (8.1.3), choosing

$$B := \bar{L}\bar{R}.$$

(8.1.3) is equivalent to

(8.1.9) $$B(x^{(i+1)} - x^{(i)}) = r^{(i)}$$

with the residual

$$r^{(i)} := b - Ax^{(i)}.$$

From (8.1.9) it then follows that

(8.1.10) $$x^{(i+1)} = x^{(i)} + u^{(i)}, \qquad u^{(i)} := \bar{R}^{-1}\bar{L}^{-1}r^{(i)}.$$

Note that $u^{(i)}$ can be simply computed by solving the triangular systems of equations

(8.1.11) $$\bar{L}z = r^{(i)}, \qquad \bar{R}u^{(i)} = z.$$

In general (if A is not too ill conditioned), the method converges extremely fast. Already $x^{(1)}$ or $x^{(2)}$ agrees with the exact solution x to machine accuracy. Since, precisely for this reason, there occurs severe cancellation in the computation of the residuals $r^{(i)} = b - Ax^{(i)}$, it is extremely important for the proper functioning of the method that the computation of $r^{(i)}$ be done in *double precision*. For the subsequent computation of z, $u^{(i)}$, and $x^{(i+1)} = x^{(i)} + u^{(i)}$ from (8.1.11) and (8.1.10), double precision is not required. [Programs and numerical examples for iterative refinement can be found in Wilkinson and Reinsch (1971), and in Forsythe and Moler (1967).]

8.2 Convergence Theorems

The iterative methods considered in (8.1.3), (8.1.4) produce from each initial vector $x^{(0)}$ a sequence of vectors $\{x^{(i)}\}_{i=0,1,\ldots}$. We now call the method *convergent* if for *all* initial vectors $x^{(0)}$ this sequence $\{x^{(i)}\}_{i=0,1,\ldots}$ converges toward the exact solution $x = A^{-1}b$. By $\rho(C)$ we again denote in the following the spectral radius (see Section 6.9) of a matrix C. We can then state the following convergence criterion:

(8.2.1) Theorem.

(a) *The method* (8.1.3) *converges if and only if* $\rho(I - B^{-1}A) < 1$.

(b) *It is sufficient for convergence of* (8.1.3) *that* $\operatorname{lub}(I - B^{-1}A) < 1$.

Here $\operatorname{lub}(\cdot)$ *can be taken relative to any norm.*

PROOF. (a): For the error $f_i := x^{(i)} - x$, from

$$x^{(i+1)} = (I - B^{-1}A)x^{(i)} + B^{-1}b,$$
$$x = (I - B^{-1}A)x + B^{-1}b,$$

we immediately obtain by subtraction the recurrence formula

$$f_{i+1} = (I - B^{-1}A)f_i,$$

or

(8.2.2) $$f_i = (I - B^{-1}A)^i f_0, \qquad i = 0, 1, \ldots.$$

(1) Assume now (8.1.3) is convergent. Then for all f_0 we have $\lim_{i\to\infty} f_i = 0$. Choosing in particular f_0 to be an eigenvector of $I - B^{-1}A$ corresponding to the eigenvalue λ, it follows from (8.2.2) that

(8.2.3) $$f_i = \lambda^i f_0,$$

and hence $|\lambda| < 1$, since $\lim_{i\to\infty} f_i = 0$. Therefore, $\rho(I - B^{-1}A) < 1$.

(2) If, conversely, $\rho(I - B^{-1}A) < 1$, it follows immediately from Theorem (6.9.2) that $\lim_{i\to\infty} (I - B^{-1}A)^i = 0$ and thus $\lim_{i\to\infty} f_i = 0$ for all f_0.

(b): For arbitrary norms one has $\rho(I - B^{-1}A) \leqslant \operatorname{lub}(I - B^{-1}A)$ [see Theorem (6.9.1)]. This proves the theorem. □

Theorem (8.2.1) suggests the conjecture that the rate of convergence is larger the smaller $\rho(I - B^{-1}A)$. The statement can be made more precise.

(8.2.4) Theorem. *For the method* (8.1.3) *the errors* $f_i = x^{(i)} - x$ *satisfy*

$$\sup_{f_0 \neq 0} \limsup_{i \to \infty} \sqrt[i]{\frac{\|f_i\|}{\|f_0\|}} = \rho(I - B^{-1}A). \tag{8.2.5}$$

Here $\|\cdot\|$ *is an arbitrary norm.*

PROOF. Let $\|\cdot\|$ be an arbitrary norm, and $\operatorname{lub}(\cdot)$ the corresponding matrix norm. By k we denote, for short, the left-hand side of (8.2.5). One sees immediately that $k \geqslant \rho(I - B^{-1}A)$, by choosing for f_0, as in (8.2.2), (8.2.3), the eigenvectors of $I - B^{-1}A$. Let now $\varepsilon > 0$ be arbitrary. Then by Theorem (6.9.2) there exists a vector norm $N(\cdot)$ such that for the corresponding matrix norm $\operatorname{lub}_N(\cdot)$,

$$\operatorname{lub}_N(I - B^{-1}A) \leqslant \rho(I - B^{-1}A) + \varepsilon.$$

According to Theorem (4.4.6) all norms on $\mathbb{C}^n$ are equivalent, and there exist constants m, $M > 0$ with

$$m\|x\| \leqslant N(x) \leqslant M\|x\|.$$

If now $f_0 \neq 0$ is arbitrary, from these inequalities and (8.2.2) it follows that

$$\begin{aligned}\|f_i\| &\leqslant \frac{1}{m} N(f_i) = \frac{1}{m} N((I - B^{-1}A)^i f_0)\\ &\leqslant \frac{1}{m}[\operatorname{lub}_N(I - B^{-1}A)]^i N(f_0)\\ &\leqslant \frac{M}{m}(\rho(I - B^{-1}A) + \varepsilon)^i \|f_0\|,\end{aligned}$$

or

$$\sqrt[i]{\frac{\|f_i\|}{\|f_0\|}} \leqslant [\rho(I - B^{-1}A) + \varepsilon] \sqrt[i]{\frac{M}{m}}.$$

Since $\lim_{i \to \infty} \sqrt[i]{M/m} = 1$, one obtains $k \leqslant \rho(I - B^{-1}A) + \varepsilon$, and since ε was arbitrary, $k \leqslant \rho(I - B^{-1}A)$. The theorem is now proved. □

Let us apply these results first to the Jacobi method (8.1.7). We continue using the notation (8.1.5)–(8.1.6) introduced in the previous section. Relative to the maximum norm $\operatorname{lub}_\infty(C) = \max_i \sum_k |c_{ik}|$, we then have

$$\operatorname{lub}_\infty(I - B^{-1}A) = \operatorname{lub}_\infty(J) = \max_i \frac{1}{|a_{ii}|} \sum_{k \neq i} |a_{ik}|.$$

If $|a_{ii}| > \sum_{k \neq i} |a_{ik}|$ for all i, we get immediately

$$\operatorname{lub}_\infty(J) < 1.$$

From Theorem (8.2.1b) we thus obtain at once the first part of the following theorem:

(8.2.6) Theorem.

(a) Strong Row Sum Criterion: *The Jacobi method is convergent for all matrices A with*

$$|a_{ii}| > \sum_{k \neq i} |a_{ik}| \quad \text{for } i = 1, 2, \ldots, n. \tag{8.2.7}$$

(b) Strong Column Sum Criterion: *The Jacobi method converges for all matrices A with*

$$|a_{kk}| > \sum_{i \neq k} |a_{ik}| \quad \text{for } k = 1, 2, \ldots, n. \tag{8.2.8}$$

A matrix A satisfying (8.2.7) [(8.2.8)] is called *strictly row-wise (column-wise) diagonally dominant.*

PROOF OF (b). If (8.2.8) is satisfied by A, then (8.2.7) holds for the matrix A^T. The Jacobi method therefore converges for A^T, and thus, by Theorem (8.2.1a), $\rho(X) < 1$ for $X := I - D^{-1}A^T$. Now X has the same eigenvalues as X^T and $D^{-1}X^T D = I - D^{-1}A$. Hence also $\rho(I - D^{-1}A) < 1$, i.e., the Jacobi method is convergent also for the matrix A. □

For irreducible matrices A the strong row (column) sum criterion can be improved. A matrix A is called *irreducible* if there is no permutation matrix P such that $P^T AP$ has the form

$$P^T AP = \begin{bmatrix} \tilde{A}_{11} & \tilde{A}_{12} \\ 0 & \tilde{A}_{22} \end{bmatrix},$$

where $\tilde{A}_{11}$ is a $p \times p$ matrix and $\tilde{A}_{22}$ a $q \times q$ matrix with $p + q = n$, $p > 0$, $q > 0$.

The irreducibility of a matrix A can often be readily tested by means of the (directed) *graph* $G(A)$ associated with the matrix A. If A is an $n \times n$ matrix, then $G(A)$ consists of n vertices $P_1, \ldots, P_n$ and there is an (oriented) arc $P_i \to P_j$ in $G(A)$ precisely if $a_{ij} \neq 0$.

EXAMPLE.

$$A = \begin{bmatrix} 1 & 2 & 0 \\ -1 & 1 & 0 \\ 3 & 0 & 1 \end{bmatrix},$$

$G(A)$: 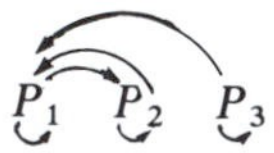

It is easily shown that A is irreducible if and only if the graph $G(A)$ is *connected* in the sense that for each pair of vertices (P_i, P_j) in $G(A)$ there is an oriented path from P_i to P_j.

For irreducible matrices one has the following:

(8.2.9) Theorem (Weak Row Sum Criterion). *If A is irreducible and*

$$|a_{ii}| \geqslant \sum_{k \neq i} |a_{ik}| \quad \text{for all } i = 1, 2, \ldots, n,$$

but $|a_{i_0, i_0}| > \sum_{k \neq i_0} |a_{i_0 k}|$ for at least one i_0, then the Jacobi method converges.

A matrix A satisfying the hypotheses of the theorem is called *irreducibly diagonally dominant.*

Analogously there is, of course, also a weak column sum criterion for irreducible A.

Proof. From the assumptions of the theorem it follows for the Jacobi method, as in the proof of (8.2.6a), that

$$\operatorname{lub}_\infty(I - B^{-1}A) = \operatorname{lub}_\infty(J) \leqslant 1,$$

and from this,

$$|J|e \leqslant e, \quad |J|e \neq e, \qquad e := [1, 1, \ldots, 1]^T. \tag{8.2.10}$$

(Absolute value signs $|\cdot|$ and inequalities for vectors or matrices are always to be understood componentwise.)

Now, with A also J is irreducible. In order to prove the theorem, it suffices to show the inequality

$$|J|^n e < e,$$

because from it we have immediately

$$[\rho(J)]^n = \rho(J^n) \leqslant \operatorname{lub}_\infty(J^n) \leqslant \operatorname{lub}_\infty(|J|^n) < 1.$$

Now, in view of (8.2.10) and $|J| \geqslant 0$,

$$|J|^2 e \leqslant |J|e \lneqq e,$$

and generally,

$$|J|^{i+1}e \leqslant |J|^i e \leqslant \cdots \lneqq e,$$

i.e., the vectors $t^{(i)} := e - |J|^i e$ satisfy

$$0 \lneqq t^{(1)} \leqslant t^{(2)} \leqslant \cdots. \tag{8.2.11}$$

We show that the number τ_i of nonvanishing components of $t^{(i)}$ increases monotonically with i as long as $\tau_i < n$: $0 < \tau_1 < \tau_2 < \cdots$. If this were not the case, there would exist, because of (8.2.11), a first $i \geqslant 1$ with $\tau_i = \tau_{i+1}$. Without loss of generality, let $t^{(i)}$ have the form

$$t^{(i)} = \begin{bmatrix} a \\ 0 \end{bmatrix} \quad \text{with a vector } a > 0,\ a \in \mathbb{R}^p,\ p > 0.$$

Then, in view of (8.2.11) and $\tau_i = \tau_{i+1}$, also $t^{(i+1)}$ has the form

$$t^{(i+1)} = \begin{bmatrix} b \\ 0 \end{bmatrix} \quad \text{with a vector } b > 0,\ b \in \mathbb{R}^p.$$

Partitioning $|J|$ analogously,

$$|J| = \begin{bmatrix} |J_{11}| & |J_{12}| \\ |J_{21}| & |J_{22}| \end{bmatrix}, \qquad |J_{11}| \text{ a } p \times p \text{ matrix},$$

it follows that

$$\begin{bmatrix} b \\ 0 \end{bmatrix} = t^{(i+1)} = e - |J|^{i+1}e \geqslant |J|e - |J|^{i+1}e$$
$$= |J|t^{(i)} = \begin{bmatrix} |J_{11}| & |J_{12}| \\ |J_{21}| & |J_{22}| \end{bmatrix} \begin{bmatrix} a \\ 0 \end{bmatrix}.$$

Because $a > 0$, this is only possible if $J_{21} = 0$, i.e., if J is reducible. Hence, $0 < \tau_1 < \tau_2 < \cdots$, and thus $t^{(n)} = e - |J|^n e > 0$. The theorem is now proved. □

The conditions of Theorems (8.2.6) and (8.2.9) are also sufficient for the convergence of the Gauss–Seidel method. We show this only for the strong row sum criterion. We have, even more precisely:

(8.2.12) Theorem. *If*

$$|a_{ii}| > \sum_{k \neq i} |a_{ik}| \quad \textit{for all } i = 1, 2, \ldots, n,$$

the Gauss–Seidel method is convergent, and furthermore [*see* (8.1.6)]

$$\operatorname{lub}_\infty(H) \leqslant \operatorname{lub}_\infty(J) < 1.$$

PROOF. Let $\kappa_H := \operatorname{lub}_\infty(H)$, $\kappa_J := \operatorname{lub}_\infty(J)$. As already exploited repeatedly, the assumption of the theorem implies

$$|J|e \leqslant \kappa_J e < e, \qquad e = [1, \ldots, 1]^T,$$

for the matrix $J = L + U$. From this, because $|J| = |L| + |U|$, one concludes

$$|U|e \leqslant (\kappa_J I - |L|)e. \tag{8.2.13}$$

Now L and $|L|$ are both lower triangular matrices with vanishing diagonal. For such matrices, as is easily verified,

$$L^n = |L|^n = 0,$$

so that $(I - L)^{-1}$ and $(I - |L|)^{-1}$ exist and

$$0 \leqslant |(I - L)^{-1}| = |I + L + \cdots + L^{n-1}|$$
$$\leqslant I + |L| + \cdots + |L|^{n-1} = (I - |L|)^{-1}.$$

Multiplying (8.2.13) by the nonnegative matrix $(I - |L|)^{-1}$, one obtains, because $H = (I - L)^{-1}U$,

$$\begin{aligned}|H|e &\leqslant (I - |L|)^{-1}|U|e \leqslant (I - |L|)^{-1}(I - |L| + (\kappa_J - 1)I)e \\ &= (I + (\kappa_J - 1)(I - |L|)^{-1})e.\end{aligned}$$

Now, $(I - |L|)^{-1} \geqslant I$ and $\kappa_J < 1$, so that the chain of inequalities can be continued:

$$|H|e \leqslant (I + (\kappa_J - 1)I)e = \kappa_J e.$$

But this means

$$\kappa_H = \mathrm{lub}_\infty(H) = \mathrm{lub}_\infty(|H|) \leqslant \kappa_J,$$

as was to be shown. □

Since $\mathrm{lub}_\infty(H) \geqslant \rho(H)$, $\mathrm{lub}_\infty(J) \geqslant \rho(J)$, this theorem may suggest the conjecture that under the assumptions of the theorem also $\rho(H) \leqslant \rho(J) < 1$, i.e., in view of Theorem (8.2.4), that the Gauss–Seidel method converges at least as fast as the Jacobi method. This, however, as examples show, is not true in general, but only under further assumptions on A. Thus, e.g., the following theorem is valid, which we state without proof [for a proof, see Varga (1962)].

(8.2.14) Theorem (Stein, Rosenberg). *If the matrix $J = L + U$ is nonnegative, then for J and $H = (I - L)^{-1}U$ precisely one of the following relations holds:*

(1) $\rho(H) = \rho(J) = 0$,
(2) $0 < \rho(H) < \rho(J) < 1$,
(3) $\rho(H) = \rho(J) = 1$,
(4) $\rho(H) > \rho(J) > 1$.

The assumption $J \geqslant 0$ is satisfied in particular [see (8.1.5), (8.1.6)] if the matrix A has positive diagonal elements and nonpositive off-diagonal elements: $a_{ii} > 0$, $a_{ik} \leqslant 0$ for $i \neq k$. Since this condition happens to be satisfied in almost all systems of linear equations which are obtained by difference approximations to linear differential operators (cf., e.g., Section 8.4), this theorem provides in many practical cases the significant information that the Gauss–Seidel method converges faster than the Jacobi method, if one of the two converges at all.

8.3 Relaxation Methods

The results of the previous section suggest looking for simple matrices B for which the corresponding iterative method (8.1.3) converges perhaps still faster than the Gauss–Seidel method, $\rho(I - B^{-1}A) < \rho(H)$. More generally, one can consider classes of suitable matrices $B(\omega)$ depending on a parameter ω and try to choose the parameter ω in an "optimal" way, i.e., so that $\rho(I - B(\omega)^{-1}A)$ as a function of ω becomes as small as possible. In the *relaxation methods* (*SOR methods*) one studies the following class of matrices $B(\omega)$:

$$B(\omega) = \frac{1}{\omega} D(I - \omega L). \tag{8.3.1}$$

Here again we use the notation (8.1.5)–(8.1.6). This choice is obtained through the following considerations.

Suppose for the $(i + 1)$st approximation $x^{(i+1)}$ we already know the components $x_k^{(i+1)}$, $k = 1, 2, \ldots, j - 1$. As in the Gauss–Seidel method (8.1.8), we then define an auxiliary quantity $\tilde{x}_j^{(i+1)}$ by

$$a_{jj}\tilde{x}_j^{(i+1)} = -\sum_{k<j} a_{jk}x_k^{(i+1)} - \sum_{k>j} a_{jk}x_k^{(i)} + b_j, \quad 1 \leqslant j \leqslant n, \quad i \geqslant 0, \tag{8.3.2}$$

whereupon $x_j^{(i+1)}$ is determined through a certain averaging of $x_j^{(i)}$ and $\tilde{x}_j^{(i+1)}$, viz.

$$x_j^{(i+1)} := (1 - \omega)x_j^{(i)} + \omega\tilde{x}_j^{(i+1)} = x_j^{(i)} + \omega(\tilde{x}_j^{(i+1)} - x_j^{(i)}). \tag{8.3.3}$$

Eliminating the auxiliary quantity $\tilde{x}_j^{(i+1)}$ from (8.3.3) by means of (8.3.2), one obtains

$$a_{jj}x_j^{(i+1)} = a_{jj}x_j^{(i)} + \omega\left[-\sum_{k<j} a_{jk}x_k^{(i+1)} - a_{jj}x_j^{(i)} - \sum_{k>j} a_{jk}x_k^{(i)} + b_j\right],$$
$$1 \leqslant j \leqslant n, \quad i \geqslant 0.$$

In matrix notation this is equivalent to

$$B(\omega)x^{(i+1)} = (B(\omega) - A)x^{(i)} + b,$$

where $B(\omega)$ is defined by (8.3.1) and

$$B(\omega) - A = \frac{1}{\omega} D((1 - \omega)I + \omega U).$$

For this method the rate of convergence, therefore, is determined by the spectral radius of the matrix

$$H(\omega) := I - B(\omega)^{-1}A = (I - \omega L)^{-1}[(1 - \omega)I + \omega U]. \tag{8.3.4}$$

One calls ω the *relaxation parameter* and speaks of *overrelaxation* if $\omega > 1$ and *underrelaxation* if $\omega < 1$. For $\omega = 1$ one exactly recovers the Gauss–Seidel method.

We begin by listing, in part without proof, a few qualitative results about $\rho(H(\omega))$. The following theorem shows that in relaxation methods only parameters ω with $0 < \omega < 2$, at best, lead to convergent methods:

(8.3.5) Theorem (Kahan). *For arbitrary matrices A one has*

$$\rho(H(\omega)) \geqslant |\omega - 1|$$

for all ω.

PROOF. $I - \omega L$ is a lower triangular matrix with 1 as diagonal elements, so that $\det(I - \omega L) = 1$ for all ω. For the characteristic polynomial $\varphi(\lambda)$ of $H(\omega)$ it follows that

$$\begin{aligned}\varphi(\lambda) &= \det(\lambda I - H(\omega)) = \det((I - \omega L)(\lambda I - H(\omega))) \\ &= \det((\lambda + \omega - 1)I - \omega\lambda L - \omega U).\end{aligned}$$

The constant term $\varphi(0)$ of $\varphi(\lambda)$ is equal to the product of the eigenvalues $\lambda_i(H(\omega))$:

$$\prod_{i=1}^{n} \lambda_i(H(\omega)) = \varphi(0) = \det((\omega - 1)I - \omega U) = (\omega - 1)^n.$$

It follows immediately that $\rho(H(\omega)) = \max_i |\lambda_i(H(\omega))| \geqslant |\omega - 1|$. □

For matrices A with $L \geqslant 0$, $U \geqslant 0$, only overrelaxation can give faster convergence than the Gauss–Seidel method:

(8.3.6) Theorem. *If the matrix A is irreducible and* $J = L + U \geqslant 0$, *and if the Jacobi method converges,* $\rho(J) < 1$, *then the function* $\rho(H(\omega))$ *is strictly decreasing on the interval* $0 < \omega \leqslant \bar{\omega}$ *for some* $\bar{\omega} \geqslant 1$.

For a proof, see Varga (1962).

One further shows:

(8.3.7) Theorem (Ostrowski, Reich). *For positive definite matrices A one has*

$$\rho(H(\omega)) < 1 \textit{ for all } 0 < \omega < 2.$$

In particular, the Gauss–Seidel method ($\omega = 1$) *converges for positive definite matrices.*

PROOF. Let $0 < \omega < 2$, and A be positive definite. Then $F = E^H$ in the decomposition (8.1.5), $A = D - E - F$, of A. For the matrix $B = B(\omega)$ in (8.3.1) one has $B = (1/\omega)D - E$, and the matrix

$$\begin{aligned}B + B^H - A &= \frac{1}{\omega}D - E + \frac{1}{\omega}D - F - (D - E - F) \\ &= \left(\frac{2}{\omega} - 1\right)D\end{aligned}$$

is positive definite, since the diagonal elements of the positive definite matrix A are positive [Theorem (4.3.2)] and $(2/\omega) - 1 > 0$.

We first show that the eigenvalues λ of $A^{-1}(2B - A)$ all lie in the interior of the right half plane, Re $\lambda > 0$. Indeed, if x is an eigenvector for λ, then

$$A^{-1}(2B - A)x = \lambda x,$$
$$x^H(2B - A)x = \lambda x^H Ax.$$

Taking the conjugate complex of the last relation gives, because $A = A^H$,

$$x^H(2B^H - A)x = \bar{\lambda} x^H Ax.$$

By addition, it follows that

$$x^H(B + B^H - A)x = \text{Re}\, \lambda x^H Ax.$$

But now, A and $B + B^H - A$ are positive definite and thus Re $\lambda > 0$. For the matrix $Q := A^{-1}(2B - A) = 2A^{-1}B - I$ one has

$$(Q - I)(Q + I)^{-1} = I - B^{-1}A = H(\omega).$$

[Observe that B is a nonsingular triangular matrix; therefore B^{-1} and thus $(Q + I)^{-1}$ exist.] If μ is an eigenvalue of $H(\omega)$ and x a corresponding eigenvector, then from

$$(Q - I)(Q + I)^{-1}x = H(\omega)x = \mu x$$

it follows, for the vector $y := (Q + I)^{-1}x \neq 0$, that

$$(Q - I)y = \mu(Q + I)y,$$
$$(1 - \mu)Qy = (1 + \mu)y.$$

Since $y \neq 0$, we must have $\mu \neq 1$, and one finally obtains

$$Qy = \frac{1 + \mu}{1 - \mu} y,$$

i.e., $\lambda = (1 + \mu)/(1 - \mu)$ is an eigenvalue of $Q = A^{-1}(2B - A)$. Hence, $\mu = (\lambda - 1)/(\lambda + 1)$. For $|\mu|^2 = \mu\bar{\mu}$ one obtains

$$|\mu|^2 = \frac{|\lambda|^2 + 1 - 2\,\text{Re}\,\lambda}{|\lambda|^2 + 1 + 2\,\text{Re}\,\lambda},$$

and since Re $\lambda > 0$ for $0 < \omega < 2$,

$$|\mu| < 1, \quad \text{i.e., } \rho(H(\omega)) < 1. \qquad \square$$

For an important class of matrices the more qualitative assertions of Theorems (8.3.5)–(8.3.7) can be considerably sharpened. This is the class of matrices with *property A* introduced by Young [see, e.g., Young (1971)], or its generalization due to Varga (1962), the class of *consistently ordered* matrices [see (8.3.10)].

(8.3.8) Definition. The matrix A has *property A* if there exists a permutation matrix P such that PAP^T has the form

$$PAP^T = \begin{bmatrix} D_1 & M_1 \\ M_2 & D_2 \end{bmatrix}, \qquad D_1,\ D_2 \text{ diagonal matrices.}$$

The most important fact about matrices with property A is given in the following theorem:

(8.3.9) Theorem. *For every $n \times n$ matrix A with property A and $a_{ii} \neq 0$, $i = 1, \ldots, n$, there exists a permutation matrix P such that the decomposition (8.1.5), (8.1.6), $\bar{A} = D(I - L - U)$, of the permuted matrix $\bar{A} := PAP^T$ has the following property: The eigenvalues of the matrix*

$$J(\alpha) := \alpha L + \alpha^{-1} U, \qquad \alpha \in \mathbb{C}, \quad \alpha \neq 0$$

are independent of α.

Proof. By Definition (8.3.8) there exists a permutation P such that

$$PAP^T = \begin{bmatrix} D_1 & M_1 \\ M_2 & D_2 \end{bmatrix} = D(I - L - U),$$

$$D := \begin{bmatrix} D_1 & 0 \\ 0 & D_2 \end{bmatrix}, \qquad L = -\begin{bmatrix} 0 & 0 \\ D_2^{-1} M_2 & 0 \end{bmatrix}, \qquad U = -\begin{bmatrix} 0 & D_1^{-1} M_1 \\ 0 & 0 \end{bmatrix}.$$

Here D_1 and D_2 are nonsingular diagonal matrices. For $\alpha \neq 0$, one now has

$$\begin{aligned} J(\alpha) = -\begin{bmatrix} 0 & \alpha^{-1} D_1^{-1} M_1 \\ \alpha D_2^{-1} M_2 & 0 \end{bmatrix} &= -S_\alpha \begin{bmatrix} 0 & D_1^{-1} M_1 \\ D_2^{-1} M_2 & 0 \end{bmatrix} S_\alpha^{-1} \\ &= S_\alpha J(1) S_\alpha^{-1} \end{aligned}$$

with the nonsingular diagonal matrix

$$S_\alpha := \begin{bmatrix} I_1 & 0 \\ 0 & \alpha I_2 \end{bmatrix}, \qquad I_1,\ I_2 \text{ unit matrices.}$$

The matrices $J(\alpha)$ and $J(1)$ are similar, and hence have the same eigenvalues. □

Following Varga (1962), a matrix A which, relative to the decomposition (8.1.5), (8.1.6), $A = D(I - L - U)$, has the property that the eigenvalues of the matrices

$$J(\alpha) = \alpha L + \alpha^{-1} U$$

for $\alpha \neq 0$ are independent of α, is called

(8.3.10) *consistently ordered.*

Theorem (8.3.9) asserts that matrices with property A can be ordered consistently, i.e., the rows and columns of A can be rearranged by a permutation P such that there results a consistently ordered matrix PAP^T.

Consistently ordered matrices A, however, need not at all have the form

$$A = \begin{bmatrix} D_1 & M_1 \\ M_2 & D_2 \end{bmatrix}, \qquad D_1, D_2 \text{ diagonal matrices.}$$

This is shown by the important example of block tridiagonal matrices A, which have the form

$$A = \begin{bmatrix} D_1 & A_{12} & & \\ A_{21} & \ddots & \ddots & \\ & \ddots & \ddots & A_{N-1,N} \\ & & A_{N,N-1} & D_N \end{bmatrix}, \qquad D_i \text{ diagonal matrices.}$$

If all D_i are nonsingular, then the matrices

$$J(\alpha) = -\begin{bmatrix} 0 & \alpha^{-1}D_1^{-1}A_{12} & & & \\ \alpha D_2^{-1}A_{21} & 0 & \ddots & & \\ & \ddots & \ddots & \ddots & \\ & & \ddots & 0 & \alpha^{-1}D_{N-1}^{-1}A_{N-1,N} \\ & & & \alpha D_N^{-1}A_{N,N-1} & 0 \end{bmatrix}$$

obey the relation

$$J(\alpha) = S_\alpha J(1) S_\alpha^{-1}, \qquad S_\alpha := \begin{bmatrix} I_1 & & & \\ & \alpha I_2 & & \\ & & \ddots & \\ & & & \alpha^{N-1} I_N \end{bmatrix},$$

that is, A is consistently ordered. In addition, block tridiagonal matrices have property A. We show this only for the special 3×3 matrix

$$A = \begin{bmatrix} 1 & b & 0 \\ a & 1 & d \\ 0 & c & 1 \end{bmatrix}.$$

One has, in fact,

$$PAP^T = \left[\begin{array}{c|cc} 1 & a & d \\ \hline b & 1 & 0 \\ c & 0 & 1 \end{array}\right] \quad \text{for } P := \begin{bmatrix} 0 & 1 & 0 \\ 1 & 0 & 0 \\ 0 & 0 & 1 \end{bmatrix}.$$

In the general case, one proceeds analogously.

We point out, however, that there are consistently ordered matrices which do not have property A. This is shown by the example

$$A := \begin{bmatrix} 1 & 0 & 0 \\ 1 & 1 & 0 \\ 1 & 1 & 1 \end{bmatrix}.$$

For irreducible $n \times n$ matrices A with nonvanishing diagonal elements $a_{ii} \neq 0$ and decomposition $A = D(I - L - U)$ it is often easy to find out whether or not A has property A by considering the graph $G(J)$ associated with the matrix $J = L + U$. For this, one examines the lengths $s_1^{(i)}, s_2^{(i)}, \ldots$ of all closed oriented paths (oriented cycles)

$$P_i \to P_{k_1} \to P_{k_2} \to \cdots \to P_{k_{s^{(i)}}} = P_i$$

in $G(J)$ which lead from P_i to P_i. Denoting by l_i the greatest common divisor of the $s_1^{(i)}, s_2^{(i)}, \ldots,$

$$l_i = \gcd(s_1^{(i)}, s_2^{(i)}, \ldots),$$

the graph $G(J)$ is called *2-cyclic* if $l_1 = l_2 = \cdots = l_n = 2$ and *weakly 2-cyclic* if all l_i are even.

The following theorem then holds, which we state without proof.

(8.3.11) Theorem. *An irreducible matrix A has property A if and only if $G(J)$ is weakly 2-cyclic.*

EXAMPLE. To the matrix

$$A := \begin{bmatrix} 4 & -1 & 0 & -1 \\ -1 & 4 & -1 & 0 \\ 0 & -1 & 4 & -1 \\ -1 & 0 & -1 & 4 \end{bmatrix}$$

there belongs the matrix

$$J := \frac{1}{4}\begin{bmatrix} 0 & 1 & 0 & 1 \\ 1 & 0 & 1 & 0 \\ 0 & 1 & 0 & 1 \\ 1 & 0 & 1 & 0 \end{bmatrix}$$

with the graph

$G(J)$: 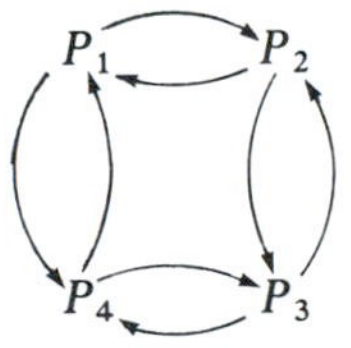

.

$G(J)$ is connected, so that J, and thus also A, is irreducible (see Section 8.2). Since $G(J)$ is evidently 2-cyclic, A has property A.

The significance of consistently ordered matrices [and therefore by (8.3.9), indirectly also of matrices with property A] lies in the fact that one can explicitly show how the eigenvalues μ of $J = L + U$ are related to the eigenvalues $\lambda = \lambda(\omega)$ of $H(\omega) = (I - \omega L)^{-1}((1 - \omega)I + \omega U)$:

(8.3.12) Theorem (Young, Varga). *Let A be a consistently ordered matrix* (8.3.10) *and $\omega \neq 0$. Then:*

(a) *With μ, also $-\mu$ is an eigenvalue of $J = L + U$.*
(b) *If μ is an eigenvalue of J and*

$$(\lambda + \omega - 1)^2 = \lambda\omega^2\mu^2, \tag{8.3.13}$$

then λ is an eigenvalue of $H(\omega)$.
(c) *If $\lambda \neq 0$ is an eigenvalue of $H(\omega)$ and* (8.3.13) *holds, then μ is an eigenvalue of J.*

PROOF. (a): Since A is consistently ordered, the matrix $J(-1) = -L - U = -J$ has the same eigenvalues as $J(1) = J = L + U$.

(b): Because $\det(I - \omega L) = 1$ for all ω, one has

$$\begin{aligned} \det(\lambda I - H(\omega)) &= \det[(I - \omega L)(\lambda I - H(\omega))] \\ &= \det[\lambda I - \lambda\omega L - (1 - \omega)I - \omega U] \\ &= \det((\lambda + \omega - 1)I - \lambda\omega L - \omega U). \end{aligned} \tag{8.3.14}$$

Now let μ be an eigenvalue of $J = L + U$ and λ a solution of (8.3.13). Then $\lambda + \omega - 1 = \sqrt{\lambda}\omega\mu$ or $\lambda + \omega - 1 = -\sqrt{\lambda}\omega\mu$. Because of (a) we can assume without loss of generality that

$$\lambda + \omega - 1 = \sqrt{\lambda}\omega\mu.$$

If $\lambda = 0$, then $\omega = 1$, so that by (8.3.14),

$$\det(0 \cdot I - H(1)) = \det(-U) = 0,$$

i.e., λ is an eigenvalue of $H(\omega)$. If $\lambda \neq 0$, it follows from (8.3.14) that

$$\begin{aligned} \det(\lambda I - H(\omega)) &= \det\left[(\lambda + \omega - 1)I - \sqrt{\lambda}\omega\left(\sqrt{\lambda}L + \frac{1}{\sqrt{\lambda}}U\right)\right] \\ &= (\sqrt{\lambda}\omega)^n \det\left[\mu I - \left(\sqrt{\lambda}L + \frac{1}{\sqrt{\lambda}}U\right)\right] \\ &= (\sqrt{\lambda}\omega)^n \det(\mu I - (L + U)) = 0, \end{aligned} \tag{8.3.15}$$

since the matrix $J(\sqrt{\lambda}) = \sqrt{\lambda}L + (1/\sqrt{\lambda})\,U$ has the same eigenvalues as $J = L + U$ and μ is eigenvalue of J. Therefore, $\det(\lambda I - H(\omega)) = 0$, and λ is an eigenvalue of $H(\omega)$.

(c): Now, conversely, let $\lambda \neq 0$ be an eigenvalue of $H(\omega)$, and μ a number satisfying (8.3.13), i.e., with $\lambda + \omega - 1 = \pm\omega\sqrt{\lambda}\mu$. In view of (a) it suffices to show that the number μ with $\lambda + \omega - 1 = \omega\sqrt{\lambda}\mu$ is an eigenvalue of J. This, however, follows immediately from (8.3.15). □

As a side result the following is obtained for $\omega = 1$:

(8.3.16) Corollary. *Let A be a consistently ordered matrix* (8.3.10). *Then for the matrix $H = H(1) = (I - L)^{-1}U$ of the Gauss–Seidel method one has*

$$\rho(H) = [\rho(J)]^2.$$

In view of Theorem (8.2.4) this means that with the Jacobi method one has to carry out about twice as many iteration steps as with the Gauss–Seidel method, to achieve the same accuracy.

We now wish to exhibit in an important special case the optimal relaxation parameter ω_b, characterized by [see Theorem (8.3.5)]

$$\rho(H(\omega_b)) = \min_{\omega \in \mathbb{R}} \rho(H(\omega)) = \min_{0<\omega<2} \rho(H(\omega)).$$

(8.3.17) Theorem (Young, Varga). *Let A be a consistently ordered matrix. Let the eigenvalues of J be real and $\rho(J) < 1$. Then*

$$\omega_b = \frac{2}{1+\sqrt{1-\rho(J)^2}}, \qquad \rho(H(\omega_b)) = \omega_b - 1 = \left(\frac{\rho(J)}{1+\sqrt{1-\rho(J)^2}}\right)^2.$$

One has, in general,

(8.3.18)

$$\rho(H(\omega)) = \begin{cases} \omega - 1 & \text{for } \omega_b \leqslant \omega \leqslant 2, \\ 1 - \omega + \frac{1}{2}\omega^2\mu^2 + \omega\mu\sqrt{1 - \omega + \frac{1}{4}\omega^2\mu^2} & \text{for } 0 < \omega \leqslant \omega_b, \end{cases}$$

where the abbreviation $\mu := \rho(J)$ is used (see Figure 29).

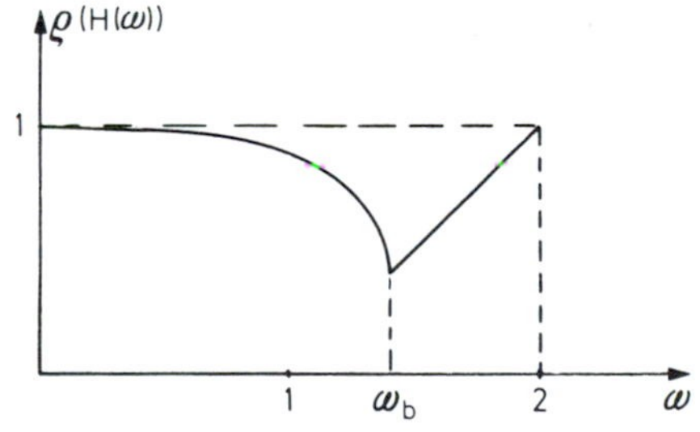

Figure 29 Spectral radius of $H(\omega)$.

Note that the left-hand differential quotient of $\rho(H(\omega))$ at ω_b is "$-\infty$". One should preferably take, therefore, as relaxation parameter ω a number that is slightly too large, rather than one that is too small, if ω_b is not known exactly.

PROOF. The eigenvalues μ_i of the matrix J, by assumption, are real, and

$$-\rho(J) \leqslant \mu_i \leqslant \rho(J) < 1.$$

For fixed $\omega \in (0, 2)$ [by Theorem (8.3.5) it suffices to consider this domain] to each μ_i there belong two eigenvalues $\lambda_i^{(1)}(\omega, \mu_i)$, $\lambda_i^{(2)}(\omega, \mu_i)$ of $H(\omega)$, which are obtained by solving the quadratic equation (8.3.13) in λ. Geometrically, $\lambda_i^{(1)}(\omega)$, $\lambda_i^{(2)}(\omega)$ are obtained as abscissae of the points of intersection of the straight line

$$g_\omega(\lambda) = \frac{\lambda + \omega - 1}{\omega}$$

with the parabola $m_i(\lambda) := \pm\sqrt{\lambda}\mu_i$ (see Figure 30). The line $g_\omega(\lambda)$ has the slope $1/\omega$ and passes through the point (1, 1). If it does not intersect the parabola $m_i(\lambda)$, then $\lambda_i^{(1)}$, $\lambda_i^{(2)}$ are conjugate complex numbers with modulus $|\omega - 1|$, as one finds immediately from (8.3.13). Evidently,

$$\rho(H(\omega)) = \max_i (|\lambda_i^{(1)}(\omega)|, |\lambda_i^{(2)}(\omega)|) = \max(|\lambda^{(1)}(\omega)|, |\lambda^{(2)}(\omega)|),$$

the $\lambda^{(1)}(\omega)$, $\lambda^{(2)}(\omega)$ being obtained by intersecting $g_\omega(\lambda)$ with $m(\lambda) := \pm\sqrt{\lambda}\mu$, where $\mu = \rho(J) = \max_i |\mu_i|$. By solving the quadratic equation (8.3.13), with $\mu = \rho(J)$, for λ, one verifies (8.3.18) immediately, and thus also the remaining assertions of the theorem. □

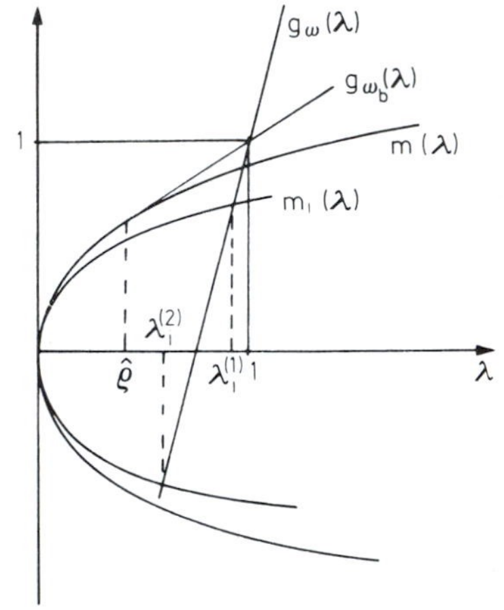

Figure 30 Determination of ω_b.

8.4 Applications to Difference Methods—An Example

In order to illustrate how the iterative methods described can be applied, we consider the Dirichlet boundary-value problem

$$(8.4.1) \qquad \begin{aligned} -u_{xx} - u_{yy} &= f(x, y), \qquad 0 < x, y < 1, \\ u(x, y) &= 0 \quad \text{for } (x, y) \in \partial\Omega, \end{aligned}$$

for the unit square $\Omega := \{x, y \mid 0 < x,\ y < 1\} \subseteq \mathbb{R}^2$ with boundary $\partial\Omega$ (cf. Section 7.7). We assume $f(x, y)$ continuous on $\Omega \cup \partial\Omega$. Since the various methods for the solution of boundary-value problems are usually compared on this problem, (8.4.1) is also called the *model problem*. To solve (8.4.1) by means of a difference method, one replaces the differential operator by a difference operator, as described in Section 7.4 for boundary-value problems in ordinary differential equations. One covers $\Omega \cup \partial\Omega$ with a grid $\Omega_h \cup \partial\Omega_h$:

$$\Omega_h := \{(x_i, y_i) \mid i, j = 1, 2, \ldots, N\},$$

$$\partial\Omega_h := \{(x_i, 0), (x_i, 1), (0, y_j), (1, y_j) \mid i, j = 0, 1, \ldots, N+1\},$$

where, for abbreviation, we put (see Figure 31)

$$x_i = ih, \quad y_j = jh, \qquad i, j = 0, 1, \ldots, N+1,$$

$$h := \frac{1}{N+1}, \qquad N \geqslant 1 \text{ an integer.}$$

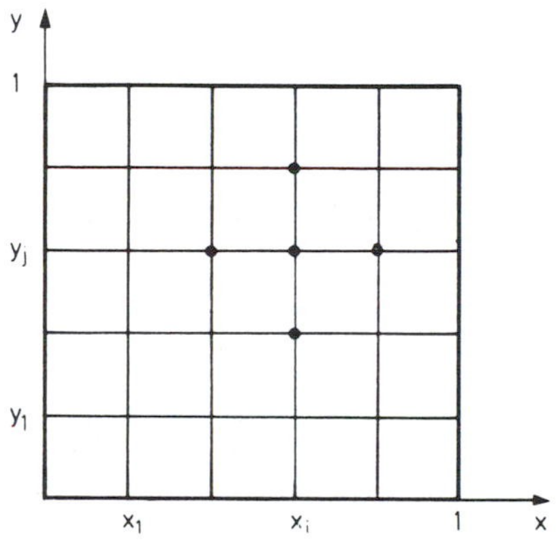

Figure 31 The grid Ω_h.

With the further abbreviation

$$u_{ij} := u(x_i, y_j), \qquad i, j = 0, 1, \ldots, N+1,$$

the differential operator

$$-u_{xx} - u_{yy}$$

can be replaced for all $(x_i, y_j) \in \Omega_h$ by the difference operator

$$\frac{4u_{ij} - u_{i-1,j} - u_{i+1,j} - u_{i,j-1} - u_{i,j+1}}{h^2} \tag{8.4.2}$$

up to an error τ_{ij}. The unknowns u_{ij}, $1 \leqslant i, j \leqslant N$ [because of the boundary conditions the $u_{ij} = 0$ are known for $(x_i, y_j) \in \partial\Omega_h$] therefore obey a system of linear equations of the form

$$
\begin{aligned}
(8.4.3)\qquad & 4u_{ij} - u_{i-1,j} - u_{i+1,j} - u_{i,j-1} - u_{i,j+1} \\
& \quad = h^2 f_{ij} + h^2 \tau_{ij}, \qquad (x_i, y_j) \in \Omega_h,
\end{aligned}
$$

with $f_{ij} := f(x_i, y_j)$. Here the errors τ_{ij} of course depend on the mesh size h. Under appropriate differentiability assumptions for the exact solution u, one shows as in Section 7.4 that $\tau_{ij} = O(h^2)$. For sufficiently small h one can thus expect that the solution z_{ij}, $i, j = 1, \ldots, N$, of the linear system of equations

$$
\begin{aligned}
& 4z_{ij} - z_{i-1,j} - z_{i+1,j} - z_{i,j-1} - z_{i,j+1} = h^2 f_{ij}, \qquad i, j = 1, \ldots, N, \\
(8.4.4)\quad & z_{0j} = z_{N+1,j} = z_{i0} = z_{i,N+1} = 0 \quad \text{for } i, j = 0, 1, \ldots, N+1,
\end{aligned}
$$

obtained from (8.4.3) by omitting the error τ_{ij}, agrees approximately with the u_{ij}. To every grid point (x_i, y_j) of Ω_h there belongs exactly one component z_{ij} of the solution of (8.4.4). Collecting the N^2 unknowns z_{ij} and the right-hand sides $h^2 f_{ij}$ row-wise (see Figure 31) into vectors

$$
\begin{aligned}
z &= [z_{11}, z_{21}, \ldots, z_{N1}, z_{12}, \ldots, z_{N2}, \ldots, z_{1N}, \ldots, z_{NN}]^T, \\
b &= h^2 [f_{11}, \ldots, f_{N1}, \ldots, f_{1N}, \ldots, f_{NN}]^T,
\end{aligned}
$$

then (8.4.4) is equivalent to a system of linear equations of the form

$$
Az = b,
$$

with the $N^2 \times N^2$ matrix

(8.4.5)

$$
A = \left[
\begin{array}{cccc|cccc|cccc|cccc}
4 & -1 & & & -1 & & & & & & & & & & & \\
-1 & \ddots & \ddots & & & \ddots & & & & & & & & & & \\
 & \ddots & \ddots & -1 & & & \ddots & & & & & & & & & \\
 & & -1 & 4 & & & & -1 & & & & & & & & \\
\hline
-1 & & & & 4 & -1 & & & \ddots & & & & & & & \\
 & \ddots & & & -1 & \ddots & \ddots & & & \ddots & & & & & & \\
 & & \ddots & & & \ddots & \ddots & -1 & & & \ddots & & & & & \\
 & & & -1 & & & -1 & 4 & & & & \ddots & & & & \\
\hline
 & & & & \ddots & & & & \ddots & & & & -1 & & & \\
 & & & & & \ddots & & & & \ddots & & & & \ddots & & \\
 & & & & & & \ddots & & & & \ddots & & & & \ddots & \\
 & & & & & & & \ddots & & & & \ddots & & & & -1 \\
\hline
 & & & & & & & & -1 & & & & 4 & -1 & & \\
 & & & & & & & & & \ddots & & & -1 & \ddots & \ddots & \\
 & & & & & & & & & & \ddots & & & \ddots & \ddots & -1 \\
 & & & & & & & & & & & -1 & & & -1 & 4
\end{array}
\right]
$$

$$
= \begin{bmatrix}
A_{11} & A_{12} & & 0 \\
A_{21} & \ddots & \ddots & \\
 & \ddots & \ddots & A_{N-1,N} \\
0 & & A_{N,N-1} & A_{NN}
\end{bmatrix}.
$$

A is partitioned in a natural way into blocks A_{ij} of order N, which are induced by the partitioning of the points (x_i, y_j) of Ω_h into horizontal rows $(x_1, y_j), (x_2, y_j), \ldots, (x_N, y_j)$.

The matrix A is quite sparse. In each row, at most five elements are different from zero. For this reason, in the execution of an iteration step $z^{(i)} \to z^{(i+1)}$ of, say, the Jacobi method (8.1.7) or the Gauss–Seidel method (8.1.8), one requires only about $5N^2$ operations (1 operation = 1 multiplication or division + 1 addition). (If the matrix A were dense, N^4 operations per iteration step would be required.) Compare this expenditure with that of a direct method for solving $Az = b$. If we were to compute a triangular decomposition $A = LL^T$ of A, say with the Cholesky method (below we will see that A is positive definite), then L would be an $N^2 \times N^2$ lower triangular matrix of the form

$L =$

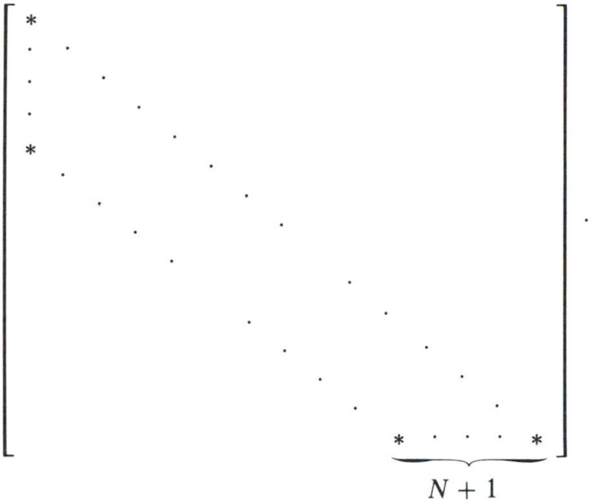

The computation of L alone requires approximately $\frac{1}{2}N^4$ operations. Since the Jacobi method, e.g., requires approximately $(N + 1)^2$ iterations (overrelaxation method: $N + 1$ iterations) [see (8.4.9)] in order to obtain a result accurate to 2 decimal places, the Cholesky method would be less expensive than the Jacobi method. The main difficulty with the Cholesky method, however, lies in the fact that with today's memory capacities the storage of the approximately N^3 nonvanishing elements of L requires too much space. (Typical order of magnitude of N: 50–100.) Here lie the advantages of iterative methods: Their storage requirement is only of the order of magnitude N^2.

To the Jacobi method belongs the matrix

$$J = L + U = \tfrac{1}{4}(4I - A).$$

The associated graph $G(J)$ (for $N = 3$),

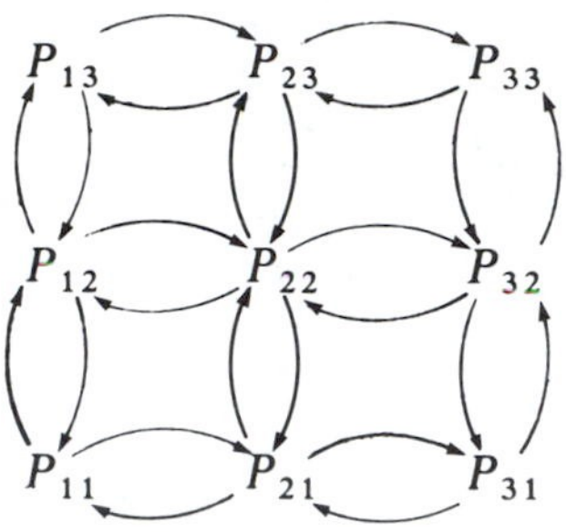

is connected and 2-cyclic. A is therefore irreducible (see Section 8.2) and has property A (Theorem 8.3.11). One easily sees, in addition, that A is already consistently ordered.

The eigenvalues and eigenvectors of $J = L + U$ can be determined explicitly. One verifies at once, by substitution, that the N^2 vectors

$$z^{(k,\,l)}, \qquad k, l = 1, 2, \ldots, N,$$

with components

$$z_{ij}^{(k,\,l)} := \sin \frac{k\pi i}{N+1} \sin \frac{l\pi j}{N+1}, \qquad 1 \leqslant i, j \leqslant N,$$

satisfy

$$Jz^{(k,\,l)} = \mu^{(k,\,l)} z^{(k,\,l)}$$

with

$$\mu^{(k,\,l)} := \frac{1}{2}\left(\cos \frac{k\pi}{N+1} + \cos \frac{l\pi}{N+1}\right), \qquad 1 \leqslant k, l \leqslant N.$$

J thus has the eigenvalues $\mu^{(k,\,l)}$, $1 \leqslant k, l \leqslant N$.

The spectral radius of J, therefore, is

(8.4.6) $$\rho(J) = \max_{k,\,l} |\mu^{(k,\,l)}| = \cos \frac{\pi}{N+1}.$$

To the Gauss–Seidel method belongs the matrix $H = (I - L)^{-1}U$ with the spectral radius [see (8.3.16)]

$$\rho(H) = \rho(J)^2 = \cos^2 \frac{\pi}{N+1}.$$

According to Theorem (8.3.17) the optimal relaxation parameter ω_b and $\rho(H(\omega_b))$ are given by

(8.4.7)
$$\omega_b = \frac{2}{1 + \sqrt{1 - \cos^2 \dfrac{\pi}{N+1}}} = \frac{2}{1 + \sin \dfrac{\pi}{N+1}},$$
$$\rho(H(\omega_b)) = \frac{\cos^2 \dfrac{\pi}{N+1}}{\left(1 + \sin \dfrac{\pi}{N+1}\right)^2}.$$

The number $\kappa = \kappa(N)$ with $\rho(J)^\kappa = \rho(H(\omega_b))$ indicates how many steps of the Jacobi method produce the same error reduction as one step of the optimal relaxation method. Clearly,

$$\kappa = \frac{\ln \rho(H(\omega_b))}{\ln \rho(J)}.$$

Now, for small z one has $\ln(1 + z) = z - z^2/2 + O(z^3)$, and for large N

$$\cos \frac{\pi}{N+1} = 1 - \frac{\pi^2}{2(N+1)^2} + O\left(\frac{1}{N^4}\right),$$

so that

$$\ln \rho(J) = -\frac{\pi^2}{2(N+1)^2} + O\left(\frac{1}{N^4}\right).$$

Likewise,

$$\begin{aligned}\ln \rho(H(\omega_b)) &= 2\left[\ln \rho(J) - \ln\left(1 + \sin \frac{\pi}{N+1}\right)\right]\\ &= 2\left[-\frac{\pi^2}{2(N+1)^2} - \frac{\pi}{N+1} + \frac{\pi^2}{2(N+1)^2} + O\left(\frac{1}{N^3}\right)\right]\\ &= -\frac{2\pi}{N+1} + O\left(\frac{1}{N^3}\right),\end{aligned}$$

so that finally, asymptotically for large N,

(8.4.8)
$$\kappa = \kappa(N) \approx \frac{4(N+1)}{\pi},$$

i.e., the optimal relaxation method is more than N times as fast as the Jacobi method. The quantities

(8.4.9)
$$\begin{aligned} R_J &:= -\frac{\ln 10}{\ln \rho(J)} \approx 0.467(N+1)^2, \\ R_{GS} &:= \tfrac{1}{2} R_J \approx 0.234(N+1)^2, \\ R_{\omega_b} &:= -\frac{\ln 10}{\ln \rho(H(\omega_b))} \approx 0.367(N+1) \end{aligned}$$

indicate the number of iterations required in the Jacobi method, the Gauss–Seidel method, and the optimal relaxation method, respectively, in order to reduce the error by a factor of $\frac{1}{10}$.

8.5 Block Iterative Methods

As the example of the previous section shows, the matrices A arising in the application of difference methods to partial differential equations often exhibit a natural block structure,

$$A = \begin{bmatrix} A_{11} & \cdots & A_{1N} \\ \vdots & & \vdots \\ A_{N1} & \cdots & A_{NN} \end{bmatrix},$$

where A_{ii} are square matrices. If now, in addition, all A_{ii} are nonsingular, it seems natural to introduce block iterative methods relative to the given partition π of A, in the following way: In analogy to the decomposition (8.1.5) of A one defines, relative to π, the decomposition

$$A = D_\pi - E_\pi - F_\pi, \qquad L_\pi := D_\pi^{-1} E_\pi, \qquad U_\pi := D_\pi^{-1} F_\pi$$

with

(8.5.1)

$$D_\pi := \begin{bmatrix} A_{11} & & & 0 \\ & A_{22} & & \\ & & \ddots & \\ 0 & & & A_{NN} \end{bmatrix}, \qquad E_\pi := -\begin{bmatrix} 0 & & & 0 \\ A_{21} & \ddots & & \\ \vdots & \ddots & \ddots & \\ A_{N1} & \cdots & A_{N,N-1} & 0 \end{bmatrix},$$

$$F_\pi := -\begin{bmatrix} 0 & A_{12} & \cdots & A_{1N} \\ & \ddots & \ddots & \vdots \\ & & \ddots & A_{N-1,N} \\ 0 & & & 0 \end{bmatrix}.$$

One obtains the *block Jacobi method* (*block total-step method*) for the solution of $Ax = b$ by choosing in (8.1.3), analogously to (8.1.7), $B := D_\pi$. One thus obtains the iteration algorithm

$$D_\pi x^{(i+1)} = b + (E_\pi + F_\pi)x^{(i)},$$

or, written out in full,

$$\text{(8.5.2)} \quad A_{jj}x_j^{(i+1)} = b_j - \sum_{k \neq j} A_{jk}x_k^{(i)}, \qquad j = 1, 2, \ldots, N, \quad i = 0, 1, \ldots.$$

Here, the vectors $x^{(s)}$, b are of course partitioned similarly to A. In each step $x^{(i)} \to x^{(i+1)}$ of the method we now must solve N systems of linear equations of the form $A_{jj} z = y$, $j = 1, 2, \ldots, N$. This is accomplished by first obtaining, by the methods of Section 4, a triangular decomposition (or a Cholesky decomposition, if appropriate) $A_{jj} = L_j R_j$ of the A_{jj}, and then reducing $A_{jj} z = y$ to the solution of two triangular systems of equations

$$L_j u = y, \qquad R_j z = u.$$

For the efficiency of the method it is essential that the A_{ii} be simply structured matrices for which the triangular decomposition is easily carried out. This is the case, e.g., for the matrix A in (8.4.5) of the model problem. Here, A_{jj} are positive definite tridiagonal $N \times N$ matrices

$$A_{jj} = \begin{bmatrix} 4 & -1 & & \\ -1 & \ddots & \ddots & \\ & \ddots & \ddots & -1 \\ & & -1 & 4 \end{bmatrix}, \qquad L_j = \begin{bmatrix} * & & & \\ * & \ddots & & \\ & \ddots & \ddots & \\ & & * & * \end{bmatrix},$$

whose Cholesky decomposition requires a very modest amount of work (number of operations proportional to N).

The rate of convergence of (8.5.2) of course is now determined by the spectral radius $\rho(J_\pi)$ of the matrix

$$J_\pi := I - B^{-1}A = L_\pi + U_\pi.$$

In the same way one can define, analogously to (8.1.8), a *block Gauss–Seidel method* (*block single-step method*), through the choice

$$B := D_\pi - E_\pi, \qquad H_\pi := (I - B^{-1}A) = (I - L_\pi)^{-1}U_\pi,$$

or, explicitly,

$$\text{(8.5.3)} \quad \begin{aligned} A_{jj}x_j^{(i+1)} &= b_j - \sum_{k<j} A_{jk}x_k^{(i+1)} - \sum_{k>j} A_{jk}x_k^{(i)}, \\ j &= 1, 2, \ldots, N, \quad i = 0, 1, \ldots. \end{aligned}$$

Again, systems of equation with the matrices A_{jj} need to be solved in each iteration step.

As in Section 8.3, one can also introduce block relaxation methods through the choice $B := B(\omega) = (1/\omega)D_\pi(I - \omega L_\pi)$. Explicitly [cf. (8.3.3)], let $\tilde{x}_j^{(i+1)}$ be the solution of (8.5.3); then

$$x_j^{(i+1)} = \omega(\tilde{x}_j^{(i+1)} - x_j^{(i)}) + x_j^{(i)}, \qquad j = 1, 2, \ldots, N.$$

Now, of course,

$$H_\pi(\omega) := I - B(\omega)^{-1}A = (I - \omega L_\pi)^{-1}[(1-\omega)I + \omega U_\pi].$$

Also the theory of consistently ordered matrices of Section 8.3 carries over, if one defines A as *consistently ordered* whenever the eigenvalues of the matrices

$$J_\pi(\alpha) = \alpha L_\pi + \alpha^{-1} U_\pi$$

are independent of α. Optimal relaxation factors are determined as in Theorem (8.3.17) with the help of $\rho(J_\pi)$.

One expects intuitively that the block methods will converge faster with increasing coarseness of the block partition π of A. This indeed can be shown under fairly general assumptions on A [see Varga (1962)]. For the coarsest partition π of A into a single block, e.g., the iterative method "converges" after just one step. It is then equivalent to a direct method. This example shows that arbitrarily coarse partitions are only of theoretical interest. The reduction in the number of iterations is compensated, to a certain extent, by the larger computational work for each individual iteration step. For the most common partitions, in which A is block tridiagonal and the diagonal blocks usually tridiagonal (this, e.g., is the case for the model problem), one can show, however, that the computational work involved in block methods is equal to that in ordinary methods. In these cases, block methods bring real advantages. For the model problem (see Section 8.4), relative to the partition given in (8.4.5), the spectral radius $\rho(J_\pi)$ can again be determined explicitly. One finds

$$\rho(J_\pi) = \frac{\cos \dfrac{\pi}{N+1}}{2 - \cos \dfrac{\pi}{N+1}} < \rho(J).$$

For the corresponding optimal block relaxation method one has asymptotically for $N \to \infty$

$$\rho(H_\pi(\omega_b)) \approx \rho(H(\omega_b))^\kappa \quad \text{with } \kappa = \sqrt{2}.$$

The number of iterations is reduced by a factor $\sqrt{2}$ compared to the ordinary optimal relaxation method (proof: see Exercise 17).

8.6 The ADI-Method of Peaceman and Rachford

Still faster convergence than in relaxation methods is obtained with the *ADI methods* (alternating-direction implicit iterative methods) for the iterative computation of the solution of a system of linear equations which arises in difference methods. From amongst the many variants of this method we describe here only the historically first method of this type, which is due to Peaceman and Rachford [see Varga (1962) and Young (1971) for an exposition of further variants]. We illustrate the method for the boundary-value problem

$$\text{(8.6.1)}\quad \begin{aligned} &-u_{xx}(x, y) - u_{yy}(x, y) + \sigma u(x, y) = f(x, y) \quad \text{for } 0 < x,\ y < 1, \\ &u(x, y) = 0 \quad \text{for } (x, y) \in \partial\Omega, \qquad \Omega := \{(x, y) \mid 0 < x,\ y < 1\} \end{aligned}$$

on the unit square, which on account of the term σu is only slightly more general than the model problem (8.4.1). We assume in the following that σ is constant and non-negative. Using the same discretization and the same notation as in Section 8.4, in place of (8.4.4) one now obtains the system of linear equations

$$\text{(8.6.2)}\quad \begin{aligned} &4z_{ij} - z_{i-1,j} - z_{i+1,j} - z_{i,j-1} - z_{i,j+1} + \sigma h^2 z_{ij} = h^2 f_{ij}, \qquad 1 \leqslant i, j \leqslant N, \\ &z_{0j} = z_{N+1,j} = z_{i0} = z_{i,N+1} = 0 \quad \text{for } 0 \leqslant i, j \leqslant N+1 \end{aligned}$$

for the approximate values z_{ij} of the values $u_{ij} = u(x_i, y_j)$ of the exact solution. To the decomposition

$$\begin{aligned} &4z_{ij} - z_{i-1,j} - z_{i+1,j} - z_{i,j-1} - z_{i,j+1} + \sigma h^2 z_{ij} \\ &\qquad \equiv [2z_{ij} - z_{i-1,j} - z_{i+1,j}] + [2z_{ij} - z_{i,j-1} - z_{i,j+1}] + [\sigma h^2 z_{ij}] \end{aligned}$$

into variables which are located on the same horizontal and vertical line (see Figure 29) of Ω_h, respectively, there corresponds a decomposition of the matrix A of the system of equations (8.6.2), $Az = b$, of the form

$$A = H + V + \Sigma.$$

Here, H, V, Σ are defined through their actions on a vector z:

$$\text{(8.6.3)}\quad \begin{aligned} w_{ij} &= 2z_{ij} - z_{i-1,j} - z_{i+1,j} && \text{if } w = Hz, \\ w_{ij} &= 2z_{ij} - z_{i,j-1} - z_{i,j+1} && \text{if } w = Vz, \\ w_{ij} &= \sigma h^2 z_{ij} && \text{if } w = \Sigma z. \end{aligned}$$

Σ is a diagonal matrix with nonnegative elements; H and V are both symmetric and positive definite. It suffices to show this for H: If the z_{ij} are ordered in correspondence to the rows of Ω_h (see Figure 29), $z = [z_{11}, z_{21}, \ldots, z_{N1}, \ldots, z_{1N}, z_{2N}, \ldots, z_{NN}]^T$, then

$$H = \left[\begin{array}{c|c|c} \begin{matrix} 2 & -1 & & \\ -1 & \ddots & \ddots & \\ & \ddots & \ddots & -1 \\ & & -1 & 2 \end{matrix} & & \\ \hline & \ddots & \\ \hline & & \begin{matrix} 2 & -1 & & \\ -1 & \ddots & \ddots & \\ & \ddots & \ddots & -1 \\ & & -1 & 2 \end{matrix} \end{array}\right].$$

But according to Theorem (7.4.7) the matrices

$$\begin{bmatrix} 2 & -1 & & \\ -1 & \ddots & \ddots & \\ & \ddots & \ddots & -1 \\ & & -1 & 2 \end{bmatrix},$$

and hence also H, are positive definite. For V one proceeds similarly.

Analogous decompositions of A are also obtained for boundary-value problems which are considerably more general than (8.6.1).

In the ADI method of Peaceman and Rachford the system of equations

$$Az = b,$$

in accordance with the decomposition $A = H + V + \Sigma$, is now transformed equivalently into

$$(H + \tfrac{1}{2}\Sigma + rI)z = (rI - V - \tfrac{1}{2}\Sigma)z + b$$

and also

$$(V + \tfrac{1}{2}\Sigma + rI)z = (rI - H - \tfrac{1}{2}\Sigma)z + b.$$

Here r is an arbitrary real number. With the abbreviations $H_1 := H + \frac{1}{2}\Sigma$, $V_1 := V + \frac{1}{2}\Sigma$, one obtains the iteration algorithm of the ADI method,

(8.6.4a) $$(H_1 + r_{i+1}I)z^{(i+1/2)} = (r_{i+1}I - V_1)z^{(i)} + b,$$

(8.6.4b) $$(V_1 + r_{i+1}I)z^{(i+1)} = (r_{i+1}I - H_1)z^{(i+1/2)} + b.$$

Given $z^{(i)}$, one first computes $z^{(i+1/2)}$ from the first of these equations, then substitutes this value into the second and computes $z^{(i+1)}$. The quantity r_{i+1} is a real parameter which may be chosen differently from step to step. With suitable ordering of the variables z_{ij} the matrices $H_1 + r_{i+1}I$ and $V_1 + r_{i+1}I$ are positive definite tridiagonal matrices (assuming $r_{i+1} \geq 0$), so that the systems of equations (8.6.4a, b) can easily be solved for $z^{(i+1/2)}$ and $z^{(i+1)}$ via a Cholesky decomposition of $H_1 + r_{i+1} I$ and $V_1 + r_{i+1} I$.

Eliminating $z^{(i+1/2)}$, one obtains from (8.6.4)

(8.6.5) $$z^{(i+1)} = T_{r_{i+1}} z^{(i)} + g_{r_{i+1}}(b)$$

with

(8.6.6) $$\begin{aligned} T_r &:= (V_1 + rI)^{-1}(rI - H_1)(H_1 + rI)^{-1}(rI - V_1), \\ g_r(b) &:= (V_1 + rI)^{-1}[I + (rI - H_1)(H_1 + rI)^{-1}]b. \end{aligned}$$

For the error $f_i := z^{(i)} - z$ it follows from (8.6.5) and the relation $z = T_{r_{i+1}} z + g_{r_{i+1}}(b)$, by subtraction, that

(8.6.7) $$f_{i+1} = T_{r_{i+1}} f_i,$$

and therefore

(8.6.8) $$f_m = T_{r_m} T_{r_{m-1}} \cdots T_{r_1} f_0.$$

As in relaxation methods, one tries to choose the parameters r_i in such a way that the method converges as fast as possible. In view of (8.6.7), (8.6.8), this means that the r_i are to be determined so that the spectral radius $\rho(T_{r_m} \cdots T_{r_1})$ becomes as small as possible.

We first want to consider the case in which the same parameter $r_i = r$ is chosen for all $i = 1, 2, \ldots$. Here one has the following:

(8.6.9) Theorem. *Under the assumption that H_1 and V_1 are positive definite one has $\rho(T_r) < 1$ for all $r > 0$.*

Note that the assumption is satisfied for our special problem (8.6.1). From Theorem (8.6.9) it follows in particular that each constant choice $r_i = r > 0$ leads to a convergent iterative method (8.6.4).

PROOF. By assumption, V_1 and H_1 are positive definite; therefore $(V_1 + rI)^{-1}$, $(H_1 + rI)^{-1}$ exist for $r > 0$, and hence also T_r of (8.6.6). The matrix

$$\begin{aligned} \tilde{T}_r &:= (V_1 + rI) T_r (V_1 + rI)^{-1} \\ &= [(rI - H_1)(H_1 + rI)^{-1}][(rI - V_1)(V_1 + rI)^{-1}] \end{aligned}$$

is similar to T_r; hence $\rho(T_r) = \rho(\tilde{T}_r)$. The matrix $\tilde{H} := (rI - H_1)(H_1 + rI)^{-1}$ has the eigenvalues

$$\frac{r - \lambda_j}{r + \lambda_j},$$

where $\lambda_j = \lambda_j(H_1)$ are the eigenvalues of H_1, which are positive by assumption. Since $r > 0$, $\lambda_j > 0$, it follows that

$$\left| \frac{r - \lambda_j}{r + \lambda_j} \right| < 1,$$

and thus $\rho(\tilde{H}) < 1$. Since with H_1 also $\tilde{H}$ is Hermitian, relative to the Euclidean norm $\|\cdot\|_2$ one has (see Section 4.4) $\operatorname{lub}_2(\tilde{H}) = \rho(\tilde{H}) < 1$. In the same way one has $\operatorname{lub}_2(\tilde{V}) < 1$, $\tilde{V} := (rI - V_1)(V_1 + rI)^{-1}$, and thus [see Theorem (6.9.1)]

$$\rho(\tilde{T}_r) \leqslant \operatorname{lub}_2(\tilde{T}_r) \leqslant \operatorname{lub}_2(\tilde{H}) \operatorname{lub}_2(\tilde{V}) < 1. \qquad \square$$

For the model problem (8.4.1) more precise statements can be made. The vectors $z^{(k,l)}$, $1 \leqslant k, l \leqslant N$, introduced in Section 8.4, with

$$\text{(8.6.10)} \qquad z_{ij}^{(k,l)} := \sin \frac{k\pi i}{N+1} \sin \frac{l\pi j}{N+1}, \qquad 1 \leqslant i, j \leqslant N,$$

as is easily verified, are eigenvectors of $H = H_1$ *and* $V = V_1$, and thus also of T_r. Therefore, the eigenvalues of T_r can be exhibited explicitly. One finds

$$\text{(8.6.11)} \qquad \begin{aligned} H_1 z^{(k,l)} &= \mu_k z^{(k,l)}, \\ V_1 z^{(k,l)} &= \mu_l z^{(k,l)}, \\ T_r z^{(k,l)} &= \mu^{(k,l)} z^{(k,l)}, \end{aligned}$$

with

$$\text{(8.6.12)} \qquad \mu^{(k,l)} = \frac{r - \mu_l}{r + \mu_l} \frac{r - \mu_k}{r + \mu_k}, \qquad \mu_j := 4 \sin^2 \frac{j\pi}{2(N+1)},$$

so that

$$\text{(8.6.13)} \qquad \rho(T_r) = \max_{1 \leqslant j \leqslant N} \left| \frac{r - \mu_j}{r + \mu_j} \right|^2.$$

By a discussion of this expression (see Exercise 20) one finally finds a result of Varga,

$$\min_{r>0} \rho(T_r) = \rho(H(\omega_b)) = \frac{\cos^2 \dfrac{\pi}{N+1}}{\left(1 + \sin \dfrac{\pi}{N+1}\right)^2},$$

where ω_b characterizes the best (ordinary) relaxation method [cf. (8.4.7)]. In other words, the best ADI method, assuming constant choice of parameters, has the same rate of convergence for the model problem as the optimal ordinary relaxation method. Since the individual iteration step in the ADI method is a great deal more expensive than in the relaxation method, the ADI method would appear to be inferior. This is certainly true if for all iteration steps one and the same parameter $r = r_1 = r_2 = \cdots$ is chosen. However, if one makes use of the option to choose a separate parameter r_i in each step, the picture changes in favor of the ADI method. For the model problem one can argue, e.g., as follows: The vectors $z^{(k,l)}$ are eigenvectors of

T_r for arbitrary r, with corresponding eigenvalue $\mu^{(k,l)}$ in (8.6.12); therefore, the $z^{(k,l)}$ are also eigenvectors of $T_{r_i} \dots T_{r_1}$ in (8.6.8). Indeed,

$$T_{r_i} \cdots T_{r_1} z^{(k,l)} = \mu^{(k,l)}_{r_i, \dots, r_1} z^{(k,l)},$$

where

$$\mu^{(k,l)}_{r_i, \dots, r_1} := \prod_{j=1}^{i} \frac{(r_j - \mu_l)(r_j - \mu_k)}{(r_j + \mu_l)(r_j + \mu_k)}.$$

Choosing $r_j := \mu_j$, $j = 1, 2, \dots, N$, we have

$$\mu^{(k,l)}_{r_N, \dots, r_1} = 0 \quad \text{for all } 1 \leqslant k, l \leqslant N,$$

so that, by the linear independence of the $z^{(k,l)}$,

$$T_{r_N} \cdots T_{r_1} = 0.$$

With this special choice of the r_j, the ADI method for the model problem terminates after N steps with the exact solution. This, of course, is a happy coincidence, which is due to the following essential assumptions:

(1) H_1 and V_1 have in common a set of eigenvectors which span the whole space.
(2) The eigenvalues of H_1 and V_1 are known.

One cannot expect, of course, that these assumptions will be satisfied in practice for problems other than (8.6.1), (8.6.2); in particular, one will hardly know the exact eigenvalues σ_i of H_1 and τ_i of V_1, but only, at best, lower and upper bounds $\alpha \leqslant \sigma_i, \tau_i \leqslant \beta$ for these eigenvalues. We would like to deal with this situation and give first a criterion for the validity of (1).

(8.6.14) Theorem. *For two Hermitian matrices H_1 and V_1 of order n there exist n linearly independent (orthogonal) vectors $z_1, \dots, z_n$, which are common eigenvectors of H_1 and V_1,*

(8.6.15) $$H_1 z_i = \sigma_i z_i, \quad V_1 z_i = \tau_i z_i, \qquad i = 1, 2, \dots, n,$$

if and only if H_1 commutes with V_1: $H_1 V_1 = V_1 H_1$.

PROOF. (1) From (8.6.15) it follows that

$$H_1 V_1 z_i = \sigma_i \tau_i z_i = V_1 H_1 z_i \quad \text{for all } i = 1, 2, \dots, n.$$

Since the z_i form a basis in $\mathbb{C}^n$, it follows at once that $H_1 V_1 = V_1 H_1$.

(2) Conversely, let $H_1 V_1 = V_1 H_1$. Let $\lambda_1 < \cdots < \lambda_r$ be the eigenvalues of V_1 with the multiplicities $\sigma(\lambda_i)$, $i = 1, \dots, r$. Then, according to Theorem (6.4.2), there exists a unitary matrix U with

$$\Lambda_V := U^H V_1 U = \begin{bmatrix} \lambda_1 I_1 & & \\ & \ddots & \\ & & \lambda_r I_r \end{bmatrix},$$

where I_j is a unit matrix of order $\sigma(\lambda_j)$. From $H_1 V_1 = V_1 H_1$ it follows immediately that $\tilde{H}_1 \Lambda_V = \Lambda_V \tilde{H}_1$, with the matrix $\tilde{H}_1 := U^H H_1 U$. We partition $\tilde{H}_1$ analogously to Λ_V:

$$\tilde{H}_1 = \begin{bmatrix} H_{11} & H_{12} & \cdots & H_{1r} \\ H_{21} & H_{22} & & H_{2r} \\ \vdots & \vdots & & \vdots \\ H_{r1} & H_{r2} & \cdots & H_{rr} \end{bmatrix}.$$

By multiplying out $\tilde{H}_1 \Lambda_V = \Lambda_V \tilde{H}_1$, one obtains $H_{ij} = 0$ for $i \neq j$, since $\lambda_i \neq \lambda_j$. The H_{ii} are Hermitian matrices of order $\sigma(\lambda_i)$. Again by Theorem (6.4.2) there exist unitary matrices $\bar{U}_i$ of order $\sigma(\lambda_i)$ such that $\bar{U}_i^H H_{ii} \bar{U}_i$ becomes a diagonal matrix Λ_i. For the unitary $n \times n$ matrix

$$\bar{U} = \begin{bmatrix} \bar{U}_1 & & \\ & \ddots & \\ & & \bar{U}_r \end{bmatrix},$$

since $H_{ij} = 0$ for $i \neq j$, there follow the relations

$$(U\bar{U})^H H_1 (U\bar{U}) = \bar{U}^H \tilde{H}_1 \bar{U} = \Lambda_H = \begin{bmatrix} \Lambda_1 & & \\ & \ddots & \\ & & \Lambda_r \end{bmatrix} \quad \Rightarrow \quad H_1(U\bar{U}) = (U\bar{U})\Lambda_H,$$

$$(U\bar{U})^H V_1 (U\bar{U}) = \bar{U}^H \Lambda_V \bar{U} = \Lambda_V \quad \Rightarrow \quad V_1(U\bar{U}) = (U\bar{U})\Lambda_V,$$

so that the columns $z_i := (U\bar{U})e_i$ of the unitary matrix $U\bar{U} = [z_1, \ldots, z_n]$ can be taken as n common orthogonal eigenvectors of H_1 and V_1. □

Unfortunately, the condition $H_1 V_1 = V_1 H_1$ is rather severe. One can show [see Varga (1962), Young (1971)] that it is satisfied only, essentially, for boundary-value problems of the type

$$-\frac{\partial}{\partial x}\left(p_1(x)\frac{\partial u(x, y)}{\partial x}\right) - \frac{\partial}{\partial y}\left(p_2(y)\frac{\partial u(x, y)}{\partial y}\right) + \sigma u(x, y) = f(x, y)$$

$$\text{for } (x, y) \in \Omega,$$

$$u(x, y) = 0 \quad \text{for } (x, y) \in \partial\Omega,$$

$$\sigma > 0 \text{ constant}, \qquad p_1(x) > 0, \quad p_2(y) > 0 \quad \text{for } (x, y) \in \Omega,$$

with rectangular domain Ω and the usual discretization (rectangular grid, etc.). Nevertheless, practical experience with the ADI method seems to suggest that the favorable convergence properties which can be proven in the commutative case frequently are present also in the noncommutative case. We therefore assume for the following discussion that H_1 and V_1 are two positive definite commuting $n \times n$ matrices with (8.6.15) and that two numbers α, β are given such that $0 < \alpha \leqslant \sigma_i, \tau_i \leqslant \beta$ for $i = 1, 2, \ldots, n$. Then

$$T_r z_i = \frac{r - \sigma_i}{r + \sigma_i} \, \frac{r - \tau_i}{r + \tau_i} \cdot z_i \quad \text{for all } r > 0, \; i = 1, 2, \ldots, n,$$

so that

(8.6.16)
$$\rho(T_{r_m} \cdots T_{r_1}) = \max_{1 \leqslant i \leqslant n} \prod_{j=1}^{m} \left| \frac{r_j - \sigma_i}{r_j + \sigma_i} \frac{r_j - \tau_i}{r_j + \tau_i} \right| \leqslant \max_{\alpha \leqslant x \leqslant \beta} \prod_{j=1}^{m} \left| \frac{r_j - x}{r_j + x} \right|^2 .$$

For given m, it is natural, therefore, to choose the parameters $r_i > 0$, $i = 1, \ldots, m$, so that the function

(8.6.17)
$$\varphi(r_1, \ldots, r_m) := \max_{\alpha \leqslant x \leqslant \beta} \prod_{j=1}^{m} \left| \frac{r_j - x}{r_j + x} \right|$$

becomes as small as possible. It can be shown that this problem for each $m > 0$ has a unique solution. For each m there are uniquely determined numbers $\bar{r}_i$ with $\alpha < \bar{r}_i < \beta$, $i = 1, \ldots, m$, such that

(8.6.18)
$$d_m(\alpha, \beta) := \varphi(\bar{r}_1, \ldots, \bar{r}_m) = \min_{\substack{r_i > 0 \\ 1 \leqslant i \leqslant m}} \varphi(r_1, \ldots, r_m).$$

The optimal parameters $\bar{r}_1, \ldots, \bar{r}_m$ can even be given explicitly, for each m, in terms of elliptic functions [see Wachspress (1966), Young (1971)]. One further knows good approximations for the $\bar{r}_i$ which are easy to compute. In the special case $m = 2^k$, however, the optimal parameters $\bar{r}_i$ can also be easily computed by recursion. For this case, the relevant results will now be presented without proof [for proofs, see, e.g., Wachspress (1966), Young (1971), Varga (1962)].

Let $r_i^{(m)}$, $i = 1, 2, \ldots, m$, denote the optimal ADI parameters for $m = 2^k$. The $r_i^{(m)}$ and $d_m(\alpha, \beta)$ can be computed recursively by means of Gauss's arithmetic-geometric mean algorithm.

It can be shown that

(8.6.19)
$$d_{2n}(\alpha, \beta) = d_n\left(\sqrt{\alpha\beta}, \frac{\alpha + \beta}{2} \right),$$

the optimal parameters of the minimax problem (8.6.18), $r_i^{(2n)}$ and $r_i^{(n)}$, being related by

(8.6.20)
$$r_i^{(n)} = \frac{r_i^{(2n)} + \alpha\beta / r_i^{(2n)}}{2}, \qquad i = 1, 2, \ldots, n.$$

Starting with this observation, one obtains the following algorithm for determining $r_i^{(m)}$. Define

(8.6.21)
$$\alpha_0 := \alpha, \qquad \beta_0 := \beta,$$
$$\alpha_{j+1} := \sqrt{\alpha_j \beta_j}, \quad \beta_{j+1} := \frac{\alpha_j + \beta_j}{2}, \qquad j = 0, 1, \ldots, k-1.$$

Then

(8.6.22)
$$d_{2^k}(\alpha_0, \beta_0) = d_{2^{k-1}}(\alpha_1, \beta_1) = \cdots = d_1(\alpha_k, \beta_k) = \frac{\sqrt{\beta_k} - \sqrt{\alpha_k}}{\sqrt{\beta_k} + \sqrt{\alpha_k}}.$$

The solution of $d_1(\alpha_k, \beta_k)$ can be found with $r_1^{(1)} = \sqrt{\alpha_k \beta_k}$. The optimal ADI parameters $r_i^{(m)}$, $i = 1, 2, \ldots, m = 2^k$ can be computed as follows:

(8.6.23).

(1) $s_1^{(0)} := \sqrt{\alpha_k \beta_k}$.

(2) For $j = 0, 1, \ldots, k-1$, determine $s_i^{(j+1)}$, $i = 1, 2, \ldots, 2^{j+1}$, as the 2^{j+1} solutions of the 2^j quadratic equations in x,

$$s_i^{(j)} = \frac{1}{2}\left(x + \frac{\alpha_{k-1-j}\beta_{k-1-j}}{x}\right), \qquad i = 1, 2, \ldots, 2^j.$$

(3) Put $r_i^{(m)} := s_i^{(k)}$, $i = 1, 2, \ldots, m = 2^k$.

The $s_i^{(j)}$, $i = 1, 2, \ldots, 2^j$, are just the optimal ADI parameters for the interval $[\alpha_{k-j}, \beta_{k-j}]$.

Let us use these formulas to study for the model problem (8.4.1), (8.4.4), with $m = 2^k$ fixed, the asymptotic behavior of $d_{2^k}(\alpha, \beta)$ as $N \to \infty$. For α and β we take the best possible bounds (see (8.6.12)):

(8.6.24)
$$\alpha = 4\sin^2\frac{\pi}{2(N+1)},$$
$$\beta = 4\sin^2\frac{N\pi}{2(N+1)} = 4\cos^2\frac{\pi}{2(N+1)}.$$

We then have

(8.6.25) $$d_m(\alpha, \beta) \sim 1 - 4\sqrt[m]{\frac{\pi}{4(N+1)}} \quad \text{as } N \to \infty, \qquad m := 2^k.$$

PROOF. By mathematical induction on k. With the abbreviation

$$c_k := \sqrt{\alpha_k/\beta_k},$$

one obtains from (8.6.21), (8.6.22)

(8.6.26a) $$d_{2^k}(\alpha, \beta) = \frac{1 - c_k}{1 + c_k},$$

(8.6.26b) $$c_{k+1}^2 = \frac{2c_k}{1 + c_k^2}.$$

In order to prove (8.6.25), it suffices to show

(8.6.27) $$c_k \sim 2\sqrt[2^k]{\frac{\pi}{4(N+1)}}, \qquad N \to \infty,$$

since it follows then from (8.6.26a) that for $N \to \infty$

$$d_{2^k}(\alpha, \beta) \sim 1 - 2c_k.$$

But (8.6.27) is true for $k = 0$, since

$$c_0 = \tan \frac{\pi}{2(N+1)} \sim \frac{\pi}{2(N+1)}.$$

If (8.6.27) is valid for some $k \geqslant 0$, then it is also valid for $k + 1$, because from (8.6.26b) we have at once

$$c_{k+1} \sim \sqrt{2c_k} \quad \text{as } N \to \infty;$$

hence the assertion. □

In practice, the parameters r_i are often repeated cyclically, i.e., one chooses a fixed m (e.g., in the form $m = 2^k$), then determines approximately the optimal ADI parameters $r_i^{(m)}$ belonging to this m, and finally takes for the ADI method the parameters

$$r_{jm+i} := r_i^{(m)} \quad \text{for } i = 1, 2, \ldots, m, \qquad j = 0, 1, \ldots.$$

If m individual steps of the ADI method are considered a "big iteration step," the quantity

$$\frac{-\ln 10}{\ln \rho(T_{r_m} \cdots T_{r_1})}$$

indicates how many big iteration steps are required to reduce the error by a factor of $\frac{1}{10}$, i.e.,

$$R_{\text{ADI}}^{(m)} = -m \frac{\ln 10}{\ln \rho(T_{r_m} \cdots T_{r_1})}$$

indicates how many ordinary ADI steps, on the average, are required for the same purpose. In case of the model problem one obtains for the optimal choice of parameters and $m = 2^k$, by virtue of (8.6.16) and (8.6.25),

$$\rho(T_{r_m} \cdots T_{r_1}) \leqslant d_m(\alpha, \beta)^2 \sim 1 - 8 \sqrt[m]{\frac{\pi}{4(N+1)}}, \qquad N \to \infty,$$

$$\ln \rho(T_{r_m} \cdots T_{r_1}) \mathrel{\dot{\leqslant}} -8 \sqrt[m]{\frac{\pi}{4(N+1)}}, \qquad N \to \infty,$$

so that

$$\text{(8.6.28)} \qquad R_{\text{ADI}}^{(m)} \mathrel{\dot{\leqslant}} \frac{m}{8} \ln (10) \sqrt[m]{\frac{4(N+1)}{\pi}} \quad \text{for } N \to \infty.$$

Comparison with (8.4.9) shows that for $m > 1$ the ADI method converges considerably faster than the optimal ordinary relaxation method. It is this convergence behavior which establishes the practical significance of the ADI method.

8.7 The Conjugate-Gradient Method of Hestenes and Stiefel

For the solution of a system of linear equations

(8.7.1) $\qquad Ax = b, \qquad A$ a (real) positive definite $n \times n$ matrix,

beginning with a vector x_0, the conjugate-gradient method of Hestenes and Stiefel (1952) produces a chain of vectors

$$x_0 \to x_1 \to \cdots \to x_n,$$

which in exact arithmetic terminates with the desired solution $x_n = x$ after at most n steps. On account of rounding errors, however, x_n will not yet be sufficiently accurate, as a rule. Therefore, further steps $x_k \to x_{k+1}$ are carried out in this method, as in true iterative methods, until a sufficiently accurate solution has been found (usually a total of $3n$ to $5n$ steps). The amount of work per step $x_k \to x_{k+1}$ about equals that of multiplying the matrix A into a vector. For this reason, the method is very advantageous for sparse, unstructured matrices A of medium size, as they occur, e.g., in network calculations, but is not recommended for dense matrices or band matrices.

The idea of the method comes from the observation that the functional $F : \mathbb{R}^n \to \mathbb{R}$,

$$\begin{aligned} F(z) &= \tfrac{1}{2}(b - Az)^T A^{-1}(b - Az) \\ &= \tfrac{1}{2}z^T Az - b^T z + \tfrac{1}{2}b^T A^{-1}b \end{aligned}$$

is minimized by the exact solution x,

$$0 = F(x) = \min_{z \in \mathbb{R}^n} F(z).$$

This follows immediately from the fact that with A also A^{-1} is positive definite [Theorem (4.3.2)] and that for a vector z the corresponding residual $r := b - Az$ vanishes only for $z := x$. This suggests the "method of steepest descent," in which (cf. Section 5.4.1) the sequence $x_0 \to x_1 \to \cdots$ is found by one-dimensional minimization of F in the direction of the gradient:

$$x_{k+1}\colon \quad F(x_{k+1}) = \min_u F(x_k + ur_k) \quad \text{with } r_k := -DF(x_k)^T = b - Ax_k.$$

In the step $x_k \to x_{k+1}$ of the conjugate-gradient method, instead, a $(k + 1)$-dimensional minimization is carried out:

$$\text{(8.7.2)} \qquad \begin{aligned} x_{k+1}\colon \qquad & F(x_{k+1}) = \min_{u_0, \ldots, u_k} F(x_k + u_0 r_0 + \cdots + u_k r_k), \\ & r_i := b - Ax_i \quad \text{for } i \leqslant k. \end{aligned}$$

It turns out that x_{k+1} can be computed rather simply. The r_i obtained are orthogonal, and thus linearly independent, as long as $r_k \neq 0$. In exact computation there is thus a first $k \leqslant n$ with $r_k = 0$, since in $\mathbb{R}^n$ at most n vectors are linearly independent. The corresponding x_k is the desired solution of (8.7.1).

We first describe the method, and then verify its properties.

(8.7.3) Conjugate Gradient Method. *Initialization: Choose* $x_0 \in \mathbb{R}^n$, *and put* $p_0 := r_0 := b - Ax_0$.
For $k = 0, 1, \ldots$:

(1) *if* $p_k = 0$, *stop*: x_k *is the solution of* $Ax = b$. *Otherwise,*
(2) *compute*

$$a_k := \frac{r_k^T r_k}{p_k^T A p_k}, \qquad x_{k+1} := x_k + a_k p_k,$$

$$r_{k+1} := r_k - a_k A p_k, \qquad b_k := \frac{r_{k+1}^T r_{k+1}}{r_k^T r_k},$$

$$p_{k+1} := r_{k+1} + b_k p_k.$$

For the realization of this method, one needs to store only four vectors, x_k, r_k, p_k, and Ap_k. In each iteration step, only one matrix multiplication, Ap_k, must be performed; the remaining work amounts to the calculation of six inner products in $\mathbb{R}^n$. The total amount of work, for sparse matrices, is therefore modest.

The most important theoretical properties of the method are summarized in the following theorem:

(8.7.4) Theorem. *Let* A *be a positive definite (real)* $n \times n$ *matrix and* $b \in \mathbb{R}^n$. *Then for each initial vector* $x_0 \in \mathbb{R}^n$ *there exists a smallest nonnegative integer* $l \leqslant n$ *such that* $p_l = 0$. *The vectors* x_k, p_k, r_k, $k \leqslant l$, *generated by the conjugate-gradient method* (8.7.3) *have the following properties:*

(a) $Ax_l = b$: *the method thus produces the exact solution of the equation* $Ax = b$ *after at most* n *steps.*
(b) $r_i^T p_j = 0$ *for* $0 \leqslant j < i \leqslant l$.
(c) $r_i^T p_i = r_i^T r_i$ *for* $i \leqslant l$.
(d) $p_i^T A p_j = 0$ *for* $0 \leqslant i < j \leqslant l$, $\quad p_j^T A p_j > 0$ *for* $j < l$.
(e) $r_i^T r_j = 0$ *for* $0 \leqslant i < j \leqslant l$, $\quad r_j^T r_j > 0$ *for* $j < l$.
(f) $r_i = b - Ax_i$ *for* $i \leqslant l$.

From Theorem (8.7.4) it follows, in particular, that the method is well defined, since $r_k^T r_k > 0$, $p_k^T A p_k > 0$ for $p_k \neq 0$. Furthermore, the vectors p_k, because of (d), are A-conjugate, which explains the name of the method.

PROOF. We begin by showing, using mathematical induction on k, that the following statement (A_k) is valid for all $k \leqslant l$, where l is the first index with $p_l = 0$:

$$(A_k)\quad \begin{aligned} &(1)\ r_i^T p_j = 0 \text{ for } j < i \leqslant k,\\ &(2)\ r_i^T r_i > 0 \text{ for } i < k, \quad r_i^T p_i = r_i^T r_i \text{ for } i \leqslant k,\\ &(3)\ p_i^T A p_j = 0 \text{ for } i < j \leqslant k,\\ &(4)\ r_i^T r_j = 0 \text{ for } i < j \leqslant k,\\ &(5)\ r_i = b - Ax_i \text{ for } i \leqslant k. \end{aligned}$$

(A_0) is trivially true. We assume, inductively, that (A_k) holds for some $0 \leqslant k < l$ and show (A_{k+1}).

(1): From (8.7.3), where $p_k^T A p_k > 0$, since A is positive definite and $p_k \neq 0$, it follows that

$$(8.7.5)\qquad r_{k+1}^T p_k = (r_k - a_k A p_k)^T p_k = r_k^T p_k - \frac{r_k^T r_k}{p_k^T A p_k} p_k^T A p_k = 0,$$

because of (A_k) (2). For $j < k$, analogously,

$$r_{k+1}^T p_j = (r_k - a_k A p_k)^T p_j = 0,$$

because of (A_k) (1), (3). This proves (A_{k+1}) (1).

(2): We have $r_k^T r_k > 0$, since otherwise $r_k = 0$ and thus, in view of (8.7.3),

$$(8.7.6)\qquad p_k = \begin{cases} r_0 & \text{if } k = 0,\\ b_{k-1} p_{k-1} & \text{if } k > 0. \end{cases}$$

Since $k < l$, we must have $k > 0$, because otherwise, $p_0 = r_0 \neq 0$. For $k > 0$, in view of $p_k \neq 0$ $(k < l)$, we get from (8.7.6) and (A_k) (3) the contradiction $0 < p_k^T A p_k = b_{k-1} p_{k-1}^T A p_k = 0$. Therefore $r_k^T r_k > 0$, so that b_k and p_{k+1} are well defined through (8.7.3). It thus follows from (8.7.3) and (8.7.5) that

$$r_{k+1}^T p_{k+1} = r_{k+1}^T (r_{k+1} + b_k p_k) = r_{k+1}^T r_{k+1}.$$

This proves (A_{k+1}) (2).

(3): From what was just proved, $r_k \neq 0$, so that a_k^{-1} is well defined. From (8.7.3), we thus get for $j \leqslant k$

$$\begin{aligned} p_{k+1}^T A p_j &= r_{k+1}^T A p_j + b_k p_k^T A p_j\\ &= a_j^{-1} r_{k+1}^T (r_j - r_{j+1}) + b_k p_k^T A p_j\\ &= a_j^{-1} r_{k+1}^T (p_j - b_{j-1} p_{j-1} - p_{j+1} + b_j p_j) + b_k p_k^T A p_j\\ &= \begin{cases} 0 & \text{for } j < k, \quad \text{because of } (A_k)\ (3) \text{ and } (A_{k+1})\ (1),\\ 0 & \text{for } j = k, \quad \text{by the definition of } a_k \text{ and } b_k,\\ & \qquad\qquad \text{and } (A_{k+1})\ (1), (2). \end{cases} \end{aligned}$$

(Here, for $j = 0$, the vector p_{-1} has to be interpreted as the zero vector $p_{-1} = 0$.) This proves (A_{k+1}) (3).

(4): By (8.7.3), and (A_{k+1}) (1), we have for $i \leqslant k$ $(p_{-1} := 0)$,

$$r_i^T r_{k+1} = (p_i - b_{i-1} p_{i-1})^T r_{k+1} = 0.$$

(5): From (8.7.3), and (A_k) (5) one gets

$$b - Ax_{k+1} = b - A(x_k + a_k p_k) = r_k - a_k A p_k = r_{k+1}.$$

With this, (A_{k+1}) is proved. Therefore, (A_l) holds true.

Because of (A_l) (2), (4) we have $r_i \neq 0$ for all $i < l$, and these vectors form an orthogonal system in $\mathbb{R}^n$. Consequently, $l \leqslant n$. From $p_l = 0$ it finally follows, by virtue of (A_l) (2), that $r_l^T r_l = r_l^T p_l = 0$, and thus $r_l = 0$, so that x_l is a solution of $Ax = b$. The proof of (8.7.4) is now completed. □

With the information in Theorem (8.7.4) we can finally show (8.7.2). To begin with, it is seen from (8.7.3) that for $k < l$ the vectors r_i, $i \leqslant k$, and p_i, $i \leqslant k$, span the same subspace of $\mathbb{R}^n$:

$$S_k := \{u_0 r_0 + \cdots + u_k r_k \,|\, u_i \in \mathbb{R}\} = \{v_0 p_0 + \cdots + v_k p_k \,|\, v_i \in \mathbb{R}\}.$$

For the function

$$\Phi(v_0, \ldots, v_k) := F(x_k + v_0 p_0 + \cdots + v_k p_k)$$

however, we have for $j \leqslant k$

$$\frac{\partial \Phi(v_0, \ldots, v_k)}{\partial v_j} = -r^T p_j,$$

where $r = b - Ax$, $x := x_k + v_0 p_0 + \cdots + v_k p_k$. With the choice

$$v_j := \begin{cases} a_k & \text{for } j = k, \\ 0 & \text{for } j < k, \end{cases}$$

one thus obtains, by (8.7.3), $x = x_{k+1}$, $r = r_{k+1}$ and by (8.7.4b), $-r_{k+1}^T p_j = 0$, so that indeed

$$\min_{v_0, \ldots, v_k} F(x_k + v_0 p_0 + \cdots + v_k p_k) = \min_{u_0, \ldots, u_k} F(x_k + u_0 r_0 + \cdots + u_k r_k) = F(x_{k+1}).$$

In exact arithmetic, we would have, at the latest, $r_n = 0$, and thus in x_n the desired solution of (8.7.1). Because of the effects of rounding errors the computed r_n is different from zero, as a rule. In practice, therefore, the method is simply continued beyond the value $k = n$ until an r_k (or p_k) is found which is sufficiently small. An ALGOL program for a variant of this algorithm can be found in Wilkinson and Reinsch (1971); an extensive account of numerical experiments, in Reid (1971) and further results in Axelsson (1976).

The conjugate-gradient method, in particular, can also be used to solve the least-squares problem for overdetermined systems,

(8.7.7) $$\text{Determine} \quad \min_x \|Bx - c\|_2,$$

where B is a sparse $m \times n$ matrix with $m \geqslant n$ and rank $B = n$. According to Section 4.8.1, indeed, the optimal solution $\bar{x}$ of (8.7.7) is also solution of the normal equations

$$Ax = b, \qquad A := B^T B, \quad b := B^T c,$$

where A is a positive definite matrix. Since even when B is sparse the matrix $A = B^T B$ can be dense, the following variant of the conjugate-gradient method (8.7.3) suggests itself for the solution of (8.7.7), and has proved useful in practice:

Initialization: Choose $x_0 \in \mathbb{R}^n$ *and compute* $s_0 := c - Bx_0$, $p_0 := r_0 := B^T s_0$.
For $k = 0, 1, \ldots$:

(1) *if* $p_k = 0$, *stop*: x_k *is the optimal solution of* (8.7.7). *Otherwise*,
(2) *compute*

$$q_k := Bp_k, \qquad a_k := \frac{r_k^T r_k}{q_k^T q_k},$$

$$x_{k+1} := x_k + a_k p_k, \qquad s_{k+1} := s_k - a_k q_k,$$

$$r_{k+1} := B^T s_{k+1}, \qquad b_k := \frac{r_{k+1}^T r_{k+1}}{r_k^T r_k},$$

$$p_{k+1} := r_{k+1} + b_k p_k.$$

The minimum property (8.7.2) can be used to estimate the speed of convergence of the conjugate-gradient method (cg method). By using the recursions of (8.7.3) for the vectors r_k and p_k, it is easy to verify

$$p_k \in \operatorname{span}[r_0, Ar_0, \ldots, A^k r_0],$$

so that

$$S_k = \operatorname{span}[p_0, \ldots, p_k] = \operatorname{span}[r_0, Ar_0, \ldots, A^k r_0] = K_{k+1}(r_0, A)$$

is the $(k+1)$-st Krylov space of A belonging to the vector r_0 (see 6.5.3). In terms of the norm $\|u\|_A := (u^T A u)^{1/2}$ associated with A, we can interpret the function $F(z) = \|z - x\|_A^2$ as a measure for the distance of the vector z from the exact solution x, and (8.7.2) can be written in the form

$$\|x_k - x\|_A = \min\{\|u - x\|_A | u \in x_0 + K_k\}.$$

If we introduce the error $e_j := x_j - x$ of x_j, then because of $r_0 = -Ae_0$, any $u \in x_0 + K_k$ satisfies

$$x - u \in e_0 + \operatorname{span}[Ae_0, A^2 e_0, \ldots, A^k e_0],$$

that is, there is a real polynomial $p(t) = 1 + \alpha_1 t + \cdots + \alpha_k t^k$ with $x - u = p(A)e_0$. Therefore

$$\|e_k\|_A = \min\{\|p(A)e_0\|_A | p \in \overline{\Pi}_k\},$$

where $\overline{\Pi}_k$ denotes the set of all real polynomials of degree $\le k$ with $p(0) = 1$. Now the positive-definite matrix A has n eigenvalues $\lambda_1 \ge \lambda_2 \ge \cdots \ge \lambda_n > 0$ and associated orthonormal eigenvectors z_i, $Az_i = \lambda z_i$, $z_i^T z_j = \delta_{ij}$ [Theorems (6.4.2) and (6.4.4)]. We may write e_0 in the form $e_0 = \rho_1 z_1 + \cdots + \rho_n z_n$, which then implies

$$\|e_0\|_A^2 = e_0^T A e_0 = \sum_{i=1}^n \lambda_i \rho_i^2,$$

$$\|p(A)e_0\|_A^2 = \sum_{i=1}^n p(\lambda_i)^2 \lambda_i \rho_i^2 \le \left(\max_i p(\lambda_i)^2\right) \cdot \|e_0\|_A^2,$$

and therefore

(8.7.8) $$\frac{\|e_k\|_A}{\|e_0\|_A} \le \min_{p \in \overline{\Pi}_k} \max_i |p(\lambda_i)| \le \min_{p \in \overline{\Pi}_k} \max_{\lambda \in [\lambda_n, \lambda_1]} |p(\lambda)|.$$

In terms of the Chebyshev polynomials

$$T_k(x) := \cos(k \arccos x) = \cos k\theta, \qquad k = 0, 1, \ldots, \text{ if } \cos\theta = x,$$

which obviously satisfy $|T_k(x)| \le 1$ for $x \in [-1, 1]$, we can construct a polynomial of $\overline{\Pi}_k$ with small $\max\{|p(\lambda)|, \lambda \in [\lambda_n, \lambda_1]\}$ in the following way (in fact, we so obtain the optimal polynomial): Consider the mapping

$$\lambda \mapsto x = x(\lambda) := (2\lambda - (\lambda_n + \lambda_1))/(\lambda_1 - \lambda_n),$$

which maps the interval $[\lambda_n, \lambda_1]$ onto $[-1, 1]$. Then, the polynomial

$$p_k(\lambda) := \frac{T_k(x(\lambda))}{T_k(x(0))}$$

belongs to $\overline{\Pi}_k$ and satisfies

$$\max_{\lambda \in [\lambda_n \lambda_1]} |p_k(\lambda)| = |T_k(x(0))|^{-1} = \left| T_k\left(\frac{c+1}{c-1}\right)\right|^{-1}.$$

Here, $c := \lambda_1/\lambda_n$ is just the condition number of the matrix A with respect to the $\mathrm{lub}_2(\cdot)$-norm [see Example (b) of Section 4.4].

It is easy to find an upper bound for $|T_k(x(0))|^{-1}$. Using that $T_k(x) = (z^k + z^{-k})/2$, if $x = (z + z^{-1})/2$, and

$$\frac{c+1}{c-1} = \frac{1}{2}\left(\frac{\sqrt{c}+1}{\sqrt{c}-1} + \frac{\sqrt{c}-1}{\sqrt{c}+1}\right),$$

we finally obtain the estimates

$$(8.7.9) \qquad \frac{\|e_k\|_A}{\|e_0\|_A} \le \left(T_k\left(\frac{c+1}{c-1}\right)\right)^{-1} \le 2\left(\frac{\sqrt{c}-1}{\sqrt{c}+1}\right)^k.$$

Thus the speed of convergence of the conjugate-gradient method for $k \to \infty$ increases if the condition number c of A decreases.

This behavior is exploited by the so-called *preconditioning techniques* in order to accelerate the conjugate-gradient method. Here, one tries to approximate as well as possible the positive-definite matrix A by another positive-definite matrix B, the *preconditioner*, so that $B^{-1}A$ is a good approximation of the unit matrix. Then the positive-definite matrix

$$A' := B^{1/2}(B^{-1}A)B^{-1/2} = B^{-1/2}AB^{-1/2},$$

which is similar to $B^{-1}A$, has a much smaller condition than A, $c' = \operatorname{cond}(A') \ll c = \operatorname{cond}(A)$. [Here, we have used that for any positive definite matrix B there exists a positive definite matrix $C =: B^{1/2}$ with $C^2 = B$. This follows easily from Theorem (6.4.2)]. Moreover, the matrix B should be chosen such that linear equations $Bq = r$ are easily solvable, which is the case, e.g., if B has Cholesky decomposition $B = LL^T$ with known sparse Cholesky factor L. After having chosen B, the vector $x' := B^{1/2}x$ solves the system

$$A'x' = b', \qquad b' := B^{-1/2}b,$$

which is equivalent to $Ax = b$. We now apply the conjugate-gradient method (8.7.3) to solve the primed system $A'x' = b'$, using $x_0' := B^{1/2}x_0$ as starting vector. Because of (8.7.9) and $c' \ll c$, the sequence x_k' generated by the cg method will converge very rapidly toward x'. But, instead of computing the matrix A and the vectors x_k' explicitly, we generate the sequence $x_k := B^{-1/2}x_k'$ associated with x_k' directly as follows: Using the transformation rules

$$A' = B^{-1/2}AB^{-1/2}, \qquad b' = B^{-1/2}b,$$

$$x_k' = B^{1/2}x_k, \qquad r_k' = b' - A'x_k' = B^{-1/2}r_k, \qquad p_k' = B^{1/2}p_k,$$

we obtain from the recursions of (8.7.3) for the primed system $A'x' = b'$ immediately the recursions of the following method.

(8.7.10) Preconditioned Conjugate Gradient Method.
Initialization: Choose $x_0 \in R^n$, *compute* $r_0 := b - Ax_0$, $q_0 := B^{-1}r_0$ *and put* $p_0 := q_0$.
For $k = 0, 1, \ldots$:

(1) *If* $p_k = 0$, *stop*: x_k *is the solution of* $Ax = b$. *Otherwise*,
(2) *compute*

$$a_k := \frac{r_k^T q_k}{p_k^T A p_k}, \qquad x_{k+1} := x_k + a_k p_k,$$

$$r_{k+1} := r_k - a_k A p_k, \qquad q_{k+1} := B^{-1} r_{k+1},$$

$$b_k := \frac{r_{k+1}^T q_{k+1}}{r_k^T q_k}, \qquad p_{k+1} := q_{k+1} + b_k p_k.$$

Essentially, the only difference, compared to (8.7.3), is that we have to solve in each step an extra linear system $Bq = r$ with the matrix B.

Now, the problem arises of finding an appropriate preconditioning matrix B, a problem similar to the problem of finding a suitable iterative method studied in Sections 8.1 to 8.3. When solving the linear equations $Ax = b$ arising from the discretization of boundary value problems for elliptic equations, say, the model problem of Section 8.4, it turned out to be useful to choose B as the SSOR matrix (cf. 8.3) defined by

(8.7.11)
$$B = \frac{1}{2-\omega}\left(\frac{1}{\omega}D - E\right)\left(\frac{1}{\omega}D\right)^{-1}\left(\frac{1}{\omega}D - E^T\right)$$

with a suitable $\omega \in (0, 2)$ [(see Axelsson (1977)]. Here, D and E are defined as in the standard decomposition (8.1.5) of A, that is, $A = D - E - E^T$.

Note that the factor $L = (1/\omega)D - E$ of B is a lower triangular matrix that is as sparse as the matrix A: Below the diagonal it is nonzero at the same positions as A.

Another more general proposal is due to Meijerink and van der Vorst (1977): They proposed to determine the preconditioner B and its Cholesky decomposition by the so-called *incomplete Cholesky factorization* of A. Slightly more general than in Section 4.3, we consider here Cholesky decompositions of the form $B = LDL^T$, where L is a lower triangular matrix with $l_{ii} = 1$ and D is a positive-definite diagonal matrix. With the incomplete Cholesky decomposition it is even possible to prescribe the sparsity structure of L: Given an arbitrary set $G \subset \{(i, j) \mid j \le i \le n\}$ of indices with $(i, i) \in G$ for all i, it is possible to find an L with

$$l_{i,j} \neq 0 \Rightarrow (i, j) \in G.$$

However, incomplete Cholesky factorization gives a decent approximation B to A only for positive-definite matrices A, which are also *M matrices*, that is, matrices A with $a_{ij} \le 0$ for $i \neq j$ and $A^{-1} \ge 0$ (see Meijerink and van der Vorst).

Fortunately, M matrices occur very frequently in applications and there are simple sufficient criteria for A to be an M matrix. For instance, any matrix $A = A^T$ with $a_{ii} > 0$, $a_{ij} \le 0$ for $i \ne j$ that also satisfies the hypotheses of Theorem (8.2.9) (weak row sum criterion) is an M matrix (e.g., the matrix A (8.4.5) of the model problem). This is proved, as in the proofs of Theorems (8.2.9) and (8.2.12), by showing the convergence of the Neumann series

$$A^{-1} = (I + J + J^2 + \cdots)D^{-1} \ge 0$$

for $A = D(I - J)$.

Given an index set G as earlier, the incomplete Cholesky factorization of a positive-definite M matrix A produces the factors D and L of a positive-definite matrix $B = LDL^T$ approximating A according to the following rules (cf. the program for the Cholesky algorithm given at the end of Section 4.3):

(8.7.12) Incomplete Cholesky Factorization.

For $i = 1, 2, \ldots$:

$$d_i := a_{ii} - \textstyle\sum_{k=1}^{i-1} d_k l_{ik}^2$$

For $j = i + 1, \ldots, n$:

$$d_i l_{ji} := \begin{cases} a_{ji} - \sum_{k=1}^{i-1} d_k l_{jk} l_{ik} & \textit{if } (i, j) \in G, \\ 0 & \textit{otherwise.} \end{cases}$$

That is, the only difference, compared to the ordinary Cholesky algorithm, is that $l_{ij} = 0$ is set equal to zero at the "forbidden" positions $(i, j) \notin G$.

8.8 The Algorithm of Buneman for the Solution of the Discretized Poisson Equation

Slightly generalizing the model problem (8.4.1), we consider the Poisson problem [compare (8.6.1)]

$$(8.8.1) \qquad \begin{aligned} -u_{xx} - u_{yy} + \sigma u &= f(x, y) \quad \text{for } (x, y) \in \Omega, \\ u(x, y) &= 0 \quad \text{for } (x, y) \in \partial\Omega \end{aligned}$$

on the rectangle $\Omega := \{(x, y) \mid 0 < x < a, 0 < y < b\} \subseteq \mathbb{R}^2$ with boundary $\partial\Omega$. Here $\sigma > 0$ is a constant and $f : \Omega \cup \partial\Omega \to \mathbb{R}$ a continuous function.

Discretizing (8.8.1) in the usual way, one obtains for the approximate values z_{ij} of $u(x_i, y_j)$, $x_i = i\,\Delta x$, $y_j = j\,\Delta y$, $\Delta x := a/(p + 1)$, $\Delta y := b/(q + 1)$, the equations

$$\frac{-z_{i-1,j} + 2z_{ij} - z_{i+1,j}}{\Delta x^2} + \frac{-z_{i,j-1} + 2z_{ij} - z_{i,j+1}}{\Delta y^2} + \sigma z_{ij} = f_{ij} = f(x_i, y_j)$$

for $i = 1, 2, \ldots, p$, $j = 1, 2, \ldots, q$. Together with the boundary values

$$z_{0,j} := z_{p+1,j} := 0 \quad \text{for } j = 0, 1, \ldots, q+1,$$
$$z_{i,0} := z_{i,q+1} := 0 \quad \text{for } i = 0, 1, \ldots, p+1,$$

one thus obtains for the unknowns

$$z = \begin{bmatrix} z_1 \\ z_2 \\ \vdots \\ z_q \end{bmatrix}, \qquad z_j = [z_{1j}, z_{2j}, \ldots, z_{pj}]^T$$

a linear system of equations which can be written in the form [cf. (8.4.5)]

$$Mz = b \tag{8.8.2a}$$

with

$$M = \begin{bmatrix} A & I & & & 0 \\ I & A & I & & \\ & \ddots & \ddots & \ddots & \\ & & \ddots & \ddots & I \\ 0 & & & I & A \end{bmatrix}, \qquad b = \begin{bmatrix} b_1 \\ b_2 \\ \vdots \\ b_q \end{bmatrix}, \tag{8.8.2b}$$

where $I = I_p$ is the $p \times p$ unit matrix, A is a $p \times p$ Hermitian tridiagonal matrix, and M consists of q block rows and columns.

In the last few years, several very effective methods for solving (8.8.2) have been proposed which are superior even to the ADI method (see Section 8.6). All these methods are reduction methods: exploiting the special structure of the matrix M, one reduces the solution of (8.8.2) recursively to the solution of systems of equations which are similarly structured, but have only half as many unknowns, and in this way successively halves the number of unknowns. Because of its simplicity we describe here only one of the first of these methods, the algorithm of Buneman (1969) [see also Buzbee, Golub, and Nielson (1970)]. For related methods see also Hockney (1969) and Swarztrauber (1977).

The following observation is essential for the reduction method of Buneman: In the system of equations (8.8.2), written out in full as

$$\begin{aligned} Az_1 + z_2 \qquad\qquad &= b_1, \\ z_{j-1} + Az_j + z_{j+1} \qquad &= b_j, \qquad j = 2, 3, \ldots, q-1, \\ z_{q-1} + Az_q &= b_q, \end{aligned} \tag{8.8.3}$$

from the three consecutive equations

$$\begin{aligned} z_{j-2} + Az_{j-1} + z_j \qquad\qquad &= b_{j-1}, \\ z_{j-1} + Az_j + z_{j+1} \qquad &= b_j, \\ z_j + Az_{j+1} + z_{j+2} &= b_{j+1}, \end{aligned}$$

one can, for all even $j = 2, 4, \ldots$, eliminate the variables z_{j-1} and z_{j+1} by subtracting A times the second equation from the sum of the others:

$$z_{j-2} + (2I - A^2)z_j + z_{j+2} = b_{j-1} - Ab_j + b_{j+1}.$$

For q odd, one thus obtains the reduced system

(8.8.4)

$$\begin{bmatrix} 2I - A^2 & I & & & 0 \\ I & 2I - A^2 & I & & \\ & I & \ddots & \ddots & \\ & & \ddots & \ddots & I \\ 0 & & & I & 2I - A^2 \end{bmatrix} \begin{bmatrix} z_2 \\ z_4 \\ \vdots \\ \\ z_{q-1} \end{bmatrix} = \begin{bmatrix} b_1 + b_3 - Ab_2 \\ b_3 + b_5 - Ab_4 \\ \vdots \\ \\ b_{q-2} + b_q - Ab_{q-1} \end{bmatrix}$$

for $z_2, z_4, \ldots$. Once a solution of (8.8.4) (i.e., the subvectors z_{2j} with even indices) is known, the vectors $z_1, z_3, \ldots$ with odd indices can be determined from the following equations, which immediately follow from (8.8.3) for $j = 1, 3, \ldots$:

$$\text{(8.8.5)} \qquad \begin{bmatrix} A & & & 0 \\ & A & & \\ & & A & \\ & & & \ddots & \\ 0 & & & & A \end{bmatrix} \begin{bmatrix} z_1 \\ z_3 \\ z_5 \\ \vdots \\ z_q \end{bmatrix} = \begin{bmatrix} b_1 - z_2 \\ b_3 - z_2 - z_4 \\ b_5 - z_4 - z_6 \\ \vdots \\ b_q - z_{q-1} \end{bmatrix}.$$

In this way, the solution of (8.8.2) is reduced to the solution of the system (8.8.4) with half the number of unknowns and subsequent solution of (8.8.5). Now (8.8.4) again has the same structure as (8.8.2):

$$M^{(1)}z^{(1)} = b^{(1)}$$

with

$$M^{(1)} = \begin{bmatrix} A^{(1)} & I & & & 0 \\ I & A^{(1)} & I & & \\ & I & \ddots & \ddots & \\ & & \ddots & \ddots & I \\ 0 & & & I & A^{(1)} \end{bmatrix}, \qquad A^{(1)} := 2I - A^2,$$

$$z^{(1)} = \begin{bmatrix} z_1^{(1)} \\ z_2^{(1)} \\ \vdots \\ z_{q_1}^{(1)} \end{bmatrix} := \begin{bmatrix} z_2 \\ z_4 \\ \vdots \\ z_{q-1} \end{bmatrix}, \qquad b^{(1)} = \begin{bmatrix} b_1^{(1)} \\ b_2^{(1)} \\ \vdots \\ b_{q_1}^{(1)} \end{bmatrix} := \begin{bmatrix} b_1 + b_3 - Ab_2 \\ b_3 + b_5 - Ab_4 \\ \vdots \\ b_{q-2} + b_q - Ab_{q-1} \end{bmatrix},$$

so that the reduction procedure just described can again be applied to $M^{(1)}$, etc. In general, for $q := q_0 := 2^{k+1} - 1$, one obtains in this way a sequence of matrices $A^{(r)}$ and vectors $b^{(r)}$ according to the following prescription:

(8.8.6).

Initialization: Put $A^{(0)} := A$; $b_j^{(0)} := b_j$, $j = 1, 2, \ldots, q_0$, $q_0 := q = 2^{k+1} - 1$.
For $r = 0, 1, 2, \ldots, k-1$: *Put*

(1) $A^{(r+1)} := 2I - (A^{(r)})^2$,
(2) $b_j^{(r+1)} := b_{2j-1}^{(r)} + b_{2j+1}^{(r)} - A^{(r)} b_{2j}^{(r)}, \qquad j = 1, 2, \ldots, 2^{k-r} - 1 =: q_{r+1}$.

For each stage $r+1$, $r = 0, 1, \ldots, k-1$, one thus obtains a system of linear equations

$$M^{(r+1)} z^{(r+1)} = b^{(r+1)},$$

or, written out in full,

$$\begin{bmatrix} A^{(r+1)} & I & & 0 \\ I & A^{(r+1)} & \ddots & \\ & \ddots & \ddots & I \\ 0 & & I & A^{(r+1)} \end{bmatrix} \begin{bmatrix} z_1^{(r+1)} \\ z_2^{(r+1)} \\ \vdots \\ z_{q_{r+1}}^{(r+1)} \end{bmatrix} = \begin{bmatrix} b_1^{(r+1)} \\ b_2^{(r+1)} \\ \vdots \\ b_{q_{r+1}}^{(r+1)} \end{bmatrix}.$$

Its solution $z^{(r+1)}$ immediately furnishes the subvectors with even indices of the solution $z^{(r)}$ of the corresponding system of equations $M^{(r)} z^{(r)} = b^{(r)}$ in stage r,

$$\begin{bmatrix} z_2^{(r)} \\ z_4^{(r)} \\ \vdots \\ z_{q_r - 1}^{(r)} \end{bmatrix} := \begin{bmatrix} z_1^{(r+1)} \\ z_2^{(r+1)} \\ \vdots \\ z_{q_{r+1}}^{(r+1)} \end{bmatrix},$$

while the subvectors with odd indices of $z^{(r)}$ can be obtained by solving the equations

$$\begin{bmatrix} A^{(r)} & & 0 \\ & A^{(r)} & \\ & \ddots & \\ 0 & & A^{(r)} \end{bmatrix} \begin{bmatrix} z_1^{(r)} \\ z_3^{(r)} \\ \vdots \\ z_{q_r}^{(r)} \end{bmatrix} = \begin{bmatrix} b_1^{(r)} - z_2^{(r)} \\ b_3^{(r)} - z_2^{(r)} - z_4^{(r)} \\ \vdots \\ b_{q_r}^{(r)} - z_{q_r - 1}^{(r)} \end{bmatrix}.$$

From the data $A^{(r)}$, $b^{(r)}$, produced by (8.8.6), the solution $z := z^{(0)}$ of (8.8.2) is thus finally obtained by the following procedure:

(8.8.7).

(0) *Initialization: Determine* $z^{(k)} = z_1^{(k)}$ *by solving the system of equations*

$$A^{(k)} z^{(k)} = b^{(k)} = b_1^{(k)}.$$

(1) *For* $r = k-1, k-2, \ldots, 0$:
  (a) *Put* $z_{2j}^{(r)} := z_j^{(r+1)}$, $j = 1, 2, \ldots, q_{r+1} = 2^{k-r} - 1$
  (b) *For* $j = 1, 3, 5, \ldots, q_r$ *compute the vector* $z_j^{(r)}$ *by solving*

$$A^{(r)} z_j^{(r)} = b_j^{(r)} - z_{j-1}^{(r)} - z_{j+1}^{(r)} \qquad (z_0^{(r)} := z_{q_r+1}^{(r)} := 0).$$

(2) *Put* $z := z^{(0)}$.

In the form (8.8.6), (8.8.7), the algorithm is still unsatisfactory, as it has serious numerical drawbacks. In the first place, the explicit computation of $A^{(r+1)} = 2I - (A^{(r)})^2$ in (8.8.6) (1) is very expensive: The tridiagonal matrix $A^{(0)} = A$, as r increases, very quickly turns into a dense matrix ($A^{(r)}$ is a band matrix with band width $2^r + 1$), so that, firstly, the computation of $(A^{(r)})^2$, and secondly, the solution of the systems of linear equations in (8.8.7) (1b) with increasing r become more and more expensive. In addition, it is easily seen that the magnitude of the matrices $A^{(r)}$ grows exponentially: For the model problem (8.4.1), e.g., one has

$$A = A^{(0)} = \begin{bmatrix} -4 & 1 & & 0 \\ 1 & -4 & \ddots & \\ & \ddots & \ddots & 1 \\ 0 & & 1 & -4 \end{bmatrix},$$

$$\|A^{(0)}\| \geqslant 4, \quad \|A^{(r)}\| \approx \|A^{(r-1)}\|^2 \geqslant 4^{2^r},$$

so that in the computation of $b_j^{(r+1)}$ in (8.8.6) (2) one incurs substantial losses of accuracy for larger values of r, since, in general, $\|A^{(r)}b_{2j}^{(r)}\| \gg \|b_{2j-1}^{(r)}\|$, $\|b_{2j+1}^{(r)}\|$, and therefore the information contained in $b_{2j-1}^{(r)}$, $b_{2j+1}^{(r)}$, when performing the addition in (8.8.6) (2), gets lost.

Both drawbacks can be avoided by a suitable reformulation of the algorithm. The explicit computation of $A^{(r)}$ is avoided if one exploits the fact that $A^{(r)}$ can be represented as a product of tridiagonal matrices:

(8.8.8) Theorem. *One has, for all* $r \geqslant 0$,

$$A^{(r)} = -\prod_{j=1}^{2^r} [-(A + 2\cos\theta_j^{(r)} \cdot I)],$$

where $\theta_j^{(r)} := (2j-1)\pi/2^{r+1}$ *for* $j = 1, 2, \ldots, 2^r$.

Proof. By (8.8.6) (1), one has, with $A^{(0)} = A$,

$$A^{(r+1)} = 2I - (A^{(r)})^2,$$

so that there exists a polynomial $p_r(t)$ of degree 2^r such that

(8.8.9) $$A^{(r)} = p_r(A).$$

Evidently, the polynomials p_r satisfy

$$p_0(t) = t,$$
$$p_{r+1}(t) = 2 - (p_r(t))^2,$$

so that p_r has the form

(8.8.10) $$p_r(t) = -(-t)^{2^r} + \cdots.$$

One now shows by mathematical induction, using the substitution $t = -2\cos\theta$, that

$$p_r(-2\cos\theta) = -2\cos(2^r\theta). \tag{8.8.11}$$

The formula is trivial for $r = 0$. If it is valid for some $r \geqslant 0$, then it is also valid for $r + 1$, since

$$\begin{aligned} p_{r+1}(-2\cos\theta) &= 2 - (p_r(-2\cos\theta))^2 \\ &= 2 - 4\cos^2(2^r\theta) \\ &= -2\cos(2\cdot 2^r\theta). \end{aligned}$$

In view of (8.8.11), the polynomial $p_r(t)$ has the 2^r distinct real zeros

$$t_j = -2\cos\left(\frac{2j-1}{2^{r+1}}\pi\right), \qquad j = 1, 2, \ldots, 2^r,$$

and therefore, by (8.8.10), the product representation

$$p_r(t) = -\prod_{j=1}^{2^r}[-(t - t_j)].$$

From this, by virtue of (8.8.9), the assertion of the theorem follows at once. □

The preceding theorem can now be exploited, in practice, to reduce the solution of the various systems of equations

$$A^{(r)}u = b$$

in (8.8.7) (1b), with matrices $A^{(r)}$, recursively to the solution of 2^r systems of equations with the tridiagonal matrices

$$A_j^{(r)} := -A - 2\cos\theta_j^{(r)}\cdot I, \qquad j = 1, 2, \ldots, 2^r,$$

as follows:

$$\begin{aligned} A_1^{(r)}u_1 &= b && \Rightarrow u_1, \\ A_2^{(r)}u_2 &= u_1 && \Rightarrow u_2, \\ &\;\vdots \\ A_{2^r}^{(r)}u_{2^r} &= u_{2^r-1} && \Rightarrow u_{2^r} \Rightarrow u := -u_{2^r}. \end{aligned} \tag{8.8.12}$$

Since, as is easily verified, the tridiagonal matrices $A_j^{(r)}$ for our discretization of problem (8.8.1) are positive definite, one can solve these systems very inexpensively by means of triangular factorizations of $A_j^{(r)}$ without pivoting (see Section 4.3).

The numerical instability which occurs in (8.8.6) (2) because of the exponential growth of the $A^{(r)}$ can be avoided, following a suggestion of Buneman, by introducing in place of the $b_j^{(r)}$ other vectors $p_j^{(r)}$, $q_j^{(r)}$, $j = 1, 2, \ldots, q_r$, which are related to the $b_j^{(r)}$ in the following way:

$$b_j^{(r)} = A^{(r)} p_j^{(r)} + q_j^{(r)}, \qquad j = 1, 2, \ldots, q_r, \tag{8.8.13}$$

and which can be computed in a numerically more stable manner than the $b_j^{(r)}$. Vectors $p_j^{(r)}$, $q_j^{(r)}$ with these properties are generated recursively as follows:

(8.8.14).

Initialization: Put $p_j^{(0)} := 0$, $q_j^{(0)} := b_j = b_j^{(0)}$, $j = 1, 2, \ldots, q_0$.
For $r = 0, 1, \ldots, k - 1$:
Compute for $j = 1, 2, \ldots, q_{r+1}$:

(1) $p_j^{(r+1)} := p_{2j}^{(r)} - (A^{(r)})^{-1}[p_{2j-1}^{(r)} + p_{2j+1}^{(r)} - q_{2j}^{(r)}]$,
(2) $q_j^{(r+1)} := q_{2j-1}^{(r)} + q_{2j+1}^{(r)} - 2p_j^{(r+1)}$.

Naturally, the computation of $p_j^{(r+1)}$ in substep (1) amounts to first determining, as just described [see (8.8.12)], the solution u of the system of equations

$$A^{(r)} u = p_{2j-1}^{(r)} + p_{2j+1}^{(r)} - q_{2j}^{(r)}$$

with the aid of the factorization of $A^{(r)}$ in Theorem 8.8.8, and then computing $p_j^{(r+1)}$ from u by means of

$$p_j^{(r+1)} := p_{2j}^{(r)} - u.$$

Let us prove by induction on r that the vectors $p_j^{(r)}$, $q_j^{(r)}$, defined by (8.8.14), satisfy the relation (8.8.13).

For $r = 0$, (8.8.13) is trivial. We assume inductively that (8.8.13) holds true for some $r \geqslant 0$. Because of (8.8.6) (2) and $A^{(r+1)} = 2I - (A^{(r)})^2$, we then have

$$\begin{aligned}
b_j^{(r+1)} &= b_{2j+1}^{(r)} + b_{2j-1}^{(r)} - A^{(r)} b_{2j}^{(r)} \\
&= A^{(r)} p_{2j+1}^{(r)} + q_{2j+1}^{(r)} + A^{(r)} p_{2j-1}^{(r)} + q_{2j-1}^{(r)} - A^{(r)}[A^{(r)} p_{2j}^{(r)} + q_{2j}^{(r)}] \\
&= A^{(r)}[p_{2j+1}^{(r)} + p_{2j-1}^{(r)} - q_{2j}^{(r)}] \\
&\quad + A^{(r+1)} p_{2j}^{(r)} + q_{2j-1}^{(r)} + q_{2j+1}^{(r)} - 2p_{2j}^{(r)} \\
&= A^{(r+1)} p_{2j}^{(r)} + (A^{(r)})^{-1}\{[2I - A^{(r+1)}][p_{2j+1}^{(r)} + p_{2j-1}^{(r)} - q_{2j}^{(r)}]\} \\
&\quad + q_{2j-1}^{(r)} + q_{2j+1}^{(r)} - 2p_{2j}^{(r)} \\
&= A^{(r+1)}\{p_{2j}^{(r)} - (A^{(r)})^{-1}[p_{2j-1}^{(r)} + p_{2j+1}^{(r)} - q_{2j}^{(r)}]\} \\
&\quad + q_{2j-1}^{(r)} + q_{2j+1}^{(r)} - 2p_j^{(r+1)} \\
&= A^{(r+1)} p_j^{(r+1)} + q_j^{(r+1)}.
\end{aligned}$$

By (8.8.13) we can now express the vectors $b_j^{(r)}$ in (8.8.7) in terms of the $p_j^{(r)}$, $q_j^{(r)}$ and obtain, for example, from (8.8.7) (1b) for $z_j^{(r)}$ the system of equations

$$A^{(r)}z_j^{(r)} = A^{(r)}p_j^{(r)} + q_j^{(r)} - z_{j-1}^{(r)} - z_{j+1}^{(r)},$$

which can be solved as follows: Determine the solution u of

$$A^{(r)}u = q_j^{(r)} - z_{j-1}^{(r)} - z_{j+1}^{(r)}$$

[one uses here again the factorization of Theorem (8.8.8)] and put $z_j^{(r)} := u + p_j^{(r)}$.

Replacing in this way the $b_j^{(r)}$ in (8.8.6), (8.8.7) systematically by $p_j^{(r)}$ and $q_j^{(r)}$, one obtains:

(8.8.15) The Algorithm of Buneman.

Assumption: Consider the system of equations (8.8.2), *with* $q = 2^{k+1} - 1$.

(0) *Initialization: Put* $p_j^{(0)} := 0$, $q_j^{(0)} := b_j$, $j = 1, 2, \ldots, q_0 := q$.
(1) *For* $r = 0, 1, \ldots, k - 1$:
For $j = 1, 2, \ldots, q_{r+1} := 2^{k-r} - 1$:
Compute the solution u of the system of equations

$$A^{(r)}u = p_{2j-1}^{(r)} + p_{2j+1}^{(r)} - q_{2j}^{(r)}$$

by means of the factorization of Theorem (8.8.8) *and put*

$$p_j^{(r+1)} := p_{2j}^{(r)} - u, \qquad q_j^{(r+1)} := q_{2j-1}^{(r)} + q_{2j+1}^{(r)} - 2p_j^{(r+1)}.$$

(2) *Determine the solution u of the system of equations*

$$A^{(k)}u = q_1^{(k)},$$

and put $z^{(k)} := z_1^{(k)} := p_1^{(k)} + u$.
(3) *For* $r = k - 1, k - 2, \ldots, 0$:
(a) *Put* $z_{2j}^{(r)} := z_j^{(r+1)}$ *for* $j = 1, 2, \ldots, q_{r+1}$
(b) *For* $j = 1, 3, 5, \ldots, q_r$ *determine the solution u of the system of equations*

$$A^{(r)}u = q_j^{(r)} - z_{j-1}^{(r)} - z_{j+1}^{(r)},$$

and put

$$z_j^{(r)} := p_j^{(r)} + u.$$

(4) *Put* $z := z^{(0)}$.

This method is very efficient: An operation count shows that for the solution of the model problem (8.4.1) ($a = b = 1, p = q = N = 2^{k+1} - 1$), with its N^2 unknowns, one requires approximately $3kN^2 \approx 3N^2 \log_2 N$ multiplications and about the same number of additions. An analysis of the numerical stability of the method can be found in Buzbee, Golub, and Nielson (1970).

In the form described, the method serves to solve the discretized Dirichlet boundary-value problem for the Poisson equation on a rectangular domain. There are variants of the method for the solution of analogous boundary-value problems for the Helmholtz equation or the biharmonic equation on rectangular domains.

Newer reduction methods with even better stability properties for the solution of such problems are given, and extensively studied, by Schröder, Trottenberg, and Reutersberg (1976). There are, in addition, more complicated versions of these methods for the respective discretized boundary-value problems on nonrectangular domains [see Buzbee and Dorr (1974); Buzbee, Dorr, George, and Golub (1971); Proskurowski and Widlund (1976); O'Leary and Widlund (1979)]. While the methods discussed are direct and noniterative, competitive iterative methods with substantially improved convergence properties have recently been developed. Examples are the *multigrid* methods and modern *domain decomposition* methods, which are rapidly being developed. The principles of multigrid methods will be briefly explained in the next section. With respect to domain decomposition methods, we refer the reader to the special literature, for instance, Chan, Glowinski, Periaux, and Widlund (1989); Glowinsky, Golub, Meurant, and Periaux (1988); and Keyes and Gropp (1987).

8.9 Multigrid Methods

Multigrid methods belong to the most efficient methods for the solution of those linear equations that result from the discretization of differential equations. As these methods are very flexible, there are many variants of them. Here, we wish to explain only the basic ideas behind these powerful methods, and do this in a rather simple situation, which, however, already reveals their typical properties. For a detailed treatment, we have to refer the reader to the special literature, for instance Brandt (1977), Hackbusch and Trottenberg (1982), and in particular to the monograph of Hackbusch (1985). Our treatment follows the elementary exposition of Briggs (1987). Instead of boundary value problems for partial differential equations, where multigrid methods have their greatest impact, we only consider their application to the boundary value problem [cf. (7.4.1)]

$$
\begin{aligned}
-y''(x) &= f(x) \qquad \text{for } x \in \Omega := (0, \pi), \\
y(0) &= y(\pi) = 0
\end{aligned}
\tag{8.9.1}
$$

for an ordinary differential equation, which can be viewed as the one-dimensional analog of the two-dimensional model problem (8.4.1). The standard discretization (see Section 7.4) with the grid size $h = \pi/n$ leads to a one-dimensional grid $\Omega_h = \{x_j = jh \mid j = 1, \ldots, n-1\} \subset \Omega$ and the following set of linear equations for a vector $u_h = (u_{h;1}, \ldots, u_{h;n-1})^T$ of approximations $u_{h;j} \approx y(x_j)$ for the exact solution y on the grid Ω_h:

$$(8.9.2) \quad A_h u_h = f_h, \quad A_h := \frac{1}{h^2}\begin{bmatrix} 2 & -1 & & 0 \\ -1 & 2 & \ddots & \\ 0 & \ddots & \ddots & -1 \\ & & -1 & 2 \end{bmatrix}, \quad f_h := \begin{bmatrix} f(x_1) \\ f(x_2) \\ \vdots \\ f(x_{n-1}) \end{bmatrix}.$$

The index h also indicates that u_h and f_h can be viewed as functions on the grid Ω_h. Therefore, we will sometimes write the jth component $u_{h;j}$ of u_h as the value of a *grid function* $u_h(x)$ for $x = x_j \in \Omega_h$, $u_{h;j} = u_h(x_j)$. The matrix A_h is a matrix of order $(n-1)$, for which the eigenvalues $\lambda_h^{(k)}$ and eigenvectors $z_h^{(k)}$ are known explicitly (cf. Section 8.4):

$$(8.9.3) \quad \begin{aligned} z_h^{(k)} &:= (\sin kh, \sin 2kh, \ldots, \sin(n-1)kh)^T, \\ \lambda_h^{(k)} &:= \frac{1}{h^2} 4 \sin^2 \frac{kh}{2} = \frac{2}{h^2}(1 - \cos kh), \qquad k = 1, 2, \ldots, n-1. \end{aligned}$$

This is easily verified by checking $A_h z_h^{(k)} = \lambda_h^{(k)} z_h^{(k)}$, $k = 1, \ldots, n-1$. The vectors $z_h^{(k)}$ have the Euclidean norm $\|z_h^{(k)}\| = \sqrt{n/2}$ and are orthogonal to each other [Theorem (6.4.2)].

If we consider for fixed k the components $\sin jkh = \sin(jk\pi/n)$ of the eigenvector $z_h^{(k)}$ at the grid points x_j of Ω_h for $j = 1, \ldots, n-1$, we see that the grid function $z^{(k)} = z_h^{(k)}$ describes a wave on Ω_k with "frequency" k and "wavelength" $2\pi/k$: The number k just gives the number of half-waves on Ω_h (see Figure 32 for $n = 6$, $k = 2$).

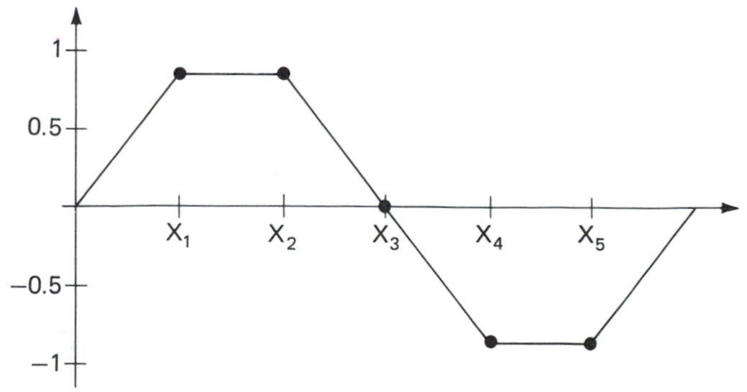

Figure 32 The grid function $z^{(2)}$.

In order to simplify the notation, we omit the index h occasionally, if it is clear from the context to which grid size h and grid Ω_h the vectors $u = u_h$, $f = f_h$ and the matrix $A = A_h$ belong.

One motivation for multigrid methods is connected with the convergence behavior of the standard iterative methods (8.1.7) and (8.1.8) for solving $Au = f$. We study this in more detail for the Jacobi method (8.1.7). The usual decomposition (8.1.5), (8.1.6)

$$A_h = D_h(I - J_h), \qquad D_h = \frac{2}{h^2} I,$$

of $A = A_h$ leads to the matrix of order $(n-1)$

$$J = J_h = I - \frac{h^2}{2} A_h = \frac{1}{2} \begin{bmatrix} 0 & 1 & & 0 \\ 1 & 0 & \ddots & \\ & \ddots & \ddots & 1 \\ 0 & & 1 & 0 \end{bmatrix}$$

and the iteration of the Jacobi method

$$v^{(i+1)} = Jv^{(i)} + \frac{h^2}{2} f.$$

The errors $e^{(i)} := v^{(i)} - u$ of the iterates $v^{(i)}$ then satisfy the recursion

$$e^{(i+1)} = Je^{(i)} = J^{i+1} e^{(0)}.$$

Clearly, the iteration matrix $J = J_h = I - (h^2/2) A_h$ has the eigenvalues

$$\mu^{(k)} = \mu_h^{(k)} = 1 - \frac{h^2}{2} \lambda_h^{(k)} = \cos kh, \qquad k = 1, \ldots, n-1,$$

but still the same eigenvectors $z_h^{(k)}$ as A_h. In order to analyze the behavior of the error e under an iteration step $e \to \bar{e} = Je$, we write e as a linear combination of the eigenvectors $z^{(k)} = z_h^{(k)}$ of $J = J_h$ (cf. Section 6.6.3),

$$e = \rho_1 z^{(1)} + \cdots + \rho_{n-1} z^{(n-1)}.$$

The weight ρ_k measures the influence of frequency k in e. Because of

$$\bar{e} = \rho_1 \mu^{(1)} z^{(1)} + \cdots + \rho_{n-1} \mu^{(n-1)} z^{(n-1)}$$

and $1 > \mu^{(1)} > \mu^{(2)} > \cdots > \mu^{(n-1)} = -\mu^{(1)} > -1$, we see that all frequencies $k = 1, \ldots, n-1$ are damped in $\bar{e}$, but to a different extent. The central frequencies $k \approx n/2$ are damped most, the extreme frequencies $k = 1$ and $n - 1$ only slightly.

The damping of the large frequencies k with $n/2 \le k \le n-1$ can be much improved by introducing a suitable relaxation factor ω into the iteration matrix. To this end, we consider a slightly more general decomposition of A [cf. (8.1.3), (8.1.4), (8.1.7)] defined by $A = A_h = B - (B - A)$ with $B := (1/\omega)D$, which leads to the *damped Jacobi method* with the iteration rule

$$v^{(i+1)} = J(\omega)v^{(i)} + \frac{\omega}{2}h^2 f \tag{8.9.4}$$

in terms of the matrix $J_h(\omega) = J(\omega) := (1 - \omega)I + \omega J$. The original Jacobi method corresponds to $\omega = 1$, $J(1) = J$. Clearly, the eigenvalues $\mu^{(k)}(\omega) = \mu_h^{(k)}(\omega)$ of $J(\omega)$ are given by

$$\mu^{(k)}(\omega) := 1 - \omega + \omega\mu^{(k)} = 1 - 2\omega \sin^2 \frac{kh}{2}, \qquad k = 1, \ldots, n-1, \tag{8.9.5}$$

and they belong to the same eigenvectors $z^{(k)} = z_h^{(k)}$ as before.

Now, an iteration step transforms the error as follows:

$$e = \sum_{k=1}^{n-1} \rho_k z^{(k)} \to \bar{e} = J(\omega)e = \sum_{k=1}^{n-1} \rho_k \mu^{(k)}(\omega) z^{(k)}. \tag{8.9.6}$$

Since $|\mu^{(k)}(\omega)| < 1$ for all $0 < \omega \le 1$, $k = 1, \ldots, n-1$, all frequencies k will be damped if $0 < \omega \le 1$. However, by a suitable choice of ω it is possible to damp the high frequencies $n/2 \le k \le n-1$ most heavily. In particular,

$$\max_{n/2 \le k \le n-1} |\mu^{(k)}(\omega)|$$

becomes minimal for the choice $\omega = 2/3$, and then $|\mu^{(k)}(\omega)| \le 1/3$ for $n/2 \le k \le n-1$: the method acts as a "smoother," as the high-frequency oscillations are smoothed out. Note that the damping factor 1/3 for the high frequencies does not depend on h, but the overall damping factor $\max_k |\mu^{(k)}(\omega)| = \mu^{(1)}(\omega) = 1 - 2\omega \sin^2(h/2) = 1 - O(h^2)$ converges to 1 as $h \downarrow 0$, so that the convergence rate of the damped Jacobi method deteriorates as h tends to zero (cf. the discussion in Section 8.4).

In any case, after relatively few steps of the damped Jacobi method one finds an iterate $v^{(i)} = v_h^{(i)}$ with an error

$$e_h^{(i)} = v_h^{(i)} - u_h = \rho_1^{(i)} z_h^{(1)} + \cdots + \rho_{n-1}^{(i)} z_h^{(n-1)}$$

containing almost no high frequencies anymore:

$$\max_{n/2 \le k < n} |\rho_k^{(i)}| \ll \max_{1 \le k < n/2} |\rho_k^{(i)}|.$$

Now there is a new consideration that comes into play: The vector $e_h^{(i)}$ is the exact solution of the system $A_h e_h^{(i)} = -r_h^{(i)}$, where $r_h^{(i)} = f_h - A_h v_h^{(i)}$ is the residual of $v_h^{(i)}$; hence the decomposition of

$$r_h^{(i)} = -\sum_{k=1}^{n-1} \rho_k^{(i)} \lambda_h^{(k)} z_h^{(k)}$$

essentially contains only contributions of the lower frequencies. But a "long wave" grid function g_h on Ω_h can be approximated quite well by a grid function g_{2h} on the coarser grid $\Omega_{2h} = \{j \cdot 2h \mid j = 1, 2, \ldots, n/2 - 1\}$ (here we assume that n is even) by means of a *projection operator* I_h^{2h}:

$$g_{2h} := I_h^{2h} g_h, \qquad I_h^{2h} := \frac{1}{4}\begin{bmatrix} 1 & 2 & 1 & & & & & \\ & & 1 & 2 & 1 & & & \\ & & & & \cdots & \cdots & & \\ & & & & & 1 & 2 & 1 \end{bmatrix}.$$

Here, I_h^{2h} is an $((n/2) - 1) \times (n - 1)$ matrix. The coarse-grid function g_{2h} on Ω_{2h} is obtained from the fine-grid function g_h on Ω_h by averaging:

$$g_{2h}(j \cdot 2h) = \frac{1}{4} g_h((2j - 1)h) + \frac{2}{4} g_h(2j \cdot h) + \frac{1}{4} g_h((2j + 1)h), \quad j = 1, \ldots, \frac{n}{2} - 1$$

(see Figure 33 for $n = 6$).

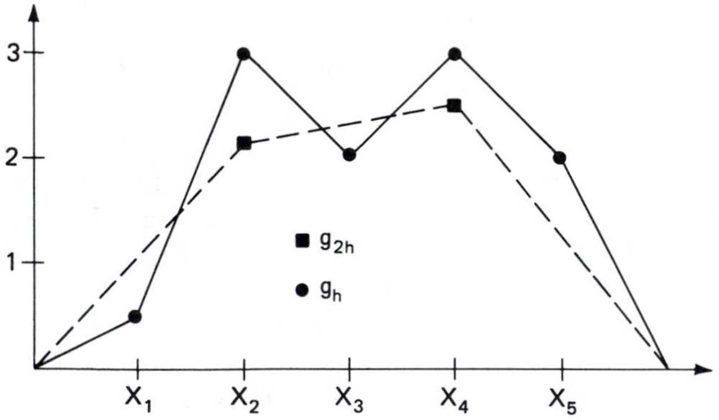

Figure 33 Projection by averaging.

Instead of forming averages, one could also use the simple restriction operator

$$I_h^{2h} = \begin{bmatrix} 0 & 1 & 0 & & & & & \\ & & 0 & 1 & 0 & & & \\ & & & & \cdots & \cdots & & \\ & & & & & 0 & 1 & 0 \end{bmatrix}$$

for the projection. Then the function $g_{2h} = I_h^{2h} g_h$ would be just the restriction of the function g_h on Ω_h to Ω_{2h},

$$g_{2h}(j \cdot 2h) := g_h(2j \cdot h), \qquad j = 1, \ldots, \frac{n}{2} - 1.$$

We do not pursue this possibility further.

Conversely, *interpolation operators* I_{2h}^h can be used to extend a grid function g_{2h} on the coarse grid Ω_{2h} to a grid function $g_h = I_{2h}^h g_{2h}$ on the fine grid Ω_h, by defining, say,

$$I_{2h}^h = \frac{1}{2}\begin{bmatrix} 1 & & & \\ 2 & & & \\ 1 & 1 & & \\ & 2 & & \\ & 1 & \vdots & \\ & & \vdots & 1 \\ & & & 2 \\ & & & 1 \end{bmatrix}.$$

Here, I_{2h}^h is an $(n-1) \times ((n/2)-1)$ matrix, and the function g_h is obtained from g_{2h} by interpolation ($g_{2h}(0) = g_{2h}(\pi) := 0$):

$$g_h(jh) := \begin{cases} g_{2h}\left(\dfrac{j}{2} \cdot 2h\right) & \text{if } j \text{ is even,} \\ \dfrac{1}{2} g_{2h}\left(\dfrac{j-1}{2} \cdot 2h\right) + \dfrac{1}{2} g_{2h}\left(\dfrac{j+1}{2} \cdot 2h\right) & \text{otherwise,} \end{cases} \qquad j = 1, \ldots, n-1$$

(see Figure 34 for $n = 6$).

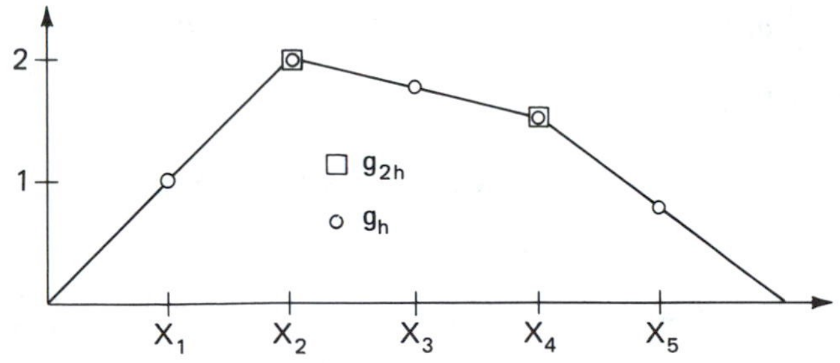

Figure 34 Extension by interpolation.

Now an elementary form of a multigrid method runs as follows: A given approximate solution $v_h^{(i)}$ of $A_h u_h = f_h$ is first transformed by a finite number steps of the damped Jacobi method (8.9.4) into a new approximate solution $w_h^{(i)}$ of $A_h u_h = f_h$ with error $e_h^{(i)}$ and residual $r_h^{(i)}$. Then the residual $r_h^{(i)}$ is projected to the coarse grid Ω_{2h}, $r_h^{(i)} \to r_{2h}^{(i)} := I_h^{2h} r_h^{(i)}$, and the linear "coarse-grid equation"

$$A_{2h} e_{2h}^{(i)} = -r_{2h}^{(i)}$$

is solved. Its solution $e_{2h}^{(i)}$ is then extended to the fine grid $e_{2h}^{(i)} \to \tilde{e}_h^{(i)} := I_{2h}^h e_{2h}^{(i)}$ by interpolation. We expect that $\tilde{e}_h^{(i)}$ is a good approximation for the exact solution $e_h^{(i)}$ of $A_h e_h^{(i)} = -r_h^{(i)}$, since $e_h^{(i)}$ is a low-frequency grid function. Therefore, $v_h^{(i+1)} := v_h^{(i)} - \tilde{e}_h^{(i)}$ will presumably be a much better approximation to u_h than $v_h^{(i)}$.

In this way, we obtain the *two-grid method*, whose basic step $v_h^{(i)} \to v_h^{(i+1)} := TGM(v_h^{(i)})$ is defined by a mapping TGM according to the following rules.

(8.9.7) Two-Grid Method. *Let v_h be a grid vector on Ω_h.*

(1) *Perform ν steps of the damped Jacobi method* (8.9.4), *with $\omega = \omega_0 := 2/3$ and the starting vector v_h, which results in the vector w_h with the residual $r_h := f_h - A_h w_h$ (smoothing step).*
(2) *Compute $r_{2h} := I_h^{2h} r_h$ (projecting step).*
(3) *Solve $A_{2h} e_{2h} = -r_{2h}$ (coarse-grid solution).*
(4) *Set $TGM(v_h) := w_h - I_{2h}^h e_{2h}$ (interpolation and fine-grid correction step).*

It is relatively easy to analyze the behavior of the error

$$e_h := v_h - u_h \to \bar{e}_h := \bar{v}_h - u_h$$

during one iteration step $v_h \to \bar{v}_h = TGM(v_h)$ of (8.9.7) in the case of our simple model problem. Because of (8.9.6), after the ν smoothing steps, the error $d_h := w_h - u_h$ of w_h satisfies

$$d_h = J(\omega_0)^\nu e_h, \qquad r_h = -A_h d_h = -A_h J(\omega_0)^\nu e_h.$$

Further, by (8.9.7),

$$A_{2h} e_{2h} = -r_{2h} = -I_h^{2h} r_h = I_h^{2h} A_h d_h,$$
$$\bar{e}_h = \bar{v}_h - u_h = d_h - I_{2h}^h e_{2h},$$

and we find the formula

$$\text{(8.9.8)} \qquad \begin{aligned} \bar{e}_h &= (I - I_{2h}^h A_{2h}^{-1} I_h^{2h} A_h)\, d_h \\ &= C_h \cdot J(\omega_0)^\nu e_h, \end{aligned}$$

where C_h is the $(n-1) \times (n-1)$ matrix

$$C_h := I - I_{2h}^h A_{2h}^{-1} I_h^{2h} A_h.$$

In order to study the propagation of the frequencies contained in e_h, we need explicit formulas for the maps $C_h z_h^{(k)}$ of the eigenvectors of A_h. Using the abbreviations $c_k := \cos^2(kh/2)$, $s_k := \sin^2(kh/2)$, and $k' := n - k$, a short direct calculation shows

(8.9.9) $$I_h^{2h} z_h^{(k)} = \begin{cases} c_k z_{2h}^{(k)} & \text{for } k = 1, \ldots, \dfrac{n}{2} - 1, \\[2ex] -s_{k'} z_{2h}^{(k')} & \text{for } k = \dfrac{n}{2}, \ldots, n - 1. \end{cases}$$

Here, the vectors $z_{2h}^{(k)}$, $1 \le k < n/2$, are just the eigenvectors of A_{2h} for the eigenvalues

$$\lambda_{2h}^{(k)} = \frac{4}{(2h)^2} \sin^2 kh = \frac{1}{h^2} \sin^2 kh$$

[see (8.9.3)], so that

$$A_{2h}^{-1} z_{2h}^{(k)} = \frac{1}{\lambda_{2h}^{(k)}} z_{2h}^{(k)}, \qquad k = 1, 2, \ldots, \frac{n}{2} - 1.$$

Again, by a simple direct calculation one verifies

(8.9.10) $$I_{2h}^h z_{2h}^{(k)} = c_k z_h^{(k)} - s_k z_h^{(k')}, \qquad k = 1, 2, \ldots, (n/2) - 1.$$

Combining these results gives for $k = 1, 2, \ldots, (n/2) - 1$

$$\begin{aligned} I_{2h}^h A_{2h}^{-1} I_h^{2h} A_h z_h^{(k)} &= \lambda_h^{(k)} I_{2h}^h A_{2h}^{-1} I_h^{2h} z_h^{(k)} \\ &= \lambda_h^{(k)} c_k I_{2h}^h A_{2h}^{-1} z_{2h}^{(k)} \\ &= \frac{\lambda_h^{(k)}}{\lambda_{2h}^{(k)}} c_k I_{2h}^h z_{2h}^{(k)} \\ &= \frac{\lambda_h^{(k)}}{\lambda_{2h}^{(k)}} c_k (c_k z_h^{(k)} - s_k z_h^{(k')}). \end{aligned}$$

Using that

$$\frac{\lambda_h^{(k)}}{\lambda_{2h}^{(k)}} = \frac{\dfrac{4}{h^2} \sin^2 kh/2}{\dfrac{1}{h^2} \sin^2 kh} = \frac{4 s_k}{\sin^2 kh}, \qquad c_k s_k = \frac{1}{4} \sin^2 kh,$$

we finally obtain for $k = 1, 2, \ldots, (n/2) - 1$

(8.9.11) $$C_h z_h^{(k)} = (I - I_{2h}^h A_{2h}^{-1} I_h^{2h} A_h) z_h^{(k)} = s_k z_h^{(k)} + s_k z_h^{(k')}.$$

Similarly, the maps of the high-frequency vectors $z_h^{(k')}$ are given by

(8.9.12) $$C_h z_h^{(k')} = c_k z_h^{(k)} + c_k z_h^{(k')}, \qquad k = 1, \ldots, \frac{n}{2}.$$

We are now able to show the following theorem.

(8.9.13) Theorem. *Let $\nu = 2$ and $\omega_0 = 2/3$, and suppose that one step of the two-grid method* (8.9.7) *transforms the vector v_h into $\bar{v}_h := TGM(v_h)$. Then the errors $e_h := v_h - u_h$, $\bar{e}_h := \bar{v}_h - u_h$ of v_h and $\bar{v}_h$ satisfy*

$$\|\bar{e}_h\|_2 \le 0.782\|e_h\|_2.$$

Thus, the two-grid method generates a sequence $v_h^{(j+1)} = TGM(v_h^{(j)})$, $j = 1, 2, \ldots$, whose errors $e_h^{(j)} = v_h^{(j)} - u_h$ converge to 0 with a linear rate of convergence that is *independent* of h,

$$\|e_h^{(j)}\|_2 \le 0.782^j \|e_h^{(0)}\|_2.$$

This is quite remarkable: The convergence rates of all iterative methods considered so far (cf. Section 8.4) depend on h and deteriorate as $h \downarrow 0$.

PROOF. We start with the decomposition of the error $e_h = v_h - u_h$,

$$e_h = \rho_1 z_h^{(1)} + \cdots + \rho_{n-1} z_h^{(n-1)}.$$

We have already seen, in (8.9.5), that the vectors $z_h^{(k)}$ are also the eigenvectors of $J(\omega_0)$ belonging to the eigenvalues $\mu_h^{(k)}(\omega_0) = 1 - 2\omega_0 s_k$. The choice of ω_0 guarantees, for $k = 1, \ldots, n/2$, $k' := n - k$, that

$$|\mu_h^{(k)}(\omega_0)| < 1, \qquad |\mu_h^{(k')}(\omega_0)| \le \frac{1}{3}.$$

Next, (8.9.6), (8.9.11), and (8.9.12) imply for $k = 1, \ldots, n/2$

$$\begin{aligned} C_h J(\omega_0)^\nu z_h^{(k)} &= (\mu_h^{(k)}(\omega_0))^\nu (s_k z_h^{(k)} + s_k z_h^{(k')}) =: \alpha_k (z_h^{(k)} + z_h^{(k')}), \\ C_h J(\omega_0)^\nu z_h^{(k')} &= (\mu_h^{(k')}(\omega_0))^\nu (c_k z_h^{(k)} + c_k z_h^{(k')}) =: \beta_k (z_h^{(k)} + z_h^{(k')}), \end{aligned}$$

where the constants α_k and β_k are estimated by

$$|\alpha_k| \le s_k \le \frac{1}{2}, \qquad |\beta_k| \le \frac{1}{3^\nu} \qquad \text{for } k = 1, \ldots, \frac{n}{2}.$$

We thus obtain the following formula for the error $\bar{e}_h$:

$$\begin{aligned} \bar{e}_h &= C_h J(\omega_0)^\nu e_h \\ &= \sum_{k=1}^{n/2} \delta_k (\rho_k \alpha_k + \rho_{k'} \beta_k)(z_h^{(k)} + z_h^{(k')}), \end{aligned}$$

where we have used the abbreviations $\delta_k := 1$ for $k < n/2$ and $\delta_{n/2} := 1/2$. Finally, the orthogonality of the vectors $z_h^{(k)}$ and $\|z_h^{(k)}\|^2 = n/2$ imply that

$$
\begin{aligned}
\|\bar{e}_h\|^2 &= n\left[\sum_{k=1}^{n/2} \delta_k(\rho_k^2\alpha_k^2 + \rho_{k'}^2\beta_k^2 + 2\rho_k\rho_{k'}\alpha_k\beta_k)\right] \\
&\le n\left[\sum_{k=1}^{n/2} \delta_k(\rho_k^2\alpha_k^2 + \rho_{k'}^2\beta_k^2 + (\rho_k^2 + \rho_{k'}^2)|\alpha_k\beta_k|)\right] \\
&\le n\left(\frac{1}{4} + \frac{1}{2\cdot 3^\nu}\right)\sum_{k=1}^{n/2} \delta_k(\rho_k^2 + \rho_{k'}^2) \\
&= \left(\frac{1}{2} + \frac{1}{3^\nu}\right)\|e_h\|^2.
\end{aligned}
$$

For $\nu = 2$ we obtain the estimate of the theorem. □

With the two-grid method there remains the problem of how to solve the linear equation $A_{2h}e_{2h} = -r_{2h}$ "on the coarse grid" Ω_{2h} in step (3) of (8.9.7). Here, the idea suggests itself to use the two-grid method again, thereby reducing this problem to the problem of solving further linear equations on the still coarser grid Ω_{4h}, etc. In this way, we obtain multigrid methods proper. From among the many variants of these methods we only describe the so-called *multigrid V-cycle*, which is an essential ingredient of all such methods. In order to solve $A_h u_h = f_h$ on the grid Ω_h, the multigrid V-cycle visits all grids

$$\Omega_h \to \Omega_{2h} \to \cdots \to \Omega_{2^j h} \to \Omega_{2^{j-1}h} \to \cdots \to \Omega_h$$

between the finest grid Ω_h and a coarsest grid $\Omega_{2^j h}$ in the indicated order: It first descends from the finest to the coarsest grid, and then ascends again to the finest grid, which also explains the name of the method. During one V-cycle, an approximate solution v_h of the fine-grid equation $A_h u_h = f_h$ is replaced by a new approximate solution

$$v_h \leftarrow MV_h(v_h, f_h)$$

of the same equation, where the function $MV_h(v_h, f_h)$ is recursively defined by the following.

(8.9.14) Multigrid V-Cycle. *Suppose v_h, f_h are given vectors on Ω_h. Put $H := h$.*

(1) *By ν steps of the damped Jacobi method* (8.9.4) *with $\omega_0 = 2/3$, transform the approximate solution v_H of $A_H u = f_H$ into another approximate solution, again denoted by v_H.*
(2) *If $H = 2^j h$ goto* (4). *Otherwise put*

$$f_{2H} := I_H^{2H}(f_H - A_H v_H), \qquad v_{2H} := MV_{2H}(0, f_{2H}).$$

(3) *Compute $v_H := v_H + I_{2H}^H v_{2H}$.*
(4) *Apply the damped Jacobi method* (8.9.4) *ν times with $\omega_0 = \frac{2}{3}$ to transform the approximate solution v_H of $A_H u = f_H$ into another approximate solution of these equations, again denoted by v_H.*

Further variants of the multigrid method are described and analyzed in the literature [see, e.g., Brandt (1977), Hackbusch and Trottenberg (1982), Hackbusch (1985), and McCormick (1987)]. The most efficient of these methods require only $O(N)$ operations to compute an approximate solution v_h of a system $A_h u_h = f_h$ with N unknowns, which is sufficiently accurate in the following sense: The error $\|v_h - u_h\| = O(h^2)$ has the same order of magnitude as the truncation error $\max_{x \in \Omega_h} \|y(x) - u_h(x)\| = \tau(h) = O(h^2)$ (cf. Theorem 7.4.10) of the underlying discretization method. Since the exact solution u_h of the discretized equation $A_h u_h = f_h$ differs from the exact solution $y(x)$ of the boundary value problem (8.9.1) by the truncation error $\tau(h)$ anyway, it makes no sense to compute an approximation v_h to u_h with $\|v_h - u_h\| \ll \tau(h)$.

For the simple two-grid method (8.9.7), Theorem (8.9.13) implies only a weaker result: Because of $N = n - 1$ and $h^2 = \pi^2/n^2$, this method requires $j = O(\log N)$ iterations to compute an approximate solution $v_h^{(j)}$ of $A_h u_h = f_h$ with $\|v_h^{(j)} - u_h\| = O(h^2)$, if we start with $v_h^{(0)} := 0$. Since the tridiagonal system in step (3) of (8.9.7) can be solved with $O(N)$ operations, the two-grid method requires altogether $O(N \log N)$ operations in order to find an approximate solution v_h of acceptable accuracy.

8.10 Comparison of Iterative Methods

In order to compare the iterative methods discussed in this chapter for solving systems of linear equations, we consider the special model problem (8.4.1),

(8.10.1)
$$\begin{aligned} -u_{xx} - u_{yy} &= 2\pi^2 \sin \pi x \sin \pi y, \\ u(x, y) &= 0 \quad \text{for } (x, y) \in \partial\Omega, \\ \Omega &:= \{(x, y) \mid 0 < x, y < 1\}, \end{aligned}$$

which has the exact solution

$$u(x, y) = \sin \pi x \sin \pi y.$$

Using the discretization described in Section 8.4, there results the system of linear equations

(8.10.2)
$$\begin{aligned} Az &= b, \\ A &\text{ as in (8.4.5)}, \\ b &= 2h^2\pi^2\bar{u}, \end{aligned}$$

with

$$\begin{aligned} \bar{u} &:= [\bar{u}_{11}, \bar{u}_{21}, \ldots \bar{u}_{N1}, \ldots, \bar{u}_{1N}, \ldots, \bar{u}_{NN}]^T, \\ \bar{u}_{ij} &:= u(x_i, y_j) = \sin i\pi h \sin j\pi h, \qquad h = 1/(N+1). \end{aligned}$$

In Section 8.4 we exhibited the eigenvectors of the Jacobi iteration matrix J associated with A. One sees immediately that the vector b of (8.10.2) is an eigenvector of J, and hence, by virtue of $J = (4I - A)/4$, also an eigenvector of A. The exact solution of (8.10.2) can therefore easily be found. We have

$$Jb = \mu b \quad \text{with } \mu = \cos \pi h.$$

For z one obtains

$$z := \frac{h^2\pi^2}{2(1 - \cos \pi h)} \bar{u}. \tag{8.10.3}$$

In the practical comparison we included the Jacobi and the Gauss–Seidel method, as well as the relaxation method, where the optimal relaxation parameter ω_b was chosen according to Theorem (8.3.17), using for $\rho(J)$ the exact value in (8.4.6). In the ADI method the optimal ADI parameters were computed by (8.6.23) for $m = 2^k$, $k = 2, 4$. For α and β we used the values given in (8.6.24). For the tests with the conjugate-gradient method we used a program of Ginsburg [in Wilkinson and Reinsch (1971)].

As a measure for the error we took the residual, weighted by $1/h^2$,

$$\bar{r}^{(i)} := \frac{1}{h^2} \|Az^{(i)} - b\|_\infty,$$

in order to make the results comparable for different values of N, and of h. The iteration was terminated as soon as $\bar{r}^{(i)}$ was reduced to the order of magnitude 10^{-4}.

In the Jacobi, Gauss–Seidel, relaxation, and ADI methods we started the iteration with the vector $z^{(0)} := 0$. The corresponding initial residual is $\bar{r}^{(0)} = 2\pi^2 \approx 20$. The results are found in the table at the top of page 634, in which, besides N, the number of iterations i is also given, as well as the terminal residual $\bar{r}^{(i)}$.

The results for the conjugate-gradient method are contained in the table at the bottom of page 634. Here, a starting vector was chosen which differs from $z^{(0)} = 0$; for $z^{(0)} := 0$ the conjugate-gradient method furnishes the exact solution $z = z^{(1)}$ in (8.9.3) after only one iteration, since b is an eigenvector of A. Furthermore, since the residuals $\bar{r}^{(i)}$ do not decrease as regularly as in the other methods, we give in the table at the bottom of page 634 not only the initial residual $\bar{r}^{(0)}$, but also the residuals $\bar{r}^{(i)}$ for additional values of i.

The results of the table at the top of page 634 are in agreement with the results on the rate of convergence in the previous sections: the Gauss–Seidel method converges twice as fast as the Jacobi method [Corollary (8.3.16)], the relaxation methods bring a further reduction of the number of iterations [Theorem (8.3.17), (8.4.9)], and the ADI method requires the fewest iterations [cf. (8.6.28)]. The conjugate-gradient method (the table at the bottom of page 634) requires more iterations than the ADI method. Its convergence behavior is somewhat "unsteady". The residual, at times, remains of the same order of magnitude over several iterations, only to become suddenly much smaller. (Similar behavior is also observed with the ADI method.)

Method	k	N	$r^{(i)}$	i
Jacobi		5	3.5×10^{-3}	60
		10	1.2×10^{-3}	235
Gauss–Seidel		5	3.0×10^{-3}	33
		10	1.1×10^{-3}	127
		25	5.6×10^{-3}	600
Relaxation		5	1.6×10^{-3}	13
		10	0.9×10^{-3}	28
		25	0.6×10^{-3}	77
		50	1.0×10^{-2}	180
ADI	2	5	0.7×10^{-3}	9
		10	4.4×10^{-3}	12
		25	2.0×10^{-2}	16
	4	5	1.2×10^{-3}	9
		10	0.8×10^{-3}	13
		25	1.6×10^{-5}	14
		50	3.6×10^{-4}	14

Since the algorithm of Buneman (see Section 8.8) is a noniterative method which (in exact arithmetic) yields the exact solution of (8.10.2) in a finite number of steps at the expense (see Section 8.8) of about $3N^2 \log_2 N$ operations (multiplications), the computational results for this algorithm will not be given here. In comparing the iterative methods with the Buneman algorithm for the solution of (8.10.2), one should keep in mind that all these methods merely compute the solution z of (8.10.2). This solution z, however,

N	$\bar{r}^{(i)}$	i	N	$\bar{r}^{(i)}$	i
5	2.6×10^{2}	0	10	9.8×10^{2}	0
	1.8×10^{1}	2		5.2×10^{1}	4
	4.0×10^{0}	4		1.6×10^{1}	8
	8.7×10^{-1}	6		3.6×10^{0}	12
	2.8×10^{-1}	8		1.3×10^{0}	16
	4.0×10^{-2}	10		7.5×10^{-2}	20
	3.6×10^{-10}	11		1.2×10^{-2}	24
				2.8×10^{-4}	28
				6.3×10^{-6}	32
				6.8×10^{-8}	36

is only a poor approximation to the desired solution $u(x, y)$ of (8.10.1). Indeed, from (8.10.3), by Taylor expansion in powers of h, we have

$$z - \bar{u} = \left(\frac{\pi^2 h^2}{2(1 - \cos \pi h)} - 1\right)\bar{u} = \frac{h^2\pi^2}{12}\,\bar{u} + O(h^4),$$

so that the error $\|z - \bar{u}\|_\infty$, inasmuch as $\|\bar{u}\|_\infty \leqslant 1$, satisfies

$$\|z - \bar{u}\|_\infty \leqslant \frac{h^2\pi^2}{12} + O(h^4).$$

Since in practice one is not interested in the solution z of (8.10.2), but in the solution $u(x, y)$ of (8.10.1), there would be little point in approximating z to higher accuracy than an error of the order of magnitude $h^2 = 1/(N + 1)^2$. Since the initial vector $z^{(0)} = 0$ has an error $\|z - z^{(0)}\| \approx 1$, the Jacobi–Gauss–Seidel and the optimal SOR method, according to (8.4.9), need the following numbers of iterations and operations (one iteration requires approximately $5N^2$ operations) in order to compute z with an error of the order h^2:

Method	Number of iterations	Number of operations
Jacobi	$0.467(N + 1)^2 \log_{10} (N + 1)^2 \approx N^2 \log_{10} N$	$5N^4 \log_{10} N$
Gauss–Seidel	$0.234(N + 1)^2 \log_{10} (N + 1)^2 \approx 0.5N^2 \log_{10} N$	$2.5N^4 \log_{10} N$
Optimal SOR	$0.36(N + 1) \log_{10} (N + 1)^2 \approx 0.72N \log_{10} N$	$3.6N^3 \log_{10} N$

To analyze the ADI method we use (8.6.28). One easily verifies, on the basis of this formula, that for given N the number $R_{\text{ADI}}^{(m)}$ is minimized for $m \approx ln[4(N + 1)/\pi]$, in which case $\sqrt[m]{4(N + 1)/\pi} \approx e$. The ADI method with optimal choice of m and optimal choice of parameters thus requires

$$R_{\text{ADI}}^{(m)} \log_{10} (N + 1)^2 \approx 3.60(\log_{10} N)^2$$

iterations to approximate the solution z of (8.10.2) with an error of the order of magnitude h^2. Per iteration the ADI method requires approximately $8N^2$ operations (multiplications), so that the total number of operations is about

$$28.8N^2(\log_{10} N)^2.$$

The Buneman method, on the other hand, according to Section 8.8 requires only

$$3N^2 \log_2 N \approx 10N^2 \log_{10} N$$

operations (multiplications) for the computation of the exact solution of (8.10.2).

A somewhat larger number of $O(N^{2.5} \log N)$ operations is required by the conjugate-gradient method (8.7.10) using the SSOR preconditioner

(8.7.11), as was shown by Axelsson (1977). Unsurpassed are certain multigrid methods, which require only $O(N^2)$ operations to find a sufficiently accurate solution [see, e.g., Hackbusch (1985)]: With these methods the number of operations grows only proportionally with the number N^2 of unknowns of (8.10.1).

In this connection, however, one should also point out the limitations of these methods, which derive from the fact that they enable one, at best, to obtain the solution $u(x, y)$ of (8.10.1) to within an error of the order $h^2 \approx 1/N^2$. The difference method thus gives a method of second order for the determination of u. In order to approximate the desired solution to an error of the order of magnitude 10^{-6}, one has to choose $h \approx (\sqrt{12}/\pi) \times 10^{-3}$, $N \approx 1000$. The system of equations (8.10.2) for this h would contain approximately 10^6 equations! To achieve higher accuracies it is more effective to use better difference methods than the one in (8.4.4), or variational methods, say of the type discussed in Section 7.5, provided they lead to methods of order higher than 2. Already a method of third order (one such method was given in Section 7.7) would in $\mathbb{R}^2$ merely require a stepsize h of the order of magnitude $h \approx 10^{-2}$, $N \approx 100$, to obtain an approximate solution with an error of about 10^{-6}. The corresponding system of linear equations in this case would have "only" 10^4 equations. The saving over the simple difference methods is so considerable that the extra work in computing the coefficients of the system of equations pays off much more than it did in boundary-value problems for ordinary differential equations (cf. Section 7.6).

Exercises for Chapter 8

1. Show: $\rho(A) < 1 \Leftrightarrow \lim_{i\to\infty} A^i = 0$.
   Hint: Use Theorem (6.9.2).

2. Let A be an $m \times m$ matrix and $S_n = \sum_{i=0}^n A^i$. Show: $\lim_{n\to\infty} S_n$ exists if and only if $\rho(A) < 1$, and then

$$\lim_{n\to\infty} S_n = (I - A)^{-1}.$$

   Hint: Use Exercise 1 and the identity $(I - A)S_n = I - A^{n+1}$.

3. In floating-point arithmetic, instead of (8.1.10), one effectively carries out the following iteration:

$$\bar{x}^{(0)} := x^{(0)}, \qquad B := \bar{L} \cdot \bar{R},$$

$$\bar{x}^{(i+1)} := \bar{x}^{(i)} + B^{-1}\bar{r}^{(i)} + a^{(i)},$$

   with

$$\bar{r}^{(i)} := b - A\bar{x}^{(i)}$$

   and

$$a^{(i)} := \mathrm{fl}(\bar{x}^{(i)} + B^{-1}\bar{r}^{(i)}) - (\bar{x}^{(i)} + B^{-1}\bar{r}^{(i)}).$$

(Using the theory in Sections 4.5, 4.6, one can estimate $\|a^{(i)}\|$ from the above.) Show:

(a) The error $\varepsilon^{(i)} := \bar{x}^{(i)} - x$, $x := A^{-1}b$ obeys the recursion

$$(*) \qquad \varepsilon^{(i+1)} = (I - B^{-1}A)\varepsilon^{(i)} + a^{(i)} = C\varepsilon^{(i)} + a^{(i)}, \qquad C := I - B^{-1}A.$$

From this, derive an explicit formula for $\varepsilon^{(i)}$.

(b) Show that the $\|\varepsilon^{(i)}\|$ remain bounded as $i \to \infty$ if $\rho(C) < 1$ and the $a^{(i)}$ remain bounded, $\|a^{(i)}\| \leqslant \eta$ for all i. Find as good an upper bound as you can for $\limsup_{i\to\infty} \|\varepsilon^{(i)}\|$.

4.

(a) Show: A is irreducible if and only if the graph $G(A)$ belonging to A is connected.
Hint: Use the fact that the graphs $G(A)$ and $G(P^TAP)$, P a permutation matrix, coincide up to the numbering of the vertices.

(b) Given

$$A_1 = \begin{bmatrix} 1 & 0 & 2 \\ 3 & 1 & 0 \\ -1 & 0 & 1 \end{bmatrix}, \qquad A_2 = \begin{bmatrix} 1 & 2 & 0 \\ -1 & 1 & 0 \\ 3 & 0 & 1 \end{bmatrix},$$

show: There exists a P such that $A_2 = P^TA_1P$, P a permutation matrix; $G(A_1)$ and $G(A_2)$ are identical up to a renaming of the vertices of the graphs.

5. Given

$$A = \begin{bmatrix} 2 & 0 & -1 & -1 \\ 0 & 2 & -1 & -1 \\ -1 & -1 & 2 & 0 \\ -1 & -1 & 0 & 2 \end{bmatrix},$$

show:

(a) A is irreducible.

(b) The Jacobi method does not converge.

6. Given

$$A = \begin{bmatrix} 2 & -1 & 0 & -1 \\ -1 & 2 & -1 & 0 \\ 0 & 0 & 2 & -1 \\ -1 & 0 & -1 & 2 \end{bmatrix},$$

show: A is irreducible and nonsingular.

7. Consider the system of linear equations $Ax = b$, A a nonsingular $n \times n$ matrix. If A is reducible, then the system of equations can always be decomposed into N systems, $2 \leqslant N \leqslant n$, of the form

$$\sum_{k=j}^{N} A_{jk}x_k = b_j, \qquad A_{jj} \text{ an } m_j \times m_j \text{ matrix}, \quad \sum_j m_j = n,$$

where all A_{jj} are irreducible.

8. (Varga 1962.) Consider the ordinary differential equation

$$-\frac{d}{dx}\left(p(x)\frac{d}{dx}y(x)\right) + \sigma(x)y(x) = f(x), \qquad a \leqslant x \leqslant b,$$

$$y(a) = \alpha_1, \qquad y(b) = \alpha_2,$$

$p(x) \in C^3[a, b]$, $\sigma(x)$ continuous and $\sigma(x) > 0$, $p(x) > 0$ on $a \leqslant x \leqslant b$. Discretize the differential equation relative to the general subdivision $a = x_0 < x_1 < x_2 < \cdots < x_N < x_{N+1} = b$, with

$$h_i := x_{i+1} - x_i,$$

using

$$(*) \qquad \frac{d}{dx}\left(p(x)\frac{d}{dx}y(x)\right)\bigg|_{x=x_i} = \frac{p_{i+1/2}\dfrac{y_{i+1}-y_i}{h_i} - p_{i-1/2}\dfrac{y_i - y_{i-1}}{h_{i-1}}}{\dfrac{h_i + h_{i-1}}{2}} + \begin{cases} O(\bar{h}_i^2) & \text{if } h_i = h_{i-1}, \\ O(\bar{h}_i) & \text{if } h_i \neq h_{i-1}, \end{cases}$$

with $\bar{h}_i = \max(h_i, h_{i-1})$ and

$$p_{i+1/2} = p(x_{i+1/2}) = p(x_i + h_i/2), \; p_{i-1/2} = p(x_{i-1/2}) = p(x_i - h_{i-1}/2).$$

Show:

(a) The validity of $(*)$ by means of Taylor expansion.
(b) If the resulting system of linear equations is written as $Ax = b$, then the following holds for A: A is real, tridiagonal with positive diagonal and negative side diagonal elements, provided the h_i are sufficiently small for all i.
(c) A is irreducible and satisfies the weak row sum criterion.
(d) The Jacobi iteration matrix J is irreducible, $J \geqslant 0$, 2-cyclic, and consistently ordered, and $\rho(J) < 1$.
(e) Do the Jacobi, Gauss–Seidel, and relaxation methods converge with the finest partitioning?

9. Consider

$$A_1 = \begin{bmatrix} 0 & 1 & 1 \\ 1 & 0 & 1 \\ 1 & 1 & 0 \end{bmatrix}, \qquad A_2 = \begin{bmatrix} 0 & 1 & 0 \\ 1 & 0 & 1 \\ 1 & 1 & 0 \end{bmatrix},$$

$$A_3 = \begin{bmatrix} 0 & 1 & 0 \\ 1 & 0 & 1 \\ 0 & 1 & 0 \end{bmatrix}, \qquad A_4 = \begin{bmatrix} 0 & 1 & 0 & 0 \\ 0 & 0 & 1 & 0 \\ 0 & 1 & 0 & 1 \\ 1 & 0 & 0 & 0 \end{bmatrix},$$

$$A_5 = \begin{bmatrix} 0 & 1 & 0 & 0 \\ 0 & 0 & 1 & 0 \\ 0 & 0 & 0 & 1 \\ 1 & 0 & 0 & 0 \end{bmatrix}, \qquad A_6 = \begin{bmatrix} 0 & 1 & 0 & 0 \\ 0 & 0 & 1 & 1 \\ 0 & 1 & 0 & 1 \\ 1 & 0 & 0 & 0 \end{bmatrix}.$$

Which graphs $G(A_i)$ are 2-cyclic?

10. Consider the 9 × 9 matrix A,

$$(*) \qquad A = \begin{bmatrix} M & -I & O \\ -I & M & -I \\ O & -I & M \end{bmatrix}, \quad \text{with } M = \begin{bmatrix} 4 & -2 & 0 \\ -1 & 4 & -1 \\ 0 & -1 & 6 \end{bmatrix}.$$

(a) Specify the matrices B, C, of the splitting $A = B - C$ which correspond to the following four iterative methods:

(1) Jacobi
(2) Gauss–Seidel } with the finest partitioning,

(3) Jacobi
(4) Gauss–Seidel } with the partitioning used in (∗).

(b) Show: A is irreducible, $G(J)$ is 2-cyclic, and A has property A.

(c) Show that methods (1) and (2) converge for A, and that (2) converges faster than (1).

(d) Show that (3) and (4) converge.

Hint: Do not compute M^{-1}, but derive a relationship between the eigenvalues of J_π (and of H_π) and M. Observe, in this connection, that for specially partitioned matrices such as

$$S = \begin{bmatrix} O & R & O \\ R & O & R \\ O & R & O \end{bmatrix},$$

the eigenvalues of S can be expressed in terms of those of R, by considering an eigenvector of S which is partitioned analogously to S.

11. For the following matrix A show that it has property A and that it is not consistently ordered, and find a permutation such that the permuted matrix $P^T AP$ is consistently ordered:

$$A = \begin{bmatrix} 4 & -1 & 0 & 0 & 0 & -1 \\ -1 & 4 & -1 & 0 & -1 & 0 \\ 0 & -1 & 4 & -1 & 0 & 0 \\ 0 & 0 & -1 & 4 & -1 & 0 \\ 0 & -1 & 0 & -1 & 4 & -1 \\ -1 & 0 & 0 & 0 & -1 & 4 \end{bmatrix}.$$

12. Show: All block tridiagonal matrices

$$\begin{bmatrix} D_1 & A_{12} & & & \\ A_{21} & D_2 & A_{23} & & \\ & \ddots & \ddots & \ddots & \\ & & \ddots & \ddots & A_{N-1,N} \\ & & & A_{N,N-1} & D_N \end{bmatrix},$$

D_i nonsingular diagonal matrices, $A_{ij} \neq 0$, have property A.

13. Show that (8.4.5) is consistently ordered.

14. Verify: A is consistently ordered but does not have property A:

$$A := \begin{bmatrix} 1 & -1 & 0 \\ 1 & 1 & 0 \\ 1 & 1 & 1 \end{bmatrix}.$$

15. Consider the Dirichlet boundary value problem on the domain Ω (see Figure 35) with mesh Ω_h for $h = 1$.

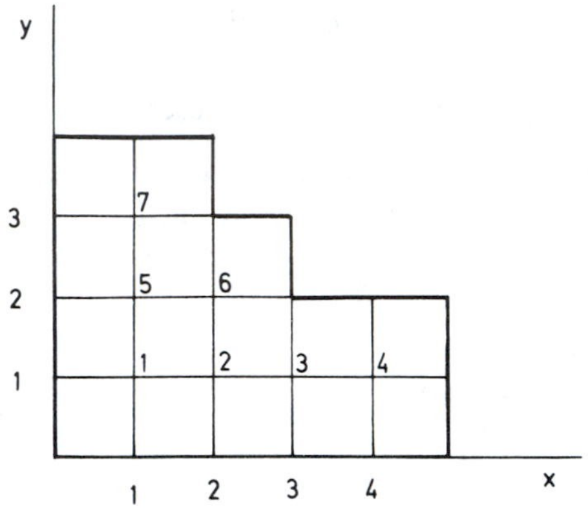

Figure 35 Domain Ω.

(a) Derive the associated system of equations $Ax = b$, using the discretization as given in (8.4.2), and specify A, assuming that the vector x is ordered in the manner shown in Figure 35, i.e., $x = [x_{11}, x_{21}, x_{31}, x_{41}, x_{12}, x_{22}, x_{13}]^T$.

(b) Show: A is irreducible and symmetric, $G(J)$ is 2-cyclic, A has property A, and $\rho(J) < 1$.

(c) Order A consistently, and indicate to which renumbering of the variables x this corresponds in Figure 35. Is this reordering unique?

16. Consider

$$A := \begin{bmatrix} 3 & -1 & 0 & 0 & 0 & -1 \\ -1 & 3 & -1 & 0 & -1 & 0 \\ 0 & -1 & 3 & -1 & 0 & 0 \\ 0 & 0 & -1 & 3 & -1 & 0 \\ 0 & -1 & 0 & -1 & 3 & -1 \\ -1 & 0 & 0 & 0 & -1 & 3 \end{bmatrix}.$$

(a) Show that the Jacobi, Gauss–Seidel, and relaxation methods converge.

(b) Which of these methods is to be preferred?

(c) Write down explicitly the iteration which belongs to A and to the method chosen in (b).

17. Show that for the model problem one has for the block relaxation method, relative to the partitioning given in (8.4.5), as $N \to \infty$,

$$\rho(H_\pi(\omega_b)) \approx \rho(H(\omega_b))^\kappa, \quad \text{with } \kappa = \sqrt{2}.$$

Hint: Observe the information on the model problem given in Section 8.5.

18. (Varga 1962.) Consider the matrix

$$A = \left[\begin{array}{cc|c} 5 & 2 & 2 \\ 2 & 5 & 3 \\ \hline 2 & 3 & 5 \end{array}\right].$$

(a) Determine the spectral radius ρ_1 of the Gauss–Seidel matrix H_1 belonging to A for the finest partitioning.
(b) Determine $\rho_2 = \rho(H_2)$, where H_2 is the matrix associated with the block Gauss–Seidel method. Use the partitioning indicated above.
(c) Show: $1 > \rho_2 > \rho_1$, i.e., the block Gauss–Seidel method does not necessarily converge faster than the point Gauss–Seidel method.
(d) Answer (a), (b) for

$$\bar{A} = \left[\begin{array}{cc|c} 5 & -2 & -2 \\ -2 & 5 & -3 \\ \hline -2 & -3 & 5 \end{array}\right];$$

do we again have $1 > \rho_2 > \rho_1$?

19. Verify (8.6.10)–(8.6.13).

20. Show:

(a)
$$\min_{r>0} \max_{0<\alpha\leqslant x\leqslant\beta} \left|\frac{r-x}{r+x}\right|^2 = \left(\frac{\sqrt{\beta}-\sqrt{\alpha}}{\sqrt{\beta}+\sqrt{\alpha}}\right)^2,$$

where the minimum is assumed precisely for $r = \sqrt{\alpha\beta}$.
(b) For the model problem with

$$\mu_j = 4 \sin^2 \frac{j\pi}{2(N+1)}$$

(cf. (8.6.12)), one then obtains

$$\min_{r>0} \rho(T_r) = \frac{\cos^2 \dfrac{\pi}{N+1}}{\left(1 + \sin \dfrac{\pi}{N+1}\right)^2}.$$

21. For (8.6.18) one can find approximate solutions which in practice, for small m, often approximate the exact solution with sufficient accuracy. Peaceman and Rachford proposed the following approximation:

$$r_j = \beta \cdot \left(\frac{\alpha}{\beta}\right)^{(2j-1)/2m}, \qquad j = 1, 2, \ldots, m.$$

Show:

(a)
$$\left|\frac{r_j - x}{r_j + x}\right| < 1 \quad \text{for all } j, \qquad \alpha \leqslant x \leqslant \beta.$$

(b) $\beta > r_1 > r_2 > \cdots > r_m > \alpha$, and, with $z := (\alpha/\beta)^{1/(2m)}$, one has

$$\left|\frac{r_k - x}{r_k + x}\right| \leqslant \frac{1 - z}{1 + z} \quad \text{for } k = m,\ \alpha \leqslant x \leqslant r_m \text{ or } k = 1,\ \beta \geqslant x \geqslant r_1,$$

$$\left|\frac{x - r_{i+1}}{x + r_{i+1}}\,\frac{x - r_i}{x + r_i}\right| \leqslant \left(\frac{1 - z}{1 + z}\right)^2, \qquad r_i \geqslant x \geqslant r_{i+1}.$$

(c) From (a) and (b) it follows for (8.6.17) that

$$\varphi(r_1, \ldots, r_m) \leqslant \frac{1 - z}{1 + z}.$$

[Details for these approximations can be found in Young (1971).]

22. Show for the sequence α_j, β_j generated in (8.6.21):

$$\alpha_0 < \alpha_j < \alpha_{j+1} < \beta_{j+1} < \beta_j < \beta_0, \qquad j \geqslant 1,$$

and

$$\lim_{j\to\infty} \alpha_j = \lim_{j\to\infty} \beta_j.$$

23. Determine the parameters $r_i^{(4)}$ from (8.6.23) for $m = 4$, $\alpha = 0.5$, $\beta = 3.5$, and compare them with the approximate values obtained from the formula of Peaceman and Rachford given in Exercise 21. In particular, compare $\varphi(r_1, \ldots, r_4)$ with the upper bound given in Exercise 21.

24. Consider the Dirichlet problem for the differential equation

$$u_{xx} + u_{yy} + \frac{1}{x} u_x + \frac{1}{y} u_y = 0$$

in a rectangular domain Ω, where Ω lies in the region $x \geqslant 1$, $y \geqslant 1$. Find a difference approximation to the problem so that the resulting system of linear equations of the form

$$(H + V)z = b$$

has the property $HV = VH$.

25. (Varga 1962.) Consider the differential equation

$$u_{xx} + u_{yy} = 0$$

on the rectangular domain $\Omega := \{(x, y) \mid 0 \leqslant x, y \leqslant 1\}$, with the boundary conditions

$$u(0, y) = 1, \quad u(1, y) = 0, \qquad 0 \leqslant y \leqslant 1,$$

$$\frac{\partial u}{\partial y}(x, 0) = \frac{\partial u}{\partial y}(x, 1) = 0, \qquad 0 \leqslant x \leqslant 1.$$

(a) In analogy to (8.4.2), discretize $u_{xx} + u_{yy}$ with mesh size $h = \frac{1}{3}$ for all grid points at which the solution is unknown. Take account of the boundary condition $u_y(x, 0) = u_y(x, 1) = 0$ by introducing fictitious grid points, e.g. $(x_i, -h)$, and approximate $u_y(x_i, 0)$ by

$$u_y(x_i, 0) = \frac{u(x_i, 0) - u(x_i, -h)}{h} + O(h), \qquad i = 1, 2.$$

One obtains a system of linear equations $Az = b$ in eight unknowns. Specify A and b.

(b) Find the decomposition $A = H_1 + V_1$ exemplified in (8.6.3) and show:

H_1 is real, symmetric, positive definite,
V_1 is real, symmetric, positive semidefinite.

(c) Show: $H_1 V_1 = V_1 H_1$.
(d) Even though the assumptions of (8.6.9) are not fulfilled, show that $\rho(T_r) < 1$ for $r > 0$ and compute r_{opt}:

$$\rho(T_{r_{\text{opt}}}) = \min_{r>0} \rho(T_r).$$

The result is $r_{\text{opt}} = \sqrt{3}$, with

$$\rho(T_{r_{\text{opt}}}) \leqslant \frac{\sqrt{3} - 1}{\sqrt{3} + 1}.$$

Hint: Use the result of Exercise 20(a) and the exact eigenvalues of H_1. Observe the hint to Exercise 10(d).

26. Consider the system of linear equations $Az = b$ which results from the problem of Exercise 15 if the unknowns form the components of z not in the order {1, 2, 3, 4, 5, 6, 7} given in Exercise 15(a) with respect to the numbering of the unknowns in Figure 30, but in the order {7, 5, 6, 1, 2, 3, 4}.

(a) Show that with this ordering A is consistently ordered.
(b) Find the decomposition analogous to (8.6.3),

$$A = H_1 + V_1$$

and show:

(α) H_1, V_1 are symmetric and real; H_1 is positive definite; V_1 has negative eigenvalues.
(β) $H_1 V_1 \neq V_1 H_1$.

References for Chapter 8

Axelsson, O.: Solution of linear systems of equations: Iterative methods. In: Barker (1977).

Barker, V. A. (Ed.): *Sparse Matrix Techniques*. Lecture Notes in Mathematics 572. Berlin, Heidelberg, New York: Springer-Verlag (1977).

Brandt, A.: Multi-level adaptive solutions to boundary value problems. *Math. of Comput*. **31**, 333–390 (1977).

Briggs, W. L.: *A Multigrid Tutorial.* Philadelphia: SIAM (1987).

Buneman, O.: A compact non-iterative Poisson solver. Stanford University, Institute for Plasma Research Report No. 294, Stanford, CA (1969).

Buzbee, B. L., Dorr, F. W.: The direct solution of the biharmonic equation on rectangular regions and the Poisson equation on irregular regions. *SIAM J. Numer. Anal.* **11**, 753–763 (1974).

———, ———, George, J. A., Golub, G. H.: The direct solution of the discrete Poisson equation on irregular regions. *SIAM J. Numer. Anal.* **8**, 722–736 (1971).

———, Golub, G. H., Nielson, C. W.: On direct methods for solving Poisson's equations. *SIAM J. Numer. Anal.* **7**, 627–656 (1970).

Chan, T. F., Glowinski, R., Periaux, J., Widlund, O. (Eds.): *Proceedings of the Second International Symposium on Domain Decomposition Methods.* Philadelphia: SIAM (1989).

Forsythe, G. E., Moler, C. B.: *Computer Solution of Linear Algebraic Systems.* Series in Automatic Computation. Englewood Cliffs, N.J.: Prentice-Hall (1967).

Glowinski, R., Golub, G. H., Meurant, G. A., Periaux, J. (Eds.): *Proceedings of the First International Symposium on Domain Decomposition Methods for Partial Differential Equations.* Philadelphia: SIAM (1988).

Hackbusch, W.: *Multigrid Methods and Applications.* Berlin, Heidelberg, New York: Springer-Verlag (1985).

———, Trottenberg, U. (Eds.): *Multigrid Methods.* Lecture Notes in Mathematics 960. Berlin, Heidelberg, New York: Springer-Verlag (1982).

Hestenes, M. R., Stiefel, E.: Methods of conjugate gradients for solving linear systems. *Nat. Bur. Standards J. Res.* **49**, 409–436 (1952).

Hockney, R. W.: The potential calculation and some applications, *Methods of Computational Physics* **9**, 136–211. New York, London: Academic Press (1969).

Householder, A. S.: *The Theory of Matrices in Numerical Analysis.* New York: Blaisdell Publ. Co. (1964).

Keyes, D. E., Gropp, W. D.: A comparison of domain decomposition techniques for elliptic partial differential equations and their parallel implementation. *SIAM J. Sci. Statist. Comput.* **8**, s166–s202 (1987).

McCormick, S.: *Multigrid Methods.* Philadelphia: SIAM (1987).

Meijerink, J. A., van der Vorst, H. A.: An iterative solution method for linear systems of which the coefficient matrix is a symmetric M-matrix. *Math. Comp.* **31**, 148–162 (1977).

O'Leary, D. P., Widlund, O.: Capacitance matrix methods for the Helmholtz equation on general three-dimensional regions. *Math. Comp.* **33**, 849–879 (1979).

Proskurowski, W., Widlund, O.: On the numerical solution of Helmholtz's equation by the capacitance matrix method. *Math. Comp.* **30**, 433–468 (1976).

Reid, J. K. (Ed.): *Large Sparse Sets of Linear Equations.* London, New York: Academic Press (1971).

———, On the method of conjugate gradients for the solution of large sparse systems of linear equations. In: Reid (1971), 231–252.

Rice, J. R., Boisvert, R. F.: *Solving Elliptic Problems Using ELLPACK.* Berlin, Heidelberg, New York: Springer (1984).

Schröder, J., Trottenberg, U.: Reduktionsverfahren für Differenzengleichungen bei Randwertaufgaben I. *Numer. Math.* **22**, 37–68 (1973).

———, ———, Reutersberg, H.: Reduktionsverfahren für Differenzengleichungen bei Randwertaufgaben II. *Numer. Math.* **26**, 429–459 (1976).

Swarztrauber, P. N.: The methods of cyclic reduction, Fourier analysis and the FACR algorithm for the discrete solution of Poisson's equation on a rectangle. *SIAM Rev.* **19**, 490–501 (1977).

Varga, R. S.: *Matrix Iterative Analysis.* Series in Automatic Computation. Englewood Cliffs, N.J.: Prentice-Hall (1962).

Wachspress, E. L.: *Iterative Solution of Elliptic Systems and Application to the Neutron Diffusion Equations of Reactor Physics*. Englewood Cliffs, N.J.: Prentice-Hall (1966).

Wilkinson, J. H., Reinsch, C.: *Linear Algebra*. Handbook for Automatic Computation, Vol. II. Grundlehren der mathematischen Wissenschaften in Einzeldarsstellungen, Bd. 186. Berlin, Heidelberg, New York: Springer (1971).

Wittmeyer, H.: Über die Lösung von linearen Gleichungssystemen durch Iteration. *Z. Angew. Math. Mech.* **16**, 301–310 (1936).

Young, D. M.: *Iterative Solution of Large Linear Systems*. Computer Science and Applied Mathematics. New York: Academic Press (1971).

General Literature on Numerical Methods

Ciarlet, P. G., Lions, J. L., Eds.: *Handbook of Numerical Analysis.* Vol. I: *Finite Difference Methods (Part 1), Solution of Equations in* $\mathbb{R}^n$ *(Part 1).* Vol. II: *Finite Element Methods (Part 1).* Amsterdam: North Holland (1990), (1991).

Conte, S. D., de Boor, C.: *Elementary Numerical Analysis, an Algorithmic Approach,* 3d edition. New York: McGraw-Hill (1980).

Dahlquist, G., Björck, Å.: *Numerical Methods.* Englewood Cliffs, N.J.: Prentice-Hall (1974).

Forsythe, G. E., Malcolm, M. A., Moler, C. B.: *Computer Methods for Mathematical Computations.* Englewood Cliffs, N.J.: Prentice-Hall (1977).

Fröberg, C. E.: *Numerical Mathematics.* Menlo Park, Calif.: Benjamin/Cummings (1985).

Gregory, R. T., Young, D. M.: *A Survey of Numerical Mathematics.* Vols. 1, 2. Reading, Mass.: Addison-Wesley (1972), (1973).

Hämmerlin, G., Hoffmann, K.-H.: *Numerical Mathematics.* Berlin, Heidelberg, New York: Springer-Verlag (1991).

Henrici, P.: *Elements of Numerical Analysis.* New York: McGraw-Hill (1964).

Hildebrand, F. B.: *Introduction to Numerical Analysis,* 2d edition. New York: McGraw-Hill (1974).

Householder, A. S.: *Principles of Numerical Analysis.* New York: McGraw-Hill (1953).

Isaacson, E., Keller, H. B.: *Analysis of Numerical Methods.* New York: John Wiley (1966).

Press, W. H., Flannery, B. P., Teukolsky, S. A., Vetterling, W. T.: *Numerical Recipes. The Art of Scientific Computing.* Cambridge: Cambridge University Press (1990).

Ralston, A., Rabinowitz, P.: *A First Course in Numerical Analysis.* New York: McGraw-Hill (1978).

Rutishauser, H.: *Lectures on Numerical Mathematics.* Edited by M. Gutknecht with the assistance of P. Henrici et al. and translated by W. Gautschi. Boston: Birkhäuser (1990).

Schaback, R., Werner, H.: *Numerische Mathematik.* Berlin, Heidelberg, New York: Springer-Verlag (1991).

Schwarz, H.-R.: *Numerical Analysis. A Comprehensive Introduction.* With a contribution by J. Waldvogel. Chichester: Wiley (1989).

Stiefel, E.: *An Introduction to Numerical Mathematics.* New York, London: Academic Press (1963).
Todd, J.: *A Survey of Numerical Analysis.* New York: McGraw-Hill (1962).
———: *Basic Numerical Mathematics. Vol. 1. Numerical Analysis.* Basel: Birkhäuser (1978) (also New York: Academic Press 1978).
———: *Basic Numerical Mathematics. Vol. 2. Numerical Algebra.* Basel: Birkhäuser (1977) (also New York: Academic Press 1977).

Index

In general, page numbers in italics refer to definitions.